SERIES IN SENSORS

STRUCTURAL SENSING, HEALTH MONITORING, AND PERFORMANCE EVALUATION

D. Huston
University of Vermont
Burlington, USA

CRC Press
Taylor & Francis Group
Boca Raton London New York

CRC Press is an imprint of the
Taylor & Francis Group, an **informa** business

A TAYLOR & FRANCIS BOOK

CRC Press
Taylor & Francis Group
6000 Broken Sound Parkway NW, Suite 300
Boca Raton, FL 33487-2742

First issued in paperback 2019

ISBN-13: 978-0-7503-0919-6 (hbk)
ISBN-13: 978-0-367-38359-6 (pbk)

Library of Congress Cataloging-in-Publication Data

Huston, D. (Dryver R.), 1958-
    Structural sensing, health monitoring, and performance evaluation / author, D. Huston.
        p. cm. -- (Series in sensors)
    "A CRC title."
    Includes bibliographical references and index.
    ISBN 978-0-7503-0919-6 (hard back : alk. paper)
    1. Structural health monitoring. 2. Machinery--Monitoring. 3. Detectors. 4. Automatic data collection systems. 5. Nondestructive testing. 6. Mechatronics. I. Title. II. Series.

TA656.6.H87 2011
624.1028'7--dc22

2009021484

Visit the Taylor & Francis Web site at
http://www.taylorandfrancis.com

and the CRC Press Web site at
http://www.crcpress.com

Contents

Preface

This book represents the author's views and impressions of structural sensing, health monitoring, and prognosis. These were derived from the vast body of work developed over the past couple of decades by highly creative and productive researchers. Although some of the citation lists are lengthy, no attempt has been made to be comprehensive in the literature reviews. The author welcomes suggestions for improvements and corrections to the book.

D. Huston
Burlington, Vermont

Acknowledgments

The author would like to thank all the people who graciously allowed their photographs and artwork to be included in the text; and all the collaborators, program managers, students, and people who helped with words of encouragement in the writing of this book: Emin Aktan, Sreenivas Allampalli, Tim Ambrose, Steve Arms, Jean-Guy Beliveau, Brent Boerger, Harold Bosch, Dylan Burns, Wai-Fah Chen, Yong Chen, Rob Church, Rick Claus, Edgardo Colon-Emeric, Hunter Cui, Ahmed Elgamal, Brian Esser, Mohammed Ettouney, Peter Fuhr, Yozo Fujino, Mack Gardner-Morse, Hamid Ghasemi, Victor Giurgiutiu, Florian Grätz, William Graves, Nenad Gucunski, Wolfgang Habel, Doug Hamilton, Zhi He, Sonja Hölzl, Jing Hu, David Hurley, Frank Jalinoos, Travis Johnston, Nick Jones, Minhaj Kirmani, Steve Kreiser, Shunli Ma, Ali Maher, Ken Maser, Scott McNulty, Paul Montane, Matt Nelson, David Ogden, Noel Pelczarski, James Plumpton, Josef Ponn, Qian Qin, Iassen Raykov, David Rosowsky, Wolfgang Sauter, Warren Schmelzer, Robert Selzer, Peter Sonntag, Mete Sozen, William Spillman Jr., Xiaoyan Sun, Hiroshi Tanaka, Peter Thiringer, Glenn Washer, Eddie Weber, William Weedon, Michael Werner, Christian Wettach, Jamie Wilsey, Haiyan Zhang, Xiangdong Zhao, and Jinyu Zheng. My apologies to anyone inadvertently overlooked.

1

Introduction

Structural sensing, structural health monitoring (SHM), structural performance assessment, and health prognosis are integral components of modern structural engineering practice. Due to recent extraordinary levels of development in sensor, communication, and signal processing technologies, it is now possible to measure structural properties and behavior with sufficient clarity to assess damage levels and to predict future courses of structural health. SHM can help the owners, builders, and designers of structures in rational decision-making [1–4]. Possible benefits include enhancing the safety of structures with warnings of impending failures, prompting more efficient use of maintenance resources, and providing information that leads to better designs. In the aerospace industry the benefits of SHM can appear as increased up-time usage rates for aerospace systems and improved designs [5]. SHM measurements can help optimize maintenance activity planning [6]. Financial entities, such as insurance companies and municipal bond markets, make use of the rational assessment of structural conditions and prognoses [7–9]. Certain high-performance structures, for example, reusable launch vehicles, may not be able to operate effectively without SHM [10–12]. Many relatively low-performance ancillary structures, such as highway signs and light poles, can also benefit from SHM [13].

Traditional SHM combines visual observations and heuristic assumptions with mathematical models of predicted behavior [14]. More modern versions include sensors and automated reasoning techniques [15]. Figure 1.1 shows a simplified hierarchical representation of a much more complicated thought process [16,17]. An alternative view of the process as an analogy to baseball appears in Figure 1.2 [18]. Effective implementation of the high-level SHM actions in Figures 1.1 and 1.2 requires a successful execution of steps that range from sensor installation and operation to data processing to data integration to integration with the needs of structural owners and society at large.

A five-level classification (derived from Rytter's original four-level classification) of SHM activities is as follows [19–21]:

Level 1: Detection—Determine if damage is present in the structure.

Level 2: Localization—Locate the site of the damage.

Level 3: Assessment—Estimate the amount of damage.

Level 4: Prognosis—Estimate the future progress of damage and the remaining life of the structure.

Level 5: Remediation—Determine, implement, and evaluate effective remediation and repair efforts.

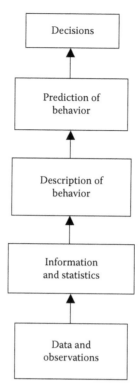

FIGURE 1.1   Layered flow of data into information into judgment/decisions. (Adapted from Ullman DG. 2003. *The Mechanical Design Process*, 3rd Ed. McGraw-Hill, Boston.)

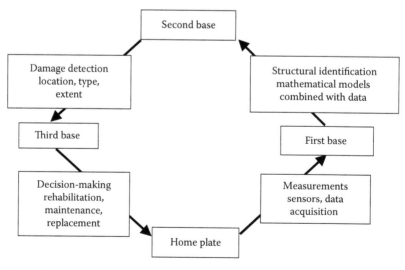

FIGURE 1.2   Baseball analogy of engineering of structural health. (Adapted from Alampalli S and Ettouney M. 2003. In *Structural Health Monitoring*, DEStech, Lancaster.)

1.1 Need for Structural and Performance Health Monitoring

Premature structural failure can have large public safety and economic consequences. SHM can play a role in preventing and mitigating the course of structural damage. SHM has a clear role in virtually all structural engineering efforts. The traditional practice of SHM has used humans as visual inspectors, possibly with the aid of simple instruments, such as plumb lines and levels. These low-tech techniques have been highly successful for millennia. Moving beyond these simple techniques to the use of modern sensing systems for SHM inevitably involves added expenses that need to be justified in terms of increased benefits to the owners and operators of structural systems. When the primary concern is to maintain the integrity of a structure, then it is relatively straightforward to make the case for expensive monitoring efforts [22]. Non-safety-critical applications of SHM include providing information for improved scheduling of maintenance, evaluating structural performance, and predicting expected lifetimes. In these cases, the justification for using SHM requires a more detailed and often more complicated comparison of the costs to benefits.

Figure 1.3 illustrates some of the potential utility of SHM [23]. The vertical axis is a health index of a structural system. Larger health index values indicate an increased state of health. A value of zero corresponds to when the system fails to operate as desired. The curves A, B, and C are possible life paths of a system. The system following path A experiences a shorter than expected lifespan. This may be due to severe loading and/or poor fabrication. Path B corresponds to what may be expected of an average system. Path C represents a system with a longer than average lifetime. Such a path may arise due to lighter than expected operational loads and/or superior fabrication. If it is unknown which path (A, B, or C) a system will follow, then operating the system and/or deciding to remove it from service inherently incurs costs related to premature failure or premature removal. Using

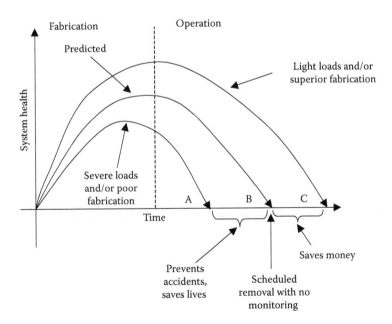

FIGURE 1.3 Potential utility of system and SHM versus operation without monitoring. (Adapted from Baker TE. 1999. *Proc 2nd Intl Workshop on Structural Health Monitoring*, Stanford. Technomic, Lancaster.)

SHM can increase the certainty as to which life path the system is following. This allows for informed case-by-case decision-making to prevent both accidents and premature removal. In many cases, in spite of persuasive arguments along the lines of Figure 1.3, SHM remains a potentially disruptive technology, especially for systems that are well understood and have established track records [24].

An example of SHM application is managing the deterioration and subsequent reconstruction of reinforced concrete highway bridge decks. Two principal mechanisms of bridge deck deterioration are (1) freeze/thaw damage to the concrete (punky concrete) and (2) corrosion-induced delamination resulting from the infiltration of chlorides introduced by winter road salting operations or by ocean spray. Overall costs due to maintaining and repairing the bridge decks, including traffic disruptions and safety hazards for motorists and repair crews, are major concerns. Chamberlin and Weyers demonstrated that the application of chloride and corrosion monitoring combined with appropriate maintenance techniques can lead to lifetimes of 35 years or more for conventional reinforced concrete bridge decks [25].

The utility of SHM is not always inherently obvious. Davison describes this as the *predictive maintenance dilemma*. When SHM data indicate that a structure is in need of repairs, but the damage is not visible to the human eye and the damage progresses slowly, then it is difficult to justify the value of SHM. If the SHM readings confirm the presence of a dangerous or damaged condition that is visible to the human eye, then it is difficult to justify the value of SHM. If the SHM readings identify a dangerous situation, but there are no good remedies, then it is difficult to justify the value of SHM. If the SHM readings are ignored and a catastrophic failure occurs, then it is difficult to justify the value of SHM [26]. In fact, the use of SHM without a clear means of structural rehabilitation or repair may increase the liability of owners and engineers without providing for any increased protection [27]. A further complication is that most structures are well designed. Unexpected failures are rare. Without an impending failure, the placement of SHM instruments on a structure may not lead to quickly tangible economic benefits [28]. Clearly, the value of SHM lies not in merely taking measurements, but in providing useful information. In some situations, it may be advantageous to continue with minimal maintenance and monitoring efforts [29].

Managing many structural conditions benefits from the use of SHM. Others do not. Four criteria for assessing whether it is useful to employ SHM techniques are as follows: (1) *The structural problem should be important*: If the problem is important, it is worth monitoring. It is wasteful to monitor unimportant problems. For example, certain cracks are problematic whereas others are benign [30]. Monitoring the potentially problematic cracks may be more useful than monitoring benign cracks. (2) *The structural problem should be fixable or preventable, if detected in a sufficiently early stage of development*: Certain failure modes happen so quickly, or so inevitably, that monitoring does not provide much opportunity for palliative efforts. Others, such as termite infestations, benefit greatly from early detection. (3) *Structural measurements and data processing should be easy*: The measurements should be relatively easy to take and interpret. Continued development of sensing, networking, and analysis technologies is expanding the scope of easy measurements. (4) *The measurements should correlate well with the underlying problem, with a minimum of confounding issues*: For example, temperature changes can produce large variations in the vibration properties of structures due to mechanical processes that have very little to do with structural health. Using vibration measurements as an SHM tool without compensating for temperature changes may be ineffective. Mitigating the impact of confounding effects is particularly important when translating laboratory measurements into field techniques.

SHM practice has matured to where many large structures use permanent sensor systems. The technical literature contains descriptions of efforts from around the globe. Yun, Park et al., and Koh and Choo describe the monitoring of several bridges in Korea [31–33]. Lau et al. and Wong report on SHM installations on three cable stayed bridges in Hong Kong [34,35]. Figueiras et al. installed a suite of fiber Bragg gratings (FBGs) and other sensors on the Sorraia River Bridge in Portugal [36]. Falkner and Henke conducted long-term strain monitoring of a concrete box girder bridge in Germany [37]. Hariri and Budelmann monitored a bascule bridge that shows early signs of distress in concrete post-tensioning strands [38]. Many European efforts appeared under the umbrella of the SAMCO Network (Structural Assessment, Monitoring, and Control) [39]. Sun et al. and Dong et al. installed SHM instruments on the Sutong Bridge in China with particular emphasis on monitoring the performance of wind–rain cable vibration countermeasures [40,41]. Ou and Zhou report on the monitoring of several bridges in China [42]. Masri et al. implemented a real-time SHM system on the Vincent Thomas Bridge in California [43]. Chompooming et al. instrumented a steel riveted bridge with a fixed bascule truss [44]. Watkins et al. used a combination of Fabry–Perot fiber optic strain gages, RFID tags, and video imaging to monitor an all-composite bridge [45]. Moe and Nouziaan monitored the health of the long prestressed-concrete boxgirder Confederation Bridge in Canada [46]. Maalej et al. used FBGs to monitor two bridges in Canada with innovative structural systems [47]. Shoukry et al. installed over 700 sensors to monitor the Star City Bridge in West Virginia [48].

1.2 Technical Challenges in SHM and Performance Assessment

A structural lifetime typically encompasses four main phases: *construction*, *usage*, *rehabilitation*, and *demolition* or *abandonment*. During each of these phases, a variety of loads (wind, gravity, floods, earthquakes, corrosion, chemical attack, insects, fire, construction, maintenance, thermal gradients, etc.) stress the structure. SHM systems can combine measurements of structural loads and responses with an understanding of structural behavior to produce an estimate of the present state and to provide a prognosis as to the future course of structural health. The process uses techniques that range from simple heuristics based on experience to advanced pattern recognition algorithms and stochastic modeling. In spite of the ready availability of powerful tools for measurement and analysis, predicting the future with certainty from uncertain information inevitably remains a difficult challenge.

Measurands, geometry, and *timing* are principal descriptive parameters of structural sensing systems. Measurands are measured physical quantities. Geometric descriptions pertain to both structural and sensing system geometries. Timing describes both the speed and time of data acquisition, as well as the structural events of interest.

The wide variety of potentially useful measurands challenges the SHM engineer to decide which ones to measure and which ones to neglect. Measurands of interest include strain, acceleration, wind pressure, chloride ion concentration, temperature, fatigue, damage, and so on.

Geometry imposes many technical challenges on SHM system design and operation. One is that many measurands of interest lie in difficult-to-access positions. Sensing often requires supplying power, transmitting data, and servicing at these remote locations. Spatial resolution and sensitivity comprise another set of geometric challenges. Many structures are large. Important events often occur in localized regions. Damage localization challenges SHM systems to provide simultaneously high-spatial-resolution-localized sensing and global coverage. Various strategies for meeting this challenge include identifying *a priori* which local regions are prone to damage or critical to stability, deploying vast arrays of inexpensive sensors with localized point-like regions of sensitivity, using distributed sensors, using mobile and redeployable systems, and using adaptive squinting or zooming sensor systems. The design of structures to facilitate SHM can be another key ingredient.

Sensor locations relative to the surface of the structure are important geometric descriptors. Sensor mounting geometries include subsurface embedment, surface attachment, and off-board remote (Figure 1.4). It is usually easier to embed sensors during construction or factory fabrication than to embed them as a retrofit into existing structures. Structures made of reinforced concrete or fiber-reinforced composites are particularly amenable to construction-phase embedment. Surface attachment has advantages of easier installation, especially with completed and in-service structures. Noncontacting stand-off sensors are another possibility. Noncontacting sensors are easy to move and can inspect broad swaths of a structure, but typically require a high degree of sensing system sophistication. Similar issues of geometric location arise when considering sensor system maintenance.

Measurement timing issues include the following: (1) *Timing of the data acquisition relative to the present life phase of the structure*: Some of the key life phases are construction, usage, performance evaluation, rehabilitation, and demolition. Each life phase may be best served by a different sensing system. (2) *Duration of the sensing effort*: The duration can range from short-term and temporary efforts to cases that last throughout the life of the structure. Long-term sensing is common with high-performance structures, such as dams, where failures are a serious threat to public safety. Long-term measurements require robust sensors and data acquisition systems. The most severe sensing conditions often occur during some of the most important measurement intervals, such as during earthquakes, hurricanes, floods, fire, blasts, and so on. Sensing during extreme events can provide invaluable information, but may require the sensor system to be more robust than the structure. (3) *Interval over which the data are acquired*: Several data acquisition timing approaches and variants are common. One is to sense continuously for an extended period of time. Continuous sensing can capture data from all events of interest, but can potentially acquire enormous amounts of data with little useful information. Intermittent-interval sensing reduces data bandwidth and storage requirements. Fixed-interval and scheduled-intermittent sensing strategies are

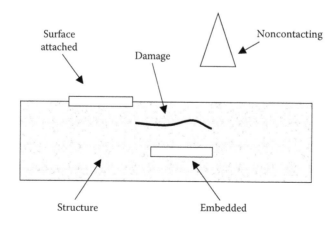

FIGURE 1.4 Sensor location relative to structure.

useful for long-term monitoring applications. An alternative is to use triggers to start the data collection when the data levels exceed preset thresholds, such as during strong earthquakes. Pretrigger buffering can enhance information capture by measuring and storing data immediately prior to the triggering event. (4) *Sampling rate*: The desired data sampling rates depend on the frequency content of data signals. Data acquisition system capabilities often limit the sampling rates. Compromises between sampling too fast and sampling too slow are common. Sampling too fast collects an enormous amount of data that quickly becomes burdensome to store, transmit, process, and comprehend. Sampling too slowly runs the risk of missing key events and mismeasuring higher-frequency data.

1.3 Measurands

Measurands are physical properties that are directly measurable with sensing instruments. For most measurands, a variety of sensors can measure and transduce the information into an easily measured property, such as voltage, optical intensity, or direct digital readout.

1.3.1 Geometric Configuration

Geometry is a principal SHM measurement. Geometric measurands include displacement, tilt, flatness, camber, and changes from a nominal configuration. Human eyesight and simple instruments, such as plumb lines or levels, can quickly measure gross geometric properties. Sophisticated modern instruments enable rapid, precise, and long-term geometric measurements. All geometric measurements require an established reference frame with an associated coordinate system. The reference frame may be fixed, such as being attached to bedrock, or it can be floating, as when measuring interstory drift in a building during an earthquake or the flexing of an airplane. For large and safety critical structures, it is often common to make specific provisions to establish fixed and stable reference points.

1.3.2 Kinematics: Displacement, Velocity, and Acceleration

Structural kinematics, that is, displacement, velocity, and/or acceleration, reflect the action of many underlying structural processes. Displacements are common structural measurements. Structural stiffness calculations and models often make use of displacement measurements. Structural velocity measurements are less common. The availability of inexpensive velocimeters has led to widespread velocity measurements in applications such as the seismic exploration of large geological structures and the monitoring of rotating machinery. Structural velocities tend to be associated with damping or other dissipative processes, or specific dynamic loading processes, such as aerodynamic and traffic loads. Accelerations are generally easier and less expensive to measure than displacements or velocities. Acceleration measurements are quite common in the assessment and control of structural dynamics. Due to the relative ease of measurement, there is considerable interest in using accelerations for SHM.

1.3.3   Strain

Strain describes the local state of deformation in continuous bodies. A full three-dimensional (3-D) characterization uses a symmetric second-order tensor with six independent components and six independent scalar measurements. In many situations it is adequate to measure only a subset of the strain components, particularly when the geometry limits access to the surface. In the majority of structural engineering applications, the deformations and rotations are sufficiently small to allow using the linearized version of the strain tensor [49,50]. Notable exceptions are tires, cables, and fabric structures, where the combination of large rotations and large in-line tensions can cause significant geometric nonlinearities that require using nonlinear strain tensor formulations for analysis [51].

1.3.4   Force, Pressure, and Stress

Directly measuring force, pressure, and stress is difficult, if not impossible [52,53]. Indirect measurements (typically based on deformation) are more common. For example, a force-measuring load cell actually measures the deformation of an internal elastic element and then uses an elastic force–deformation model, combined with a precise calibration, to infer the amount of force traversing the load cell. Force is a 3-D vector quantity and requires three independent measurements for full characterization. Stress is a second-order symmetric tensor and requires six independent scalar measurements for full characterization [49]. Most stress measurements examine only a subset of the six components. Modern prepackaged integrated transducers measure forces, stresses, and pressures without requiring much interaction between the user and the inner workings of the transduction mechanisms.

Several situations arise that require specialized force and stress measurements. One situation (fairly uncommon in structural mechanics) is the distinction between thermodynamic and mechanical notions of pressure [49]. Another is to measure generalized forces that are associated with concepts of analytical mechanics [54,55]. Large-deformation situations may benefit from interpreting the measurements in terms of specialized Piola–Kirchof stress tensors [49]. In cases of mixed solid–fluid media, such as water-saturated concrete or soil, it is often of interest to measure the fluid pressure in a manner that is independent of the stresses in the solid.

1.3.5   Strength

Strength describes the ability of a structure or a structural element to withstand or sustain a load without yielding, collapse, or serious damage. Predicting the failure load of a particular structure, especially one with complicated subsystem interactions, remains a severe technical challenge. Uncertainty as to the failure mode and the properties of materials under severe loading conditions limits the utility of submaximal load tests in ultimate strength assessments.

1.3.6   Temperature

Structural temperatures often reflect much useful information about underlying conditions, that is, thermal performance, the presence of thermoelastic stresses, and underlying structural faults, such as delaminations. The large thermal inertias of massive structures, coupled with low-frequency seasonal and annual variations of ambient temperatures, can require temperature measurements over long periods of time—possibly several years or even decades—for a full thermal characterization.

1.3.7 Cracks

Cracks may or may not indicate structural problems. Typical crack characteristics are size, location, and dimensional changes. Determining if a crack is growing requires repeated measurements. When cracks are deemed to be potentially dangerous, then automated or routine crack inspections may be warranted or, in more severe cases, immediate removal of the structure from service.

1.3.8 Fatigue

Fatigue causes failure in structural components. The mechanics of fatigue processes are a complicated interplay of microcracking and mechanical loading [56]. Presently available fatigue measurement methods are largely inadequate. Some of the modestly successful approaches include those based on fatigue fuses and indirect measurements, such as load history measurements combined with models of fatigue behavior.

1.3.9 Chemical State

All structural elements rely on chemical bonding for strength and stability. Chemical changes—including corrosion—are major factors in the damage and damage resistance of structural elements [57]. For example, hydrogen sulfide can severely damage concrete sewer pipes. Most of the presently available tools for measuring structural chemistry remain underdeveloped.

1.3.10 Air/Moisture Permeability and Content

Concrete, masonry, and other structural materials are typically porous and permeable to the ingress and egress of moisture and other fluids. The amount of permeability and the associated air entrainment affects freeze/thaw performance. High levels of moisture inside a structure can cause damage and may be a symptom of other problems. Specialized instruments for air/moisture permeability and content measurements are available.

1.3.11 Electrical, Magnetic, and Electromagnetic Properties

The electrical properties of structural materials often strongly correlate with underlying structural properties. SHM efforts exploit this degree of correlation with a panoply of electrical, magnetic, and electromagnetic (EM) property measurements of materials to infer the structural properties. These include electrical resistance, inductance, capacitance, magnetic susceptibility, dielectric properties, and the propagation of eddy currents.

1.3.12 Scour Meters

Scour is the erosion of the riverbed around bridge piers and is a common cause of bridge collapse. Scour processes can occur due to normal riverbed erosion. Floods accelerate the process. Scour measurements including those prior to, during, and after floods can be crucial components of efforts to improve bridge survivability.

1.4 Sensor Performance Benchmarks

The amount and quality of the information contained in SHM data depend on many factors. Sensor performance is one of the principal factors. The following describe some common sensor performance benchmarks.

1.4.1 Sensor Dimensionality and Spatial Dependence

Sensors measure through physical processes that span a multidimensional space–time environment. In most cases, it is reasonable to approximate these processes as occurring over single, multiple, and/or fractional dimensions (Figure 1.5). *Point sensors* have approximately zero-dimensional (0-D) spatial measurement sensitivity. Examples of point sensors are foil strain gages, thermocouples, and accelerometers. When the spatial sensitivity of the sensor extends beyond a point, then the sensor is called a *distributed sensor*. Examples of distributed sensors include linear sensors based on activity along a cable, such as time domain reflectometry, aerial imaging, and volume averaging systems such as x-ray tomography and ground-penetrating radar. One-dimensional (1-D) sensors measure along a line or a curve. An example is a long strain gage that measures strain in concrete by spanning several grains of aggregate. Two-dimensional (2-D) sensors measure over a distributed area or surface. Many 2-D sensors use standoff imaging techniques. Three-dimensional volumetric sensors measure properties over volumes, often with the aid of penetrating radiation and imaging techniques. Certain classes of sensors have dimensions that are effectively fractional. Between zero and one dimension are the sensors that require measurements at two or more

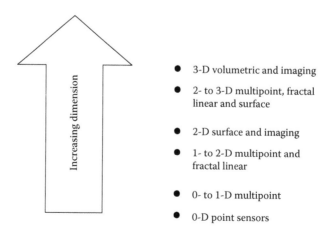

FIGURE 1.5 Relative dimensionality of various sensor types.

points to operate. Between zero, one, two, and three dimensions are distributed linear sensors and arrays of point sensors that can provide substantial coverage of an area or volume. Sensors with fractal geometry, such as the branching nervous system in vertebrates, are intriguing possibilities for SHM system design.

1.4.2 Data Format

Data come in a wide variety of analog, digital, and qualitative formats. Modern sensing systems often simultaneously acquire multiple channels with a heterogeneous mix of data types. Combining data from large numbers of sensors with location information can form image data. *Features* are higher-level data derived from preliminary statistical analyses that represent an information-dense condensation of raw low-level data.

1.4.3 Precision and Accuracy

Precision and *accuracy* describe measurement quality. At least three different definitions appear with these terms. The first definition uses *precision* and *accuracy* synonymously to describe the correctness of the measurement and the smallest measurable difference. Lay practitioners of the technical arts commonly use this first definition. The second definition uses *precision* as the size of the smallest quantity that can be distinguished by a system. For example, a meter stick marked with millimeter gradations has a precision of 1 mm. In contrast, *accuracy* describes the quality and/or confidence in a given measurement. An example of the distinction between precision and accuracy is the use of a ruler to measure length. If the ruler is 100 m long, with increments of 1 mm, it has a precision of 1 mm. If, on a hot day, the ruler expands due to heating by 5 mm, then it is accurate to within 5 mm or one part in 20,000, but retains a precision of 1 mm. International and national standards agencies adopt a third definition of accuracy based on the concept of a *standard uncertainty* and generally avoid using the term *precision*.

1.4.4 Linearity

Sensors and sensing systems convert measurands into numerical values. The relation between the true value of the measurand and the value reported by the sensor invariably depends on a complicated interplay of physical effects and data processing techniques. Since there is uncertainty in all physical measurements and concepts, the term *conventional true value* describes an ideal best representation of physical reality. A functional relation, F, between the conventional true value, x, the sensor output, V, and random noise or uncertainty in the measurement, N, is

$$V = F(x, N). \tag{1.1}$$

F can be a nonlinear functional operator and may involve convolutions (possibly a Volterra series) over time and space. A standard method of linearizing the functional is to ignore temporarily the noise and convolution effects and to expand the function in a first-order Taylor series:

$$V = ax + b. \tag{1.2}$$

The successful use of linearization requires sufficient linearity and smoothness in the sensor output as a function of measurand value. The basis of many sensors is an underlying physical phenomenon that is linearizable, such as with a resistive foil strain gage. High-accuracy measurements with linear transducers often involve correcting for small nonlinearities. The basis of other sensors is inherently nonlinear. An example is a Pitot tube that measures fluid velocity with a square root relation based on Bernoulli's principle. Processing the signals with nonlinear inversion relations, polynomial curvefits, or lookup tables can linearize inherently nonlinear outputs.

1.4.5 Dynamic Range

Dynamic range quantifies measurable amplitude range signals. It is the ratio of the largest to smallest measurable value and is often expressed in terms of decibels (dB). Sensors and sensing systems with large dynamic ranges are usually preferable to those with smaller dynamic ranges, but tend to be more expensive and require more digits to represent the data.

1.4.6 Rise Time and Frequency Dependence

Rise time quantifies the speed with which a system will respond to a transient load. A standard definition is the time required for a system to move from 10% to 90% of the amplitude of the steady-state response to a step input (Figure 1.6) [58]. The rise time imposes a limitation on the minimum temporal (or spatial) resolution of a system (Figure 1.7).

Frequency dependence quantifies the frequency domain abilities of a sensor to measure time-varying phenomena. All sensors have a limited frequency range. An example is a piezoelectric inertial-mass accelerometer. Piezoelectric charge leakage limits low-frequency measurement capabilities. Inertial-mass resonance and mounting support flexibility limit high-frequency measurements.

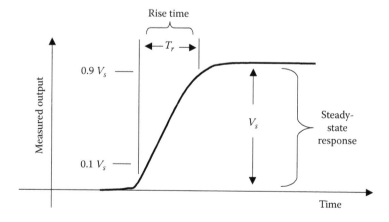

FIGURE 1.6 Rise time defined in terms of step response of the measurement system. (Derived from Andrews JR. 1989. TDR, step response and "S" parameter measurements in the time domain. *Application Note AN-4*, Picosecond Pulse Labs, Boulder.)

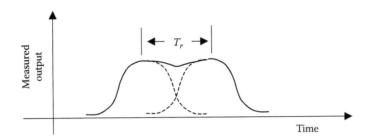

FIGURE 1.7 Minimum temporal resolution in terms of the rise time T_r. (Derived from Andrews JR. 1989. TDR, step response and "S" parameter measurements in the time domain. *Application Note AN-4*, Picosecond Pulse Labs, Boulder.)

1.4.7 Cross Talk

Cross talk is a loosely defined term that describes the contamination of measurements by unwanted signals. Undesired physical effects contaminating and confounding measurements, such as thermal drift in a strain gage, are one form of cross talk. The contamination of signals in one channel from another data channel, such as due to inductive coupling or software bugs, is another form of cross talk. Statistical correlation techniques can help detect cross-talk problems.

1.4.8 Durability, Maintainability, and Redundancy

Sensor *durability* can be vital to the success of the sensing effort. The sensor may have to be more durable than the structure being monitored, especially during extreme event or long-term monitoring applications. For many of these applications, the sensor must not only survive, but it must also sustain a stable and/or accurate calibration. Sensor *maintainability* is another important factor. Ideally, once a sensor is installed, it will not require any maintenance. However, if a sensor fails, it is desirable to replace or repair it. Embedded sensors without access for repair can be troublesome in this regard. Redundancy is an alternative to durability. Redundancy uses multiple sensors to take a measurement. Heterogeneity among the sensor types tends to increase the redundancy and overall robustness of the sensing effort beyond that which may be possible with redundant homogeneous sensor arrays.

1.4.9 Calibration and Self-Calibration

Calibrating sensors against a known standard helps to obtain accurate measurements. Some sensors have an inherently stable calibration and may only require an initial calibration in the factory. Other sensors are less stable and need to be calibrated on a routine basis. Sensors based on the IEEE 1451.2 (TEDS) standard have calibration information stored onboard and accessible with embedded microprocessors. Some sensors have onboard self-calibration circuitry and hardware. Measuring the shunt resistance in a strain gage circuit is an example of a partial onboard calibration. The continued development of microelectromechanical system (MEMS) technology will likely lead to more creative designs of sensors with self-calibration capabilities.

1.4.10   Registration

Registration identifies measurements with position. Most structural measurements require some form of registration. Simple manual registration techniques include recording sensor locations on drawings and/or relative to fiduciary marks on the structure. Registration with electronic instruments and/or imaging systems is becoming more common. Combining information from heterogeneous sensors requires a cross-registration of the data [59].

1.4.11   Power

All sensors require *power* to operate. This can be a major constraint and cost factor. Simple mechanical sensors, such as plumb bobs or liquid levels, require human power for setup and reading the sensor. Providing human power may be easy, on it may be very expensive and dangerous. Power issues become particularly acute in embedded or inaccessible sensing situations. Running power cables to the sensor may be impractical. Instead, battery power, with attendant battery replacement issues, or energy harvesting techniques, such as solar power, is required.

1.4.12   Ease of Use and Cost

Ease of use and *cost* are important. The decision to employ an SHM system on a structure requires that the system provide a payback to the owners or to society in terms of reduced maintenance costs, overall lifetime costs, and/or increased structural safety. Establishing the cost versus benefit of a sensor system can be a nontrivial exercise.

1.5   Data Collection, Storage, and Transmission

Modern SHM systems can rapidly collect massive amounts of data in short periods of time. The data need to be collected, stored, and transmitted for use in further analyses. Data transmission networks continue to increase in diversity of architectures and capability. Wireless, wired, fiber optic, and hybrid systems all appear in routine SHM practices.

1.6   Data Processing, Decision Making, and Prognosis

Collected data need to be processed to assess the state of health of the structure and to make prognoses about future health states. This is a complex multistage and multiscale process that makes extensive use of modern digital data processing and statistical analyses techniques and combines the results with human judgment. For example, Fraser et al. describe the layout for a complex multilevel SHM system for bridges [60].

1.7   Future of SHM

The future of SHM and structural prognosis is that of a rapidly growing field of research and engineering practice. Advanced sensing, information processing, and networking technologies are pushing the field. The continued development of nano- and microelectromechanical systems technologies are leading to smaller, less expensive—yet better performing—integrated sensor systems [61]. Engineers, owners, and various financial entities that want to make better use of their structures are pulling the development of even more advanced technologies. A key to success will be in developing technologies and designs that make both engineering and economic success.

References

1. Housner GW, Bergman LA, Caughey TK, Chassiakos AG, Claus RD, Masri SF, Skelton RE, Soong TT, Spencer BF, and Yao J. 1997. Structural control: Past, present and future. *J Eng Mech*, 123(9), 897–971.
2. Aktan AE, Catbas FN, Grimmelsman KA, and Tsikos CJ. 2000. Issues in infrastructure health monitoring for management. *J Eng Mech*, 126(7), 711–724.
3. Fujino Y and Abe M. 2004. Structural health monitoring—current status and future. *Proc 2nd European Workshop on Structural Health Monitoring*, DEStech, Munich.
4. Hess III PE. 2007. Structural health monitoring for high-speed naval ships. In *Structural Health Monitoring*, FK Chang (ed.). DEStech, Lancaster.
5. Schmidt HJ, Telgkamp J, and Schmidt-Brandecker B. 2004. Application of structural health monitoring to improve efficiency of aircraft structure. *Proc 2nd European Workshop on Structural Health Monitoring*, DEStech, Munich.
6. Furuta H, Hirokane M, Dogaki M, and Frangopol DM. 2004. Optimal bridge maintenance using health monitoring. *Proc 2nd European Workshop on Structural Health Monitoring*, DEStech, Munich.
7. Wong F and Ettouney M. 2003. Engineering tools for infrastructure condition forecasting. In *Structural Health Monitoring 2003*, FK Chang (ed.). DEStech, Lancaster.
8. Ellingwood BR. 2005. Risk-informed condition assessment of civil infrastructure: State of practice and research issues. *Struct Infrastruct Eng*, 1(1), 7–18.
9. PB Consult, Inc. PricewaterhouseCoopers, and LLC Cambridge Systematics, Inc. 2004. A review of DOT compliance with GASB 34 requirements. *NCHRP Report 522*, TRB, Washington, DC.
10. Renson LF. 2003. Health monitoring system for future reusable launcher: From operational aspects to mission risk management impacts. *Smart Nondestructive Evaluation and Health Monitoring of Structural and Biological Systems II*, SPIE 5047.
11. Calabro AM and Sellitto M. 2003. Intelligent and integrated health management system (I2H) on reusable launch vehicles RLV: Concepts and instruments. In *Structural Health Monitoring*, FK Chang (ed.). DEStech, Lancaster.
12. Degang C, Xinmao G, Baoqi T, Hongdong H, and Huaiqiang X. 1999. The investigation of health monitoring and strength of adaptive composite structures. *Proc 2nd Intl Workshop on Structural Health Monitoring*, Stanford, Technomic, Lancaster.
13. Fish P. 2000. Inspection of ancillary structures, a public safety responsibility. In *Structural Materials Technology IV*, S Allampalli (ed.). Technomic, Lancaster.
14. Happold E. 1980. *Appraisal of Existing Structures*. The Institute of Structural Engineers, London.
15. Liu SC, Tomizuka M, and Ulsoy G. 2005. Strategic issues in sensors and smart structures. *Struct Control Health Monit*, 13, 946–957.
16. Ullman DG. 2003. *The Mechanical Design Process*, 3rd Ed. McGraw-Hill, Boston.
17. Elgamal A, Conte JP, Masri S, Fraser M, Fountain T, Gupta A, Trivedi M, and El Zarki M. 2003. Health monitoring framework for bridges and civil infrastructure. In *Structural Health Monitoring*, FK Chang (ed.). DEStech, Lancaster.
18. Alampalli S and Ettouney M. 2003. Summary of the workshop on engineering structural health. In *Structural Health Monitoring*, FK Chang (ed.). DEStech, Lancaster.

19. Doebling SW, Farrar CR, and Prime MB. 1998. A summary review of vibration-based damage identification methods. *Shock Vib Dig*, 30(2), 91–105.
20. Rytter A. 1993. Vibration based inspection of civil engineering structures. PhD dissertation, Department of Building Technology and Structural Engineering, Aalborg University, Aalborg.
21. Worden K, Manson G, and Allman D. 2003. Experimental validation of a structural health monitoring methodology: Part I. Novelty detection of a laboratory structure. *J Sound Vib*, 259(2), 323–343.
22. Stepinski T. 2003. NDE of copper canisters for long term storage of spent nuclear fuel for the Swedish nuclear power plants. *Nondestructive Detection and Measurement for Homeland Security*, SPIE 5048.
23. Baker TE. 1999. Condition monitoring innovations for U.S. army rotorcraft. *Proc 2nd Intl Workshop on Structural Health Monitoring*, Stanford. Technomic, Lancaster.
24. Gunn III LC. 1999. Operational experience with health monitoring of the Delta II Program. *Proc 2nd Intl Workshop on Structural Health Monitoring*, Stanford. Technomic, Lancaster.
25. Chamberlin WP and Weyers RE. 1994. Field performance of latex-modified and low-slump dense concrete bridge deck overlays in the United States. *Amer Conc Inst*, ACI SP 151-1.
26. Davison G. 2003. The dilemma—to call or not to call. *Sound Vib*, (October), 5–6.
27. Velde HV. 1998. Condition evaluation and methods of corrosion protection for unbonded post-tension cables. *Proc Intl Conf on Corrosion and Rehabilitation of Reinforced Concrete Structures*, Orlando.
28. Box RA, McCullough PJ, and Bistline RS. 2000. Introduction to the Delaware River Port Authority's "SmartBridges" Initiative. *Nondestructive Evaluation of Highways, Utilities, and Pipelines IV*, SPIE 3995.
29. Dubey B, Bagchi A, and Alkass S. 2007. Selection of optimal alternative for bridge monitoring: Weighted objective decision analysis. In *Structural Health Monitoring*, FK Chang (ed.). DEStech, Lancaster.
30. Broede J and Koehl M. 1999. Aero engine life usage monitoring including safe crack propagation. *Proc 2nd Intl Workshop on Structural Health Monitoring*, Stanford. Technomic, Lancaster.
31. Yun CB. 2004. Recent R&D activities on structural health monitoring for bridge structures in Korea. *Proc IABMAS'04 Bridge Maintenance, Safety, Management and Cost*, Kyoto. Taylor & Francis, London.
32. Park JC, Park CM, and Chang SP. 2004. Long-term behaviors of a cable-stayed bridge based on full scale measurements. *Proc IABMAS'04 Bridge Maintenance, Safety, Management and Cost*, Kyoto. Taylor & Francis, London.
33. Koh HM and Choo JF. 2004. Preparing for the future: National research programs for the next generation of bridge design and maintenance in Korea. *Proc IABMAS'04 Bridge Maintenance, Safety, Management and Cost*, Kyoto. Taylor & Francis, London.
34. Lau CK, Mak WP, Wong KY, Chan WY, and Man KL. 1999. Structural health monitoring of three cable-supported bridges in Hong Kong. *Proc 2nd Intl Workshop on Structural Health Monitoring*, Stanford. Technomic, Lancaster.
35. Wong KY. 2004. Instrumentation and health monitoring of cable-supported bridges. *Struct Control Health Monit*, 11, 91–124.
36. Figueiras J, Felix C, Matos JC, and Sousa H. 2004. An integrated system for structural health monitoring—application to the Sorraia River Bridge. *Proc IABMAS'04 Bridge Maintenance, Safety, Management and Cost*, Kyoto. Taylor & Francis, London.
37. Falkner H and Henke V. 2004. Better bridge assessment by monitoring. *Proc IABMAS'04 Bridge Maintenance, Safety, Management and Cost*, Kyoto, Taylor & Francis, London.
38. Hariri K and Budelmann H. 2004. Monitoring of the Bridge Herrenbrucke in Lubeck: Motivation, procedures, results and data evaluation. *Proc 2nd European Workshop on Structural Health Monitoring*, DEStech, Munich.
39. Wenzel H and Pardi L. 2004. The European Funded Project SAMCO—structural assessment, monitoring and control. *Proc IABMAS'04 Bridge Maintenance, Safety, Management and Cost*, Kyoto. Taylor & Francis, London.

40. Sun L, Zhang Q, Chen A, and Lin Z. 2004. Cable vibration control countermeasures and structural health monitoring system design of Sutong Bridge. *Proc IABMAS'04 Bridge Maintenance, Safety, Management and Cost*, Kyoto. Taylor & Francis, London.
41. Dong X, Zhang Y, Xu H, and Ni YQ. 2005. Research and design of structural health monitoring system for the Sutong Bridge. In *Structural Health Monitoring*, FK Chang (ed.). DEStech, Lancaster.
42. Ou JP and Zhou Z. 2004. Techniques of optical fiber Bragg grating smart sensors and intelligent monitoring systems of infrastructures. In *Advanced Smart Materials and Smart Structures Technology*, FK Chang, CB Yun, and BF Spencer, Jr. (eds.). DEStech, Lancaster.
43. Masri SF, Sheng LH, Wahbeh M, Caffrey J, Nigbor R, and Abdel-Ghaffar A. 2004. Design and implementation of a web-enabled real-time monitoring system for civil infrastructures. *Proc IABMAS'04 Bridge Maintenance, Safety, Management and Cost*, Kyoto. Taylor & Francis, London.
44. Chompooming K, Sirimontri S, Prasarnklieo W, and Sirisak S. 2003. Structural health monitoring system of a riveted steel truss bridge. In *Structural Health Monitoring*, FK Chang (ed.). DEStech, Lancaster.
45. Watkins SE, Unser JF, and Nanni A. 2001. Instrumentation and manufacture of a smart composite bridge for short-span applications. *Smart Systems for Bridges, Structures, and Highways*, SPIE 4330.
46. Moe MS and Noruziaan B. 2004. Health monitoring system of the confederation bridge—a living lab for research, education and furthering the engineering practice. In *Advanced Smart Materials and Smart Structures Technology*, FK Chang, CB Yun, and BF Spencer, Jr. (eds.). DEStech, Lancaster.
47. Maalej M, Karasaridis A, Pantazopoulou S, and Hatzinakos D. 2002. Structural health monitoring of smart structures. *Smart Mater Struct*, 11, 581–589.
48. Shoukry SN, Riad MY, and William GW. 2006. Remote health monitoring and modeling of Star City Bridge, WV. Paper no. 06-2843, 86th Annual Meeting, TRB, Washington, DC.
49. Malvern LE. 1969. *Introduction to the Mechanics of a Continuous Medium*. Prentice-Hall, Englewood Cliffs, NJ.
50. Novozhilov W. 1953. *Foundations of the Nonlinear Theory of Elasticity*. Dover Publications, Mineola, NY.
51. Irvine HM. 1981. *Cable Structures*. MIT Press, Cambridge, MA.
52. Lindsay RB. 1969. *Concepts and Methods of Theoretical Physics*. Dover Publications, New York, NY.
53. Ciufoloini I and Wheeler JA. 1995. *Gravitation and Inertia*. Princeton University Press, Princeton.
54. Goldstein H. 2002. *Classical Mechanics*, 3rd Ed. Prentice-Hall, Englewood Cliffs, NJ.
55. Meirovitch L. 1970. *Methods of Analytical Dynamics*, McGraw-Hill, New York, NJ.
56. Hertzberg RW. 1989. *Deformation and Fracture Mechanics of Engineering Materials*, 3rd Ed. Wiley, New York.
57. Fuhr PL and Huston DR. 1998. Embedded fiber optic sensors for bridge deck chloride penetration measurements. *Opt Eng*, 37, 1221–1229.
58. Andrews JR. 1989. TDR, step response and "S" parameter measurements in the time domain. *Application Note AN-4*, Picosecond Pulse Labs, Boulder.
59. Hajnal JV, Hill DL, and Hawkes DJ. 2001. *Medical Image Registration*. CRC Press, Boca Raton.
60. Fraser M, Elgamal A, Conte JP, Masri S, Fountain T, Gupta A, Trivedi M, and El Zarki M. 2003. Elements of an integrated health monitoring framework. *Smart Nondestructive Evaluation and Health Monitoring of Structural and Biological Systems II*, SPIE 5047.
61. Varadan VK. 2003. Nanotechnology, MEMS and NEMS based complex adaptive devices and systems. In *Structural Health Monitoring*, FK Chang (ed.). DEStech, Lancaster.

2

Point Sensors

Sensors transduce physical quantities. Measurements occur over finite regions of space. The size, geometry, and effective dimension of the measurement region have a large effect on the nature of the measurement and also serve as a coarse means of sensor classification. *Point sensors* measure in a region that is sufficiently small to be approximated as a point. *Distributed sensors* measure over larger regions with an effective dimension greater than zero. This chapter describes the operating principles and uses of many of the point sensors commonly used for SHM. Chapter 3 covers distributed sensors. The organization of this chapter is somewhat arbitrary with an alignment that follows the underlying operating mechanisms of the sensors, that is, mechanical, electrical, electromagnetic (EM), optical, modern physics, and/or chemical effects.

2.1 Mechanical Sensors

Mechanical sensors rely primarily on mechanical effects for transduction. Simplicity and reliability are common characteristics of mechanical sensors. Some mechanical sensors date back to the earliest of SHM endeavors.

2.1.1 Accelerometers

Inertial accelerometers measure acceleration relative to an inertial reference frame. A typical principle of operation measures the relative displacement of a spring-mounted proof mass with respect to a moving base (Figure 2.1). Under suitable circumstances, the relative displacement is essentially proportional to the acceleration of the moving base.

The equation of motion for the proof mass relative to the inertial reference frame of the vibrating base is

$$m\ddot{x} = -ky - c\dot{y}, \tag{2.1}$$

where m is the inertia of the proof mass and c and k are the damping and stiffness of the mount between the proof mass and base support. $x(t)$, $y(t)$, and $z(t)$ are the proof mass inertial displacement, proof mass displacement relative to the base, and base displacement relative to the inertial reference frame, respectively. Using the kinematic relation,

$$x = y_0 + y + z \tag{2.2}$$

with y_0 as a fixed offset, produces an equation of motion

$$\ddot{y} + 2\varsigma\omega_0\dot{y} + \omega_0^2 y = -\ddot{z}, \tag{2.3}$$

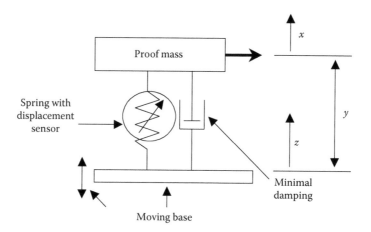

FIGURE 2.1 Inertial accelerometer.

where $\zeta = c/2\,(k/m)^{1/2}$ is the damping ratio and $\omega_0 = (k/m)^{1/2}$ is the circular natural frequency. Assuming harmonic motions at frequency ω,

$$y(t) = Y e^{i\omega t}, \tag{2.4}$$

$$z(t) = Z e^{i\omega t}, \tag{2.5}$$

$$\ddot{z} = -\omega^2 Z e^{i\omega t}. \tag{2.6}$$

Defining the dimensionless frequency ratio $\Omega = \omega/\omega_0$ gives the frequency relation

$$[(1 - \Omega^2) + 2i\zeta\Omega]Y = \Omega^2 Z. \tag{2.7}$$

When the spring is lightly damped and sufficiently stiff for the vibrations to occur at a frequency less than about 20% of the proof mass natural frequency, then the coefficient of the first term in Equation 2.7 approximates unity and the relative displacement of the proof mass and base is approximately proportional to the base acceleration, that is,

$$\ddot{z} \approx -\omega_0^2 y. \tag{2.8}$$

The linear relation in Equation 2.8 forms the basis of many accelerometer designs. A common design approach collocates stiffness and sensing with springs made from piezoelectric materials, such as quartz. Strain-induced piezoelectric charges are proportional to acceleration. Piezoelectric accelerometers are robust, inexpensive, and widely used, but have difficulty in measuring low-frequency vibrations. This can be problematic when attempting to measure the vibrations of large structures, and in using static gravity acceleration measurements for calibration or tilt measurement.

The servo-accelerometer is an alternative accelerometer designed with low-frequency sensitivity that replaces the stiff spring proof mass mount with a soft spring and an electromechanical servomechanism. MEMS accelerometers use silicon cantilever structures, and capacitive or semiconductor-strain-gage displacement measurements [1]. Most MEMS accelerometers measure static as well as transient accelerations. Loh et al. developed a MEMS accelerometer with an onboard optical interferometric transducer [2]. Plaza et al.

used evanescent wave coupling in proof mass springs as the transducer in an optical MEMS accelerometer [3]. Through careful design, it is now possible to fabricate MEMS accelerometers with performance capabilities that approach those of more macrosized devices [4,5]. Dau et al. replaced the solid proof mass with a fluid variant that uses convective heat transfer in a hermetic MEMS chamber to transduce acceleration [6].

Similar to the MEMS accelerometer is the MEMS vibratory gyroscope. Such devices typically combine Coriolis effects with a vibrating structure (often in a ring geometry) to measure angular rotation [1,7,8]. While these devices have yet to attain much attention from the structural sensing community, they have been demonstrated as viable dynamic angular measurement transducers [9].

2.1.2 Dial Indicators, Whittemore and Scratch Gages

The *dial indicator* measures relative displacement between two solid surfaces. A spring-loaded piston attached to a mechanical drive amplifies and converts relative displacements into needle rotations on an indicator dial. Typical piston stroke ranges are 12 mm with resolutions as small as 1–2 μm. Dial indicators are very reliable and easy to use, especially if the structural surfaces are easily accessible. Short-term deflection measurements are a common application of dial indicators. Automated electronic instruments may be more practical for longer-term measurements.

A *Whittemore gage* (a.k.a. *Demec gage*) measures the distance between two reference points that are fixed onto a structural element (Figure 2.2). Whittemore gages provide very reliable and low-cost long-term strain measurements with a minimum of maintenance. Common methods of establishing the reference points are to drill two or more small conical holes directly into the structure, or to surface-bond inserts with predrilled holes. High-precision dial or displacement indicators measure the relative motion of the reference points. The distance separating the reference holes is often nominally 254 mm (10 in). Length accuracies on the order of 13 μm are reported, with a corresponding strain resolution of about 50 με [10].

An interesting, though now obsolete, device is the *mechanical scratch gage* [10]. The scratch gage records mechanical deformations as scratches on a brass disk by using a deformation-powered linkage. The cyclic motion of the strains activates the recording disk and scratching mechanism. This is an early example of an ambient energy-harvesting sensor system.

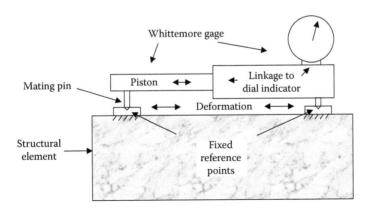

FIGURE 2.2 Operating principle of Whittemore gage.

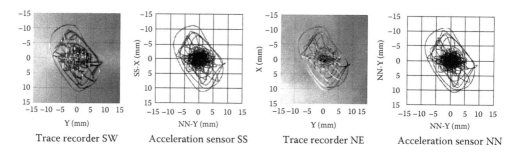

FIGURE 2.3 Comparison of mechanically actuated and electronic measurements of the interstory drift of a building during an earthquake. (Courtesy of Wada A, et al. 2006. *Proc 4th China–Japan–US Symp on Structural Control and Monitoring*, Hangzhou, China.)

Wada et al. also used a scratch technique to measure 2-D interstory drift in buildings during earthquakes (Figure 2.3) [11].

2.1.3 Tiltmeters

Tiltmeters measure tilt, that is, rotation relative to the vertical axis defined by gravity. High-sensitivity tiltmeters are relatively inexpensive and easy to use. One operating principle transduces the gravity and tilt-induced motion of a fluid inside of a small chamber. Electrical transduction techniques are common. Optical methods are also possible. Typical fluid-filled tiltmeter sensitivities are less than 0.5 arcsecond. An alternative design uses an accelerometer with a low-frequency steady-state sensitivity. Since tiltmeters use the acceleration of gravity to actuate the transducer, they are also sensitive to other forms of acceleration. Tiltmeters are useful for applications such as monitoring bridge pier tilt [12].

2.1.4 Vibrating-Wire Strain Gages

A vibrating-wire strain gage transduces by using strain to stretch a wire. This alters the fundamental resonance vibration frequency, f_0, according to

$$f_0 = \frac{1}{2L}\left[\frac{T}{m}\right]^{1/2},\qquad(2.9)$$

where L is the length of the wire, T is the wire tension, and m is the mass per unit length [5]. The wire mounts to the structure at two points. Structural strains cause a relative displacement of the mounts and stretch the wire. Electromechanical induction is a common method of measuring the resonance frequency. The vibrating wire gage derives long-term stability from the resonance frequency being sensitive to tension and insensitive to most confounding effects. The disadvantages of vibrating wire gages are that the gages tend to be relatively large, require bulky cables for power supply and signal transmission, and are fairly expensive. Common uses are in critical applications that require stable long-term measurements and justify the added expense, such as dams. As an example of a successful severe environment application, Wells et al. monitored 10 months of post-construction strains in concrete pavements with vibrating wire gages [13].

Measuring temperature-induced frequency shifts can serve to measure tension in large cables or external prestressing strands [14,15]. A tap with a hammer or ambient loads readily

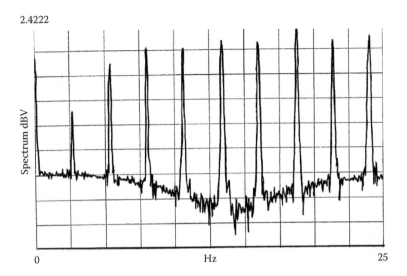

2.4222

Spectrum dBV

0   Hz   25

FIGURE 2.4 Acceleration frequency spectrum measured on a high-tension solid bar construction stay cable on Cornish–Windsor Bridge (VT and NH, USA). [Redrawn from Huston DR, Beliveau JG, and Durham DR. 1990. In *Proc 6th US–Japan Bridge Engineering Workshop*, UJNR, U.S.–Japan Panel on Wind and Seismic Effects, Lake Tahoe.]

excites the vibrations. A complication is motions of higher modes. Accelerometers amplify signals from higher frequency modes as the square of the frequency. If the cable tension is sufficiently high for the dynamics to approximate that of a string, that is, the geometric stiffening due to tension dominates the bending or shear stiffness, the modal frequencies are integer multiples of the fundamental frequency (Figure 2.4) [16,17]. Measuring the frequency difference of the higher harmonic modes indicates the fundamental frequency and the cable tension.

2.1.5 Wire Tension Meter

A wire tension meter measures the tension in wires and cables by pseudostatic mechanical actuation (Figure 2.5). The tension meter applies a three-point lateral load to the wire (Figure 2.6). If the wire mechanics approximate that of a string, then a static analysis produces the force–displacement relation:

$$T = \frac{aF}{2d}. \tag{2.10}$$

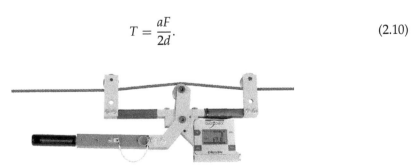

FIGURE 2.5 Quick Check™ three-point wire tension meter. (Photograph courtesy of Dillon Force Measurement Products, a member of the Avery Weigh-Tronix Group.)

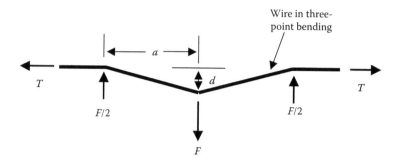

FIGURE 2.6   Three-point wire bend tension meter mechanics.

When the bending stiffness of the wire is significant, then models that combine beam and string mechanics may be more appropriate. The bending stiffness of many wires and cables, particularly multistrand varieties, is difficult to determine from first principles. Such situations generally require the use of empirical calibrations. If the cables are large, vibration testing may be a more practical approach to determining tension [18].

2.1.6   Crack and Joint Width Measurement Systems

A variety of instruments can measure structural crack location, length, and rate of growth [19]. Monitoring structural joint behavior often makes use of similar instruments. One approach is to use displacement gages that span the cracks and joint gaps [5]. Carlson patented a wire strain-gage instrument for crack width measurement in 1934 [20]. Figure 2.7 shows an optical device that uses overlaid grids and Moiré fringe enhancement to measure crack width. Spring-loaded vibrating wire gages, linear variable differential transformers (LVDTs), and similar electronic displacement transducers can automate the process [5,21].

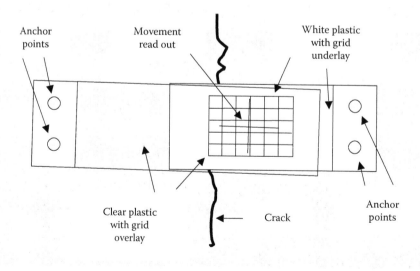

FIGURE 2.7   Manually readable optical crack width monitoring gage. (Adapted from Avongard Products USA Ltd.)

When multiple cracks are present, it may be more convenient to use a crack-monitoring transducer that spans several cracks [22]. Monitoring cracks with strain gages is nontrivial due to the uncertainties in the resulting strain gradients. However, strain gages can be effective if the crack-induced strain distributions are well understood [23]. Measuring the crack length in addition to width and detecting the presence of incipient cracks usually requires an imaging technique. A close-up examination of cracks can provide additional information concerning age and underlying crack mechanisms [24]. The presence of carbonation in concrete cracks may indicate that the crack is old. The converse is not necessarily true [25].

2.1.7 Schmidt Hammer

The Schmidt hammer estimates the strength of a brittle material, for example, concrete or masonry, by measuring the amount of energy absorbed during a standardized hammer impact. An ideal elastic impact and rebound will not transfer energy between the hammer and the material. The result is a hammer rebound back to the original height. Inelastic impacts on a strong material will cause a small amount of damage that transfers a small amount of energy from the hammer to the site of the damage. The consequence is a slightly reduced hammer rebound height. Impact of the tip causes localized yielding or crushing of the material at the impact point. Weaker materials suffer more damage and absorb more energy on impact with a resulting smaller hammer rebound. Several types of Schmidt hammers are available. The operating principle of a pendulum version appears in Figure 2.8. The amount of energy absorbed, E_a, is

$$E_a = U_1 - U_2 = WL\left(\cos\theta_2 - \cos\theta_1\right), \tag{2.11}$$

where W and L are the weight and length of the hammer. Pendulum versions are useful for testing light materials, such as plaster. Spring-loaded variants are more common for testing heavier materials, such as concrete (Figure 2.9).

Penetration tests measure the strength of materials by assessing the capability of the material to resist mechanical penetration by a probe. The Windsor probe is a standard penetration test for concrete as specified in ASTM C803 [26]. The Windsor probe uses a small precisely measured explosive charge to drive a probe into the surface. A similar, though more qualitative, penetration test uses a pick or knife to probe into wood and other materials. Sound material resists penetration. It is often easy to shove a probe into the heart of a rotted wood member. Shotcrete construction quality control measures use electrically actuated penetrometers [27].

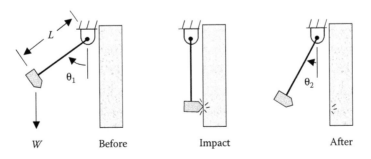

FIGURE 2.8   Operating principle of pendulum-style Schmidt hammer.

FIGURE 2.9 Spring loaded Schmidt impact hammer with recording capability. (Photograph courtesy of Proceq SA, Switzerland.)

2.1.8 Concrete Force Extraction and Pulloff Testers

Pulloff testers measure the strength of concrete and masonry surfaces, interfaces, and anchors. A typical tester pulls on an anchored surface connection with a load that increases until the connection fails or reaches a specified level (Figure 2.10) [26]. An electronic load cell measures and records the peak load. A common configuration uses a manually actuated mechanical screw drive to provide the pullout force.

2.1.9 Permeability Testers

Permeability quantifies the ease with which fluids flow through porous materials. Permeability affects the long-term endurance and performance of structural materials. Testing for permeability uses chemical diffusion, pressure, or other means, such as electric fields to drive the fluid through the material, and then measures the corresponding amount of flow [28]. Figure 2.11 is a schematic of a Torrent permeability tester that uses pressure as the driving force. The *Rapid Chloride Permeability Test* uses electric fields to drive chloride ions through samples of permeable concrete [29,30].

2.1.10 Stress Meters and Pressure Gages

Direct measurements of stress and pressure are difficult, if not impossible. Indirect measurements based on stress-induced deformations are standard. Field representations of mechanical stress use a symmetric second-order tensor with six independent components. Full characterization of the stress tensor at a point requires measuring six independent scalars. Measuring the individual stress tensor components involves either analog transduction devices to isolate the effect of the individual components on individual sensor channels, or multichannel measurements with subsequent signal processing to isolate the individual components.

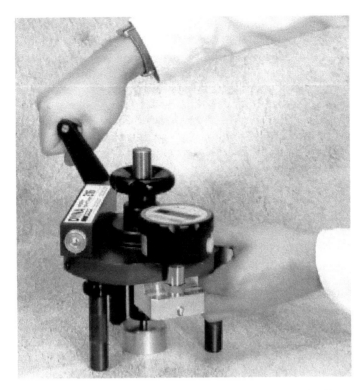

FIGURE 2.10   Concrete pulloff tester. (Photograph courtesy of Proceq SA, Switzerland.)

The use of embedded stress transducers requires careful mechanical design. A key issue is that an embedded sensor may significantly alter the stress field in a region near to the sensor and cause spurious readings. In 1936, Carlson patented a widely used stress meter [31,32]. The Carlson meter uses a stress-field-matching design based on a relatively large plate and a thin mercury layer to isolate the stress component in question. A wire-driven resistance-based Wheatstone bridge measures the deformations. Typical uses of the Carlson stress gage lie in the monitoring of stresses in large structures, such as dams. The notion of matching

FIGURE 2.11   Torrent permeability tester mechanics. (Photograph courtesy of Proceq SA, Switzerland.)

the geometry and stiffness of an embedded stress sensor has reemerged in the context of piezoelectric sensing elements [33]. It is possible to measure the stress in tight locations with embedded thin-film sensors. An example by Anderson et al. used embedded PVDF thin films to measure dynamic stress field distributions in adhesive joints [34]. Javidinejad and Joshi demonstrated embeddable MEMS pressure and temperature sensors that could survive the rigors of autoclave composites manufacturing [35].

The Gloetzel pressure gage measures pressure at remote sites, such as the bottom of a dam [5]. The measurand and reference pressure are applied to opposite sides of a flexure-mounted diaphragm (Figure 2.12). Raising the reference pressure until it exceeds the measurand pressure causes a bypass valve to open and to indicate the level of the measurand pressure. The Gloetzel gage is stable, long-term reliable, and relatively inexpensive, but requires installing and maintaining fluid conductors. The Osterberg Cell or O-Cell™ is a similar device for measuring the strength of buried shafts, piers, and other geotechnical structures [36,37].

2.1.11 Mine Posts and Mine Flags

Underground mining is a hazardous undertaking. A major danger is unexpected mine roof collapses. The sagging and downward movement of the roof is often a reliable precursor of mine roof collapses. Several methods and instruments can measure mine roof deflections and can serve as safety monitoring instruments. A traditional method is the jamming of a wooden post between the mine roof and floor. When the roof sags, the roof loads transfer to the post and the post produces cracking sounds (audible acoustic emissions, AEs) as a warning of imminent collapse. Modern versions of this concept are known as *tell tales*. The tell tales bolt into holes drilled into the roof of the mine. The tell tales span stable and unstable portions of the rock (about 1 m). The tell tale measures the differential settlement due to the motion of unstable rocks and gives a warning when the settlement exceeds a preset threshold [38].

2.1.12 Peak Strain Sensors and Fatigue Fuses

Recording and analyzing load histories can provide information for estimating the amount of damage and fatigue in a structure. A historical datum of interest is the maximum strain

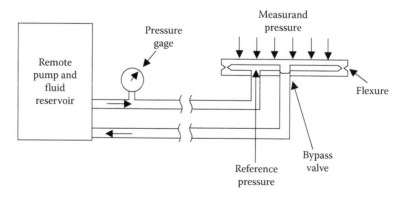

FIGURE 2.12 Operating principle of a Gloetzel pressure gage. (Adapted from US Army Corps of Engineers. 1980. Engineering and design instrumentation for concrete structures. EM 1110-2-4300, Department of the Army, Office of the Chief of Engineers, Washington, DC.)

experienced by a structural member, that is, the *peak strain*. One method of transducing peak strain uses the strain to induce a permanent phase change in the material. Specific alloys of austenitic stainless steel [transformation-induced plasticity (TRIP) steel] change from a nonmagnetic to a ferromagnetic form as a function of plastic deformation. Magnetic susceptibility measurements are a convenient method of measuring peak strain with these materials [39]. Surface-attached transducer elements and structural elements, such as concrete reinforcing bars made of TRIP steel, provide potentially useful peak strain sensing opportunities [40,41]. The state of magnetization of many ordinary steel alloys reflects the stress and external magnetic history [42]. An alternative is to use a miniature ratchet mechanism [43].

Another load-history item of interest is a measure of the cyclic fatigue loads experienced by a structure. One technique is to mount custom-shaped sacrificial coupons to the structure so as to undergo the same service loading as the structure [44]. The coupon geometry includes specialized stress concentration notches (Figure 2.13). The intent is for the notches to induce fatigue in the coupon more quickly than the base structure [45]. Visual or automated techniques can detect the fatigue cracking in the coupon before similar cracks develop in the structure. Fujimoto et al. compared conductive film sensors, conductive paint sensors, plastic optical fiber, glass optical fiber, and carbon fiber bundles as possible fatigue crack sensing mechanisms in ship structures [46]. The various sensors worked well when the cracks propagated from the structure and through the sensing elements. Inamura et al. developed miniaturized prepackaged fatigue fuses for placement on structural fatigue hot spots [47].

2.1.13 Scour Meters

The hydraulic action of flowing water, often during floods, can *scour* and remove the soil supporting bridge piers and foundations. Scour is the principal cause of many bridge

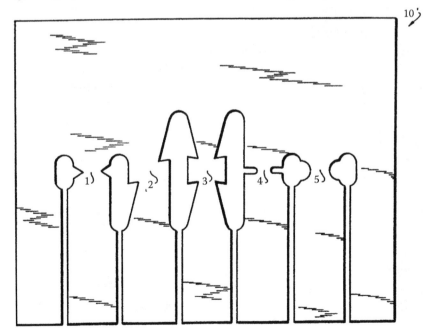

FIGURE 2.13 Metal fatigue detector with specialized notches. (Data from de la Veaux RC. 1993. Metal fatigue detector. US Patent 5,237,875.)

collapses. A difficulty with scour monitoring is that eroded regions of the riverbed often refill with silt. The silt does not provide much structural support, easily rescours in repeat flood events, but nonetheless confounds simple observational-type measurements. Harsh river environments place extreme demands on the design of successful scour meters. One approach is to bury mechanical sensors that float or sink during scour events [48]. *Float-outs* are buoyant riverbed-buried plastic objects. Recovery of a float-out indicates the occurrence of a scour event. More sophisticated scour transducers use devices such as magnetic collars that slide along tubes and sink as the riverbed scours, river-flow-sensitive mini-cantilevers, and toughened multipoint pressure sensor arrays [49–51]. Measuring pier tilt with tiltmeters or clinometers can also provide useful information concerning scour effects [52].

2.2 Electrical Sensors

Electrical sensors use electrical and/or EM phenomena for transduction. While most electrical sensors ultimately rely on the measurement of current, the development of miniaturized high-impedance circuits enables measuring a variety of quantities, such as charge, capacitance, inductance, resistance, and voltage, without regard to the details of the measurement circuit.

Many electronic sensors rely on approximations that coalesce complex 3-D EM behaviors into equivalent scalar lumped circuit elements, that is, resistance (R), capacitance (C), and inductance (L). Ohm's law is the lumped relation defining resistance in terms of voltage (V) and current (i):

$$V = iR. \tag{2.12}$$

The lumped voltage–current relation for a capacitor is

$$\frac{dV}{dt} = \frac{i}{C}. \tag{2.13}$$

The lumped voltage–current relation for an inductor is

$$V = L\frac{di}{dt}. \tag{2.14}$$

The time-varying nature of Equations 2.13 and 2.14 require using time-varying signals for capacitance and inductance measurements. In the context of sinusoidal oscillations, the impedance, Z, forms a combined representation of the action of resistance, capacitance, and inductance. For example, monitoring the response to high-frequency EM oscillations can measure the impedance of granular materials, which correlates with moisture content [53]. Figure 2.14 shows a microwave moisture sensor that helps to control the quality of hot-mix asphalt [54].

There are situations, such as the operation of circuits at GHz frequencies, when lumped scalar circuit models provide an inadequate description. Such situations may require using a 3-D model at the expense of significantly increased complexity. In other cases, simpler models that make use of lumped scalar elements with frequency-dependent impedance properties work well in lieu of full 3-D EM analyses.

FIGURE 2.14 Microwave moisture probe used in hot-mix asphalt quality control. (From West RC and Turner P. 2006. Evaluation of automated process control testing for hot mix asphalt production. Paper no. 06-2072, 86th Annual Meeting, TRB, Washington, DC. With permission.)

2.2.1 Electrical Resistance Sensors

Electrical resistance sensors transduce changes in the resistance of a sensor element. Resistance is a macroscopic quantity that depends on the material resistivity, ρ, and the current flow geometry. Resistance is generally relatively easy to measure and varies with a variety of physical effects. An example is to consider the resistance of a wire, R, with a uniform cross section, homogeneous properties, and a uniform pattern of electric current flow, as in Figure 2.15. For the geometry,

$$R = \frac{\rho L}{A}. \tag{2.15}$$

The change in resistance, dR, due to small changes in the resistivity, length, and cross-sectional area ($d\rho$, dL, and dA), is approximately [55]

$$dR = \frac{L}{A}d\rho + \frac{\rho}{A}dL - \frac{\rho L}{A^2}dA. \tag{2.16}$$

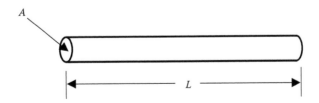

FIGURE 2.15 Long wire geometry.

For linear elastic isotropic materials, with Poisson's ratio, v, and an area strain, dA/A, the changes in resistance in terms of the axial strain and material resistivity change are

$$\varepsilon_x = \frac{dL}{L}, \tag{2.17}$$

$$\frac{dA}{A} \approx -2v\varepsilon_x, \tag{2.18}$$

$$\frac{dR}{R} = \frac{d\rho}{\rho} + (1 + 2v)\varepsilon_x. \tag{2.19}$$

The resistivity change may depend linearly on change in temperature T, axial strain ε_x, and transverse strain, ε_y. A linearized representation is

$$d\rho = K_T\, dT + K_x\varepsilon_x + K_y\varepsilon_y \tag{2.20}$$

with K_T, K_x, and K_y being constants. Combining Equations 2.15, 2.19, and 2.20 yields

$$dR = \frac{L}{A}\{K_T\, dT + [(K_x - vK_y) + \rho(1 + 2v)]\varepsilon_x\}. \tag{2.21}$$

Commonly used sensor materials include specialized alloys, such as Constantan and doped semiconductors [1,55]. Conductive piezoresistive polymers, such as *poly 3,4-ethylenedioxythiophene*, may prove to be useful for strain gaging due to favorable properties that enable micron-scale (or smaller) patterning with microelectronic manufacturing techniques [56]. It is possible to fabricate materials out of carbon nanotubes with custom-resistive strain sensing capabilities [57,58].

Li et al. found that shape memory alloys (SMAs), such as nitinol, undergo large changes in resistivity that correspond to the large phase-change shape deformations characteristic of the materials [59–61]. Although SMAs are common in actuation applications, it is also possible to use SMAs as a combination of high-performance structural elements and sensing materials [62,63]. Embeddable humidity and moisture sensors can make use of the water-sensitive resistance or capacitance of certain polymers (Figure 2.16) [64,65]. Similar to SMAs, electro-active polymers exhibit large changes in electrical properties with respect to deformation [66].

2.2.2 Resistive Strain Gages

Strain is a local measure of material deformation. A complete field characterization capable of describing length, angular, and volumetric changes uses a second-order symmetric strain tensor [67]. The diagonal terms of the tensor correspond to axial strains and length changes. The off-diagonal terms correspond to shear strains and angular changes. The strain tensor may be nonlinear or linear or depending on whether the deformation gradients are sufficiently large for product terms to be significant.

The surface attachment geometry of most strain gages restricts the measurements to only one or a few tensor components. The resistive strain gage is common. Traditional variants transduce the longitudinal strain of an elongated wire or foil conductor into a resistance change using the effects described in Equation 2.21 [68]. Since resistance changes are small for metallic resistors operating within elastic deformation limits, it is usually necessary

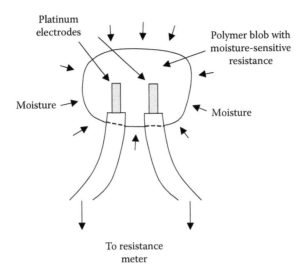

FIGURE 2.16 Embeddable polymer-based electrical resistance moisture sensor. (Adapted from McGovern ST et al. 2005. *Advanced Sensor Technologies for Nondestructive Evaluation and Structural Health Monitoring*, SPIE 5770.)

to amplify the resistance-change signal by both mechanical and electrical means. Typical mechanical amplification techniques include extending the effective length of the resistor element by folding or wrapping, placing the gage at high-strain locations, and/or using mechanical structures that specifically isolate and amplify the strain. Early versions used a wrapped-wire configuration for mechanical strain amplification. Since the geometry of wrapped wire is awkward for a point sensor, modern versions replace the wire with a multibend foil equivalent (Figure 2.17). Mass production of foil strain gages at low cost with a wide variety of micropatterned foil geometries is routine [55].

Semiconductor strain gages are more sensitive to strain than foil gages. The signals tend to not require as much amplification, but are highly sensitive to confounding effects, such as transverse strains and temperature variations [1]. It is possible to miniaturize semiconductor gages and/or build them directly into MEMS devices [69]. An example

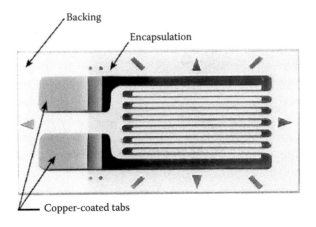

FIGURE 2.17 Resistive strain gage. (Courtesy of Vishay Intertechnology, Inc.)

is the *dual bond stress and temperature* (DBST) measurement sensor that can detect debonding of materials, such as solid fuel rocket fuels [70]. The DBST uses semiconductor strain gages on a thin ring-supported plate for transduction.

A typical resistive strain gage application uses axial strain in the structure to induce strain in the gage by the transfer of shear stress across a bonded mounting interface. The induced strain causes a small, but measurable, resistance change in the gage. Direct calibration is the typical procedure to compensate for the *shear lag* difference between the strain on the surface of the structure and the gage [71]. Wheatstone bridge circuits combined with signal amplifiers are common methods of measuring the resistance changes. The Wheatstone bridge circuit acts as an analog computer to add and subtract resistances (Figure 2.18). It is possible to design a Wheatstone bridge configuration to be sensitive to certain effects, such as bending, and insensitive to other effects, such as temperature shifts.

The Wheatstone bridge applies an excitation voltage, V_{AC}, across two opposing corners (A and C) and reads a signal voltage, V_{BD}, from the other two opposing corners (B and D). If the circuit contains only resistive elements

$$V_{BD} = V_{AC} \left[\frac{R_1 R_3 - R_2 R_4}{(R_1 + R_2)(R_3 + R_4)} \right]. \tag{2.22}$$

The bridge is balanced when the output voltage $V_{BD} = 0$, that is,

$$\frac{R_1}{R_4} = \frac{R_2}{R_3}. \tag{2.23}$$

When $R_1, R_2, R_3, R_4,$ and V_{BD} undergo small changes, that is,

$$R_1 \rightarrow R_1 + \Delta R_1,$$
$$R_2 \rightarrow R_2 + \Delta R_2,$$
$$R_3 \rightarrow R_3 + \Delta R_3, \tag{2.24}$$
$$R_4 \rightarrow R_4 + \Delta R_4,$$
$$V_{BD} \rightarrow V_{BD} + \Delta V_{BD},$$

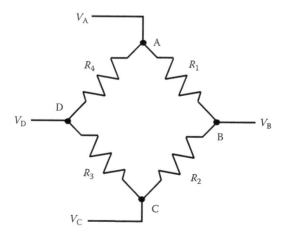

FIGURE 2.18 Wheatstone bridge circuit.

the following approximate relation holds:

$$\Delta V_{BD} = \left[\frac{\Delta R_1 R_3 + R_1 \Delta R_3 - \Delta R_2 R_4 - R_2 \Delta R_4}{(R_1 + R_2)(R_3 + R_4)} \right]. \tag{2.25}$$

If the nominal strain gage resistances R_1, R_2, R_3, and R_4 are equal,

$$\Delta V_{BD} = G(\Delta R_1 - \Delta R_2 + \Delta R_3 - \Delta R_4), \tag{2.26}$$

where G is a constant that incorporates strain gage properties, along with amplifier and data acquisition gains [55].

The voltage versus resistance change relation of Equation 2.26 can serve as a guide in the design of strain gage sensor systems. For example, Figure 2.19 shows a strain gage bridge configuration that is sensitive to bending in a beam, but insensitive to axial and torsional deformations, and resistance changes due to homogeneous temperature shifts. The resistance change for each strain gage is

$$\begin{aligned}
\Delta R_1 &= \Delta R_B + \Delta R_A + \Delta R_T + \Delta R_\theta, \\
\Delta R_2 &= -\Delta R_B + \Delta R_A + \Delta R_T + \Delta R_\theta, \\
\Delta R_3 &= \Delta R_B + \Delta R_A + \Delta R_T + \Delta R_\theta, \\
\Delta R_4 &= -\Delta R_B + \Delta R_A + \Delta R_T + \Delta R_\theta,
\end{aligned} \tag{2.27}$$

where ΔR_B, ΔR_A, ΔR_T, and ΔR_θ are the changes in gage resistance due to bending, axial deformation, temperature, and torsion, respectively. For this case, the bridge voltage output

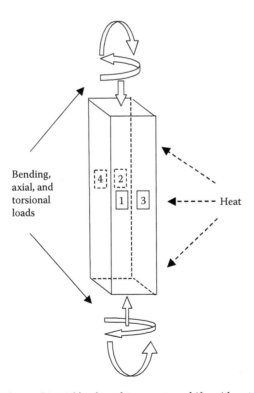

Bending, axial, and torsional loads

Heat

FIGURE 2.19 Beam subjected to multi-axial loads and temperature shifts with a strain gage Wheatstone bridge.

sums to

$$V_{BD} = 4G\,\Delta R_B. \tag{2.28}$$

A principal assumption in the above derivation is that all of the strain gages experience the same temperature change. Unanticipated nonhomogeneous temperature gradients can cause spurious readings, especially with semiconductor strain gages. Using gages with temperature-induced resistance changes that compensate for the thermal expansion of the base material can reduce the effects of temperature gradients. Custom temperature-compensating strain gages are available for measuring the strain in many common engineering materials.

For cases where it is impractical to use four strain gages in a full Wheatstone bridge, *quarter bridge*, and *half bridge*, variants replace the active strain gages with dummy resistors or gages. Using one active and three dummy gages forms a quarter bridge. Using two active and two dummy gages forms a half bridge. Interpreting the bridge output of quarter and half bridges requires suitable modifications of the linearized bridge output (Equation 2.26). A key detail in the use of quarter and half bridges is to prevent the dummy resistors from introducing spurious content into the readings, such as through the use of precision resistors. Typical fabrication techniques for precision resistors include using wire wraps of alloys with minimal temperature sensitivity. A disadvantage of wire-wrapped precision resistors is the difficulty with miniaturization. Another option is to use unloaded dummy strain gages that compensate for temperature shifts. As early as 1959, Murray and Stein described *chevron* Wheatstone bridges that have multiple arms [68]. These chevron bridges enable interesting possibilities for creating more complex analog computers for strain gage measurements, including load cells, but are relatively rare.

Mounting strain gages onto structures is a labor and skill-intensive process [72]. The technology is mature. Vendors offer a wide variety of adhesives and bonding techniques for different applications. Yost and Assis report that the installation of strain gages on a steel bridge takes about 5 min per gage [73]. Strain gages embed into composites and reinforced concrete structures with relative ease (Figure 2.20) [74,75].

FIGURE 2.20 Embedment of vibrating wire, reinforcement attached resistive and embeddable resistive strain gages in bridge deck prior to concrete placement. (Photograph courtesy of Cuelho E et al. 2006. Paper no. 06-2302, 86th Annual Meeting, TRB, Washington, DC.)

The lack of long-term bond stability and difficulties with corrosion of electrical connections renders foil strain gages to be somewhat unreliable for many applications that extend beyond a year or two. Moisture attack and debonding are major sources of the problems. The skilled application of the gages and the use of suitable protective coatings can extend strain gage lifetimes indefinitely. For example, Prine and Socie found that foil strain gages could survive in the harsh environment of a ferry hull subjected to winter weather for at least a year [76]. Ware et al. report that 60% of foil strain gages lasted 10–13 years in a fleet-wide aircraft SHM effort [77]. Joas used foil strain gages over a 20-year period for monitoring the stress in pipelines [78].

Alternate strain gage mounting approaches use prepackaged weldable and clamp-on mounts. Weldable strain gages are prepackaged units with metal-plate-mounted prebonded and hermetically sealed resistive strain gages. Tack welding the metal plate to the structure under test makes for a quick and reliable strain gage mount. Tack welding occasionally raises concerns about the introduction of stress risers in the base structure. Clamp-on strain gages are a similar less permanent mounting technique. Clamp-on prebonded gages tend to use longer plates, often with a Z-pattern geometry (Figure 2.21). The rationale for the Z-pattern geometry is that the central region is in approximately pure shear. Placing four gages in a Wheatstone bridge configuration rotated 45° relative to the principal axes of the Z-plate creates a sensor that is sensitive to longitudinal strains and insensitive to other strain

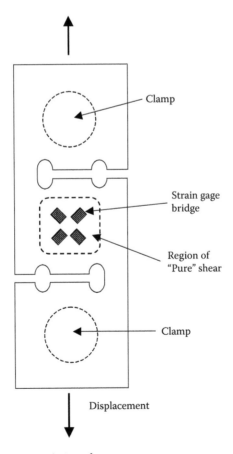

FIGURE 2.21 Z-geometry clamp-on strain transducer.

components. Kim et al. and Nam et al. developed specialized box and enclosure techniques for installing strain gages in concrete bridge decks [79,80].

The amplitude of the signal from a resistive-strain-gage Wheatstone bridge usually is on the order of millivolts, with microvolt resolutions required for many applications. Such signal levels are usually too small to be measured with standard analog-to-digital converters (ADCs), and suffer from noise contamination when transmitted in a raw analog form over any appreciable distance. Amplification and conditioning before measurement and/or transmission can help to ameliorate noise contamination. Several amplification techniques are possible. One uses specialized strain gage and signal conditioning amplifiers. Commercially available strain gage amplifiers are quite capable, with adjustable gains, offsets, and filtering and other features, but may be too bulky and power hungry for many field applications. Alternatives are to use smaller, but less capable, operational amplifiers, packaged modules, or integrated circuits. Adding an adjustable potentiometer, as shown in Figure 2.22, is a simple method for removing a direct current (DC) offset from an unbalanced Wheatstone bridge.

An easy method for detecting potential problems in a Wheatstone bridge built with four nominally equal resistors is to disconnect the bridge circuit from the amplifier and to measure the resistance across the diagonals (Figure 2.22, A to C or B to D). The diagonal resistance will equal the nominal resistance of an individual arm when all of the contacts and gages are intact and properly wired. If the diagonal resistance differs from that of the nominal gage resistance, the circuit has a fault.

A *star bridge circuit* can establish a zero reference in a strain gage circuit (Figure 2.23) [81]. The star bridge mimics the electrical loads of a strain gage Wheatstone bridge, but produces a zero output signal. This zeroing circuit is sufficiently stable to establish a reliable reference between different data acquisition systems.

It is often desirable to minimize strain gage circuit power consumption. One reason is the tight power budgets of battery and remotely powered instruments. Another reason is that the passage of electric currents through strain gages causes Joule heating and can lead to thermal drift problems, particularly when the substrate is a poor thermal conductor and variable convective fluid heat transfer effects are at play. Some techniques to reduce power consumption in strain gage circuits are as follows: (1) *Use strain gages with high nominal resistances*: Increasing the resistance of the strain gages reduces the current flow

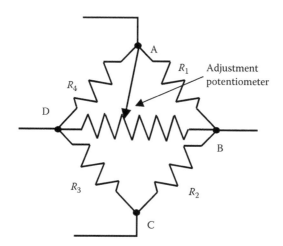

FIGURE 2.22 Wheatstone bridge circuit with potentiometer for offset compensation.

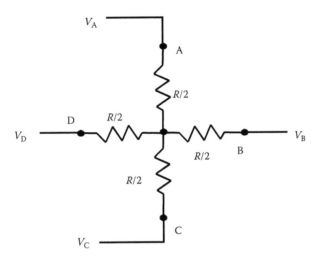

FIGURE 2.23 Star bridge circuit. (Redrawn from Wright C. 1993. *Personal Eng Instrum News,* 10(10), October.)

and power consumption for a given excitation voltage and sensitivity. Manufacturing the strain gages with thinner metal lines can increase the resistance while retaining the same strain sensitivity (gage factor). Typical strain gages have nominal resistances of 120, 350, and 1000 Ω. A Wheatstone bridge built with 1000 Ω strain gages uses only 8.3% of the power as that of an equivalent 120 Ω gage bridge. Higher resistance strain gages also allow for the use of larger excitation voltages while reducing thermal problems due to Joule heating. (2) *Cycle the power*: If the requirements for the strain measurements are such that the sampling interval is fairly long (on the order of 0.1 s or more), then it is practical to cycle the excitation voltage on and off. Measurements need be taken only while the circuit is on. It is relatively easy to achieve duty cycles of 1 ms on and 99 ms off and a resulting 99% reduction in power consumption without incurring voltage transient problems [82]. (3) *Use constant current excitation and measure the voltage across a single strain gage*: Constant current excitation is particularly useful in miniature integrated devices where semiconductor-based electronics are used. The *Kelvin 4-point probe,* Figure 2.24, is a standard circuit for these measurements [83]. Graichen and Bergmann used such a circuit with a cyclic duty cycle in an artificial hip wireless strain telemetry system [84]. Constant current circuits using more than one gage in a differential configuration are also possible [85]. (4) *Use a high-resolution ADC*: 24-bit ADCs can measure the voltage directly from a Wheatstone bridge circuit with sufficient resolution to circumvent the need for secondary amplification and signal processing. The 24-bit ADCs are somewhat limited in sampling frequency. This is a minimal constraint for many structural applications.

2.2.3   2-D and 3-D Strain Gage Rosettes and Spiders

Strain gage rosettes and spiders are arrays of strain gages with configurations that measure individual components of strain tensors. In 2-D, a symmetric 2×2 matrix (tensor) with three independent components describes the strain field at a point as

$$[\varepsilon] = \begin{bmatrix} \varepsilon_{xx} & \varepsilon_{xy} \\ \varepsilon_{xy} & \varepsilon_{yy} \end{bmatrix}. \tag{2.29}$$

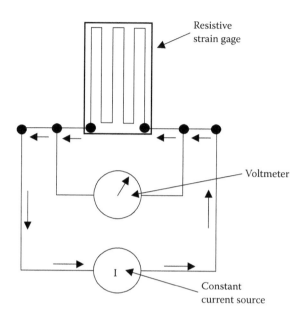

FIGURE 2.24   Kelvin 4-point circuit for measurement of resistance with constant current.

In 3-D, a symmetric 3×3 matrix (tensor) with six independent components describes strain as

$$[\varepsilon] = \begin{bmatrix} \varepsilon_{xx} & \varepsilon_{xy} & \varepsilon_{zx} \\ \varepsilon_{xy} & \varepsilon_{yy} & \varepsilon_{yz} \\ \varepsilon_{zx} & \varepsilon_{yz} & \varepsilon_{zz} \end{bmatrix}. \tag{2.30}$$

The diagonal terms of the strain tensor correspond to axial or longitudinal strains, for example,

$$\varepsilon_{xx} = \frac{\partial u}{\partial x} \approx \frac{\Delta L}{L}, \tag{2.31}$$

where u is the displacement in the x direction, ΔL is the change in length of a member, and L is the original length. The off-diagonal terms represent the shear strain,

$$\varepsilon_{xy} = \frac{1}{2}\left(\frac{\partial u}{\partial y} + \frac{\partial v}{\partial x}\right) = \frac{1}{2}\gamma_{xy}, \tag{2.32}$$

where u and v are the displacements in the x and y directions, respectively. γ_{xy} is the *engineering shear strain*, which represents angle changes from perpendicular due to deformation.

    The design of strain gage rosettes exploits the notion that the measurement of a 2-D strain tensor with three components requires three independent measurements (three gages). Similarly a 3-D tensor requires six independent measurements (six gages). A straightforward algebraic manipulation of the strain gage readings often yields the strain tensor components. A required feature in the analysis is a knowledge of the strain gradients. A uniform strain gradient-free state simplifies the analysis. A common 2-D rosette design uses three gages attached at angles of 0°, 45°, and 90° (Figure 2.25). The unit vectors, $\{u_i\}$, for the gage

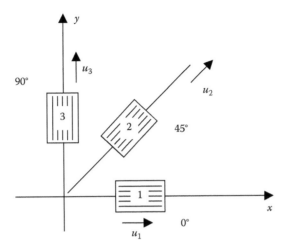

FIGURE 2.25 0°, 45°, and 90° strain gage rosette.

directions are

$$\{u_1\} = \begin{bmatrix} 1 \\ 0 \end{bmatrix}, \quad \{u_2\} = \begin{bmatrix} \sqrt{2}/2 \\ \sqrt{2}/2 \end{bmatrix}, \quad \{u_3\} = \begin{bmatrix} 0 \\ 1 \end{bmatrix}. \tag{2.33}$$

The axial strains in the gages are

$$
\begin{aligned}
\varepsilon_1 &= \{u_1\}^T[\varepsilon]\{u_1\} = \varepsilon_{xx}, \\
\varepsilon_2 &= \{u_2\}^T[\varepsilon]\{u_2\} = \frac{1}{2}\varepsilon_{xx} + \varepsilon_{xy} + \frac{1}{2}\varepsilon_{yy}, \\
\varepsilon_3 &= \{u_3\}^T[\varepsilon]\{u_3\} = \varepsilon_{yy}.
\end{aligned}
\tag{2.34}
$$

Solving the simultaneous linear equations in Equation 2.34 produces the strain tensor components ε_{xx}, ε_{xy}, and ε_{yy}. The above formulation becomes more complicated when using gages with transverse and/or shear sensitivity, as is often the case in semiconductor-strain-gage-based MEMS sensors. For strain gages that are sensitive to axial strains and insensitive to shear strains, the rosette design uses the relation between the axial strain in a particular direction, ε_n, defined by a double scalar product of a unit vector $\{u_n\}$, and the strain tensor [67].

$$\varepsilon_n = \{u_n\}^T[\varepsilon]\{u_n\}. \tag{2.35}$$

Stress relaxation is a semidestructive method of determining intrinsic stress. The method removes a small amount of material and observes the resulting deformations due to the relaxation of the intrinsic stress. Hole drilling may be the most common of these techniques [86]. The method is to drill into the material at a predetermined point central to a triangular three-gage rosette. The removal of the material by drilling causes the stress pattern around the hole to change in a predictable manner that allows for the estimation of the prestress. Sabaté used a focused ion beam to mill micron-scale slots in thin films to determine intrinsic stress distributions [87]. A similar approach known as the *doorstopper* technique can determine the prestress on the interior of concrete and rock structures [88].

Measuring components of a 3-D strain tensor is a straightforward, extension of the 2-D strain gage rosette method. 3-D rosettes are sometimes termed *spiders*. Perhaps the most common application of strain spiders is in geotechnical applications [5].

The above discussion used the assumption that the strain field corresponds to the small gradient linear version of the strain tensor. If the deformation gradients are large enough to require nonlinear strain tensor representations or the deformations are plastic, then it may still be possible to use rosettes to calculate the strain field, provided that the nonlinear and inelastic behaviors are reasonably well understood. For example, Gran and Seaman used piezoresistive strain gages in a three-gage array to measure plastic deformation [89]. The method requires that the gages be aligned with the principal directions of the plastic deformation.

When deformations are large, particle-tracking methods may be a convenient strain field mapping option. It is the relative motion of the particles that corresponds to strains, not gross rigid body motions. Estimating the components of a 2-D strain tensor requires three measurements of relative motion and the tracking of three noncolinear particles. Similarly, estimating the components of a 3-D strain tensor requires six measurements of relative motion and the tracking of four noncoplanar particles. The analysis proceeds with Equation 2.35 in a manner similar to the strain gage rosette. Kusaka used such an approach for the real-time tracking of crack growth in concrete [90]. Strain data from particle image tracking indicated the location of cracks. The principal strains indicated the direction of the crack, and whether it was an opening crack.

2.2.4   Load Cells

Load cells measure forces and other structural loads. The design of most load cells is to measure a single component of force and to be insensitive to force components in other directions and to bending moments. These designs are executed with a variety of specialized, and often proprietary, internal geometries [91]. Resistive and piezoelectric sensor elements are common in load cell transducers. Measuring transient loads requires using transducers with high-frequency sensitivity, such as quartz piezoelectric versions.

Successful use of load cells requires careful consideration of the mounting configuration. Statically determinate configurations are desirable, although not always achievable. A standard construction monitoring application is a single-axis load cell with a large center hole that permits the concentric passage and loading of a rod with a jack (Figure 2.26). Measuring with load cells when the load reverses requires preloaded attachment details to avoid issues of unwanted nonlinear hysteresis effects due to backlash. Multi-axis load cells are available, but are somewhat uncommon in structural sensing applications [92].

A pressure gage is a variant of the load cell. Pressure gages come in many forms and use a variety of transduction mechanisms. A common design transduces pressure by measuring the flexural strain due to a pressure differential across a diaphragm with a foil strain gage in a specialized radial gage pattern. MEMS-based pressure transducers are common [1]. Some of these can measure shear stresses on surfaces for applications including wind tunnel model testing [93].

Issues of dynamic range pose a bit of a challenge in the measurement of pressure loads on structures during windstorms. Large low-frequency shifts in barometric pressure during a storm aggravate measures of high-frequency pressure fluctuations. Differential pressure transducers have difficulty in establishing reliable reference pressures. Absolute pressure

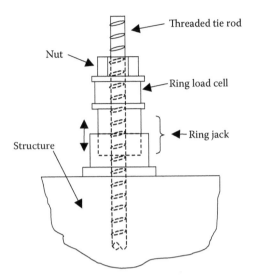

FIGURE 2.26 Statically determinate ring load cell configuration used to monitor jacking loads on tie rods.

transducers may lack the resolution to measure Bernoulli-type pressures along with large barometric pressure changes.

Exploiting measurand-induced loading on a load cell enables building a variety of specialized sensors. As an example, Saafi and Romine developed a MEMS-based humidity transducer. The transduction mechanism used a material that swells with moisture [94].

2.2.5 Resistive Temperature Gages

Resistive thermometers harness temperature-induced resistance changes for transduction. Resistance temperature *detectors* (RTDs) are metallic variants. In metals, the resistance increases linearly with temperature over fairly wide temperature ranges [95]. The long-term stability of platinum makes it a preferred metal for RTDs. A disadvantage of platinum is the small coefficient of resistance change with temperature (0.0039/°C). Measuring the resistance change in an RTD usually requires specialized circuits, such as a Wheatstone bridge or a Kelvin 4-point probe, and special considerations to compensate for stray capacitances [96]. *Thermistors* are semiconductor-based RTDs. The temperature versus resistivity behavior of semiconductors differs from that of metals. The resistivity of semiconductors drops as temperature increases. An advantage of thermistors is that the temperature sensitivity is sufficiently large to minimize the need for specialized resistance measurement circuits. Packaging thermistors in small containers with a small thermal inertia enables a fairly rapid dynamic response. Thermistors tend to be inexpensive. Some disadvantages of thermistors are as follows: (1) the temperature ranges for measurement are fairly limited; (2) the measuring process can induce confounding temperature changes; and (3) microcracking-type failures are common. Nonetheless, through the use of proper techniques, thermistors can be quite effective. As an example, Hwang and Lee conducted long-term measurements (in excess of two years) of the interior temperature of a concrete bridge with a multithermistor sensor array that can measure internal gradients [97].

2.2.6 Thermoelectric Sensors

Joining two dissimilar materials produces a very repeatable temperature-dependent voltage difference and is the basis of stable high precision and accuracy temperature sensors, that is, thermocouples. A disadvantage of thermocouples is that the output relation between temperature and voltage is small and nonlinear. Measurement requires using microvolt-resolution data acquisition systems and nonlinear lookup tables. Standardized thermocouple junctions go by the letters B, E, J, K, N, R, S, and T. These thermocouples come with published polynomial coefficients for calculating the nonlinear voltage–temperature relations [98].

2.2.7 Capacitive Sensors

A wide variety of sensing applications use measurand-induced capacitance changes [99]. In structural measurements, capacitance change provides the ability to measure small displacements inside transducers, such as accelerometers, and to measure material property changes. Capacitance measurements can be highly sensitive. An example is the measurement of nanoscale motions in ultrahigh precision machines [100]. Capacitive sensors scale well with reduced size. They are fairly common in MEMS sensors. Examples are accelerometers and AE transducers [101,102].

The physical mechanism of a capacitor is that the application of a voltage across the terminal causes negative and positive charge concentrations to accumulate in separate locations. The capacitance, C, quantifies the relation between the applied voltage, V, and charge buildup, q,

$$V = \frac{q}{C}. \tag{2.36}$$

It should be noted that capacitance appears as a reciprocal in Equation 2.36, which is unlike similar relations for resistance, R, and inductance, L, which are linear terms.

Capacitance depends on both dielectric material properties and geometry. Practical capacitive sensor design uses specialized circuits and gap geometries. A simple configuration is the parallel plate capacitor as shown in Figure 2.27. If the gap dimension, x, is considerably smaller than the lateral dimensions of the plate, then the capacitance is approximately

$$C = \frac{\varepsilon_0 \varepsilon_r A}{x}. \tag{2.37}$$

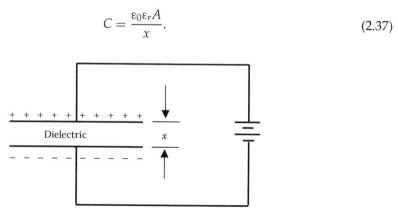

FIGURE 2.27 Parallel plate capacitor.

ε_0 is the electrical permittivity of vacuum or air, where $\varepsilon_0 = 8.854 \times 10^{-12}$ F/m. ε_r is the relative electrical permittivity of the material placed in the gap. $\varepsilon_r > 1$ for virtually all passive materials. A is the area of the plate surface.

A standard capacitance measurement technique applies a harmonic oscillating voltage, $V(t) = V_0 \sin(2\pi ft)$, across the capacitor and measures the resulting current, $i(t) = I_0 \cos(2\pi ft)$, where

$$C = \frac{I_0}{2\pi f V_0}. \tag{2.38}$$

Capacitors are integral elements of many analog electronic oscillator circuits. Measurand-induced capacitance changes transduce the oscillatory behaviors of these circuits. Wireless inductive pickups can readily detect the effect of capacitance changes. Examples include wireless humidity and resistance sensors [103,104]. If all that is required of a measurement is the transduction of a single bit of information, such as the presence of corrosion or cracking, then a frequency-shifting sensor with a sensing capacitor in parallel with another capacitor and in series with an inductor is an option [105,106]. The presence of the measurand, such as a crack, breaks the circuit continuity to the sensing capacitor and causes an easily measured resonant frequency shift (see Chapter 4 for more details).

2.2.8 Magnetic Inductance and Reluctance-Based Sensors

The LVDT, also known as the *Linear Variable Displacement Transducer*, is a common high-performance displacement transducer [5,107]. LVDTs are capable of submicron resolution, are accurate ($\pm 0.5\%$), and can be long-term stable. The LVDT transduces displacement with a piston-like motion of a core cylinder inside of a hollow cylinder. The hollow cylinder houses three cylindrical coils of wire as shown in Figure 2.28. The center coil is the primary coil. The upper and lower coils, that is, the secondary coils, are nominally identical, but wrapped in opposite directions. The core cylinder uses soft magnetic materials (low magnetic hysteresis but mechanically rigid) that partially spans all three coils and forms two transformer circuits, each with the same primary coil, but with a different secondary coil. Driving an oscillating current through the primary coil induces voltages and currents in the secondary coils. Displacing the core alters the relative amount of inductive coupling between the primary and two secondary coils. The induced secondary coil voltages are

$$V_A = L_A \sin(2\pi ft - \phi_A),$$
$$V_B = L_B \sin(2\pi ft - \phi_B). \tag{2.39}$$

L_A and L_B represent the core-position-dependent mutual inductances of the primary and secondary coils. For a good approximation, L_A and L_B are proportional to the core displacement. If the secondary coils are nearly identical, then

$$\phi_A \approx \phi_B = \phi \tag{2.40}$$

and the differential voltage, V_{AB}, between the secondary coils becomes

$$V_{AB} = V_A - V_B = (L_A - L_B) \sin(2\pi ft - \phi). \tag{2.41}$$

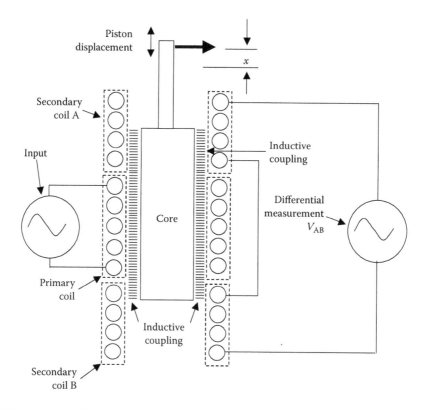

FIGURE 2.28   Schematic of LVDT.

Most LVDTs use a demodulator circuit to convert V_{AB} into an equivalent analog or digital displacement signal. The demodulator takes advantage of V_{AB} being an oscillating sine wave with the same frequency as the primary excitation. The amplitude is proportional to absolute value of the core displacement from a zero offset. The demodulator detects that the phase angle shifts 180° when the displacement changes the sign.

Reluctance quantifies the ease with which magnetic fields flow through magnetic circuits. Magnetic fields flow through ferromagnetic materials much more readily than through air gaps. Using a variable air gap in series with a ferromagnetic circuit element causes the overall reluctance of the circuit to depend on the gap size. The relation is nonlinear and decreases with gap distance. A differential configuration that uses two air gaps can eliminate much of the nonlinear relation between inductance (reluctance), air gap, and displacement [107]. Advantages of variable reluctance transducers are miniaturization and use in high-precision (fine-scale resolution) transducers (Figure 2.29) [108]. Some concrete rebar location and cover meters use reluctance for measurements (Figure 2.30). Difficulties with confounding effects have generally led to the use of similar, but more reliable instruments that use transient pulse techniques to induce detectable eddy currents in the rebars.

2.2.9   Electromechanical Sensors

Electromechanical sensors couple moving mechanical parts with EM fields as the basis of transduction. *Voice-coil velocimeters* and *force-balance servo-accelerometers* are common examples [107,109]. The voice-coil velocimeter, also known as a *geophone*, is an inexpensive and

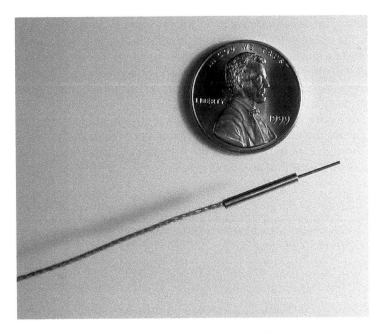

FIGURE 2.29 Miniature differential variable reluctance transducer for displacement measurement. (Photograph courtesy of Microstrain, Inc.)

easy-to-use device for measuring dynamic motions, usually velocities. It is a standard to employ large arrays of voice coils in dynamic geotechnical experiments, such as oil exploration with explosive-seismic techniques, and in the impedance testing of drilled piles [110]. Huber et al. found that velocimeters were particularly useful in measuring low-frequency structural vibrations [111]. Lemke, however, found that the large inertial masses required for low-frequency sensitivity limited the utility of geophones in certain large-array applications that required sensitivities for frequencies less than 10 Hz [112]. Velocimeters are quite

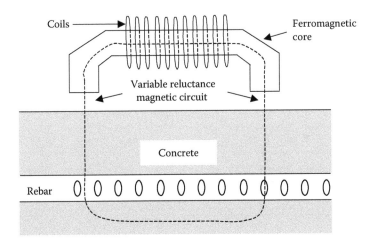

FIGURE 2.30 Variable reluctance rebar detector and concrete cover meter.

common in rotating machinery monitoring measurements. In addition to low cost, a primary reason for the use of velocimeters is that the stress in a machine component vibrating in a linear mode is proportional to velocity [113].

The voice-coil velocimeter uses a permanent magnet proof mass that moves on a low-resonant-frequency mount inside a conductive coil (Figure 2.31). Relative motions of the proof mass magnet and the coil induce measurable voltages that are approximately proportional to the relative velocity. Attaching the coil to a high-impedance circuit, so that very little current flows, causes the mechanics of the proof mass to match that of the accelerometer (Equations 2.1 through 2.3). In the case of the velocimeter, however, the natural frequency, ω_0, is set very low so that Equation 2.3 becomes

$$-\ddot{z} \approx \ddot{y}. \tag{2.42}$$

The coil voltage, V, is proportional to the relative velocity between the magnet and the coil, which mounts rigidly to a moving base

$$V = C_V \dot{y}. \tag{2.43}$$

The inductive coupling constant, C_V, depends on the magnetic field strength, the number of turns in the coil, and geometric factors. Integrating Equation 2.42 and combining with Equation 2.43 yields the relation between the velocity of the base and the coil voltage

$$V = -C_V \dot{z}. \tag{2.44}$$

The proof mass in a servo-accelerometer is also spring mounted to the base. When the base accelerates, the proof mass moves relative to the base. A transducer converts the relative displacement into a voltage. An amplifier converts the voltage into a current that activates a voice coil to move the proof mass back to a neutral position. The amount of feedback current required is proportional to the base acceleration. The resulting small relative motions of the proof mass allow for mounting it on a low-stiction flexure mount in an effort to

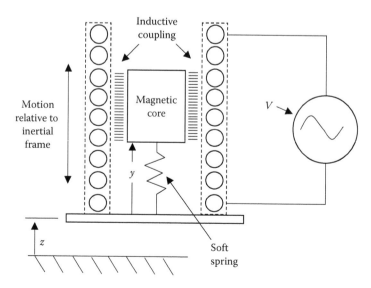

FIGURE 2.31 Voice coil velocimeter.

eliminate mechanical hysteresis. Servo-accelerometers are sensitive to low-frequency or DC oscillations and can also be used as tiltmeters.

While the servo-accelerometer is similar to the velocimeter, it is fairly expensive. A primary application is low-frequency and low-amplitude acceleration measurements, as in seismic studies. The disadvantages of servo-accelerometers include their relatively large power consumption, size, and weight. MEMS versions of servo-accelerometers have yet to become commercially available for low-cost low-acceleration and low-frequency measurements.

2.2.9.1 Piezoelectric and Magnetostrictive Sensors

The *piezoelectric effect* is the strain-induced voltage or the voltage-induced strain in solids. Piezoelectric materials find use in a wide variety of sensors, such as load cells, accelerometers, pressure gages, ultrasonic sources and sensors, and AE sensors. Three main classes of materials appear in piezoelectric transducers: (1) *single crystal*, for example, quartz (natural or fused); (2) *polycrystalline piezoceramic*, for example, lead zirconate titanate (PZT); and (3) *piezopolymers*, for example, polyvinylidene fluoride (PVDF).

The basis of the piezoelectric effect lies in a material micro- and nanostructure that is inherently anisotropic with a nonhomogeneous charge distribution. When strained, positive and negative charges move in opposite directions. The net effect is a macroscopically measurable voltage [114,115]. A linear relation between strain and voltage in indicial tensor notation is

$$V_i = C_{ijk}\varepsilon_{jk}, \tag{2.45}$$

or in matrix form is

$$
\begin{bmatrix} V_x \\ V_y \\ V_z \end{bmatrix} =
\begin{bmatrix}
d_{11} & d_{12} & d_{13} & d_{14} & d_{15} & d_{16} \\
d_{21} & d_{22} & d_{23} & d_{24} & d_{25} & d_{26} \\
d_{31} & d_{32} & d_{33} & d_{34} & d_{35} & d_{36}
\end{bmatrix}
\begin{bmatrix}
\varepsilon_{xx} \\ \varepsilon_{yy} \\ \varepsilon_{zz} \\ \gamma_{xy} \\ \gamma_{yz} \\ \gamma_{zx}
\end{bmatrix}, \tag{2.46}
$$

where V_i, C_{ijk}, and ε_{jk} are the voltage vector, piezoelectric constitutive property tensor, and the strain tensor, respectively. The 3×6 matrix $[d_{ij}]$ is equivalent to piezoelectric tensor C_{ijk}. Crystal and strain tensor symmetries reduce the number of independent piezoelectric coupling parameters. For example, a hexagonal crystal, such as quartz, has three independent piezoelectric parameters. The d_{13} and d_{16} parameters tend to be the most important for use in piezoelectric transducers. Since it is often easier to produce uniform shear strains than a uniform axial strains, many piezoelectric transducers employ shear effects in the transducer element with a sensitivity determined by d_{16}.

Building a piezoelectric transducer poses a couple of challenges. One is that the output voltage with most materials is small. Another is that the piezoelectric sensor acts as a capacitor. Measuring piezo-induced voltage requires sampling of a small amount of charge. Since the available charge is small, even a high-impedance voltage measurement draws down the charge and appreciably affects the measurement. Piezoelectric sensors have difficulty in measuring steady-state or low-frequency signals. Piezoelectric transducers are better suited for transient or oscillating signals, including those at high frequencies.

Specialized charge amplifiers and collocated amplifier circuits measure piezoelectric charge (voltage). The charge amplifier is an off-board signal conditioning instrument. Using charge amplifiers is somewhat difficult due to the need for maintaining high-impedance interconnects between the transducer and the amplifier. Collocating the charge amplifier in a miniaturized form near to or integrated with the piezoelectric transducer avoids some of the cabling and connector problems. A standard collocation technique is to drive the amplifier on the transducer with a constant current source. The amplifier regulates the voltage to be proportional to the piezoelectric charge and transduced signal (Figure 2.32) [116,117]. Such a configuration allows for a simple two-wire (often coaxial for high-frequency fidelity) connection to the transducer.

Polymer piezoelectric materials, such as PVDF, produce larger voltages than ceramic or single crystal materials, and are somewhat more convenient for building custom and non-production transducers. Gu et al. demonstrated a network of wireless PVDF displacement sensors [118]. Park et al. and Kim et al. detected low-velocity impacts and resulting damage on composite panels with surface-attached PVDF patches [119,120].

An advantage of piezoelectric devices is that many of the effects scale nicely at small dimensions. Embeddable piezoelectric sensors and MEMS devices using piezoelectric effects are common [121,122]. Thielicke et al. describe techniques for high-reliability manufacturing of PZT piezoelectric sensor patches [123]. Two different piezoelectric sensor types performed an excess of 10^7 strain measurements with amplitudes of 0.12% without any appreciable signs of degradation. *Lead magnesium niobate–lead titanate* (PMN–PT) is an alternative crystalline piezoelectric material with potential for high performance in harsh environments [124].

Since the piezoelectric effect is reversible, an applied time-varying voltage induces controllable mechanical distortions over a wide frequency range. Embedding or attaching piezoelectric elements to structures forms an electromechanical system. Measuring the impedance of such an electromechanical system provides useful information [125]. Oshima et al. measured the electromechanical impedance (EMI) of a curing-composite-embedded piezoelectric patch [126]. The impedance shifted as the composite went from a state of resin-impregnated fiber matrix to a gel to a cured solid. Chen et al. correlated EMI shifts of a piezoelectric disc embedded in concrete with stress [127]. The shifts appeared in both the reactive (real) and dissipative (imaginary) impedance components. The ability to manufacture small geometry piezoelectric elements creates the opportunity for embedded active sensing applications, such as testing bolted joints [128,129].

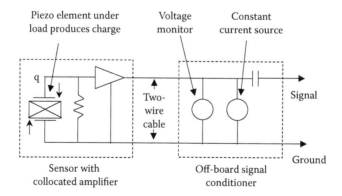

FIGURE 2.32 Integrated Circuit-Piezoelectric (ICP) collocated amplifier circuit. (Adapted from PCB Piezotronics, Inc. 2002. *General Piezoelectric Theory*. Depew, New York.)

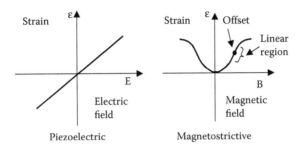

FIGURE 2.33 Qualitative behavior of linear piezoelectric and nonlinear magnetoelastic effects.

Magnetostriction is similar to the piezoelectric effect. A magnetic field induces a strain in a solid or vice versa [130–132]. A principal distinction is that piezoelectric effects tend to be linear and magnetostrictive are nonlinear (Figure 2.33). Reversing the electric field applied to a piezoelectric solid changes the sign of the induced strain. Reversing the magnetic field applied to a magnetostrictive material does not cause the strain to reverse. The strain instead depends on the absolute value of the magnetic field. The magnetic field–strain relation is nonlinear, but reversible. Applying a mechanical prestress can move the material into a linear range. Certain *Galfenol* alloys appear to have an inherent prestress that causes the material to operate in a linear range without the need for external prestress [133]. In spite of the inconvenience of the nonlinear behavior, magnetostrictive devices have certain advantages over piezoelectric counterparts. Magnetoelastic strains are an order of magnitude larger than piezoelectric strains. Magnetoelastic strains are capable of sustaining static displacements or readings in DC fields. *Giant magnetostrictive alloys* can extend these performance parameters even farther [134]. Sammal and Ranniku developed miniaturized Carlson stress meters for concrete based on magnetoelastic transducers [135].

An additional class of magnetostrictive sensors transduces strain-induced changes in magnetization hysteresis of ferromagnetic materials, such as steel [136]. An attractive feature of this approach is that it is possible to detect prestresses without requiring unstressed reference states. Wang et al. developed steel cable tension transducers based on measurements of relative magnetic permeability and hysteresis [137]. Han et al. demonstrated noncontacting magnetostrictive strain sensors for rotating shafts [138]. Murai et al. measured stress-induced anisotropy in the magnetization properties of steel bridge members with a four-pole probe [139]. Ogawa and Takahashi used a similar technique to measure the stress in steel pipes in pipeline bridges [140].

2.3 Fiber Optic Sensors

Fiber optic sensors use light for both transduction and signal transmission. There are many similarities between fiber optic and electrical sensors. Electrical sensors transduce by modifying voltage, current, frequency, or phase of an electrical signal. Fiber optic sensors transduce by modifying the intensity, fast frequency (wavelength), slow frequency (time-modulated intensity), polarization, phase, and coherence of optical signals for transduction. Optical effects with proven utility in fiber optic sensors include intensity modulation, modal domain interference, Bragg grating wavelength selective transmission and reflection, noncoherent white light interference, Brillouin backscatter, and fluorescence.

An *extrinsic* fiber optic sensor uses a measurand to modify optical properties at a point. An optical cable carries the modified optical signal to a remote sensor for further transduction. *Intrinsic* fiber optic sensors use the measurand to modify light as it propagates along the length of the fiber. Intrinsic fiber optic sensors have distributed sensing capabilities.

Examples of where fiber optic sensors may be superior to similar electrical sensors include the following: spectroscopic chemical measurements; distributed load measurements over very large distances; measurements in EMI harsh environments such as structures prone to lightning strikes; embedded measurements, and long-term strain measurements [141–143]. Bridges, pavements, and airplanes are all potential candidates for miniaturized embeddable fiber optic sensors [144–148]. Issues of concern with fiber optic sensors is the faithful transfer of the strain fields from the base material to the sensor and the actions of confounding effects [149]. For example, more than one component of the strain tensor can contribute to the output of a single-channel fiber optic strain sensor [150,151]. Due to the relative novelty of the technology, the long-term reliability of fiber optic sensors remains to be determined, but the prognosis appears to be favorable [152–156].

2.3.1 Intensity-Based Fiber Optic Sensors

Intensity-based fiber optic sensors use measurand-induced changes in the intensity of the transmitted or reflected light. These can be made of low-cost light sources and detectors, that is, phototransistors, photoresistors, and photodiodes. Intensity-based sensors are fairly simple to build and demonstrate. A disadvantage is that many effects besides the measurand, such as connector infidelities, can confound the intensity readings. This is the reason why On/Off 1-bit and relative-intensity sensors are more common than absolute-intensity sensors. An industrial application is the use of intensity sensors to measure the presence and proximity of objects. Referencing schemes, such as using optical bridge configurations with controlled transient reference signals, combined with careful design and packaging enables making absolute intensity sensors with utility for SHM [157–163]. Asanuma developed intensity-based sensors that operate by an *in situ* fiber cracking [164]. This technique allows for the embedment of fiber optic sensors in active composites and other advanced materials.

2.3.2 Interference-Based Point Sensors

Optical interference is a consequence of the wave nature of light. A common case occurs with the splitting of a light beam into two parts, sending both down a separate path, rejoining the beams, and then projecting the result onto a detector. Depending on the relative phase, the rejoined light beams either add or subtract to form fringe patterns of varying intensities. Changes to the effective relative path lengths alter the phase angles at the detector and cause the fringe patterns to shift. Measuring interference pattern shifts can provide highly precise and accurate information about changes in physical parameters.

Interferometric sensing techniques can use either coherent (usually laser light) or incoherent (white) light (see Appendix A for some more details). The measured intensity of monochromatic coherent light depends on the relative path length difference, Δx_i. A maximum intensity difference occurs when the path length differential equals an integer multiple of the wavelength. Counting fringes gives a resolution equal to the wavelength of the light ($1\,\mu$m approximately). A disadvantage of coherent light sensors is that the phase angle change is ambiguous when the differential length change exceeds that of a wavelength of light. Another disadvantage is that the direction of motion is not obvious. Fringe counting and dual-beam phase quadrature techniques can resolve some of these issues. Commercially available interferometers using dual wavelength and subfringe intensity interpolation techniques can achieve nanometer resolutions. Incoherent or white light interferometers use analog statistical averaging. The maximum detected intensity occurs when the path lengths are equal. Multiple fringe ambiguities are avoided. White light interferometric fiber optic displacement sensors may have a long-term stability that rivals the vibrating wire strain gage.

Interferometers come in different forms with a variety of path geometries, several of which are useful in fiber optic sensors [165]. The Fabry–Perot interferometer, Figure 2.34, uses two parallel surfaces with an alignment perpendicular to the light beam. The incident surface is a partial reflector. The second surface is a complete reflector [166,167]. The first surface reflects part of the beam back towards the source and transmits the remainder to the second surface. The second surface reflects the remainder back towards the source. Both of the reflected beams rejoin and interfere at a detector located at the source end of the fiber.

The Fabry–Perot sensor seems to be particularly well suited for precise pressure sensing, high-temperature and harsh environment sensing, and embedment applications [168–172]. Chen et al. successfully embedded 45 Fabry–Perot sensors to monitor stresses in a concrete bridge [173]. Benmokrane et al. embedded Fabry–Perot sensors to monitor fiber-reinforced polymer (FRP) bridge decks [174]. Herrero and Casas monitored the tension in concrete bridge post-tensioning strands with Fabry–Perot strain readings in a specialized application that required measuring strains up to 20,000 $\mu\varepsilon$ [175]. Lawrence et al. measured process-induced strains in the manufacture of composites [176]. Stewart et al. embedded Fabry–Perot fiber optic thermal sensors into composites for thermographic damage detection [177]. Valla et al. developed Fabry–Perot strain gage rosettes [178]. Specialized signal demodulation instruments can read Fabry–Perot interferometer sensors. For example, Yu developed an improved detector system based on a digital demodulation principle [179]. The Fabry–Perot interferometer has a nonlinear periodic phase-dependent output for large deformations that cover multiple wavelengths. Kim et al. developed phase-tracking techniques that overcome this limitation [180].

2.3.3 Polarization-Based Sensors

Polarization-based sensors use measurand modification of polarization [181]. One technique uses photoelastic materials to transduce strain by changing the polarization phase

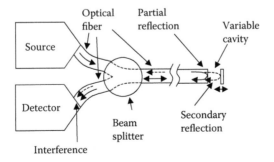

FIGURE 2.34 Fiber optic Fabry–Perot interferometer.

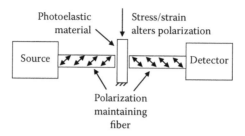

FIGURE 2.35 Schematic of photoelastic polarization-based fiber optic strain sensor.

of transmitted light (Figure 2.35). It is possible to use polarization in distributed fiber optic sensors. Sensing hydrostatic pressure is a possible application [182]. Polarization-based fiber optic sensors have, to date, not achieved much usage beyond laboratory demonstrations. The primary reason seems to be that many confounding effects can alter polarization, as well as an annoying 2π phase angle ambiguity.

2.3.4 Bragg Grating-Based Fiber Optic Strain Gages

FBG strain gages are the strain transducers of choice for many applications. The FBG uses the Bragg effect for transduction. An optical medium with a periodically varying index of refraction will selectively reflect and transmit waves in a narrow wavelength band, with a center wavelength band, λ_B, equal to the spatial wavelength of the index of refraction variations, λ_S [183]

$$\lambda_B = \lambda_S. \tag{2.47}$$

The operating principle of the FBG is that strain and temperature changes alter the effective Bragg grating spacing and the wavelength of the reflected signal (Figure 2.36) [184–187]. A linear representation is

$$\frac{\Delta\lambda_B}{\lambda_B} = (1 - p_e)\varepsilon + (\alpha_\Lambda + \alpha_n)\,\Delta T, \tag{2.48}$$

where ε is the mechanical strain, ΔT is a temperature shift, and p_e, α_Λ, and α_n are strain-optic, fiber thermal expansion, and thermal-optic coefficients, respectively. Guo found that a quadratic curve performs a good fit to the temperature-dependent behavior of FBGs down to cryogenic temperatures of 125 K [188]. Dong et al. compensated for temperature shifts in FBG outputs by a combination of triangular cantilever beam mounts, broad banded light sources, and reflected power measurements (instead of wavelength) [189]. Giaccari et al. used a coupled micromechanical and optical model to ferret out the effects of local strain gradients on FBG readings [190].

A typical sensing configuration uses the FBG as a strain transducer. Temperature compensation, along with simultaneous strain and temperature sensing are possible [191,192]. Specialized coatings that protect the FBG fiber, such as nickel plating, can alter the temperature sensitivity [193]. A principal advantage of FBGs is that the reflected light is a wavelength-modulated signal. The signal is insensitive to intensity-based noise disturbances, in a manner similar to frequency-modulated (FM) radio. The ability of an FBG

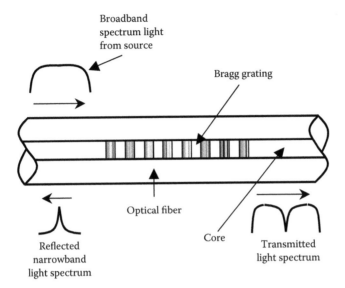

Broadband
spectrum light
from source

Bragg grating

Optical fiber

Core

Reflected
narrowband
light spectrum

Transmitted
light spectrum

FIGURE 2.36 Operating principle of fiber optic Bragg grating. (Adapted from Dunphy J, Meltz G, and Morey W. 1991. *Fiber Optic Smart Structures*. Wiley, New York, NY.)

to selectively filter optical signals makes it very useful in a variety of optoelectronic applications that extend well beyond strain and temperature sensing. An example is to use an FBG to fabricate a refractive index sensor to monitor the state of cure in thermoset resins [194].

The standard FBG manufacturing technique writes the grating onto the transmission core of a germanium-doped optical fiber with spatially patterned ultraviolet (UV) light (usually from an interference pattern). The UV light causes a small, localized pattern of permanent shifts in the refractive index of the glass core. Alternative manufacturing processes include lithographic masking techniques. Some sensors use specialized alternative approaches, such as etching the side of D-shaped cores for high-temperature sensing applications [195].

Since FBGs selectively reflect narrow bandwidths of light, it is possible to place multiple FBGs with different center wavelengths on the same cable [197–199]. An additional advantage is that since the transduction mechanism is robust with respect to connection changes, it is possible to use the same readout instruments with multiple installations that can reduce the overall cost of ownership [200]. Figure 2.37 shows the load testing of a composite highly reliable advanced grid structure developed by Takeya et al. that contains an array of embedded FBGs [201].

FBGs can be a cost-competitive strain transducer, especially with the development of affordable readout instrumentation [202,203]. Sirkis laid out many of the issues related to the practical and effective use of FBGs in construction materials and bridges [204]. This includes stress gradients, long-term reliability, and temperature compensation. As with foil strain gages, debonding can be an issue [205]. FBG strain sensors are beginning to be used routinely [206,207]. Martinez et al., Yamakawa et al., and Li and Zhu compared the industrial and structural testing utility of FBGs with that of foil strain gages [208–210]. They concluded that FBGs were superior because of the long-term stability of the readings, immunity to EM interference, ease of multiplexing on a single cable, test instrument portability, and ease of surface mounting. FBGs are finding routine use in certain demanding applications, such as embedded rebar strain sensing, dynamic measurements of bridge

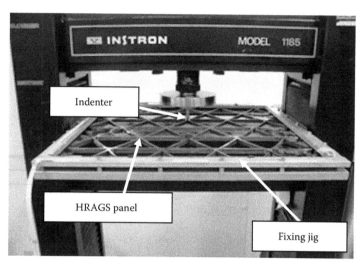

FIGURE 2.37   Composite grid panel with array of embedded FBGs undergoing load test. (Photograph courtesy of N. Takeda.)

structural components, strain measurements in complex airplane box structures, composite bridge, ship hull and mast strain measurements, the root of a helicopter blade, real-time measurements of cryogenic fuel tanks, train pantograph electrical contact force sensing, and composite cure monitoring [211–230]. Specialized packaging for harsh environments is available. An example is undersea pipeline vibration measurements [231,232]. Schmidt-Hattenberger et al. measured the construction loads on large piles with FBGs [233]. The results showed excellent agreement with foil strain gage measurements. Hudson and Little monitored solid fuel rocket motors with embedded multi-axis FBGs along with embedded pressure and temperature sensors [234]. Figure 2.38 shows data from the FBGs as the bridge underwent load testing [235].

The narrow diameter (typically 100–200 µm) of FBGs makes them attractive for embedment applications [236]. These include monitoring the performance of composite patches on aluminum aircraft structures, parachutes, metallic–intermetallic laminates, composite lap joints, composite load cells for bridge bearings, filament wound composite pressure vessels, and the process monitoring of resin flow in the molds of the vacuum-assisted resin transfer molding (VARTM) composite manufacturing process [185,237–246]. Embedding the sensor has the added advantage of avoiding confounding readings due to inconsequential surface microcracking and other defects [247]. Smaller diameter FBGs can be used when necessary [248]. Hongo et al. developed FBGs with diameters less than 50 µm for embedment applications [249]. Mizutani et al. report FBGs with outside diameters less than 6.5 µm [250]. The small diameters enabled quantitative transverse crack measurement by an analysis of optical spectral peak widths and shapes. Lin et al. developed metallic coatings for long-term durable FBG packaging [251]. FBGs produce very little heat, which is attractive for measuring the deformation of materials that are poor heat conductors, such as wood [252]. However, there are cases where embedding FBGs can produce measurable changes in the dynamics of structures [253]. Ahmad et al. demonstrated an FBG-based pressure sensor that embedded the fiber in a polymer jacket with one end of the jacket cylinder exposed to the measurand pressure [254]. Hybrid sensing systems, such as those that combine the sensitivity and EM immunity of FBGs for sensing with piezoelectric patch elements for

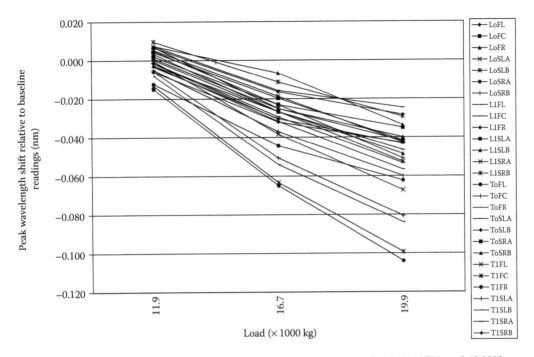

FIGURE 2.38 Spectral shift of 24 fiber grating strain sensors for weights of 11,900, 16,700, and 19,900 kg on Horsetail Falls Bridge in Oregon, USA. (From Seim J, Udd E, Schulz W, and Laylor HM. 1999. Smart Systems for Bridges, Structures, and Highways, SPIE 3746. With permission.)

the generation of probing elastic waves, or those that combine FBGs with Fabry–Perot sensors for combined temperature and strain sensing, are possible solutions for demanding applications, such as aircraft structural monitoring or strain sensing over large temperature swings [255–257].

Long-term static measurements require temperature compensation [258]. Using calibrated mounts enables configuring FBGs to act as temperature gages. Ecke et al. measured the temperature on the skin of helium balloons with an array of 20 FBGs [259]. Player et al. developed an FBG-based humidity sensor that used moisture-induced polymer swelling as the transduction mechanism [260]. Specialized multi-axis FBG strain and temperature sensors are possible [261–263]. Zhang et al. used FBG-based geophones for explosive-seismic oil exploration [264].

Readout instrumentation for FBG sensors continues to show steady development in terms of improved performance and reduced cost. Koh et al. used a Mach–Zehnder interferometer to measure FBG signals at frequencies up to 200 kHz [265]. Nguyen used a lock-in technique to measure gated ultrasound waves at 200 kHz [266]. Hsu et al. developed an FBG interrogation system based on swept frequency techniques that can measure at rates of up to 250 kHz [267]. Coppola et al. and Thomson et al. obtained similar performance using an etalon for FBG interrogation and by using gas cell wavelength references [268,269]. Komatsuzaki used an *arrayed wavelength grating* (AWG) filter to obtain FBG measurements with speeds up to 300 kHz [270]. Qiao et al. used a two-wave mixing technique to create an FBG system with a dynamic sensitivity of at least 500 kHz [271]. Ye and Tatam measured continuous ultrasonic solid elastic waves with a resolution for out-of-plane displacements of 3×10^{-15} m Hz$^{-1/2}$ at a frequency of 470 kHz with a path-length-unbalanced readout Mach–Zehnder interferometer [272]. Abdi et al. demonstrated and calibrated a 3-FBG strain sensing system with coherent optical frequency domain reflectometry [273]. Adamovsky et al. developed an unbalanced interferometer system for interrogating FBGs [274]. The instrument can measure temperature over wide dynamic ranges. MEMS-based readout instrumentation is a recent further development in the trend towards miniaturization [275]. FBG sensors readily integrate into other transducers, such as accelerometers and load cells [276,277]. Continued improvements in micro- and nanoscale manufacturing technologies are enabling the fabrication of a variety of strain-sensitive photonic devices. One example is the photonic ring resonator [278,279]. The operating principle of this sensor is that a microscale fiber optic loop forms a high-Q optical resonator. Strain deformation of the resonator affects the resonant wavelength in a readily detectable manner.

2.4 Chemical and Corrosion Sensors

Changes in the chemical state of materials can dramatically affect the strength and overall health of structures. Since most cases of chemical attack on structures occur over extended time periods, early detection and intervention can lead to positive outcomes [280].

Electrical and optical transduction mechanisms are the basis of most field-deployable chemical sensors. Electrical chemical sensors generally use electrochemical reactions to produce voltages or to induce changes in the electrical properties of materials. Optical chemical sensors often use a chemical reaction to produce a measurable change in the optical properties of a material as the transduction mechanism. A variety of transmission and reflection methods couple light into and out of an optical chemical sensor, often with the aid of a fiber optic waveguide. Evanescent surface waves can be particularly useful since they can measure chemical reactions on surfaces.

Sensitivity, specificity, response time, and *long-term stability* are important chemical sensor performance parameters. The sensitivity is the size of the sensor output in terms of the concentration of the chemical reagent under analysis. Specificity is the ability of the sensor to distinguish the chemical reagents. Long-term stability is a particularly important concern for chemical sensors that carry onboard reagents.

Sensing of color, or the equivalent but more detailed optical spectrum, is useful for corrosion and other chemical detection applications. High-resolution optical spectrometers are available, with impressive progress being made in terms of reduced cost and size [281].

Direct spectroscopy is an examination of the spectrum of broad-banded (white) light that reflects off of a surface, or transmits through a material (Figure 2.39). Rust colors can indicate the presence of corrosion, especially for steel, and somewhat less so for aluminum [282–284]. Water and corrosion products, such as aluminum hydroxide, will absorb radiation at infrared (IR) wavelengths that are readily detectable with methods such as *Fourier transform infrared* (FTIR) spectroscopy [285,286]. Huang et al. used a similar fiber arrangement to that of Figure 2.39, but measured the amount of *scattering* due to surface roughness as an indicator of corrosion [287]. Palmer et al. ascertained the health of electric power transformer fluids with an optical absorption measurement in the UV range [288]. Fair and Parthasarathy developed a temperature history sensor that ascertains whether specialized polymers have exceeded certain temperature thresholds by an examination of the IR absorption spectra [289].

Indicator spectroscopy uses a chemical indicator to indicate the presence of a particular chemical species or pH [290–292]. Indicator spectroscopy can be quite effective, but some of the dyes tend to bleach and experience long-term stability problems. Wiese et al. measured

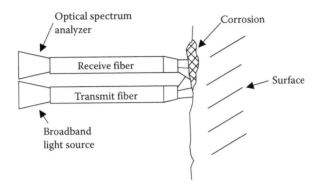

FIGURE 2.39 Direct spectroscopy of surface for corrosion.

moisture content in curing concrete by indicator spectroscopy [293]. The indicator was pyridinium-*N*-phenolate betaine dye embedded in a polymethylmethacrylonitrile polymer matrix. The presence of moisture caused the primary spectral peak to shift from 602 nm down to as low as 562 nm. Dantan et al. developed a fiber optic pH sensor based on indicator spectroscopy for the specific application of detecting corrosion precursors in reinforced concrete [294]. Srinivasan et al. found *Clayton Yellow* to be a reliable indicator of pH in concrete [295]. Pidaparti et al. used *fluorescein of sodium* to indicate the pH of aluminum corrosion with UV light [296]. Ghandehari et al. developed a fiber optic evanescent wave oxygen luminescent gas leak detector for pipeline monitoring applications [297]. Figure 2.40 shows a fiber optic chloride detector mounted into a bridge deck among the rebars before the concrete is poured. This sensor uses indicator spectroscopy to measure the presence of chloride ions in concrete. Such a sensor could assess concrete rebar corrosion potential. Distinguishing troublesome-free chloride ions versus more benign chemically bound ions is an important consideration [298]. Kanada et al. developed an improved reagent-free concrete chloride sensor that makes use of absorbance spectroscopy (Figure 2.41) [299].

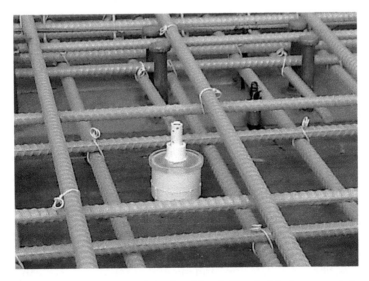

FIGURE 2.40 Fiber optic chloride sensor module placed in a bridge deck before concrete is poured. (Adapted from Fuhr PL and Huston DR. 1998. *J Opt Eng*, 37(4), 1221–1228.)

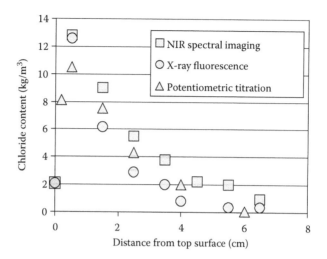

FIGURE 2.41   Comparison of fiber optic NIF, x-ray fluorescence and potentiometric titration chloride detection in concrete sample versus depth. (Adapted from Kanada H et al. 2006. *NDE Conf on Civil Engineering,* American Society for Nondestructive Testing, St. Louis.)

The sensor exploits the absorbance of *near-infrared* (NIR) radiation at 2,266 nm of chloride-laden cement. A fiber optic probe can readily detect the presence of chlorides. Analyzing the radiation pattern indicates the presence and concentration of ions and elements of interest, such as chlorides or sulfur in concrete [300]. Grubb et al. report that taking a small sample by sanding, and determining the pH with indicator paper, is superior to other available pH test procedures for concrete [301].

Nonlinear spectroscopy uses one of several optical effects to measure the presence of chemical species. These include Raman, Stokes, Brillouin, and other phenomena. A difficulty with nonlinear optical spectroscopy is that the effects tend to be weak, and the instruments are expensive and bulky. It is likely that with increased technical development the performance will increase as the size and cost decrease [303]. *Laser-induced breakdown spectroscopy* (LIBS) directs a high-intensity laser pulse at a material. The energy in the laser pulse breaks down chemical bonds and forms a plasma that radiates in a manner characteristic of the elemental species in the material.

Corrosion can occur through one of several distinct electrochemical processes. The result is that it is generally necessary to use a variety of techniques for comprehensive corrosion sensing. *Visual inspection* can identify gross regions of corrosion. Access limitations prevent easy visual inspection in many applications. The *half-cell* is a quantitative electrochemical technique that is fairly easy to implement on accessible structural members (see Chapter 3). A disadvantage of the half-cell method is that the corrosion processes must be active during the measurement. Unexpected chemical species and electrochemical reactions can confound the measurements. *Sacrificial member sensing* monitors corrosion processes in a sample coupon. The standard approach uses a coupon made of an alloy that matches the structure in question and undergoes an environmental exposure that also matches the structure. Onboard instruments, such as vibrating beams, fiber optic strain gages, wireless inductive couplers, fiber optic detectors, and integrated MEMS devices, can detect corrosion in a sacrificial member [304–310]. A serious challenge to successful sacrificial coupon corrosion sensing is that corrosion processes are often so localized that they may aggressively attack the primary structure, but not be detected with a nearby sacrificial member, or vice versa.

Corrosion processes that form localized electrochemical cells can be particularly troublesome because the local environment can change from a passive state to the one that rapidly accelerates corrosion, while remaining largely hidden from view. Examples of structural situations of concern are crevice corrosion in aircraft and drilled rock anchors. Crevices, such as lap joints, trap water, set up a localized environment with depleted oxygen and shift the pH towards acid levels. The presence of aggressive chemical species, such as chloride ions, further aggravates the situation. Detecting crevice corrosion at an early stage enables economically viable corrections and repairs. The result is that there is a need for effective *in situ* corrosion sensors. Cooper et al. sensed the potential for lap joint corrosion activity with a long-gage-grating fiber optic moisture sensor [311]. Flachsbarth et al. developed a moisture-detecting fiber optic sensor [312]. The technique embeds Reichart's dye in custom-formulated polymethacrylnitrile. In an example of combined multisensor corrosion detection, Cusson et al. monitored corrosion processes in a reinforced concrete bridge with an array of relative humidity, temperature, electrochemical potential, and strain sensors as an assessment of the performance of several different proprietary anticorrosion systems [313]. Fink and Payer used electrical time of wetness, pH, and temperature sensors to evaluate corrosion proclivities in heat exchangers [314]. Braunling and Dietrich combined galvanic sensing washers and bolt caps with environmental sensors to form a corrosion sensing system for helicopters [315]. Gostautas et al. examined multisensor systems for corrosion detection in suspension bridge cables [316]. Similar to aircraft crevice corrosion, detecting corrosion in large cables without embedded sensors is difficult because of the need for expensive partial disassemblies. In an early (ca. 1995) development, Cosentino et al. developed fiber optic chloride sensors that used silver nitrate as an indicator and evanescent wave coupling of light into the fiber [317].

References

1. Hsu TR. 2002. *MEMS & Microsystems Design and Manufacture*. McGraw-Hill, Boston.
2. Loh NC, Schmidt MA, and Manalis SR. 2002. Sub-10 cm^3 interferometric accelerometer with nano-g resolution. *J Microelectromech Syst*, 11(3), 182–187.
3. Plaza JA, Llobera A, Dominguez C, Esteve J, Salinas I, Garcia J, and Berganzo J. 2004. BESOI-based integrated optical silicon accelerometer. *J Microelectromech Syst*, 13(2), 355–364.
4. Abdolvand R, Amini BV, and Ayazi F. 2007. Sub-micro-gravity in-plane accelerometers with reduced capacitive gaps and extra seismic mass. *J Microelectromech Syst*, 16(5), 1036–1043.
5. Han KH and Cho YH. 2003. Self-balanced navigation-grade capacitive microaccelerometers using branched finger electrodes and their performance for varying sense voltage and pressure. *J Microelectromech Syst*, 12(1), 11–20.
6. Dau VT, Dao DV, and Sugiyama S. 2007. A 2-DOF convective microaccelerometer with a low thermal stress sensing element. *Smart Mater Struct*, 16, 2308–2314.
7. Fox C, Rourke A, Eley R, Fell C, and McWilliam S. 2003. Multi-channel and multi-axis inertial sensor concepts based on vibrating structures. *Smart Sensor Technology and Measurement Systems*, SPIE 5050.
8. Asokanthan SF, Ariaratnam ST, Cho J, and Wang T. 2006. MEMS vibratory angular rate sensors: Stability considerations for design. *Struct Control Health Monit*, 13, 76–90.
9. Masuda A, Sone A, and Morita S. 2004. Continuous damage monitoring of civil structures using vibratory gyroscopes. *Sensors and Smart Structures Technologies for Civil, Mechanical and Aerospace Systems*, SPIE 5391.

10. US Army Corps of Engineers. 1980. Engineering and design instrumentation for concrete structures. EM 1110-2-4300, Department of the Army, Office of the Chief of Engineers, Washington, DC.

11. Wada A, Kasai K, Takeuchi T, Yamanaka H, Sakata H, Yamada S, Morikawa H, and Ooki Y. 2006. Structural design and comprehensive monitoring of seismic isolated tall steel building. *Proc 4th China–Japan–US Symp on Structural Control and Monitoring*, Hangzhou, China.

12. Schuyler JN and Gularte F. 2004. Automated tiltmeter monitoring of bridge response to compaction grouting. *Nondestructive Evaluation of Highways, Utilities, and Pipelines IV*, SPIE 3995.

13. Wells SA, Phillips BM, and Vandenbossche JM. 2006. Characterizing strain induced by environmental loads in jointed plain concrete pavements immediately after paving and throughout the first ten months. Paper no. 06-2812, 86th Annual Meeting, TRB, Washington, DC.

14. Riad KH and Fehling E. 2004. Determination of the effective tensile force in external prestressing cables on the basis of vibration measurements. *Proc Structural Materials Technology VI, an NDT Conference*, Buffalo.

15. Geier R, Hoffmann S, and Distl J. 2007. ISyS—a project for the development of intelligent systems for cable force measurements. In *Structural Health Monitoring*, FK Chang (ed.). DEStech, Lancaster.

16. Huston DR, Beliveau JG, and Durham DR. 1990. Wind-induced failure of bar-cable on the Cornish–Windsor cable-stayed bridge. In *Proc 6th US–Japan Bridge Engineering Workshop*, AM Abdel-Ghafar (ed.). UJNR, U.S.–Japan Panel on Wind and Seismic Effects, Lake Tahoe.

17. Gardner-Morse M and Huston D. 1993. Modal identification of a cable-stayed pedestrian bridge. *J Struct Eng*, 119(11), 3384–3404.

18. Zheng G, Ko JM, and Ni YQ. 2001. Multimode-based evaluation of cable tension force in cable-supported bridges. *Smart Systems for Bridges, Structures, and Highways*, SPIE 4330.

19. Hertzberg RW. 1989. *Deformation and Fracture Mechanics of Engineering Materials*, 3rd Ed. Wiley, New York.

20. Carlson RW. 1936. Telemetric device. US Patent 2,036,458.

21. Sundquist H and James G. 2004. Monitoring of shear cracks and the assessment of strengthening on two newly-built light-rail bridges in Stockholm. *Proc IABMAS'04 Bridge Maintenance, Safety, Management and Cost*, Kyoto. Taylor & Francis, London.

22. Kim YY, Fischer G, and Li VC. 2004. Performance of bridge deck link slabs designed with ductile engineered cementitious composite. *ACI Struct J*, 101(1), 42–49.

23. Rahman MS, Oshima T, Mikami S, Yamazaki T, and Tamba I. 2004. Diagnosis of aged bridge by intelligent monitoring system. *Proc 2nd Intl Workshop on Structural Health Monitoring*, Stanford. Technomic, Lancaster.

24. Jana D and Erlin B. 2007. Carbonation as an indicator of crack age. *Concrete Intl*, 29(5), 39–42.

25. Neville A. 2004. Can we determine the age of cracks by measuring carbonation? Part 2. *Concrete Intl*, (January).

26. American Concrete Institute. 2003. *In-place Methods to Estimate Concrete Strength*, ACI 228.1R-03.

27. Jolin M, Beaupre D, and Mindess S. 2002. Quality control of dry-mix shotcrete during construction. *Concrete Intl*, 24(10), 69–74.

28. Martys NS. 1995. Survey of concrete transport properties and their measurement. *NIST*, IR 5592, (February).

29. Luping T and Nilsson LO. 1992. Rapid determination of the chloride diffusivity in concrete by applying an electrical field. *ACI Mater J*, 89-M6, 89(1), 49–53.

30. Feldman RF, Chan GW, Brousseau RJ, and Tumidajski PJ. 1994. Investigation of the rapid chloride permeability test. *ACI Mater J*, 91-M23, 91(3), 246–255.

31. Carlson RW. 1936. Stress meter. US Patent 2,148,013.

32. Carlson RW. 1970. Meter for measuring compressive stress in earthen masses and the like. US Patent 3,529,468, September 22.

33. Wakha K, Majed MA, Dasgupta A, and Pines DJ. 2003. Multifunctional piezoelectric stiffness/energy sensor for monitoring the health of structures. *Smart Structures and Integrated Systems*, SPIE 5056.

34. Anderson GL, Robertson RC, Peterson BL, and Dillard DA. 1994. Embedded piezoelectric sensors to measure peel stresses in adhesive joints. *Exp Mech*, 34(3), 194–201.

35. Javidinejad A and Joshi SP. 2001. Autoclave reliability of MEMS pressure and temperature sensors embedded in carbon fiber composites. *J Electron Packag*, 123, 79–82.

36. Fellenius BH. 2000. The O-Cell—a brief introduction to an innovative engineering tool. *Väg-och Vattenbyggaren*, Stockholm, 47(4), 11–14.

37. Nusairat J, Engel R, Liang RY, and Yang K. 2006. Vertical and lateral load testing of drilled shafts socketed into rock at the Pomeroy–Mason Bridge over the Ohio River. Paper no. 06-2683, 86th Annual Meeting, TRB, Washington, DC.

38. Iannachione AT, Prosser LJ, Grau R, Oyler DC, Dolinar DR, Marshall TE, and Compton CS. 2000. Roof monitoring helps prevent injuries in stone mines. *Min Eng*, 52(11), 32–37.

39. Kaneko Y, Kirikoshi K, Onishi K, Suzuki T, Miyamoto N, and Sumitro S. 2007. Material characteristics of TRIP steel with self-diagnosis and application to structural systems. *Smart Mater Struct*, 16, 2464–2476.

40. Thompson L and Westermo B. 1994. A new strain measurement technology for material damage assessment. *Smart Sensing, Processing, and Instrumentation*, SPIE 2191.

41. Adachi Y and Unjoh S. 2001. Seismic damage sensing of bridge structures with TRIP reinforcement steel bars. *Smart Systems for Bridges, Structures, and Highways*, SPIE 4330.

42. Yang E and Li L. 2005. Magnetization changes induced by low cycle fatigue both in the geomagnetic field and the magnetic-free environment. *Nondestructive Evaluation and Health Monitoring of Aerospace Materials, Composites and Civil Infrastructure IV*, SPIE 5767.

43. Arms SW, Guzik DC, and Townsend CP. 1998. Microminiature high-resolution linear displacement sensor for peak strain detection in smart structures. *Sensory Phenomena and Measurement Instrumentation for Smart Structures and Materials*, SPIE 3330.

44. de la Veaux RC. 1993. Metal fatigue detector. US Patent 5,237,875.

45. Abe M, Komon K, Narumoto A, Sugidate M, Mori T, and Miki C. 2000. Monitoring of railway bridges in Japan. *Nondestructive Evaluation of Highways, Utilities, and Pipelines IV*, SPIE 3995.

46. Fujimoto Y, Shintaku E, Kim SC, and Takamoto T. 1995. A study of health monitoring of ship structures. *Proc MARIENV'95*, Tokyo.

47. Inamura F, Muragishi O, Nihei K, Kobayashi T, Ohgaki K, Kawaguchi Y, and Umeda A. 2004. Remaining life evaluation by fatigue detecting sensor. *Proc IABMAS'04 Bridge Maintenance, Safety, Management and Cost*, Kyoto. Taylor & Francis, London.

48. Raghavendrachar M, Murugesh G, Ng S, and Shepard RW. 2004. Health monitoring of California bridges. *Proc IABMAS'04 Bridge Maintenance, Safety, Management and Cost*, Kyoto. Taylor & Francis, London.

49. Davis J, Nassif H, and Ertekin AO. 2004. Field evaluation of methods for monitoring bridge scour. *Proc IABMAS'04 Bridge Maintenance, Safety, Management and Cost*, Kyoto. Taylor & Francis, London.

50. Chern CC, Tsai IC, Chang KC, and Lin YB. 2004. The safety improvement and inspection study for bridges with scoured foundations in Taiwan. *Proc IABMAS'04 Bridge Maintenance, Safety, Management and Cost*, Kyoto. Taylor & Francis, London.

51. Mercado EJ and Rao JR. 2006. The pneumatic scour detection system. *NDE Conf on Civil Engineering*, American Society for Nondestructive Testing, St. Louis.

52. Marron D. 2000. Remote monitoring of structural stability using electronic clinometers. In *Structural Materials Technology IV*, S Allampalli (ed.). Technomic, Lancaster.

53. Assenheim JG. 1997. Apparatus and method for measurements of moisture concentration in granular materials. US Patent 5,666,061.

54. West RC and Turner P. 2006. Evaluation of automated process control testing for hot mix asphalt production. Paper no. 06-2072, 86th Annual Meeting, TRB, Washington, DC.

55. Dally JW and Riley WF. 1991. *Experimental Stress Analysis*. McGraw-Hill, New York.

56. Mateiu R, Lillemose M, Hansen TS, Boisen A, and Geschke O. 2007. Reliability of poly-3,4-ethylenedioxythiophene strain gauge. *Microelectron Eng*, 84, 1270–1273.

57. Loh KJ, Lynch JP, and Kotov NA. 2005. Conformable single-walled carbon nanotube thin film strain sensors for structural monitoring. In *Structural Health Monitoring*, FK Chang (ed.). DEStech, Lancaster.

58. Park CS, Kang BS, Lee DW, Choi TY, and Choi YS. 2007. Fabrication and characterization of a pressure sensor using a pitch-based carbon fiber. *Microelectron Eng*, 84, 1316–1319.

59. Li H, Mao C, Ou J, and Li Z. 2003. Advanced health monitoring and damage repair technologies by using shape memory alloys. In *Structural Health Monitoring*, FK Chang (ed.). DEStech, Lancaster.

60. Zhang Y and Brown DJ. 2003. NiTi-fiber reinforced concrete with damage monitoring capability. In *Structural Health Monitoring*, FK Chang (ed.). DEStech, Lancaster.

61. Nagai H and Oishi R. 2006. Shape memory alloys as strain sensors in composites. *Smart Mater Struct*, 15, 493–498.

62. Srinivasan AV and McFarland DM. 2001. *Smart Structures Analysis and Design*. Cambridge University Press, Cambridge.

63. Ogisu T, Nomura M, Kikiukawa H, and Takeda N. 1999. Development of a health monitoring system using embedded SMA foils in CFRP laminates. *Proc 2nd Intl Workshop on Structural Health Monitoring*, Stanford. Technomic, Lancaster.

64. McGovern ST, Spinks GM, and Wallace GG. 2005. The use of embedded sensors for the monitoring of adhesive joints in marine environments. *Advanced Sensor Technologies for Nondestructive Evaluation and Structural Health Monitoring*, SPIE 5770.

65. Dokmeci M and Najafi K. 2001. A high-sensitivity polyimide capacitive relative humidity sensor for monitoring anodically bonded hermetic micropackages. *J Microelectromech Syst*, 10(2), 197–204.

66. Bonomo C, Fortuna L, Giannone P, Graziani S, and Strazzeri S. 2006. A model for ionic polymer metal composites as sensors. *Smart Mater Struct*, 15, 749–758.

67. Malvern LE. 1969. *Introduction to the Mechanics of a Continuous Medium*. Prentice-Hall, Englewood Cliffs, NJ.

68. Murray WM and Stein PK. 1959. *Strain Gage Techniques*. MIT Press, Cambridge.

69. Mohammed AA, Moussa WA, and Lou E. 2007. A novel MEMS strain sensor for structural health monitoring applications under harsh environmental conditions. In *Structural Health Monitoring*, FK Chang (ed.). DEStech, Lancaster.

70. Chelner H and Buswell J. 2005. Embedded sensors for structural health monitoring of rocket motors. In *Structural Health Monitoring*, FK Chang (ed.). DEStech, Lancaster.

71. Hautamaki C, Cao L, Zhou J, Mantell SC, and Kim TS. 2003. Calibration of MEMS strain sensors fabricated on silicon: Theory and experiments. *J Microelectromech Syst*, 12(5), 720–727.

72. Sharma V, Gamble WL, Choros J, and Reinschmidt AJ. 1994. Moving load tests on three precast pretensioned concrete railway bridges. *New Experimental Techniques for Evaluating Concrete Material and Structural Performance*, American Concrete Institute, ACI SP-143.

73. Yost JR and Assis G. 2000. Load testing of the Washington & Main Street Railroad Bridge. In *Structural Materials Technology IV*, S Allampalli (ed.). Technomic, Lancaster.

74. Cuelho E, Stephens J, Smolenski P, and Johnson J. 2006. Evaluating concrete bridge deck performance using active instrumentation. Paper no. 06-2302, 86th Annual Meeting, TRB, Washington, DC.

75. Frosch RJ and Matamoros AB. 1999. Monitoring and assessment of reinforced concrete structures. *Proc 2nd Intl Workshop on Structural Health Monitoring*, Stanford. Technomic, Lancaster.

76. Prine DW and Socie D. 2000. Continuous remote monitoring of the merrimac free ferry. In *Structural Materials Technology IV*, S Allampalli (ed.). Technomic, Lancaster.

77. Ware R, Reams R, Woods A, and Selder R. 2005. Sensor reliability in fielded C-17 aircraft strain gages. In *Structural Health Monitoring*, FK Chang (ed.). DEStech, Lancaster.

78. Joas HD. 2000. 20 years of life integrity of a high-pressure natural gas pipeline using strain ages. *Nondestructive Evaluation of Highways, Utilities, and Pipelines IV*, SPIE 3995.

79. Kim J, Jeong JH, Kwon SM, and Lee JH. 2006. Development of methods and devices for installation of strain gauges in concrete pavements. Paper no. 06-1378, 86th Annual Meeting, TRB, Washington, DC.

80. Nam JH, Kim SM, and Won MC. 2006. Measurement and analysis of early-age concrete strains and stresses in continuously reinforced concrete pavement under environmental loading. Paper no. 06-0852, 86th Annual Meeting, TRB, Washington, DC.

81. Wright C. 1993. For obtaining valid results, good operational procedures are as important as a good design. *Personal Eng Instrum News*, 10(10), (October).

82. Kirkham H and Jackson SP. 1997. Low-power, microprocessor-controlled strain-gauge circuit. *NASA Tech Brief*, 21(1), Item #74, *JPL New Technology Report*, NPO-19750, January.

83. Keithley Instruments. 2001. *Data Acquisition and Control Handbook*. Cleveland, OH.

84. Graichen F and Bergmann G. 1991. Four-channel telemetry system for in vivo measurement of hip joint forces. *J Biomed Eng*, 13, 370–374.

85. Bergmann G, Graichen F, Siraky J, Jendrzynski H, and Rohlmann A. 1988. Multichannel strain gage telemetry for orthopaedic implants. *J Biomech*, 21(2), 169–176.

86. Shenton III HW, Gallagher K, and Chajes MJ. 2004. Applicability of the ASTM hole-drilling method for measuring dead load stresses. *Proc Structural Materials Technology VI, an NDT Conference*, Buffalo.

87. Sabaté N, Vogel D, Keller J, Gollhardt A, Marcos J, Gràcia I, Cané C, and Michel B. 2007. FIB-based technique for stress characterization on thin films for reliability purposes. *Microelectron Eng*, 84, 1783–1787.

88. Leite M and Corthesy R. 2001. Stress measurements in concrete structures with modified doorstopper technique. *ACI Struct J*, 98-S59, 98(5), 619–628.

89. Gran JK and Seaman L. 1997. Analysis of piezoresistance gauges for stress in divergent flow fields. *J Eng Mech*, 123(1), 36–44.

90. Kusaka T. 2005. Real-time monitoring of crack growth behavior in concrete structures using the grid method with a crack detection algorithm. In *Structural Health Monitoring*, FK Chang (ed.). DEStech, Lancaster.

91. Measurements Group, Inc. 1988. *Strain Gage Based Transducers*, 2nd Ed. Rayleigh, NC.

92. Huston DR, Fuhr PL, Rosowsky DV, Chen WF, and Kirmani M. 1999. Shoring measurements at museum towers. *Structural Engineering in the 21st Century, Proc ASCE Struct Cong*.

93. Xu Y, Tai YC, Huang A, and Ho CM. 2003. IC-integrated flexible shear-stress sensor skin. *J Microelectromech Syst*, 12(5), 740–747.

94. Saafi M and Romine P. 2003. Health monitoring of concrete structures using embedded MEMS. In *Structural Health Monitoring*, FK Chang (ed.). DEStech, Lancaster.

95. Bolton W. 2000. *Instrumentation & Measurement Pocket Book*, 3rd Ed. Newnes, Oxford.

96. Agilent Technologies. 2000. Practical temperature measurements. *Application Note* 290.

97. Hwang ES and Lee JJ. 2004. Temperature loads for concrete slab bridges. *Proc IABMAS'04 Bridge Maintenance, Safety, Management and Cost*, Kyoto. Taylor & Francis, London.

98. Croarkin MC, Guthrie DC, Burns GW, Kaeser M, and Strouse GF. 1993. Temperature-electromotive force reference function and tables for letter-designated thermocouple types based on the ITS-90. *NIST Monograph 175*, Gaithersburg.

99. Baxter LK. 1997. *Capacitive Sensors Design and Applications*. IEEE Press, New York.

100. Slocum AH. 1992. *Precision Machine Design*. Society of Manufacturing Engineers, Dearborn.

101. Ozevin D, Pessiki S, Jain A, Greve DW, and Oppenheim IJ. 2003. Development of a MEMS device for acoustic emission testing. *Smart Systems and Nondestructive Evaluation for Civil Infrastructures*, SPIE 5057.

102. Acar C and Shkel AM. 2003. Comparative characterisation of low-g capacitive MEMS accelerometers. In *Structural Health Monitoring*, FK Chang (ed.). DEStech, Lancaster.

103. Harpster TJ, Hauvespre S, Dokmeci MR, and Najafi K. 2002. A passive humidity monitoring system for in situ remote wireless testing of micropackages. *J Microelectromech Syst*, 11(1), 61–67.

104. Spillman Jr WB and Durkee SR. 1995. Contactless interrogation of sensors for smart structures. US Patent 5,433,115.

105. Novak LJ, Grizzle KM, Wood SL, and Neikirk DP. 2003. Development of state sensors for civil engineering structures. *Smart Systems and Nondestructive Evaluation for Civil Infrastructures*, SPIE 5057.

106. Finkenzeller K. 2003. *RFID Handbook*, 2nd Ed. Wiley, Chichester.

107. Bolton W. 2000. *Instrumentation & Measurement Pocket Book*, 3rd Ed. Newnes, Oxford.

108. Arms S and Townsend CP. 1996. Differential variable reluctance transducer. US Patent 5,497,147.

109. Pierson JG. 1999. *The Art of Practical and Precise Strain Based Measurement*, 2nd Ed. Pierson and Associates, Rockaway, NJ.

110. Finno RJ and Chao HC. 2000. Nondestructive evaluation of selected drilled shafts at the central artery/tunnel project. In *Structural Materials Technology IV*, S Allampalli (ed.). Technomic, Lancaster.

111. Huber MS, Bay JA, Halling MW, and Womack KC. 2000. Use of velocity transducers in low-frequency modal testing of structures. *Nondestructive Evaluation of Highways, Utilities, and Pipelines IV*, SPIE 3995.

112. Lemke J. 2000. A remote vibration monitoring system using wireless internet data transfer. *Nondestructive Evaluation of Highways, Utilities, and Pipelines IV*, SPIE 3995.

113. Gaberson HA. 2007. Conditions under which displacement, velocity, or acceleration should be used for diagnostic monitoring. *Proc 31st Annual Meeting Vibration Institute*, San Antonio.

114. Eringen AC. 1980. *Mechanics of Continua*. Krieger, New York.

115. Nowacki W. 1979. In *Foundations of Linear Piezoelectricity Electromagnetic Interactions in Elastic Solids*, H. Parkus (ed.). Springer, Wien.

116. PCB Piezotronics, Inc. 2002. *General Piezoelectric Theory*. Depew, New York.

117. Endevco Corp. 2001. *The Handbook of Dynamic Force, Pressure and Acceleration Measurement*. San Juan Capistrano.

118. Gu H, Jin P, Zhao Y, Lloyd GM, and Wang M. 2004. Design and experimental validation of a wireless PVDF displacement sensor for structure monitoring. *Nondestructive Detection and Measurement for Homeland Security II*, SPIE 5395.

119. Park CY and Lee JH. 2004. Low velocity impact monitoring for composite sandwich panels using PVDF sensor. *Proc 2nd European Workshop on Structural Health Monitoring*. DEStech, Munich.

120. Kim IG, Lee HY, Kim JW, and Park CY. 2004. Impact damage monitoring of composite laminates using PVDF sensor signals. *Proc 2nd European Workshop on Structural Health Monitoring*. DEStech, Munich.

121. Blanas P, Rigas E, and Das-Gupta DK. 1999. Health monitoring of composite structures using composite piezoelectric transducers. *Proc 2nd Intl Workshop on Structural Health Monitoring*, Stanford. Technomic, Lancaster.

122. Varadan VK and Varadan VV. 1999. Wireless remotely readable and programmable microsensors and MEMS for health monitoring of aircraft structures. *Proc 2nd Intl Workshop on Structural Health Monitoring*, Stanford. Technomic, Lancaster.

123. Thielicke B, Gesang T, and Wierach P. 2003. Reliability of piezoceramic patch sensors under cyclic mechanical loading. *Smart Mater Struct*, 12, 993–996.

124. Aggarwal MD, Kochary F, Penn BG, and Miller J. 2007. Bulk crystal growth of piezoelectric PMN-PT crystals using a gradient freeze technique for improved SHM sensors. In *Structural Health Monitoring*, FK Chang (ed.). DEStech, Lancaster.

125. Miyashita T, Abe M, and Fujino Y. 2003. Identification of stress field using piezoelectric patch as actuators and sensors. In *Structural Health Monitoring*, FK Chang (ed.). DEStech, Lancaster.

126. Oshima N, Inoue K, Motogi S, and Fukuda T. 2003. Constructing of cure monitoring system with piezoelectric ceramics for composite laminate. *Smart Structures and Integrated Systems*, SPIE 5056.

127. Chen Y, Wen Y, and Li P. 2005. Characterization of concrete stress by measuring dissipation factors of embedded piezoelectric ceramic disc. *Sensors and Smart Structures Technologies for Civil, Mechanical, and Aerospace Systems*, SPIE 5765.

128. Ritdumrongkul S, Abe M, and Fujino Y. 2004. Quantitative health monitoring of bolted joints using a piezoceramic actuator-sensor and laser Doppler vibrometers. *Proc 2nd European Workshop on Structural Health Monitoring*. DEStech, Munich.

129. Xie W, Zhang B, Du S, and Dai F. 2007. Experimental investigation of bolt loosening detection in thermal protection panels at high temperature. In *Structural Health Monitoring*, FK Chang (ed.). DEStech, Lancaster.

130. Moon F. 1984. *Magneto-solid Mechanics*. Wiley, New York.

131. McMasters OD. 1986. Method of forming magnetostrictive rods from rare earth-iron alloys. US Patent 4,609,402.

132. Snodgrass JD and McMasters OD. 2001. High performance rare earth-transition metal magnetostrictive materials. US Patent 2,273,966.

133. Slaughter JC, Raim J, Wun-Fogle M, and Restorff JB. 2005. Effect of cyclic stresses and magnetic fields on stress-annealed galfenol alloys. *Active Materials: Behavior and Mechanics*, SPIE 5761.

134. Quandt E. 1997. Giant magnetostrictive thin film materials and applications. *J Alloys Comp*, 258, 133–137.

135. Sammal O and Ranniku E. 1998. Direct stress measurement in concrete medium with small-sized magnetoelastic stress transducer estodic M20. In *Structural Engineering Worldwide*. GL Fenves, NK Srivastava, AH Ang, and RG Domer (eds.). Elsevier, Amsterdam, T213-5.

136. Bartels KA, Dynes C, Lu Y, and Kwun H. 1999. Evaluation of concrete reinforcements using magnetostrictive sensors. *Nondestructive Evaluation of Bridges and Highways III*, SPIE 3587.

137. Wang ML, Wang G, and Yim J. 2006. Smart cables for cable-stay bridge. *NDE Conf on Civil Engineering*, American Society for Nondestructive Testing, St. Louis.

138. Han SW, Kim YY, and Lee HC. 2003. Non-contact wave sensing for damage detection in a rotating shaft using magnetostrictive sensors. *Smart Sensor Technology and Measurement Systems*, SPIE 5050.

139. Murai R, Yanagisawa E, Oka S, Nakaya E, and Sakate M. 2004. Experimental study on stress measurement of steel bridges using magnetostriction effect. *Proc IABMAS'04 Bridge Maintenance, Safety, Management and Cost*, Kyoto. Taylor & Francis, London.

140. Ogawa Y and Takahashi A. 2004. Maintenance for steel pipelines attached bridges. *Proc IABMAS'04 Bridge Maintenance, Safety, Management and Cost*, Kyoto. Taylor & Francis, London.

141. Culshaw B and Michie WC. 1993. The OSTIC programme—its achievements and their impact on instrumentation in civil engineering. In *Applications of Fiber Optic Sensors in Engineering Mechanics*, F Ansari (ed.). ASCE, New York.

142. Brown JC. 1995. A system to measure lightning-induced transients. *NASA Tech Briefs*, LEW-16095.

143. Grattan KT. 1995. Fiber optic techniques for temperature measurement. In *Optical Fiber Sensor Technology*, K Grattan and B Meggitt (eds.). Chapman & Hall, London.

144. Livingston RA. 1998. Federal highway administration research program in fiber optics for the infrastructure. In *Fiber Optic Sensors for Construction Materials and Bridges*, F Ansari (ed.). Technomic, Lancaster.

145. Elster J, Trego A, Catterall C, Averett J, Jones M, Evans M, and Fielder B. 2003. Flight demonstration of fiber optic sensors. *Smart Sensor Technology and Measurement Systems*, SPIE 5050.

146. Kister G, Winter D, Badcock RA, Fernando GF, Gebremichael Y, Boyle W, Meggitt B, and Keats MA. 2004. Structural health monitoring of an all-composite bridge using fibre Bragg grating sensors. *Proc 2nd European Workshop on Structural Health Monitoring*. DEStech, Munich.

147. Doyle C, Staveley C, and Henderson P. 2003. Structural health monitoring using optical fibre strain sensing systems. In *Structural Health Monitoring*, FK Chang (ed.). DEStech, Lancaster.

148. Kim KS, Chung C, and Kim HJ. 2003. Fiber optic structural monitoring system of concrete beam retrofitted by composite patches. *Industrial and Commercial Applications of Smart Structures Technologies*, SPIE 5054.

149. Li Q, Li G, and Wang G. 2003. Elasto-plastic bond mechanics of embedded fiber optic sensors in concrete under uniaxial tension with strain localization. *Smart Mater Struct*, 12, 851–858.

150. Sirkis JS. 1993. Optical and mechanical isotropies in embedded fiber optic sensors. *Smart Mater Struct*, 2, 255–259.

151. Calero J, Wu SP, Pope C, Chuang SL, and Murtha JP. 1994. Theory and experiments on birefringent optical fibers embedded in concrete structures. *J Lightwave Tech*, 12(6), 1081–1091.

152. Xu Z, Bassam A, Jia H, Tennant A, and Ansari F. 2005. Fiber optic sensor reliability issues in structural health monitoring. *Smart Sensor Technology and Measurement Systems*, SPIE 5758.

153. Inaudi D. 2005. Long-term reliability testing of packaged strain sensors. *Smart Sensor Technology and Measurement Systems*, SPIE 5758.

154. Udd E, Winz M, Kreger S, and Heider D. 2005. Failure mechanisms of fiber optic sensors placed in composite materials. *Smart Sensor Technology and Measurement Systems*, SPIE 5758.

155. Berghmans F. 2005. Reliability of components for fiber optic sensors. *Smart Sensor Technology and Measurement Systems*, SPIE 5758.

156. Ott M. 2005. Validation of commercial fiber optic components for aerospace environments. *Smart Sensor Technology and Measurement Systems*, SPIE 5758.

157. Murtaza G and Senior JM. 1995. Schemes for referencing of intensity-modulated optical sensor systems. In *Optical Fiber Sensor Technology*, K Grattan and B Meggitt (eds.). Chapman & Hall, London.

158. Grossman B, Cosentino P, Doi S, Kumar G, and Vergese J. 1994. Development of microbend sensors for pressure, load, and displacement measurements in civil engineering. *Smart Sensing, Processing, and Instrumentation*, SPIE 2191.

159. Lessing R. 1990. Light wave conductor-bending sensor with sliding rails for monitoring bridge structures or the like. US Patent 4,972,073.

160. Braustein JB, Ruchala J, and Hodac B. 2002. Smart structures: Fiber-optic deformation and displacement monitoring. *First Intl Conf on Bridge Maintenance, Safety and Management*. IABMAS, Barcelona.

161. Donlagic D and Culshaw B. 1998. Mode selective/filtering microbend sensor structure for use in distributed and quasi distributed sensor systems. *Sensory Phenomena and Measurement Instrumentation for Smart Structures and Materials*, SPIE 3330.

162. Dansich LA. 1992. Bend-enhanced fiber optic sensors. *Fiber Optic and Laser Sensors X*, SPIE 1795.

163. Dubaniewicz TH, Heasley KA, and DiMartino MD. 1994. Fiber optic stress sensor development. *Time Domain Reflectometry in Environmental, Infrastructure, and Mining Applications*, US Bureau of Mines, SP 19-94, September.

164. Asanuma H. 2004. Formation of innovative sensors in polymer and metal based materials and composites for health monitoring. *Proc 2nd European Workshop on Structural Health Monitoring*. DEStech, Munich.

165. Udd E. 1991. *Fiber Optic Sensors*. Wiley, New York.

166. Claus RO, Gunther MF, Wang AB, Murphy KA, and Sun D. 1993. Extrinsic Fabry–Perot sensor for structural evaluation. In *Applications of Fiber Optic Sensors in Engineering Mechanics*, F Ansari (ed.). ASCE, New York.

167. Lee CE, Alcoz JJ, Yeh Y, Gibler WN, Atkins RA, and Taylor HF. 1992. Optical fiber Fabry–Perot sensors for smart structures. *Smart Mater Struct*, 1, 123–127.

168. Gloetzel R, Hofmann D, Basedau F, and Habel W. 2005. Geotechnical pressure cell using a long-term reliable high-precision fibre optic sensor head. *Smart Sensor Technology and Measurement Systems*, SPIE 5758.

169. Lopex-Anido R and Fifield S. 2003. Experimental methodology for embedding fiber optic strain sensors in fiber reinforced composites fabricated by the vartm/scrimp process. In *Structural Health Monitoring*, FK Chang (ed.). DEStech, Lancaster.

170. Masri SF, Agbabian MS, Abdel-Ghaffar AM, Higazy M, Claus RO, and de Vries MJ. 1994. Experimental study of embedded fiber-optic strain gauges in concrete structures. *J Eng Mech*, 120(8), 1696–1717.

171. Choquet P, Leroux R, and Juneau F. 1997. New Fabry–Perot fiber-optic sensors for structural and geotechnical monitoring applications. *Trans Res Rec*, 1596, 39–44.

172. Bonfiglioli B and Pascale G. 2003. Internal strain measurements in concrete elements by fiber optic sensors. *J Mater Civil Eng*, 15(2), 125–133.

173. Chen WM, Zhu Y, Fu YM, and Huang S. 2004. Study on the embedment of fiber Fabry–Perot strain sensor in prestressed reinforced concrete bridges. *Sensors and Smart Structures Technologies for Civil, Mechanical and Aerospace Systems*, SPIE 5391.

174. Benmokrane B, Quiriron M, El-Salakawy E, Debaikey A, and Lackey T. 1993. Fabry–Perot sensors for the monitoring of FRP reinforced bridge decks. *Nondestructive Evaluation and Health Monitoring of Aerospace Materials and Composites III*, SPIE 5393.

175. Herrero VV and Casas JR. 2004. Long-term monitoring of prestressing strands in post-tensioned concrete bridges and structures. *Proc IABMAS'04 Bridge Maintenance, Safety, Management and Cost*, Kyoto. Taylor & Francis, London.

176. Lawrence CM, Nelson DV, Spingarn JR, and Bennett TE. 1996. Measurement of process-induced strains in composite materials using embedded fiber optic sensors. *Smart Sensing, Processing and Instrumentation*, SPIE 2718.

177. Stewart A, Carman G, and Richards L. 2003. Thermography technique for graphite/epoxy composite structures utilizing embedded thermal fiber optic sensors. *Smart Structures and Integrated Systems*, SPIE 5056.

178. Valla T, Hogg D, and Measures RM. 1992. Fiber-optic Fabry–Perot strain rossettes. *Smart Mater Struct*, 1, 227–232.

179. Yu M. 2005. Digital phase demodulation technique based on low coherence fiber optic interferometry and phase measurement interferometry. *Smart Sensor Technology and Measurement Systems*, SPIE 5758.

180. Kim DH, Chang YH, Han JH, and Lee I. 2005. Optical phase estimation for a patch-type extrinsic Fabry–Perot interferometer sensor system and its application to flutter suppression. *Smart Mater Struct*, 14, 696–706.

181. Frazer R. 1992. Polarization-rotating sensors connected to optical fibers. *NASA Tech Brief*, 16(5), 64.

182. Bock WJ and Eftimov TA. 1993. Single- and few-mode fiber optic pressure sensors. In *Applications of Fiber Optic Sensors in Engineering Mechanics*, F Ansari (ed.). ASCE, New York, NY.

183. Dunphy J, Meltz G, and Morey W. 1991. Optical fiber Bragg grating sensors: A candidate for smart structures applications. *Fiber Optic Smart Structures*, Chapter 10, E Udd (ed.). Wiley-Interscience, New York, NY.

184. Maher MH, Chen B, Prohaska JD, Nawy EG, and Snitzer E. 1994. Fiber optic sensor for measurement of strain in concrete structures. *New Experimental Techniques for Evaluating Concrete Material and Structural Performance*, American Concrete Institute, ACI SP-143.

185. Budtsev Y, Gorbatov N, Tur M, Green AK, Kressel I, Ben-Simon U, Ghilai G, Shafir E, Berkovic G, and Gali S. 2004. Smart bonded composite repairs for aging aircraft. *Proc 2nd European Workshop on Structural Health Monitoring*. DEStech, Munich.

186. Chehura E, Ye CC, Staines SE, James SW, and Tatam RP. 2004. Characterization of the response of fibre Bragg gratings fabricated in stress and geometrically induced high birefringence fibres to temperature and transverse load. *Smart Mater Struct*, 13, 888–895.

187. Bosia F, Giaccari P, Botsis J, Facchini M, Limberger HG, and Salathé RP. 2003. Characterization of the response of fibre Bragg grating sensors subjected to a two-dimensional strain field. *Smart Mater Struct*, 12, 925–934.

188. Guo ZS. 2007. Cryogenic temperature characteristics of the fiber Bragg grating sensors. In *Structural Health Monitoring*, FK Chang (ed.). DEStech, Lancaster.

189. Dong X, Yang X, Zhao CL, Ding L, Shum P, and Ngo NQ. 2005. A novel temperature-insensitive fiber Bragg grating sensor for displacement measurement. *Smart Mater Struct*, 14, N7–N10.
190. Giaccari P, Dunkel GR, Humbert L, Botsis J, Limberger HG, and Salathé RP. 2005. On direct determination of non-uniform internal strain fields using fibre Bragg gratings. *Smart Mater Struct*, 14, 127–136.
191. Ferguson S, Graver TW, and Mendez A. 2005. Advances in temperature compensation of fiber-optic strain sensors. In *Structural Health Monitoring*, FK Chang (ed.). DEStech, Lancaster.
192. Yoon HJ, Costantini DM, Michaud V, Limberger HG, Manson JA, Salathé PR, Kim GK, and Hong GS. 2005. In-situ simultaneous strain and temperature measurement of adaptive composite materials using a fiber Bragg grating based sensor. *Smart Sensor Technology and Measurement Systems*, SPIE 5758-8.
193. Xie JF, Zhang H, Zhu Z, Xu JN, Hu RH, and Song LF. 2007. A study of the temperature sensitivity of fiber Bragg gratings after metallization. *Smart Mater Struct*, 16, 1837–1842.
194. Giordano M, Laudati A, Russo M, Nasser J, Cutolo A, and Cusano A. 2003. Full cure monitoring by fiber optic dual functionality sensing system. In *Structural Health Monitoring*, FK Chang (ed.). DEStech, Lancaster.
195. Smith KH, Ipson BL, Lowder TL, Hawkins AR, Selfridge RH, and Schultz SM. 2005. Etched in-fiber Bragg gratings for temperature sensing at high temperatures. *Smart Sensor Technology and Measurement Systems*, SPIE 5758.
196. Udd E. 1995. *Fiber Optic Smart Structures*. Wiley, New York.
197. Measures RM, Melle S, and Liu K. 1992. Wavelength demodulated Bragg grating fiber optic sensing system for addressing smart structure critical issues. *Smart Mater Struct*, 1, 39–44.
198. Grant J, Kaul R, Taylor S, Myer G, Jackson K, and Sharma A. 2003. Distributed sensing of carbon-epoxy composites and composite wound pressure vessels using fiber-Bragg gratings. *Smart Sensor Technology and Measurement Systems*, SPIE 5050.
199. Breglio G, Cusano A, Irace A, and Cutolo A. 2003. A new method for the multiplexing of fiber optic Bragg sensor arrays. In *Structural Health Monitoring*, FK Chang (ed.). DEStech, Lancaster.
200. Magne S, Boussoir J, Rougeault S, Marty-Dewynter V, Ferdinand P, and Bureau L. 2003. Health monitoring of the Saint-Jean Bridge of Bordeaux, France using fiber Bragg grating extensometers. *Smart Sensor Technology and Measurement Systems*, SPIE 5050.
201. Takeya H, Ozaki T, and Takeda N. 2005. Structural health monitoring of advanced grid structure using multipoint FBG sensors. *Industrial and Commercial Applications of Smart Structures Technologies*, SPIE 5762.
202. Measures RM. 1993. Smart structure technology. In *Applications of Fiber Optic Sensors in Engineering Mechanics*, F Ansari (ed.). ASCE, New York.
203. Neubauer S, Hemphill D, Phares BM, Wipf TJ, Doornink JD, Greimann LF, and Monk C. 2004. Use of fiber Bragg gratings for the long-term monitoring of a high performance steel bridge. *Proc Structural Materials Technology VI, an NDT Conference*, Buffalo.
204. Sirkis JS. 1998. Using Bragg grating sensor systems in construction materials and bridges: Perspectives and challenges. In *Fiber Optic Sensors for Construction Materials and Bridges*, F Ansari (ed.). Technomic, Lancaster.
205. Matrat J, Levin K, and Jarlas R. 1999. Effect of debonding on strain measurement of embedded Bragg grating sensors. *Proc 2nd Intl Workshop on Structural Health Monitoring*, Stanford. Technomic, Lancaster.
206. Tanaka N, Okabe Y, and Takeda N. 2003. Temperature-compensated strain measurement using fiber Bragg grating sensors embedded in composite laminates. *Smart Mater Struct*, 12, 940–946.
207. Sennhauser U, Broennimann R, Mauron P, and Nellen PM. 1998. Reliability of optical fibers and Bragg grating sensors for bridge monitoring. In *Fiber Optic Sensors for Construction Materials and Bridges*, F Ansari (ed.). Technomic, Lancaster.
208. Martinez F, Martinez M, Obieta G, Sanchez F, and Santiago J. 2004. Structural monitoring with FBG sensors: Industrial applications. *Proc 2nd European Workshop on Structural Health Monitoring*. DEStech, Munich.

209. Yamakawa H, Iwaki H, Mita A, and Takeda N. 1999. Health monitoring of steel structures using fiber Bragg grating sensors. *Proc 2nd Intl Workshop on Structural Health Monitoring*, Stanford. Technomic, Lancaster.

210. Li SG and Zhu J. 2006. Application study on surface FBG sensor in structure experiment. *Proc 4th China–Japan–US Symp on Structural Control and Monitoring*, Hangzhou.

211. Spammer SJ, Fuhr PL, Nelson M, and Huston D. 1988. Rebar-epoxied optical fiber Bragg gratings for civil structures. *Microwave Opt Technol Lett*, 18(3), 214–219.

212. Davis MA, Bellemore DG, Kersey AD, Putnam MA, Friebele EJ, Idriss RL, and Kodinduma M. 1996. High sensor count Bragg grating instrumentation system for large-scale structural monitoring applications. *Smart Sensing, Processing and Instrumentation*, SPIE 2718.

213. Ou JP and Zhou Z. 2004. Techniques of optical fiber Bragg grating smart sensors and intelligent monitoring systems of infrastructures. In *Advanced Smart Materials and Smart Structures Technology*. FK Chang, CB Yun, and BF Spencer, Jr. (eds.). DEStech, Lancaster.

214. Idriss RL, Kodindouma MB, Kersey AD, and Davis MA. 1998. Multiplexed Bragg grating optical fiber sensors for damage evaluation in highway bridges. *Smart Mater Struct*, 7, 209–216.

215. Baukat DM and Gallon A. 2002. Performance of fiber Bragg grating sensors during testing of FRP composite bridge decks. In *Proc Structural Materials Technology V, an NDT Conference*, S Allampalli and G Washer (eds.). American Society of Nondestructive Testing, Cincinnati.

216. Sundaram R, Kamath GM, Gupta N, and Rao MS. 2005. Structural health monitoring of co-cured composite structures using FBG sensors. *Smart Structures and Integrated Systems*, SPIE 5764.

217. Cheng LK, Liu SY, Guan BO, Chung WH, Chan TH, Schaarsberg JJ, Oostijk BW, and Tam HY. 2004. Dynamic load monitoring of the Tsing Ma Bridge using a high-speed FBG sensor system. *Proc 2nd European Workshop on Structural Health Monitoring*. DEStech, Munich.

218. Kawiecki G, Pena J, and Guemes A. 2004. Strain monitoring using FBG optical transducers in a mechanical test of an aeronautical structure. *Proc 2nd European Workshop on Structural Health Monitoring*. DEStech, Munich.

219. Takeda N and Mizutani T. 2004. Real-time strain measurement of a composite LH2 tank mounted on a reusable rocket test vehicle using an onboard FBG measurement system. *Proc 2nd European Workshop on Structural Health Monitoring*. DEStech, Munich.

220. Johnson GA, Pran K, Wang G, Havsgard GB, and Vohra ST. 1999. Structural monitoring of a composite hull air cushion catamaran with a multi-channel fiber Bragg grating sensor system. *Proc 2nd Intl Workshop on Structural Health Monitoring*, Stanford. Technomic, Lancaster.

221. Read IJ, Foote PD, Roberts D, Zhang L, Bennion I, and Carr M. 1999. Smart carbon-fibre mast for a super-yacht. *Proc 2nd Intl Workshop on Structural Health Monitoring*, Stanford. Technomic, Lancaster.

222. Read IJ and Foote PD. 2001. Sea and flight trials of optical fibre Bragg grating strain sensing system. *Smart Mater Struct*, 10, 1085–1094.

223. Foedinger R, Rea D, Sirkis J, Troll J, Grande R, and Vandiver TL. 1999. Structural health monitoring and impact damage detection for filament wound composite pressure vessels. *Proc 2nd Intl Workshop on Structural Health Monitoring*, Stanford. Technomic, Lancaster.

224. Idriss RL, White KR, Pate JW, Vohra ST, Chang CC, and Danver BA. 1998. Monitoring of an Interstate Highway Bridge using a network of optical fiber sensors. In *Structural Engineering Worldwide*. GL Fenves, NK Srivastava, AH Ang, and RG Domer (eds.). Elsevier, Amsterdam, T208-6.

225. Zhao H, Zhang B, Wang R, Wu Z, and Wang D. 2007. Monitoring of composite pressure vessel using two kinds of fiber optic sensors. In *Structural Health Monitoring*, FK Chang (ed.). DEStech, Lancaster.

226. Fernandez-Lopez A, Wagner W, and Guemes A. 2007. Embedded sensors at the root of a helicopter blade. *Structural Health Monitoring*, FK Chang (ed.). DEStech, Lancaster.

227. Zonta D, Bursi OS, and Pozzi M. 2004. Development of an FBG-based dynamic measurement system for real-time monitoring of RC elements. *Proc 2nd European Workshop on Structural Health Monitoring*. DEStech, Munich.

228. Maurin L, Ferdinand P, Laffont G, Roussel N, Boussoir J, and Rougeault S. 2007. High speed real-time contact measurements between a smart train pantograph with embedded fibre Bragg grating sensors and its overhead contact line. In *Structural Health Monitoring*, FK Chang (ed.). DEStech, Lancaster.

229. Saouma VE, Anderson DZ, Ostrander K, Lee B, and Slowik V. 1998. Application of fiber Bragg grating in local and remote infrastructure health monitoring. *Mater Struct*, 31, 259–266.

230. Montanini R and D'Acquisto L. 2007. Simultaneous measurement of temperature and strain in glass fiber/epoxy composites by embedded fiber optic sensors: I. Cure monitoring. *Smart Mater Struct*, 16, 1718–1726.

231. Wnuk VP, Méndez A, and Graver T. 2005. Process for mounting and packaging of fiber Bragg grating strain sensors for use in harsh environment applications. *Smart Sensor Technology and Measurement Systems*, SPIE 5758.

232. Ren L, Li HN, Sun L, and Li DS. 2005. Development of tube-packaged FBG strain sensor and application in the vibration experiment of submarine pipeline model. *Advanced Sensor Technologies for Nondestructive Evaluation and Structural Health Monitoring*, SPIE 5770.

233. Schmidt-Hattenberger C, Straub T, Naumann M, Borm G, Laurerer R, Beck C, and Schwartz W. 2003. Strain measurements by fiber Bragg grating sensors for in-situ pile loading tests. *Smart Sensor Technology and Measurement Systems*, SPIE 5050.

234. Hudson HL and Little RR. 2005. Installation and demonstration of embedded sensors for solid rocket motor health monitoring. In *Structural Health Monitoring*, FK Chang (ed.). DEStech, Lancaster.

235. Seim J, Udd E, Schulz W, and Laylor HM. 1999. Health monitoring of an Oregon Historical Bridge with fiber grating strain sensors. *Smart Systems for Bridges, Structures, and Highways*, SPIE 3671.

236. Pan XW, Liang DK, and Li D. 2006. Optical fiber sensor layer embedded in smart composite material and structure. *Smart Mater Struct*, 15, 1231–1234.

237. Takeda S, Yamamoto T, Okabe Y, and Takeda N. 2004. Debonding monitoring of a composite repair patch using small-diameter FBG sensors. *Smart Electronic, MEMS, BioMEMS, and Nanotechnology*, SPIE 5389.

238. Li HC, Davis CE, Herszberg I, Mouritz AP, and Galea SC. 2004. Application of fiber optic Bragg grating sensors for the detection of disbonds in bonded composite ship joints. *Proc 2nd European Workshop on Structural Health Monitoring*. DEStech, Munich.

239. Guemes A, Pintado JM, Frovel M, and Menendez JM. 2004. FBG based SHM systems in monolithic composite structures for aerospace. *Proc 2nd European Workshop on Structural Health Monitoring*. DEStech, Munich.

240. Kosmatka J. 2003. Detecting impact damage and location in multifunctional metallic–intermetallic (MIL) composites. In *Structural Health Monitoring*, FK Chang (ed.). DEStech, Lancaster.

241. Mueller UC, Raffaelli L, Reutlinger A, Latka I, Ecke W, Tumino G, and Baier H. 2004. Integration and operation of fiber optic sensors in cryogenic composite tank structures. *Proc 2nd European Workshop on Structural Health Monitoring*. DEStech, Munich.

242. Udd E, Schulz WL, Seim J, Corona-Bittick K, Dorr J, Slattery K, Laylor H, and McGill G. 1999. Fiber optic smart bearing load structure. *Nondestructive Evaluation of Bridges and Highways III*, SPIE 3587.

243. El-Sherif M, Li M, El-Sherif D, Hidayet M, Rahman A, and Lee C. 2003. Fiber optic system for 2-D strain measurements in parachutes. In *Structural Health Monitoring*, FK Chang (ed.). DEStech, Lancaster.

244. Connelly MJ, Moloney S, and Butler P. 2005. Tunable laser based carbon composite strain sensing system using wavelength division multiplexed fiber Bragg grating sensors. *Smart Sensor Technology and Measurement Systems*, SPIE 5758.

245. Kang D, Park S, and Kim C. 2005. In-situ health monitoring of filament wound pressure tanks using embedded fiber Bragg grating sensors. In *Structural Health Monitoring*, FK Chang (ed.). DEStech, Lancaster.

246. Eum S, Kageyama K, Maurayama H, Uzawa K, Ohsawa I, Kanai M, and Igawa H. 2007. Process monitoring for composite structures fabricated by VARTM using fiber Bragg grating sensors. In *Structural Health Monitoring*, FK Chang (ed.). DEStech, Lancaster.

247. Ling HY, Lau KT, Zhou LM, Cheng L, Teng JG, and Jin W. 2004. Ultilisation of embedded fibre-optic sensors for infrastructure applications. In *Advanced Smart Materials and Smart Structures Technology*. FK Chang, CB Yun, and BF Spencer, Jr. (eds.). DEStech, Lancaster.

248. Satori K, Ikeda Y, Kurosawa Y, Hongo A, and Takeda N. 2000. Development of small-diameter optical fiber sensors for damage detection in composite laminates. *Sensory Phenomena and Measurement Instrumentation for Smart Structures and Materials*, SPIE 3986.

249. Hongo A, Fukuchi K, Kojima S, and Takeda N. 2003. Embedded small-diameter fiber Bragg grating sensors and high speed wavelength detection. *Smart Sensor Technology and Measurement Systems*, SPIE 5050.

250. Mizutani T, Okabe Y, and Takeda N. 2003. Quantitative evaluation of transverse cracks in carbon fiber reinforced plastic quasi-isotropic laminates with embedded small-diameter fiber Bragg grating sensors. *Smart Mater Struct*, 12, 898–903.

251. Lin YB, Chang KC, Chern JC, and Wang LA. 2005. Packaging methods of fiber-Bragg grating sensors in civil structure applications. *IEEE Sens J*. 5(3), 419–424.

252. Falciai R, Trono C, Lanterna G, and Castelli C. 2003. FBG sensors for painted wood panel deformation monitoring. *Smart Sensor Technology and Measurement Systems*, SPIE 5050.

253. Yoon Y, Chung S, Kim M, Lee WI, and Lee B. 2003. Vibration detection for a composite smart structure embedded with a fiber grating sensor. *Smart Sensor Technology and Measurement Systems*, SPIE 5050.

254. Ahmad H, Harun SW, Chong WY, Zulkifli MZ, Thant MM, Yusof Z, and Poopalan P. 2008. High-sensitivity pressure sensor using a polymer-embedded FBG. *Microwave Opt Technol Lett*, 50(1), 60–61.

255. Gonzalez IF, Chang FK, Qing PX, Kumar A, and Zhang D. 2005. High speed hybrid active system. *Sensors and Smart Structures Technologies for Civil, Mechanical, and Aerospace Systems*, SPIE 5765.

256. Dai J, Zhang W, Du Y, and Sun B. 2006. Study on meantime measurement of the strain and the temperature based on fiber optic sensing. *Proc 4th China–Japan–US Symp on Structural Control and Monitoring*, Hangzhou.

257. Chang YH, Kim DH, Lee I, and Han JH. 2006. A patch-type smart self-sensing actuator. *Smart Mater Struct*, 15, 667–677.

258. Moerman W, Taerwe L, De Waele W, Degrieck J, and Baets R. 1999. Remote monitoring of concrete elements by means of Bragg gratings. *Proc 2nd Intl Workshop on Structural Health Monitoring*, Stanford. Technomic, Lancaster.

259. Ecke W, Latka I, Hoefer B, Reutlinger A, and Willsch R. 2003. Optical fiber grating sensor network monitors helium gas temperature profile in airship. *Smart Sensor Technology and Measurement Systems*, SPIE 5050.

260. Player J, Slade J, Kristoff SB, Kokkins SJ, Lusignea R, and Livingston R. 2005. Fiber optic Bragg grating sensor for in situ measurement of relative humidity in concrete. In *Structural Health Monitoring*, FK Chang (ed.). DEStech, Lancaster.

261. Kunzler M, Udd E, Johnson M, and Mildenhall K. 2005. Use of multidimensional fiber grating strain sensors for damage detection in composite pressure vessels. *Smart Sensor Technology and Measurement Systems*, SPIE 5758.

262. Udd E, Kreger S, Calvert S, Kunzler M, and Davol K. 2003. Usage of multi-axis fiber grating strain sensors to support nondestructive evaluation of composite parts and adhesive bond lines. In *Structural Health Monitoring*, FK Chang (ed.). DEStech, Lancaster.

263. Michie C, Culshaw B, Thursby G, Jin W, and Kostantaki M. 1996. Optical sensors for temperature and strain measurement. *Smart Sensing, Processing and Instrumentation*, SPIE 2718.

264. Zhang Y, Yin Z, Chen B, and Cui HL. 2005. A novel fiber Bragg grating based seismic geophone for oil/gas prospecting. *Sensors and Smart Structures Technologies for Civil, Mechanical, and Aerospace Systems*, SPIE 5765.

265. Koh JI, Bang HJ, Kim CG, and Hong CS. 2005. Simultaneous measurement of strain and damage signal of composite structures using a fiber Bragg grating sensor. *Smart Mater Struct*, 14, 658–663.

266. Nguyen AD. 2007. High performance fiber optic strain and ultrasonic wave testing. In *Structural Health Monitoring*, FK Chang (ed.). DEStech, Lancaster.

267. Hsu K, Haber T, Mock J, Volcy J, and Graver TW. 2003. High-speed swept laser interrogation system for vibration monitoring. In *Structural Health Monitoring*, FK Chang (ed.). DEStech, Lancaster.

268. Coppola G, Ferraro P, Iodice M, Rendina I, Rocco A, Inserra S, and Camerlingo FP. 2003. A new approach for detecting strain in FBG sensor by a silicon based etalon filter. In *Structural Health Monitoring*, FK Chang (ed.). DEStech, Lancaster.

269. Thomson DJ, Rivera E, Brown B, and Mukhi SN. 2003. High speed strain measurements with fiber Bragg sensors and gas cell wavelength references. In *Structural Health Monitoring*, FK Chang (ed.). DEStech, Lancaster.

270. Komatsuzaki S, Kojima S, Hongo A, Takeda N, and Sakurai T. 2005. Development of high-speed optical wavelength interrogation system for damage detection in composite materials. *Smart Sensor Technology and Measurement Systems*, SPIE 5758.

271. Qiao Y, Zhou Y, and Krishnaswamy S. 2005. Adaptive two-wave mixing wavelength demodulation of fiber Bragg grating sensor for monitoring dynamic strains. *Smart Sensor Technology and Measurement Systems*, SPIE 5758.

272. Ye C-C and Tatam RP. 2005. Ultrasonic sensing using Yb^{3+}/Er^{3+}-codoped distributed feedback fibre grating lasers. *Smart Mater Struct*, 14(1), 170–176.

273. Abdi AM, Suzuki S, Schülzgen A, and Kost AR. 2005. Fiber Bragg grating array calibration. *Sensors and Smart Structures Technologies for Civil, Mechanical, and Aerospace Systems*, SPIE 5765.

274. Adamovsky G, Juergens J, and Floyd B. 2005. Processing of signals from fiber Bragg gratings using unbalanced interferometers. *Smart Sensor Technology and Measurement Systems*, SPIE 5758.

275. Okabe Y, Minakuchi S, and Takeda N. 2005. Identification of impact damage in sandwich structures by application of high speed MEMS-OSA to FBG sensors. *Smart Sensor Technology and Measurement Systems*, SPIE 5758.

276. Nogueira MM, Morikawa SR, Kato CC, Valente LC, and Braga AM. 2005. Fiber Bragg grating accelerometer for structural monitoring of transmission lines. In *Structural Health Monitoring*, FK Chang (ed.). DEStech, Lancaster.

277. Jiang DS and Liu SC. 2005. The application and research of FBG sensors based load cell to Bowstring Arch Bridge health monitoring system. In *Structural Health Monitoring*, FK Chang (ed.). DEStech, Lancaster.

278. Bhola B, Song HC, Tazawa H, and Steier WH. 2005. Polymer microresonator strain sensors. *IEEE Photon Technol Lett*, 17(4), 867–869.

279. Taillaert D, Van Paepegem W, Vlekken J, and Baetsa R. 2007. A thin foil optical strain gage based on silicon-on-insulator microresonators. *Third European Workshop on Optical Fibre Sensors*, SPIE 6619.

280. Huston D and Fuhr PL. 1997. Fiber optic bridge deck chloride detection. *Building to Last*, ASCE Structures Congress, pp. 974–979.

281. Hewlett-Packard. 1996. Optical spectrum analysis. *Application Note* 1550-4.

282. Boisde G and Harmer A. 1996. *Chemical and Biochemical Sensing with Optical Fibers and Waveguides*. Artech House, Norwood.

283. Norris JO. 1995. Multimode optical fiber chemical sensors. In *Optical Fiber Sensor Technology*, K Grattan and B Meggitt (eds.). Chapman & Hall, London.

284. Fuhr PL and Huston DR. 1998. Corrosion detection in reinforced concrete roadways and bridges via embedded fiber optic sensors. *Smart Mater Struct*, 7, 217–228.

285. Ghandehari M and Vimer C. 2004. Optical spectroscopy for detection of moisture ingress in civil infrastructure materials. *Proc Structural Materials Technology VI, an NDT Conference*, Buffalo.

286. Namkung JS, Hoke M, and Rogowski RS. 1994. Optical fiber FTIR remote detection of aluminum hydroxide. *Smart Sensing, Processing, and Instrumentation*, SPIE 2191.

287. Huang H, Tata US, and Majumdar A. 2007. A novel all-fiber surface roughness sensor based on laser scattering. In *Structural Health Monitoring*, FK Chang (ed.). DEStech, Lancaster.

288. Palmer JA, Wang X, Shoureshi RA, Mander A, and Torgerson D. 2000. Optical methodology for the health assessment of power transformers. *Nondestructive Evaluation of Highways, Utilities, and Pipelines IV*, SPIE 3995.

289. Fair GE and Parthasarathy TA. 2006. Polymeric thermal history sensor. US Patent 7,080,939.

290. Beckmann M, Kroener M, Makendowski P, Wiese S, Johannes HH, and Kowalsky W. 2003. New designs of fiber-optical sensors in structural health monitoring. In *Structural Health Monitoring*, FK Chang (ed.). DEStech, Lancaster.

291. Vimer CS, Ghandehari M, and Spellane P. 2003. Monitoring degradation chemistries of infrastructural materials. In *Structural Health Monitoring*, FK Chang (ed.). DEStech, Lancaster.

292. Fuhr PL and Huston DR. 2000. Fiber optic chloride threshold detectors for concrete structures. *J Struct Control*, 7(1), 77–102.

293. Wiese S, Kowalsky W, Wichern J, and Grahn W. 1999. Fiberoptical sensors for on-line monitoring of moisture in concrete structures. *Proc 2nd Intl Workshop on Structural Health Monitoring*, Stanford. Technomic, Lancaster.

294. Dantan N, Habel WR, and Wolfbeis OS. 2005. Fiber optic pH sensor for early detection of danger of corrosion in steel-reinforced concrete structures. *Smart Sensor Technology and Measurement System*, SPIE 5758.

295. Srinivasan R, Phillips TE, Bargeron CB, Carlson MA, Schemm ER, and Saffarian HM. 2000. Embedded micro-sensor for monitoring pH in concrete structures. *Smart Systems for Bridges, Structures, and Highways*, SPIE 3988.

296. Pidaparti RM, Neblett EB, Miller SA, and Alvarez JC. 2008. Monitoring the corrosion process of Al alloys through pH induced fluorescence. *Smart Mater Struct*, 17, 015001.

297. Ghandehari M, Khalil G, and Kimura F. 2005. Detection of gas leaks in the subsurface environment. *Nondestructive Detection and Measurement for Homeland Security III*, SPIE 5769.

298. Pauri M, Monosi S, Alvera I, and Collepardi M. 1989. Assessment of free and bound chloride in concrete. *Mater Eng*, 1(2), 497–501.

299. Kanada H, Ishikawa Y, and Uomoto T. 2006. Measurement of chloride content in concrete using near-infrared spectroscopy and X-ray fluorescence analysis. NDE *Conf on Civil Engineering*, American Society for Nondestructive Testing, St. Louis.

300. Wilsch G, Weritz F, Schaurich D, Taffe A, and Wiggenhauser H. 2004. Laser-induced breakdown spectroscopy for on-site determination of sulfur and concrete. *Proc Structural Materials Technology VI, an NDT Conference*, Buffalo.

301. Grubb JA, Limaye HS, Kakade AM. 2007. Testing pH of concrete. *Concrete Intl*, 29(4), 78–83.

302. Fuhr PL and Huston DR. 1998. Embedded fiber optic sensors for bridge deck chloride penetration measurements. *J Opt Eng*, 37(4), 1221–1228.

303. Washer G and Brooks T. 2006. Application of Raman spectroscopy for nondestructive evaluation of composite materials. *NDE Conf on Civil Engineering*, American Society for Nondestructive Testing, St. Louis.

304. Simonen JT, Andringa MM, Grizzle KM, Wood SL, and Neikirk DP. 2004. Wireless sensors for monitoring corrosion in reinforced concrete members. *Sensors and Smart Structures Technologies for Civil, Mechanical and Aerospace Systems*, SPIE 5391.

305. Udd E, Hague ED, and Trego A. 2000. Fiber optic grating corrosion and chemical sensor. US Patent 6,144,026.

306. Dickerson NP, Simonen JT, Andringa MM, Wood SL, and Neikirk DP. 2005. Wireless low-cost corrosion sensors for reinforced concrete structures. *Sensors and Smart Structures Technologies for Civil, Mechanical, and Aerospace Systems*, SPIE 5765.

307. Rutherford P, Ikegami R, and Shrader J. 1996. Novel NDE fiber optic corrosion sensor. *Smart Sensing, Processing and Instrumentation*, SPIE 2718.
308. Mugambi EM, Kwan K, Laskowski BC, Ooi TK, and Corder A. 2007. MEMS based strain and corrosion sensors for structural health monitoring. In *Structural Health Monitoring*, FK Chang (ed.). DEStech, Lancaster.
309. Li XM, Chen W, Huang ZQ, Huang S, and Bennett KD. 1998. Fiber optic corrosion sensor fabricated by electrochemical method. *Sensory Phenomena and Measurement Instrumentation for Smart Structures and Materials*, SPIE 3330.
310. Dong S, Peng G, and Luo Y. 2007. Preparation techniques of metal clad fibres for corrosion monitoring of steel materials. *Smart Mater Struct*, 16, 733–738.
311. Cooper KR, Ma Y, Wikswo JP, and Kelly RG. 2004. Simultaneous monitoring of the corrosion activity and moisture inside aircraft lap joints. *Sensors and Smart Structures Technologies for Civil, Mechanical and Aerospace Systems*, SPIE 5391.
312. Flachsbarth J, Brandes M, Kowalsky W, and Johannes HH. 2004. Humidity sensor based on Reichart's dye prepared with eximer-laser. *Proc 2nd European Workshop on Structural Health Monitoring*. DEStech, Munich.
313. Cusson D, Taylor D, Glazer R, and Arnott M. 1998. Remote sensing of the performance of a repaired concrete highway bridge. *Proc Intl Conf on Corrosion and Rehabilitation of Reinforced Concrete Structures*, Orlando.
314. Fink K and Payer JH. 1991. Corrosion sensors for condensate corrosivity in heat exchangers. Paper no. 563, *Proc NACE Annual Conference*, Houston, TX.
315. Braunling R and Dietrich P. 2005. Corrosion and corrosivity monitoring system. *Sensors and Smart Structures Technologies for Civil, Mechanical, and Aerospace Systems*, SPIE 5765.
316. Gostautas R, Finlayson R, Betti R, Smyth A, Khazem D, Mahmoud K, and Aktan AE. 2005. Corrosion monitoring research for suspension bridge cables. In *Structural Health Monitoring*, FK Chang (ed.). DEStech, Lancaster.
317. Cosentino P, Grossman B, Shieh C, Doi S, Xi H, and Erbland P. 1995. Fiber optic chloride sensor development. *J Geotech Eng*, 121(8), 610–617.
318. Seim J, Udd E, Schulz W, and Laylor HM. 1999. Smart Systems for Bridges, Structures, and Highways, SPIE 3746.

3

Distributed and Noncontact Sensors

3.1 Introduction

Distributed sensors measure over lines, curves, areas, and/or volumes and have capabilities and measurement modalities that extend beyond those of isolated point sensors or arrays of point sensors. Figure 3.1 shows a combinatorial classification of distributed sensors based on geometry, dimension, imaging modality, placement technique, and physical operating principle. Distributed sensing typically requires more technical sophistication than point sensors. However, in many cases, the superior performance outweighs the increased costs.

Quantitative descriptions of the action of distributed sensors consider the conversion of a distributed measurand $m(x, y, z, t)$ into a measurement, $V(x, y, z, t)$, that is,

$$V(x, y, z, t) = S[G(x, y, z, t), m(x, y, z, t)]. \tag{3.1}$$

The variable S can take on a variety of forms, such as a direct mapping with suitable gains, a convolution integral, and even more complex functional forms. $G(x, y, z, t)$ is the *antenna gain* [1].

3.2 Noncolocated Point Sensors

The collective action of an array of *noncolocated point sensors* often acts to form a distributed sensor. Many noncolocated point sensors use separation distance to integrate and amplify physical effects for ease of measurement. Examples of noncolocated point sensors are gravity-operated devices, such as plumb lines and liquid-level meters, optical line-of-sight devices, and electrochemical half-cells.

3.2.1 Plumb Bobs and Inclinometers

The plumb bob and line provides a stable and reliably vertical orientation for most earth-bound structures. The conventional plumb bob and line system is a heavy weight suspended on a flexible yet inextensible line and indicator markings on the structure (Figure 3.2). In the absence of lateral disturbances, the plumb line remains vertical. A primary usage is to assess the verticality of structures and elements.

Plumb lines are often parts of permanently installed sensor suites in large dams [2]. A recommendation for dams is to use a plumb bob of 11 kg (25 lb) and to place it in a shaft 200 mm (8 in) in diameter for shafts 60 m long, and to use a shaft with 300 mm diameter for lengths between 200 and 300 m. Placing the plumb bob in a bath of light oil will dampen lateral pendular vibrations.

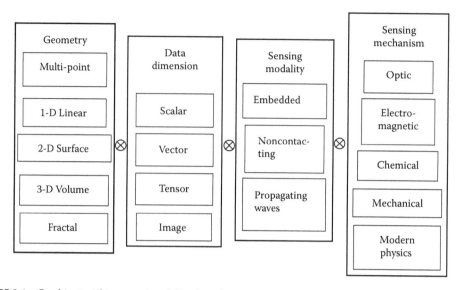

FIGURE 3.1 Combinatorial taxonomies of distributed sensors.

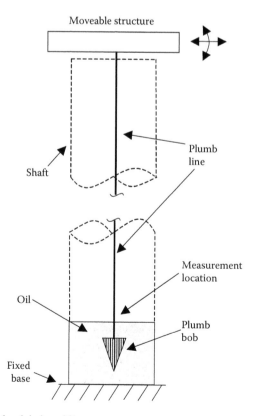

FIGURE 3.2 Conventional plumb bob and line.

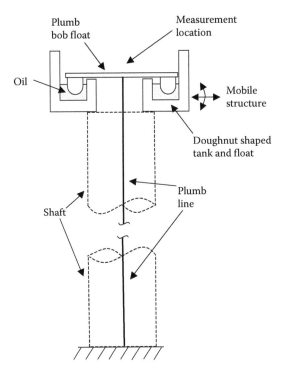

Plumb
bob float

Measurement
location

Oil

Mobile
structure

Doughnut shaped
tank and float

Plumb
line

Shaft

FIGURE 3.3 Inverted pendulum plumb bob and line with doughnut float.

When geometric constraints make it inconvenient to use traditional pendulum-style plumb bobs and lines, inverted versions are useful alternatives. The inverted bob floats in a liquid bath at the top, usually with a doughnut configuration (Figure 3.3). As an example, Romera and Hernández monitored the Belesar Arch Dam in northwest Spain for excessive displacements due to *alkali aggregate reaction* (AAR) or *alkali silica reaction* (ASR) with five inverted pendula, each at five locations along the length of the cable for a total of 25 measurements [3]. The pendula can track displacements due to daily temperature changes, reservoir height, and AAR.

Inclinometers measure tilt. These measurements require two fixed points to establish a tilt line. The angle between the tilt line and reference line (often a vertical line established by gravity) forms the *tilt angle*. A common application of inclinometers monitors movement in geotechnical structures with a hollow pipe fixed in the structure that contains an internal grove track on which rides the inclinometer probe (Figure 3.4).

3.2.2 Liquid-Level Meter

A *liquid-level meter* measures relative vertical displacements in earthbound structures [2]. The instrument uses two or more interconnected fluid tanks and the quasi-static gravity-induced movement of liquids for transduction. A fluid-carrying conductor connects the tanks and allows for gravity to equalize the heights of the fluids in both tanks (Figure 3.5). A comparison of the local fluid height in both tanks detects the relative vertical displacement of the tanks. Liquid-level meters can be very effective, but the need for running long fluid-conducting tubes between the tanks can be inconvenient. Terzaghi often receives attribution

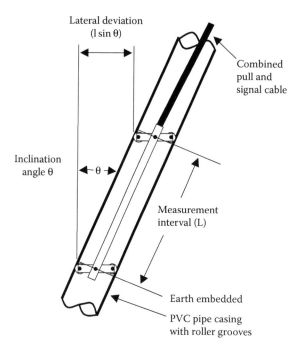

FIGURE 3.4   Mobile inclinometer probe inside of pipe track. (Picture derived from Durham Geo Slope Indicator, Inc., literature.)

for developing the liquid-level meter, but it appears that Takahasi had developed a system at an earlier date [4].

3.2.3   Half-Cell Corrosion and Resistance Meters

A half-cell meter detects and measures the amount of electrochemical activity in a path between two points on a structure (Figures 3.6 and 3.7). A predominant SHM application is multipoint readings that map areas of active corrosion in reinforced concrete [5–7]. The half-cell technique with reinforced concrete typically attaches a copper electrode to steel reinforcing bars and presses a porous wet $Cu/CuSO_4$ electrode to the concrete surface. This arrangement sets up a half-cell that indicates the level of corrosion activity by the difference in oxidation–reduction potentials of iron and copper ions [8–11]. If the half-cell potential

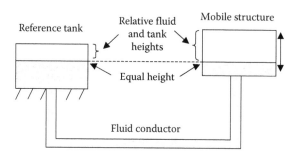

FIGURE 3.5   Liquid-level meter.

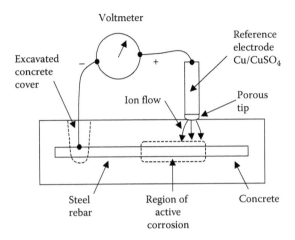

FIGURE 3.6 Half-cell corrosion potential measurement configuration.

is less than $-0.35\,V$, then there is a 90% probability of corrosion being active. If the half-cell potential is in the range of -0.2–$0\,V$, there is a 90% probability of no active corrosion being present. If the half-cell potential is positive, no reliable conclusions can be drawn from the measurement. The *linear polarization resistance* (LPR) method is a three-electrode variant that applies an external potential to the structure and measures the resistance and

FIGURE 3.7 Half-cell concrete corrosion potential measurement instrument. (Photograph courtesy of the Proqec SA, Switzerland.)

instantaneous rates of corrosion [12]. The *Stearn–Geary* theory of corrosion underlies the operating principle of LPR [13].

The half-cell method works well if chloride ion concentrations around the rebar are the source of the corrosion. Confounding factors include a decrease in oxygen concentration at the reinforcing steel, carbonation, the use of corrosion inhibitors, epoxy-coated and galvanized rebars, pavement patch repairs, and cathodic protection systems. A disadvantage of the half-cell method is the need for excavating the rebars for electrode attachment, usually with power tools [14]. The half-cell method has also been criticized for not guaranteeing the location and extent of the region of measurement. An electrically conductive guard ring and/or a linear polarizing device attached to the half-cell electrodes can improve the location resolution [15–17]. While the half-cell method may effectively detect active corrosion processes, it has difficulty with detecting corrosion that has already progressed significantly and/or may rapidly progress in the future, but is at present in a passive state.

Additional half-cell configurations appear in specialized applications. Watters et al. developed a small embeddable chloride detector that measures the voltage developed between silver/silver chloride and copper/copper sulfate electrodes [18]. A similar system using steel and titanium or steel and titanium with a metal oxide mixture coating also measures active corrosion. Corrosion in the steel sets up electrochemical currents to the titanium member. Sorensen et al. installed a suite of these corrosion sensors in bridges in Denmark [19]. Hansson et al. used an array of embedded sensors with half-cells to detect chloride penetration into concrete [20]. Schoess and Havey patented a smart fastener that senses aircraft corrosion with miniaturized electrochemical sensors that fit into a bolt or rivet [21]. Laird and Li developed a half-cell variant *Electrochemical fatigue sensor* that measures the early growth of fatigue cracks in metals [22].

Concrete electrical resistivity measurements can also assess corrosion proclivities [23]. The measurement systems typically use a four-point Kelvin probe, also known as a *Wenner probe*, to drive a known current through the concrete with two contacts and measures the voltage drop across two additional contacts [24]. Figure 3.8 shows a commercially available concrete resistance meter.

An integrated multisensor system that measures electrochemical potentials of pH and chloride ions as well as moisture and temperature may be a highly effective means of assessing corrosion activity. As examples, Agarwala and Fabiszewski miniaturized and packaged full-cell galvanic current sensors into self-contained sensor units, and Kelly et al. developed a multisensor embeddable MEMS sensor that measures and transmits relevant corrosion data [25,26]. Rolling variants of the electrochemical sensors find occasional usage [27].

3.2.4 Line-of-Sight Optical Instruments

Since light travels along predictable ray paths for macroscopic dimensions, it is possible to build line-of-sight optical instruments for measuring structural deformation. Line of sight can be accurate for measuring cross-range angles, but tends to be less accurate for measuring down range distances. In this context, traditional surveying instruments can readily measure gross structural deformations or movements. The present resolutions are in the range of 125 μm. More sophisticated and automated instruments can measure smaller deflections and motions [28–30]. Wahbeh et al. tracked motions of the Vincent Thomas suspension bridge near Los Angeles with optical instruments that measured the motions of two LEDs attached near the mid-span of the bridge [31]. Kanda et al. used optical tracking to

FIGURE 3.8 Concrete resistance meter used to assess potential for corrosion. (Photograph courtesy of the Proqec SA, Switzerland.)

measure motions and the collapse of building structures in the laboratory [32]. An interesting example of line-of-sight techniques is the use of slight orbital trajectory perturbations to deduce the location of air leaks in the International Space Station by a momentum transfer analysis [33].

Ray optics predicts an infinite resolution at the fine scale for line-of-sight optics. However, ray optics is only an approximation to more rigorous optical theories, including wave models (see Appendix A for some more details). The wave nature of light limits fine-scale resolution [34]. Based on scalar wave diffraction considerations, Rayleigh's criterion gives an approximate expression for the lateral resolution, $d_{\min}$, of a set of optics as

$$d_{\min} = \frac{0.61\lambda}{n\sin(\delta)} = \frac{0.61\lambda}{\text{NA}}, \tag{3.2}$$

where λ is the wavelength of light. $\text{NA} = n\sin(\delta)$ is the *numerical aperture* of the optical system [35]. Boehler et al. have described methods of determining laser scanner accuracy [36].

3.2.5 Radio-Based, Global Positioning System, and Time-of-Flight Instruments

Radio-based and *Global Positioning System* (GPS) locating instruments are similar to the line-of-sight optical instruments. The basis of these systems is signal propagation timing between two or more points. The SHM literature contains a modest number of publications on the use of these instruments. Steltzer et al. developed custom wireless position sensing systems with update rates in excess of 1000 Hz [37]. The combination of reduced cost and increased performance of GPS systems technology has progressed to where it is practical to conduct long-term multipoint measurements on large structures. Breuer et al. measured the displacement of high-rise towers subjected to wind loading with GPS instruments [38]. Xu et al. used a 12-channel GPS system to measure ambient vibrations on the Humen Bridge in Guangdong [39]. Fujino et al. report a long-term GPS displacement monitoring effort on the Akashi Kaikyo Bridge [40]. The system used a reference station and a fiber optic interconnect between the GPS receivers to enable measurements with a resolution of 5 mm for horizontal displacements and 10 mm for vertical displacements. Chan et al. measured the dynamic accuracy of GPS displacement instruments [41]. Accuracies of 1.3–3.6% were measured for motions with frequencies less than 1 Hz. The accuracy degraded significantly at 1.8 Hz. Meng et al. used real-time kinetic GPS to measure the motions of four large bridges in the United Kingdom [42]. Huang et al. used a combination of 12 dual-frequency GPS systems to monitor the health of the Nanpu Bridge in Shanghai [43]. Liu et al. used 12 dual-frequency GPS receivers and two reference receivers to monitor the Donghai Great Bridge near Shanghai [44]. The timing of signal propagation also forms the basis of elastic-wave distance measuring systems. Mandracchia demonstrated an ultrasonic strain measuring system based on wave propagation timing [45].

3.2.6 Long-Gage-Length Fiber Optic Strain Sensors

Long-gage-length strain sensors use measuring length to advantage. A typical foil gage technique uses a gage with sufficient length to span local strain inhomogeneities to produce an average strain measurement. A related technology uses fiber optic sensors to form long-gage sensors with lengths that range from tens of mm to several km.

Another application of long gages is to increase the sensitivity of measurements of homogeneous strain fields. The differential displacements increase with gage length. Applying a long gage mount can amplify the overall delta displacement due to strain. Li and Wu developed specialized mounting packages to form long-gage FBG strain sensors [46]. Schulz et al. used this technique to form long-gage FBG sensors on the Horsetail Falls Bridge in Oregon [47,48].

The long-gage *white light interferometer* uses a noncoherent light source with a moderately short coherence length. The interferometer splits the light source into two beams that follow separate paths and then recombine at a detector. The detected intensity is maximum only when the effective lengths of both paths are equal. When the path lengths differ from equality, the intensity drops off. This differs from coherent-light laser-based interferometry that has a detected intensity maximum when the path lengths are equal plus or minus a wavelength multiple. The white light interferometer has the advantage of avoiding the 2π phase ambiguity characteristic of coherent light interferometers. A disadvantage of white light interferometers is that the different path lengths have to be equal to within a wavelength of light. Figure 3.9 shows a fiber optic version of a white light Michelson interferometer. The difference in the path lengths, ΔL, is

$$\Delta L = L_1 - L_2. \tag{3.3}$$

The detected light intensity is maximum when $\Delta L = 0$ (Figure 3.10).

In terms of practical measurements, the system in Figure 3.9 is a bit too simple. The addition of secondary interferometer arms and a piezoelectric-stage-mounted mirror creates a

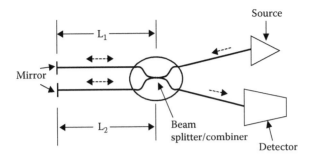

FIGURE 3.9   Simple version of fiber optic noncoherent white light interferometer.

practical measurement tool. Figure 3.11 shows the SOFO fiber optic long-gage white light interferometric displacement/strain measurement system [49]. Sampling rates of at least 1 kHz are possible with this system [50].

SHM monitoring systems use SOFO long-gage strain sensors on a wide variety of bridges, dams, and other structures. To date, the sensors have demonstrated long-term absolute strain reading stability and a laboratory-verified performance [51,52]. Figure 3.12 shows the embedded installation of a sensor along with a resistive strain gage in a concrete box girder [53]. A layer of grout covers the strain gages and leaves a barely visible installation. Figure 3.13 shows a sensor installation on the Lutrive Bridge. Figure 3.14 compares SOFO strain gage data and bridge pier tilt data collected on the Lutrive Bridge. Leung et al. used white light interferometry to measure the debonding action of optical fiber sensors in concrete [54]. Heaton et al. wrapped a long-gage white light interferometer sensing cable into an FRP composite motor case. The sensor measured bulk case deformations [55]. Donlagic and Hanc detected the presence of vehicle axles with roadway-embedded fiber optic interferometers [56].

Pavement-mounted weigh-in-motion sensors occasionally use long-gage fiber optic sensors. For example, Téral et al. used a single-mode polarization-sensitive fiber optic strain sensor to measure truck axle weights [57]. A polyurethane matrix held the fiber and transferred the load from the tire to the transverse axis of the fiber.

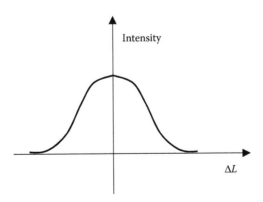

FIGURE 3.10   Intensity versus path length difference in the white light interferometer.

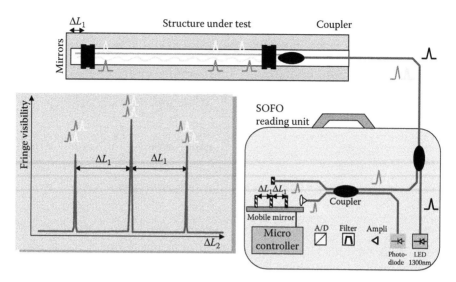

FIGURE 3.11 Schematic of the operating principle of the SOFO© noncoherent interferometric fiber optic strain sensing system. (Picture courtesy of the Smartec SA, Switzerland.)

3.2.7 Multipoint Extensometer Sensing Systems

Integrated multipoint arrays of extensometers can measure multicomponent structural deformations. Static deformations combined with favorable mounting geometries are amenable to the use of these systems. Crack movement measurement is an example application. The motion of a crack in a plane has at least three degrees of freedom: opening, sliding,

FIGURE 3.12 Embedded installation of long-gage fiber optic strain sensor along with resistive strain sensor in concrete box girder at the Berlin Lertner Hbf train station [53]. (From Niemann J, Habel W, and Hille F. 2005. Keynote lecture at the *2nd Intl Conf Reliability, Safety and Diagnostics of Transport Structures and Means* at the University of Pardubice/Czech, July 7–8, 2005, pp. 248–256. With permission.)

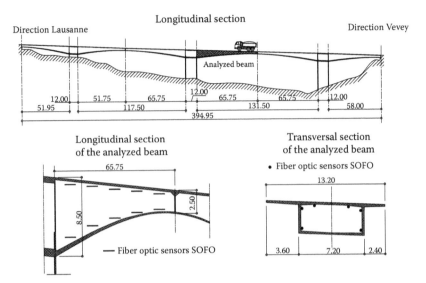

FIGURE 3.13 Schematic of SOFO strain gage sensor installation on the Lutrive Bridge in Switzerland. (Photograph courtesy of the Smartec SA, Switzerland.)

and rotation. Figure 3.15 shows a rosette of three LVDTs that measure all three in-plane crack motions simultaneously. Long-term monitoring of the crack with this particular arrangement generally requires placing a protective cover over the array. A geotechnical multipoint application uses an extensometer probe to measure borehole motions (Figure 3.16).

3.2.8 Modern Physics Sensors

With the notable exception of lasers, the operating basis of the majority of sensors used in SHM lies within the realm of classical physics, that is, Newton's equations for mechanics and Maxwell's equations for electromagnetism. The following is a brief listing of some SHM sensing techniques with operating principles based on modern physics, that is, principles of quantum mechanics and particle physics.

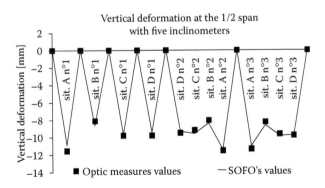

FIGURE 3.14 Comparison of SOFO strain gage data with that taken from inclinometers. (Photograph courtesy of the Smartec SA, Switzerland.)

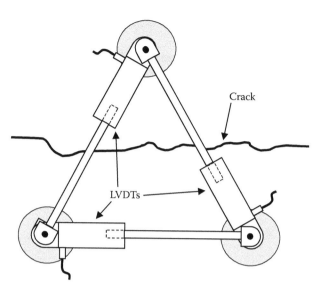

FIGURE 3.15 Monitoring crack width, sliding, and rotation motions with a three-point rosette of LVDTs. (Adapted from Germann Instruments literature, http://www.germann.org/Products/CMD/CMD.pdf)

Quadrupole resonance measures the resonant oscillations of the atomic nuclei of specific isotopes (usually nitrogen 14). These nuclei have internal quadrupole electrical gradients that interact with external electrical fields. Such nuclei vibrate measurably in response to an appropriate electrodynamic stimulus. If nitrogen atoms are part of a crystal, deformations of the crystal affect the quadrupole resonance vibrations. This effect is the basis of embedded sensing elements that use nitrogen-laden crystals [58].

Launching subatomic particles of various types into materials and observing the resulting interactions can indicate subsurface conditions. Gamma-rays are high-energy photons that penetrate materials and interact with atoms in a manner characteristic of the specific element. Minchin found that gamma-rays were effective in identifying water in concrete post-tensioning ducts [59]. Collico et al. used a radioisotope neutron source to identify the elemental constituents of concrete samples including the presence and concentration of chlorides [60]. Srikar et al. used lower energy photons in the IR range to stimulate Raman scattering for measurements of stress in silicon with a spatial resolution down to 1 μm [61]. Positrons are antimatter versions of electrons. When a positron encounters an electron, the particles annihilate one another and release detectable gamma-rays. This reaction forms the basis of a nondestructive evaluation (NDE) technique for metals known as *positron annihilation* [62]. Radioactive sources, such as ^{22}Na, ^{68}Ge, or ^{58}Co, generate suitable positrons. When the positrons penetrate and interact with a metallic solid, the positrons tend to become trapped in dislocations and then react to annihilate free electrons. Positron annihilation can detect dislocation-based fatigue damage in metals at an early stage.

The *electron work function* (EWF) is the amount of energy required to remove an electron from the surface of a metallic solid. The EWF decreases in the presence of dislocations [63]. Balachov et al. measured the EWF as a function of plastic deformation and concluded that it may be possible to use the EWF as an indicator of damage before it emerges into a crack [64]. Electrons will also *tunnel* across nanometer-scale gaps in a manner that depends highly on

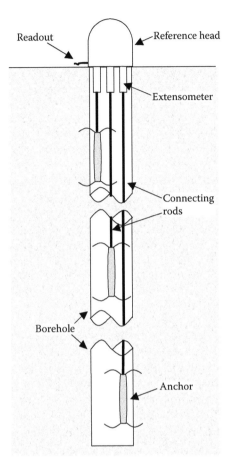

Readout

Reference head

Extensometer

Connecting rods

Borehole

Anchor

FIGURE 3.16 Multipoint extensometer system for measuring borehole motions. (Picture derived from Durham Geo Slope Indicator, Inc., literature.)

the gap dimension. Liu and Kenny used this effect to build an MEMS servo accelerometer with a rather high resolution of 20 nano-g$/\sqrt{Hz}$ and a 5 Hz–1.5 kHz bandwidth [65].

Many atomic species have naturally occurring radioactive isotopes. Detecting these isotopes with radiation detectors can be indicative of the concentration of a particular chemical species. Livingston et al. measured the radiation level of ^{40}K isotope decay in concrete to assess the overall levels of potassium present and the propensity for the concrete to be subject to the deleterious formation of *ettrigite* [66].

Giant magneto resistance (GMR) is a physical effect that causes certain solids with thin nanostructures to be highly sensitive to external magnetic fields [67]. Thin layers of chromium between iron layers are common materials in GMR-active solids. The explanation for this behavior lies in the spin of the electrons in two nearby iron layers. An external magnetic field aligns the spins on opposite sides of the iron layer. The spin alignment facilitates the passage of electrons through the chromium layer and reduces the macroscopic electrical resistance. Zero external magnetic fields leave the spins antialigned and produce the opposite effect. GMR materials can form the sensing component in highly sensitive magnetic detectors. Computer hard drive read-write heads are common commercial GMR applications. Eddy current crack detectors are an SHM application.

3.3   1-D Distributed Sensors

1-D distributed sensors use curvilinear paths for both the sensor and signal transmission. The collocation of sensing and communication media enables a distributed measurement of physical effects along the path of the conductor via position-dependent analog signal processing. The basis for a 1-D sensor is virtually any physical mechanism that transduces a signal as it propagates through a conductor along a linear or curvilinear geometry (Figure 3.17). Electrical, optical, hydraulic, and pneumatic conductors are all possibilities.

The speed of signal transmission and transduction is a distinguishing feature of 1-D distributed sensors. In *quasi-static* sensors the speed of information transmission through the sensor is higher than the time constants of signal transduction. The signal transmission is effectively instantaneous. Spatial resolution usually requires the aid of antenna gain techniques. *Time-domain* sensors use the time-of-flight of a signal as it propagates through the sensor as a measurand that provides spatial resolution.

3.3.1   Quasi-Static Sensors

Signal attenuation is the basis of many quasi-static sensors. An event detector may rely on signal transmission disruption as the signal. Electric, hydraulic, pneumatic, and fiber optic conductors all have the potential to serve as signal disruption event detectors.

Modulating the intensity of light transmission through a fiber optic cable is a viable signal attenuation transduction mechanism. Intensity-based sensors are relatively simple to build and use as benchtop technology demonstrators. They are a bit more difficult to implement in practical applications. Intensity-based fiber optic sensors have a serious disadvantage. Spurious conditions, such as connector failures, can attenuate the signal and confound data interpretation. This is analogous to amplitude-based noise appearing as static in AM radio

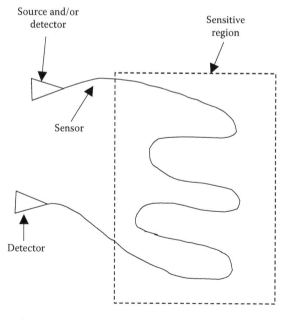

FIGURE 3.17   1-D distributed sensor.

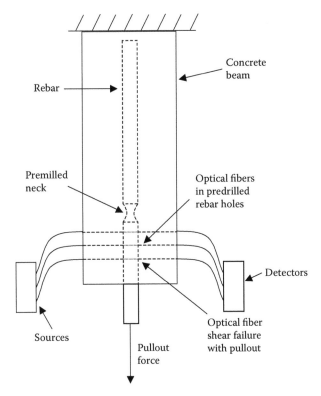

FIGURE 3.18 Schematic layout of intensity-based fiber optic sensors for measuring rebar debonding during a pullout test.

communication. A simple intensity-based event-detecting fiber optic transducer modulates the light intensity transmission by the fracture and subsequent separation of fiber optic cable [68]. Large deformation short of fracture can also be detectable [69]. Dry used a similar approach [70]. Structural fractures cause cracks in fluid-filled tubes to actuate self-healing structural systems. Cracking in the tubes causes the fluid to flow, react, and alter the intensity of the transmitted light.

Arrays of quasi-static distributed sensors can potentially detect and localize damage. As an example, Figure 3.18 shows measurements of pullout failure in a concrete reinforcing bar. The rebar contained a set of predrilled holes that traversed the center of the rebar cross section. The optical fibers ran through both the concrete and the rebar. Pulling the rebar out of the concrete caused the individual fibers to crack, separate, and alter the transmitted intensity [71]. The intensity plots appear in Figure 3.19. Similar methods using embedded fiber optics produce real-time crack detectors in concrete members and composite plates [72,73]. Tennyson developed a system for detecting hypervelocity micrometeorite impact damage on spacecraft with a grid array of interferometric long-gage fiber optic sensors [74]. Wooddell et al. developed a fiber optic sensor with randomly distributed holes perpendicular to the axis of the fiber for distributed chemical and photovoltaic sensing applications [75].

The wide variety of available optical sensing techniques enables a variety of additional optical 1-D distributed sensors. Tennyson et al. used long-gage FBGs to monitor strain in pipelines [76]. Takeda et al. developed chirped FBGs where the spacing of the grating varies along the length of the fiber [77]. The variable grating spacing can sense strain gradients,

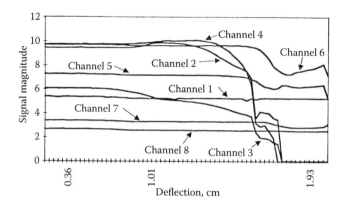

FIGURE 3.19 Fiber optic transmitted intensity versus rebar pullout displacement. (Data from Huston DR et al. 1994. *Time Domain Reflectometry in Environmental, Infrastructure, and Mining Applications*, US Bureau of Mines, SP 19-94.)

such as due to composite delaminations. Ling et al. detected composite delaminations with FBGs [78]. Stress gradients across the FBG attributed to the delaminations caused distortions of the spectral peaks of the signal reflected from the grating. Polymer optical fibers can be superior to glass fibers in damage-detecting distributed sensors because of tougher mechanical properties [79,80]. An example is the use of polymer optical fibers as embedded wear monitors for wood staves in large shipboard propeller shaft bearings [81,82]. Dakai et al. used transmitted intensity sensing in vascular self-healing composites to assess the occurrence of damage and fluid transport [83].

Electrical and piezoelectric phenomena are also the basis of a variety of distributed 1-D sensors. One approach is for the measurand to change the resistivity along the length of the sensor. Examples include the Carlson meter and other long-gage resistive sensors. Datta et al. formed distributed sensors with interdigitated piezoelectric PZT [84]. The *Picostrain* system is an electrical quasi-static distributed sensor based on the changes in the electrical properties of cables as they deform [85]. This system has been demonstrated as a means of weighing trucks at highway speeds [86]. Takagi et al. developed carbon fiber-reinforced polymer (CFRP) panels with embedded piezoelectric fibers that act as both sensors and actuators [87]. Damage to the panel damages the sensory fibers with a loss of signal indicating the occurrence of damage.

A plastic fluid tube can be a combined distributed fire detection and safety actuation device for large natural gas storage tanks (Figure 3.20). Normal operation requires a pilot pressure to activate the tank valve into an open position. Using the sensing tube to supply the pilot pressure creates a safety device. If a fire occurs near the tank, the sensing fluid tube melts. The line loses pressure along with the pilot valve actuator. This causes the valve to shut off the gas flow out of the tank in an attempt to prevent a conflagration.

The use of continuity in hydraulic or pneumatic conductors as a signal readily extends to more complicated configurations. *Comparative vacuum monitoring* (CVM) detects crack formation by measuring the differential pressure in segregated fluid conductor circuits (Figure 3.21) [88]. When a crack is of sufficient size to breach both of the segregated circuits, differential pressure measurements between vacuum lines, and those at atmospheric pressure can detect the leak [89,90]. Since it is relatively easy to mass produce the conductor circuits in a variety of patterns, that is, interdigitated arrays with submillimeter pitches, and since the circuits are lightweight, the CVM technique has garnered considerable attention for potential usage in aircraft [91].

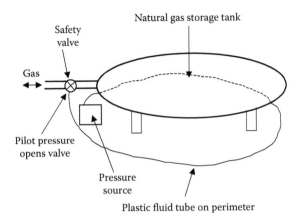

FIGURE 3.20 Plastic tube on the perimeter of a natural gas tank as a fire safety device.

3.3.2 Operational Model of Quasi-Static 1-D Sensors

An operational approach can explain certain 1-D distributed sensing phenomena [1]. The process initiates with the launching of a signal into a conductor, followed by the alteration (transduction) of the signal along the path by measurands along the path, and measurement at the end. Figure 3.22 is a close-up of the first active segment. The action of the measurand on this segment operates on the signal S_0 with G_1 and produces an output S_1, that is,

$$S_1 = G_1 S_0. \tag{3.4}$$

The cascading action of multiple sensitive segments multiplies. For two segments

$$S_2 = G_2 S_1 = G_2 G_1 S_0 \tag{3.5}$$

and for N segments

$$S_N = G_N S_{N-1} = G_N G_{N-1} \ldots G_2 G_1 S_0 \tag{3.6}$$

or

$$S_N = \left[\prod_{i=1}^{N} G_i \right] S_0. \tag{3.7}$$

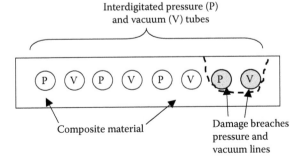

FIGURE 3.21 Operational principle of CVM.

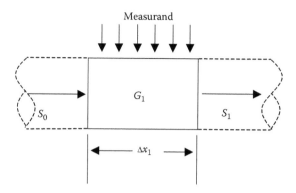

FIGURE 3.22   Operation of measurand on signal on the first segment.

As an example, consider an intensity-modulating sensor, such as a multimode optical fiber subjected to distributed microbending [92]. The transduction mechanism attenuates the amount of light in each segment based on the local curvature of the fiber. The fractional reduction in transmission, r_i, per segment i depends on the local radius of curvature so that

$$G_i = (1 - r_i).\tag{3.8}$$

The exit intensity, I, as a function of curvature and the input intensity, I_0, is

$$I = \left[\prod_{i=1}^{n} G_i\right] I_0 = \left[\prod_{i=1}^{n}(1 - r_i)\right] I_0.\tag{3.9}$$

Taking the natural logarithm converts the products into sums

$$\ln(I) = \ln(I_0) + \sum_{i=1}^{n} \ln(1 - r_i).\tag{3.10}$$

If it is assumed that the localized measurand-induced attenuation per unit length is proportional to the measurand, m_i, and the segment length, Δx_i, then

$$r_i = m_i \Delta x_i\tag{3.11}$$

and

$$\ln(I) = \ln(I_0) + \sum_{i=1}^{n} \ln(1 - m_i \Delta x_i).\tag{3.12}$$

If $m_i \Delta x_i$ is small, then

$$\ln(1 - m_i \Delta x_i) \approx -m_i \Delta x_i.\tag{3.13}$$

In the limit as Δx_i goes to zero and n goes to infinity

$$\ln[I(x)] = \ln[I_0] - \int_{0}^{x} m(x)\, dx\tag{3.14}$$

or

$$I(x) = I_0 - \exp\left[\int_0^x m(x)\,dx\right].$$ (3.15)

Modulating the phase angle of a signal can create a 1-D distributed sensor. Distributed interferometric, polarimetric, and elliptical core fiber optic sensors are all examples [93–97]. The operational approach provides a simplified model of distributed 1-D phase angle sensing, where the signal transmitted and transduced is the phase angle, ϕ, of the lightwave, that is,

$$S = e^{j\phi}.$$ (3.16)

The measurand modifies the phase in a segment of length Δx_i so that

$$G_i = \exp\left[jm(x_i, t)\Delta x_i\right].$$ (3.17)

Here, $m(x_i, t) = m_i$ is the influence of the measurand. The output signal, S, is

$$S = e^{j\phi} = e^{j\phi_0}\prod_{i=1}^{N} e^{jm_i\Delta x_i} = \exp\left[j\left(\phi_0 + \sum_{i=1}^{N} m_i\Delta x_i\right)\right].$$ (3.18)

Since ϕ is the parameter of interest,

$$\phi = \sum_{i=1}^{N} m_i\Delta x_i + \phi_0$$ (3.19)

or in integral form

$$\phi = \int_0^L m(x, t)\,dx + \phi_0.$$ (3.20)

3.3.3 Spatial Sensitivity, Antenna Gain, and Tuning of 1-D Sensors

A distributed 1-D sensor takes advantage of nonhomogeneous spatial sensitivity effects to allow for configurations that are sensitive to some desired physical effects and insensitive to others. Borrowing a term from radio engineering, this technique is known as tuning the *antenna gain* of the sensor [1]. Demonstrations of antenna gain tuning include point load location, vehicle type identification, and vibration mode shape identification [92,98–104].

One method of creating a spatially tuned distributed 1-D sensor uses a phase-shifting fiber optic modal domain interferometric sensor. The measurand of interest modulates multiple coherent laser-based light modes in a large core (50–150 µm) optical fiber. Upon exiting the fiber, the light wave modes project onto a surface to form a seemingly random interference speckle pattern. Flexing the fiber alters the propagation paths. The modes and speckles move accordingly (Figure 3.23) [93,105,106]. In certain configurations, modal domain fiber optic sensors are inherently nonlinear [107]. A somewhat similar distributed

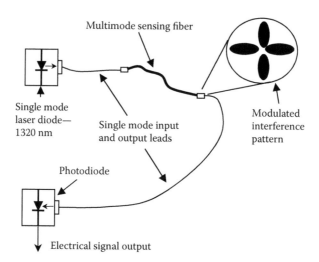

FIGURE 3.23   Distributed multimode fiber optic sensor.

sensor measures Doppler shifts in the wavelength of the light traversing a fiber that are due to dynamic motions, such as vibrations [108].

Changes to the geometry affect the output of modal domain fiber optic sensors. A linearized model of the output due to fiber curvature change is

$$V(t) = \int_0^L K \left(\frac{\partial^2 w}{\partial s^2} \right) ds, \tag{3.21}$$

where $V(t)$ is the sensitivity, L is the length of the active sensing region, K is a calibration factor, $w(x, y, t)$ is the amplitude of the transverse out-of-plane plate vibrations, and s is a distance parameter along the length of the fiber. This spatial sensitivity allows for fabricating modal domain sensors with spatial antenna gains that are sensitive to a particular type of motion, such as a natural mode of vibration [101,109].

Figure 3.24 shows data from an amplitude sensitivity calibration of a modal domain fiber optic sensor with a quartz accelerometer with an output processed into displacements. The test setup involved attaching both transducers to a simply-supported I-beam with a 25 Hz dominant bending vibration mode. Tapping the beam with a small hammer sets up vibrations with amplitudes in the micron range.

Since the plate mechanics are largely linear, the plate motion is the superposition of vibrating independent modes

$$w(x, y, t) = \sum_{i=1}^{\infty} \varphi_i(x, y) q_i(t), \tag{3.22}$$

where $\phi_i(x, y)$ is the ith mode shape, and $q_i(t)$ is the modal displacement. In terms of coordinates along the length of the fiber

$$w(s, t) = \sum_{i=1}^{\infty} \varphi_i(s) q_i(t). \tag{3.23}$$

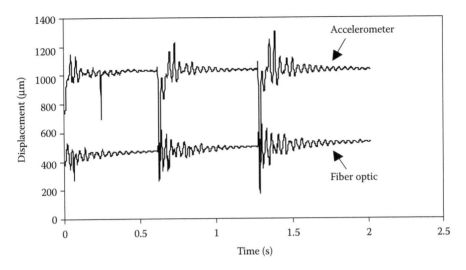

FIGURE 3.24 Sensitivity comparison of the accelerometer and distributed fiber optic sensors. (Data courtesy of W. Spillman, Jr.).

Placing a fiber sensor along a nodal line for a particular mode produces a sensor with an output that is insensitive to the vibrations of that mode. If instead the fiber placement was orthogonal to the nodal line, it will flex as the mode vibrates and be sensitive to those vibrations.

The *Chladni sand technique* determined the mode shapes and natural frequencies. The first step spread sand particles uniformly over the plate. An inertial mass shaker set the plate into vibration with harmonic excitation at a single frequency. When the vibration frequency coincided with a modal resonance, the sand migrated to the nodal lines of that mode. This technique identified 16 distinct modes between 50 and 1400 Hz. Figure 3.25 shows sand patterns for modes at 92 and 492 Hz. Results from spatially weighted fiber optic sensor tests with spectra from two different spatial antenna configurations on a vibrating steel plate are shown in Figure 3.26. One sensor antenna gain configuration was sensitive to the 92 Hz mode. The second antenna gain configuration was insensitive to the 92 Hz mode.

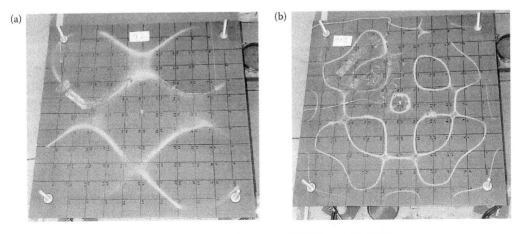

FIGURE 3.25 Chladni sand patterns for a plate vibrating at (a) 92 Hz and (b) 492 Hz.

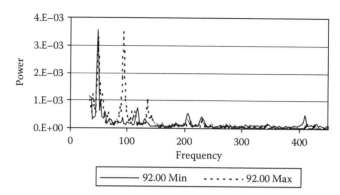

FIGURE 3.26 Spectra from distributed fiber optic sensors with differing spatial sensitivities on a plate vibrating in resonance with a 92 Hz mode. (Adapted from Huston D et al. 1999. *Smart Systems for Bridges, Structures and Highways,* SPIE 3671.)

The principles of antenna gain and spatial sensitivity tuning apply to a variety of other sensing technologies. Foil strain gages is an example. The lay-up of the foil pattern tunes the strain gage sensitivity to the needs of specific applications. Structures built of nonhomogeneous materials, such as concrete or fiber reinforced composites, have nonhomogeneous strain fields. Figure 2.21 shows a foil strain gage with a custom aggregate-spanning and strain-field-averaging sensitivity for embedded concrete strain sensing. Delepine-Lesoille et al. developed a similar mounting arrangement for embedding fiber optic sensors into concrete [110]. The vibrating wire and fiber optic white light interferometric technique strain gages are capable of measuring over much larger gage lengths. Wu et al. used the electrical resistance of CFRP composite reinforcing bars to act as long-gage damage sensors for concrete beams [111,112]. Nanni et al. produced similar damage sensing rods with a combination of CFRP and glass fibers, where the carbon fibers usually break first at specific loads [113]. It is possible to place piezoelectric materials in patterns that tune to specific behaviors, such as modal vibrations of a plate or point impacts [114,115]. Tzou developed a series of sophisticated piezoelectric pattern sensor and analysis techniques for shell structures [116]. Deraemaeker and Preumont used an array of point-sensitive accelerometers to form distributed spatial filters with collective sensitivities to particular modes and the emergence of damage [117]. Linear combinations of the individual accelerometer signals performed the spatial weighting. Lanzara et al. developed large-scale integrated silicon patterns that combine antenna gain with networking for embedded SHM [118]. White et al. used 1-D piezoelectric quartz elements as roadway weigh-in-motion sensors [119]. The quartz elements showed a potential for increased durability and performance over competing sensor technologies.

Integrating optical fibers into the winding of filament-wound composite structures during manufacturing forms long-gage embedded sensors. Wrapping the sensing fiber in radial or longitudinal directions produces a distributed sensor with circumferential and/or axial loading sensitivity. Dong et al. used a radio frequency (RF) subcarrier modulation of the laser light in the fiber for sensing and monitoring filament-wound high-pressure composite gas storage tanks [120]. Bastianini et al. used a similar technique with Brillouin backscatter measurements to detect strains in composite pipe sections [121].

A variety of more complicated distributed 1-D sensor architectures are possible. A homogeneous pattern is relatively simple to design and build. Figure 3.27 shows a possible smart composite configuration with a homogeneous distribution of sensors with localized damage that is near but does not span the sensor. Figure 3.28 is a similar smart composite

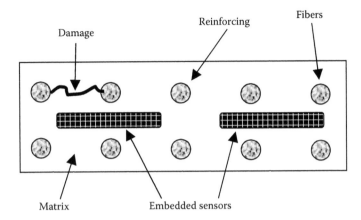

FIGURE 3.27 Smart composite with homogeneous distribution of sensors and with local damage.

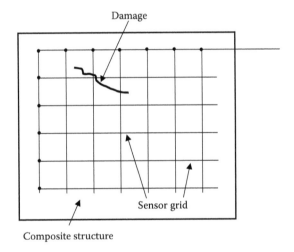

FIGURE 3.28 Smart composite with grid-like distribution of sensors and with local damage.

configuration with a grid-like distribution of sensors and with localized damage that spans the sensor. Grid patterns are a convenient method of providing a comprehensive coverage for a composite sensing system [122]. Figure 3.29 shows a smart composite structure with a hierarchical branching architecture and with local damage that also spans the sensors. Tree-like sensor geometries appear as the architectures for nervous systems in animals. Advantages of tree architectures include simplicity, scalability, and possibilities for self-assembly. A disadvantage is the potential for information bottlenecks around the trunk lines and the inability to pass information between neighboring branches on distinct trunk lines. An interesting possibility would be to create tree-like geometry SHM sensor systems with automated *Lichtenberg patterns* or similar self-forming branching techniques.

3.3.4 Time Domain Reflectometry

Time domain reflectometers (TDRs) use signal propagation timing for spatial resolution. The TDR process first sends short pulse signals down a waveguide. The pulses interact with

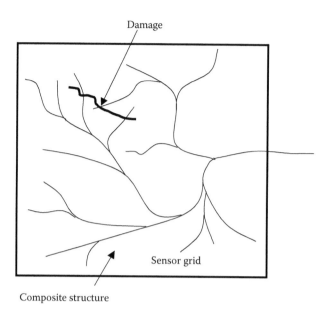

FIGURE 3.29 Smart composite structure with branching sensor architecture and with local damage.

measurands along the way. Pulse–measurand interactions produce characteristic reflections and transmissions. Analyzing reflected or transmitted signals in the time domain gives an indication of measurand properties and location. A schematic of a TDR-based 1-D distributed sensor appears in Figure 3.30. Most TDR signals travel very quickly and require high-speed instruments for time-domain spatial resolution. EM, optical, and elastic waveguides can all serve in TDR systems. *Electrical TDRs* (ETDRs) measure EM waves propagation and reflection in conductive cables [123]. Specialized variants can measure corrosion in pipes [124]. *Optical TDRs* (OTDRs) measure propagating and reflecting signals in optical cables. A *pile dynamic analyzer* (PDA) assesses the propagation of stress waves in piles as they are hammered into the Earth to detect soil properties and possible pile damage [125–127]. Ultrasound, acoustic, and elastic wave equivalents occasionally find use [128–131]. An interesting example of elastic wave timing is the *acoustic strain gage* [132]. *Pulse discharge diagnosis* assesses the integrity of insulation on electrical wiring [133]. The technique used stored charges due to insulation breakdown to produce detectable high-frequency discharge transients.

An explanation of the behavior of TDR systems derives from considering the behavior of a short-duration signal pulse with shape $I(t) = I_0 p(t)$. As the pulse propagates through a waveguide, small portions of the pulse reflect back towards the source to form the measurement signal, $d(t)$. The following assumptions are often reasonable and simplify the analysis.

1. The pulse travels through the waveguide without dispersion, at a constant speed, c, and with minimal losses.
2. All reflections are sufficiently small so that secondary reflections are negligible.
3. No nonlinear wave–wave interactions occur.
4. The source produces a narrow pulse with a characteristic length that is smaller than the desired spatial resolution.

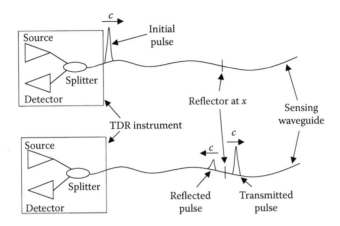

FIGURE 3.30 TDR-based sensor with initial, reflected, and transmitted pulses.

5. The detector and the source have the same location.
6. The detector has sufficient temporal resolution and amplitude range to measure the reflected pulses.

An isolated reflector located a distance x_i from the source and detector (as in Figure 3.30) with reflection coefficient R_i reflects a pulse that arrives back at the detector with a total (back and forth) travel time τ_i as

$$\tau_i = \frac{2x_i}{c}. \tag{3.24}$$

The detected signal, $d(t)$, for a single reflection at x_1, is

$$d(t) = R_1 I_1 p(t - \tau) = R_1 I_1 p\left(t - \frac{2x_1}{c}\right). \tag{3.25}$$

The pulse transmitted through the first reflector, $I_T(t) = I_1 p(t)$, has the same shape as the initial pulse, but has a reduced amplitude due to the loss of the reflected signal, that is,

$$I_T = I_2 = I_1(1 - R_1). \tag{3.26}$$

For a series of multiple reflectors, the detected signal is the sum of a set of reflected pulses, $d_i(t)$

$$d(t) = d_1(t) + d_2(t) + \cdots + d_n(t), \tag{3.27}$$

$$d_i(t) = I_i p(t - \tau_i), \tag{3.28}$$

and

$$I_i = I_{i-1}(1 - R_{i-1}) = I_0(1 - R_1)(1 - R_2)\ldots(1 - R_{i-1}). \tag{3.29}$$

If the pulse lengths are shorter than the separation distance between the reflectors, the individual pulses do not overlap and a sequence of diminishing amplitude pulses forms

$$d(t) = \sum_i d_i(t) = I_0 \sum_i \left[p(t - \tau_i) \prod_{j=1}^{i} (1 - R_1)^j \right]. \tag{3.30}$$

Often the waveguide reflects a small amount of light per unit length due to scattering processes, such as the Rayleigh, Brillouin, and Raman effects [134]. Continuum models of scattering are useful in these cases. A time-varying convolution can represent the signal output $d(t)$ from a detector.

$$d(t) = \int_t^{t+\Delta t_d} R(t)D(t,\tau)\,d\tau, \tag{3.31}$$

where $R(t)$ is the signal received at the detector, $D(t, \tau)$ is a kernel operator for the detector sensitivity and Δt_d is the time constant of the detector. A reasonable approximation to detected signal is often

$$d(t) = DR(t)\Delta t_d = DI(t,x)\rho\Delta x_d, \tag{3.32}$$

where D is a detection sensitivity constant and ρ is the amount of signal reflected per unit length. $I(t, x)$ is the intensity of the signal pulse at position x, corresponding to a received signal at t, that is,

$$I(t,x) = I_0 \prod_{j=1}^{i} (1 - \rho\Delta x_j). \tag{3.33}$$

The detected signal is

$$d(t) = D\rho\Delta x_d I_0 \prod_{j=1}^{i} (1 - \rho\Delta x_j). \tag{3.34}$$

Taking the logarithm

$$\ln[d(t)] = \ln(D\rho\Delta x_d I_0) + \sum_{j=1}^{i} \ln(1 - \rho\Delta x_j), \tag{3.35}$$

$$\ln[d(t)] \approx A - \sum_{j=1}^{i} \rho\Delta x_j = A - \sum_{j=1}^{i} 2\rho c \Delta t_j, \tag{3.36}$$

where A is a constant. In the limit as Δt_j goes to zero, the sum becomes an integral,

$$\ln[d(t)] = A - 2\rho ct. \tag{3.37}$$

This result indicates that uniformly distributed backreflections produce a received signal that decays linearly with time on a semilog plot.

A reflected TDR signal usually contains multiple distributed and discrete components [135]. Figure 3.31 is an idealized TDR signal that includes a portion where the amount of Rayleigh scattering backreflection increases due to a temperature rise [136]. Waveguide lengths of hundreds, if not thousands, of meters are required for attaining the idealized signal of Figure 3.31. Shorter length waveguides can be useful, provided that the TDR instruments are properly tuned. Limitations on the speed of the initial optical pulse are usually the limiting constraint. The use of prepositioned reflectors along the length of

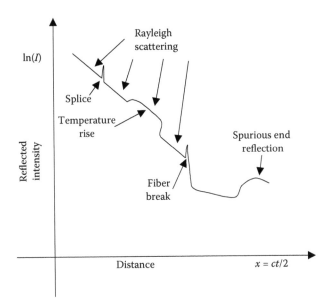

FIGURE 3.31 Nominal OTDR signal from an optical fiber with a splice joint, temperature variation, and break.

the waveguide can enhance spatial registration [137]. Lyöri et al. and Zimmerman and Claus developed long-gage strain sensors based on the relative motion of multiple discrete reflectors [138,139]. Prepositioning optical reflectors along the length of a fiber waveguide can form registration points for long-gage strain gages [140]. Crack sensing requires good mechanical coupling between the material under test and the waveguide [141]. Leung et al. developed an OTDR-based crack sensor along with a corresponding mechanical model [142]. Reentrant loops are another possibility for enhancing OTDR strain measurements [143]. OTDR techniques can also interrogate a set of discrete sensor elements along a network. Voet et al. demonstrated an array of up to eight phase shifting fiber optic pore water pressure sensors for geotechnical applications [144].

Figure 3.32 shows a diagram of an experimental test rig that produces controlled levels of shear damage in an optical fiber [71]. Figure 3.33 shows the corresponding change in the OTDR signal in the fiber with shear. These curves were taken on a ca. 1990 model OTDR instrument with significant saturation and ringing. Nonetheless, the OTDR clearly indicated the progression of the damage. Gu et al. report similar results [145].

The above described OTDR phenomena use elastic scattering and reflection effects, that is, the wavelength of the light does not change. Inelastic and nonlinear effects scatter light at wavelengths that differ from the launch signal. This includes Raman and Brillouin scattering, and underlies multiple distributed fiber optic sensor designs [146–150].

Brillouin backscatter results from light interactions with acoustic phonons. It is possible to control the spatial arrangement of phonons with acousto-optical interactions and to use this technique to form a sensor. One approach uses intense light to cause an electrostrictive mechanical deformation of an optical fiber waveguide. Due to thermodynamic considerations, this interaction produces light with lower energy and longer wavelength known as a Stokes wave. The electrostriction due to coherent light forms a forward-propagating elastic wave in the fiber that acts as a Bragg grating. The Stokes wave interacts with the elastic wave Bragg grating, and due to a Doppler effect, produces a frequency shifted backscattered Brillouin wave. Brillouin scattering is sensitive to both temperature and strain [151].

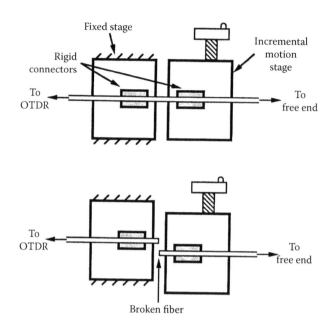

FIGURE 3.32 Diagram of optomechanical stage test setup for inducing controlled shear damage in optical fibers. (Data from Huston DR et al. 1994. *Time Domain Reflectometry in Environmental, Infrastructure, and Mining Applications*, US Bureau of Mines, SP 19-94.)

Although the frequency shifts of Brillouin scattering are small (10 GHz approximately), the relatively large amplitudes of the return signals can make it suitable for distributed temperature sensing applications, especially when combined with time-domain sensors to make a Brillouin optical TDRs (BOTDR) [152,153].

Multiple applications have demonstrated distributed strain and temperature BOTDR sensors with lengths in excess of 1 km and spatial resolutions to less than a meter [151, 154–159]. Shorter distances may also be viable [160]. Bernini et al. detected damage in a 7.0 m beam in bending with BOTDR [161]. Nishio et al. used a system with nominal 100 mm spatial resolution to measure the deflection of a cantilever beam using embedded fiber optic cables [162]. Specialized fiber optic cable packaging methods that enhance the strain transfer from the structure to the cable and specialized data analysis techniques can improve sensitivity [163,164]. Ordinary optical communications fiber optic cable is specifically designed to minimize such strain transfer. BOTDR temperature sensing has been used in pipeline leak detection [165]. Figure 3.34 shows temperature data collected from the bottom of Lake Geneva in Switzerland using an ordinary fiber optic communications cable configured as a Brillouin backscatter sensor. Coupling water-swellable polymers to a BOTDR fiber forms a distributed water sensor, possibly for use in corrosion condition monitoring [166]. TDR can interrogate arrays of FBGs along an optical fiber [167,168]. It is possible to combine BOTDRs with FBG strain sensors on the same optical waveguide [169]. Fiber optic techniques that combine Brillouin sensing with long FBGs may produce a sensing system that is superior for large civil structures [170,171]. Sensor systems capable of simultaneously measuring strain and temperature with sufficient spatial resolution to be applicable to monitoring modest-sized composite elements are possible [172]. Instrument speed, rather than transduction speed, seems to impose the biggest constraint on the use of BOTDR for dynamic testing. Texier et al. conducted BOTDR dynamic measurements on a vibrating beam with natural frequency of 4.2 Hz and a maximum strain of 500 $\mu\varepsilon$ [173].

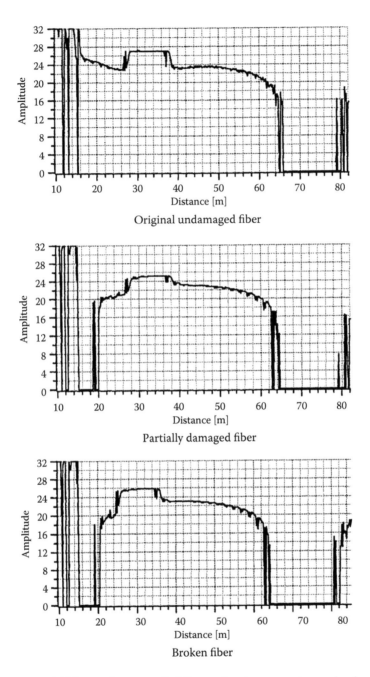

Original undamaged fiber

Partially damaged fiber

Broken fiber

FIGURE 3.33 Change in OTDR trace from an optical fiber as shearing causes damage by shearing. (Data from Huston DR et al. 1994. *Time Domain Reflectometry in Environmental, Infrastructure, and Mining Applications*, US Bureau of Mines, SP 19-94.)

Installation over long distances and signal registration with position on the structure are nontrivial technical challenges that affect the practical use of distributed sensors. Jin et al. developed an interesting and highly effective method for installing fiber optic cables in long conduits for undersea pipeline monitoring applications [174]. The technique uses

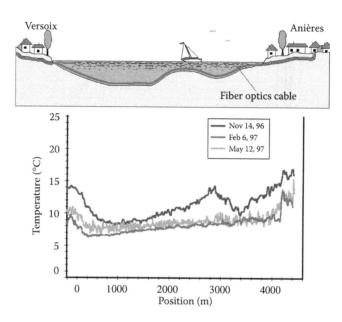

FIGURE 3.34 Temperature profile recorded from the bottom of Lake Geneva with a distributed Brillouin backscatter sensor. (Reproduced from Selker JS et al. 2006. *Water Resour. Res.*, 42, W12202. Copyright 2006 American Geophysical Union. With permission.)

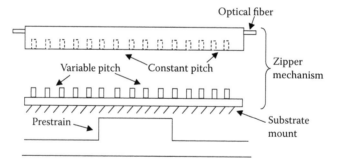

FIGURE 3.35 Neubrex variable pitch zipper device for distributed optical fiber mounting combined with fiber prestrain for registration. (Adapted from Kishida K et al. 2008. US Patent 7,356,238.)

compressed air to drag the fiber cable while providing lubrication between the cable and a smooth steel conduit. Installation distances of 1–3 km are practical. Figure 3.35 shows the operating principle of the Neubrex zipper device that aids in the mounting of distributed optical fiber sensors while also introducing a prestrain in the fiber. The combination of prestrain and mounting location registers the fiber on the structure.

Similar to OTDR is *optical frequency domain reflectometry* (OFDR) where frequency sweeping of the light from a tunable laser source interrogates an optical fiber in a manner similar to step frequency or FM radar. Sweeping the laser light frequency allows for a determination of the strain in the individual gratings to produce a distributed sensor. Spillman demonstrated an OFDR system based on strain-induced birefringence and polarization along the length of the fiber [175].

ETDR can also detect damage and deformation in structures. One application is to monitor mine roofs with embedded coaxial cables [137,176]. It is possible to precrimp the cable at specific locations to provide spatial registration capabilities [177]. Chajes et al. monitored pre- and post-tensioning strands in concrete for corrosion and other defects with ETDR by converting the strands into an electrical waveguide with a short-circuited end [178]. Chen et al. developed cables with spiral wrapped foil conductors, which have an improved spatial resolution for resolving damage with ETDR in concrete beams [179,180]. Frequency sweeping of high-frequency signals can synthesize ETDR signals. Many vector network analyzers have this capability. Budelmann et al. used frequency-sweeping methods to locate cracks in elongated conductors, such as prestressing strands [181]. Liang et al. monitored subsurface moisture levels in highway pavements with ETDR [182]. Dennis et al. detected shearing soil motions due to shallow slope failures with ETDR [183]. Francke et al. measured rock motions in a nuclear waste isolation pilot plant with ETDR [184]. Daita et al. measured soil conductivity to determine the effects of admixtures on quality with ETDR [185].

Elastodynamic TDR methods are also practical. Anastasi and Madaras used ultrasonic 1-D waveguide transmission to measure wire insulation losses [186]. Flexural modes were easier to launch and receive than axial modes, but suffered from dispersion problems. Embedded rods with elastic moduli that differ from the surrounding elastic medium form elastic waveguides. Reis et al. detected corrosion in reinforcing bars by propagating ultrasonic waves along the bar [187]. Railroad rails and steel post-tensioning strands may also be suitable for guided wave SHM tests [188,189]. Gong et al. evaluated the condition of soil nails and the surrounding grout by using the nails as waveguides in a sonic-echo TDR system [190]. Loveday used railroad rails as elastic waveguides with reliable transmission distances of 2.5 km [191].

Washer and Fuchs developed sensors that measure tension in steel post-tensioning strands for concrete by using the strand as an elastic waveguide and by using magnetoelastic effects to both launch and analyze propagating elastic waves [192]. Na et al. used a similar EM acoustic wave technique where magnetostrictive and/or Lorentz EM forces launched elastic waves to inspect concrete filled steel pipes [193]. Kwun and Teller II developed a magnetoelastic system for actively and passively monitoring cable structures [194]. Kim and Kim combined magnetostrictive sensing and wavelet analysis of dispersive bending waves to identify damage in steel beams [195]. Waveguides can also combine with magnetostrictive effects to form position transducers. Figure 3.36 show the principle of operation of the Temposonics™ magnetostrictive elastic waveguide displacement transducer. Anderegg et al. used magnetostrictive elastic waveguide displacement transducer as part of a long-term highway deformation monitoring effort [196].

3.3.5   Elastic Waves: Sonar and Ultrasound

Propagating elastic and acoustic waves are the underpinning mechanisms of a variety of SHM sensor systems. These systems generally use four types of wave–material interactions: (1) *Interactions with boundaries between media with different wave speeds and the dimensions of the region are large compared with a wavelength*: The dimensions of the regions enclosed by the boundary are much larger than the wavelength. The timing and amplitude of reflections from the boundaries indicate the location of the boundary and change in material properties. (2) *Interactions with boundaries and the transverse dimensions of the region are on the order of a wavelength*: Elastic scattering off of small inclusions, cracks, and delaminations are typical in these cases. An analysis of the scattered waves can indicate the size and nature of the inhomogeneity. (3) *Interaction of waves with a distribution of scatterers*: An analysis of the bulk

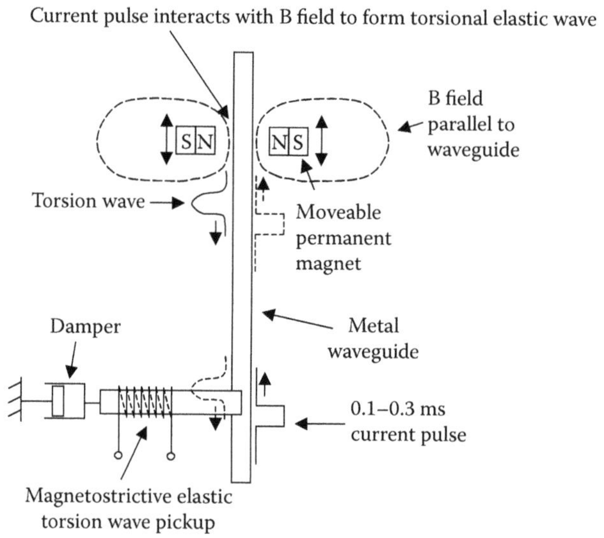

Current pulse interacts with B field to form torsional elastic wave

B field parallel to waveguide

Torsion wave →

Moveable permanent magnet

Damper

Metal waveguide

0.1–0.3 ms current pulse

Magnetostrictive elastic torsion wave pickup

FIGURE 3.36 Principle of operation of Temposonics™ magnetostrictive waveguide position transducer. (Picture derived from Nevius SG. 1996. US Patent 3,249,854 and MTS Temposonics™ literature.)

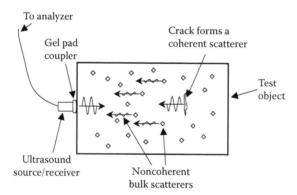

To analyzer

Gel pad coupler

Crack forms a coherent scatterer

Test object

Ultrasound source/receiver

Noncoherent bulk scatterers

FIGURE 3.37 Ultrasonic testing of solid with a coherent crack scatterer and an incoherent heterogeneous bulk scatterer.

scattering of waves can indicate the nature and distribution of the scatterers (Figure 3.37). (4) *Measurement of wave speeds*: This can help indicate macroscopic material properties, such as bulk modulus. The waves used in the tests typically travel through bulk solid cross sections, along surfaces or channels that guide the waves in specially constructed waveguides. Wavelength-dependent wave speeds, that is, dispersion, can provide addition insight into material properties.

The composition of ultrasound return signals typically consists of *ballistic* and *scattering* components. Ballistic signals are coherent reflections off of material discontinuities. Scattering signals are the diffuse reflections off of bulk inhomogeneities with noncoherent configurations in the material. Bulk ultrasonic testing works well for the characterization of certain homogeneous media, such as steel, or those with favorable transmission and scattering characteristics, that is, the soft tissue in the human body [197]. Modern instruments are now capable of rapidly examining complicated parts and the imaging of small defects.

Possibilities include corrosion cracking in metals or the potential for percolation damage in concrete [198,199].

The application of ultrasound to SHM is common in situations where the failure of an individual part has severe consequences and the failure mode is a subsurface defect in an otherwise homogeneous solid. Ultrasonic thickness measurements using ballistic signals on metal parts allow for assessments of corrosion, cracking, layers of impurities, wear, and as-built dimensions [200,201]. An example is fracture-critical steel hanger pins on older bridges. The failure of one of these pins can cause the bridge to collapse. Subsurface cracks often form in the pins. These cracks often grow slowly, but have the potential to reach dangerous dimensions. Ultrasonic inspection can detect the cracks at early stages of development [202]. Blackshire and Cooney imaged hidden corrosion defects with dimensions as small as 100 μm, as well as surface cracks [203,204]. Materials with complicated microstructures are more difficult to examine with conventional ultrasound signal processing methods. Statistical pattern recognition approaches, along with the acousto-ultrasonic method, may be useful in applications that involve bulk scattering, such as the cure monitoring of particle board manufacture [205–207]. Nonlinear techniques that generate short-wavelength harmonics can enable high-resolution ultrasound imaging techniques [208].

Elastic wave speed, scattering, and absorption measurements can provide useful information for estimating bulk material properties [209]. Nonhomogeneous materials, such as concrete, attenuate ultrasound waves with size- and wavelength-dependent scattering processes being principal mechanisms [210,211]. If propagating waves through the medium for the purpose of probing underlying features is the intent, then it may be necessary to mitigate the effects of scattering and attenuation by a careful tuning of the transmission wavelength [212]. If the intent is to determine bulk material properties, then it can be useful to measure the scattering and attenuation directly. Popovics et al. correlated ultrasonic wave speed and attenuation with the state of cure in fresh concrete [213]. Gallo and Popovics correlated wave speed with concrete ASR activity [214]. Takatsubo et al. found that subwavelength voids cause a delay and broadening of ultrasound pulses [215]. The delay correlates well with the volume fraction of the voids. Tang correlated microcracking in concrete with an increase in ultrasonic wave attenuation [216]. Krause et al. found that good condition concrete enables imaging of multiple layers of steel and polystyrene subsurface features using a frequency range of 50–100 kHz [217]. Drabkin and Kim showed that severely damaged concrete heavily attenuates ultrasonic waves in specific frequency bands, that is, 40–80 kHz [218]. Benedetti assessed levels of fire damage in reinforced concrete structures with ultrasound [219]. Ramamoorthy et al. and Le Marrec et al. suggest using concrete ultrasound scattering as a method to assess the depth of cracks in concrete [220,221]. Aktan et al. correlated ultrasonic pulse velocity with permeability in early age concrete [222]. Mirmiran and Wei correlated ultrasound pulse velocity with damage in FRP-encased concrete columns [223]. Ross and Pellerin give an extensive review of the use of ultrasound and other NDE methods on wood structures in ca. 1994 [224].

The guided nature of certain ultrasonic and elastic waves enables using panels, plates, pipes, overhead transmission lines, and similar elongated elements as waveguides for at-a-distance structural evaluations [225–230]. Lamb waves propagate in plate-like structures and are perhaps the most common of the guided waves used in damage detection (see Appendix A). Modeling high-frequency waves in beams often requires supplementing the Bernoulli–Euler formulation with shear and rotary inertia terms, as in the Timoshenko beam model [231,232].

The SHM literature contains multiple examples of procedures that exploit the attenuation, scattering, and reflection of Lamb waves by damage. Early works include that of Firestone and Ling (ca. 1946) and Worlton (ca. 1965) that convert plane waves into Lamb waves by a subsurface delamination to detect the presence of damage [233,234]. Neary et al. detected damage in composite plates with embedded piezoelectric transducers during signal attenuation techniques [235]. Tseng et al. and Engberg and Ooi used ultrasonic impedance measurements in composite plates at cryogenic temperatures [236,237]. Swenson and Crider detected "hot-spot" damage in relatively inaccessible regions of aircraft bulkheads with Lamb waves launched by piezoelectric transducers [238]. Yang et al. used piezoelectric sensing washers in high-temperature panels to detect fastener failures (a severe sensing challenge) [239]. Berger and Baier used Lamb wave reflections on aluminum and CFRP beams to detect damage [240]. Cawley measured corrosion in pipes with Lamb waves [241]. Thomas and Welter detected corrosion damage by Lamb wave distortion measurements [242]. Guy et al. measured damage-induced energy leakage between different Lamb wave modes in composite plates [243]. Osegueda et al. detected cracks near rivet holes in flat panels by observing Lamb wave scattering [244]. Jain et al. used MEMS transducers with interdigitated sensors to provide a directional sensitivity for ultrasound damage detection [245]. Murphy et al. detected composite plate delaminations with embedded Fabry–Perot fiber optic sensors [246]. Pierce et al. measured damage-dependent changes in Lamb wave propagation with a fiber optic Mach–Zehnder interferometer [247]. Thursby et al. used Bragg grating and polarimetric fiber optic sensors to detect Lamb waves, both of which have a directional dependence that is useful in location algorithms [94]. Jung et al. detected damage in the form of buried inclusions that simulated damage in concrete with Lamb waves [248]. Luangvilai et al. measured composite concrete repair patch performance with laser ultrasound-induced Lamb waves [249]. Kudryavtsev et al. measured the prestress in steel welds by determining slight changes in ultrasonic wave speeds due prestress-induced acousto-elastic effects [250]. Calomfirescu and Herrmann used Lamb wave attenuation as an indicator of damage in composites [251]. Chellapilla et al. developed formulations that describe the interaction of Lamb waves with cylindrical fasteners in aircraft structures [252]. Tian et al. used Lamb wave mode conversion for crack detection [253]. Petculescu et al. used the modal group velocity change and time delay change of A_0 modes to both locate and quantify delaminations and damage in composite plates [254]. Takahashi et al. examined Lamb wave detection of delaminations in composite repair patches with a laser-scanning system [255]. The technique uses a thermoelastic Nd:YAG laser source and a piezoelectric patch pickup. Scanning the laser source over the patch and using concepts of wave reversibility enable the synthesis of Lamb wave images with sufficient detail to detect damage.

Shear horizontal (SH) waves are similar to Lamb waves, except that the particles move in the direction of the plane of a plate. This causes SH wave propagation to be particularly sensitive to surface cracks. Yamada et al. demonstrated through visualization techniques that SH wave scattering is sensitive to both the location and size of surface cracks [256].

Thick solids with flat surfaces guide *surface acoustic waves* (SAW), that is, *Rayleigh waves* along the surface. Surface features, such as cracks, will reflect, scatter, and attenuate SAW waves [257,258]. Marzani et al. generated SAW waves and used mobile acoustic pickups to detect cracks in pavements [259]. Kuhr et al. developed specialized interdigitated piezoelectric transducers for launching and receiving SAW waves in complex geometry aerospace structures [260]. Similarly, Wen and Zhu developed an analog wavelet array processor using specialized interdigitated piezoelectric transducers for launching and receiving SAW waves [261]. A set of highly sensitive MEMS-based chemical and strain sensors uses SAW waves for sensing [262,263]. SAW waves are useful in geotechnical applications for determining the strength of soils without resorting to coring or penetration tests [264,265].

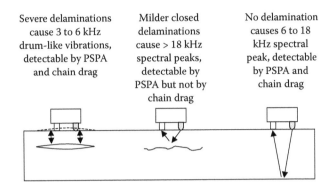

Severe delaminations cause 3 to 6 kHz drum-like vibrations, detectable by PSPA and chain drag

Milder closed delaminations cause > 18 kHz spectral peaks, detectable by PSPA but not by chain drag

No delamination causes 6 to 18 kHz spectral peak, detectable by PSPA and chain drag

FIGURE 3.38 Detection of delaminations in concrete with a Portable Seismic Pavement Analyzer (PSPA). (Adapted from Gucunski N and Maher A. 2000. Bridge deck evaluation using portable seismic pavement analyzer (PSPA). Final Report, FHWA, NJ 2000-05.)

This method is sometimes called the *surface acoustic stress wave* (SASW) technique. The acronym SASW also corresponds to *spectral analysis of surface waves*. The measurements use an impulsive source that generates the stress waves and transducers that measure the timing and amplitude of the waves (Figure 3.38). Figure 3.39 show versions of portable elastic wave analyzers for use in pavement testing [266–268]. Frequency domain analysis can be quite useful in sorting out the different elastic wave effects and designing the appropriate test protocol [269]. Falling weight deflectometer (FWD) measurements can provide estimates of pavement layer stiffness, thickness, and distress [270–274]. FWD testing is in many respects a quasi-static version of SASW testing.

Launching and receiving ultrasound waves requires transducers with suitable frequency and energy transduction characteristics. Piezoelectric and embedded transducers, laser pulse heating thermoelastic, magnetostrictive transducers, magnetostrictive interactions with ferromagnetic materials, EM eddy-current Lorentz force methods, air-coupling, water jet coupling, and spray-on sol–gel coatings are viable options for launching and detecting

FIGURE 3.39 Portable Seismic Pavement Analyzer.

the guided ultrasound waves [277–286]. Point sources can launch waves, but will excite multiple wave modes and complicate the data analysis. Specialized transducers that launch more of the source energy into a particular wave mode can alleviate some of these difficulties. Possibilities include interdigitated transducers with a spatial configuration tuned to excite a particular Lamb wave mode, synthetic focusing ultrasound Lamb wave arrays scanning laser thermoelastic methods, and Lamb wave crack-induced mode conversion transducers [287–291]. Receivers include most of the same transducers as the launchers, along with embeddable receive-only transducers, such as FBGs [292]. Laser ultrasound sends and receives ultrasound signals with laser techniques [210,293–297]. Local transient thermal heating by a laser pulse converts into an elastodynamic pulse that propagates throughout a structure. Interferometric and similar laser-based methods measure the resulting ultrasound surface motions. Hybrid laser–piezoelectric ultrasound systems are also common. It is possible to fabricate sensors that use shear waves and focusing lenses to identify thin debonding defects [298]. Situations involving bulk temperature changes can influence the nature of propagating waves and may require compensation techniques to distinguish temperature-induced changes from damage-induced changes [299].

An often advantageous approach combines different source and receiver technologies to form hybrid elastic wave test systems. Ogisu et al. inspected aircraft with a hybrid PZT source and FBG pickup technique [300]. Lanza di Scalea et al. formed ultrasonic waveguides for damage detection with lap shear joints [301]. Yuan et al. detected damage in composite honeycomb plates with wideband Lamb wave methods [302]. Wideband methods have the potential for sensitivity over a wider range of conditions. Vinh recommends using Stoneley waves to detect delaminations [303]. Divekar and Vizzini detected delaminations in tapered composite structures using embedded PZT ultrasound source transducers and fiber optic pickups [304]. Malkin measured debonding in composite repair patches with guided waves [305]. Bescond et al. measured surface stresses in metals due to shot-peening and similar effects with laser-generated surface P-waves [306]. The P and S wave speeds in metals depend on prestress, in a material-specific manner. In steel, the speed change in the P-waves is often of the opposite sign of the S-wave speed change. Thien et al. developed a system for pipeline inspection that uses Lamb waves propagating along the axis of the pipe surface to identify cracks, corrosion and through-wall thickness; and impedance measurements of pipe section joint integrity with the same set of *macro-fibre composite* (MFC) transducers [307].

Triangulation techniques using the time-of-flight of elastic waves with an array of multiple point sensors can locate objects, features, and defects. Barnoncel et al. propagated flexure waves from an array of piezoelectric transducers on the perimeter of a plate to detect damage [308]. Li et al. improved the arrival time estimations of complicated Lamb waves with time correlation techniques [309]. Sonar is a similar technique for measuring the distance to underwater objects. A structural application of sonar monitors the distance to the riverbed next to bridge piers with potential scour problems [310–312]. Measuring wave direction as well as timing can improve the location estimates. Betz et al. and Matt and Lanza di Scalea developed sensor rosettes that determine the direction of Lamb wave propagation [313,314]. The sensor uses three fiber optic FBG sensors on a single fiber with the sensors laid up into a triangle. Processing the signals from the individual FBGs indicates the direction of propagation, as well as the time of signal arrival.

3.3.6 Impact-Echo Testing

Impact-echo (IE) testing is similar to ultrasonic methods. IE launches elastic waves into a solid structural element and analyzes the reflected and transmitted elastic waves for information concerning subsurface conditions. The distinction between IE and ultrasonic testing lies primarily in the signal source. IE uses a mechanical impact to generate the elastic waves that are impulsive and broad banded. Ultrasonic testing uses ultrasonic transducers with signals that are typically gated sinusoids and possibly with chirping. IE has the advantage of a robust and convenient excitation source at the expense of producing a somewhat less-controllable waveform. The technique dates back to at least 1962 with a pole testing system due to Harris [315]. Standard IE testing techniques are relatively slow, require good contact between the transducers and the surface being tested, and require careful control and tuning of the IE source pulse parameters. Detecting the arrival times of IE pulses can be challenging due to weak return signals, noise contamination, and wave scattering. It is possible to determine the depth of the asphalt concrete overlay as well as the location and depth of features such as post-tensioning strands, voids, bridge pier foundations delaminations, and other defects with IE methods [316–324]. Using special provisions for access often aids in data acquisition. An example is the construction of large concrete piers and piles. The method is to place an array of plastic pipes with diameters of 20–30 mm parallel to the axis of the pier or pile and near the perimeter. These pipes allow for the insertion of sonic-cross hole logging instruments that can detect voids and other gross defects (Figure 3.40).

IE testing calculates the depth, D, of a reflector with the formula

$$D = \frac{\beta C_P}{2f},$$

(3.38)

where β is a geometric correction factor, C_P is the P-wave speed and f is the frequency of the wave (vibration mode). The geometric correction factor β accounts for modal vibration effects on the propagation of waves. If the solid is infinite or semiinfinite, plane waves can propagate and $\beta = 1$ (but no signals will return). If the solid is a slab with finite depth and infinite horizontal extent, then $\beta = 0.96$. Sansalone and Carino initially determined this value empirically [317]. Numerical finite element and zero-order symmetric Lamb waves provide analytical confirmation [325]. Different geometries vibrate with different mode shapes and require different β values. Gassman and Zein determined that $\beta = 0.87$ for a particular concrete T-beam [326]. Supplementary SAW (Rayleigh wave) techniques can

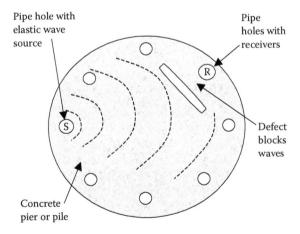

FIGURE 3.40 Cross section of concrete pipe or pier with cross-hole sonic logging detection of void defect.

determine the wave speed C_P for use in Equation 3.35 [327]. Similar techniques known as *sonic echo* or *pile integrity* testing can determine scour and damage in bridge piles [328]. The method is to launch elastic waves from the top of the pile and then deduce subsurface conditions from an analysis of the reflected waves. Accounting for the dispersive nature of elastic wave propagation is often a necessary improvement to the data analysis [329]. *Parallel seismic* testing can measure the length of a pile by sending and receiving elastic waves from a small shaft bored parallel to the pile. IE testing can also provide a means of determining C_P when the relative geometries of the reflectors are known. An example application is the cure monitoring of fresh concrete [330].

Many IE tests produce high-quality data that do not require extensive signal processing for interpretation. Nonetheless, signal processing can enhance the data interpretation by providing additional information, such as improved estimates of wave arrival times. Bayard and Joshi processed acoustic reflection waves with Weiner filters [331]. Frequency domain signal processing can also be effective. The primary measured parameter is usually the fundamental frequency of vibration. Ata and Ohtsu used frequency domain analysis of multipoint IE tests on post-tensioned concrete beams in an effort to detect grouting voids [332]. Finno and Chao measured the mobility of drilled piles with the spectral analysis of velocimeter readings [333]. Deviations from expected values of the mobility can correspond to defects in the piles. IE testing can also measure subsurface layers stiffnesses. Since the impacts produce waves that radiate from the impact point, accounting for wave shape and near field effects by mathematical devices, such as the cone model, provides estimates of underlying stiffness that are superior to those derived from simple single degree of freedom models [334].

Similar to IE methods is the diagnostic method of applying a small mechanical tap to a structure and then analyzing the emanating sound. Tapping structures with cracks, delaminations, voids, and other defects often produces sounds that differ from that of healthy solid structures. Tapping sound analysis with the human ear as the sensing system has been an SHM staple for millennia. Modern instrumentation can enhance these techniques. Canning et al. analyzed hammer tap responses in asphalt cylinders to develop an improved testing method for moisture damage [335]. Kweon and Kim used IE testing to determine complex elastic moduli of asphalt samples [336]. Butt et al. showed that acoustic waves attenuate more rapidly in damaged concrete bridge decks [337]. It is possible for automated instruments to create and analyze tap sounds [338]. Savalli et al. developed an automatic three-finger piezoelectric cantilevers tapping system for SHM [339]. Two of the fingers tap and the third acts as a pickup.

3.3.7 EM Waves: 1-D Ground Penetrating Radar

GPR testing is the EM analog of ultrasonic or elastic wave testing. Reports of the use of GPR in bridge and highway inspections date back to at least the early 1980s [340]. An antenna launches EM waves or pulses into the structure. Measuring and analyzing the reflected waves can identify subsurface features and properties. 1-D GPR testing treats the EM waves and subsurface structural features as having a planar and layered structure. Ray optics and Snell's law of reflection and refraction are often sufficient for the interpretation of 1-D GPR data. The timing, strength, and phase of the waves or pulses that reflect off of subsurface features have simple and direct correlations with the feature type and location. GPR applications in the testing of reinforced concrete, asphalt, other pavement materials, masonry, and geotechnical structures are common.

A common SHM GPR application measures the bulk EM properties of concrete and pavements. Material deterioration changes EM properties. These changes may be in the fine structure of the material, such as the formation of additional interfaces and dielectric discontinuities created by delaminations, microcracking, and freeze-thaw damage; or they may occur in the bulk properties of the material, such as an increase in the dielectric constant (decreased velocity) or an increase in the attenuation due to higher moisture content, chloride content, and porosity. GPR-based bulk property measurements can provide bridge decks deterioration prognostic on a par with chain drag and half-cell methods [341]. GPR can also identify and locate features, such as reinforcing bars, post-tensioning strands, pavement and masonry layer depths, railroad bed ballast condition, delaminations, and voids [342–345]. Jones et al. compared 3-D GPR imaging techniques with Schmidt hammer and coring techniques to assess the condition of a cylindrical reinforced concrete ringwall [346]. The 3-D GPR data provided sufficient information to location regions of moisture ingress and early-stage reinforcing bar corrosion activities.

Table 3.1 lists the wavelengths of GPR waves in various media. The majority of GPR systems for concrete bridge inspection operate near 1 GHz, but higher frequency concrete tests show promise [347]. Millimeter-wavelength testing shows promise for the inspection of composite and foam structures [348]. The type of EM waves used in the testing is largely hardware dependent. Ultra-wideband (UWB) pulses, gated sinusoids, and swept measurements have all seen some degree of success. The assumed wavefront shape is planar or spherical, but is usually more complex in the near field.

Figure 3.41 shows the primary components of a GPR system. Radar systems use either a single antenna (monostatic) or separate antennas (bistatic) to send and receive signals. For pavement testing, air launched antennas are more convenient than ground coupled antennas because they can be used at highway speeds. However, air launched antennas couple less energy into the ground [349]. Systems with arrays of antennas that use advanced imaging techniques are viable [350].

An antenna produces EM waves by a complex interaction of varying electric and magnetic fields [351]. Figure 3.42 shows a horn antenna specifically designed to transmit and receive signals in the 0.5–6.0 GHz range. A key feature of this antenna is the integrated balun feed connector (Figure 3.43). This connector, if properly tuned, can produce a feedpoint with minimal backreflections, while effectively incorporating a balun to balance the load on opposite sides of the antenna [352,353].

TABLE 3.1

EM Wavelengths as a Function of Frequency for Media

	Wavelength (m)			
Frequency (GHz)	Air $\epsilon_r = 1.0$ $c = 3E+8$ m/s	Asphalt $\epsilon_r = 5.0$ $c = 1.34E+8$ m/s	Concrete $\epsilon_r = 7.5$ $c = 1.09E+8$ m/s	Water $\epsilon_r = 78.0$ $c = 3.39E+7$ m/s
0.5	0.600	0.268	0.218	0.068
1	0.300	0.134	0.109	0.034
2	0.150	0.067	0.055	0.017
4	0.075	0.034	0.027	0.009
8	0.038	0.017	0.014	0.004
16	0.019	0.009	0.007	0.002

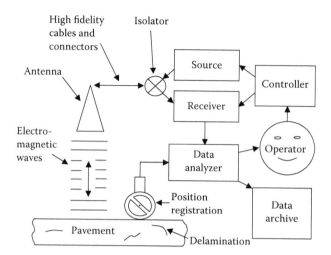

FIGURE 3.41 Schematic of information flow in a GPR system.

GPR data processing uses a variety of techniques that range from color and graphical enhancements of the plots of the raw data to complicated systems that use synthetic aperture radar (SAR) and migration techniques to reconstruct underlying features from the data [354–358]. Cylindrical reinforcement bars reflect incident waves from most angles. Reflections from bars that are not directly under the antenna are also received due to off-axis beam spreading. The effect is to produce a reflection at an apparent depth that differs from the true depth and needs to be resolved by calibration. Arrays of reinforcing bars are differentially sensitive to the polarization of the GPR waves [359].

Frequency dependence constrains GPR system performance. High-frequency rolloff limits fine-scale resolution. Low-frequency rolloff limits penetration depths. Since material

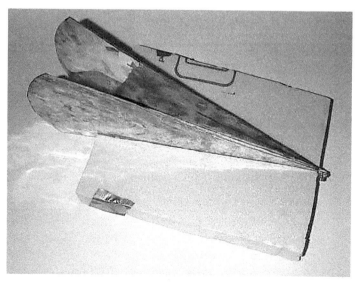

FIGURE 3.42 Integrated-balun, rounded-end horn antenna. (Adapted from Huston DR et al. 2001. *Smart Systems for Bridges, Structures, and Highways*, SPIE 4330.)

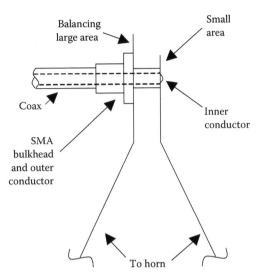

FIGURE 3.43 Integrated-balun apex feed detail for horn antenna. (Adapted from Huston D et al. 2000. *Nondestructive Evaluation of Highways, Utilities, and Pipelines IV*, SPIE 3995.)

properties are often frequency dependent, it can be advantageous to use a GPR system with tunable frequency ranges [361,362]. The voltage standing wave ratio (VSWR) is a standard performance measure of radio and microwave system components such as antennas. The VSWR is the ratio of the maximum to minimum voltage vector amplitude, that is,

$$\text{VSWR} = \frac{1 + |\Gamma(f)|}{1 - |\Gamma(f)|}, \tag{3.39}$$

where $\Gamma(f)$ is the reflection coefficient as a function of frequency [363]. Values of the VSWR larger than one indicate the presence of backreflections and suboptimal performance.

A simple and effective GPR testing technique infers bulk material dielectric constants by surface reflectivity measurements. The *reflection coefficient*, R_{12}, between media 1 and 2, is the amplitude ratio of the incoming and reflected waves. R_{12} depends on the dielectric constants (ε_1 and ε_2) of the two layers

$$R_{12} = \frac{\sqrt{\varepsilon_1} - \sqrt{\varepsilon_2}}{\sqrt{\varepsilon_1} + \sqrt{\varepsilon_2}}. \tag{3.40}$$

Reflection amplitude measurements from the surface of concrete, A_c, and a metal plate placed on the surface, A_{pl}, provide the information needed to calculate the relative dielectric constant of the concrete ε_c. This formula inspires simple reflection-type measurements of the bulk dielectric constant by comparing the amplitude of the surface reflection, A_c, to that reflected from a metal plate, A_{pl} [364]. The metal plate reflection amplitude corresponds to that of an incident wave with the sign reversed so that $R_{12} = -A_c/A_{pl}$ and

$$\varepsilon_c = \left[\frac{1 + (A_c/A_{pl})}{1 - (A_c/A_{pl})} \right]^2. \tag{3.41}$$

Surface reflection dielectric measurements are standard tools in the GPR assessment of concrete bridge decks. Xing et al. developed an alternative pavement dielectric measurement technique using two antennas with evanescent wave coupling [365]. When the surface has an overlay, such as asphalt on top of concrete, the reflection coefficients from both the air–asphalt and the asphalt–concrete interfaces combine to yield expressions for the dielectric coefficient of the concrete, ε_c, and the asphalt overlay, ε_a,

$$\varepsilon_c = \varepsilon_a \left[\frac{F - R_2}{F + R_2} \right]^2,$$
(3.42)

where

$$F = \frac{4\sqrt{\varepsilon_a}}{1 - \varepsilon_a}.$$
(3.43)

An examination of Equation 3.40 indicates that transmitting a wave from a high dielectric layer into a low dielectric layer causes a positive amplitude reflection. Conversely, transmitting a wave from a region of low dielectric to high dielectric inverts and changes the sign of the reflected pulse. Placing a thin layer of material of differing dielectric value perpendicular to a wavefront can act as an analog wave differentiator.

The dielectric constant of normal concrete ranges from 7.5 to 10.5 with deteriorated concrete generally having higher values [364,366,367]. Baker-Jarvis et al. describe in detail methods for measuring the EM properties of building materials [368]. Grivas et al. measured dielectric properties with a relative percentage error of 6% using a GPR system operating between 1 and 2.4 GHz [369].

An example of using surface reflectivity to indicate the integrity of the surface of a roadway is a series of reflection measurements taken on a highly distressed bridge (Bostwick Rd, Shelburne, VT). The measurements traversed paths corresponding to sections of undistressed, partially distressed, and highly distressed with an asphalt patch repair pavement regions (Figures 3.44 and 3.45). GPR can also measure asphalt and pavement thicknesses, especially when the pavement thickness is less than about 180 mm [370–372]. GPR is an

GIMA1, the first test, the first line 24 to 28 feet, Bostwick bridge, Shelburne, Vermont

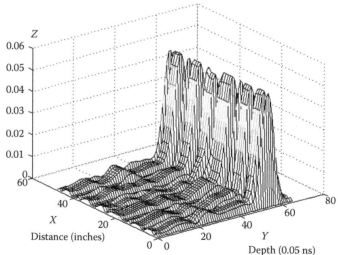

FIGURE 3.44 GPR reflectivity from a relatively undistressed portion of the deck of the Bostwick Bridge (Shelburne, VT).

GIMA1, the first test, the first line 8 to 12 feet, Bostwick bridge, Shelburne, Vermont

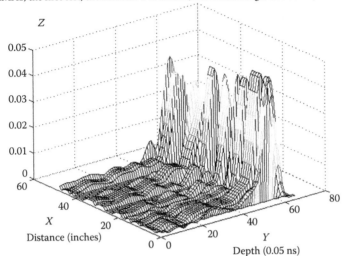

FIGURE 3.45 GPR reflectivity from a distressed portion on the deck of the Bostwick Bridge (Shelburne, VT).

effective tool for geotechnical applications. For example, Millard et al. detected leaks in water pipes by measuring changes in the dielectric properties of the clay soils surrounding buried pipes [373].

An important, but difficult, GPR measurement is the detection of air-filled delaminations in concrete bridge decks. One possibility is to test with high frequency, that is >1 GHz signals. Figure 3.46 shows the setup of a test of high-frequency UWB GPR (1–16 GHz). The purpose of these tests was to assess the detectability of 1 mm air gaps between concrete slabs [352]. Figure 3.47 shows the pulse response, which indicates that the delaminations are detectable. These tests were conducted under ideal laboratory conditions, but are indicative of the possibility of measuring thin air-filled delaminations. Real delaminations have more complicated fracture surfaces and should produce a more scattered response, with a weaker on-axis return amplitude [374]. It may be interesting to look for polarization-dependent Brewster angle effects as an additional means of detecting air-filled delaminations. Park et al. suggest using three-dipole antennas in a triangular rosette arrangement to exploit polarization effects for determining void shapes [375].

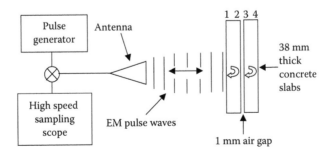

FIGURE 3.46 Geometry of the high-frequency radar test. (Adapted from Huston DR et al. 2000. *J Appl Geophys*, 43, 139–146.)

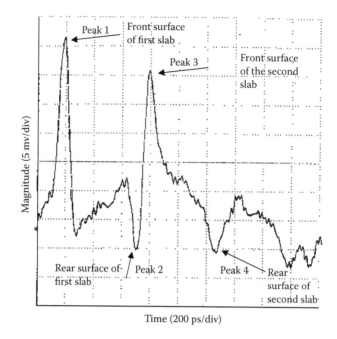

FIGURE 3.47 Reflected signal versus time for the high-frequency radar test. Peaks 2 and 3 are approximately derivative of peak 1. (Adapted from Huston DR et al. 2000. *J Appl Geophys*, 43, 139–146.)

3.3.8 Magnetic Induction and Circuits

Defects, such as voids or cracks, distort the flow of electricity and resulting magnetic fields through materials (Figure 3.48). Defect-induced magnetic field distortion has been a tool for inspecting railroad rails since the 1920s. The original versions used brush contacts to create the electric current and required manual operation. Automated mobile vehicular versions using noncontact EM induction appeared in the 1960s [376]. Similar systems inspect wire ropes for loss of cross sectional area [377,378]. Krieger et al. detected broken

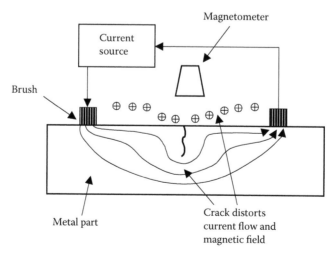

FIGURE 3.48 Magnetic induction crack sensor.

steel reinforcing and prestressing strands in concrete structures by measuring deviations in externally applied magnetic fields with superconducting quantum interference detectors (SQUIDs) [379]. Xiong and Wang also used field distortion techniques to detect damage in steel pipes by magnetic flux leakage detection [380]. NASA developed an angular position and rate transducer based on the inductive coupling between magnets in a central core and coils in an outer race rotating element [381]. This arrangement enables wireless data transmission.

3.4 2-D Sensors

2-D distributed sensors can operate in either nonimaging or imaging modes. Nonimaging sensors produce scalar outputs corresponding to the collective action of measurands. Imaging sensors resolve spatially localized measurand variations to form images.

3.4.1 Nonimaging 2-D Sensors

Many nonimaging 2-D sensors operate by attaching active materials to a surface. Piezoelectric films, such as PVDF, are candidate active materials. A wide variety of flexure and delamination-sensitive strain transducers, including some with pattern-specific antenna gain designs, are possible [382–385]. Specialized sensing paints are another approach. Mixing chemicals into paints, such as phenolphthalein, that react in a measurable manner in the presence of corrosion byproducts is a viable corrosion detection scheme [386]. Specialized paints can form electrochemical sensor electrodes for the corrosion monitoring of critical surfaces [387]. Pressure-sensitive paints are useful in wind tunnel testing and miniaturize nicely for microchannel testing [388–390].

3.4.2 Piezoelectric Transducer Networks

Arrays of point-geometry piezoelectric transducers can form sensor networks with the transducers serving as the nodes and elastic waves as the interconnects. The sensor network measures perturbations and changes to the transmissibility of ultrasound waves between transducers (Figure 3.49) [391–398]. This technology is maturing to resolve manufacturing and reliability issues for integrated systems, such as the SMART (Stanford Material Receive and Transmit) Layer, to be commercially available [399–402]. Relatively straightforward techniques can extend the piezoelectric sensor network method to more complicated structures, such as honeycomb composite structures, stiffened panels, masonry panels, and trusses [403–406].

An advantage of piezoelectric sensor networks is that the same transducers can operate in different modes for different sensing applications. These include the detection of delaminations in composites with ultrasound pitch and catch techniques, the detection of bulk fatigue effects with EMI tests, and vibration testing of plate-like structures, such as liquid storage tanks [407,408]. Piezoelectric frequency-domain testing can also assess plate-like structures. Deviations from nominal impedances are indicative of structural changes, which may correspond to damage [409–411].

A principal difficulty with piezoelectric sensor networks is the nontriviality of the requisite signal processing. Under certain circumstances, it is possible to identify the effects of

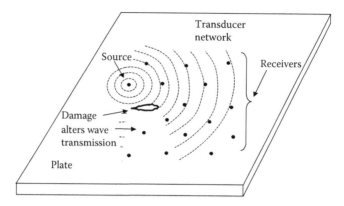

FIGURE 3.49   Piezoelectric sensor network detecting damage by transmission obstruction.

damage on elastic wave propagation from first principles [412–414]. Many wavelengths are of the same size as the damage. Diffraction is complication. Fritzen et al. processed the sensor array data by forming a Hankel matrix of intersensor correlations and comparing the matrices before and after damage by singular value decompositions [415]. The method detected fairly small and localized damage on a satellite structure. Derveaux et al. showed that Kirchoff migration methods of identifying damage were unreliable, without assumptions as to the nature of the damage, such as number of damaged locations [416]. A variant approach collects and interprets the data in the frequency domain, that is, measures the EMI [417]. It is possible to monitor the state of cure in composites by frequency-shift tracking [418]. More sophisticated statistical pattern recognition approaches, such as neural networks and genetic algorithms, provide an additional set of signal processing options [419,420]. Royer et al. developed a tomographic technique that detects corrosion in complicated aircraft components with a piezoelectric sensor array that launches and receives guided Lamb waves [421].

Under appropriate circumstances, synthetic methods will convert the impedance measurements into equivalent pulse waves and images (see Appendix A). Such techniques enable using sensor arrays as passive detectors to determine the time and location of impact events [422,423].

Huang and Peumans developed mass production techniques for surface deployable sensor networks that machine an array of interconnected sensors from monolithic silicon wafers using standard microelectronic lithographic techniques [424]. The interconnections are flexible spirals that unwrap to allow for placing the sensor modules at distant locations by stretching the network mesh.

3.4.3   2-D Surface Profile Measurements: Mechanical, Laser Scanning, and Interferometric Instruments

Measuring the geometry of a 2-D surface (planar or nonplanar) requires instruments to contact and establish surface point positions in 3-D. A variety of instruments can take such measurements. Some use mechanical contact with the surface to establish the geometry, whereas others use optical techniques. Figure 3.50 shows an instrument that measures the flatness of floors in accordance with ASTM E-1155-96 (2001) and ACI-117 specifications. The instrument operates by measuring surface elevations as it is pulled along the floor.

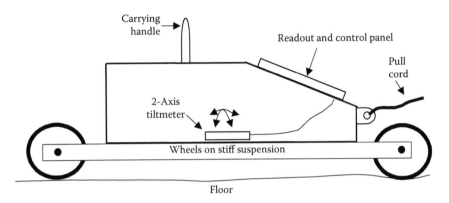

FIGURE 3.50 F-meter floor flatness measuring instrument. (Adapted from Allen Face Co, LLC. At http://www.allenface.com/fmeter.php)

Sampling rates of up to 5000 points per hour with rated accuracies on the order of ±50 μm are possible.

Laser scanning methods that measure surface geometry and kinematics use techniques such as triangulation, speckle pattern analysis, and Doppler frequency shifts. Jacobs indicates that a laser scanning instrument should have the following characteristics: (1) High point density, (2) Ultra-fast data capture, (3) Remote, reflectorless measurement, (4) 3-D visualization, (5) Informative imagery, and (6) Unattended, robotic operation [425]. Pai et al. assessed damage in vibrating airplane wings with a scanning laser vibrometer [426]. Leong et al. measured cracks with a combination of Lamb waves and scanning laser vibrometry [427]. Sundaresan et al. detected cracks in airplane panels subjected to vibration test signals with a scanning laser vibrometer [428]. Buckner et al. used laser scanning to detect surface defects in Inconel metal alloy parts that were precursor indicators of fatigue damage [429]. Triangulation methods and stereo digital image correlation techniques continue to improve in performance where high-resolution 3-D profiles of surfaces can be measured quickly [430,431].

Optical instruments can measure deflections at remote parts of structures. As an example, Mehrabi measured the natural frequencies of stay cables with a laser vibrometer in order to monitor the tension in a cable-stayed bridge [432]. Chen et al. used laser vibrometry to evaluate portable bridges [433]. Bennett et al. measured interstory drift in buildings by detecting intensity shifts of projections of Gaussian laser beams [434]. Markov et al. used a multiple beam laser Doppler velocimeter to assess damage in composite plates [435]. 3-D laser scanning instruments revealed glyphs carved into the monoliths at Stonehenge in the United Kingdom that are too weathered to be detected by the human eye [436]. Fuchs et al. measured the deflections of several highway bridges with similar instruments [437].

Moiré interferometry can measure micron-scale deformations on surfaces. Li et al. used pre-etched lines for interferometric Moiré pattern strain sensing in microstructures. Easley et al. observed fiber debonding in concrete with Moiré interferometry [438]. Wang et al. detected interlaminar strain gradients in AS4 carbon fiber reinforced PEEK composites using Moiré interferometry and a wavelet analysis of the image data [439]. Jensen et al. measured material properties and test structure nonidealities in MEMS at micron scales with interferometry [440]. *Acoustography* is a somewhat related technique where liquid crystals combine with ultrasound to form a real-time ultrasound image scanning system [441]. Likewise radar-based interferometric techniques can detect changes in large-scale objects.

These methods are variants of SAR. Shinozuka et al. used SAR to estimate damage in urban areas following an earthquake [442]. Shearography uses the entire surface to form an interferogram as an object deforms [443,444]. Leviton invented a 2-D high-resolution optical encoder for precision surface location and orientation measurement [445].

It is possible to detect the presence of particular chemical species by the appearance of specific optical properties. An example is an airborne laser system that detects leaks emanating from natural gas pipelines [446]. The system uses a 3-line tunable laser to detect the presence of methane and ethane from an airplane.

3.4.4   Visual Inspection

The human eye is the most common sensor for image-based structural inspection. The human eye can quickly scan large areas of structural surfaces and detect damaged or suspicious areas. Trained inspectors can be very effective at diagnosing dangerous situations, but have shortcomings. (See Chapter 8 for more details of human aspects of visual inspection.) Wavelength, visual acuity, human subjectivity, and the routine need for testing in remote and dangerous environments are all difficulties with using humans as imaging systems.

3.4.5   2-D Imaging Sensors

Cameras and specialized sensors can extend visual inspection capabilities to other parts of the EM spectrum. UV light, x-rays, and gamma-rays lie at wavelengths below the visible range. Safety concerns and requirements for positioning the source and sensor on opposite sides of the structure often limit the practicality of using x-rays and other penetrating nonreflecting radiations for inspection. A common application of UV light imaging is *in situ* chemical analysis, such as detecting ASR in concrete. Above the visible light range lie IR, terahertz, microwave, and radio wave radiation. IR imaging methods can identify features interior to a structure that induce surface temperature changes, such as delaminations. At longer wavelengths, microwave radiation (GPR) reflects and scatters off features underlying the surface of dielectric (nonmetallic) solids, such as concrete and geotechnical structures. In between IR and microwave lies the terahertz regime, which is largely unexplored and is just beginning to be used as a potential SHM tool [447].

3.4.6   Photographic and Video Image Processing

Modern digital cameras and machine vision systems can capture, store, and transmit images of structures at low cost and create entirely new modalities of visual inspection. Collecting, archiving, and analyzing massive amounts of image data are possible. For example, Mraz et al. used digital image analyzers to detect and quantify cracks in asphalt pavements with a technique that was largely independent of vehicle, but was sensitive to lighting conditions [448].

Satellite-based Earth-imaging technology has astonishing capabilities. These systems use visible, IR, and multispectral wavelength imagers. Images with meter resolutions are publicly available. The processing of such satellite images can reveal and locate large-scale infrastructure damage following natural and manmade catastrophes, such as earthquakes and hurricanes [449]. The damaged buildings are clearly visible. Adams et al. developed a

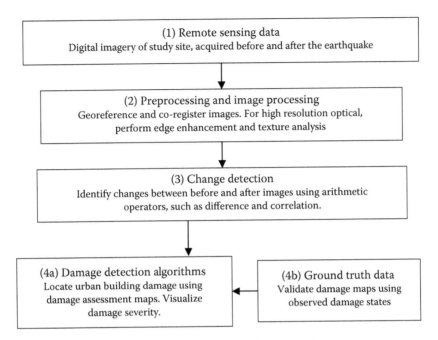

FIGURE 3.51   VIEWS algorithm for detecting damage by combining satellite imagery and ground truth data. (Adapted from Adams BJ et al. 2004. *MCEER Earthquake Investment Report*, Bam 12-26-03. With permission.)

system for *Visualizing impacts of earthquakes with satellites* (VIEWS) [450]. The system combines satellite imagery with ground truth verification (Figure 3.51). One method of analysis is to overlay before and after images to provide an assessment of the extent and location of damage.

A careful comparison of high-resolution before and after images often can provide sufficient information to estimate structural deformations. The tracking of the relative motions of features or landmarks is a possibility. Active optical methods can enhance the displacement tracking. One method of high resolution is to project interferometric speckles from a laser onto the surface of the structure. As the structure deforms, the speckles move in well-understood manners. An analysis of the speckle motions produces a structural deformation map. Subpixel interpolation methods extend the resolution to levels sufficient for estimating the corresponding strain fields. Example applications include the inspection of welds and measurement of curing-induced strains in concrete, and the determination of residual stresses in microscopic thin films by focused ion beam milling [451–454].

Video equipment can record dynamic SHM data. Perhaps the most famous dynamic SHM footage is Farquharson's recording of the Tacoma Narrows Bridge collapse. Machine-based image processing methods can capture dynamic data and use the motions for SHM and system identification purposes [455,456]. Photogrammetry techniques can provide sufficient information to construct 3-D geometrical renderings by analyzing image data from multiple cameras or a single camera at multiple positions [457]. Tyson et al. measured strain and 3-D displacement on the surface of objects under mechanical loading with two cameras and a random spot correlation technique [458]. Doer et al. developed video image capturing techniques to track the motions of buildings in seismic events [459]. Chang and Ji used videogrammatic techniques to measure the motion of vibrating structures [460]. Poudel et al. performed vibration modal analysis on beams with damage with high-resolution image processing techniques [461].

Image analysis can indicate incipient damage. *Lüders lines* are visible striations due to localized yielding that appear in certain metallic specimens when loaded slightly above the yield stress. Ichinose et al. assessed damage in steel members under cyclic loading by detecting Lüders lines [462]. Tateishi and Hanji detected incipient low-cycle fatigue damage in steel plate welded joints with a large-strain field optical imaging system [463]. Yoshida et al. observed a similar phenomenon with speckle interferometry where white bands formed and moved synchronously with Lüders fronts [464]. A detailed examination of the phenomenon indicated that the lines of plastic yielding were too small to be imaged directly and instead collectively appeared as white bands. Tracking the white bands indicated the location of future fractures.

3.4.7 Temperature Sensing and IR Thermography

The temperature field on the surface of a structure reflects the combined interactions of external thermal loading with surface and subsurface features. IR thermography, temperature indicating paints, and surface attachable sensors are all viable methods of measuring surface temperatures. It is possible to use measurements of surface temperature distributions to identify the presence of subsurface features, such as delaminations.

IR thermography captures and analyzes the spectra of temperature-dependent thermal radiation to estimate surface temperature. The Planck formula for the radiation spectrum of a blackbody radiator is [465]

$$E(\lambda, T) = \frac{3.7415 \times 10^{-16}}{\lambda^5 [\exp(1.4388 \times 10^{-2}/\lambda T) - 1]} \frac{W}{m^3}, \tag{3.44}$$

where λ is the wavelength in meters and T is the temperature in degree Kelvin. The spectrum in Equation 3.44 is a smooth curve, with a single maximum at wavelength, λ_{max}, where

$$\lambda_{max} = \frac{2.898 \times 10^{-3} \, mK}{T}. \tag{3.45}$$

Surface conditions on real materials alter this result somewhat. Most of the temperatures encountered in structural engineering have peak radiation wavelengths in the IR range.

Due to the spectral breadth of thermal radiation, radiation-based temperature measurements typically rely on one or a few detectors with coarse wavelength resolution. Since the detectors also radiate thermal energy according to Equation 3.45, cooling the detector may be necessary for low temperature or high-resolution measurements. Variable surface emissivities can confound quantitative IR temperature measurements, especially in field applications [466]. Aside from the detectors, standard principles of visible light optics usually apply to the design and use of IR imaging systems. Fiber optic blackbody temperature sensors are possible [467].

Active thermal imaging systems apply controlled thermal loads to the surface of a structure and measure the ensuing transients and resulting gradients. Thermal transients that interact with subsurface features that in turn affect the surface temperature distribution and provide an enhanced thermal imaging capability. These active techniques include the application of pulse, step, convective, and gradient-based thermal loads [468–472]. Active methods work best when conduction from the source (usually on the surface) to the subsurface dominates the heat flow [473,474]. When the conduction is weak, the active thermal methods are less successful. For example, Aktas et al. found that active IR thermography was only a weak indicator of crushing damage in the foam cores of sandwich composite wings [475].

Transient 1-D thermal conduction models describe many of the important aspects of surface to subsurface heat transfer. The transient temperature distribution, $T(z, t)$, following the application of the temperature load corresponds to a diffusion equation with the depth, z, time, t, and thermal diffusivity, α

$$\frac{\partial^2 T}{\partial z^2} - \frac{1}{\alpha}\frac{\partial T}{\partial t} = 0. \tag{3.46}$$

When the system is linear, the solutions are exponential. A plot of the logarithm of surface temperature versus time produces a straight line. Deviations from the ideal behavior of Equation 3.46 cause the plot to deviate from a straight line at a time corresponding to the depth of the deviation.

SHM applications of IR thermography include detecting subsurface delaminations in concrete roadways and composite repair patches, detecting the breakdown of refractory insulation in large boilers, monitoring the state of cure in large concrete structures, maintaining quality control in concrete construction and hot-mix asphalt paving, moisture ingress detection in masonry, damage and delamination detection in Si/SiC composites, and the thermal performance/efficiency of structures [476–485]. Highly sensitive instruments can detect small thermoelastic temperature changes in stressed materials that may correlate with damage, including subsurface damage that is not visible on the surface [486,487]. Temperature measurements can also identify active damage processes, such as fatigue, microcracking, and yielding [488–490]. Subjecting a structure to relatively high-frequency, for example, 15 kHz, excitation, induces friction-type energy dissipative processes corresponding to cracks. Thermography instruments can detect the resulting localized heating

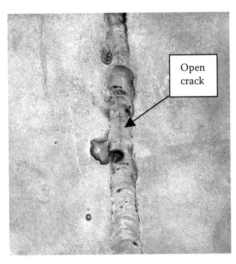

Visible   Infrared ultrasound

FIGURE 3.52 Visible and ultrasound IR imaging of weld defects. Both images show open crack. IR image shows poor weld bond that is not in visible image. (From Han X et al. 2005. *Sensors and Smart Structures Technologies for Civil, Mechanical, and Aerospace Systems*, SPIE 5765. With permission.)

(Figure 3.52) [491]. A potential advantage of thermography is that the measurements can be quicker than competing techniques, such as ultrasound [492].

Specialized thermally sensitive paints enable sensing structural temperatures with optical techniques in the visible light range. Temperature-induced color changes may be permanent or reversible. Permanent color changes are useful in situations where excessive temperatures are indicative of dangerous conditions, such as in a boiler. Other applications, such as teapots or mood rings, benefit from reversible temperature-induced color changes. *Thermoluminescent paints* produce temperature-sensitive luminescent decays. A standard technique uses UV light to excite atoms and molecules in the paint to emit luminescent radiation. The rate of decay of intensity of the luminescent radiation is temperature dependent. Thermoluminescent paints are potential substitutes for IR thermography in situations where access is difficult [493]. Chen et al. used lithium oxide as an indicator for emission spectroscopy SHM of yttria stabilized zirconia thermal barrier coatings [494]. NASA reports on a specialized rare earth luminescent layer (YSZ doped with erbium, neodymium, or praseodymium) that can indicate delaminations in thermal barrier coatings [495]. The technique works with the aid of evanescent wave coupling across the delamination enhancing the luminescent visibility.

3.4.8 Magnetic Particle Detection

Magnetic particle detection can identify cracks and other defects in ferromagnetic materials, that is, steel. The principle of operation is that defects, such as cracks in ferromagnetic materials, distort static magnetic fields flowing through a structural member. The technique proceeds by magnetizing a piece of steel, with an externally applied field, and then spraying iron filings on the steel. Cracks on the surface distort the magnetic field so that the iron filings conglomerate near the cracks. This method can identify tight cracks, particularly those that are perpendicular to the magnetic field lines. Magnetic particle crack and defect detection does not require highly skilled operators, but is limited to detecting near-surface defect.

3.4.9 Dye Penetrant and Soap Examination

Liquids with low surface tension and low viscosity readily penetrate into and out of tight surface cracks. The *dye penetrant* procedure applies a photoactive penetrating fluid to a surface under examination. The fluid penetrates into surface cracks. The next step wipes off the excess penetrant. This leaves the penetrant in the surface cracks. The next step sprays a developer onto the surface. Penetrant fluid seeps out of the cracks and interacts with the developer to produce a visible crack line. A similar technique uses soap to indicate the presence of cracks in structural members undergoing cyclic loads.

3.4.10 Electrical Impedance Tomography

Electrical impedance tomography (EIT) maps localized electrical impedance variations in an object (usually in 2-D) from a set of cross impedance measurements taken at points on the boundary. This is an inverse problem. The number of cross impedance measurements limits the maximum number of locations (pixels) at which the impedance can be determined. In general, issues of ill-conditioning tend to require substantially more measurements than locations for property estimation. EIT can detect and locate damage and delaminations in CFRP composite plates [497,498]. Wang and Chung found that oblique diagonal through-thickness electrode arrangements were superior to straight through and lateral electrode configurations for resistance-based damage detection and localization in CFRP composite plates [499].

3.5 3-D Sensing

Sensing the presence and nature of features inside of 3-D objects inherently requires some sort of probing of the internal structure. One method probes the interior with endoscopes or drilling out of core samples. Another approach uses NDE methods to launch energy into an object and to infer the internal structure by measuring the reflection, transmission, and scattering of waves. The retrieved data can be in a nonimaging format, such as the measurement of bulk material properties and layer thicknesses, or can be more detailed and assembled into an image. An *A-scan* is an individual trace of the received signal versus time (or depth). If the signal has a large oscillatory component, the A-scan is typically the envelope of the signal. The *B-scan*, that is, brightness scan, is a common method of displaying 3-D information. A B-scan is an assembled collection of the data from reflection measurements. Each reflected trace is plotted along a vertical line where the amplitude of the reflected signal is converted into pixel brightness (or color). For oscillatory source signals, such as ultrasound or radar, the envelope of the return signal is used as the intensity. The Hilbert transform is convenient for calculating the signal envelope (see Chapter 5). Demonstrations of the ultrasound imaging of defects appeared as early as the 1980s with B-scan imaging and is now a common practice [500,501]. Somewhat less common, but also quite useful, are C-scans [502–504]. These are images of horizontal slices through the object, also known as the *coronal planes*. In a study by Good et al., C-scan imaging proved useful in displaying ultrasonic bolt inspection results, especially for the identification of small (0.51 mm) thread root cracks [505].

3.5.1 Nonimaging 3-D Systems

Bulk material properties can indicate overall structural health. Common bulk properties of interest are elastic moduli, electrical resistance, capacitance, and magnetostriction. The measurement of bulk electrical material properties does not require overly sophisticated instrumentation. Bulk resistance measurements are practical assessment methods. Many physical effects alter bulk resistivity. For example, Zhou et al. developed a fatigue sensor based on changes in the electrical resistance of a metal coupon. Fatigue-generated microcracks cause a measurable increase in electrical resistance [506]. Carbon-based resistive bulk sensors have found a fairly wide level of usage in prototype and demonstration applications. Inada et al. installed carbon-resistance sensors in large concrete piles for foundations [507]. A variant is a flexion transducer based on conductive ink and patented by Gentile et al. [508]. Guan et al. developed specialized embedded electrode sensor nets to use with carbon fibers to form both bulk sensors and reinforcing in concrete [509]. Wang et al. demonstrated that the electrical properties of the interlaminar interfaces in cross-ply-layer constructed CFRPs are sensitive to multiple effects including temperature, moisture, thermal damage, and stress [510]. The cross-ply nature of the construction enables an addressable location of measurand properties. Park et al. developed a network model of the changes in bulk conductivity of CFRP composites due to microcracking [511]. Hirano and Todoroki and

Todoroki et al. developed a composite delamination detection system by measuring changes in electrical resistance [512,513]. Resistivity correlates with the size and depth of cracks in reinforced concrete as well as with damage in CFRP composites [514,515]. Thostenson and Chou and Loh et al. used distributed networks of carbon nanotubes as resistance-based damage, strain, and pH detectors [516,517]. A potential issue with resistivity and conductivity measurements is that the measurement process drives current through the material. The current can cause Joule heating of the material and confound the measurements. Nokken and Hooton propose alleviating Joule heating problems associated with the ASTM C 1202 concrete electrical conductivity test by using a 60 V excitation for 1 min [518].

It is possible to extend bulk electrical property measurement techniques by the custom fabrication of materials [519]. Embedded metal wires in a grid pattern in a composite plate at an offset from the centroidal axis forms a distributed sensor that transduces flexure into a measurable resistance change [520]. Adjusting the configuration of steel wire reinforcing mesh can form a capacitive bulk strain sensor in a rubber tire [521]. Custom configurations of carbon fiber and carbon particle-embedded composites can form bulk electrical resistance damage sensors [522–525]. Hou and Hayes used an electrically isolated anisotropic carbon fiber layer to detect and locate damage in CFRP plates from simple edge resistance measurements [526]. The resistance change with strain and damage is typically fairly large, but tends to be nonlinear and hysteretic. The recent development of engineered self-assembled materials at the molecular scale opens the possibility for interesting custom high-performance sensors. Lalli et al. used a self-assembly of polymers and metals to create *metal rubber* strain sensors with a reversible strain-sensitive change in resistance in excess of 150% [527]. Carbon nanotubes are another potentially promising approach [528]. Kang et al. fabricated distributed piezoresistive strain sensors using carbon nanotubes [529].

Magnetic fields flow through elongated ferromagnetic structures in patterns similar to the flow of electric current through conductors. Lumped circuit element methods using quantities such as reluctance, can provide a simplified yet often adequate description of the phenomena. Nakamura et al. developed the *main flux method* corrosion sensor for bridge cables [530]. The method sends a magnetic flux longitudinally through a cable section. A cable with a reduced cross section changes the reluctance of the circuit. An automated gondola used the main flux method to inspect vertical hanger cables on the Innoshima and Ohnaruto Bridges in Japan. Schempf built a pipe-crawling inspection and repair robot that uses magnetic circuits to detect cracks in cast iron pipes [531].

Magnetoelasticity is the interaction of the magnetic properties and deformations in ferromagnetic materials [532]. One approach measures changes in magnetization hysteresis curves with inductive sensors. If the magnetoelastic properties of the material are known, then magnetoelastic testing can be quite effective. Evaluating materials of unknown composition may require on-site calibrations. Singh et al. measured corrosion-induced cross section losses in bridge cables with magnetoelastic transducers [533]. Sakai et al. measured friction coefficients and tensions in concrete bridge box girder post-tensioning strands with magnetoelastic effects [534]. Budelmann et al. describe similar results, as well as the possibility of using high-frequency EM oscillations to detect cracks and defects, such as due to corrosion [181]. Bruns et al. assessed the fatigue life of steel welds with magnetoelastic effects [535].

The *Barkhausen effect* is the rapid switching of micromagnetic domains in ferromagnetic materials to align with external and internal magnetic fields. The micromagnetic domain jumps cause detectable signals that appear as in external EM pickups and is known as *Barkhausen noise* [536]. Stresses in ferromagnetic materials cause Barkhausen noise effects that are useful in NDE evaluations of welds and other high-stress structural elements [537].

Bulk electrical properties depend on both the effects of the present state and the past loading history of a material. Okuhara et al. used changes in electrical resistance to transduce past loading in composite materials [538]. The study examined carbon fiber, carbon particle, and titanium nitride fibers as sensing elements. The carbon fiber materials exhibited loading-based sensitivity. Fiber preloading, as may be required in post- or pretensioned concrete constructions, enhanced the sensitivity. Dispersing the carbon particles in the resin was difficult and increased the brittleness of the composite material. Nonetheless, based on a presumed percolation effect, the material showed a strong increase in resistivity with loading and damage. The titanium nitride wires were perhaps the most promising technique. Since the nitride layer thickness is controllable, it may be possible to optimize fibers for load history sensitivity. Quattrone et al. examined the utility of mixing sub 45 μm magnetostrictive particles in polymer composites as a means of measuring strain and detecting damage [532]. The results were largely successful, except that mixing the particles uniformly into the polymer matrix proved to be a challenge. Lee et al. measured the *dielectrostrictive* behavior of composites, that is, the change in dielectric properties due to mechanical strain in polymeric solids [539]. Pethrick et al. correlated the 0–6 GHz frequency-dependent dielectric behavior in polymer adhesive joints with moisture content [540]. Moisture and freeze-thaw action are major degradation mechanisms in adhesive joints.

3.5.2 Chain Drag and Rotary Percussion Sounding

The *chain drag* test (ASTM D 4580) is a standard method of identifying medium and severe delaminations in exposed concrete bridge decks. The procedure is to listen to the sounds produced while dragging a set of chain links across the surface of a bridge deck. Delaminated concrete sounds distinctly different. The chain drag technique is generally only useful on bare concrete with no asphalt overlays. The chain drag technique can be difficult to use in noisy environments, such as that due to nearby heavy traffic.

A potential improvement to the chain drag test is *rotary percussion sounding* [541]. The method produces small repeated impacts with a pair of rotating discs [542]. The discs have teeth that are similar to those found on industrial chain drive sprockets. The rotating discs are mounted on extender arms for the testing of vertical or overhead surfaces. Inspection times with the rotary percussion sounding instrument are reported to be as low as 20% of that required for an equivalent chain drag test.

3.5.3 Eddy Current Sensing

Metallic structures reflect and repel EM oscillations by the induction of induced internal eddy currents that act to counter the changing magnetic fields. The eddy currents follow structural geometries and the external fields. Cracks in the structure can interrupt the flow of the eddy currents. Measuring disturbances in the magnetic fields produced by eddy currents can indicate the presence of cracks. It is, in principle, possible to use physics-based inversion techniques to identify the nature of subsurface features [543–545]. Inspecting plates sandwiched by rivets in airplane skins for cracks is a common application of eddy current sensing. A typical readout instrument plots the input voltage versus the detected voltage on an *x–y* oscilliscope [546]. A skilled operator can correlate the shape of the *x–y* curve with subsurface defects.

A key feature of eddy current testing is the short depth of penetration of an oscillating EM field into metallic structures. Higher frequency oscillations penetrate less deeply than the fields of lower frequency oscillations. A consequence is that the *penetration depth* (a.k.a. the *skin depth*), δ, is a key factor affecting the performance of eddy current sensors

$$\delta = \left[\frac{2\rho}{\mu_0 \omega} \right]^{1/2}, \tag{3.47}$$

where ρ is the bulk resistivity of the metal and ω is the circular frequency of oscillation [547].

Commercial eddy current instruments oscillate at frequencies on the order of 10 kHz and have penetration depths of a couple of mm. Ferromagnetic components confound the measurement and penetration depth estimates. Measuring at deeper depths with eddy currents requires specialized techniques that combine penetrating low or multifrequency waves with highly sensitive inductive pickups. Liu et al. used multifrequency eddy current excitations to compensate for the different penetration depths [548]. Buckley et al. used a square wave as the excitation source to provide a transient broad banded input in an attempt to circumvent issues related to the tuning of the frequency source [549]. GMR sensors are a class of highly sensitive magnetic field sensors that can enhance eddy current sensing [550–557].

The small size of the magnetic fields caused by eddy currents interacting with subsurface features has prompted the development of specialized magnetic pickups. Goldfine et al. developed a *meandering winding magnetometer* that measures early stage bulk fatigue cracking in metals with eddy currents [558]. The device uses an electrode with primary and secondary circuits that track in parallel in a serpentine zigzag manner on the metal surface. An impedance analyzer measures the level of inductive coupling between the primary and secondary circuits. Rakow and Chang developed specialized meandering and serpentine sensors for the *in situ* fatigue damage detection of in bolted joints [559]. Magneto-optical effects using garnet crystals as pickups are a possible eddy current imaging method [560].

The utility and relative simplicity of eddy current testing inspires extensions of the technique. High-frequency testing is one possibility. Qaddoumi et al. used excitation frequencies in the range of 36.5 GHz to detect fatigue cracks in metals using open ended waveguides to transmit the EM waves [561]. Nishimura et al. and Lemistre and Placko extended the eddy current method to graphite composite inspection [562,563]. Graphite composites being conductive will also sustain eddy currents similar to metals, but have an inherent anisotropy due to the fiber lay-up. Sun et al. remotely induced eddy currents in metal parts and then detected defects as the parts move past a pair of magnetic poles [564].

3.5.4   Imaging GPR and Ultrasound

GPR and ultrasound are reflective NDE testing methods. A source antenna or transducer launches waves that propagate into and through a transmissive medium. The waves reflect and scatter off subsurface features. A receiving antenna detects the reflected and/or scattered waves. GPR uses EM waves. Ultrasound uses elastic waves. The analysis usually uses scalar wave models. Although the bulk of the discussion in this section describes GPR imaging methods, most of these principles also apply directly to ultrasound or other scalar wave testing techniques.

The source waves can be any of a variety of signal types. This includes sinusoidal, gated sinusoids, pulses, or even more complicated waveforms. The choice largely depends on a combination of source and detector capabilities, along with the need for sufficient frequency domain and temporal resolution. Since most media are linear, Fourier techniques

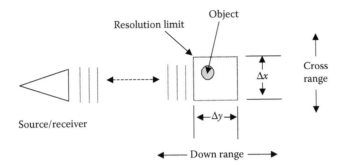

FIGURE 3.53 Position resolution limits of a radar scatterer.

are convenient methods of synthesizing and analyzing the waves. Wavelength-dependent wave speeds, that is, dispersion, can complicate the analysis. Describing the oscillating field produced by a source antenna in the near field can be quite complicated and often requires resort to numerical methods for detailed analysis. In the far field, the wave descriptions are simpler. A reasonable far-field approximation is to assume that the antenna produces circular (spherical in 3-D) waves that appear as plane waves to reflectors and scatterers.

A standard measurement determines the distance to a reflector from the time that it takes for a wave to travel back and forth between the antenna and the reflective feature. If r is the distance between the reflector and the antenna, c is the wave speed, and Δt is the time-of-flight, then

$$r = \frac{1}{2}c\Delta t. \tag{3.48}$$

The factor of $1/2$ accounts for the round trip.

The time-of-flight calculation in Equation 3.48 is an idealization that does not include resolution limits on measurements imposed by uncertain frequency–wavelength interdependencies (Figure 3.53). The smallest dimension, Δy, that can be resolved in the down range, y, direction is approximately

$$\Delta y = \frac{c}{4B_0} = \frac{c}{2N\Delta f}, \tag{3.49}$$

where B_0 is the bandwidth of the system [565]. N is the number of steps in a step-frequency radar system [566]. Δf is the step frequency. For a system with a bandwidth of 5.5 GHz operating in concrete with a relative dielectric constant of 5, the down range resolution is 6 mm. The smallest resolvable dimension, Δx, in the cross range, x, direction is

$$\Delta x = \frac{cx}{Lf_{max}}, \tag{3.50}$$

where f_{max} is the maximum frequency, and L is the antenna aperture. In a non-far-field application Millard reports on resolution values in tests with bowtie antennas on concrete where the maximum down range resolution is one-half the center wavelength and the maximum cross-range resolution is the −3 dB beamwidth of the antenna [567].

Raising the frequency to reduce the wavelengths is a natural approach for increasing spatial resolution. Practical issues of higher attenuation, possibly weaker reflections, and the need for more expensive instruments tend to limit the upper range of frequencies.

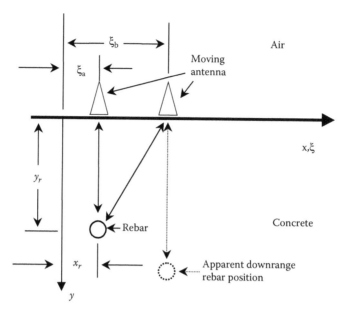

FIGURE 3.54 2-D geometry of moving monostatic antenna and hyperbolic nonlinear distortion. (Adapted from Huston D et al. 2000. *Nondestructive Evaluation of Highways, Utilities, and Pipelines IV*, SPIE 3995.)

Micowave-induced oscillations of molecules can explain many aspects of the frequency-dependent attenuation of microwaves. Water molecules resonate at 2.4 GHz. This causes water-saturated pavements and soils to be strong absorbers of microwaves near 2.4 GHz. Many GPR instruments operate near 900 MHz. The 2.4 GHz water resonance absorption has traditionally been viewed as being a barrier beyond which meaningful GPR measurements could not be made. However, water is a weaker microwave absorber in the 3 and 12 GHz range. There has been some activity in testing in these higher frequencies. Kim et al. used a 5.2 GHz system with synthetic aperture methods to image steel and voids in concrete specimens [568]. Pieraccini et al. used a 10 GHz center band step-frequency SAR system for imaging subsurface features in masonry walls [569]. Khanfar and Qaddoumi report that the C-band between 3 and 4 GHz is good for detecting delaminations in concrete [570]. Huston et al. detected mm-scale air-filled gaps between concrete slabs with a 0–20 GHz impulse radar system [347].

When a material has a layered structure, 1-D ray tracing models are often sufficient to interpret GPR data. Considering that 2-D and 3-D effects improve the quality of the interpretations, especially in the presence of features that vary with horizontal (cross-range) position, will confound 1-D layered media models [571]. A next step in increasing model complexity is to assume that the antenna acts as a point source (and receiver) that launches spherical waves with a corresponding spherically symmetric ray pattern. The notion of using time-of-flight calculations as in Equation 3.48 remains valid with the use of the diagonal distance from the antenna to the reflector as the travel distance, r (Figure 3.54).

$$r = [(x_r - \xi)^2 + y_r^2]^{1/2}. \tag{3.51}$$

Ray-tracing time-of-flight analysis generally gives only one piece of information—the distance to a reflecting feature (based on an assumed wave speed). Locating an isolated object in a 2-D, or 3-D, half-space requires more information, such as moving the antenna across the surface of an object and collecting more data or using multiantenna testing arrays. In the 2-D geometry shown in Figure 3.54, the time for a signal to travel round-trip from the antenna to the reflector is

$$\Delta t = \frac{2}{c_m}[(x_r - \xi)^2 + y_r^2]^{1/2}. \tag{3.52}$$

For a fixed reflector position, Equation 3.52 forms a hyperbola that varies with feature depth (Figure 3.55). The deeper the reflector, the flatter the hyperbola appears.

Reconstructing subsurface features from surface measurements is an inverse problem [572,573]. An analysis of B-scan data often reliably determines the depth of reinforcing bars in roadways and bridge decks [574]. Independent developments in the contexts of geophysical exploration and airborne radar with the names *migration* and SAR led to solution procedures for this type of inverse problem [565,575–577]. One of the earlier reported applications of the SAR technique to the GPR imaging of subsurface features in concrete is due to Mast [578]. Johansson and Mast further extended this work into 3-D imaging of subsurface features [579]. The performance of such algorithms can be quite striking. The system is able to image rebars quite nicely. These methods have been further extended with the development of multichannel automated moving roadway test systems [580–582]. Registering the GPR position with respect to the roadway requires additional instruments [583]. Imaging defects, such as cracks and delaminations, remains a largely unsolved technical challenge. Wang and Yuan used migration methods to detect damage in plates with ultrasound signals [584]. Alver and Ohtsu used stacked IE signals to detect voids in concrete [585].

Traditional GPR systems point downward and are not configured to test nonhorizontal members, such as columns and retaining walls. Specialized hand-held GPR testing instruments are available. Common applications are locating and determining the amount of concrete cover over reinforcing bars [586]. Figure 3.56 shows an early prototype hand-held system. Some building studies also use GPR systems [587]. Sarri et al. used a handheld four-bowtie antenna system for measuring subsurface feature in walls with 1.6 GHz pulses [588]. Jackson et al. found that a 2 GHz hand-held GPR system combined with IR thermography was an effective approach for detecting both shallow and deep defects in FRP composite-reinforced concrete bridge columns [589].

Most of the above discussion concerns the GPR testing of materials with a geometry that is essentially a layered half-space. Curved and more complicated 3-D geometries pose

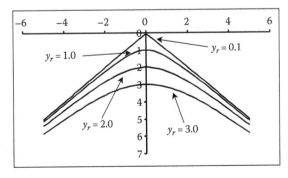

FIGURE 3.55 Hyperbolic nonlinearity that results from moving an antenna relative to a fixed reflector at various depths. (Adapted from Huston D et al. 2000. *Nondestructive Evaluation of Highways, Utilities, and Pipelines IV,* SPIE 3995.)

FIGURE 3.56 Prototype handheld GPR system (ca. 2001). (Data from Huston DR et al. 2001. *Smart Systems for Bridges, Structures, and Highways,* SPIE 4330.)

additional measurement and data interpretation challenges. Feng et al. developed a system for inspecting FRP-wrapped concrete structures with microwaves [590]. Wolfe and Wahbeh found that 3-D geometric effects can seriously confound the ultrasonic testing of steel welds and curved girders [591]. Careful consideration of the particular measurement details alleviates many of these problems.

3.5.5 GPR Regulatory Environment

GPR systems must comply with regulations that govern the emission of EM radiation. Until recently, the regulatory environment for GPR systems was relatively unstructured, with most countries banning the routine use of GPR systems, while simultaneously allowing operation under an experimental classification.

GPR systems use a variety of waveforms such as pulses, gated sine waves, and step-frequency sweeping of sine waves. Many GPR systems use very short (~1–1000 ps) EM pulses with a corresponding very broad frequency band content (~1–10 GHz). These are known as UWB pulses [593]. The history of UWB technology dates back to the earliest days of radio communication with spark gap generators. Technical advances have replaced the spark gap generators, largely with narrow band signal modulators. In the 1970s, UWB technology reemerged in a variety of defense-related applications and then permeated into other fields, for example, GPR and short-range high-speed data transmission for consumer products. Since UWB signals span such a broad portion of the EM spectrum, the possibility arises of UWB systems interfering with other users of the EM spectrum.

The term UWB appears to have originated in 1990 at an OSD/DARPA workshop [594]. OSD/DARPA defined UWB as transmitted EM waves with a fractional bandwidth (FBW)

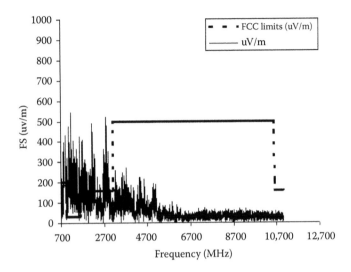

FIGURE 3.57 Radiated emissions measurements from PERES II GPR system operating in high-power mode versus allowable FCC 02-48 levels. (Adapted from Huston DR. 2006. *Project Report, Federal Highway Administration,* DTFH61-03-H0012.)

greater than 0.25, regardless of the center frequency. The FBW is given by

$$FBW = \frac{2(f_H - f_L)}{f_H + f_L},$$ (3.53)

where f_H and f_L are the upper and lower frequencies, respectively, of the $-20\,dB$ emission rolloff point. The center frequency, f_c, is the average of the upper and lower frequencies

$$f_c = \frac{f_H + f_L}{2}.$$ (3.54)

The OSD/DARPA UWB definition is practical for RF emissions under 2 GHz. Above 2 GHz the US Federal Communications Commission (FCC) modified the OSD/DARPA definition to include devices with an FBW greater than 0.20 or a $-10\,dB$ bandwidth occupying more than 500 MHz of spectrum.

FCC 02-48 is perhaps the most clearly defined regulation of UWB radiated emissions [595–598]. FCC 02-48 highly restricts radiated emission levels in the 960 MHz to 3.1 GHz frequency range. Below 960 MHZ and between 3.1 GHz and 10.6 GHz the allowable levels for radiated emissions are much higher. These regulations govern the emissions that radiated upward from the back side of the antenna and do not govern the amount of radiation directed downward to the ground. This leaves open the possibility of using highly directional antennas. As an example of noncompliance, Figure 3.57 shows radiated emissions measurements from a micropower impulse radar GPR system (PERES II) operating in its high-power mode versus the allowable FCC 02-48 levels [599].

3.5.6 Acoustic Emission Monitoring

Acoustic emissions (AE) are short-duration high-frequency elastic waves in solids caused by incipient microfractures and other localized events, such as small pressure-driven fluid leaks and SMA Martensitic phase transformations [600,601]. Most AE source events produce broadband localized pulse-like elastic stresses that excite many of the possible elastic wave modes in the solid, that is, pressure, shear, Lamb, Rayleigh, and so on. Unlike ultrasound testing, AE monitoring is a passive technique. A simple form of AE monitoring is the measurement and counting of events. More sophisticated implementations attempt to locate the source event by triangulation, as in Figure 3.58, and/or classify the event by an analysis of the elastic waveform [602,603]. The complexities of the signal processing may warrant use of neural networks or other advanced pattern recognition techniques [604]. Automated systems with sophisticated user interfaces have appeared [605]. Event counting is routinely used for acceptance testing of high-performance structures. The testing applies a submaximal load to the part. High event count levels can be an indicator of deteriorating structural health and the progression of damage in a structure. The occurrence of the number of AE

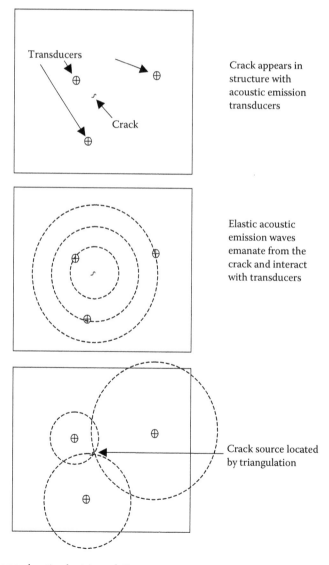

FIGURE 3.58   AE source location by triangulation.

events exceeding a preset threshold warrants the rejection of the part [606]. Measuring AE emissions using simple online statistics, such as root mean square (RMS), can assist with real-time cutting process monitoring [607].

Piezoelectric transducers, fiber optic sensors, and laser interferometers are common AE transducers. Laser interferometers have the advantage of being noncontacting, but can be more expensive than competing piezoelectric systems [608,609]. An attractive feature of fiber optic sensors is the potential for very high frequency measurements [610]. The range of possible fiber optic AE sensor configurations includes arrays of serially multiplexed Bragg grating sensors; Fabry–Perot interferometric sensors; or distributed fiber optic sensors, such as a modal domain or Mach–Zehnder interferometric sensor [611,612]. Duke Jr and Cassino used Fabry–Perot sensors to measure in-plane motions that indicate failures in FRP patches [613]. Butler et al. used a tapered fiber optic coupler as an AE sensor [614].

AE methods can locate and assess microcracking damage in concrete, asphalt pavements and composites and can also facilitate more fundamental fracture mechanics studies [615–624]. Figure 3.59 shows an array of AE monitoring instruments on a reinforced concrete bridge girder. Maji and Kratochvil measured AE events in steel bridges in the laboratory and then compared the results with field tests on a steel bridge in New Mexico [625,626]. The steel girders were particularly amenable to propagating AE signals in the form of Lamb waves. Paulson measured the fracture of individual cable strands in the Bronx-Whitestone Bridge in New York City with AE techniques [627]. Since the cables are under high tension, a cable strand releases considerable energy into AE waves as it fractures. This high-energy release creates an easy to identify acoustic signal [628]. NCHRP recommends conducting AE monitoring on suspension bridge cables for a period of at least 12–18 months, if the cables are in a stage 3 (out of 4) level of corrosion [629]. Fumo and Worthington found that AE events in prestressed concrete cylinder pipes propagated nicely through water in the pipes and allowed for using hydrophones as detectors, especially of circumferential prestressing wire breaks [630]. AE methods can detect incipient aircraft damage that is too small to be detected with the present NDE systems [631–633]. Huang and Nissen used AE monitoring

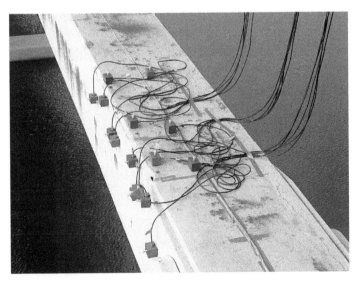

FIGURE 3.59 Array of AE sensors on reinforced concrete bridge girder with corrosion-induced cracking. (Photograph courtesy of Physical Acoustics Corp.)

to detect cryogenic insulation debonding in an LH_2 tank during flight on a reusable launch vehicle [634]. Triangulating location methods in anisotropic materials, such as fiber reinforced composite plates, require compensating for directionally dependent elastic wave speeds that correspond to elliptical wavefronts [635]. Hashida et al. extended AE emission testing to the field testing of concrete structures with an active *in situ* material break off technique [636]. The method machines the surface of a concrete structure to form a small subsurface cantilever beam and then loads the cantilever while observing AE behavior.

AE test instruments are convenient for both real-time and long-term monitoring [637]. Moving machinery, joints in structures, helicopter rotor hubs, and *thermal protection systems* (TPSs) are all situations where a direct attachment of transducers to the structural members of interest may be difficult. Since AE waves propagate through tight bearings and joints, it is often possible to monitor parts at convenient remote positions [638]. Developments in wireless AE sensor and transmission systems are under way [639].

AE signals are often sufficiently information rich to enable crack-type identification from relatively simple signal processing. A standard technique extracts and analyzes both amplitude and frequency content to distinguish shear cracks from tension cracks [640,641]. Shear cracks tend to produce low-frequency and high-amplitude signals. Tension cracks instead produce high-frequency and low-amplitude signals [642,643]. A distinguishing characteristic of delaminations in composites is the production of high-frequency flexure waves combined with extensional waves [644]. Conversely, matrix cracking usually does not produce extensional waves. AE waveform signature analysis can also identify different types of fiber breakage mechanisms in composites [645].

AE techniques can be useful in load testing of new structures. Many structures and materials exhibit an irreversible and memory-dependent break-in behavior known as the *Kaiser effect* [646–648]. A pristine structure may produce many AE events during the first loading sequence, even if the load is a fraction of the failure load. Unloading the structure and then reloading to the same level will again produce AE events. The number of events depends on whether the reload sequence is causing damage. If the reloading does not damage the structure, the number of new AE events will be much smaller than the first cycle (Figure 3.60). If the reloading damages the structure, the number of events is higher, particularly after reaching a threshold load level. The ratio of the present threshold load to the previous maximum load is the *load ratio* (also known as the *felicity ratio*). The ratio of the amount of AE activity on the unloading cycle to the previous maximum loading cycle is the *calm ratio*.

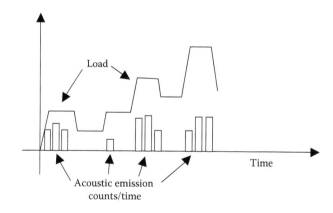

FIGURE 3.60 Kaiser effect in AE monitoring. (Adapted from Lim MK and Koo TK. 1989. *Mag Concr Res*, 41, 189.)

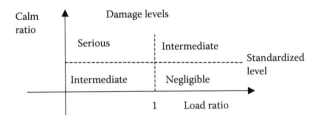

FIGURE 3.61 AE damage indication as a function of Caim and Load Ratios. (Adapted from Ohtsu M et al. 2002. *ACI Struct J*, 99(4), 411–417.)

Comparing load ratio with the calm ratio can quantify the Kaiser effect (Figure 3.61) [649]. Successful verification of this technique on reinforced concrete beams has led to the adoption of the Japanese standard NDIS-2421 [650]. Colombo et al. compared the ratios of levels of AE events in the loading and unloading phases on a concrete structure [651]. Likewise, Scheerer et al. demonstrated that the AE event rates in composites undergoing high cycle fatigue loading initially decrease following a break-in period and then increase significantly with fatigue damage [652].

A primary criticism of AE testing is difficulty in detecting preexisting damage when the damage does not actively produce AE signals. There are circumstances, however, where structural materials with preexisting damage exhibit sufficiently distinctive AE behavior and can be distinguished from pristine materials. For example, Yoon et al. identified distinctive AE signals corresponding to preexisting damage and corrosion in concrete specimens [653].

Another difficulty with AE monitoring occurs when the material tends to scatter and/or attenuate AE elastic waves, such as in concrete. Henkel and Wood measured attenuation rates of 40 dB/m in concrete [654]. Such circumstances may require dense arrays of sensors for reliable AE event detection. Liang et al. developed a system using serially multiplexed Bragg grating fiber optic AE sensors to monitor post-tensioning FRP rods in concrete beams [655]. If event counting and not event location is important, then certain types of distributed sensors may be useful. Possibilities include multimode fiber optic sensors, or a series arrangement of piezoelectric point sensors [656,657]. Chen and He used a similar arrangement with an array of surface attached elastic waveguides for transmission of AE signals to remote locations for detection [658]. Caneva et al. embedded PVDF piezoelectric film into fiber reinforced composites in an attempt to reduce some of the localization issues associated with AE sensing [659]. The transmission of AE signals over large distances and through joints in structures can also confound localization efforts. The use of guard sensors placed around the perimeter of the region in question can help with identifying and alleviating the influence of these external AE signals.

3.5.7 Radiography

Radiography uses penetrating radiation, such as x-rays or gamma-rays, to examine interior features in an object from differential absorption and transmission patterns. Radiography has several advantages as a nondestructive method of evaluation, including penetration depth and internal feature resolution. Despite these advantages, geometric problems of positioning source and detector on opposite sides of the structure and safety concerns limit the practical uses of radiography for field testing. Factory-based applications of radiography resolve most of these issues and are quite routine for the subsurface inspection of critical high-performance components. Since penetrating radiation travels through objects with a minimum of scattering and secondary reflections, it is possible to reconstruct internal feature geometries using tomographic techniques without resorting to complicated migration reconstructions. Multiple examples of x-ray tomography appear in the SHM literature. Abdel-Ghaffar inspected asphalt cylinders for damage with used x-ray tomography, Speck et al. inspected solder joints and other features of electric circuit boards with tomography and Saleh et al. used high-energy (6 MeV) x-rays to examine concrete bridge post-tensioning tendons [660–662]. Bentz et al. achieved 20 μm resolution with x-ray tomography in an effort to track water movements in concrete cylinders [663]. Ruderman inspected solid-fuel rocket motors, and Liu monitored cracking processes in elastomers with x-ray imaging [664,665]. Thieberger et al. developed algorithms for tomography of concrete with gamma-rays [666].

While most x-rays penetrate through an object, *Compton scattering* interactions cause detectable backscattered and off-axis scattered waves. Dugan et al. exploited this effect to develop an imaging system for the damage detection in space shuttle thermal insulating foam [668]. A primary advantage of this *lateral migration radiography* technique is that it requires placing a test instrument on only one side of the structure.

References

1. Spillman Jr W and Huston D. 1995. Scaling and antenna gain in integrating fiber optic sensors. *IEEE J Lightwave Technol*, 13(7), 1222–1230.
2. US Army Corps of Engineers. 1980. *Engineering and design instrumentation for concrete structures*, EM 1110-2-4300, Department of the Army, Office of the Chief of Engineers, Washington, DC.
3. Romera LE and Hernández S. 2002. Modeling an arch dam suffering from alkali–aggregate reaction. *Concrete Intl*, 24(12), 34–38.
4. Ladner M. 1985. Unusual methods for deflection measurements. *American Concrete Institute*, Detroit, ACI-SP-88-1.
5. AASHTO. 2003. *AASHTO Manual for Condition Evaluation of Bridges*, 1994, 2nd Ed., with Rev. through 2002, American Association of State Highway and Transportation Officials, Washington, DC.
6. Johnsen TH, Geiker MR, and Faber MH. 2003. Quantifying condition indicators for concrete structures. *Concrete Intl*, 25(12), 47–54.
7. Berke NS, Hicks MC, and Hoopes RJ. 1994. Condition assessment of field structures with calcium nitrite. *American Concrete Institute*, ACI SP 151-1.
8. Scannell WT, Sohanghpurwala AA, and Islam M. 1996. Assessment of physical condition of concrete bridge components. *FHWA-SHRP Showcase*.
9. Gu P and Beaudoin JJ. 1997. Obtaining effective half-cell potential measurements in reinforced concrete structures. *Construction Technology Update*, No. 18, Inst for Research in Construction, NRC Canada.
10. ASTM 1999. Standard test method for half-cell potentials of uncoated reinforced concrete. ASTM C876-91, ASTM International, West Conshohocken, PA, DOI: 10.1520/C0876-91R99.
11. Fontana MG and Greene ND. 1978. *Corrosion Engineering*, 2nd Ed. McGraw-Hill, New York.
12. So HS, Millard SG, and Law DW. 2006. Environmental influences on corrosion rate measurements of steel in reinforced concrete. *NDE Conf on Civil Engineering*, American Society for Nondestructive Testing, St. Louis.
13. Stearn M and Geary AL. 1952. Electrochemical polarization. *J Electrochem Soc*, 104(1), 56–63.
14. Michael AP, Hamilton III HR, Green PS, and Boyd AJ. 2004. Long term monitoring of the FRP repair and corrosion of steel reinforcement at the University Boulevard Bridge in Jacksonville, Florida. *Proc Structural Materials Technology VI, an NDT Conference*, Buffalo.

15. Broomfield JP. 1996. Field measurement of the corrosion rate of steel in concrete using a micro-processor controlled unit with a monitored guard ring for signal confinement. *Techniques to Assess the Corrosion Activity of Steel Reinforced Concrete Structures*, ASTM STP 1276.

16. Broomfield JP, Rodriguez J, Ortega LM, and Garcia AM. 1994. Corrosion rate measurements in reinforced concrete structures by linear polarization device. *American Concrete Institute*, ACI SP 151-1.

17. Feliu S, Gonzalez JA, and Andrade A. 1996. Electrochemical methods for on-site determinations of corrosion rates of rebars. *Techniques to Assess the Corrosion Activity of Steel Reinforced Concrete Structures*, ASTM STP 1276.

18. Watters DG, Jayaweera P, Bahr AJ, Huestis DL, Priyantha N, Meline R, Reis R, and Parks D. 2003. Smart pebble™: Wireless sensors for structural health monitoring of bridge decks. *Smart Systems and Nondestructive Evaluation for Civil Infrastructures*, SPIE 5057.

19. Sorensen RE, Buhr B, and Frolund T. 2004. Sensoring corrosion—the Danish way. *Proc IABMAS'04 Bridge Maintenance, Safety, Management and Cost*, Kyoto. Taylor & Francis, London.

20. Hansson CM, Seabrook PT, and Marcotte TD. 2004. In-place corrosion monitoring. *Concrete Intl*, 26(7), 59–65.

21. Schoess JN and Havey GD. 1996. Smart fastener. US Patent 5,549,803.

22. Laird C and Li Y. 2000. Methods and devices for electrochemically determining metal fatigue status. US Patent 6,026,691.

23. Camoes A, Cruz PJ, Jalali S, and Ferreira RM. 2004. Durability performance of concrete bridge piers made with metakaolin and latex mixes. *Proc IABMAS'04 Bridge Maintenance, Safety, Management and Cost*, Kyoto. Taylor & Francis, London.

24. Laurens S, Sbartaï ZM, Kacimi S, Balayssac JP, and Arliguie G. 2006. Prediction of concrete electrical resistivity using artificial neural networks. *NDE Conf on Civil Engineering*, American Society for Nondestructive Testing, St. Louis.

25. Agarwala VS and Fabiszewski A. 1991. Thin film microsensors for integrity of coatings, composites and hidden structures. Paper no. 342, *Proc NACE Annual Conference*, National Association for Corrosion Engineering, Houston, TX.

26. Kelly RG, Yuan J, Jones SH, Wang W, Hudson K, Sime A, Schneider, and Clemena GG. 1999. Embeddable sensor for corrosion measurement. *Nondestructive Evaluation of Bridges and Highways III*, SPIE 3587.

27. John G, Hladky K, Gaydecki P, and Dawson J. 1992. Recent developments in inspection techniques for corrosion damaged structures. In *Corrosion Forms and Control for Infrastructure*, ASTM STP 1137, V Chakker (ed.)., 246–257.

28. Merkle WJ and Myers JJ. 2004. Use of the total station for load testing of retrofitted bridges with limited access. *Sensors and Smart Structures Technologies for Civil, Mechanical and Aerospace Systems*, SPIE 5391.

29. Tyrsa V, Burtseva L, Lopez MR, and Tyrsa V. 2004. Monitoring of civil engineering structures. *Health Monitoring and Smart Nondestructive Evaluation of Structural and Biological Systems III*, SPIE 5394.

30. Foltz LB. 2000. 3D laser scanner provides benefits for PennDOT bridge and rockface surveys. *Professional Surveyor*, 20(5), 24–28.

31. Wahbeh AM, Caffrey JP, and Masri SF. 2003. A vision-based approach for the direct measurement of displacements in vibrating systems. *Smart Mater Struct*, 12, 785–794.

32. Kanda K, Miyamoto Y, Kondo A, and Oshio M. 2005. Monitoring of earthquake induced motions and damage with optical motion tracking. *Smart Mater Struct*, 14, S32–S38.

33. Ames B. 2003. Software designed to find pinhole leaks in International Space Station. *Military and Aerospace Electronics*, (November).

34. Pauli W. 1973. *Optics and the Theory of Electrons*. Dover Publications, Mineola, NY.

35. Levinson HJ and Arnold WH. 1997. Optical lithography. In *Handbook of Microlithography, Micromachining and Microfabrication*, vol. 1, P Choudury (ed.), SPIE, Bellingham.

36. Boehler W, Bordas M, and Marbs A. 2003. Investigating laser scanner accuracy. *Proc CIPA XIX Intl Symposium*, Sept 30–Oct 4, Antalya, Turkey, pp. 696–702.

37. Steltzer A, Fischer A, Weinberger F, and Vossiek M. 2003. RF-sensor for a local position measurement system. *Nondestructive Detection and Measurement for Homeland Security*, SPIE 5048.

38. Breuer P, Chmielewski T, Gorski P, and Konopka E. 2002. Application of GPS technology to measurements of displacements of high-rise structures due to weak winds. *J Wind Eng Ind Aerodyn*, 90(3), 223–230.

39. Xu L, Guo JJ, and Jiang JJ. 2002. Time–frequency analysis of a suspension bridge based on GPS. *J Sound Vib*, 254(1), 105–116.

40. Fujino Y, Murata M, Okano S, and Takeguchi M. 2000. Monitoring system of the Akashi Kaikyo bridge and displacement measurement using GPS. *Nondestructive Evaluation of Highways, Utilities, and Pipelines IV*, SPIE 3995.

41. Chan WS, Xu YL, Ding XL, Xiong YL, and Dai WJ. 2005. Dynamic displacement measurement accuracy of GPS for monitoring large civil engineering structures. *Sensors and Smart Structures Technologies for Civil, Mechanical, and Aerospace Systems*, SPIE 5765.

42. Meng X, Roberts GW, and Dodson AH. 2005. GNSS for structural deflection monitoring: Implementation and data analysis. In *Structural Health Monitoring*, FK Chang (ed.). DEStech, Lancaster.

43. Huang W, Wang R, Meng X, Yao L, and Yang B. 2007. Identification studies on a prototype structural health monitoring system for the Nanpu bridge in Shanghai, P.R. China. In *Structural Health Monitoring*, FK Chang (ed.). DEStech, Lancaster.

44. Liu C, Meng X, and Yao L. 2007. A real-time kinematic GPS positioning based structural health monitoring system for the 32 km Donghai Bridge in China. In *Structural Health Monitoring*, FK Chang (ed.). DEStech, Lancaster.

45. Mandracchia EA. 1996. Ultrasonic testing of highway bridges. *Nondestructive Evaluation of Bridges and Highways*, SPIE 2946.

46. Li S and Wu Z. 2005. Characterization of long-gage fiber optic sensors for structural identification. *Sensors and Smart Structures Technologies for Civil, Mechanical, and Aerospace Systems*, SPIE 5765.

47. Schulz WL, Conte JP, and Udd E. 2001. Long gage fiber optic bragg grating strain sensors to monitor civil structures. *Smart Systems for Bridges, Structures, and Highways*, SPIE 4330.

48. Seim J, Udd E, Schulz W, and Laylor HM. 1999. Health monitoring of an Oregon historical bridge with fiber grating strain sensors. *Smart Systems for Bridges, Structures, and Highways*, SPIE 3671.

49. Inaudi D, Elamari A, Pflug L, Gisin N, Breguet J, and Vurpillot S. 1994. Low-coherence deformation sensors for the monitoring of civil-engineering structures. *Sens Actuat A*, 44, 125–130.

50. Lienhart W. 2005. Experimental investigation of the performance of the SOFO measurement system. In *Structural Health Monitoring*, FK Chang (ed.). DEStech, Lancaster.

51. Kurokawa S, Shimano K, Sumitro S, and Suzuki M. 2004. Global concrete structure monitoring by utilizing fiber optic sensor. *Proc IABMAS'04 Bridge Maintenance, Safety, Management and Cost*, Kyoto. Taylor & Francis, London.

52. Inaudi D, Glisic B, and Posenato D. 2004. High-speed demodulation of long-gauge fibre optic strain sensors for dynamic structural monitoring. *Proc 2nd European Workshop on Structural Health Monitoring*, DEStech, Munich.

53. Habel W, Kohlhoff H, Knapp J, Helmerich R, Hänichen H, and Inaudi D. 2002. Monitoring system for long-term evaluation of prestressed railway bridges in the new Lehrter Bahnhof in Berlin. *Proc 3rd World Conf on Structural Control*, April 7–12, 2002, Como.

54. Leung CK, Wang X, and Olson N. 2000. Debonding and calibration shift of optical fiber sensors in concrete. *J Eng Mech*, 126(3), 300–307.

55. Heaton LC, Kranz M, and Williams J. 2004. Embedded fiber optics for structural health monitoring of composite motor cases. *Nondestructive Evaluation and Health Monitoring of Aerospace Materials and Composites III*, SPIE 5393.

56. Donlagic D and Hanc M. 2003. Vehicle axle detector for roadways based on fiber optic interferometer. *Smart Sensor Technology and Measurement Systems*, SPIE 5050.

57. Téral SR, Larcher SJ, Caussignac JM, and Barbachi M. 1996. Fiber optic weigh-in-motion sensor: Correlation between modeling and practical characterization. *Smart Sensing, Processing and Instrumentation*, SPIE 2718.

58. Vierkoetter SA, Ward CR, Gregory DM, Menon SM, and Roach DP. 2003. NDE of composites *via* quadrupole resonance spectroscopy. *Nondestructive Evaluation and Health Monitoring of Aerospace Materials and Composites II*, SPIE 5046.

59. Minchin Jr RE. 2006. Identification and demonstration of a technology adaptable to locating water in post-tensioned bridge tendons. Florida Department of Transportation, Final Report, BD545-45.

60. Collico Savio DL, Mariscotti MA, and Ribeiro Guevara S. 1995. Elemental analysis of a concrete sample by capture gamma rays with a radioisotope neutron source. *Nucl Instrum Methods Phys Res B*, 95, 379–388.

61. Srikar VT, Swan AK, Ünlü MS, Goldberg BB, and Spearing SM. 2003. Micro-Raman measurement of bending stresses in micromachined silicon flexures. *J Microelectromech Syst*, 12(6), 779–787.

62. Akers DW and Denison AB. 2001. Nondestructive examination using neutron activated positron annihilation. US Patent 6,178,218.

63. Li W and Li DY. 2002. Effects of dislocation on electron work function of metal surface. *Mater Sci Technol*, 18, 1057–1060.

64. Balachov II, Sannikov AA, and Ustenko I. 2003. Detection and monitoring of incipient cracks. In *Structural Health Monitoring*, FK Chang (ed.). DEStech, Lancaster.

65. Liu CH and Kenny TW. 2001. A high-precision, wide-bandwidth micromachined tunneling accelerometer. *J Microelectromech Syst*, 10(3), 425–433.

66. Livingston RA, Saleh HH, Ceary MS, and Amdes MA. 2004. Development of an image plate system for NDT potassium measurement in concrete. *Proc Structural Materials Technology VI, an NDT Conference*, Buffalo.

67. Baibich MN, Broto JM, Fert A, Van Dau FN, Petroff F, Eitenne P, Creuzet G, Friedrich A, and Chazelas J. 1988. Giant magnetoresistance of (001)Fe/(001)Cr magnetic superlattices. *Phys Rev Lett*, 61(2), 2472–2475.

68. Fuhr P, Huston D, Kajensk P, and Snyder D. 1993. Curing and stress monitoring of concrete beams with embedded optical fiber sensors. *J Struct Eng*, 119(7), 2263–2269.

69. Wolff R, Miesseler HJ, and Weiser M. 1991. Method for monitoring deformations with light waveguides. US Patent 5,044,205.

70. Dry C. 1996. Crack and damage assessment in concrete and polymer matrices using liquids released internally from hollow optical fibers. *Smart Sensing, Processing and Instrumentation*, SPIE 2718.

71. Huston DR, Fuhr PL, and Ambrose TP. 1994. Damage detection in structures using OTDR and intensity measurements. *Time Domain Reflectometry in Environmental, Infrastructure, and Mining Applications*, US Bureau of Mines, SP 19-94.

72. Ansari F. 1992. Real-time condition monitoring of concrete structures by embedded optical fibers. In *Nondestructive Testing of Concrete Elements and Structures*, F Ansari and S Sture (eds.). ASCE, New York.

73. Hayes S, Brooks D, Liu T, Vickers S, and Fernando GF. 1996. *In-situ* self-sensing fibre reinforced composites. *Smart Sensing, Processing and Instrumentation*, SPIE 2718.

74. Tennyson R. 2004. SHM system for detecting foreign object impact damage on spacecraft. *Proc 2nd European Workshop on Structural Health Monitoring*, DEStech, Munich.

75. Wooddell M, Pickrell G, and Ooi TK. 2007. Development of stochastic optical fiber sensors for structural health monitoring applications. In *Structural Health Monitoring*, FK Chang (ed.). DEStech, Lancaster.

76. Tennyson RC, Morison WD, and Manuellpillai G. 2003. Intelligent pipelines using fiber optic sensors. *Smart Sensor Technology and Measurement Systems*, SPIE 5050.

77. Takeda S, Okabe Y, and Takeda N. 2003. Application of chirped FBG sensors for detection of local delamination in composite laminates. *Smart Sensor Technology and Measurement Systems*, SPIE 5050.

78. Ling HY, Lau KT, Cheng L, and Au HY. 2005. Evaluation of embedded fibre Bragg grating sensor for delamination detection in composite structures. In *Structural Health Monitoring*, FK Chang (ed.). DEStech, Lancaster.

79. Fuhr PL, Huston DR, and Ambrose TP. 1994. Polymer optical fiber sensors for structural sensing. In *Applications of Photonic Technology*, R Measures (ed.), Plenum Publishing, New York.

80. Kuang KS, Quek ST, and Maalej M. 2005. Polymer-based optical fiber sensors for health monitoring of engineering structures. *Sensors and Smart Structures Technologies for Civil, Mechanical, and Aerospace Systems*, SPIE 5765.

81. Cohen EI, Mastro SA, Nemarich CP, Korczynski JF, Jarrett AW, and Jones WC. 1999. Recent developments in the use of plastic optical fiber for an embedded wear sensor. *Sensory Phenomena and Measurement Instrumentation for Smart Structures and Materials*, SPIE 3670.

82. Cohen EJ. 2000. Embedded wear sensor. US Patent 6,080,982.

83. Dakai L, Mingshun H, Baoqi T, and Hao Q. 1999. Research on some problems about optical fiber embedded in carbon fiber smart structures. In *Proc 2nd Intl Workshop on Structural Health Monitoring*, FK Chang (ed.), Stanford. Technomic, Lancaster.

84. Datta S, Hause J, Hurd D, Kirikera GR, Schulz MJ, Ghoshal A, and Sundaresan MJ. 2003. An active fiber continuous sensor. In *Structural Health Monitoring*, FK Chang (ed.). DEStech, Lancaster.

85. Latta BM. 2003. Picostrain engineering data acquisition system. US Patent 6,556,927.

86. Sener JC, Latta BM, and Ross JD. 2004. Development and installation of picostrain sensors in structural systems. *Sensors and Smart Structures Technologies for Civil, Mechanical and Aerospace Systems*, SPIE 5391.

87. Takagi K, Sato H, and Saigo M. 2005. Damage detection and gain-scheduled control of CFRP smart board mounting the metal core assisted piezoelectric fiber. *Modeling, Signal Processing, and Control*, SPIE 5757.

88. Davey KJ. 1998. Monitoring apparatus for monitoring impending faults in the integrity of a component or structure. US Patent 5,770,794.

89. Wheatley G, Kollgaard J, Register J, and Zaidi M. 2003. Comparative vacuum monitoring as an alternative means of compliance. In *Structural Health Monitoring*, FK Chang (ed.). DEStech, Lancaster.

90. Wishaw M and Barton DP. 2001. Comparative vacuum monitoring: A new method of *in-situ* real-time crack detection and monitoring. *AINDT Conference*, Brisbane.

91. Stehmeier H and Speckmann H. 2004. Comparative vacuum monitoring (CVM): Monitoring of fatigue cracking in aircraft structures. *Proc 2nd European Workshop on Structural Health Monitoring*, DEStech, Munich.

92. Udd E. 1991. *Fiber Optic Sensors*. Wiley, New York.

93. Spillman Jr WB. 1989. Multi-mode optical fiber sensor and method. US Patent 4,863,270.

94. Thursby G, Sorazu B, Betz D, Staszewski W, and Culshaw B. 2004. Comparison of point and integrated fiber optic sensing techniques for ultrasound detection and location of damage. *Smart Sensor Technology Measurement Systems*, SPIE 5384.

95. Paolozzi A, Felli F, and Caponero MA. 1999. Global temperature measurements of aluminum alloy specimens with embedded optical fibers. *Proc 2nd Intl Workshop on Structural Health Monitoring*. Stanford. Technomic, Lancaster.

96. Fuerstenau N, Janzen DD, and Schmidt W. 1993. In-flight strain measurements on structurally integrated composite plates using fiber-optic interferometric strain gauges. *Smart Mater Struct*, 2, 147–156.

97. Lisboa O, Tsukahara Y, Neron C, and Jen CK. 1994. A quasi-distributed optical fiber sensor. *Smart Mater Struct*, 3, 157–163.

98. Spillman W Jr and Huston D. 1996. Impact detection, location and characterization using spatially weighted distributed fiber optic sensors. *Distributed Fiber Optic Sensors*, SPIE 2838.

99. Spillman W Jr and Huston D. 1997. Detection, location and characterization of point perturbations over a two dimensional area using two spatially weighted distributed fiber optic sensors. *Smart Sensing, Processing, and Instrumentation*, SPIE 3042.
100. Huston DR, Spillman WB Jr, Claus RO, and Ayra V. 1996. Vehicle classification by pattern matching gage sensors. *Smart Sensing, Processing and Instrumentation*, SPIE 2718.
101. Huston D, Spillman W Jr, Sauter W, and Pelczarski N. 1999. Monitoring micro floor vibrations with distributed fiber optic sensors. *Smart Systems for Bridges, Structures and Highways*, SPIE 3671.
102. Spillman WB Jr, Huston DR, and Wu J. 2001. Seismic event monitoring using very long gauge length integrating fiber optic sensors. In *Selected Papers on Distributed Fiber Optic Sensors and Measuring Networks*, YN Kulchin (ed.), SPIE 4357.
103. Bourquin F. 2004. Towards modal fibre optic sensors for slender beams. *Proc 2nd European Workshop on Structural Health Monitoring*, DEStech, Munich.
104. Tanaka N, Snyder SD, and Hansen CH. 1996. Distributed parameter modal filtering using smart sensors. *J Vib Acous*, 118, 630–640.
105. Sawyer J, Ruffin P, and Lofts C. 1994. Investigation of a fiber specklegram sensor for structural fatigue monitoring. *Smart Sensing, Processing, and Instrumentation*, SPIE 2191.
106. Pan K, Uang CM, Cheng F, and Yu FT. 1994. Multimode fiber sensing by using mean-absolute speckle-intensity variation. *Appl Opt*, 33(10), 2095–2098.
107. Lindner DK, Zvonar GA, Baumann WT, and Delos PL. 1993. Nonlinear effects of a modal domain optical fiber sensor in a vibration suppression control loop for a flexible structure. *J Vib Acous*, 115, 120–128.
108. Kageyama K, Ohsawa I, Kanai M, Machijima Y, Matsumura F, and Nagata K. 2003. Development of a new fiber-optic vibration sensor and its application to structural health monitoring of composite structures. *Nondestructive Evaluation and Health Monitoring of Aerospace Materials and Composites II*, SPIE 5046.
109. Huston D, Fuhr P, Beliveau JG, and Spillman Jr W. 1991. Structural member vibration measurements using a fiber optic sensor. *J Sound Vib*, 149(2), 348–353.
110. Delepine-Lesoille S, Merliot E, Nobili M, Dupont J, and Caussignac JM. 2004. Design of a new optical fiber sensor body meant for embedding into concrete. *Proc 2nd European Workshop on Structural Health Monitoring*, DEStech, Munich.
111. Wu Z, Yang C, and Takahasi T. 2003. Self-diagnosis of concrete beams reinforced with hybrid CFRP rods. In *Structural Health Monitoring*, FK Chang (ed.). DEStech, Lancaster.
112. Wu ZS, Yang CQ, Harada T, and Ye LP. 2005. Self-diagnosis of structures strengthened with hybrid carbon-fiber-reinforced polymer sheets. *Smart Mater Struct*, 14, S39–S51.
113. Nanni F, Ruscito G, Forte G, and Gusmano G. 2007. Design, manufacture and testing of self-sensing carbon fibre–glass fibre reinforced polymer rods. *Smart Mater Struct*, 16, 2368–2374.
114. Friswell MI. 2003. Modal actuators and sensors for beam and plate structures. *Smart Sensor Technology and Measurement Systems*, SPIE 5050.
115. Bray BC and Washington GN. 2003. Positional impact sensing with spatially shaded polyvinylidene flouride film. *Smart Sensor Technology and Measurement Systems*, SPIE 5050.
116. Tzou HS. 1993. *Piezoelectric Shells: Distributed Sensing and Control of Continua*. Kluwer Academic, Dordrecht.
117. Deraemaeker A and Preumont A. 2004. Modal filters for vibration based damage detection. *Proc 2nd European Workshop on Structural Health Monitoring*, DEStech, Munich.
118. Lanzara G, Feng J, Huang K, Dinyari R, Kim JY, Peumans P, and Chang FK. 2007. Stretching of monolithic silicon-based sensor network for large area embedded structural health monitoring. In *Structural Health Monitoring*, FK Chang (ed.). DEStech, Lancaster.
119. White R, Song J, Haas C, and Middleton D. 2006. An evaluation of quartz piezo-electric weigh-in-motion sensors. Paper no. 06-2591, 86th Annual Meeting, TRB, Washington, DC.
120. Dong F, Thursby G, and Culshaw B. 2003. RF sub-carrier based fibre strain sensor for monitoring composite gas storage tanks. *Smart Sensor Technology and Measurement Systems*, SPIE 5050.

121. Bastianini F, Cargnelutti M, Di Tommaso A, and Toffanin M. 2003. Distributed brillouin fiber optic strain monitoring application in advanced composite materials. *Smart Systems and Nondestructive Evaluation for Civil Infrastructures*, SPIE 5057.

122. van Schoor M, Lengyel A, Muller GJ, and Radighieri B. 2004. Method and sheet like sensor for measuring stress distribution. US Patent 6,802,216.

123. Sun S, Pommerenke D, Drewniak JL, and Chen G. 2004. Signal loss, spatial resolution, and sensitivity of long coaxial crack sensors. *Sensors and Smart Structures Technologies for Civil, Mechanical and Aerospace Systems*, SPIE 5391.

124. Kuo JT and Burnett GD. 2005. Pipe testing apparatus and method using electrical or electromagnetic pulses transmitted into the pipe. US Patent 6,194,902.

125. Webster S and Teferra W. 1996. Pile damage assessments using the pile driving analyzer. *Proc 5th Intl Conf on the Application of Stress-wave Theory to Piles (STRESSWAVE '96)*, Orlando, FL.

126. Alvarez C, Zuckerman B, and Lemke J. 2006. Dynamic pile analysis using CAPWAP and multiple sensors. *ASCE GEO Congress*, Atlanta.

127. Hertlein B and Davis A. 2006. *Nondestructive Testing of Deep Foundations*. Wiley, Chichester.

128. Gnaedinger RJ and Gnaedinger JP. 1972. Method and apparatus for determining structural characteristics. US Patent 3,641,811.

129. Oyadiji SO and Feroz KT. 1999. Applications of PZT sensors and finite element analysis in defect detection in bars. *Proc 2nd Intl Workshop on Structural Health Monitoring*, Stanford. Technomic, Lancaster.

130. Wu J and Spillman Jr WB. 1994. Distributed defect detection system using a unidirectional acoustic waveguide. *Smart Sensing, Processing, and Instrumentation*, SPIE 2191.

131. Todorovska MI and Trifunac MD. 2008. Impulse response analysis of the Van Nuys 7-storey hotel during 11 earthquakes and earthquake damage detection. *Struct Control Health Monit*, 15, 90–116.

132. Hugentobler MK and Mandracchia EA. 1998. Acoustic strain gauge and assembly and method for measuring strain. US Patent 5,750,900.

133. Yan W, Goebel KF, and Evers N. 2005. Algorithms for partial discharge diagnostics applied to aircraft wiring. *Proc Aging Aircraft Wiring Conference*, Palm Springs.

134. Born M and Wolf E. 1999. *Principles of Optics*, 7th Ed. Cambridge University Press, Cambridge.

135. Smith C. 1994. OTDR technology and optical sensors time domain reflectometry. *Time Domain Reflectometry in Environmental, Infrastructure, and Mining Applications*, US Bureau of Mines, SP 19-94.

136. Wanser KH, Haselhuhn M, and LaFond M. 1993. High temperature distributed strain and temperature sensing using OTDR. In *Applications of Fiber Optic Sensors in Engineering Mechanics*, F Ansari (ed.), ASCE, New York.

137. O'Connor KM and Dowding CH. 1999. *GeoMeasurements by Pulsing TDR Cables and Probes*. CRC Press, Boca Raton, FL.

138. Lyöri V, Mantyniemi A, Kilpela A, Duan Q, and Kostamovaara J. 2003. A fibre-optic time-of-flight radar with a sub-metre spatial resolution for the measurement of integral strain. *Smart Sensor Technology and Measurement Systems*, SPIE 5050.

139. Zimmerman BD and Claus RO. 1993. Spatially multiplexed optical fiber time domain sensors for civil engineering applications. In *Applications of Fiber Optic Sensors in Engineering Mechanics*, F. Ansari (ed.), ASCE, New York.

140. Lou KA, Zimmermann B, and Yaniv G. 1994. Combined sensor system for process and in-service health monitoring of composite structures. *Smart Sensing, Processing, and Instrumentation*, SPIE 2191.

141. Diaz de Leon A, Cruz PJ, and Leung CK. 2004. An innovative fiber optic sensor for cracking detection and monitoring. *Proc IABMAS'04 Bridge Maintenance, Safety, Management and Cost*, Kyoto. Taylor & Francis, London.

142. Leung CK, Olson N, Wan KT, and Meng A. 2005. Theoretical modeling of signal loss versus crack opening for a novel crack sensor. *J Eng Mech*, 131(8), 777–790.

143. Kercel SW and Muhs JD. 1993. Repeatable sensitivity of OTDR-based reentrant loop strain measurement using single-mode and graded-index fibers. *Smart Mater Struct*, 2, 249–254.
144. Voet MR, Bonne AA, Caufriez V, and Lebrun C. 1994. Fiber optic pore water pressure field measurements at the Hades underground research facility and the L'Eau D'Heure Dam. *Time Domain Reflectometry in Environmental, Infrastructure, and Mining Applications*, US Bureau of Mines, SP 19-94.
145. Gu X, Chen Z, and Ansari F. 2000. Embedded fiber optic crack sensor for reinforced concrete structures. *ACI Struct J*, 97(3), 468–476.
146. Bibby GW. 1989. Temperature measurement. US Patent 4,859,065.
147. Billington R. 1999. Measurement methods for stimulated Raman and Brillouin scattering in optical fibres. *NPL Report*, COEM 31.
148. Basnett T and Barber SJ. 1989. Distributed temperature measurement with optical fibres. *Fiber Optics '89*, SPIE 1120.
149. Rogers A. 1995. Nonlinear effects in optical fibers. In *Optical Fiber Sensor Technology*, K Grattan and B Meggitt (eds.). Chapman & Hall, London.
150. Kleinerman M and Kelleher PW. 1991. A distributed force-sensing optical fiber using forward time division multiplexing. *Distributed and Multiplexed Fiber Optic Sensors*, SPIE 1586.
151. Horiguchi T, Kurashima T, and Koyamada Y. 1992. Measurement of temperature and strain distribution by Brillouin frequency shift in silica optical fibers. *Distributed and Multiplexed Fiber Optic Sensors II*, SPIE 1797.
152. Dakin JP. 1995. Distributed optical fiber sensors. In *Fiber Optic Smart Structures*, E Udd (ed.), Wiley, New York.
153. Cho SB, Jung-Ju Lee JJ, and Kwon IB. 2006. Pulse base effect on the strain measurement of a Brillouin-scattering-based distributed optical fiber sensor. *Smart Mater Struct*, 15, 315–324.
154. Bao X, Webb DJ, and Jackson DA. 1994. Combined distributed temperature and strain sensor based on Brillouin loss in an optical fiber. *Opt Lett*, 19, 141.
155. Hartog AH. 2003. Optical time domain reflectometry method and apparatus. US Patent 6,542,228.
156. Kwon IB and Kim CY. 2003. Distributed strain and temperature measurement of a beam using fiber optic BOTDA sensor. *Smart Systems and Nondestructive Evaluation for Civil Infrastructures*, SPIE 5057.
157. Zhang W, Shi B, Gao JQ, Ding Y, and Zhu H. 2005. Nondestructive testing of bridges strengthened by external prestressing using distributed fiber optic sensing. *Proc IABMAS'04 Bridge Maintenance, Safety, Management and Cost*, Kyoto. Taylor & Francis, London.
158. Yamaura T, Inoue Y, Kino H, and Nagai K. 1999. Development of structural health monitoring system using Brillouin optical time domain reflectometer. *Proc 2nd Intl Workshop on Structural Health Monitoring*. Stanford. Technomic, Lancaster.
159. Kishida K, Zhang H, Li CH, Guzik A, Suzuki H, and Wu Z. 2005. Diagnosis of corrosion based thinning in steam pipelines by means of Neubrescope high precision optical fiber sensing system. In *Structural Health Monitoring*, FK Chang (ed.). DEStech, Lancaster.
160. Li C, Sun Y, Zhao YG, Liu H, Gao LM, Zhang ZL, and Qiu HT. 2006. Monitoring pressure and thermal strain in the second lining of a tunnel with a Brillouin OTDR. *Smart Mater Struct*, 15, N107–N110.
161. Bernini R, Fraldi M, Minardo A, Minutolo V, Carrannante F, Nunziante L, and Zeni L. 2005. Damage detection in bending beams through Brillouin distributed optic-fiber sensor. In *Structural Health Monitoring*, FK Chang (ed.). DEStech, Lancaster.
162. Nishio M, Mizutani T, and Takeda N. 2007. Structural shape identification of composite structures using embedded optical fiber sensors. In *Structural Health Monitoring*, FK Chang (ed.). DEStech, Lancaster.
163. Oka K, Ohno H, Kurashima T, Matsumoto M, Kumagai H, Mita A, and Sekijima K. 1999. Fiber optic distributed sensor for structural monitoring. In *Proc 2nd Intl Workshop on Structural Health Monitoring*. Stanford. Technomic, Lancaster.

164. Kim SH, Lee JJ, and Kwon IB. 2002. Structural monitoring of a bending beam using Brillouin distributed optical fiber sensors. *Smart Mater Struct*, 11, 396–403.

165. Nikles M, Vogel B, Briffod F, Grosswig S, Sauser F, Luebbecke S, Bals A, and Pfeiffer T. 2004. Leakage detection using fiber optics distributed temperature monitoring. *Smart Sensor Technology Measurement Systems*, SPIE 5384.

166. Anastasio S, Pamukcu S, and Pervispour M. 2007. Chemical selective BOTDR sensing for corrosion detection on structural systems. In *Structural Health Monitoring*, FK Chang (ed.). DEStech, Lancaster.

167. Gifford DK, Childers BA, Duncan RG, Jackson AC, Shaw S, Schwienberg B, and Mazza J. 2003. Structural integrity monitoring of aircraft panels using a distributed Bragg grating sensing technique. *Smart Sensor Technology and Measurement Systems*, SPIE 5050.

168. Wang A, Zhang P, Cooper KL, and Pickrell GR. 2003. Highly multiplexed fiber optic sensors for civil infrastructure monitoring. In *Structural Health Monitoring*, FK Chang (ed.). DEStech, Lancaster.

169. Davis MA and Kersey AD. 1996. Separating the temperature and strain effects on fiber Bragg grating sensors using stimulated Brillouin scattering. *Smart Sensing, Processing and Instrumentation*, SPIE 2718.

170. Wu Z. 2006. Recent developments in fiber optic based distributed monitoring technologies. *Proc 4th China–Japan–US Symp on Structural Control and Monitoring*, Hangzhou.

171. Lee JW, Huh YC, Nam YY, Lee GH, Lee YB, Kim YS, and Kim JY. 2007. Damage detection using distributed optical fiber strain sensor. In *Structural Health Monitoring*, FK Chang (ed.). DEStech, Lancaster.

172. Yari T, Nagai K, Shimizu T, and Takeda N. 2003. Overview of damage detection and damage suppression demonstrator and strain distribution measurement using distributed BOTDR sensors. *Industrial and Commercial Applications of Smart Structures Technologies*, SPIE 5054.

173. Texier S, Pamukcu S, Toulouse J, and Ricles J. 2005. Brillouin scattering fiber optic strain sensor for distributed applications in civil infrastructure. In *Structural Health Monitoring*, FK Chang (ed.). DEStech, Lancaster.

174. Jin WL, Shao JW, and Wu YH. 2006. A new health monitoring technique on submarine pipeline. *Proc 4th China–Japan–US Symp on Structural Control and Monitoring*, Hangzhou.

175. Spillman Jr WB. 1991. Composite integrity monitoring. US Patent 4,983,034.

176. McKown TO and Eilers DD. 1994. Explosive-array performance measurements using time domain reflectometry. *Time Domain Reflectometry in Environmental, Infrastructure, and Mining Applications*, US Bureau of Mines, SP 19-94.

177. Pierce CE, Bilaine C, Huang FC, and Dowding CH. 1994. Effects of multiple crimps and cable length on reflection signatures from long cables. *Time Domain Reflectometry in Environmental, Infrastructure, and Mining Applications*, US Bureau of Mines, SP 19-94.

178. Chajes M, Hunsperger R, Liu W, Li J, and Kunz E. 2002. Nondestructive evaluation of pre- and post-tensioning strands using time domain reflectometry. In *Proc Structural Materials Technology V, an NDT Conference*, S. Allampalli and G Washer (eds.). Cincinnati.

179. Chen G, Mu H, Pommerenke D, and Drewniak JL. 2003. Continuous coaxial sensors for monitoring of RC structures with electrical time domain reflectometry. *Smart Systems and Nondestructive Evaluation for Civil Infrastructures*, SPIE 5057.

180. Chen GD, Sun SS, Pommerenke D, Drewniak JL, Greene GG, McDaniel RD, Belarbi A, and Mu HM. 2005. Crack detection of a full-scale reinforced concrete girder with a distributed cable sensor. *Smart Mater Struct*, 14, S88–S97.

181. Budelmann H, Hariri K, Holst A, and Wichmann HJ. 2004. New monitoring techniques for bridge tendons. *Proc IABMAS'04 Bridge Maintenance, Safety, Management and Cost*, Kyoto. Taylor & Francis, London.

182. Liang RY, Al-Akhras K, and Rabab'ah S. 2006. Field monitoring of moisture variations under flexible pavement. Paper no. 06-1683, 86th Annual Meeting, TRB, Washington, DC.

183. Dennis ND, Ooi CW, and Wong VH. 2006. Monitoring shallow slope failures in residual soils using time domain reflectometry. Paper no. 06-2738, 86th Annual Meeting, TRB, Washington, DC.

184. Francke JL, Terrill LJ, and Francke CT. 1994. Time domain reflectometry study at the waste isolation pilot plant. *Time Domain Reflectometry in Environmental, Infrastructure, and Mining Applications*, US Bureau of Mines, SP 19-94.

185. Daita RK, Drnevich VP, Kim D, and Chen R. 2006. Quality assessment of lime kiln dust (LKD) modified soils by electrical conductivity measured with time domain reflectometry (TDR). Paper no. 06-1480, 86th Annual Meeting, TRB, Washington, DC.

186. Anastasi RF and Madaras EI. 2003. Aging wire insulation assessment by phase spectrum examination of ultrasonic guided waves. *Nondestructive Evaluation and Health Monitoring of Aerospace Materials and Composites II*, SPIE 5046, Bellingham, WA.

187. Reis H, Ervin B, Kuchma DA, and Bernhard J. 2003. Evaluation of corrosion damage in steel reinforced concrete. In *Structural Health Monitoring*, FK Chang (ed.). DEStech, Lancaster.

188. Zumpano G and Meo M. 2004. A new damage detection technique for rails base on wave propagation. *Sensors and Smart Structures Technologies for Civil, Mechanical and Aerospace Systems*, SPIE 5391.

189. Laguerre L, Bouhelier M, and Grimault A. 2004. Applications of ultrasonic guided waves to the evaluation of steel members integrity identification to structural health monitoring. *Proc 2nd European Workshop on Structural Health Monitoring*, DEStech, Munich.

190. Gong J, Jayawickrama PW, and Tinkey Y. 2006. Non-destructive evaluation of installed soil nails. Paper no. 06-1364, 86th Annual Meeting, TRB, Washington, DC.

191. Loveday PW. 2000. Development of piezoelectric transducers for a railway integrity monitoring system. *Smart Systems for Bridges, Structures, and Highways*, SPIE 3988.

192. Washer G and Fuchs P. 2004. Ultrasonic health monitoring of prestressing tendons. *Proc Structural Materials Technology VI, an NDT Conference*, Buffalo.

193. Na WB, Kundu T, Ryu YS, and Kim JT. 2005. Concrete filled steel pipe inspection using electro magnetic acoustic transducer (EMAT). *Sensors and Smart Structures Technologies for Civil, Mechanical, and Aerospace Systems*, SPIE 5765.

194. Kwun H and Teller II CM. 2005. Nondestructive evaluation of ferromagnetic cables and ropes using magnetostrictively induced acoustic/ultrasonic waves and magnetostrictively detected acoustic emissions. US Patent 5,456,113.

195. Kim IK and Kim YY. 2005. Damage size estimation by the continuous wavelet ridge analysis of dispersive bending waves in a beam. *J Sound Vib*, 287, 707–722.

196. Anderegg P, Brönnimann R, Raab C, and Partl M. 2000. Long-term monitoring of highway deformation. *Nondestructive Evaluation of Highways, Utilities, and Pipelines IV*, SPIE 3995.

197. Lerski RA. 1988. *Practical Ultrasound*. IRL Press, Oxford.

198. Kacmar C and Blackshire JL. 2003. *In-situ* measurements of stress-corrosion crack growth using laser ultrasonics. *Testing, Reliability, and Application of Micro- and Nano-Material Systems*, SPIE 5045.

199. Nemati KM and Gardoni P. 2005. Microstructural and statistical evaluation of interfacial zone percolation in concrete. *Strength, Fract Complex*, 3, 191–197.

200. Firestone FA. 1942. Flaw detecting device and measuring instrument. US Patent 2,280,226.

201. Severns DA and Ceccarelli C. 2004. Utilization of NDE technologies in in-service and fabrication inspection for the Nevada department of transportation. *Proc Structural Materials Technology VI, an NDT Conference*, Buffalo.

202. Graybeal BA, Walthier RA, and Washer GA. 2000. Ultrasonic inspection of bridge hanger pins. In *Structural Materials Technology IV*, S. Allampalli (ed.), Technomic Press, Lancaster.

203. Blackshire JL, Hoffmann J, Kropas-Hughes C, and Tansel I. 2003. Microscopic NDE of hidden corrosion. *Testing, Reliability, and Application of Micro- and Nano-Material Systems*, SPIE 5045.

204. Cooney AT and Blackshire JL. 2004. Characterization of microscopic surface breaking cracks using the near-field intensification of non-destructive laser generated surface waves. *Sensors and Smart Structures Technologies for Civil, Mechanical and Aerospace Systems*, SPIE 5391.

205. Vary A. 1998. The acousto-ultrasonic approach. In *Acousto-Ultrasonics, Theory and Applications*, JC Duke Jr (ed.), Plenum Press, New York.

206. Cosgriff LM, Martin RE, and Baaklini GY. 2003. Acousto-ultrasonic characterisation of C/SiC composites under different loading configurations. *Nondestructive Evaluation and Health Monitoring of Aerospace Materials and Composites II*, SPIE 5046, Bellingham, WA.

207. Beall FC and Chen LH. 2000. Ultrasonic monitoring of resin curing in a press for the production of particle board and similar materials. US Patent 6,029,520.

208. Maev RG. 2006. New generation of high resolution ultrasonic imaging technique for material characterization and NDT. *Proc ECNDT 2006—Fr.1.6.1*.

209. Washer G, Fuchs P, Rezai A, and Ghasemi H. 2005. Ultrasonic measurement of the elastic properties of ultra high performance concrete (UHPC). *Nondestructive Evaluation and Health Monitoring of Aerospace Materials, Composites and Civil Infrastructure IV*, SPIE 5767.

210. Owino JO and Jacobs LJ. 1999. Attenuation measurments in cement-based materials using laser ultrasonics. *J Eng Mech*, 126(6), 637–647.

211. Becker J, Jacobs LJ, and Qu J. 2003. Characterization of cement-based materials using diffuse ultrasound. *J Eng Mech*, 129(12), 1478–1484.

212. Fisk P, Holt R, Sargent D, and El Beik A. 2002. Detection and evaluation of voids in tendon ducts by sonic resonance and tomography techniques. In *Proc Structural Materials Technology V, an NDT Conference*, S. Allampalli and G Washer (eds.). Cincinnati.

213. Popovics S, Silva-Rodriguez R, Popovics JS, and Martucci V. 1994. Behavior of ultrasonic pulses in fresh concrete. *New Experimental Techniques for Evaluating Concrete Material and Structural Performance*, ACI SP-143-12.

214. Gallo G and Popovics JS. 2006. Application of ultrasonic surface waves to characterize ASR damage in concrete. *NDE Conf on Civil Engineering*, American Society for Nondestructure Testing, St. Louis.

215. Takatsubo J, Urabe K, Tsuda H, Toyama N, and Wang B. 2004. Ultrasonic technique for nondestructive characterization of void inclusion. In *Advanced Smart Materials and Smart Structures Technology*, FK Chang, CB Yun, and BF Spencer, Jr. (eds.). DEStech, Lancaster.

216. Tang FF. 1994. Energy characterization for brittle material response under uniaxial compression through ultrasonic scanning. *New Experimental Techniques for Evaluating Concrete Material and Structural Performance*, ACI SP-143-13.

217. Krause M, Mielentz F, Milmann B, Streicher D, and Mayer K. 2006. Ultrasonic reflection properties at interfaces between concrete steel and air: Imaging and modelling. *NDE Conference on Civil Engineering*, Americen Society Nondestructure Testing, St. Louis.

218. Drabkin S and Kim DS. 1994. Fatigue monitoring of concrete specimens using frequency characteristics of ultrasonic waves. *New Experimental Techniques for Evaluating Concrete Material and Structural Performance*, ACI SP-143-14.

219. Benedetti A. 1998. On the ultrasonic pulse propagation into fire damaged concrete. *ACI Struct J*, 953., S24.

220. Ramamoorthy SK, Kane Y, and Turner JA. 2004. Ultrasound diffusion for crack depth determination in concrete. *J Acoust Soc Am*, 1152, 523–529.

221. Le Marrec L, Abraham O, Chekroun M, Leparoux D, Laguerre L, Derode A, and Campillo M. 2006. Towards multidiffusive ultrasonic propagation for nondestructive evaluation of concrete: Theoretical overview. *NDE Conference on Civil Engineering*, American Society for Nondestructure Testing, St. Louis.

222. Aktan HM, Yaman IO, Udegbunam O, and Hearn N. 2000. Early age durability assessment of cast-in-place RC bridge deck. *Nondestructive Evaluation of Highways, Utilities, and Pipelines IV*, SPIE 3995.

223. Mirmiran A and Wei Y. 2001. Damage assessment of FRP-encased concrete using ultrasonic pulse velocity. *J Eng Mech*, 127(2), 126–135.

224. Ross RJ and Pellerin RF. 1994. Nondestructive testing for assessing wood members in structures a review. *USDA Forest Service Technical Report*, FPL-GTR-70.

225. Graff KF. 1973. *Wave Motion in Elastic Solids*. Ohio State University Press, Columbus, OH.

226. Lee BC and Staszewski WJ. 2003. Lamb wave interaction with structural defects—Modelling and simulations. *Modeling, Signal Processing and Control*, SPIE 4049.

227. Pretorius JV, Van Shoor MC, Muller GJ, and Jessiman AW. 2005. Piezoelectric structural acoustic leak detection for pressurized pipelines. *Industrial and Commercial Applications of Smart Structures Technologies*, SPIE 5762.

228. Wilson MS and Hurlebaus S. 2007. Monitoring of overhead transmission lines. In *Structural Health Monitoring*, FK Chang (ed.). DEStech, Lancaster.

229. Guo D and Kundu T. 2000. Lamb wave sensors for detecting wall defects in pipes. *Nondestructive Evaluation of Highways, Utilities, and Pipelines IV*, SPIE 3995.

230. Kannan E, Maxfield BW, and Balasubramaniam K. 2007. SHM of pipes using torsional waves generated by *in situ* magnetostrictive tapes. *Smart Mater Struct*, 16, 2505–2515.

231. Yan W, Lim CW, Cai JB, and Chen WQ. 2007. An electromechanical impedance approach for quantitative damage detection in Timoshenko beams with piezoelectric patches. *Smart Mater Struct*, 16, 1390–1400.

232. Hu Y and Yang Y. 2007. Wave propagation modeling of the PZT sensing region for structural health monitoring. *Smart Mater Struct*, 16, 706–716.

233. Firestone FA and Ling DS Jr. 1951. Method and means for generating and utilizing vibrational waves in plates. US Patent 2,536,126.

234. Worlton DC. 1965. Method of applying Lamb waves in ultrasonic testing. US Patent 3,165,922.

235. Neary TE, Huston DR, Wu JR, and Spillman WB Jr. 1996. *In-situ* damage monitoring of composite structures. *Smart Sensing, Processing and Instrumentation*, SPIE 2718.

236. Tseng KK, Tinker ML, Lassiter JO, and Eckel JT. 2003. Impedance-based structural health monitoring for composite laminates in cryogenic environments. *Nondestructive Evaluation and Health Monitoring of Aerospace Materials and Composites II*, SPIE 5046.

237. Engberg RC and Ooi TK. 2004. Methods and piezoelectric imbedded sensors for damage detection in composite plates under ambient and cryogenic conditions. *Smart Sensor Technology Measurement Systems*, SPIE 5384.

238. Swenson ED and Crider JS III. 2007. Damage detection using Lamb waves for structural health monitoring on an aircraft bulkhead. In *Structural Health Monitoring*, FK Chang (ed.). DEStech, Lancaster.

239. Yang J, Chang FK, and Derriso MM. 2003. Design of built-in health monitoring system for bolted thermal protection panels. *Nondestructive Evaluation and Health Monitoring of Aerospace Materials and Composites II*, SPIE 5046.

240. Berger U and Baier H. 2004. Experimental and analytical investigations in structural health monitoring with LAMB-waves. *Proc 2nd European Workshop on Structural Health Monitoring*. DEStech, Munich.

241. Cawley P. 1997. Quick inspection of large structures using low frequency ultrasound. In *Structural Health Monitoring: Current Status and Perspectives*, F.-K. Chang, (ed.). Technomic Publishing, Lancaster.

242. Thomas D and Welter J. 2004. Corrosion damage detection with piezoelectric wafer active sensors. *Health Monitoring and Smart Nondestructive Evaluation of Structural and Biological Systems III*, SPIE 5394.

243. Guy P, Jayet Y, and Goujon L. 2003. Guided waves interaction with complex delaminations application to damage detection in composite structures. *Smart Nondestructive Evaluation and Health Monitoring of Structural and Biological Systems II*, SPIE 5047.

244. Osegueda R, Kreinovich V, Nazarian S, and Roldan E. 2003. Detection of cracks at rivet holes in thin plates using Lamb-wave scanning. *Smart Nondestructive Evaluation and Health Monitoring of Structural and Biological Systems II*, SPIE 5047.

245. Jain A, Greve DW, and Oppenheim IJ. 2003. Experiments in ultrasonic flaw detection using a MEMS transducer. *Smart Systems and Nondestructive Evaluation for Civil Infrastructures*, SPIE 5057.

246. Murphy KA, Schmid CA, Tran TA, Carman G, Wang A, and Claus RO. 1994. Delamination detection in composites using optical fiber techniques. *Smart Sensing, Processing, and Instrumentation*, SPIE 2191.

247. Pierce SG, Philp WR, Culshaw B, Gachagan A, McNab A, and Hayward G. 1996. Ultrasonic inspection of CFRP plates using surface bonded optical fibre sensors. *Smart Sensing, Processing and Instrumentation*, SPIE 2718.

248. Jung YC, Na WB, Kundu T, and Ehsani M. 2000. Damage detection in concrete using Lamb waves. *Nondestructive Evaluation of Highways, Utilities, and Pipelines IV*, SPIE 3995.

249. Luangvilai K, Jacobs LJ, and Qu J. 2003. Propagation of guided Lamb waves in concrete repaired with composite plates. *Smart Systems and Nondestructive Evaluation for Civil Infrastructures*, SPIE 5057.

250. Kudryavtsev YF, Kleiman JI, and Gustcha OI. 2000. Ultrasonic measurement of residual stresses in welded railway bridge. In *Structural Materials Technology IV*, S. Allampalli (ed.), Technomic, Lancaster.

251. Calomfirescu M and Herrmann AS. 2007. Attenuation of Lamb waves in composites: Models and possible applications. In *Structural Health Monitoring*, FK Chang (ed.). DEStech, Lancaster.

252. Chellapilla SS, Aldrin JC, and Jata KV. 2007. Interaction of guided waves with fastener sites in aircraft structures. In *Structural Health Monitoring*, FK Chang (ed.). DEStech, Lancaster.

253. Tian T, Chiu WK, and Rajic N. 2007. Health monitoring of repaired structures: A numerical analysis. In *Structural Health Monitoring*, FK Chang (ed.). DEStech, Lancaster.

254. Petculescu G, Krishnaswamy S, and Achenbach JD. 2008. Group delay measurements using modally selective Lamb wave transducers for detection and sizing of delaminations in composites. *Smart Mater Struct*, 17, 015007.

255. Takahashi I, Ito Y, Takeda SI, Iwahori Y, Yashiro S, Takasubo J, and Takeda N. 2007. Debonding detection in scarf-repaired CFRP laminates using Lamb-wave visualization generation laser scanning method. In *Structural Health Monitoring*, FK Chang (ed.). DEStech, Lancaster.

256. Yamada M, Hiramoto K, Nakahata K, and Kitahara M. 2004. Visualization of SH ultrasonic testing for surface breaking crack. *Proc IABMAS'04 Bridge Maintenance, Safety, Management and Cost*, Kyoto, Taylor & Francis, London.

257. Tittmann BR. 2003. Recent results and trends in health monitoring with surface acoustic waves. *Testing, Reliability, and Application of Micro- and Nano-Material Systems*, SPIE 5045.

258. Palmer SB, Dixon S, Edwards RS, and Jian X. 2005. Transverse and longitudinal crack detection in the head of rail tracks using Rayleigh wave-like wideband guided ultrasonic waves. *Nondestructive Evaluation and Health Monitoring of Aerospace Materials, Composites and Civil Infrastructure IV*, SPIE 5767.

259. Marzani A, Rizzo P, Lanza di Scalea F, and Benzoni G. 2004. Mobile acoustic system for the detection of surface-breaking cracks in pavement. *Smart Sensor Technology Measurement Systems*, SPIE 5384.

260. Kuhr S, Blackshire JL, Martin SA, and Na JK. 2007. Design, fabrication and testing of thin-film, surface-wave sensors for crack detection in complex geometry aerospace structures. In *Structural Health Monitoring*, FK Chang (ed.). DEStech, Lancaster.

261. Wen C and Zhu C. 2006. Time synchronous dyadic wavelet processor array using surface acoustic wave devices. *Smart Mater Struct*, 15, 939–945.

262. Sherrit S, Bao XQ, Bar-Cohen Y, and Chang X. 2003. BAW and SAW sensors for *in-situ* analysis. *Smart Sensor Technology and Measurement Systems*, SPIE 5050.

263. Sachs T, Grossman R, Michel J, and Schrueffer E. 1996. Remote sensing using quartz sensors. *Smart Sensing, Processing and Instrumentation*, SPIE 2718.

264. Ong CK, Chen S, Galloway C, and Delatte N. 2004. Correlating spectral analysis of surface wave (SASW) technique to anchor holding capacity. *Proc Structural Materials Technology VI, an NDT Conference*, Buffalo.

265. Joh SH, Kang TH, Kwon SA, and Won MC. 2006. Accelerated stiffness profiling of aggregate bases and subgrades for quality assessment of field compaction. Paper no. 06-1683, 86th Annual Meeting, TRB, Washington, DC.

266. Celaya M and Nazarian S. 2006. Seismic testing to determine quality of hot-mix asphalt. Paper no. 06-2654, 86th Annual Meeting, TRB, Washington DC.

267. Mallick RB, Bradley JE, and Nazarian S. 2006. Evaluation and recommendation of a fast nondestructive test and analysis method for in-place determination of stiffness of subsurface reclaimed layers in thin surface hot mix asphalt (HMA) pavements. Paper no. 06-0635, 86th Annual Meeting, TRB, Washington, DC.

268. Gucunski N and Maher A. 2000. Bridge deck evaluation using portable seismic pavement analyzer (PSPA). *Final Report, FHWA*, NJ 2000-05.

269. Simonin JM and Abraham O. 2006. Two complementary seismic investigation methods for the detection and characterisation of delamination in road structures. *NDE Conf on Civil Engineering*, American Society for Nondestructure Testing, St. Louis.

270. Vermont Agency of Transportation Pavement Design Committee. 2006. A study of *in situ* pavement material properties determined from FWD testing. Paper no. 06-2243, 86th Annual Meeting, TRB, Washington, DC.

271. Hossain MM and Yang WS. 2000. Determination of concrete pavement surface layer modulus and thickness using nondestructive deflection testing. In *Structural Materials Technology IV*, S Allampalli (ed.), Technomic, Lancaster.

272. Tabrizi K, Ganji V, and Sauber R. 2000. Project level application of falling weight deflectometer. In *Structural Materials Technology IV*, S Allampalli (ed.). Technomic, Lancaster.

273. Kumapley RK and Kumapley NK. 2000. NDT approach to monitoring PCC deterioration due to D-cracking in highway pavements. In *Structural Materials Technology IV*, S Allampalli (ed.). Technomic, Lancaster.

274. Ceylan H, Gopalakrishnan K, Guclu A, and Bayrak MB. 2006. Structural characterization of flexible airfield test pavement under repeated traffic loading. *NDE Conf on Civil Engineering*, American Society for Nondestructure Testing, St. Louis.

275. Ganji V, Tabrizi K, and Vittilo N. 2000. Project level application of portable seismic pavement analyzer. In *Structural Materials Technology IV*, S Allampalli (ed.). Technomic, Lancaster.

276. Nazarian S, Baker MR, and Crain K. 1997. Movable seismic pavement analyzer. US Patent 5,614,670.

277. Wang CH, Rose JT, and Chang FK. 2003. A computerized time-reversal method for structural health monitoring. *Nondestructive Evaluation and Health Monitoring of Aerospace Materials and Composites II*, SPIE 5046.

278. Kessler SS and Dunn CT. 2003. Optimization of Lamb wave actuating and sensing materials for health monitoring of composite structures. *Smart Structures and Integrated Systems*, SPIE 5056.

279. Lanza di Scalea F, Bartoli I, Rizzo P, and McNamara J. 2004. *On-line High-speed Rail Defect Detection*, US Department of Transportation, DOT/FRA/ORD-04/16.

280. Kotyaev O, Shimada Y, and Hashimoto K. 2006. Laser-based non-destructive detection of inner flaws in concrete with the use of Lamb waves. *Proc ECNDT 2006*.

281. Park G, Wait JR, Sohn H, and Farrar CR. 2003. Sensing system development for damage prognosis. In *Structural Health Monitoring*, FK Chang (ed.). DEStech, Lancaster.

282. Clark AV, Schramm RE, Schaps SR, and Filla BJ. 1995. Safety assessment of railroad wheels through roll-by detection of tread cracks. *Nondestructive Evaluation of Aging Railroads*, SPIE 2458.

283. Sorazu B, Culshaw B, and Pierce B. 2005. Optical technique for examining materials' elastic properties. *Smart Sensor Technology and Measurement Systems*, SPIE 5758.

284. Zhu J and Popovics JS. 2007. Imaging concrete structures using air-coupled impact-echo. *J Eng Mech*, 133(6), 628–640.

285. Roye W and Schieke S. 2006. Ultrasonic probes for special applications. *Proc ECNDT 2006—Mo.2.7.4*.

286. Kobayashi M, Jen CK, Moisan JF, Mrad N, and Nguyen SB. 2007. Integrated ultrasonic transducers made by the sol–gel spray technique for structural health monitoring. *Smart Mater Struct*, 16, 317–322.
287. Quek ST, Jin J, and Tua PS. 2004. Comparison of plain piezoceramics and inter-digital transducer for crack detection in plates. *Smart Electronic, MEMS, BioMEMS, and Nanotechnology*, SPIE 5389.
288. Gordon GA and Braunling R. 2005. Quantitative corrosion monitoring and detection using ultrasonic Lamb waves. *Sensors and Smart Structures Technologies for Civil, Mechanical, and Aerospace Systems*, SPIE 5765.
289. Arias I and Achenbach J. 2003. Thermoelastic generation of ultrasound by line-focused laser irradiation. *Intl J Solids Struct*, 40, 6917–6935.
290. Kearns J, Pena-Macias J, Criado-Abad A, Southward T, Evans D, and Malkin M. 2007. Development and flight demonstration of a piezoelectric phased array damage detection system. In *Structural Health Monitoring*, FK Chang (ed.). DEStech, Lancaster.
291. Kim SB and Sohn H. 2007. Simultaneous crack detection in thin metal plates and aircraft panels. In *Structural Health Monitoring*, FK Chang (ed.). DEStech, Lancaster.
292. Tsuda H. 2005. Response of FBG sensors to Lamb wave propagation and its application to active sensing. *Nondestructive Evaluation and Health Monitoring of Aerospace Materials, Composites and Civil Infrastructure IV*, SPIE 5767.
293. Dokun OD, Jacobs LJ, and Haj-Ali RM. 2000. Ultrasonic monitoring of material degradation in FRP composites. *J Eng Mech*, 126(7), 704–710.
294. Atique SH, Culshaw B, Thursby G, Dong F, Sorazi B, and Park HS. 2004. Generation of ultrasound for material testing using low power diode laser. *Proc 2nd European Workshop on Structural Health Monitoring*, DEStech, Munich.
295. Jacobs LJ and Whitcomb RW. 1997. Laser generation and detection of ultrasound in concrete. *J Nondestruct Eval*, 162., 57–65.
296. Voillaume H, Simonet D, Brousset C, Barbeau P, Arnaud JL, Dubois M, Drake T, and Osterkamp M. 2006. Analysis of commercial aeronautics applications of laser ultrasonics for composite manufacturing. *Proc ECNDT 2006—We.1.1.1.*
297. Blodgett DW and Baldwin KC. 2005. Laser-based ultrasonics: Applications at APL. *Johns Hopkins APL Tech Dig*, 26(1), 36–45.
298. Owens SE, Miyasaka C, and Shull PJ. 2004. Application of a high numerical aperture lens to visualize debonding between metallic films and polymer substrates with SAM. *Nondestructive Evaluation and Health Monitoring of Aerospace Materials and Composites III*, SPIE 5393.
299. Croxford AJ, Wilcox PD, Drinkwater BW, and Konstantinidis G. 2007. Temperature sensitivity limitations for guided wave structural health monitoring. In *Structural Health Monitoring*, FK Chang (ed.). DEStech, Lancaster.
300. Ogisu T, Shimanuki M, Kiyoshima S, Okabe Y, and Takeda N. 2004. Development of damage monitoring system for aircraft structure using a PZT actuator/FBG sensor hybrid system. *Industrial and Commercial Applications of Smart Structures Technologies*, SPIE 5388.
301. Lanza di Scalea FL, Rizzo P, and Marzani A. 2004. Propagation of ultrasonic guided waves in lap-shear adhesive joints. *Nondestructive Evaluation and Health Monitoring of Aerospace Materials and Composites III*, SPIE 5393.
302. Yuan S, Wang L, Wang X, and Chang MA. 2004. New damage signature for composite structural health monitoring. *Proc 2nd European Workshop on Structural Health Monitoring*, DEStech, Munich.
303. Vinh T. 1993. Dynamic damage indicators for composite based structures. In *Safety Evaluation Based on Identification Approaches*, HG Natke, GR Tomlinson and JTP Yao (eds.). Vieweg, Braunschweig.
304. Divekar AA and Vizzini AJ. 1994. Delamination detection in tapered laminates using embedded ultrasonics. *Smart Sensing, Processing, and Instrumentation*, SPIE 2191.
305. Malkin M. 2005. Structural health monitoring for bonded repairs: Complex structure testing. In *Structural Health Monitoring*, FK Chang (ed.). DEStech, Lancaster.

306. Bescond C, Monchalin JP, Lévesque D, Gilbert A, Talbot R, and Ochiai M. 2005. Determination of residual stresses using laser-generated surface skimming longitudinal waves. *Nondestructive Evaluation and Health Monitoring of Aerospace Materials, Composites and Civil Infrastructure IV*, SPIE 5767.

307. Thien AB, Chiamori HC, Ching JT, Wait JR, and Park G. 2008. The use of macro-fibre composites for pipeline structural health assessment. *Struct Control Health Monit*, 15, 43–63.

308. Barnoncel D, Osmont D, and Dupont M. 2003. Health monitoring of sandwich plates with real impact damages using PZT devices. In *Structural Health Monitoring*, FK Chang (ed.). DEStech, Lancaster.

309. Li F, Su Z, Ye L, and Meng G. 2006. A correlation filtering-based matching pursuit (CF-MP) for damage identification using Lamb waves. *Smart Mater Struct*, 15, 1585–1594.

310. Hunt B. 2003. Pier pressure. *Bridge Design and Engineering*, 9(30), 48–50.

311. Davis J, Nassif H, and Ertekin AO. 2004. Field evaluation of methods for monitoring bridge scour. *Proc IABMAS'04 Bridge Maintenance, Safety, Management and Cost*, Kyoto. Taylor & Francis, London.

312. Weissmann J and Diaz M. 2004. Bridge scour monitoring in Texas. *Proc IABMAS'04 Bridge Maintenance, Safety, Management and Cost*, Kyoto. Taylor & Francis, London.

313. Betz DC, Thursby G, Culshaw B, and Staszewski W. 2003. Lamb wave detection and source location using fiber Bragg grating rosettes. *Smart Sensor Technology and Measurement Systems*, SPIE 5050.

314. Matt HM and Lanza di Scalea F. 2007. Macro-fiber composite piezoelectric rosettes for acoustic source location in complex structures. *Smart Mater Struct*, 16, 1489–1499.

315. Harris WT. 1962. Pole testing apparatus. US Patent 3,066,525.

316. Gutierrez M and Enger PF. 1979. Void detector system. US Patent 4,163,393.

317. Sansalone M and Carino NJ. 1986. Impact-echo: A method for flaw detection in concrete using transient stress wave analysis. *NBSIR* 86-3452, Maryland, September.

318. Sansalone M. 1997. Impact-echo: The complete story. *ACI Struct J*, 94(6), 777–786.

319. Olson LD. 1992. Sonic NDE of structural concrete. In *Nondestructive Testing of Concrete Elements and Structures*, F Ansari and S Sture (eds.). ASCE, New York.

320. Ghorbanpoor A, Virmani YP, and Fatemi GR. 1992. Evaluation of concrete bridges by impact-echo. In *Nondestructive Testing of Concrete Elements and Structures*, F Ansari and S Sture (eds.). ASCE, New York.

321. Krause M, Wiggenhauser H, and Krieger J. 2002. NDE of a post tensioned concrete bridge girder using ultrasonic pulse echo and impact echo. *Proc Structural Materials Technology V, an NDT Conference*, S Allampalli and G Washer (eds.). Cincinnati.

322. Abraham O and Cote P. 2002. Impact-echo thickness frequency profiles for detection of voids in tendon ducts. *ACI Struct J*, 993, S25.

323. Ryden N and Park CB. 2006. A combined multichannel impact echo and surface wave analysis scheme for non-destructive thickness and stiffness evaluation of concrete slabs. *NDE Conf on Civil Engineering*, American Society for Nondestructure Testing, St. Louis.

324. Pristov E, Dalton W, Piscsalko G, and Likins G. 2006. Comparison of impact-echo with broadband input to determine concrete thickness. *NDE Conf on Civil Engineering*, American Society for Nondestructure Testing, St. Louis.

325. Gibson A and Popovics JS. 2005. Lamb wave basis for impact-echo method analysis. *J Eng Mech*, 131(4), 438–443.

326. Gassman SL and Zein L. 2004. Condition assessment of concrete bridge T-shaped girders under fatigue loading. *Proc Structural Materials Technology VI, an NDT Conference*, Buffalo.

327. Kim DS, Seo WS, and Choi YK. 2004. IE-SASW method for nondestructive evaluation of concrete structure. In *Advanced Smart Materials and Smart Structures Technology*, FK Chang, CB Yun, and BF Spencer, Jr. (eds.). DEStech, Lancaster.

328. Borg SL. 2000. Nondestructive testing for length determination of piles for five long island bridges. In *Structural Materials Technology IV*, S Allampalli (ed.). Technomic, Lancaster.

329. Holt JD and Slaughter SH. 2000. Mississippi's approach to unknown bridge foundations. In *Structural Materials Technology IV*, S Allampalli (ed.). Technomic, Lancaster.

330. Pessiki S and Johnson MR. 1994. In-place evaluation of concrete strength using the impact-echo method. *New Experimental Techniques for Evaluating Concrete Material and Structural Performance*, ACI SP-143-15.

331. Bayard, DS and Joshi SS. 1995. Detecting structural failures via acoustic impulse responses. *NASA JPL New Technology Report*, NPO-19167.

332. Ata N and Ohtsu M. 2004. Void detection in grouted sheath in post-tensioning prestressed concrete by SIBIE. *Proc IABMAS'04 Bridge Maintenance, Safety, Management and Cost*, Kyoto. Taylor & Francis, London.

333. Finno RJ and Chao HC. 2000. Nondestructive evaluation of selected drilled shafts at the central artery/tunnel project. In *Structural Materials Technology IV*, S Allampalli (ed.). Technomic, Lancaster.

334. Jackson H, Gucunski N, and Maher A. 2002. Application of cone model to the nondestructive testing of rigid pavements. In *Proc Structural Materials Technology V, an NDT Conference*, S Allampalli and G Washer (eds.). Cincinnati.

335. Canning JS, Niezrecki C, and Birgisson B. 2004. Diagnosis of moisture damage in asphalt pavement. *Sensors and Smart Structures Technologies for Civil, Mechanical and Aerospace Systems*, SPIE 5391.

336. Kweon G and Kim YR. 2006. Determination of the complex modulus of asphalt concrete using the impact resonance test. Paper no. 06-0464, 86th Annual Meeting, TRB, Washington, DC.

337. Butt SD, Limaye V, Mufti AA, and Bakht B. 2004. Acoustic transmission technique for evaluating fatigue damage in concrete bridge deck slabs. *ACI Struct J*, 101(1), 3–10.

338. Hwang JS and Kim SJ. 2003. Validation of tapping sound analysis for damage detection of composite structures. *Nondestructive Evaluation and Health Monitoring of Aerospace Materials and Composites II*, SPIE 5046.

339. Savalli N, Baglio S, Muscato G, Lemistre MB, and Balageas DL. 2003. Hybrid tactile probe for damage detection and material recognition. *Smart Nondestructive Evaluation and Health Monitoring of Structural and Biological Systems II*, SPIE 5047.

340. Alongi AV, Cantor TR, Kneeter CP, and Alongi A Jr. 1982. Concrete evaluation by radar theoretical analysis. *Transportation Research Board Record*, No. 853.

341. Barnes CL and Trottier JF. 2000. Effectiveness of ground penetrating radar for preparing pre-tender deterioration estimates on asphalt covered reinforced concrete bridge decks. In *Structural Materials Technology IV*, S Allampalli (ed.). Technomic, Lancaster.

342. Romero FA and Clark TM. 2004. Ground penetrating radar assessment comparing construction design specifications to contractor performance for thickness of hot-mix asphalt pavement. *Proc Structural Materials Technology VI, an NDT Conference*, Buffalo.

343. Kind T and Maierhofer C. 2006. Railway ballast fouling inspection by monitoring the GPR propagation velocity with a multi-offset antenna array. *11th Intl Conf on Ground Penetrating Radar*, Columbus.

344. Xie W, Al-Qadi IL, Roberts R, Tutumluer E, and Boyle J. 2006. Quantification of railroad ballast condition using ground penetrating radar data. *NDE Conf on Civil Engineering*, American Society for Nondestructure Testing, St. Louis.

345. Cardimona S, Willeford B, Wenzlick J, and Anderson N. 2000. Investigation of bridgedecks utilizing ground penetrating radar. *Intl Conf on the Application of Geophysical Tech to Planning, Design, Construction, and Maintenance of Trans Facilities*, St. Louis.

346. Jones B, Carnevale M, and Hager J. 2006. Comparison of high-resolution 3D imaging radar survey with material testing data in an overall structural integrity assessment of a concrete ringwall. *11th Intl Conf on Ground Penetrating Radar*, Columbus.

347. Huston DR, Hu JQ, Maser K, Weedon W, and Adam C. 2000. GIMA ground penetrating radar system for infrastructure health monitoring. *J Appl Geophys*, 43, 139–146.

348. Abou-Khousa M, Kharkovsky S, and Zoughi R. 2005. Embedded flaw spatial distribution retrieval from microwave and millimeter wave images. In *Structural Health Monitoring*, FK Chang (ed.). DEStech, Lancaster.

349. Smith SS. 1995. Detecting pavement deterioration with subsurface interface radar. *Sensors,* 12(9), 29–37.

350. Scott M, Rezaizadeh A, and Moore M. 2001. Phenomenology study of HERMES ground penetrating radar technology for detection and identification of common bridge deck features. FHWA-RD-01-090, USDOT Federal Highway Administration.

351. Balanis CA. 1989. *Advanced Engineering Electromagnetics.* Wiley, New York.

352. Aurand J. 1998. Personal communication to the author.

353. Andrews JR. 1999. Ultra-wideband differential measurements using PSPL BALUNs. *Application Note* AN-8, Picosecond Pulse Labs, Boulder.

354. Russ JC. 1999. *The Image Processing Handbook,* 3rd Ed. CRC Press, Boca Raton.

355. Hunt LD, Massie D, and Cull JP. 2000. Standard pallettes for GPR data analysis. *Eighth Intl Conf on Ground Penetrating Radar,* SPIE 4084.

356. Saarenketo T and Roimela P. 1998. Ground penetrating radar technique in asphalt pavement density quality control. *Proc 7th Intl Conf on Ground Penetrating Radar.* Radar Systems and Remote Sensing Laboratory, University of Kansas, Lawrence, KS, Volume 2, pp. 461–466.

357. Johansson EM and Mast JE. 1994. Three-dimensional ground penetrating radar imaging using synthetic aperture time-domain focusing. *Advanced Microwave and Millimeter-wave Detectors,* SPIE 2275.

358. Mast JE. 1998. Automated position calculating imaging radar with low-cost synthetic aperture sensor for imaging layered media. US Patent 5,796,363.

359. Helmerich R, Niederleithinger E, and Wiggenhauser H. 2006. A toolbox with non-destructive testing methods for the condition assessment of railway bridges. Paper no. 06-2143, 86th Annual Meeting, TRB, Washington, DC.

360. Huston D, Pelczarski N, Esser B, Maser K, and Weedon W. 2000. Damage assessment in roadways with ground penetrating radar. *Nondestructive Evaluation of Highways, Utilities, and Pipelines IV,* SPIE 3995.

361. Huston DR, Fuhr PL, Maser K, and Weedon WH. 2002. *Nondestructive Testing of Reinforced Concrete Bridges Using Radar Imaging Techniques,* NETC 94-2 New England Transportation Consortium.

362. Morey RM. 2001. Innovative GPR for pavement inspection. *Smart Systems for Bridges, Structures and Highways,* SPIE 4330.

363. Balanis CA. 1989. *Advanced Engineering Electromagnetics.* Wiley, New York.

364. AASHTO. 1993. Standard test method for evaluating asphalt-covered concrete bridge decks using pulsed radar. *TP36-93,* Edition 1A.

365. Xing H, Li J, Chen X, Liu CR, Michalk B, Oshinski E, Chen H, Bertrand C, and Claros G. 2006. A new method for pavement dielectric constant measurement using ground penetrating radar. Paper no. 06-0280, 85th Annual Meeting, TRB, Washington, DC.

366. Carter CR, Chung T, Holt FB, and Manning DG. 1986. An automated signal processing system for the signature analysis of radar waveforms from bridge decks. *Can Electr Eng J,* 11(3), 128–137.

367. Maser KR and Kim Roddis WM. 1990. Principles of thermography and radar for bridge deck assessment. *J Transp Eng,* 1165., 583–601.

368. Baker-Jarvis J, Janezic MD, Riddle RF, Johnk RT, Kabos P, Holloway C, Geyer RG, and Grosvenor CA. 2005. Measuring the permittivity and permeability of lossy materials: Solids, liquids, metals, building materials, and negative-index materials, NIST Technical Note 1536.

369. Grivas DA, Braijawa F, and Shin H. 2004. Use of ground penetrating radar testbed to validate pavement base and subbase properties. *Proc Structural Materials Technology VI, an NDT Conference,* Buffalo.

370. Choubane B, Fernando E, Ross SC, and Dietrich BT. 2003. Use of ground penetrating radar for asphalt thickness determination. *Testing, Reliability, and Application of Micro- and Nano-Material Systems,* SPIE 5045.

371. Maser KR, Roberts R, Kutrubes D, and Holland J. 2002. Technology for quality assurance of new pavement thickness. In *Proc Structural Materials Technology V, an NDT Conference*, S Allampalli and G Washer (eds.). Cincinnati.

372. Williams RR, Martin TJ, Maser KR, and McGovern G. 2006. Evaluation of network level ground penetrating radar effectiveness. Paper no. 06-2243, 86th Annual Meeting, TRB, Washington, DC.

373. Millard SG, Shaw M, and Bungey JH. 2004. The use of ground penetrating radar for investigating leakage for water pipes beneath highway pavements, in clay sub-strata. *Proc Structural Materials Technology VI, an NDT Conference*, Buffalo.

374. Sato M and Miwa T. 2000. Polarimetric borehole radar system for fracture measurement. *Subsurface Sensing Technol Appl*, 1(1), 161–175.

375. Park SK, Uomoto T, and Park SB. 2004. Estimation of the shape of subsurface voids using three-dipole radar advanced. In *Smart Materials and Smart Structures Technology*, FK Chang, CB Yun, and BF Spencer, Jr (eds.). DEStech, Lancaster.

376. Fitzgerald CS. 1995. Inspection for rail defects by magnetic induction. *Nondestructive Evaluation of Aging Railroads*, SPIE 2458.

377. Hall DR. 2002. Electromagnetic inspection of wire ropes in vertical lift bridges. In *Proc Structural Materials Technology V, an NDT Conference*, S Allampalli and G Washer (eds.). Cincinnati.

378. Bergamini A and Christen R. 2003. A simple approach to the localization of flaws in large diameter steel cables. *Smart Nondestructive Evaluation and Health Monitoring of Structural and Biological Systems II*, SPIE 5047.

379. Krieger J, Krause HJ, Gampe U, and Sawade G. 1999. Magnetic field measurements on bridges and development of a mobile SQUID-system. *Nondestructive Evaluation of Bridges and Highways III*, SPIE 3587.

380. Xiong EG and Wang SL. 2006. Magnetic flux leakage testing mechanism and its application. *Proc 4th China–Japan–US Symp on Structural Control and Monitoring*, Hangzhou.

381. NASA. 2007. Wireless measurement of rotation and displacement rate. *NASA Tech Briefs*, LAR-16848-1.

382. Shah DK, Chan WS, and Joshi SP. 1994. Delamination detection and suppression in a composite laminate using piezoelectric layers. *Smart Mater Struct*, 3, 263–301.

383. Georgiades G, Oyadiji SO, and Liu Z. 2005. Design of a distributive load sensor using PZT. *Smart Sensor Technology and Measurement Systems*, SPIE 5758.

384. Lee CK, Hsu YH, Hsiao WH, and Wu JW. 2003. Electrical and mechanical field interactions of piezoelectric systems: Foundation of smart structure-based piezoelectric sensors and actuators, piezo-transformers, and free-fall sensors. *Industrial and Commercial Applications of Smart Structures Technologies*, SPIE 5054.

385. Zhang Y. 2003. Dynamic strain measurement using piezoelectric paint. In *Structural Health Monitoring*, FK Chang (ed.). DEStech, Lancaster.

386. Zhang J and Frankel GS. 1999. Investigation of the corrosion-sensing behavior of an acryllic-based coating system. *Corrosion*, 5, 957.

387. Davis GD, Dacres CM, Shook M, and Wenner BS. 1998. Electrochemical *in-situ* sensors for detecting corrosion on aging aircraft. *Proc Intelligent NDE Science for Aging and Futuristic Aircraft*, El Paso.

388. Abbitt JD, Fuentes CA, and Carroll BF. 1996. Film-based pressure-sensitive-paint measurements. *Opt Lett*, 212(2), 1797–1799.

389. Liu T, Guille M, and Sullivan JP. 2001. Accuracy of pressure sensitive paint. *AIAA J*, 39(1), 103–112.

390. Huang C, Gregory JW, and Sullivan JP. 2007. Microchannel pressure measurements using molecular sensors. *J Microelectromech Syst*, 16(4), 281–288.

391. Lu Y and Su Z. 2004. Lamb wave-based crack identification for aluminum plate using active sensor network. *Proc 2nd European Workshop on Structural Health Monitoring*, DEStech, Munich.

392. Dugnani R and Malkin M. 2003. Damage detection on a large composite structure. In *Structural Health Monitoring*, FK Chang (ed.). DEStech, Lancaster.

393. DuPont M, Osmont D, Gouyon R, and Balageas DL. 1999. Permanent monitoring of damaging impacts by a piezoelectric sensor based integrated system. *Proc 2nd Intl Workshop on Structural Health Monitoring*. Stanford. Technomic, Lancaster.

394. Wang CS and Chang FK. 1999. Built-in diagnostics for impact damage identification of composite structures. *Proc 2nd Intl Workshop on Structural Health Monitoring*. Stanford. Technomic, Lancaster.

395. Tseng KK and Wang L. 2005. Impedance-based method for nondestructive damage identification. *J Eng Mech*, 131(1), 58–64.

396. Lin M, Kumar A, Qing X, Beard SJ, Russell SS, Walker JL, and Delay TK. 2003. Monitoring the integrity of filament wound structures using built-in sensor networks. *Industrial and Commercial Applications of Smart Structures Technologies*, SPIE 5054.

397. Lu Y, Ye L, and Su Z. 2006. Crack identification in aluminium plates using Lamb wave signals of a PZT sensor network. *Smart Mater Struct*, 15, 839–849.

398. Hay TR, Royer RL, Gao H, Zhao X, and Rose JL. 2006. A comparison of embedded sensor Lamb wave ultrasonic tomography approaches for material loss detection. *Smart Mater Struct*, 15, 946–951.

399. Chang FK and Lin M. 2002. Diagnostic layer and methods for detecting structural integrity of composite and metallic materials. US Patent 6,370,964.

400. Ihn JB and Chang FK. 2004. Detection and monitoring of hidden fatigue crack growth using a built-in piezoelectric sensor/actuator network: II. Validation using riveted joints and repair patches. *Smart Mater Struct*, 13, 621–630.

401. Mook G, Pohl J, and Michel F. 2003. Non-destructive characterization of smart CFRP systems. *Smart Mater Struct*, 12, 997–1004.

402. Kusaka T and Qing PX. 2003. Characterization of loading effects on the performance of smart layer embedded or surface-mounted on structures. In *Structural Health Monitoring*, FK Chang (ed.). DEStech, Lancaster.

403. Park J and Chang FK. 2003. Built-in detection of impact damage in multi-layered thick composite structures. In *Structural Health Monitoring*, FK Chang (ed.). DEStech, Lancaster.

404. Seydel RE and Chang FK. 1999. Implementation of a real-time impact identification technique for stiffened composite panels. *Proc 2nd Intl Workshop on Structural Health Monitoring*. Stanford. Technomic, Lancaster.

405. Park G, Cudney HH, and Inman DJ. 1999. Impedance-based health monitoring techniques for civil structures. *Proc 2nd Intl Workshop on Structural Health Monitoring*. Stanford. Technomic, Lancaster.

406. Zhao X, Gao H, Zhang G, Ayhan B, Yan F, Kwan C, and Rose JL. 2007. Active health monitoring of an aircraft wing with embedded piezoelectric sensor/actuator network: I. Defect detection, localization and growth monitoring. *Smart Mater Struct*, 16 1208–1217.

407. Bois C, Hochard C, de Vadder D, and Mazerolle F. 2005. Fatigue crack monitoring in glass fiber reinforced laminate with piezoelectric transducers. In *Structural Health Monitoring*, FK Chang (ed.). DEStech, Lancaster.

408. Williams SM. 2005. The feasibility of using piezoelectric transducers to generate and detect microscopic structural vibrations for monitoring storage tanks. In *Structural Health Monitoring*, FK Chang (ed.). DEStech, Lancaster.

409. Allen DW, Peairs DM, and Inman DJ. 2004. Damage detection by applying statistical methods to PZT impedance measurements. *Smart Electronic, MEMS, BioMEMS, and Nanotechnology*, SPIE 5389.

410. Rutherford AC, Park G, Sohn H, and Farrar CR. 2004. Nonlinear feature identification of impedance-based structural health monitoring. *Smart Electronic, MEMS, BioMEMS, and Nanotechnology*, SPIE 5389.

411. Islam AS and Craig KC. 1994. Damage detection in composite structures using piezoelectric materials. *Smart Mater Struct*, 3, 318–328.

412. Holnicki-Szulc J and Zielinski TG. 1999. Damage identification method based on analysis of perturbation of elastic waves propagation. *Proc 2nd Intl Workshop on Structural Health Monitoring*. Stanford. Technomic, Lancaster.
413. Keilers CH Jr. 1997. Search strategies for identifying a composite plate delamination using built-in transducers. In *Structural Health Monitoring: Current Status and Perspectives*, F.-K. Chang, (ed.). Technomic, Lancaster.
414. Prasad SM, Balasubramaniam K, and Krishnamurthy CV. 2004. Structural health monitoring of composite structures using Lamb wave tomography. *Smart Mater Struct*, 16, 1489–1499.
415. Fritzen CP, Mengelkamp G, Dietrich G, Richter W, Cuerno C, Lopez-Diez J, and Guemes A. 2004. Structural health monitoring of the ARTEMIS satellite antenna using a smart structure concept. *Proc 2nd European Workshop on Structural Health Monitoring*. DEStech, Munich.
416. Derveaux G, Papanicolaou G, and Tsogka C. 2005. Time reversal imaging for sensor networks with optimal compensation in time. In *Structural Health Monitoring*, FK Chang (ed.). DEStech, Lancaster.
417. Park G, Inman DJ, and Farrar CR. 2003. Recent studies in piezoelectric impedance-based structural health monitoring. In *Structural Health Monitoring*, FK Chang (ed.). DEStech, Lancaster.
418. Macias JP, Guemes A, and Kawiecki G. 2003. Cure monitoring using networks of piezoceramic transducers. In *Structural Health Monitoring*, FK Chang (ed.). DEStech, Lancaster.
419. Su Z and Ye L. 2003. A signal processing and interpretation technique for Lamb-wave-based damage diagnosis using digital damage fingerprints extracted from an actuator/sensor network. In *Structural Health Monitoring*, FK Chang (ed.). DEStech, Lancaster.
420. Coverley PT and Stazewski WJ. 2003. Impact damage location in composite structures using optimized sensor triangulation procedure. *Smart Mater Struct*, 12, 795–203.
421. Royer Jr RL, Zhao X, Owens SE, and Rose JL. 2007. Large area corrosion detection in complex aircraft components using Lamb wave tomography. In *Structural Health Monitoring*, FK Chang (ed.). DEStech, Lancaster.
422. Joshi SP and Munir NI. 1996. Impact location, energy and damage detection in a smart structure. *Smart Sensing, Processing and Instrumentation*, SPIE 2718.
423. Okafor AC, Chandrashekara K, and Jiang YP. 1996. Location of impact in composite plates using waveform-based acoustic emission and Gaussian cross-correlation techniques. *Smart Sensing, Processing and Instrumentation*, SPIE 2718.
424. Huang K and Peumans P. 2005. Large sensor networks built from monolithic silicon. In *Structural Health Monitoring*, FK Chang (ed.). DEStech, Lancaster.
425. Jacobs G. 2005. High definition scanning: Forensic and damage assessment surveys. *Professional Surveyor*, 25(11), 14–17.
426. Pai PF, Kim BS, and Chung JH. 2003. Dynamics-based damage inspection of an aircraft wing panel. *Nondestructive Evaluation and Health Monitoring of Aerospace Materials and Composites II*, SPIE 5046.
427. Leong WH, Lee BC, Staszewski WJ, and Scarpa F. 2004. Crack detection in metallic structures using Lamb waves and laser scanning vibrometry. *Proc 2nd European Workshop on Structural Health Monitoring*. DEStech, Munich.
428. Sundaresan MJ, Ghoshal A, Schulz MJ, Ferguson F, Pai PF, and Chung JH. 1999. Crack detection using a scanning laser vibrometer. *Proc 2nd Intl Workshop on Structural Health Monitoring*, FK Chang (ed.). Stanford. Technomic, Lancaster.
429. Buckner BD, Markov V, Earthman JC, and Angeles J. 2003. A laser scanning technique for fatigue damage precursor detection and monitoring. In *Structural Health Monitoring*, FK Chang (ed.). DEStech, Lancaster.
430. Lavalle JP, Schuet SR, and Schuet DJ. 2004. High-speed 3D scanner with real-time 3D processing. *Nondestructive Evaluation and Health Monitoring of Aerospace Materials and Composites III*, SPIE 5393.
431. Helfrick M and Niezrecki C. 2007. An investigation of the use of 3-D optical measurements to perform structural health monitoring. In *Structural Health Monitoring*, FK Chang (ed.). DEStech, Lancaster.

432. Mehrabi AB and Ciolko AT. 2001. Health monitoring of aging cable structures. *Proc ASCE Structures Congress and Exposition*. Sponsored by Structural Engineering Institute of ASCE, PC Chang (ed.). May 21–23, 2001, Washington, DC

433. Chen SE, Petro S, Venkatappa S, Ramamoody V, Moody J, Gangarao H, and Culkin A. 2000. Automated full-scale laser vibration sensing system. In *Structural Materials Technology IV*, S Allampalli (ed.). Technomic, Lancaster.

434. Bennett KD, Hoover III CW, Chen RZ, and Plone MD. 1994. Gaussian beam displacement sensor for monitoring interstory drift in smart buildings. *Smart Sensing, Processing, and Instrumentation*, SPIE 2191.

435. Markov V, Trolinger J, Kilpatrick J, Webster J, Pardoen G, and Keene D. 2003. Remote opto-acoustic sensors for structural integrity and health monitoring of complex systems. In *Structural Health Monitoring*, FK Chang (ed.). DEStech, Lancaster.

436. Kincade K. 2003. Lasers shed light on Stonehenge secrets. *Laser Focus World*.

437. Fuchs PA, Washer GA, Chase SB, and Moore M. 2004. Laser-based instrumentation for bridge load rating. *J Performance Constructed Facilities*, 18(4), 213–219.

438. Easley TC, Faber KT, and Shah SP. 2001. Moire interferometry analysis of fiber debonding. *J Eng Mech*, 127(6), 625–629.

439. Wang SB, Tong JW, Yue C, Li LA, and Shen M. 2002. New method on experimental damage study of AS4/PEEK composite laminates. *Nondestructive Evaluation and Health Monitoring of Aerospace Materials and Civil Infrastructures*, SPIE 4704.

440. Jensen BD, de Boer MP, Masters ND, Bitsie F, and LaVan DA. 2001. Interferometry of actuated microcantilevers to determine material properties and test structure nonidealities in MEMS. *J Microelectromech Syst*, 10(3), 336–346.

441. Sandhu JS, Wang H, Sonpatki MM, and Popek WJ. 2003. Real-time full-field ultrasonic inspection of composites using acoustography. *Nondestructive Evaluation and Health Monitoring of Aerospace Materials and Composites II*, SPIE 5046.

442. Shinozuka M, Ghanem R, Houshmand B, and Mansouri B. 2000. Damage detection in urban areas by SAR imagery. *J Eng Mech*, 126(7), 769–777.

443. Andhee A, Gryzagoridis J, and Findeis D. 2005. Comparison of normal and phase stepping shearographic NDE. *Nondestructive Evaluation and Health Monitoring of Aerospace Materials, Composites and Civil Infrastructure IV*, SPIE 5767.

444. Fulton JP, Namkung M, and Melvin LD. 1992. Practical estimates of the errors associated with the governing shearography equation. *Rev Prog Quant Nondestruct Eval*, 11, 427–434, DO Thompson and DE Cimenti (eds.). La Jolla, CA.

445. Leviton DB. 2004. Method and apparatus for two-dimensional absolute optical encoding. US Patent 6,765,195.

446. Kalayeh HM, Paz-Pujalt GR, and Spoonhower JP. 2006. System and method for remote quantitative detection of fluid leaks from a natural gas or oil pipeline. US Patent 6,995,846.

447. Anastasi RF and Madaras EI. 2005. Application of Hilbert–Huang transform for improved defect detection in terahertz NDE of shuttle tiles. *Nondestructive Evaluation and Health Monitoring of Aerospace Materials, Composites and Civil Infrastructure IV*, SPIE 5767.

448. Mraz A, Gunaratne M, Nazef A, and Choubane B. 2006. Experimental evaluation of the pavement imaging system of the FDOT multi-purpose survey vehicle. Paper no. 06-1124, 86th Annual Meeting, TRB, Washington, DC.

449. Shinozuka M and Rejaie A. 2001. Damage assessment from remotely sensed images using PCA. *Smart Systems for Bridges, Structures, and Highways*, SPIE 4330-01.

450. Adams BJ, Huyck CK, Mio M, Cho S, Ghosh S, Chung HC, Eguchi RT, Houshmand B, Shinozuka M, and Mansouri B. 2004. The Bam (Iran) earthquake of December 26, 2003: Preliminary reconnaissance using remotely sensed data and VIEWS system. *MCEER Earthquake Investment Report*, Bam 12-26-03.

451. Holstein D, Hartmann HJ, and Jueptner W. 1998. Investigation of laser welds by means of digital speckle photography. *Laser Interferometry IX: Techniques and Analysis*, SPIE 3478.

452. Steinchen W and Yang L. 2003. *Digital Shearography*. SPIE Press, Bellingham.

453. Sassenfeld NC, Crull MM. 1992. Concrete surface characterization using optical metrology. In *Nondestructive Testing of Concrete Elements and Structures*, F Ansari and S Sture (eds.). ASCE, New York.

454. Sabaté N, Vogel D, Keller J, Gollhardt A, Marcos J, Gràcia I, Cané C, and Michel B. 2007. FIB-based technique for stress characterization on thin films for reliability purposes. *Microelectron Eng*, 84, 1783–1787.

455. Shinozuka M, Chung HC, Ichitsubo M, and Liang J. 2001. System identification by video image processing. *Smart Systems for Bridges, Structures, and Highways*, SPIE 4330.

456. Choi MY, Heo G, Park K, and Oh S. 2001. Evaluation of structural dynamic characteristics using CCD measurement principle. *Smart Systems for Bridges, Structures, and Highways*, SPIE 4330.

457. Jáuregui DV, Tian Y, and Jiang R. 2006. Photogrammetry applications in routine bridge inspection and historic bridge documentation. Paper no. 06-0575, 86th Annual Meeting, TRB, Washington, DC.

458. Tyson J, Schmidt T, and Galanulis K. 2003. 3-D measurements enhance mechanical studies. *Biophotonics Intl*, 10(10), 50–53.

459. Doer KU, Hutchinson TC, and Kuester F. 2005. A methodology for image-based tracking of seismic-induced motions. *Smart Sensor Technology and Measurement Systems*, SPIE 5758.

460. Chang CC and Ji Y. 2005. High precision videogrammetric technique for structural dynamic response measurement. *Sensors and Smart Structures Technologies for Civil, Mechanical, and Aerospace Systems*, SPIE 5765.

461. Poudel UP, Fu G, and Ye J. 2005. Structural damage detection using digital video imaging technique and wavelet transformation. *J Sound Vib*, 286, 869–895.

462. Ichinose K, Yoshida S, Gomi K, Taniuchi K, and Fukuda K. 2003. Monitoring of surface state of carbon steel bar under cyclic loading. In *Structural Health Monitoring*, FK Chang (ed.). DEStech, Lancaster.

463. Tateishi K and Hanji T. 2004. Low cycle fatigue strength of welded joints in extreme large strain field. *Proc IABMAS'04 Bridge Maintenance, Safety, Management and Cost*, Kyoto. Taylor & Francis, London.

464. Yoshida S, Ishii H, Ichinose K, Gomi K, and Taniuchi K. 2004. Observation of optical interferometric band structure representing plastic deformation front under cyclic loading. *Jpn J Appl Phys*, 43(8A), 5451–5454.

465. Kreith F and Bohn MS. 1993. *Principles of Heat Transfer*, 5th Ed. West Publishing, St. Paul.

466. Meca FJ, Sanchez FJ, Quintas M, Calvo JA, Rodriguez D, and Sainz P. 2000. Main limitations in infrared temperature measurement of train's hot points. *Nondestructive Evaluation of Highways, Utilities, and Pipelines IV*, SPIE 3995.

467. Shifflett P. 1992. Blackbody fired on silica fiber. US Patent 5,164,999.

468. Brown JR, Hamilton III HR. 2004. Infrared thermography inspection procedures for the non-destructive evaluation of FRP composites bonded to reinforced concrete. *Proc Structural Materials Technology VI, an NDT Conference*, Buffalo.

469. Miller BC, Maser KR, and McGrath L. 2004. Active infrared thermography for bridge deck delamination detection. *Proc Structural Materials Technology VI, an NDT Conference*, Buffalo.

470. Lesniak JR, Bazile DJ, and Boyce BR. 1999. Coating tolerant thermography inspection system. *Nondestructive Evaluation of Bridges and Highways III*, SPIE 3587.

471. Duke Jr JC, Miceli M, Horne MR, and Mehl NJ. 2006. Infrared thermal imaging for NDE of FRP bridge beams and decks. *NDE Conf on Civil Engineering*, American Society for Nondestructure Testing, St. Louis.

472. Cramer KE and Winfree WP. 2000. Thermographic imaging of material loss in boiler water-wall tubing by application of scanning line source. *Nondestructive Evaluation of Highways, Utilities, and Pipelines IV*, SPIE 3995.

473. Shepard SM, Wang D, Lhota JR, Rubaduex BB, and Ahmed T. 2002. Reconstruction and enhancement of thermographic sequence data. *Nondestructive Evaluation and Health Monitoring of Aerospace Materials and Civil Infrastructures*, SPIE 4704.

474. Martin RE and Gyekenyesi A. 2002. Pulsed thermography of ceramic matrix composites. *Nondestructive Evaluation and Health Monitoring of Aerospace Materials and Civil Infrastructures*, SPIE 4704.

475. Aktas E, Seaver M, Nichols JM, and Trickey T. 2007. Detecting delamination and core crushing in a sandwich composite wing. In *Structural Health Monitoring*, FK Chang (ed.). DEStech, Lancaster.

476. Fuchs PA and Washer GA. 2004. Infrared thermographic bridge deck inspection on the poplar street bridge, St. Lois, missouri. *Proc Structural Materials Technology VI, an NDT Conference*, Buffalo.

477. Wenzlick JD and Fuchs PA. 2006. Infrared system field testing on poplar street bridge. *NDE Conf on Civil Engineering*, American Society for Nonstructure Testing, St. Louis.

478. Henault JW and Larsen DA. 2006. Thermal imaging of hot mix asphalt paving projects in Connecticut. Paper no. 06-0185, 86th Annual Meeting, TRB, Washington, DC.

479. Del Grande NK and Durbin PF. 1999. Delamination detection in reinforced concrete using thermal inertia. *Nondestructive Evaluation of Bridges and Highways III*, SPIE 3587.

480. Snell LM and Cheek MA. 2006. Estimating fresh concrete temperatures infrared thermometers provide rapid check. *Concrete Intl*, (June).

481. Hawkins GF, Johnson EC, and Nokes JP. 1999. Detecting manufacturing flaws in composite retrofits. *Nondestructive Evaluation of Bridges and Highways III*, SPIE 3587.

482. Sebesta S, Scullion T, Liu W, and Harrison G. 2006. Thermal imaging hot-mix paving operations for quality assessment: The state of the practice in texas. Paper no. 06-2467, 86th Annual Meeting, TRB, Washington, DC.

483. Casler C. 2005. Infrared thermography a tool for locating moisture infiltration in masonry walls. *Concrete Intl*, (August).

484. Cosgriff LM, Bhatt R, Choi SR, and Fox DS. 2005. Thermographic characterization of impact damage in SiC/SiC composite materials. *Nondestructive Evaluation and Health Monitoring of Aerospace Materials, Composites and Civil Infrastructure IV*, SPIE 5767.

485. Phillips L, Willoughby K, and Mahoney J. 2004. Infrared thermography revolutionizes hot-mix asphalt paving. *Proc Structural Materials Technology VI, an NDT Conference*, Buffalo.

486. Cunningham PR, Dulieu-Barton JM, and Shenoi RA. 2002. Damage location and identification using infra-red thermography and thermoelastic stress analysis. *Nondestructive Evaluation and Health Monitoring of Aerospace Materials and Civil Infrastructures*, SPIE 4704.

487. Han X, Lu J, Islam MS, Li W, Zeng Z, Favro LD, Newaz G, and Thomas RL. 2005. Sonic infrared imaging NDE. *Sensors and Smart Structures Technologies for Civil, Mechanical, and Aerospace Systems*, SPIE 5765.

488. Bremond P and Potet P. 2001. Lock-in thermography: A tool to analyse and locate thermomechanical mechanisms in materials and structures. *Thermosense XXIII*, SPIE 4360-76.

489. Songling H, Luming L, Haiqing Y, and Keren S. 2003. NDE of composites delamination by infrared thermography. *Nondestructive Evaluation and Health Monitoring of Aerospace Materials and Composites II*, SPIE 5046.

490. Birman V and Byrd LW. 1999. Theoretical foundations using thermography for nondestructive detection of matrix cracks in woven ceramic matrix composites. *Proc 2nd Intl Workshop on Structural Health Monitoring*. Stanford. Technomic, Lancaster.

491. Han X, Islam M, Li W, Loggins V, Lu J, Zeng Z, Favro LD, Newaz G, and Thomas RL. 2003. Acoustic chaos in sonic infrared imaging of cracks in aerospace components. In *Structural Health Monitoring*, FK Chang (ed.). DEStech, Lancaster.

492. Bates D, Lu D, Smith GF, and Hewitt J. 2000. Rapid NDT of composite aircraft components using lock-in ultrasonic and halogen lamp thermography. *Nondestructive Evaluation of Aging Materials and Composites IV*, SPIE 3993.

493. Bencic TJ and Eldridge JI. 2005. Smart coatings for health monitoring and nondestructive evaluation. *Advanced Sensor Technologies for Nondestructive Evaluation and Structural Health Monitoring*, SPIE 5770.

494. Chen G, Lee KN, and Tewari SN. 2006. Emission spectroscopy analysis for the non-destructive evaluation of the health of thermal barrier coatings. *J Mater Sci*, 41, 6855–6860.

495. NASA. 2005. *Delamination-Indicating Thermal Barrier Coatings*, NASA Technical Briefs LEW-17929-1/30-1.

496. Arnaud L and Houel A. 2006. Fatigue damage and cracking of asphalt pavement on orthotropic steel bridge deck. Paper no. 06-0534, 86th Annual Meeting, TRB, Washington, DC.

497. Schueller R, Joshi SP, and Schulte K. 1997. Damage detection in CFRP by resistance measurements at the boundary. In *Structural Health Monitoring: Current Status and Perspectives*, F.-K. Chang, (ed.). Technomic Publishing, Lancaster.

498. Bois C and Hochard C. 2004. Damages measurements and predictive model of damages for a complete monitoring system. *Proc 2nd European Workshop on Structural Health Monitoring*. DEStech, Munich.

499. Wang D and Chung DD. 2006. Comparative evaluation of the electrical configurations for the two-dimensional electric potential method of damage monitoring in carbon fiber polymer–matrix composite. *Smart Mater Struct*, 15, 1332–1344.

500. Brophy JW. 1988. *Automated Imaging System for Bridge Inspection*, FHWA/RD-87/090, US Department of Transportation, Federal Highway Administration.

501. Roenelle P, Valade M, and Abraham O. 2006. Impact echo B-scan for the detection of void in tendon duct: An experimental parametric study on a real size test site. *NDE Conf on Civil Engineering*, American Society for Nondestructure Testing, St. Louis.

502. Iyer S, Schokker A J, and Sinha SK. 2003. Ultrasonic C-scan imaging preliminary evaluation for corrosion and void detection in posttensioned tendons. *Transportation Research Record 1827 Highway Pavements and Structures Maintenance and Security*, Paper no. 03-2667.

503. Breen C, Guild F, and Pavier M. 2006. Impact damage to thick carbon fibre reinforced plastic composite laminates. *J Mater Sci*, 41, 6718–6724.

504. Streicher D, Algernon D, Wöstmann J, Behrens M, and Wiggenhauser H. 2006. Automated NDE of post-tensioned concrete bridges using radar, ultrasonic echo and impact-echo. *NDE Conf on Civil Engineering*, American Society for Nondestructure Testing, St. Louis.

505. Good MS, Harris Jr RV, Skorpik JR, Pardini AF, Smith CM, Bowey RE, Diaz AA, Burghard BJ, Judd KM, and Adamson JD. 2000. Development of enhanced ultrasonic imaging for *in situ* inspection of tension-stressed, threaded fastener. *Nondestructive Evaluation of Highways, Utilities, and Pipelines IV*, SPIE 3995.

506. Zhou Z, Duan Z, Jia Z, and Ou J. 2004. New kind of structural fatigue life prediction smart sensor. *Smart Sensor Technology Measurement Systems*, SPIE 5384.

507. Inada H, Okuhara Y, and Kumagai H. 2004. Experimental study on structural health monitoring of RC columns using self-diagnosis materials. *Sensors and Smart Structures Technologies for Civil, Mechanical and Aerospace Systems*, SPIE 5391.

508. Gentile CT, Wallace M, Avalon TD, Goodman S, Fuller R, and Hall T. 1992. Angular displacement sensors. US Patent 5,086,785.

509. Guan X, Ou J, and Han B. 2003. Carbon fiber reinforced cement sensor and its produce technology. In *Structural Health Monitoring*, FK Chang (ed.). DEStech, Lancaster.

510. Wang S, Kowalik DP, and Chung DD. 2004. Self-sensing attained in carbon-fiber-polymer–matrix structural composites by using the interlaminar interface as a sensor. *Smart Mater Struct*, 13, 570–592.

511. Park JB, Okabe T, Yoshimura A, Takeda N, and Curtin WA. 2002. Quantitative evaluation of the electrically conductive internal network in CFRP composites. *Nondestructive Evaluation and Health Monitoring of Aerospace Materials and Civil Infrastructures*, SPIE 4704.

512. Hirano Y and Todoroki A. 2004. Damage monitoring of woven CFRP laminates using electrical resistance change. *Proc 2nd European Workshop on Structural Health Monitoring*. DEStech, Munich.

513. Todoroki A, Tanaka Y, and Shimamura Y. 1999. Response surface for delamination monitoring of graphite/epoxy composites using electric resistance change. *Proc 2nd Intl Workshop on Structural Health Monitoring*. Stanford. Technomic, Lancaster.

514. Lataste JF, Breysse D, Sirieix C, and Frappa M. 2004. Nondestructive evaluation of damage and cracks in reinforced concrete by electrical resistivity measurement. *Proc IABMAS'04 Bridge Maintenance, Safety, Management and Cost*, Kyoto. Taylor & Francis, London.

515. Angelidis N and Irving PE. 2003. Self-sensing of impact damage in carbon fiber reinforced polymers by means of electrical potential changes. In *Structural Health Monitoring*, FK Chang (ed.). DEStech, Lancaster.

516. Thostenson ET, Chou TW. 2006. Carbon nanotube networks: Sensing of distributed strain and damage for life prediction and self healing. *Adv Mater*, 18, 2837–2841.

517. Loh KJ, Kim J, Lynch JP, Kam NW, and Kotov NA. 2007. Multifunctional layer-by-layer carbon nanotube–polyelectrolyte thin films for strain and corrosion sensing. *Smart Mater Struct*, 16, 429–438.

518. Nokken MR and Hooton RD. 2006. Electrical conductivity testing. *Concrete Intl*.

519. Chung D. 2003. *Composite Materials: Science and Applications*. Springer, London, UK.

520. Poh S and Baz A. 1996. Distributed sensor for rectangular plates. *Smart Sensing, Processing and Instrumentation*, SPIE 2718.

521. Matsuzaki R and Todoroki A. 2004. Passive wireless strain monitoring of tire using capacitance change. *Sensors and Smart Structures Technologies for Civil, Mechanical and Aerospace Systems*, SPIE 5391.

522. Chen PW and Chung DD. 1993. Carbon fiber reinforced concrete for smart structures capable of non-destructive flaw detection. *Smart Mater Struct*, 2, 22–30.

523. Muto N, Yanagida H, Nakatsuji T, Sugita M, Ohtsuka Y, and Aral Y. 1992. Design of intelligent materials with self-diagnosing function for preventing fatal fracture. *Smart Mater Struct*, 1, 324–329.

524. Kumagai H, Shiba K, Suzuki M, Sugita M, and Matsubara H. 2003. Fiber reinforced composites as self-diagnosis materials for concrete structures. *Smart Systems and Nondestructive Evaluation for Civil Infrastructures*, SPIE 5057.

525. Yoshitake K, Shiba K, Suzuki M, Sugita M, and Okuhara Y. 2004. Damage evaluation for concrete structures using fiber reinforced composites as self-diagnosis materials. *Smart Sensor Technology Measurement Systems*, SPIE 5384.

526. Hou L and Hayes SA. 2002. A Resistance-based damage location sensor for carbon-fibre composites. *Smart Mater Struct*, 11, 966–969.

527. Lalli JH, Hill A, Subrahmanyan S, Davis B, Mecham J, Goff RM, and Claus RO. 2005. Metal rubber sensors. *Smart Sensor Technology and Measurement Systems*, SPIE 5758.

528. Park JM, Jung JK, Kim SJ, and Lee JR. 2005. Nondestructive damage sensitivity of functionalized carbon nanotube and nanofiber/epoxy composites using electrical resistance measurement. *Smart Sensor Technology and Measurement Systems*, SPIE 5758.

529. Kang I, Schulz MJ, Kim JH, Shanov V, and Shi D. 2006. A carbon nanotube strain sensor for structural health monitoring. *Smart Mater Struct*, 15, 737–748.

530. Nakamura M, Saito T, and Sugimoto T. 2004. Soundness investigation of suspender ropes of the suspension bridge which has passed 20 years. *Proc IABMAS'04 Bridge Maintenance, Safety, Management and Cost*, Kyoto. Taylor & Francis, London.

531. Schempf H, Mutschler E, Goltsberg V, and Crowley W. 2003. GRISLEE: Gasmain repair and inspection system for live entry environments. *Intl J Robot Res*, 22, 7–8.

532. Quattrone RF, Berman JB, Trovillion JC, Feickert CA, Kamphaus JM, White SR, Giurgiutiu V, and Cohen GL. 2000. *Investigation of Terfenol-D for Magnetostrictive Tagging of Fiber-Reinforced Polymer Composites*, ERDC/CERL TR-00-46, US Army Core of Engineers.

533. Singh V, Lloyd GM, and Wang ML. 2003. Quantitative validation testing of magnetoelastic corrosion sensing for bridge cables. *Smart Systems and Nondestructive Evaluation for Civil Infrastructures*, SPIE 5057.

534. Sakai H, Uesugi T, Yasumori H, Sumitro S, and Wang ML. 2004. Long-term monitoring on external post-tensioned box girder bridges by utilizing EM sensory technology. *Proc IABMAS'04 Bridge Maintenance, Safety, Management and Cost*, Kyoto. Taylor & Francis, London.

535. Bruns M, Dilger K, Nitschke-Pagel T, and Wohlfahrt H. 2004. Combined application of different measurement techniques for the characterisation of cyclically loaded steel weldments. *Proc 2nd European Workshop on Structural Health Monitoring*, DEStech, Munich.

536. Feynman RP, Leighton RB, and Sands M. 1963. *The Feynman Lectures on Physics Mainly Electromagnetism and Matter*, Vol II. Addision-Wesley, Reading, MA.

537. Lachmann C, Nitschke-Pagel T, and Wohlfahrt H. 1999. Nondestructive characterization of fatigue processes in cyclically loaded welded joints by the Barkhausen noise method. *Proc 2nd Intl Workshop on Structural Health Monitoring*. Stanford. Technomic, Lancaster.

538. Okuhara Y, Jang BK, Matsubara H, and Sugita M. 2003. Fiber reinforced composites as self-diagnosis materials for memorizing damage histories. *Smart Systems and Nondestructive Evaluation for Civil Infrastructures*, SPIE 5057.

539. Lee HY, Peng Y, and Shkel YM. 2005. Monitoring liquid and solid polymers through electroactive response. *Sensors and Smart Structures Technologies for Civil, Mechanical, and Aerospace Systems*, SPIE 5765.

540. Pethrick RA, Hayward D, McConnell BK, and Crane RL. 2005. Use of high and low frequency dielectric measurements in the NDE of adhesively bonded composite joints. *Nondestructive Evaluation and Health Monitoring of Aerospace Materials, Composites and Civil Infrastructure IV*, SPIE 5767.

541. Clark P. 2004. Concrete evidence: Sampling peformance of non-destructive rotary percussion sounding for delaminations in steel reinforced suspended concrete structures. *Proc Structural Materials Technology VI, an NDT Conference*, Buffalo.

542. Clark P. 2001. Device for detecting delaminations and methods for use thereof. US Patent 6,220,097.

543. Norton SJ and Bowler FL. 1993. Theory of eddy current inversion. *J Appl Phys*, 73(2), 501–512.

544. Harfield N, Yoshida Y, and Bowler JR. 1996. Low-frequency perturbation theory in eddy-current. *J Appl Phys*, 80(7), 4090–4100.

545. Burke SK. 1994. Eddy-current inversion in the thin-skin limit: Determination of depth and opening for a long crack. *J Appl Phys*, 76(5), 3072–3080.

546. Rao BP. 2007. *Practical Eddy Current Testing*. Alpha Science International, Oxford.

547. Moon F. 1984. *Magneto-Solid Mechanics*. Wiley, New York.

548. Liu Z, Forsyth DS, Safizadeh MS, Lepine BA, and Fahr A. 2003. Quantitative interpretation of multi-frequency eddy current data by using data fusion approaches. *Nondestructive Evaluation and Health Monitoring of Aerospace Materials and Composites II*, SPIE 5046.

549. Buckley JM, Smith RA, and Skramstad JA. 2003. Transient eddy currents for aircraft structure inspection—An introduction. *Proc 2nd Intl NDT Symp and Exhibition*, Istanbul.

550. Shay I, Zilberstein V, Washabaugh A, and Goldfine N. 2003. Remote temperature and stress monitoring using low frequency inductive sensing. *Nondestructive Evaluation and Health Monitoring of Aerospace Materials and Composites II*, SPIE 5046.

551. Bajjuri S, Hoffman J, Siddojou A, and Meyendorf N. 2004. Development of GMR eddy current sensors for high temperature applications and imaging of corrosion in thick multi-layer structures. *Testing, Reliability, and Application of Micro- and Nano-Material Systems II*, SPIE 5392.

552. Perry AH. 2002. Near DC Eddy current measurement of aluminum multilayers using MR sensor and commodity low cost computer technology. *Nondestructive Evaluation and Health Monitoring of Aerospace Materials and Civil Infrastructures*, SPIE 4704.

553. Jander A, Smith C, and Schneider R. 2005. Magnetoresistive sensors for nondestructive evaluation. *Advanced Sensor Technologies for Nondestructive Evaluation and Structural Health Monitoring*, SPIE 5770.

554. Rempt RD. 2005. Using magnetoresistive (MR) sensors for a sensitivity quantum leap in aircraft nondestructive evaluation (NDE). *Advanced Sensor Technologies for Nondestructive Evaluation and Structural Health Monitoring*, SPIE 5770.

555. Na JK and Franklin MA. 2006. Magnetoresistive sensor based eddy current crack finder. US Patent Application 20060290349.

556. Wincheski RA, Namkung M, and Simpson JW. 2005. Magnetoresistive flux focusing eddy current flaw detection. US Patent 6,888,346.

557. Na JK and Franklin MA. 2006. Magnetoresistive sensor based eddy current crack finder. US Patent Application 20060290349.

558. Goldfine NJ, Clark DC, Walrath KE, Volker W, and Chepolis WM. 2002. Method of detecting widespread fatigue and cracks in a metal structure. US Patent 6,420,867.

559. Rakow A and Chang FK. 2007. *In-situ* sensor design for monitoring fatigue damage in bolted joints. In *Structural Health Monitoring*, FK Chang (ed.). DEStech, Lancaster.

560. Pinnasaud J, Joubert PY, and Lemistre M. 2005. Quantitative magneto optical imager for nondestructive evaluation. *Health Monitoring and Smart Nondestructive Evaluation of Structural and Biological Systems IV*, SPIE 5768.

561. Qaddoumi N, Mirshahi R, Ranu E, Ostahevich V, Zoughi R, McColsky JD, and Livingston R. 1997. Analysis of stress induced fatigue crack detection using microwave open-ended waveguide sensors using higher-order mode approach. In *Structural Health Monitoring: Current Status and Perspectives*. F.-K. Chang, (ed.). Technomic, Lancaster.

562. Nishimura H, Sugiyama T, Okuhara Y, Shin SG, Matsubara H, and Hiroaki Y. 2000. Application of self-diagnosis FRP to concrete pile for health monitoring. SPIE 3985 *Smart Structures and Integrated Systems*, SPIE 3985.

563. Lemistre MB and Placko D. 2004. Evaluation of the performances of an electromagnetic shm system for composite, comparison between numerical simulation, experimental data and ultrasonic investigation, *Health Monitoring and Smart Nondestructive Evaluation of Structural and Biological Systems III*, SPIE 5394.

564. Sun Y, Udpa S, Lord W, and Udpa L. 2000. Numerical modeling and experimental study of motion induced remote field current effect and its application to online inspection and quality examination of rolling metallic strips. *Nondestructive Evaluation of Highways, Utilities, and Pipelines IV*, SPIE 3995.

565. Soumekh M. 1999. *Synthetic Aperture Radar Signal Processing*. Wiley-Interscience, New York.

566. Daniels DJ. 2004. *Ground Penetrating Radar*, 2nd Ed. Institution of Electrical Engineers, London.

567. Millard SG, Shaari A, and Bungey JH. 2002. Resolution of GPR bowtie antennas. *Proc 9th Intl Conf on Ground Penetrating Radar*, Santa Barbara.

568. Kim YJ, Jofre L, De Flaviis F, and Feng MQ. 2004. Microwave subsurface imaging technology for damage detection. *J Eng Mech*, 130(7), 858–866.

569. Pieraccini M and Luzi G. 2002. A high frequency penetrating radar for masonry investigation. *Proc 9th Intl Conf on Ground Penetrating Radar*, Santa Barbara.

570. Khanfar A and Qaddoumi N. 2002. Microwave near-field nondestructive detection and characterization of disbonds in concrete structures. In *Proc Structural Materials Technology V, an NDT Conference*, S Allampalli and G Washer (eds.). Cincinnati.

571. Belli K, Wadia-Fascetti S, and Rappaport C. 2008. Model based evaluation of bridge decks using ground penetrating radar. *Comput-Aided Civ Infrastruct Eng*, 231, 3–16.

572. Cheney M. 1997. Inverse boundary-value problems. *Am Sci*, 85, 448–455.

573. Isakov V. 1998. Inverse problems for partial differential equations, In *Applied Mathematical Science* Vol. 127, JE Marsden and L Sirovich (eds.). Springer, New York.

574. Perkins AD, Amrol JJ, Romero FA, and Roberts RL. 2000. DOT specification development based on ground penetrating radar system performance in measuring concrete cover (reinforcement depth) on new bridge deck construction. In *Structural Materials Technology IV*, S Allampalli, (ed.). Technomic, Lancaster.

575. Franceshetti G and Lanari R. 1999. *Synthetic Aperture Radar Processing*. CRC Press, Boca Raton, FL.

576. Robinson EA. 1982. Spectral approach to gephysical inversion by Lorentz, Fourier and Radon transforms. *Proc IEEE*, 70(9), 1039–1054.

577. Stolt RH. 1978. Migration by Fourier transform. *Geophysics*, 43(1), 23–48.

578. Mast JE. 1993. Microwave pulse-echo radar imaging for the nondestructive evaluation of civil structures, PhD Dissertation, Electrical Engineering, University of Illinois at Urbana-Champaign.
579. Johansson EM and Mast JE. 1994. Three-dimensional ground penetrating radar imaging using synthetic aperture time-domain focusing. *Advanced Microwave and Millimeter-wave Detectors*, SPIE 2275.
580. Scott ML. 1999. Automated characterization of bridge deck distress using pattern recognition analysis of ground penetrating radar data, PhD Dissertation, Civil and Environmental Engineering, Virginia Polytechnic Institude and State University.
581. Davidson NC and Chase SB. 2001. Initial testing of advanced ground penetrating radar technology for the inspection of bridge decks—The HERMES and PERES bridge inspectors. *Smart Systems for Bridges, Structures and Highways*, SPIE 4330.
582. Al-Qadi IL, Lahouar S, and Jiang K. 2004. Measuring rebar cover-depth in rigid pavements using GPR. *Proc Structural Materials Technology VI, an NDT Conference*, Buffalo.
583. Doersken K. 2002. Improved optical positioning for GPR based structural mapping. *Proc 9th Intl Conf on Ground Penetrating Radar*, Santa Barbara.
584. Wang L and Yuan FG. 2003. Imaging of multiple damages in a composite plate by prestack reverse-time migration technique. In *Structural Health Monitoring*, FK Chang (ed.). DEStech, Lancaster.
585. Alver N and Ohtsu M. 2006. Visual identification of voids in concrete by SIBIE procedure. *NDE Conf on Civil Engineering*, American Society for Nondestructure Testing, St. Louis.
586. Riley GJ. 2003. Rebar locators. *Structure*, (July/August), 228–230.
587. Xie X, Zhao Y, and Wang C. 2002. A case study of shanghai historical bank building. *Proc 9th Intl Conf on Ground Penetrating Radar*, Santa Barbara.
588. Sarri A, Manacorda G, and Miniati M. 2002. A novel GPR system for high resolution inspection of walls and structures. *Proc 9th Intl Conf on Ground Penetrating Radar*, Santa Barbara.
589. Jackson DA, Islam M, and Alampalli S. 2000. Feasibility of evaluating the performance of fiber reinforced plastic (FRP) wrapped reinforced concrete columns using ground penetrating radar (GPR) and infrared (IR) thermography techniques. In *Structural Materials Technology IV*, S Allampalli (ed.). Technomic, Lancaster.
590. Feng MQ, De Flaviis F, and Kim YJ. 2002. Use of microwaves for damage detection of FRP-wrapped concrete structures. *J Eng Mech*, 128(2), 172–183.
591. Wolfe RW and Wahbeh AM. 2000. Non-destructive testing challenges for the california toll-bridge seismic retrofit program. In *Structural Materials Technology IV*, S Allampalli (ed.). Technomic, Lancaster.
592. Huston DR, Pelczarski N, and Esser B. 2001. Inspection of bridge columns and retaining walls with electromagnetic waves. *Smart Systems for Bridges, Structures, and Highways*, SPIE 4330.
593. Pattan B. 2003. A brief exposure to ultra-wideband signaling. *Microw J*, 46(12), 104.
594. Barrett TW. 2001. History of ultra wideband communications and radar: Part I, UWB communications. *Microw J*, 44(1), 22–54.
595. Federal Communications Commission. 2002. *Revision of Part 15 of the Commision's Rules Regarding Ultra-Wideband Transmission Systems*, FCC 02-48, Washington, DC.
596. Olhoeft GR. 2002. The new ground penetrating radar regulatory environment. *Proc 9th Intl Conf on Ground Penetrating Radar*, Santa Barbara.
597. Huston D, Miller J, and Esser B. 2004. Regulation of ultrawideband ground penetrating radar inspection systems. *ASNT Structural Materials Technology (SMT): NDT/NDE for Highways and Bridges 2004 Conference*, Buffalo.
598. Jones SK. 2004. *A Methodology for Measuring Ultra-Wideband (UWB) Emissions to Assess Regulatory Compliance*, ITU-R Study Group: Task Group 1/8 Document no. USTG 1/8-31.
599. Huston DR. 2006. Measurement of electromagnetic characteristics of ground penetration radars. *Project Report, Federal Highway Administration*, DTFH61-03-H0012.
600. Nordstrom R. 1990. Direct tank bottom leak monitoring with acoustic emission. *Mater Eval*, 48, 251–254.

601. Kato H, Nakao S, and Sasaki K. 2007. Temperature- and load-sensing *via* acoustic emission of shape memory alloy particles. *Smart Mater Struct*, 16, 1989–1996.

602. Satpathi D and Wang ML. 1998. Acosutic emission source location in plate girders using Lamb waves. In *Structural Engineering Worldwide*, GL Fenves, NK Srivastava, AH Ang, and RG Domer (eds.). Elsevier, Amsterdam, T-200-4.

603. Luo Y, Zhao GQ, Gu JZ, Zhu JG, Gong RR, and Liu ZT. 2005. New method of planar location of acoustic emission source and its application. In *Structural Health Monitoring*, FK Chang (ed.). DEStech, Lancaster.

604. Sachse W and Grabec I. 1992. Intelligent processing of acoustic emission signals. *Mater Eval*, 50(7), 826–834.

605. Martin T, Read I, and Foote P. 2003. Automated notification of structural damage for structural health management using acoustic emission detection. In *Structural Health Monitoring*, FK Chang (ed.). DEStech, Lancaster.

606. Miller RK. 1990. Acoustic emission testing of storage tanks. *TAPPI J*, 73(12), 105–109.

607. Stephan M, Fröhlich K, Frankenstein B, and Hentschel D. 2005. Statistical signal parameters of acoustic emission for process monitoring. *Testing, Reliability, and Application of Micro- and Nano-Material Systems III*, SPIE 5766.

608. Zarroli J, Karchnak M, and Goodenow T. 2004. Performance characteristics of the robust laser interferometer (RLI) with respect to health monitoring needs for civil, mechanical and aerospace infrastructure elements. *Sensors and Smart Structures Technologies for Civil, Mechanical and Aerospace Systems*, SPIE 5391.

609. Murphy J, Glass JT, Majerowicz S, and Green Jr RE. 1990. Laser interferometric probe for detection of acoustic emission. *Mater Eval*, 48, 714–720.

610. Liu K, Ferguson SM, and Measures RM. 1990. Fiber optic interferometric sensor for the detection of acoustic emission within composite materials. *Opt Lett*, 15(2), 1255–1257.

611. Berthold JW and Roman GW. 2001. Fiber optic acoustic emission sensor. US Patent 6,289,143.

612. Zheng SX, McBride R, Barton SJ, Jones JD, Hale KF, and Jones BE. 1992. Intrinsic optical fibre sensor for monitoring acoustic emission. *Sens Actuat* A, 31, 110–114.

613. Duke JC and Cassino C. 2002. Health monitoring of FRP strengthening applications. In *Proc Structural Materials Technology V, an NDT Conference*, Cincinnati.

614. Butler T, Chen R, Krishnamurthy S, Badcock RA, and Fernando GF. 2004. Damage monitoring of composite materials using a novel fibre optic acoustic emission sensor. *Proc 2nd European Workshop on Structural Health Monitoring*. DEStech, Munich.

615. Ouyang C, Landis E, and Shah SP. 1992. Damage assessment in concrete using acoustic emission. In *Nondestructive Testing of Concrete Elements and Structures*, F Ansari and S Sture (eds.). ASCE, New York.

616. Li X, Marasteanu MO, and Labuz JF. 2006. Observation of crack propagation in asphalt mixtures using acoustic emission. Paper no. 06-1903, 86th Annual Meeting, TRB, Washington, DC.

617. Landis EN and Baillon L. 2002. Experiments to relate acoustic emission energy to fracture energy of concrete. *J Eng Mech*, 128, 6.

618. Mal A, Shih F and Banerjee S. 2003. Acoustic emission waveforms in composite laminates under low velocity impact. *Nondestructive Evaluation and Health Monitoring of Aerospace Materials and Composites II*, SPIE 5046.

619. Johnson DE, Shen HW, and Finlayson RD. 2001. Acoustic emission evaluation of reinforced concrete bridge beam with graphite composite laminate. *Smart Systems for Bridges, Structures, and Highways*, SPIE 4330.

620. Mihashi H, Nomura N, and Niiseki S. 1991. Influence of aggregate size on fracture process zone of concrete detected with three dimensional acoustic emission technique. *Cement Concrete Res*, 21, 737–744.

621. Rossi P, Robert JL, Gervais JP, and Bruhat D. 1990. The use of acoustic emission in fracture mechanics applied to concrete engineering. *Fract Mech*, 35(4/5), 751–763.

622. Nguyen BN, Tucker BJ, Korolev VN, and Khaleel MA. 2005. Damage analysis, experimental characterisation and optimization of discontinuous fiber polymer composites. In *Structural Health Monitoring*, FK Chang (ed.). DEStech, Lancaster.

623. Ziehl PH and Ridge AR. 2006. Evaluation of strengthened reinforced concrete beams: Cyclic load test and acoustic emission methods. *ACI Struct J*, 103(6), 832–841.

624. Šajna A, Kovaè J, Stipanoviæ I, and Mikuliæ D. 2006. Determination of bond and flexural strength of reinforced concrete by acoustic emission. *NDE Conf on Civil Engineering*, American Society of Nondestructure Testing, St. Louis.

625. Maji AK and Kratochvil T. 1994. Acoustic emission monitoring of steel bridges. *Proc ASCE Structures Congress.'94*, NC Baker and BJ Goodno (eds.). April 24–28, Atlanta, Georgia.

626. Maji A K, Satpathi D, and Kratochvil T. 1997. Acoustic emission source location using Lamb wave modes. *J Eng Mech*, 123(2), 154–161.

627. Paulson PO. 1999. Practical continuous acoustic monitoring of suspension bridge cables. Paper no. 990893, TRB, Washington, DC.

628. Le Diouron T and Sumitoro S. 2004. Acoustic monitoring and stress measurements for deterioration assessment in concrete and steel structures. *Proc IABMAS'04 Bridge Maintenance, Safety, Management and Cost*, Kyoto. Taylor & Francis, London.

629. Mayrbaural RM and Camo S. 2004. Guidelines for inspection and strength evaluation of suspension bridge parallel wire cables. *NCHRP Report*, 534, TRB, Washington, DC.

630. Fumo J and Worthington W. 2000. Satellites and solid state electronics test concrete pressure water pipelines. *Nondestructive Evaluation of Highways, Utilities, and Pipelines IV*, SPIE 3995.

631. Searle I, Ziola S, and May S. 1997. Damage detection experiments and analysis for the F-16 In *Structural Health Monitoring: Current Status and Perspectives*, F.-K. Chang, (ed.). Technomic, Lancaster.

632. O'Brien E. 2004. Structural health monitoring of damage tolerant civil aircraft structures. *Proc 2nd European Workshop on Structural Health Monitoring*. DEStech, Munich.

633. Paget CA and Atherton KJ. 2004. Damage assessment in a full-scale aircraft wing by modified acoustic emission. *Proc 2nd European Workshop on Structural Health Monitoring*. DEStech, Munich.

634. Huang Q and Nissen GL. 1997. Structural health monitoring of DC-XA LH2 tank using acoustic emission. In *Structural Health Monitoring: Current Status and Perspectives*, F.-K. Chang, (ed.). Technomic, Lancaster.

635. Paget CA, Atherton K, and O'Brien EW. 2003. Triangulation algorithm for damage location in aeronautical structures. In *Structural Health Monitoring*, FK Chang (ed.). DEStech, Lancaster.

636. Hashida T, Takahashi H, Kobayashi S, and Fukugawa Y. 1990. Fracture toughness determination of concrete by use of breakoff tester and acoustic emission technique. *Cement Concrete Res*, 20, 687–701.

637. Carlos MR, Finlayson RD, Miller RK, Friesel MA, and Klokus LL. 1999. Acoustic emission on-line monitoring systems (AEOLMS). *Proc 2nd Intl Workshop on Structural Health Monitoring*. Stanford. Technomicl, Lancaster.

638. Schoess JN, Busch D, and Menon S. 1999. CH-46 rotor acoustic monitoring system (RAMS). *Proc 2nd Intl Workshop on Structural Health Monitoring*. Stanford. Technomic, Lancaster.

639. Ying L, Jing C, Zuting L, and Xiong W. 2007. Design and development of wireless AE sensor unit applied to ultrasonic test. In *Structural Health Monitoring*, FK Chang (ed.). DEStech, Lancaster.

640. Ohtsu M, Okamoto T, and Yuyama S. 1998. Moment tensor analysis of acoustic emission for cracking mechanisms in concrete. *ACI Struct J*, 95, S9.

641. Nkrumah F, Grandhi G, Sundaresan MJ, and Derriso M. 2005. Identification of failure modes in composite materials. *Nondestructive Evaluation and Health Monitoring of Aerospace Materials, Composites and Civil Infrastructure IV*, SPIE 5767

642. Ohtsu M, Tanaka M, and Tomoda Y. 2004. Maintenance of reinforced concrete by acoustic emission monitoring of corrosion. *Proc IABMAS'04 Bridge Maintenance, Safety, Management and Cost*, Kyoto. Taylor & Francis, London.

643. Ohtsu M, Uchida M, Okamoto T, and Yuyama S. 2002. Damage assessment of reinforced concrete beams qualified by acoustic emission. *ACI Struct J*, 99, S42.

644. Marantidis C, Van Way CB, and Kudva JN. 1994. Acoustic emission sensing in an on-board smart structural health monitoring system for military aircraft. *Smart Sensing, Processing, and Instrumentation*, SPIE 2191.

645. Ma BT, Schadler LS, and Laird C. 1990. Acoustic emission in single filament carbon/polycarbonate and Kevlar/polycarbonate composites under tensile deformation. *Polym Compos*, 11(4), 211–216.

646. Lim MK and Koo TK. 1989. Acoustic emission from reinforced concrete beams. *Mag Concr Res*, 41, 189.

647. Weng CC, Tam MT, and Lin GC. 1992. Acoustic emission characteristics of mortar under compression. *Cement Concrete Res*, 22, 641–652.

648. Kroggel O and Wilhelm T. 2006. 'Stress memory' of concrete—ultrasonic investigations. *NDE Conf on Civil Engineering*, American Society of Nondestructure Testing, St. Louis.

649. Ohtsu M, Uchida M, Okamato T, and Yuyama S. 2002. Damage assessment of reinforced concrete beams qualified by acoustic emission. *ACI Struct J*, 994, 411–417.

650. Shigeishi M and Ohtsu M. 2004. Applicability of damage estimation based on acoustic emission activity under loading to practical reinforced concrete bridge. *Proc IABMAS'04 Bridge Maintenance, Safety, Management and Cost*, Kyoto. Taylor & Francis, London.

651. Colombo S, Main I G, Shigeishi M, and Forde MC. 2004. NDT integrity and load carrying assessment of concrete bridges. *Proc IABMAS'04 Bridge Maintenance, Safety, Management and Cost*, Kyoto. Taylor & Francis, London.

652. Scheerer M, Goss T, Hahn PA, Karlovsky A, Ladstaetter E, and Poenninger A. 2005. Monitoring of failure development during fatigue loading of RTM laminates by means of acoustic emission analysis. In *Structural Health Monitoring*, FK Chang (ed.). DEStech, Lancaster.

653. Yoon DJ, Weiss WJ, and Shah SP. 2000. Assessing damage in corroded reinforced concrete using acoustic emission. *J Eng Mech*, 126(3).

654. Henkel DP and Wood JD. 1991. Monitoring concrete reinforced with bonded surface plates by the acoustic emission method. *NDT&E Intl*, 24(55), 259–264.

655. Liang Y, Sun A, and Ansari F. 2004. Acoustic emission characterization of damage in hybrid fiber-reinforced polymer rods. *J Composites for Construction*, 8(1), 70–78.

656. Vandenplas S, Papy J, Wevers M, and Van Huffel S. 2004. Acoustic emission monitoring using a multimode optical fiber sensor. *Sensors and Smart Structures Technologies for Civil, Mechanical and Aerospace Systems*, SPIE 5391.

657. Sundaresan MJ, Kemerling J, Nkrumah F, Grandhi G, and Shulz MJ. 2004. Evaluation of a scalable structural health monitoring system based on acoustic emission testing. *Proc 2nd European Workshop on Structural Health Monitoring*. DEStech, Munich.

658. Chen HL and He Y. 2001. Analysis of acoustic surface waveguide for AE monitoring of concrete beams. *J Eng Mech*, 127(1), 1–10.

659. Caneva C, Domenichini F, and Sarasini F. 2004. Structural integrity monitoring of composite laminates using embedded acoustic emission polymeric transducers. *Proc 2nd European Workshop on Structural Health Monitoring*. DEStech, Munich.

660. Abdel-Ghaffar AM, Leahy RM, Masri SF, and Synolakis CE. 1992. A feasibility study for a concrete core tomographer. In *Nondestructive Testing of Concrete Elements and Structures*, F Ansari and S Sture (eds.). ASCE, New York.

661. Speck M, Wolter KJ, Daniel D, and Danszcak M. 2004. Application of computer tomography in microelectronic packaging. *Testing, Reliability, and Application of Micro- and Nano-Material Systems II*, SPIE 5392.

662. Saleh H, Goni JJ, and Washer G. 2002. Radiographic inspection of post-tensioning tendons using high energy x-ray machine. In *Proc Structural Materials Technology V, an NDT Conference*, S Allampalli and G Washer (eds.). Cincinnati.

663. Bentz DP, Halleck PM, Grader AS, and Roberts JW. 2006. Water movement during internal curing. *Concrete Intl*, 28(10), 39–45.

664. Ruderman GA. 2005. Health management issues and strategy for air force missiles. *First Intl Forum on Integrated System Health Engineering and Management in Aerospace*, Napa.
665. Liu CT. 1999. Application of real-time x-ray technique to monitor damage process in a particulate filled elastomer. In *Proc 2nd Intl Workshop on Structural Health Monitoring*, Stanford, FK Chang (ed.), Stanford, Technomic Press, Lancaster.
666. Thieberger P, Mariscotti MA, and Ruffolo M. 2006. Simulation program for reinforced concrete tomography with gamma-rays. *NDE Conf on Civil Engineering*, American Society of Nondestructure Testing, St. Louis.
667. Dugan ET, Jacobs AM, Shedlock D, and Ekdahl D. 2004. Detection of defects in foam thermal insulation using lateral migration backscatter X-ray radiography. *Penetrating Radiation Systems and Applications VI*, SPIE 5541.
668. Niemann J, Habel W, and Hille F. 2005. Complex monitoring system for long-term evaluation of prestressed bridges in the new Lehrter Bahnhof in Berlin. Keynote lecture at the *2nd Intl Conf Reliability, Safety and Diagnostics of Transport Structures and Means* at the University of Pardubice/Czech, July 7–8, 2005, pp. 248–256.
669. Selker, JS, Thévenaz L, Huwald H, Mallet A, Luxemburg W, van de Giesen N, Stejskal M, Zeman J, Westhoff M, and Parlange MB. 2006. Distributed fiber-optic temperature sensing for hydrologic systems, *Water Resour. Res.*, 42, W12202.
670. Kishida K et al. 2008. US Patent 7,356,238.
671. Nevius SG. 1996. US Patent 3,249,854.

4

Data Acquisition, Conditioning, and Sensor Networking

Acquiring and processing SHM sensor data is a multistep process. Correctly implementing steps at both the detail and systemic level is crucial to the success of the overall SHM effort. Figure 4.1 shows a typical and relatively simple data processing sequence. Multichannel and heterogeneous multisensor data generally involve more complicated sequences and architectures [1]. Real-time monitoring systems require performing data processing steps rapidly and in an automated fashion that removes most of the detail-level operations from the end user [2–5]. This chapter discusses some of the initial and intermediate steps in the overall process, including analog signal conditioning, multiplexing, ADC, data logging, and data transmission networks.

4.1 Amplitude Range and Quantization

Most SHM data originate in analog form. ADC samples and converts analog data into a digital form, usually as binary integers. Digitization facilitates exploiting the convenience and power of modern computer hardware and software.

The initial digital format of sampled data is usually a binary integer. The most common format is *2's Complement*. An advantage of the 2's Complement format is that it eases rapid integer arithmetic. Other binary integer formats occasionally appear in practice due to specialized capabilities. An example is the *Gray code* format, which is useful for binary counting operations. Only one bit changes between counts when using Gray code.

As with almost all experimental sampling techniques, ADC is imperfect. ADC limitations in the ranges and resolutions of the amplitude, along with the time and/or spatial domains, constrain the amount of collectible information. Using ADC to capture most of the information in the original analog data, without distortion, requires understanding and managing these limitations. Improperly applied ADC techniques can produce faulty and/or superfluous data.

The ratio of the largest measurable amplitude of the signal to the smallest measurable amplitude is a primary ADC performance parameter, known as the *dynamic range*. For example, if a signal contains data with a minimal measurable resolution of 0.001 V over a ± 5 V range, then the dynamic range is 10,000 or 80 dB. The exponent of the binary representation, that is, the number of bits, provides digital quantification of the dynamic range. ADCs are typically 8-bit (256 bins), 12-bit (4096 bins), 16-bit (65,536 bins), or 24-bit (16,777,216 bins). The dynamic range of a 16-bit ADC is 65,536 or 96 dB. Adding 1-bit of dynamic range to an ADC increases the quantization *signal to noise ratio* (SNR) by approximately 6 dB [6]. As an example of the impact of dynamic range on system performance, Casciati et al. used

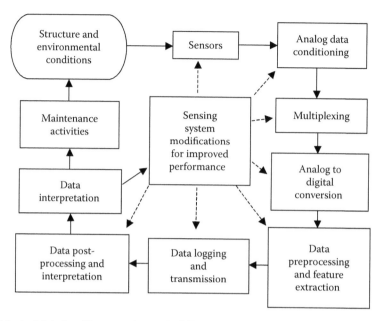

FIGURE 4.1   Principal data handling steps in a typical SHM system.

10-bit ADCs on a wireless telemetry system for long-span bridge vibration monitoring [7]. 10-bit resolution proved to be inadequate and severely limited the performance of the overall data acquisition and telemetry system. The appearance of inexpensive ADCs with 16-bit and 24-bit resolutions alleviates many difficulties with dynamic range limitations. Wang et al. demonstrated a wireless bridge monitoring system with 16-bit ADCs and a top sampling rate of 100 kHz [8]. These ADCs seemed to be sufficient for typical structural monitoring applications. Some specialized systems (telephones for example) use a nonlinear mapping of analog amplitude versus digital bin value to extend the effective dynamic range of a low-bit count ADC.

The selection of ADC hardware is often a cost/performance compromise. ADCs with larger numbers of bits offer more precision and dynamic range, but operate at slower speeds, tend to cost more, and produce data streams that require more storage space. *Successive approximation* ADCs compare the input voltage with an internally generated digital variable voltage. The measured datum is the value of the internal voltage that equals the input voltage. *Flash conversion* ADCs also use a comparison principle, but instead use a parallel array of comparators operating at preset voltages. This technique allows for very high speed data acquisition, but tends to limit the resolution. *Integrating* ADCs use the time required to charge and discharge a capacitor as an indication of voltage. Integrating ADCs can be made with a high (24-bit) resolution, but tend to be slower than other ADCs. *Sigma–delta* ADCs use an over-sampling technique.

Acquiring high-quality digital data requires a good match between the amplitude and frequency ranges of the analog data with the capabilities of the ADC. When the match to the raw data is inadequate, it is often possible to transform the signals to more suitable forms by analog signal conditioning. These techniques include amplification, DC offset shifting, DC offset removal by high-pass filtering (AC coupling), filtering to remove unwanted frequency components, and occasionally selectively amplifying frequency components by a process known as *prewhitening*. As an example, Figure 4.2 shows a typical digitized time history that

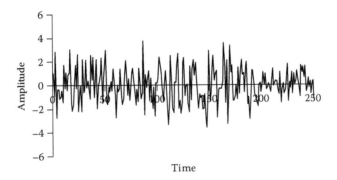

FIGURE 4.2 Random time history with acquisition error producing a hashy appearance due to inadequate amplitude quantization.

appears to have minimal acquisition errors. When the amplitude range of the data exceeds that of the ADC, data clipping and possibly damage to the ADC may occur. Figure 4.3 shows a time history with amplitude clipping errors. If the amplitude range of the data is too small, then the digital resolution of the ADC can be a concern. Amplitude quantization errors may result. Figure 4.4 shows a time history with a hashy appearance due to insufficient amplitude resolution.

Amplitude histograms can assess ADC amplitude performance. Figure 4.5a–c shows histograms of the time histories from Figures 4.2 through 4.4. The histogram of a clipped signal has large values at the clipping edges. When quantization errors occur, the histograms show concentrated values at points corresponding to ADC amplitude quanta. Amplitude histograms can also guide improving the appearance of image data by mapping the dynamic range of the image data to values that match the perception range of the human eye [9].

When the data range is known a priori, it is fairly straightforward to modify the analog data to match the input requirements of the ADC. Collecting data during extreme events, such as hurricanes, floods, overweight vehicles, and earthquakes, can be problematic due to a lack of reliable a priori range information. Strategies to improve dynamic range matching are the following: (1) use multiple independent data acquisition systems and sensors that have different sensitivity ranges; (2) monitor and adjust the range of the ADC as needed; (3) use autoranging of the analog input signal conditioning to adjust automatically the ADC

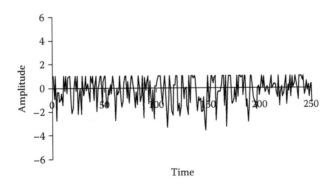

FIGURE 4.3 Random time history with acquisition error due to amplitude clipping.

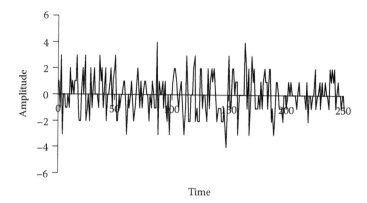

FIGURE 4.4   Random time history with acquisition error due to inadequate amplitude quantization.

resolution; (4) use high-resolution ADCs, for example, 16- or preferably 24-bit; (5) use peak detect electrical circuits, possibly with logarithm-amplitude scaling that blunts the effect of large-amplitude excursions in the raw data [10].

4.2   Digitization in Time and Space

The user typically has control over many temporal (or spatial) ADC options that affect sampling performance, such as sampling rate, duration, and resolution. Methods for quantifying and reducing ADC errors are reasonably well understood [6,11–13]. Information in parts of the signal that lie between or outside the range of the overall sampling intervals can be lost.

The simplest and by far most common ADC technique samples at equal intervals in time, Δt (or space, Δx). The data sampling rate, f_s, is

$$f_s = \frac{1}{\Delta t}. \tag{4.1}$$

The sampling rate and other considerations limit the upper frequency, f_{max}, of collectable information. It is often possible to use the *Nyquist criterion* to select the appropriate sampling rate, that is, the *Nyquist frequency* f_N, to achieve a desired f_{max}. The criterion states that f_s should be more than twice f_{max}, that is,

$$f_s \geq 2f_{max} = f_N. \tag{4.2}$$

It is usually preferable to sample at rates several times faster than f_N. Sampling at a rate less than f_N is usually undesirable. One difficulty with sub-f_N sampling is that a Fourier analysis of the data cannot faithfully report values for the high-frequency signal components. A second, and perhaps more important, issue is that the ADC samples the high-frequency components at a reduced rate and causes high-frequency data to appear falsely as low-frequency signals in a process known as *aliasing*. An example of aliasing occurs in video or motion pictures of a rotating spoked wheel or propeller. If the wheel spins faster than the video frame-sampling rate, a playback of the footage often shows the wheel spinning slowly and possibly in the wrong direction.

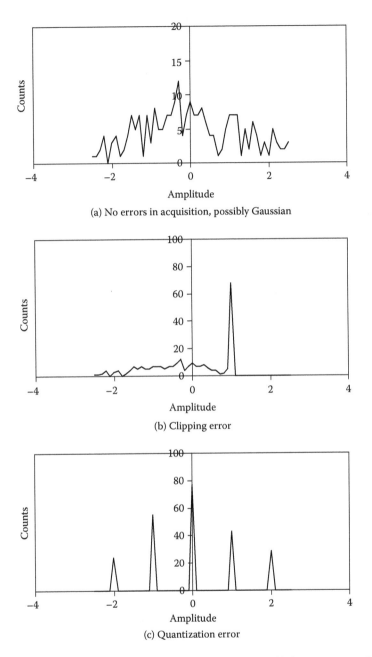

(a) No errors in acquisition, possibly Gaussian

(b) Clipping error

(c) Quantization error

FIGURE 4.5 Histograms of data with (a) no clipping or quantization error, (b) clipping error, and (c) quantization error.

Removing high-frequency content from analog signals prior to ADC can prevent aliasing. One method of high-frequency signal content removal uses an analog low-pass filter with a transmissibility that is flat at low-frequencies and drops off rapidly at higher frequencies. Standard performance specifications for low-pass filters are the *cutoff frequency*, f_c^*, and the *rolloff* (Figure 4.6) [14]. The cutoff frequency is the frequency at which the transmissibility has dropped off by a preset level (usually 20 dB). The rolloff specifies the steepness of the

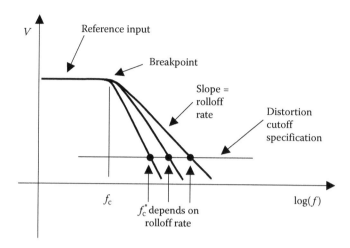

FIGURE 4.6 Rolloff rate and cutoff frequency f_c^* of a low-pass filter. (Adapted from Taylor JL. 1986. Computer-Based Data Acquisition Systems Design Techniques. Inst Soc Amer, Research Triangle Park.)

transmissibility decline with frequency. The standard units for rolloff are the dB reduced per octave, that is,

$$\text{Rolloff (dB)} = 20 \log_{10}\left(\frac{\Delta V}{V_0}\right), \tag{4.3}$$

where ΔV is the difference in the amplitude of transmitted Fourier components (often volts) with a frequency separation equal to an octave (doubling) of frequency. V_0 is a reference amplitude. One octave is a doubling (or halving) of frequency. High-quality *anti-alias filters* have rolloffs in excess of 80 dB/octave. Such filters incorporate high-precision analog electronics and tend to be expensive. A second, often more economical, approach uses a hybrid analog and digital over-sampling system that combines an inexpensive slow rolloff low-pass filter with a high-speed ADC. The signal first passes through the low-pass filter. Next, the ADC samples the filtered signal several times faster than f_N followed by a digital filter/smoother to eliminate the high-frequency data that get past the inexpensive slow rolloff analog filter. A final step often down-selects an interleaved subset of the data to reduce the amount of samples. A third and even less expensive approach combines over-sampling with digital filtering and does not use an analog filter. This can be effective when the higher frequency components of the raw signal are fairly small and tend to rolloff towards the higher frequencies. However, if the high-frequency components are large and do not rolloff, then over sampling without an analog filter may not prevent aliasing.

A difficulty with anti-alias filters is distortion of the time domain appearance of signals, especially transient signals. An idealized example of the difficulty appears in the response from feeding a sharp impulse, that is, $x(t) = \delta(t)$, into a low-pass filter with a perfectly sharp rolloff at a frequency of 0.5. The filter output, $y(t)$, is the convolution of the input with the impulse response of the filter, $h(\tau)$ (Figure 4.7).

$$y(t) = \int_{-\infty}^{\infty} x(t - \tau)h(\tau)\,d\tau. \tag{4.4}$$

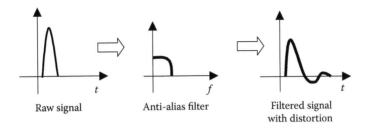

Raw signal   Anti-alias filter   Filtered signal
with distortion

FIGURE 4.7   Distortion of transient signal in time domain by an anti-alias filter.

Transforming the transients in Figure 4.7 into the frequency domain, $Y(f)$, with the Fourier transform and the convolution theorem, takes the form

$$Y(f) = X(f)H(f),\tag{4.5}$$

where

$$X(f) = \int_{-\infty}^{\infty} x(t)\exp(-j2\pi ft)\,dt = \int_{-\infty}^{\infty} \delta(t)\exp(-j2\pi ft)\,dt = 1.\tag{4.6}$$

A filter with a perfectly sharp rolloff at 0.5 has a frequency response function (FRF), $H(f)$, of the form

$$H(f) = \begin{cases} 0 & 0.5 \le f \le -0.5 \\ 1 & -0.5 < f < 0.5 \end{cases}.\tag{4.7}$$

Converting the inverse Fourier transform back to the time domain gives the filter output

$$y(t) = \int_{-\infty}^{\infty} Y(f)\exp(j2\pi ft)\,df,\tag{4.8}$$

$$y(t) = \int_{-0.5}^{.05} \left[\cos(2\pi ft) + j\,\sin(2\pi ft)\right]\,df,\tag{4.9}$$

$$y(t) = \frac{\sin(\pi t)}{\pi t} = \mathrm{sinc}(t).\tag{4.10}$$

The output $y(t)$ is a distorted version of the input $\delta(t)$, that is, $y(t) = \mathrm{sinc}(t)$. In this case, the output also has the curious noncausal property of starting before the application of the input (Figure 4.8). An additional component of time domain distortion by anti-alias filters is the action of frequency-dependent phase shifts. *Bessel or Bessel–Thompson* filters are good choices for minimizing phase distortion and retaining time domain fidelity [15].

Spatial aliasing can also occur with multipoint measurements of distributed quantities. An example is the vibration monitoring of a bridge deck with an array of accelerometers [16]. Increasing the number of sensors to reduce spatial aliasing may not be as practical as switching to scanning instruments with high spatial measurement point densities.

The requirements of high-fidelity time history reconstruction following ADC tend to be more stringent than those imposed by the Nyquist criterion for Fourier component identification. For time history reconstruction, a rule of thumb is that the sampling rate should be at least ten times that of the highest frequency of the time domain features in the data under scrutiny. This is typically several times the Nyquist frequency.

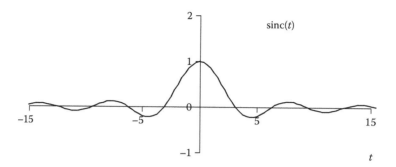

FIGURE 4.8 Normalized sinc function with noncausal behavior.

Just as the smallest sampling increment Δt constrains the ability to sense high-frequency data, the overall sampling period, $T = N \Delta t$, places limits on the measurement of low-frequency information. The minimum frequency sensing limit, f_{min}, is

$$f_{min} = \frac{1}{T}. \tag{4.11}$$

Measuring low-frequency data with confidence can be important in long-term monitoring applications. Monitoring thermal deformations in large structures is a case in point. Daily and seasonal influences require multiyear measurements [17]. The presence of cracks in monumental structures, such as Brunelleschi's dome in Florence (Italy) or the Tomb of the Unknowns in Arlington (USA), is of concern. Determining whether the cracks are stable requires separating daily and seasonal fluctuations in crack width from long-term trends [18].

Certain situations lead to sampling at nonuniform intervals. One is the measurement of very fast transients where direct ADC measurement is not possible with present-day instruments. Radar and time domain reflectometer (TDR) pulses are examples. When the high-speed transient signals repeat with minimal phase jitter, it is possible to use a *sampling oscilloscope* to take advantage of the repeated waveform structure to perform high-speed ADC. The sampling oscilloscope takes only one sample per time history. Each individual data sample covers only a very short time interval. The averaging of repeated samples with the same time lag produces an improved single data point estimate. Sweeping the sample time lag over the duration of the transient while averaging at each point reconstructs the time history. Sampling oscilloscopes with effective measurement frequencies in excess of 50 GHz are available at low cost. Nonuniform sampling also arises in the context of position estimation from inertial accelerometer or gyroscope measurements. Integration of the raw data signals induces low-frequency (LF) drift. Intermittent high-accuracy position measurements, such as with GPS signals or docking at a port, can remove the drift uncertainty. Navigation systems often combine intermittent information with real-time repetitively sampled information by using Kalman filters [19].

4.3 Transducer Registration and Smart Sensors

Registration links measurements with locations on the structure and the time of data collection. Proper registration is an obvious, but nontrivial, requirement for transducers and data channels to provide useful information. Manual registration techniques are fairly simple to implement, but can quickly become tedious and can be prone to error, for even modest sensor number counts. Imaging systems and heterogeneous sensor suites pose additional measurement challenges. Automated and/or semiautomated registration techniques can be useful alternatives.

In addition to registration, quantitative measurements require reliable transducer calibrations. Some transducers have stable calibrations. A single initial calibration may be sufficient. Other transducers require periodic calibration. Self-calibrating sensors are attractive options, but are relatively rare [20]. The integration of MEMS-based sensor technology with networked communications opens up the possibility of smart sensors with embedded calibration, registration, data acquisition, signal conditioning, and communication capabilities. Wireless *ad hoc* sensor networks provide additional capabilities, but also pose calibration challenges due to the need for minimizing power consumption [21].

4.3.1 Transducer Registration

Making effective use of data streams usually requires identifying with certainty the location and timing of data collection. Transducer registration is a key feature of this process. A variety of methods can provide a unique identification for each sensor. One method places a unique serial number on each sensor, and then manually registers the serial number in the registration process. Semiautomated processes, such as laser bar code readers or radio frequency identification (RFID) tags, can help to register the sensor serial number information. A more fully automated approach uses smart sensors that include sensor identification information as part of a header in a data stream, such as with the IEEE 1451 protocol [22].

4.3.2 Spatial Registration

Manual sensor position registration techniques record sensor locations with drawings and/or photographs. Photographs of sensor installations can be surprisingly useful. Photographs form permanent records of installation details, some of which may not be obvious to record a priori. Digital photogrammetry techniques can help with automating registration [23]. Some distributed sensing systems (such as TDR) have inherently built-in location capabilities. Nonetheless, it is still necessary to register the configuration of the distributed sensor relative to the structure. Placing crimps into coaxial TDR cables at known locations, or similar techniques with fiber optic sensors, can provide position registration marks on the data [24]. Global positioning system (GPS) and other EM range finding techniques can also provide a relatively coarse sensor registration [25].

Using image data raises additional issues of spatial registration, especially when attempting to overlay and/or use image data from different sources. Modern medical image processing techniques make quite sophisticated use of various image registration and mapping transformations to achieve high-accuracy overlays from different instruments; such as CAT scans, ultrasound and MRI images [26]. Most of these methods assume that a relatively simple geometric transformation can cause one image to overlay and register with another image. Landmark registration and/or feature correlation techniques determine the parameters required by the geometric transformation to optimize the overlay registration. Liu and Forsyth outline a series of techniques for multisensor image registration of aircraft skins [27]. The rivets served as landmarks for alignment. It is likely that similar techniques will be of use in the overlay registration of SHM data processing.

4.3.3 Temporal Registration

Time registration or *time stamping* requires (1) an accurate clock to determine the correct time, (2) a means of synchronizing clocks, and (3) a means of recording the time. When collecting data at uniform intervals, recording only the start time, sampling interval, and number of samples may be sufficient for time stamping. Intermittent data acquisition requires continual time stamping.

Multichannel data acquisition systems raise additional time registration issues due to the need for across-channel synchronization. *Time division multiplexing* (TDM) alleviates some of the synchronization issues by acquiring multiple analog data channels with a single ADC and a multiplexer. An important operational consideration is that most ADCs run at a fixed speed. The maximum sampling rate per channel for TDM data acquisition is the maximum ADC sampling rate divided by the number of channels. Since the ADC samples the data sequentially, concerns occasionally arise about the temporal sampling delay between channels. When the data acquisition rate is slower than the maximum capability of the ADC, *burst sampling* can reduce interchannel sampling temporal delays. The technique samples each channel sequentially at the highest speed of the ADC for one cycle. The system then waits without sampling data until it is time to initiate another sampling sequence (Figure 4.9). It is possible to enhance data acquisition performance by *smart multiplexing*. For example, Prosser and Perey developed an event-driven multiplexer for acoustic emission (AE) sensing. A smart multiplexer switches between channels in an adaptive pattern that depends on the nature of a trigger event [28]. The concept is that since the transducers physically nearest to the triggering transducer tend to capture the most important spatial information concerning the AE event and the near event, data should be collected with a priority over other more distant transducers.

Certain applications require truly simultaneous measurements. For example, synchronous time fidelity can be crucial to obtaining accurate vibration modal test results. *Sample and hold amplifiers* are one possibility. A sample and hold amplifier captures and holds the analog voltage of a specific data channel at a particular instant for subsequent multiplexing and ADC. For those applications where the synchronization and timing registration is not critical, for example, as in temperature or static deflection measurements, then it may be practical to let the sensor nodes run for longer periods of time without synchronization.

Using multiple ADCs exacerbates data channel synchronization issues. The clock accuracy on many low-cost microprocessors is usually not sufficient to provide long-term accurate time-stamped measurements across an array of autonomous ADCs. Timing signal broadcasting is a viable means of achieving global synchronization. One approach connects all of the data acquisition systems onto a single wired data bus, as with the IEEE 1451.3 protocol or with internet protocols [22]. Ishikawa and Mita found that time synchronization jitters were as high as $1.53\,\mu s$ for sensor separation distances of 190 m when using the fast Ethernet 100BASE-TX protocol and routing hardware [29]. Wireless techniques include broadcasting a local timing signal, using GPS signals or using some of the long-range timing signals that are broadcast by government standards organizations [30–32]. Luo et al. developed a wireless acceleration sensor system where a single source broadcasted a synchronization signal to an array of sensor nodes, each with its own internal clock [33]. The initial synchronization errors were as large as $20\,\mu s$ for transmission ranges of up to 100 m. Subsequent drifts in synchronization among the individual sensor nodes ran as high as 5 ms over a 6 minute sampling interval. Keefer et al. used wireless sensor network (WSN) systems with a custom-designed bidirectional communication to synchronize precisely the sensor nodes for vibration modal analysis [30]. Occasionally there are phenomena with sufficient inherent temporal regularity to serve as unintended synchronization sources.

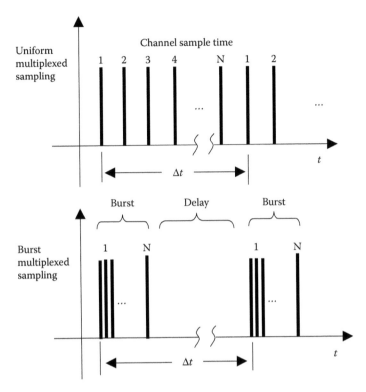

FIGURE 4.9 Uniform and burst multiplexed sampling modes.

An example is a structure that vibrates with highly stable modal frequencies and phases [34]. Whang et al. avoided broadcasting a global synchronization signal by correcting for local processor time skew with a model of the propagation delays in the system [35].

Fiber optic sensing offers several opportunities for sensor multiplexing, including time division, frequency division (Bragg grating), code division, wavelength division, coherence, and polarization division methods [36]. FBGs are often touted for their analog optical multiplexing capabilities. It is possible to multiplex several, if not many, FBG sensors on a single fiber optic cable [37]. This technique exploits the property that FBGs reflect a single wavelength of light from each sensor. Multiple FBGs, each with a different center reflection wavelength, can coexist on a single optical fiber. FBG and AE fiber optic sensor suites can be good candidates for embedded sensing networks in composite elements of airframes [38]. Seo et al. demonstrated an intensity-modulated FBG multiplexing technique [39]. Connection and splice issues continue to hamper the widespread use of too many FBGs on a single cable. Tam et al. measured train track loads and railcar body strains during routine railroad operations with arrays of 6–10 FBGs [40].

An extreme case of multiplexing occurs in the acquisition of 2-D and 3-D image data. Each pixel or voxel is essentially a separate data channel. Avoiding image distortions, particularly with time-varying interlaced image data, requires specialized techniques [9,41].

4.3.4 Smart Sensors and IEEE 1451

Smart sensors have onboard signal processing capabilities. Using smart sensors offloads some of the computational and information logging effort away from centrally located

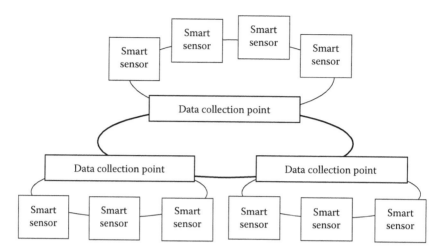

FIGURE 4.10 Hierarchical ring architecture for a smart sensor network. (Adapted from Hunter GW et al. 2005. *First Intl Forum on Integrated System Health Engineering and Management in Aerospace,* Napa, CA.)

processors onto distributed local processors. Smart sensors are particularly well suited for sensor condition monitoring, sensor number registration, triggering, and low-level signal processing. Smart sensors underlie the formation of intelligent sensors networks with architectures that provide sophisticated features, such as sensor self-diagnosis and network reconfiguration. Figure 4.10 shows an example of a smart sensor network with a hierarchical-ring architecture. In an early approach to smart sensing, Mahajan and Figueroa developed the Dynamic Across Time Autonomous-Sensing, Interpretation Model Learning and Maintenance (DATA-SIMLAMT) theory [42].

The following is a brief overview of the IEEE 1451 standards and potential applications. The IEEE 1451 standards comprise a framework for building intelligent distributed sensor systems and cover detailed issues related to the transmission of sensor data and information [22]. As with all technical standards, up-to-date revisions should be checked for detailed information.

IEEE Std 1451.1-1999 *Standard for a Smart Transducer Interface for Sensors and Actuators - Network Capable Application Processor (NCAP) Information Model*—This standard covers the software interface between a variety of transducer types and arbitrary networks.

IEEE Std 1451.2-1997 *Transducer to Microprocessor Communication protocols and Transducer Electronic Data Sheet (TEDS) Formats*—This standard sets the protocol for the storage and transmission of pertinent transducer information, including sensor identification such as manufacturer, model number, and serial number; transducer performance information, such as range limits, physical units, warm-up time, and uncertainty; data conversion information including sampling rates, read and write times, and trigger accuracy; and transducer calibration information. A distinction is the local storage of transducer information. The standard implements an efficient method for encoding the physical units of a transducer in terms of angle measures and the seven fundamental units of length, mass, time, current, temperature, luminous intensity, and amount of substance as a 10-digit integer [44,45].

IEEE 1451.3-2003 *Distributed Multidrop System for Interfacing Smart Transducers*—This standard describes a bus system that places many, possibly hundreds of, transducers on a two-wire bus. The bus simultaneously supplies power to both the sensors and transmits the data. The bus network can sustain sensor bandwidths in hundreds of kilohertz with nanosecond synchronization.

IEEE 1451.4-2004 *Mixed-mode Transducer and Interface*—This is a standard that uses analog sensors with digital TEDS in a single two-wire or four-wire package.

IEEE P1451.5 *Wireless Communication Protocols and TEDS Formats*—This is a proposed standard for wireless transmission of sensor information and data.

IEEE P1451.6 *A High-Speed CANopen-Based Transducer Network for Intrinsically Safe and Non-Intrinsically Safe Applications*—This is a proposed standard for sensor interfaces within the CANopen protocols for interfacing multiple microcontroller applications.

4.4   Data Transmission

Data transmission is an integral component of most SHM systems. Many SHM data transmission practices tend to follow standard data communication techniques, with the use of specialized variants as needed. Analog, digital, optical, acoustic, wired, and wireless techniques are all used in SHM. Figure 4.11 shows a generic communication system with a transmitter, transmission channel, and receiver. Transducers send SHM information to the transmitter. The transmitter encodes information into a suitable format and then sends the information as a signal over the transmission channel. The receiver decodes the signal information and converts it into an understandable format.

It should be noted that imperfections arise in each of the steps of transduction, transmission, reception, and state estimation and degrade the total amount of conveyed information. Nonetheless, it is possible to increase the quality and density of information. Everyone is familiar with qualitative notions of information. Each new bit of data contains information. Data bits with unexpected values carry more information than bits with expected values. For example, the observation that a bridge is standing upright is usually expected. This observation generally carries little new information. The observation of an unexpected bridge collapse carries considerably more information. Quantifying these concepts is particularly important when analyzing the performance limits in data compression, transmission, and error checking.

Information theory quantifies some of the fundamental constraints imposed on all data transmission techniques. These constraints govern data transmission rates, information content, and the integrity of signals sent over a particular channel. Fisher, Shannon, and Weiner discovered many of these concepts in the early middle of the twentieth century [46,47]. A primary notion is that there is a tradeoff between the amount of information transmitted versus the use of redundant signal content to reduce the error rate. Similar concepts

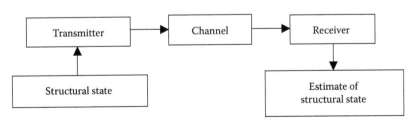

FIGURE 4.11   Generic SHM communication system. (Adapted from Reza FM. 1994. *An Introduction to Information Theory*. Dover publications, Mineola, NY.)

are also useful in statistical decision-making (see Chapter 6). An example of SHM application is the inspection of steel pipes for corrosion. Many steel pipes corrode at predictable rates. Measuring the rate of corrosion with an instrument, such as an ultrasonic thickness probe, usually confirms that the thickness of the pipe conforms to general norms of erosion and corrosion. Each additional measurement usually confirms the previous observation and usually provides little new information. The detection of unexpected accelerated corrosion is an improbable event. Such data provide a high-level of information in the signal. Reducing the rate of inspection can cause an increase in the amount of information in the signal due to the longer sampling intervals causing an increase in the uncertainty of the outcome. In the pipeline corrosion example, avoiding corrosion measurements for several years can lead to unexpected and high-information signal content.

Shannon accounts for noise and uncertainty in transmitted data with the information entropy, S. For discrete distributions of transmitted signal outcomes,

$$S = -\sum_{i=1}^{N} p_i \ln(p_i) \tag{4.12}$$

and for continuous distributions of transmitted signal outcomes

$$S = -\int p(x) \ln[p(x)] \, dx, \tag{4.13}$$

where p_i and $p(x)$ are the probability of occurrence of the ith bit of data. A key feature of information entropy is that it quantifies the notion that transmitting signals with redundancy increases the reliability of correct signal transmission at the expense of transmission rate. The entropy reflects the level of redundancy. Signals with more redundancy have a high-level of predictability, an unbalance in the probability of possible outcomes, and a lower entropy content. Signals with a lower level of predictability contain less redundancy. The probability distribution between the various possible signals is more balanced and the entropy is larger.

It has been noted that the Shannon entropy measure of information theory has the same mathematical form as the entropy measure of statistical mechanics when using the appropriate probability distribution functions (PDFs), for example, the Boltzmann distribution [47,48]. Many believe that the concept of entropy in information theory is essentially the same as that of statistical mechanics. However, the interpretation is different. Increased entropy in statistical mechanics corresponds to increased randomness or disorder in a system. In information theory, an entropy increase corresponds to an increase in the randomness of a signal and the amount of information that it contains. Signals that contain large amounts of redundant information have lower entropy, lower information density, and lower data transmission rates. However, redundant signals are more robust in terms of communication and transmission in the presence of noise than signals with high entropy, minimal redundancy, and high information density [46]. Human language is an example. Human language can be understood, even in the presence of large amounts of noise, other conversations, and moderate levels of garbling. The reason is that human language is highly redundant. It has a submaximal entropy and information density. Completing many words and phrases without hearing the entire word or phrase is often easy. Repeating a sequence of words or phrases reduces the amount of conveyed information, along with the entropy. In spite of language being redundant, using humans as information transmitters and receivers is inherently error prone.

A fundamental result of information theory is a constraint on the rate at which information can be transmitted over a given channel. The *Shannon–Hartley* rule quantifies this result:

$$C = B \log_2 \left(1 + \frac{S}{N} \right), \tag{4.14}$$

where C is the channel capacity, B is the occupied bandwidth, and S/N is the signal to noise ratio (S/N) in terms of a linear power ratio, not dB [49]. In the limiting case where the S/N ratio approaches zero, the noise overwhelms the signal and the channel capacity approaches zero. Conversely, when the noise levels go to zero, the signal can transmit data with unlimited resolution and the channel capacity goes to infinity. (The seemingly anomalous is result of infinite channel capacity is a consequence of the analog signal transmission and a continuum hypothesis that allows for infinite resolution of the signal amplitude.)

4.4.1 Analog Wired Data Transmission

A relatively simple data transmission technique sends analog data over an electrically conductive wire. The analog signal encoding is often voltages, but can also be currents, the frequency of oscillating waves, or more complicated modulation schemes. Using a common ground to connect the receiver and transmitter simplifies the wiring. The system requires only a single signal cable plus a common ground to connect the transmitters and receivers (Figure 4.12).

Single-wire data transmission works well when the signals are large compared to noise levels. When the signals are small, strong noise levels prevent effective communication. Possible solutions range from amplifying the analog signal—but not the noise—before transmission to encoding with error correction schemes to using alternative signal transmission modalities. Using *low-noise amplifiers* (LNAs) between the pickup antenna and transmission cables in microwave communication and radar signal processors is an example of pretransmission amplification. Digital data transmission methods routinely use built-in schemes that can detect and correct many types of errors.

Many noisy environment situations favor the use of differential signal transmission over two parallel wires. *Common mode* noise arises when the physical noise processes cause the same noise to appear on two channels. An example is inductive EM pickup on long cable runs where both cables experience essentially the same fields. A pure differential system, as shown in Figure 4.13, removes common mode noise by analog subtraction and floats independent of a ground (Figure 4.14). Excessive drift of the floating grounds can be a concern with a pure differential system. A practical recommendation is to connect the

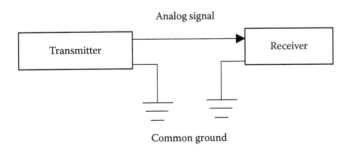

FIGURE 4.12 Single-wire analog data transmission with a common ground.

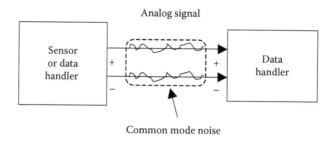

Analog signal

Common mode noise

FIGURE 4.13   Differential data handling transmission connection.

signal lines to the low-level ground of the receiving data handler with a resistor that is about 2000 times larger than the source resistance [50].

The *common mode rejection ratio* (CMMR) quantifies the ability of a differential transmission system to reject common mode noise as the ratio of the power in the signal passed through the differential amplifier to the power of the common mode noise (usually in dB) [51].

Taking full advantage of differential data transmission and acquisition requires the differencing to occur with synchronized data that have the same instantaneous noise amplitude on each line. The traditional approach of using an operational amplifier virtually assures synchronized differencing in the analog domain. Analog differencing is generally more robust than digital. Many commercially available data acquisition systems (particularly low-cost PC-based versions) perform differencing on digital data by a paired sequence of ADC operations on each line. High-frequency (HF) and computer-bus-based noise may not be rejected with such a post-ADC differencing approach. Another issue with post-ADC differencing is that common mode DC offsets cut into the dynamic range of the ADC. It should be noted that since digital data transmission virtually always relies on an analog physical transmission layer, differential transmission of digital data is often superior to single-ended digital transmission.

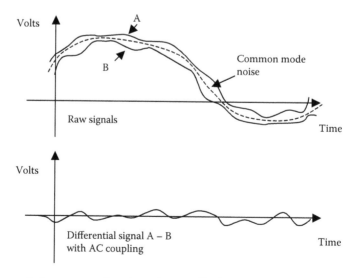

FIGURE 4.14   Removal of common mode noise and mean offset by differential and AC coupling.

The inductive coupling of transient and EM fields into the cables is a principal source of cable noise. Possible remedies to inductive noise pickup on cables are the following: (1) Differential transmission. (2) Spiral wrapping the two wires to form a *twisted pair*. (3) Using coaxial cables. (4) Running a third conductor, that is, a *shield*, parallel to and enveloping the two data lines. (5) Avoiding leaving excess cable lying around in coils, since coils are natural inductive pickups. If a length of excess wire is necessary, then it is best to fold the excess portion in the middle and to coil the doubled strand to create two identical coils with inductances of opposite sign. The opposite signs of the cable inductances add to cancel out the inductive noise pickup. (6) Avoid connectors and connection configurations that act as antennas. Sloppy solder joints with sharp-pointed solder blobs that can act as antennas or be a source for shorts should also be avoided. The use of lead-free *Restriction of Hazardous Substances* (RoHS) compliant solders exacerbates these problems. High-quality, for example, military style, connectors can be bulky and expensive, but can be a worthwhile investment. Smaller high-performance and high-channel count connectors are beginning to appear [52].

An example of the use of high-fidelity cables for long runs of analog signals to data acquisition systems is the long term for wind load and dynamic performance monitoring efforts on the Deer Isle and Luling bridges. Figure 4.15 shows the accelerometer and anemometer layouts on the Deer Isle Bridge in Maine. These systems have successfully collected data under extreme weather conditions for over 20 years [53].

A potential difficulty with using analog voltage as the data signal is that changes in connectors and/or cable resistance can cause spurious voltage changes. Using current rather than voltage as the signal is one method of avoiding this difficulty. Current signals can be more robust than voltage signals. This is a physical consequence of charge conservation in insulated transmission lines. Current-based signals form the basis of the 4–20 mA signal protocol commonly used in manufacturing and process plants, often as the standardized HART (Highway Addressable Remote Transducer) protocol.

FM encodes data as a low-frequency shift in a higher frequency harmonic carrier signal. A primary reason for the success of FM data transmission is that most noise sources affect the amplitude, but not the frequency content of the original signal.

Digital data transmission methods offer distinct advantages over analog methods and have become the dominant format for both the recording and transmitting data. A primary

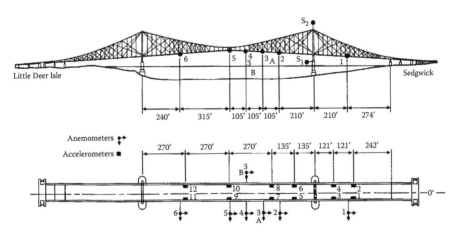

FIGURE 4.15 Accelerometer and anemometer layout on Deer Isle—Sedgwick Bridge, ME (USA). (Picture courtesy Harold Bosch, US Federal Highway Administration.)

advantage is the possibility of introducing redundancy into the signals for the detection and correction of transmission errors [46,54]. Signal redundancy facilitates the transmission of data over noisy channels, and reduces the need for high-fidelity instrument cables. Another advantage of digital data transmission is the ease of multiplexing several data channels onto a single physical transmission channel. A primary disadvantage of digital data transmission is that ADC produces errors and causes the loss of information. Another disadvantage is that digital systems cannot process and transmit data at very high-frequencies. This is not a severe restriction for most standard SHM applications. In this regard, it should be noted that digital and analog transmissions use different definitions of frequency. For analog systems, "frequency" is the inverse of the period of one single harmonic oscillation. In digital systems, "frequency" is the frequency of switching states from high to low or vice versa. Switching from low to high and then back to low is two complete cycles. The analog equivalent is one complete cycle. In principle, the frequency of a digital system is twice that of an equivalent analog system. Practical considerations of harmonic distortion generally prevent digital switching as fast as harmonic oscillations and the ratio factor of two is more of an upper limit for a particular transmission link.

Table 4.1 lists some of the standard wired data buses in use at the time of this writing. Additional buses with increased capabilities that accommodate highly integrated devices, such as the CANbus, will most likely continue to appear and find SHM usage in the future [55,56]. The majority of the standard data buses evolved for use in local area networks, laboratory, and industrial environments. No industry-standard data bus has yet to emerge specifically for SHM. Devices based on the standard digital communication protocols may be the best option because of the wide availability of the various data processing and transmission components and the universality of the associated software [58]. An IEEE 1451 smart sensor network is also an option. As an example configuration, Fu et al. describe a multichannel multisensor system for monitoring the Dafosi Bridge across the Yangtze River [59]. A heterogeneous set of fiber optic and electrical lines communicates the data to a remote monitoring station.

4.4.2   Acoustic and Ultrasonic Data Transmission

In some situations, acoustic or ultrasonic waves are a practical means of transmitting data. Gao et al. used ultrasonic waves to transmit pressure data from isolated sensors that were embedded inside an injection molding machine [60].

4.4.3   Radio Wireless Telemetry

It is often possible to replace wires with radio waves as the data transmission physical layer. Increased flexibility in terms of sensor positioning and rapid sensor deployment, as well as potentially decreased overall SHM system costs favor wireless data transmission. Prior to the advent of miniaturized digital circuitry, most wireless sensor telemetry systems used analog techniques for data modulation. Nachemson and Elfström developed an early (ca. 1971) mechanical load sensing system with analog wireless telemetry [61]. This specialized application measured forces in steel Harrington rods (surgically implanted steel devices that provide mechanical support to the human spine). Wireless telemetry eliminated the need for running wires through the patient's skin. The sensing method used mechanical loads to alter the relative position and mutual inductance of two wire coils. Near-field transformer mode techniques transferred power and signals

TABLE 4.1

Standard Wired Data Buses

Name	Maximum Distance (m)	Maximum Data Rates (kb/s)	Number of Talkers/ Listeners	Comments
4–20 mA		Analog	1 talker, 1 listener	2-, 3-, and 4-wire versions. Many industrial uses
RS-232 Serial	15	115	1 talker	Standard serial port. Cable connectors ambiguous
RS-422 Serial	1,220	115	1 listener 1 talker	Less common than RS-232 and RS-485
RS-485 Serial	1,220	115	10 listeners 32 talkers and listeners	Standard multichannel serial interface
Universal Serial Bus (USB)	5 per cable 15 total	1200	1 talker 127 listeners	Convenient interface, permits hot-plugging
Parallel printer port	15	100	1 talker 1 listener	Drives PC peripheral devices. Can be used as a digital I/O port
GPIB IEEE-488	30	1000	1 talker 15 listeners	Standard lab instrument data bus
Firewire IEEE-1394	4.5	4000	1 talker 63 listeners	Used in image data transfers and Mac platforms
Ethernet	925	100,000	Virtually unlimited	Standard high-speed network protocol

Source: Data from Harrold D. 1998. 4–20 mA Alive and Kicking. *Control Eng*, 13, 109–114.

through the skin. In similar applications, Novak et al. developed wireless analog state sensors that use the frequency shifting of a resonant inductor capacitor (LC) circuit due to the interruption of one arm in a parallel capacitor circuit for crack and corrosion sensing, and Wong and Kim developed wireless strain sensors based on interdigitated capacitive electrodes [62,63].

A wireless data transmission system typically consists of several interconnected subsystems including a physical transmission layer, data protocols, power supply, and network architectures (Figure 4.16). EM wave propagation is the most common physical layer. Other modalities, such as elastic wave propagation, appear in specialized applications. Physical data transmission converts, that is, modulates, data into a physical form for transmission to a remote receiver, transmitting the signal, receiving the physical signal, and demodulating the received signal back into a data format. Typical antenna configurations include those that are printed directly onto the circuit board, attached to the circuit board, or externally mounted. The *antenna gain*, G, is a standard antenna performance parameter that incorporates power conversion, directivity, efficiency, and impedance matching effects [64]. G is the product of the antenna directivity D, the efficiency η, and the impedance mismatch

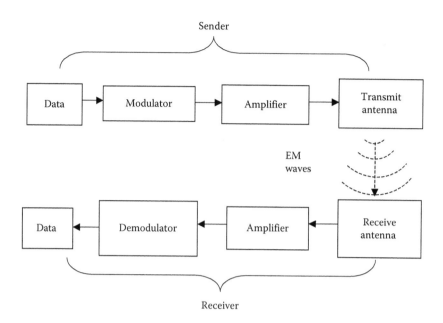

FIGURE 4.16   Schematic of a basic wireless data send and receive system.

between the source and the antenna:

$$G = D\eta\mathrm{M} = \frac{P_{d,\max}4\pi r^2}{P_{\mathrm{rad}}} \cdot \frac{P_{\mathrm{rad}}}{P_{\mathrm{accept}}} \cdot \frac{P_{\mathrm{accept}}}{P_{\mathrm{avail}}} = D\eta\left[1 - |s_{11}|^2\right], \qquad (4.15)$$

where $P_{d,\max}$ is the maximum radiation density, in any direction. r is the distance from the antenna to the observation point. P_{rad} is the total radiated power. P_{accept} is the power accepted from the source. P_{avail} is the power available from the source. s_{11} is the reflection coefficient s-parameter (approximately the equivalent of an FRF). Antennas that are significantly smaller than a wavelength tend to be inefficient, but easier to package than larger antennas.

Practical implementations of standalone digital wireless sensor systems require a tight integration of sensing, processing, telemetry, and power management (Figure 4.17). The technology of wireless sensor data transmission and networking is presently undergoing rapid growth. Lynch and Koh discuss the state-of-the-art in the design of these systems ca. 2006 [65]. Most of these design efforts integrate off-the-shelf components to form sensing nodes and systems [66].

Multiple methods and protocols are available for modulating and demodulating EM waves for use in data transmission (Table 4.2). Common among all EM wireless data transmission systems is the physics of propagating EM waves. With the exception of ultrawideband (UWB) data transmission, the majority of EM data transmission systems operate by transmitting an oscillating sinusoidal wave, known as the *carrier,* and then varying the amplitude, frequency, or phase.

Many of the early digital wireless sensor applications appeared in the aerospace industry. The development and publication of most of the basic concepts occurred by the mid-1980s [68]. Some early embedded digital wireless telemetry sensing applications also appeared in the orthopedic biomechanics literature. In 1989, Schneider et al. used a remotely

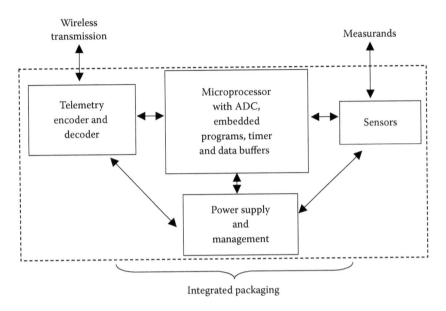

FIGURE 4.17 Elements of a standalone wireless sensor system.

powered wireless digital data acquisition and telemetry unit to measure and transmit loads on interlocking nails used in the repair of femur fractures [69]. Graichen and Bergmann measured the strains in the head of an artificial hip implant with a similar system [70]. Rohlmann et al. measured loads on spinal fixation implants [71]. The system used transformer-style remote wireless power transmission at 4 kHz, a low-power constant-current strain gage data acquisition system, and pulse-modulated radio data telemetry at 150 MHz. Neuzil et al. developed a general-purpose embeddable transducer system that used a transformer-mode remote power supply operating at 130 kHz and used differential phase shift keying of 65 kHz across the transmit coil for data telemetry [72]. Troyk et al. developed a general-purpose bidirectional telemetry implantable sensor for biomedical sensing applications using a 450 kHz carrier frequency [73]. Aoki et al. used a MEMS accelerometer-based system

TABLE 4.2

Comparison of Wireless Communications Protocols

Parameter	Wi-Fi (IEEE 802.11)	Bluetooth (IEEE 802.15.1)	ZigBee (IEEE 802.15.4)
Range	≈100 m	≈10–100 m	≈10 m
Bandwidth/throughput	868 MHz–20 kb/s	2.4 GHz–1 Mb/s	868 MHz–20 kb/s
	915 MHz–40 kb/s		915 MHz–40 kb/s
	2.4 GHz–250 kb/s		2.4 GHz–250 kb/s
Power (current)	400 mA (Tx on)	40 mA (Tx on)	30 Ma (Tx on)
Consumption	20 mA (standby)	0.2 mA (standby)	1 μA (standby)
Battery life	Minutes-hours	Hours-days	Days-years
Protocol stack size	100 kbyte	100 kbyte	32 kbyte
Relative node physical size	Large	Medium	Small
Relative cost/complexity	High	Moderate	Low

Source: Data from Allan R. 2005. Wireless sensors land anywhere and everywhere. *Electron Des*, ED Online ID #10710 (July 21).

for the dynamic monitoring of light poles [74]. Lynch et al. used a similar system for bridge monitoring [75]. Pakzad et al. demonstrated the use of a network of 12 wireless MEMS accelerometers to measure the dynamics of a pedestrian bridge with 36 channels of data, each sampled at 200 Hz [76]. Heo et al. showed that wireless MEMS accelerometers could determine mode shapes and frequencies in vibrating structures [77]. A head-to-head comparison of wireless to hard-wired telemetry using forced vibration testing of the Alamosa Canyon Bridge in New Mexico (USA) indicated that the wireless system performed adequately, but not as well as the competing hard-wired system. The reduced dynamic range resolution of both the MEMS accelerometers and the data acquisition system degraded the performance of the wireless system, especially at lower frequencies. Liu and Yuan indicate that the present state-of-the-art (ca. 2004) in wireless systems is such that the data acquisition rate is insufficient for ultrasound SHM testing [78]. Newer wireless systems with improved capabilities are under development [79,80]. Overly et al. demonstrated wireless nodes with piezoelectric impedance monitoring capability [81].

4.4.4 Eye Diagrams

In spite of most modern data storage and transmission techniques being digital, the physical transmission of signals is almost invariably analog. The *eye diagram* is a standard tool for assessing the ability of a system to transmit digital signals in an analog format (Figure 4.18). The eye diagram plots the measured ability of an analog transmission system to switch back and forth repeatedly from high to low states and vice versa. A large separation distance in the eye corresponds to robust signal transmission. Smaller separation distances indicate difficulty in distinguishing high and low states, and correspond to increased rates of transmission errors.

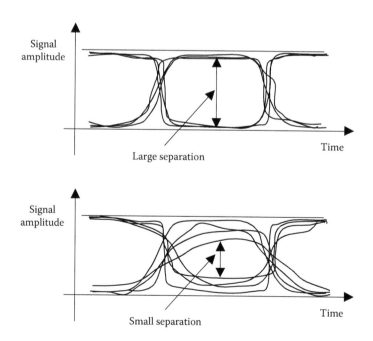

FIGURE 4.18 Eye diagrams for communication systems with large and small signal separation and discrimination.

4.4.5 Power Supply and Consumption

Power supply and consumption are important SHM system design considerations [82]. Most SHM systems use electricity as the basic medium of power transfer. Line power is an excellent option, if available. When the intent is to collect data during extreme events, such as hurricanes or earthquakes, then backup power supplies need to be available in the event of a line power dropout or loss. Remote and geometrically intractable sensing situations often prevent using line power. Possible power supplies for situations when line power is not available include harvesting energy from the ambient environment, capturing remotely transmitted energy, or using onboard power supplies, such as batteries. Energy-efficient designs and operational procedures can ease many issues of power supply. For example, sensors with wireless data transmission typically consume considerably more energy for data transmission and reception than they do for internal operations. A sensor that wakes up intermittently and transmits data and then sleeps again can provide reductions in power consumption of an order of magnitude or more, but still requires circuitry that can readily deliver peak power levels as needed [64]. Deciding when to wake up and transmit is a nontrivial issue. An issue is to manage and avoid simultaneous transmission on a single channel by multiple sensor nodes. A variety of channel sense, arbitrate, and transmit protocols may alleviate many of these issues, with an inevitable tradeoff between channel capacity and overall sensing system power management.

4.4.6 Battery Power Supply

Electrical batteries can be good sources of power. However, the need for recharging and/or replacing batteries is a limitation. Many onboard distributed and wireless sensing systems with onboard batteries operate with very tight energy budgets [83]. Design considerations for battery-powered sensor nodes include considerations of battery internal resistance, the need for recharging, low-remaining energy detection, gravimetric energy density, and cost [64]. D'Souza et al. demonstrated that wireless sensor nodes operating with a 1% On to 99% Off Sleep duty cycle can operate for several months with onboard battery power [84]. Local data preprocessing to increase the data information density reduces data transmission costs, but increases the power consumed by the local processor [85]. Sumners and Champaigne developed a low-power consumption wireless distributed piezoelectric sensor array for impact detection in spacecraft [86]. A key component is the ability to operate in a low-power mode between transient impact events, while retaining the ability to trigger, capture, and transmit data from important transient events.

4.4.7 Ambient Energy Harvesting

Harvesting energy directly from the ambient environment can be an attractive power supply option. Thermodynamic considerations for energy harvesting require available energy in the form of differentials or gradients in energy levels [48]. Thermodynamic efficiency and energy availability increase with the size of the energy differential. Ambient available energy sources may include measurand-derived power, wind, sunlight, ambient vibration, thermal gradients, and raindrops [64,87–90]. Since available ambient energy is often of low quality, some authors refer to these technologies as *energy scavenging* rather than energy harvesting.

Measurand-derived power is perhaps the simplest and most reliable method of powering a sensor. Examples of measurand-powered sensors include thermal-expansion-based thermometers, liquid-level vertical displacement gages, propeller-driven wind velocity gages, ratcheting peak displacement gages, piezoelectric patches, the mechanical scratch gage, and opti-mechanical smart bolts [91,92]. Currano developed a measurand-powered MEMS-based resettable latching shock threshold sensor (Figure 4.19) [93]. Hansen et al. developed a similar system that uses snap-through Mises truss mounts on a proof mass for latching [94].

Direct mechanical actuation of piezoelectric materials to produce electrical energy can be a straightforward harvesting technique, when conditions permit. Gao et al. used a mechanically actuated piezoelectric stack to power a pressure transducer system inside an injection molding machine [60]. Zhou et al. developed an acceleration-powered accelerometer using d_{33} piezoelectric coupling [95].

Ambient vibrations may be a good source of available energy [96–98]. A typical approach uses a spring-mounted proof mass oscillator with a piezoelectric element attached to the spring (Figure 4.20). Casciati et al. and Elvin et al. describe many of the pertinent design issues for optimal energy harvesting performance with electromechanical systems [99,100]. A key issue is to match the electromechanical receptance of the system to that of the available vibration energy [101]. Clark et al. generated approximately 0.4 mA at a rectified 15 V using a 270 Hz 0.2 g vibration source and a frequency-tuned harvester [102]. Challa et al. developed a variable-frequency energy harvester that used variable magnetic stiffness elements [103]. Churchill et al. demonstrated ambient vibration energy harvesting as a means to power wireless networks [104]. The system used a vibrating cantilever beam with a sandwich piezoelectric transducer coupled to a diode bridge to convert mechanical energy into an electrical form. Approximately one minute of vibrations at strain levels of 150 με in the piezoelectric element was sufficient to power data transmission across a wireless data link at a range of 1/2 km. Arms et al. describe a subsequent generation of this technology, which used a combination of vibration energy harvesting with wireless sensing links to measure and transmit sensor information from rotating helicopter blade hubs to a central processor

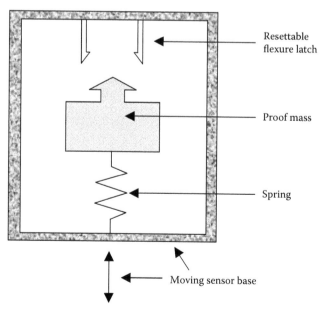

FIGURE 4.19 Operating principle of measurand-powered MEMs-based resettable latching shock or acceleration threshold sensor. (Data from Curano LA. 2005. In *Structural Health Monitoring*, DEStech, Lancaster.)

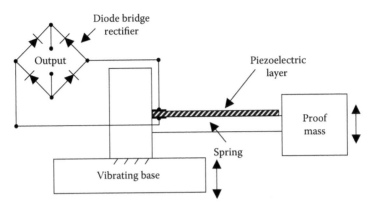

FIGURE 4.20 Vibration energy harvester using a cantilever with a piezoelectric layer and a diode bridge.

[105]. Ayers et al. achieved similar results by cyclic loading of a piezoelectric stack with strains of $70\,\mu\varepsilon$ and a battery storage of 0.86 J at 2.45 V in a device with nominal dimensions of $10 \times 10 \times 1\,mm^3$ [106]. In sea trials lasting several months, Discenzo et al. found that a piezoelectric cantilever vibration energy harvester could generate power in the order of hundreds of microwatts for pump condition monitoring [107]. Since vibration energy harvesting systems are intimately coupled to the dynamics of the base structure, Wang et al. propose using the power flows in a vibration-based energy harvesting system as a measurand in an SHM effort [108].

Solar and other light sources can be excellent supplies of harvestable ambient energy [109]. Many permanently installed SHM systems can justify the capital expense of the solar panels. For example, a sonar-based bridge pier scour-monitoring instrument on the Woodrow Wilson Bridge across the Potomac River in Washington, DC uses solar power [110].

Ambient temperature gradients provide available energy to power thermoelectric generators, usually with solid-state heat engines [48,111]. Thin-film micromanufacturing combined with multistage miniaturized integrated packaging forms the basis of specialized thermoelectric generators as energy harvesters for wireless sensing applications. Figure 4.21 shows a thermoelectric generator that produces $130\,\mu W$ and 3 V at a temperature gradient of only 5 K. Babic et al. fabricated a hydrogen electrothermal heat engine for ambient energy harvesting using modified hydrogen fuel cell technology [112].

FIGURE 4.21 Thermoelectric power generator for wireless sensing applications. (Photograph courtesy of the Thermo Life® Energy Corporation.)

4.4.8 Wireless Power Transmission

Wireless EM techniques can supply power to remote systems. Several different transmission modes are possible. LF oscillations can inductively supply power to a coil on a sensor system with EM fields that are essentially quasi-static. Higher frequency systems use EM waves. Very HF radiation, that is, light waves, uses photon energy as the energy transmission mechanism [113].

Typical transmission frequencies are less than 100 kHz. LF power transmission involves EM transmission wavelengths that are long relative to the overall dimensions of the physical components. Quasi-static inductive coupling models that largely ignore wave effects can describe the energy transfers. Due to the similarity of the energy transfer process with that of a transformer, LF power transmission is often called *transformer mode* power transmission.

Transformer mode power transmission systems are useful for situations that are amenable to bringing the sender and receiver into close proximity with one another, but nonetheless require a wireless noncontacting transmission link. Near-field quasi-static EM interactions can transmit data along with power. An analog technique includes a sensor in the inductive circuits. Changes in the measurand create measurable changes in the inductive circuit. Similar digital approaches use the switching of the properties of the inductive coupling. *Load shift coding* is a common technique. Examples of application of transformer mode telemetry are a MEMS-based system that incorporates capacitive sensing to detect and transmit pressure, temperature, and relative humidity by DeHennis and Wise [114]. Figure 4.22 is a schematic of a system developed by Microstrain Inc. (Burlington, VT) and reported by Esser et al. that uses low-frequency (kHz) waves to power a microprocessor-controlled sensor module [115]. One variant of system transmits data from the sensor to a host receiver via a 916.5 MHz telemetry link. The host receiver then detects the analog circuit changes. A second variant transmits the data by load shift coding. Harpster et al. used a sensor based on resonance changes in an LC circuit that range from 4 to 5 MHz [116]. Spillman and

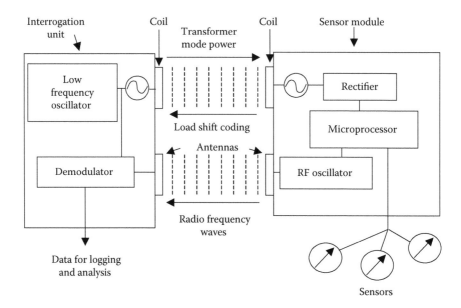

FIGURE 4.22 Schematic of wireless inductive powering and data telemetry unit using both radio telemetry and load shift coding.

Durkee used a resistive strain gage to induce a detectable change in the resonance quality of an analog circuit [117]. Shinoda et al. developed a capacitance-based wireless sensor for robotic tactile sensing [118]. Sohn et al. used inductive coupling to transfer sufficient power to drive a piezoelectric ultrasonic guided wave SHM system [119].

Rotating machines are attractive candidates for inductive power transmission because the wireless connection alleviates the need for onboard batteries or slip rings, and the close proximity of rotor and stator allows for a show transmission distance. The typical approach uses electromechanical energy harvesting with inductive elements that rotate in magnetic fields to extract energy from the rotating shaft to provide power to a rotor-mounted sensor system. Joshi et al. and Jia et al. used a wireless inductive power and data telemetry system to monitor the temperature in roller bearings [120,121]. French et al. patented a bearing that senses bearing state and transmits the state to an external monitor via a wireless link [122]. Renwen et al. demonstrated a wireless system for transmitting data through a helicopter hub [123].

A disadvantage of inductive coupling power transmission is the requirement of bringing the data sender and receiver close to one another (10 mm approximately). Other methods of remotely powering systems have appeared. One example is the directional beaming of microwaves to rectennas [124–129]. Another example is the excitation of resonant cavities by remotely supplied microwave power. Such devices are rumored to have been used as eavesdropping devices, with invention attributed to L.S. Termin (inventor of the Theremin). A similar device can remotely power and transmit strain sensor data [130]. High-frequency EM waves in the form of light can also provide remote power to sensor systems. Spillman and Andresen developed a fiber optic power transmission system for a network of capacitive sensors [131]. Feng demonstrated an optically powered electromechanical accelerometer that simultaneously sent the power supply and data signals over fiber optic cables [132].

4.4.9 Passive Wireless Sensors: RFID

In many respects, passive wireless sensor technology overlaps that of RFID systems. The original intent of RFID system development was for object identification and inventory control. Figure 4.23 shows a typical RFID system. Increases in the capability of RFID systems, including the use of onboard microprocessors and sensors, convert the systems into passive wireless sensors. For a detailed description of RFID technology (ca. 2003), see Finkenzeller [133].

The operating principles of RFID systems are frequency dependent. Typical RFID system frequencies are (1) LF 30–300 kHz, (2) HF 3–30 MHz, (3) *ultra high frequency* (UHF) 300 MHz–3 GHz, and (4) *microwave* >3 GHz. Coupling ranges are (1) *close*, with ranges of about 1 cm and frequencies of DC to 30 MHz, (2) *remote*, with ranges up to 1 m using inductive radio signals, and (3) *long range*, with ranges of at least 3 m possible using backscatter techniques. Information processing techniques are (1) *low end*, unidirectional and a small amount of information, (2) *mid-range* with writable data storage, and (3) *high end* with an onboard microprocessor.

A fundamental issue associated with passive wireless sensing and RFID technology is the method of sending and receiving the signal. The source signal is typically strong. The return signal is typically weak. Using the same frequency and modulation for the source and return signals poses difficulties. Several successful workaround solutions appear in the design of RFID systems.

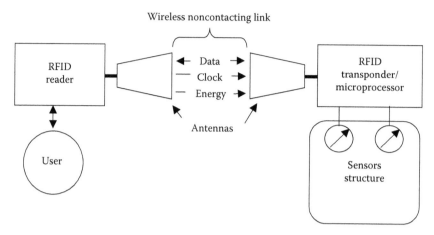

FIGURE 4.23 Typical RFID system with reader and transponder. (Adapted from Finkenzeller K. 2003. *RFID Handbook*, 2nd Ed. Wiley, Chichester.)

One bit of information adequately describes many situations. A typical 1-bit RFID transponder uses oscillatory resonance as the signal. The design is that the signal source and RFID tag inductively couple and form a single resonance system (Figure 4.24). The feedback impedance of the signal generator shifts with the presence of an RFID tag. The RFID tag normally encodes one bit of information, such as whether a particular circuit element is open or closed. These systems require large frame antennas to surround the RFID tag with an inductive field. The need for a fairly tight inductive coupling between the source and RFID tags in resonance-based systems limits the usable range to a meter or two. Such systems commonly appear at the doorways of retail stores for inventory control. Dickerson et al. used a 1-bit resonant circuit sensor system for corrosion sensing in reinforced concrete [134]. A sacrificial corrosion member served as the switch element.

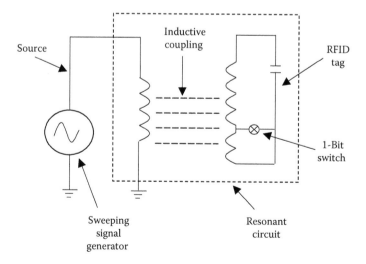

FIGURE 4.24 1-bit resonant circuit RFID wireless system. (Adapted from Finkenzeller K. 2003. *RFID Handbook*, 2nd Ed. Wiley, Chichester.)

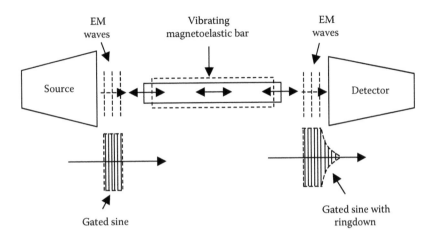

FIGURE 4.25 1-bit magnetoelastic resonant ringdown RFID wireless system. (Adapted from Finkenzeller K. 2003. *RFID Handbook*, 2nd Ed. Wiley, Chichester.)

Sending return signals at times or frequencies that differ from the source enables the detection of much weaker return signals. One technique combines a source that switches On and Off with a magnetostrictive oscillator. When the source is On, a magnetostrictive material in the RFID tag goes into resonance. When the source switches to Off, the magnetostrictive element continues to oscillate and rings down. Since magnetostriction is a reversible process, the magnetostrictive element emits EM waves during oscillation and ring-down. These waves are detectable and form the basis of a 1-bit transducer (Figure 4.25). Nonlinear superharmonic systems can transmit information over longer distances, perhaps up to a couple of meters. These devices typically operate at microwave frequencies and use nonlinear elements, such as diodes, to generate superharmonics and sidebands (Figure 4.26). Superharmonic and sideband transmission enables sending more than 1-bit on the return signal.

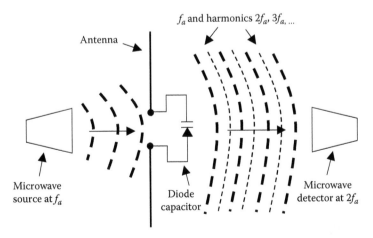

FIGURE 4.26 1-bit nonlinear superharmonic circuit RFID wireless system. (Adapted from Finkenzeller K. 2003. *RFID Handbook*, 2nd Ed. Wiley, Chichester.)

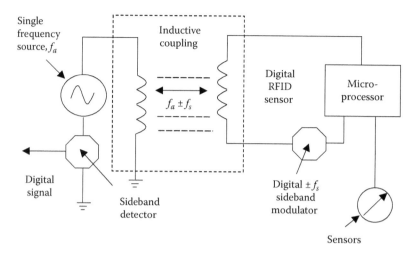

FIGURE 4.27   Digital RFID sensor system using sideband load modulation. (Adapted from Finkenzeller K. 2003. *RFID Handbook*, 2nd Ed. Wiley, Chichester.)

A variety of other similar noncontacting wireless RFID-type sensors are possible (Figure 4.27). Bowles et al. used a stress-sensitive amorphous alloy as the core material in an inductive solenoid to form a wireless battery-free stress sensor [135]. Andringa et al. describe some of the design details pertaining to resistive-sensing-based passive wireless sensors [136]. Examples of applications include tire strain sensors, sacrificial corrosion sensors, resonant cavity strain sensors, embeddable chloride sensors for bridge decks, magnetic temperature sensors, and displacement/contact indicators [130,137–142].

4.5 Sensor Networks

A sensor network integrates sensors, data processors, network controllers, and data links. The capability of a given sensor network can far exceed the sum of the capabilities of individual components. The miniaturized collocation of data collection, processing, and networking capabilities into sensor modules facilitates the creation of sensor networks. The sensor modules acquire analog and/or digital data. The network passes the sensor data from the transducers onto data handlers that process the data to increase the information density, or may act as transponders that pass the information along without any supplement processing. The communications can use a variety of wired or wireless layers (Figure 4.28).

Wired networks typically run electrical cables between the sensing nodes for both power supply and communication. When implemented at the early design and fabrication stage, embedded sensors can often provide an effective sensing solution. *Fiber-reinforced polymer composite* (FRPC) structural elements are often considered as excellent candidates for embedded sensors since the layup manufacturing process allows easy embedment. The performance requirements of many FRPC structures are sufficiently high to justify the cost of embedding sensors [143].

WSNs are potentially a convenient, readily reconfigurable, and inexpensive sensing solution [144,145]. A driving enabler is the seemingly continuous and rapid pace of advances in wireless technology, standards, hardware, and software. Most of the wireless data

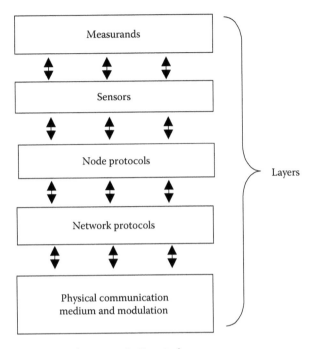

FIGURE 4.28 Generic sensor network communication stack.

transmission systems use propagating EM fields, that is, microwaves, IR, and visible light, to transmit information. Occasionally, acoustic (ultrasound) waves and subatomic particles form the physical medium for information transmission.

4.5.1 Network Architectures

A wide variety of architectures can prove useful in SHM WSNs. A simple network architecture uses a transmitter and receiver with one-way transmission (Figure 4.29). A limitation of the one-way transmission architecture is the minimal remote control over the transmitters and data acquisition system. An additional limitation is that the range and power of the telemetry units are insufficient for reliable transmission. A repeater may be necessary (Figure 4.30). Transceivers are enhanced, but more expensive, systems that act as both talkers and listeners (Figure 4.31). Protocols that allow for using multiple senders and receivers facilitate constructing a multitude of different network configurations (Figure 4.32).

A variety of competing sensor–multiplexer–data bus configuration architectures are in use or under consideration. These include (1) bus network, (2) star network, (3) hierarchical or

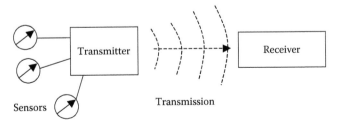

FIGURE 4.29 One-way transmission.

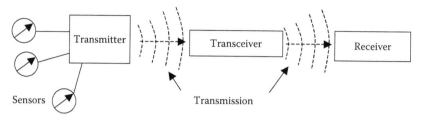

FIGURE 4.30   Transmitter, repeater (transceiver), and receiver configuration.

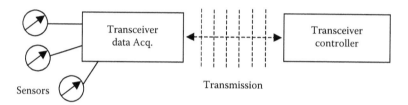

FIGURE 4.31   Bidirectional transceiver pair with data acquisition and control.

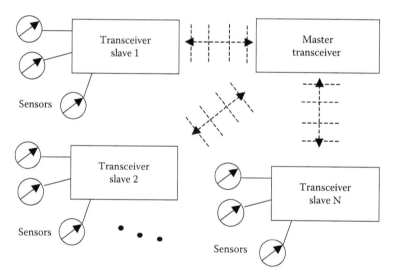

FIGURE 4.32   Two-stage sensor network architecture with multiple slave data acquisition units with a master controller and transceivers.

tree network, (4) nearest neighbor or mesh, or (5) complex network. A *data dependence graph* represents the topological linkage of a sensor network (Figure 4.33 and Figure 4.34) [146]. In many respects, the data dependence graph is an ordinary node and line graph [147]. The nodes (vertices) of the graph represent either transducers or data handlers. Possible node activities include sensing, signal conditioning, digitizing, calculating statistics, and saving data; and transmitting data, trigger, and other control signals. The lines (edges) connecting the nodes indicate a communication link between the nodes. The configuration of the graph may be static, as in a fixed hard wired data acquisition system, or the configuration can be dynamic and change with time as with an *ad hoc* WSN. If the network structure is dynamic,

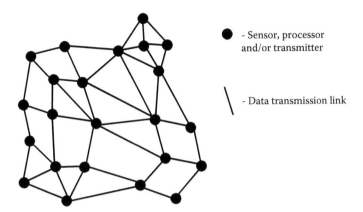

FIGURE 4.33   Data dependence graph for a system with nearest neighbor mesh topology.

or *ad hoc* self-assembling, then the nodes must periodically broadcast network configuration information to those nodes with which it is communicating. When it is desired to retain more detailed information about the network, such as energy flows, then a *bond graph* may be a useful representation [148].

The abstraction of the data dependence graph enables applying mathematical results from graph theory to the design and analysis of sensor networks. The following is a small sampling of a rich mathematical topic with literature that dates back at least to Euler. Of particular interest are results from the theory of random graphs, many of which appeared in the latter half of the twentieth century [149–152]. These techniques can estimate items such as the probability of a continuous path of links between two nodes, the average number of links per node, and the growth (or decay) of these values through the addition (or deletion) of nodes. A graph $G_{n,r}$ has n nodes and r edges (links). A common (but not necessary) assumption is that there are no double edges between nodes and no edges that self link back to a node. The nodes are typically labeled sequentially with integers. The *degree* of a node is the number of edges connecting to a node. The total number of distinct graph

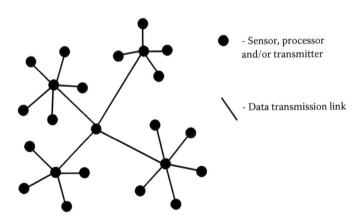

FIGURE 4.34   Data dependence graph for hierarchical cluster tree topology.

configurations $C_{n,r}$ of n nodes and r edges is

$$C_{n,r} = \left(\binom{\binom{n}{2}}{r} \right). \tag{4.16}$$

An *adjacency matrix* $[A]$ tabulates the edge configuration and forms the basis of many results that describe the connectedness of graphs. (Many of the following results derive from Tsai [153].) The elements A_{ij} equal one if there is an edge connecting nodes i and j, and equal zero if there is no connecting edge. The diagonals A_{ii} automatically equal zero. For the partially connected graph $G_{5,4}$ and the connected subgraph shown in Figure 4.35, the adjacency matrices $[A_G]$ and $[A_S]$ are, respectively,

$$[A_G] = \begin{bmatrix} 0 & 1 & 1 & 0 & 0 \\ 1 & 0 & 1 & 0 & 0 \\ 1 & 1 & 0 & 1 & 0 \\ 0 & 0 & 1 & 0 & 0 \\ 0 & 0 & 0 & 0 & 0 \end{bmatrix} \tag{4.17}$$

and

$$[A_S] = \begin{bmatrix} 0 & 1 & 1 & 0 \\ 1 & 0 & 1 & 0 \\ 1 & 1 & 0 & 1 \\ 0 & 0 & 1 & 0 \end{bmatrix}. \tag{4.18}$$

A *walk* is a sequence of alternating nodes and edges that begins and ends with a node [153]. Walks include the possibility of backtracking. *Trails* are walks that do not repeat edges. A *circuit* is a walk with no repeated nodes, except for the first and last node. A *connected graph* has a walk between every node. A *tree* is a connected graph that contains no circuits. A *spanning tree* is a tree that connects all of the nodes in a graph. It can be shown that the element (i, j) of $[A]^n$ equals the number of walks of length n from node i to node j. For example, the elements of $[A_S]^2$ describe the number of walks of length 2 between two

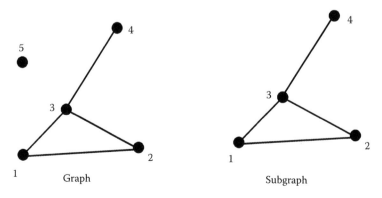

FIGURE 4.35 Partially connected graph and connected subgraph component.

nodes, that is,

$$[A_S]^2 = \begin{bmatrix} 2 & 1 & 1 & 1 \\ 1 & 2 & 1 & 1 \\ 1 & 1 & 3 & 0 \\ 1 & 1 & 0 & 1 \end{bmatrix}. \tag{4.19}$$

Global measures of connectedness are a rich topic in graph theory [154]. A typical simple result derived from replacing the diagonal elements of $-[A]$ with node degrees produces the matrix $[M]$. An interesting result is that all of the cofactors of $[M]$ are equal. The value of the cofactors equals the number of spanning trees. For the subgraph with adjacency matrix $[A_S]$, $[M_S]$ is

$$[M_S] = \begin{bmatrix} 2 & -1 & -1 & 0 \\ -1 & 2 & -1 & 0 \\ -1 & -1 & 3 & -1 \\ 0 & 0 & -1 & 1 \end{bmatrix}. \tag{4.20}$$

The cofactors of $[M_S]$ all equal 3. A visual inspection of the subgraph in Figure 4.35 confirms this result.

Random graph theory extends many of the classic results from deterministic graph theory into the stochastic domain. The random variables include the number of nodes and the edges connecting them. If the random variables are independent, and the number of nodes is finite, then many of the statistical results follow binomial distributions (as would be expected from Equation 4.16). In the case of large numbers of nodes, independent random variables lead to Poisson-type statistics [150]. If the random variables are dependent, then the statistical properties can change dramatically. Important cases are those where connected nodes tend to become more highly connected as the network grows. Power law distributions of random graph properties may result [149,151,152]. Analyzing situations with large numbers of entities that interact with one another may benefit from the abstraction of random graph theory. This is especially the case where the configuration and topology of the interconnections affect overall behavior. There are numerous examples in the physical and social sciences. The percolation of fluids through solids is an example where the existence of spanning trees (superelements) of fluid paths governs whether the fluid will flow. In the social sciences, organizational network models of "who knows who" can be highly informative. WSNs are a potential SHM application of random graph theory [155].

Many of the sensor networks encountered in SHM practice are only moderately complicated in form, such as the two-stage hierarchical network shown in Figure 4.32. Casciati et al. found it convenient to use a hybrid system with different frequencies and different protocols on different legs of the network [156]. Hedman et al. developed strain sensing microprocessors that contribute to building two-tier distributed hierarchical networks to monitor airplane fatigue cracks [157]. More complicated better performing hierarchical schemes are possible. One possibility configures the system to minimize power consumption [158]. Civera et al. used wireless sensor chains of MEMS inclinometers for geotechnical applications [159]. Ghosal et al. developed a sensor network architecture for acoustic signal detection that mimics the tree-like structure and hierarchical trigger signaling processes of many biological nervous systems [160]. Sharma et al. developed a sensor network architecture, known as Whirlpool [161]. A sector-type division of the network allows for the querying of the system in a manner that optimizes signal transmission and the identification of changes in structural behavior. Rye et al. compared the merits of global bus, nearest neighbor, and tree architectures for sensor networks embedded into composites [162]. Practical advantages of simplicity and implementation favored the global bus architecture for early developments.

4.5.2 Integrated and Real-time Systems

Large-scale structures, that is, dams, bridges, buildings, power plants, and so on, tend to be fairly fixed in form and operation. As such, fixing SHM systems into and onto fixed structures may be desirable. The architecture of most fixed SHM systems tends to fall along the lines of a heterogeneous mix of sensors with a hierarchical organization. Many systems send data on a near-real-time basis to interested parties [163].

Multiple examples of fixed SHM systems appear in the literature. The ISIS system in Canada has sensors networked on about 20 bridges, with several of the networks providing real-time condition monitoring [164]. Bakker et al. installed multiple internet-capable monitoring systems on structures in the Netherlands [165]. Chen and Gong used a two-tier system where a Fieldbus exerted local control over the data acquisition instruments and then transmitted the data to a central host with a TCP/IP network connection [166]. Zhu et al. used a two-stage hierarchical network to connect to the internet for further processing of SHM data from the Dafosi Bridge—a cable-stayed bridge across the Yangtze River in western China [167]. Three low-speed sensor types (fiber optic strain, displacement, and temperature) connected to a local computer for data preprocessing. A different local computer collected and preprocessed higher speed dynamic data. Both local computers connected to a third computer for overall system control and communication on the internet. A remote computer downloaded, processed, and stored the data.

Aerospace SHM applications can benefit from embedded lightweight integration. As an example, Trego developed and installed a data acquisition system specifically designed for airplane SHM [168]. The system acquires eight channels of electrical data and 16 channels of fiber optic data. The human interface is important. Custom software packages convert the sensor data into images for quick damage and condition assessments [169].

Local and automated processing techniques can enhance the performance of a real-time system. For example, Zhang et al. developed a lossy wavelet compression scheme for analyzing the hysteretic behavior of structures in earthquakes [170]. Changes to the hysteresis curves correspond to damage. The data compression reduces the required bandwidth and eases the development of a regional network of structural monitoring systems. Zhang and Li developed a lossless vibration data compression method based on a linear predictor algorithm [171].

4.5.3 Data Protocols: Senders, Receivers, and Transceivers

Transmitting data requires physical transmission media, physical transmission mechanisms, data modulation techniques, and data interface protocols. Both the transmitter and receiver must coordinate via an accepted protocol. SHM often uses a variety of custom data transmission protocols. An example is a system developed by Milos and Karunaratne that interrogates thermocouples on the space shuttle and other aerospace thermal protection system (TPS) applications with a wireless link [172].

Several wireless networking protocol standards are presently used by the communications industry and are potentially useful for networking sensors for SHM. The IEEE 802.11 family of protocols forms a ubiquitous set of wireless data interchange networks, including

WiFi, for computers. Typical maximum ranges are on the order of 1 km. Ranges of up to 10 km with Yagi antennas operating under ideal conditions may be possible. The IEEE 802.15 standards cover *wireless personal area networks* (WPANs). IEEE 802.15.1 and 802.15.2 cover the Bluetooth standard, which finds some usage in WSNs. IEEE 802.15.4 is a standard for low-speed and low-power-consumption data networks that specifies physical and medium access control layers for wireless network communications [64]. The standard allows for considerable flexibility in the topology and operational parameters of the network with communications over either a lower band (868.0–868.6 MHz in Europe and 902–928 MHz in the US and Pacific rim) or an upper band (2.400–2.485 GHz). The network can address up to 64,516 logical devices. The standard also incorporates the routine transmission of link quality information and controlled topological reconfiguration. *Zigbee* is a WSN network standard that uses IEEE 802.15.4 protocols [173]. Figure 4.36 shows a state-of-the-art IEEE 802.15.4 wireless sensor node. This digitally addressable node is battery powered, can acquire analog sensor data at rates up to 2,048 samples/s, and has an onboard data storage capacity of up to 1,000,000 measurements. Data transmission distances of 300 m are attainable with the use of high-gain antennas.

Since effective communication requires the coherent transmission of data, the simultaneous data transmission by network nodes may cause problematic data collisions that degrade or defeat effective communication. One method of avoiding collisions preassigns the

FIGURE 4.36 IEEE 802.15.4-compatible wireless sensor node. (Photograph courtesy of the Microstrain, Inc.)

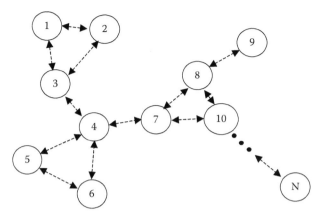

FIGURE 4.37   Self-assembled mesh network.

timing of transmission and reception for each node. Preassigning the timing simplifies control issues and has the potential to optimize bandwidth consumption by the network, but lacks flexibility. More free-form and flexible protocols are possible. Time division multiple access (TDMA) allows multiple nodes to operate on the same frequency by coordinating the timing of each individual transmission, usually in a sequential multiplexed manner. TDMA exerts a tight control of the communication timing sequence, but has the flexibility of being able to change the sequence as needed. Many early cellular telephone systems used TDMA protocols to coordinate multiple callers on the same frequency band. Adaptive transmission schemes that tolerate and recover from data collisions can be more robust, at the expense of certainty as to how the network will behave and the degree of efficient bandwidth utilization. A simple collision recovery scheme has each transmitter stop transmitting when a collision occurs and then retransmit at an individually different time.

4.5.4   Self-Assembly *Ad Hoc* Networks

Ad hoc networks can reconfigure their topology by adding and removing nodes and by changing the interconnection paths automatically as needed (Figure 4.37). Many *ad hoc* networks can self-assemble from scratch. A possible application that until recently was in the realm of science fiction is to randomly disperse a set of sensors throughout a structure. The sensors then self-assemble into a viable WSN. Mixing small wireless sensors into concrete is one possibility. There are a couple of examples of modestly self-assembling sensor systems that are already in use. The 802.11 wireless networking standards are an interesting, although not entirely identical, analog to the self-assembling network or the cellular wireless system. Cellular telephone systems are another example. Biological neural systems are very complex, yet quite effective, self-assembled sensor and transmission networks, with a largely hierarchical structure [174].

One self-assembly strategy for a network is the so-called *parent–child* technique [175]. Some of the key points of this strategy are:

- The sensor modules have an embedded self-assembly protocol.
- Each sensor has a unique identifying number or tag.
- Each sensor module has *n* neighbors.

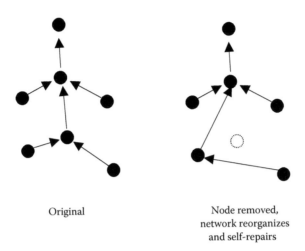

Original

Node removed,
network reorganizes
and self-repairs

FIGURE 4.38 Self-repair of *ad hoc* tree network.

- The sensors communicate with one another in discrete frames of time through a TDMA multiple access protocol.
- During each broadcast, a wave of conversations is undertaken. A sensor broadcasts a call to its neighbors. The sensor talks to its neighbors, the neighbors talk to neighbors, and a wave of communication and identification of configuration occurs.

There are some disadvantages of self-assembled networks. The autonomous (wireless) nature of the sensor modules places severe demands on the power budget for each sensor module [176]. Registering the position of the individual sensor modules relative to the structure requires an additional step, such as registering the location and identity of each sensor during assembly, or by sending some sort of separate locating signal throughout the system. Basheer et al. extended the concept of *ad hoc* self-organizing sensor networks with localized tree structures [177]. Figure 4.38 shows self-repair in an *ad hoc* network.

At the time of this writing, the development of wireless sensing nodes and network hardware is proceeding quickly. Glaser deployed and used a set of small wireless sensor nodes, known as *Motes*, in a densely packed arrangement for structural system identification [178,179]. In ca. 2004 technology, Motes could reliably self-assemble into a network, and sense and transmit data at a rate of 3 kB in 1.2 s over a distance of 30 m using Bluetooth technology. Mastroleon et al., Ou et al., and Chen et al. describe low-power-consumption designs with two-tier hybrid star topologies [180–182].

4.5.5 Distributed Computations

Modern microprocessor technology enables performing computations and statistical analyses at nodes distributed throughout a sensor network. Distributed computing can reduce the data transmission and bandwidth requirements by transmitting data with high-information content. Statistical feature selection is an important design parameter. If the sensor network measures and transmits dynamic structural information, autoregressive (AR) parameter estimation appears to be an effective feature for local calculation and for rapid identification of changes in structural behaviors [183]. In a different approach, Hashimoto et al. suggest using wavelet transforms for efficient calculation of the local statistical features of acceleration signals before transmission [184].

WSNs can also distribute computational burdens across the network so that the network operates as a distributed parallel processing computer [185]. As an example, Chintalapudi et al. developed a modal vibration WSN that coordinates distributed calculations and statistical estimates [186]. The local nodes calculate possible resonant frequencies from local vibration measurements and then transmit the information to a central controller. The central controller forms a master list of global resonant frequencies from the local frequency information and then sends the master list to the local nodes. The local nodes calculate magnitude and phase information for each mode from the local vibration measurements and then transmit the result to the central controller. The central controller calculates the global mode shapes. Similarly, Zimmerman and Lynch used a distributed *ad hoc* WSN to calculate mode shapes and shape changes with simulated annealing [187].

Self-assembling sensors networks are in many respects complicated examples of cellular automata where the behavior and operation of the entire system depends on both local and global interactions [188]. In certain circumstances it is possible to exploit local versus global behaviors to advantage. Channel degradation and loss are possible structural failure sensing techniques [189]. Wang et al. proposed a system that senses the organizational structure of an SHM WSN as a reflection of the state of damage in a structure [190,191]. Foreman et al. describe the action of self-organizing networks using an information theory representation of phase transitions and entropy to assess the location and amount of energy in structural impacts [192].

4.6 Data Storage

Proper data storage is an integral component of SHM data processing, which continues to improve at an astonishing rate. Standardized methods of structuring data would be useful in this regard. Certain applications require the use of ruggedized high-performance digital data acquisition systems [193]. Aircraft flight data recorders with crash surviving designs are an example. Long-term archiving requires a physical medium for storing the data, along with sufficient information to resurrect and understand the data several years hence.

References

1. Marecos V, Santos LO, and Branco F. 2004. Bridges safety control in real time. *Proc IABMAS'04 Bridge Maintenance, Safety, Management and Cost*, Kyoto. Taylor & Francis, London.
2. Masri SF, Sheng LH, Wahbeh M, Caffrey J, Nigbor R, and Abdel-Ghaffar A. 2004. Design and implementation of a web-enabled real-time monitoring system for civil infrastructures. *Proc IABMAS'04 Bridge Maintenance, Safety, Management and Cost*, Kyoto. Taylor & Francis, London.
3. Sereci AM, Radulescu D, and Radulescu C. 2003. Recent installation of real-time structural monitoring systems. In *Structural Health Monitoring*, FK Chang (ed.). DEStech, Lancaster.
4. Zhang S, Wang CF, and Yen M. 2005. Sensor data management in structural health monitoring systems. In *Structural Health Monitoring*, FK Chang (ed.). DEStech, Lancaster.

5. Yu P. 2007. Real-time impact detection system for thermal protection system. In *Structural Health Monitoring*, FK Chang (ed.). DEStech, Lancaster.
6. Oppenheim AV and Schafer RW. 1975. *Digital Signal Processing*. Prentice-Hall, Englewood Cliffs, NJ.
7. Casciati F, Faravelli L, and Borghetti F. 2003. Wireless links between sensor-device control stations in long span bridges. *Smart Systems and Nondestructive Evaluation for Civil Infrastructures*, SPIE 5057.
8. Wang Y, Loh KJ, Lynch JP, Fraser M, Law K, and Elgamal A. 2006. Vibration monitoring of the Voigt bridge using wired and wireless monitoring systems. *Proc 4th China–Japan–US Symp on Structural Control and Monitoring*, Hangzhou.
9. Russ JC. 1999. *The Image Processing Handbook*, 3rd Ed. CRC Press, Boca Raton, FL.
10. Bozeman Jr RJ. 1996. System for memorizing maximum values. US Patent 5,539,402.
11. Blackman RB and Tukey JW. 1958. *The Measurement of Power Spectra*. Dover Publications, Mineola, NY.
12. Box GE and Jenkins GM. 1976. *Time Series Analysis Forecasting and Control*. Holden-Day, San Francisco.
13. Bendat JS and Piersol AG. 2000. *Random Data: Analysis and Measurement Procedures*, 3rd Ed. Wiley-Interscience, New York.
14. Taylor JL. 1986. *Computer-based Data Acquisition Systems Design Techniques*. Instrument Society of America, Research Triangle Park.
15. Andrews JR. 1999. Low-pass risetime filters for time domain applications. *Application Note AN-7a*, Picosecond Pulse Labs, Boulder.
16. Stubbs N and Park S. 1996. Optimal sensor placement for mode shapes via Shannon's sampling theorem. *Microcomput Civ Eng*, 11, 411–419.
17. Mondal P and DeWolf JT. 2004. Long-term monitoring of temperatures in a segmental concrete box-girder bridge in Connecticut. *Proc Structural Materials Technology VI, an NDT Conference*, Buffalo.
18. Bartoli G, Chiarugi A, and Gusella V. 1996. Monitoring systems on historic buildings: The Brunelleschi dome. *J Struct Eng*, 122(6), 663–673.
19. Stengel RF. 1994. *Optimal Control and Estimation*. Dover Publications, Mineola, NY.
20. Lueck DE. 2006. Self-calibrating pressure transducer. US Patent 7,043,960.
21. Feng J and Potkonjak M. 2005. Sensor calibration using density estimation-based error model. *Modeling, Signal Processing, and Control*, SPIE 5757.
22. Lee K. 2000. IEEE 1451: A Standard in support of smart transducer networking. *IEEE Instrumentation and Measurement Technology Conference*, Baltimore.
23. Dillon MJ, Bono RW, and Brown DL. 2004. Use of photogrammetry for sensor location and orientation. *Sound Vib*, (November), 23–27.
24. O'Connor KM and Dowding CH. 1999. *GeoMeasurements by Pulsing TDR Cables and Probes*. CRC Press, Boca Raton, FL.
25. Zhao M and Shi X. 2005. A new developed wireless sensor based on GPRS & GPS for dynamic signals of structures. In *Structural Health Monitoring*, FK Chang (ed.). DEStech, Lancaster.
26. Hajnal JV, Hill DL, and Hawkes DJ. 2001. *Medical Image Registration*. CRC Press, Boca Raton, FL.
27. Liu Z and Forsyth DS. 2004. Registration of multi-modal NDI images for aging aircraft. *Res Nondestruct Eval*, 15, 1–17.
28. Prosser W and Perey D. 2003. Multiplexing technology for acoustic emission monitoring of aerospace vehicles. In *Structural Health Monitoring*, FK Chang (ed.). DEStech, Lancaster.
29. Ishikawa K and Mita A. 2007. Fine time synchronization system for sensor grid. In *Structural Health Monitoring*, FK Chang (ed.). DEStech, Lancaster.
30. Kiefer KF, Swanson B, Krug E, Ajupova G, and Walter PL. 2003. Wireless sensors applied to modal analysis. *J Sound Vib*, (November), 10–15.
31. Cottineau LM, Cam VL, and Ducros DM. 2003. Implementation of a low-cost, distributed absolute time reference in an application dedicated to damage detection. In *Structural Health Monitoring*, FK Chang (ed.). DEStech, Lancaster.

32. Maróti M, Kusy B, Simon G, and Lédeczi Á. 2004. The flooding time synchronization protocol. *SenSys'04*, Baltimore.

33. Luo YZ, Shen YB, Wang B, Rao L, Tong RF, Zhang F, Yuan FG, and Liu L. 2006. Development of a wireless sensor system potentially applied to large-span spatial structures. *Proc 4th China–Japan–US Symp on Structural Control and Monitoring*, Hangzhou.

34. Lei Y, Kiremidjian AS, Nair KK, Lynch JP, and Law KH. 2003. Time synchronization algorithms for wireless monitoring system. *Smart Systems and Nondestructive Evaluation for Civil Infrastructures*, SPIE 5057.

35. Whang DH, Xu N, Rangwala S, Chintalapud K, Govindan R, and Wallace JW. 2004. Development of an embedded networked sensing system for structural health monitoring. In *Advanced Smart Materials and Smart Structures Technology*, FK Chang, CB Yun, and BF Spencer, Jr. (eds.), DEStech, Lancaster.

36. Kersey AD. 1995. Fiber optic sensor multiplexing techniques. In *Fiber Optic Smart Structures*, E Udd (ed.). Wiley, New York.

37. Udd E. 1991. *Fiber Optic Sensors*. Wiley-Interscience, New York.

38. Takeda N, Tajima N, Sakurai T, and Kishi T. 2003. Structural health monitoring issues in Japanese Composite Fuselage Demonstrator Program for damage detection and suppression. In *Structural Health Monitoring*, FK Chang (ed.). DEStech, Lancaster.

39. Seo JM, Kim SH, Kwon IB, Lee JJ, and Yoon DJ. 2005. Intensity-modulated multiplexing of fiber bragg grating sensors. *Smart Mater Struct*, 14, 177–182.

40. Tam HY, Lee T, Ho SL, Haber T, Graver T, and Méndez A. 2007. Utilization of fiber optic Bragg grating sensing systems for health monitoring in railway applications. In *Structural Health Monitoring*, FK Chang (ed.). DEStech, Lancaster.

41. Soumekh M. 1999. *Synthetic Aperture Radar Signal Processing*. Wiley, New York.

42. Mahajan A and Figueroa F. 1995. Dynamic across time autonomous—Sensing, interpretation, model learning and maintenance theory (DATA-SIMLAMT). *Mechatronics*, 5(6), 665–693.

43. Hunter GW, Oberle LG, Baakalini G, Perotti J, and Hong T. 2005. Intelligent sensor systems for integrated system health management in exploration applications. *First Intl Forum on Integrated System Health Engineering and Management in Aerospace*, Napa, CA.

44. Hamilton B. 1996. A compact representation of units. *Hewlett-Packard Technical Report* HPL-96-61, May.

45. Eccles LH. 1999. The representation of physical units in IEEE 1451.2. *Sensors*, 16(4), 30–35.

46. Reza FM. 1994. *An Introduction to Information Theory*. Dover Publications, Mineola, NY.

47. Shannon CE and Weaver W. 1949. *The Mathematical Theory of Communication*. University of Illinois Press, Urbana and Chicago.

48. Callen HB. 1960. *Thermodynamics*. Wiley, New York.

49. Agilent Technologies, Inc. 2005. Ultra-wideband communication RF measurements. *Application Note* 1488.

50. Keithley Instruments. 2001. *Data Acquisition and Control Handbook*. Cleveland.

51. Carr JJ. 1993. *Sensors and Circuits*. Prentice-Hall, Englewood Cliffs, NJ.

52. Tolmie BR and Wittemeyer RH. 2001. Extruded metallic electrical connector assembly and method of producing same. US Patent 6,283,792.

53. Bosch HR and Miklofsky HA. 1993. Monitoring the aerodynamic performance of a suspension bridge. *Proc 7th US National Conf on Wind Engineering*, Los Angeles.

54. Hamming RW. 1998. *Digital Filters*. Dover Publications, Mineola, NY.

55. He L, Ou J, Yao W, and Tang W. 2005. Real-time health monitoring system of offshore platform structure based on Fieldbus CAN and OPC. *Sensors and Smart Structures Technologies for Civil, Mechanical, and Aerospace Systems*, SPIE 5765.

56. Frankenstein B, Fröhlich KJ, Hentschel D, and Reppe G. 2005. Microsystem for signal processing applications. *Advanced Sensor Technologies for Nondestructive Evaluation and Structural Health Monitoring*, SPIE 5770.

57. Harrold D. 1998. 4–20 mA Alive and Kicking. *Control Eng*, 13, 109–114.

58. Ballard CM and Chen SS. 1997. An internet structural monitoring system. In *Intelligent Civil Engineering Materials and Structures*, F Ansari (ed.). ASCE, New York.

59. Fu Y, Zhu Y, Chen W, and Huang S. 2004. Remote health monitoring system for Dafosi Yangtze river bridge. *Sensors and Smart Structures Technologies for Civil, Mechanical and Aerospace Systems*, SPIE 5391.

60. Gao RX, Kazmer DO, Zhang L, Theurer C, and Cui Y. 2004. Self-powered sensing for mechanical system condition monitoring. *Sensors and Smart Structures Technologies for Civil, Mechanical and Aerospace Systems*, SPIE 5391.

61. Nachemson A and Elfstrom G. 1971. Intravital wireless telemetry of axial forces in Harrington distraction rods in patients with idiopathic scoliosis. *J Bone Joint Surg*, 53-A(April), 3.

62. Novak LJ, Grizzle KM, Wood SL, and Neikirk DP. 2003. Development of state sensors for civil engineering structures. *Smart Systems and Nondestructive Evaluation for Civil Infrastructures*, SPIE 5057.

63. Wong ZJ and Kim CG. 2007. Development of a passive, wireless and embeddable strain sensor for structural health monitoring application. In *Structural Health Monitoring*, FK Chang (ed.). DEStech, Lancaster.

64. Callaway Jr EH. 2004. *Wireless Sensor Networks*. Auerbach, Boca Raton, FL.

65. Lynch JP and Koh KJ. 2006. A summary review of wireless sensors and sensor networks for structural health monitoring. *Shock Vib Dig*, 38(2), 91–128.

66. Chung H, Park C, Xie Q, Chou P, and Shinozuka M. 2005. Duranode: wireless-networked sensing system for structural safety monitoring. *Nondestructive Detection and Measurement for Homeland Security III*, SPIE 5769.

67. Allan R. 2005. Wireless sensors land anywhere and everywhere. *Electron Des*, ED Online ID #10710 (July 21).

68. Strock OJ. 1987. *Introduction to Telemetry*. Inst Soc Amer, Research Triangle Park.

69. Schneider E, Genge M, Michel M, Bram W, and Perren SM. 1989. Instrumented interlocking nail for telemetrized load determination in the human femur. 35th Annual Meeting, Orthopaedic Research Society, Las Vegas.

70. Graichen F and Bergmann G. 1991. Four-channel telemetry system for *in vivo* measurement of hip joint forces. *J Biomed Eng*, 13, 370–374.

71. Rohlmann A, Bergmann G, and Graichen F. 1994. A spinal fixation device for *in vivo* load measurements. *J Biomech*, 27(7), 961–967.

72. Neuzil P, Serry FM, Krenek O, and Maclay GJ. 1997. An integrated circuit to operate a transponder with embeddable MEMS microsensors for structural health monitoring. In *Structural Health Monitoring Current Status and Perspectives*. Technomic, Lancaster.

73. Troyk PR, Schwan MA, DeMichele GA, Loeb GE, Schulman J, and Strojnik P. 1996. Microtelemetry techniques for implantable smart sensors. *Smart Sensing, Processing, and Instrumentation*, SPIE 2718.

74. Aoki S, Fujino Y, and Abe M. 2003. Intelligent bridge maintenance system using MEMS and network technology. *Smart Systems and Nondestructive Evaluation for Civil Infrastructures*, SPIE 5057.

75. Lynch JP, Sundararajan A, Law KH, Kiremidjian AS, Carryer E, Sohn H, and Farrar CR. 2003. Field validation of a wireless structural monitoring system on the alamosa canyon bridge. *Smart Systems and Nondestructive Evaluation for Civil Infrastructures*, SPIE 5057.

76. Pakzad SN, Kim S, Fenves GL, Glaser SD, Culler DE, and Demmel JW. 2005. Multi-purpose wireless accelerometers for civil infrastructures monitoring. In *Structural Health Monitoring*, FK Chang (ed.). DEStech, Lancaster.

77. Heo G, Lee WS, Lee G, and Lee D. 2005. Identifying dynamic characteristics of structures to estimate the performance of a smart wireless MA System. *Smart Sensor Technology and Measurement Systems*, SPIE 5758.

78. Liu L and Yuan FG. 2004. Development of wireless piezoelecric sensor on the MICA platform. In *Advanced Smart Materials and Smart Structures Technology*, DEStech, Lancaster.

79. Ruiz-Sandoval M, Spencer BF, and Kurata N. 2003. Development of a high sensitivity accelerometer for the mica platform. In *Structural Health Monitoring*, FK Chang (ed.). DEStech, Lancaster.

80. Mitchell K, Banerjee S, and Sholy B. 2007. Wireless acquisition system for automated near real-time structural monitoring using ultrasonic sensors. In *Structural Health Monitoring*, FK Chang (ed.). DEStech, Lancaster.

81. Overly TG, Park G, and Farrar CR. 2007. Development of impedance-based wireless active-sensor node for structural health monitoring. In *Structural Health Monitoring*, FK Chang (ed.). DEStech, Lancaster.

82. Galbreath JH, Townsend CP, Mundell SW, Hamel MJ, Esser B, Huston D, and Arms S. 2003. Civil structure strain monitoring with power-efficient, high-speed wireless sensor networks. *Proc 3rd Intl Workshop on Structural Health Monitoring*, Stanford University, Stanford, CA.

83. Sazonov E, Janoyan K, and Jha R. 2004. Wireless intelligent sensor network for autonomous structural health monitoring. *Smart Sensor Technology Measurement Systems*, SPIE 5384.

84. D'Souza R, Allen A, Andric O, Pham M, Chiou W, and Hester L. 2003. Using low-cost low-power wireless sensor devices to monitor the health of structures. In *Structural Health Monitoring*, FK Chang (ed.). DEStech, Lancaster.

85. Tanner NA, Farrar CR, and Sohn H. 2002. Structural health monitoring using wireless sensing systems with embedded processing. *Nondestructive Evaluation and Health Monitoring of Aerospace Materials and Civil Infrastructures*, SPIE 4704.

86. Sumners J and Champaigne KD. 2007. Wireless data acquisition system for impact detection and structural monitoring. In *Structural Health Monitoring*, FK Chang (ed.). DEStech, Lancaster.

87. Sodano HA, Park G, Leo DJ, and Inman DJ. 2003. Use of piezoelectric energy harvesting devices for charging batteries. *Smart Sensor Technology and Measurement Systems*, SPIE 5050.

88. Connor RJ, Santosuosso BJ, and Pessiki SP. 2002. Long-term wireless remote monitoring of the Lehigh river bridge. *Proc Structural Materials Technology V, an NDT Conference*, S Allampalli and G Washer (eds). Cincinnati.

89. Guigon R, Chaillout JJ, Jager T, and Despesse G. 2008. Harvesting raindrop energy: Theory. *Smart Mater Struct*, 17, 015038.

90. Guigon R, Chaillout JJ, Jager T, and Despesse G. 2008. Harvesting raindrop energy: Experimental. *Smart Mater Struct*, 17, 015038.

91. Katsuura H, Arakawa T, Yamagami T, and Fumoto K. 2004. Development of the low power consumption dynamic stress-frequency monitoring system by the piezo film sensor. *Proc IABMAS'04 Bridge Maintenance, Safety, Management and Cost*, Kyoto. Taylor & Francis, London.

92. Popenoe CH. 1971. Opti-mechanical stress-strain indicator. US Patent 3,602,186.

93. Curano LA. 2005. No-power MEMS shock threshold sensor. In *Structural Health Monitoring*, FK Chang (ed.). DEStech, Lancaster.

94. Hansen BJ, Carron CJ, Jensen BD, Hawkins AR, and Schultz SM. 2007. Plastic latching accelerometer based on Bistable compliant mechanisms. *Smart Mater Struct*, 16, 1967–1972.

95. Zhou W, Liao WH, and Li WJ. 2005. Analysis and design of a self-powered piezoelectric microaccelerometer. *Smart Electronics, MEMS, BioMEMS, and Nanotechnology*, SPIE 5763.

96. Peano F and Tambosso T. 2005. Design and optimization of a MEMS electret-based capacitive energy scavenger. *J Microelectromech Syst*, 14(3), 429–435.

97. du Plessis AJ, Huigsloot MJ, and Discenzo FD. 2005. Resonant packaged piezoelectric power harvester for machinery health monitoring. *Industrial and Commercial Applications of Smart Structures Technologies*, SPIE 5762.

98. Lallart M, Monnier T, Guy P, Guyomar D, Jayet Y, Lefeuvre E, Petit L, and Richard C. 2007. Self-powered structural health monitoring: Autonomous wireless sensors and actuators featuring piezoactive microgenerators. In *Structural Health Monitoring*, FK Chang (ed.). DEStech, Lancaster.

99. Casciati F, Faravelli L, and Rossi R. 2005. Design of a power harvester for wireless sensing applications. In *Structural Health Monitoring*, FK Chang (ed.). DEStech, Lancaster.

100. Elvin NG, Lajnef N, and Elvin AA. 2006. Feasibility of structural monitoring with vibration powered sensors. *Smart Mater Struct*, 15, 977–986.

101. Richter B and Tweifel J. 2007. On the need of modeling the interdependence between piezoelectric generators and their environmental excitation source. In *Structural Health Monitoring*, FK Chang (ed.). DEStech, Lancaster.

102. Clark WW, Romeiko JR, Charnegie DA, Kusic G, and Mo C. 2007. A case study in energy harvesting for powering a wireless measurement system. In *Structural Health Monitoring*, FK Chang (ed.). DEStech, Lancaster.

103. Challa VR, Prasad MG, and Fisher FT. 2007. Resonant frequency tunable vibration energy harvesting device. In *Structural Health Monitoring*, FK Chang (ed.). DEStech, Lancaster.

104. Churchill DL, Hamel MJ, Townsend CP, and Arms SW. 2003. Strain energy harvesting for wireless sensor networks. *Smart Electronics, MEMS, BioMEMS, and Nanotechnology*, SPIE 5055.

105. Arms SW, Townsend CP, Churchill DL, Hamel MJ, Augustin M, Yeary D, and Phan N. 2007. Optimization of energy harvesting wireless sensors with application to flight loads monitoring of helicopter components. In *Structural Health Monitoring*, FK Chang (ed.). DEStech, Lancaster.

106. Ayers JP, Greve DW, and Oppenheim IJ. 2003. Energy scavenging for sensor applications using structural strains. *Smart Systems and Nondestructive Evaluation for Civil Infrastructures*, SPIE 5057.

107. Discenzo FM, Chung D, and Loparo KA. 2006. Pump condition monitoring using self-powered wireless sensors. *Sound Vib*, (May), 12–15.

108. Wang J, Tan CA, and Ashebo D. 2007. Power flows of energy harvesting as measurands in structural health onitoring. In *Structural Health Monitoring*, FK Chang (ed.). DEStech, Lancaster.

109. Spooncer RC and Philp GS. 1995. Hybrid optical fiber sensors. In *Optical Fiber Sensor Technology*, K Grattan and B Meggitt (eds). Chapman & Hall, London.

110. Hunt B. 2003. Pier pressure. *Bridge Design and Engineering*, First Quarter.

111. Grisso BL, Kim J, Farmer JR, Ha DS, and Inman DJ. 2007. Autonomous impedance-based SHM utilizing harvested energy. In *Structural Health Monitoring*, FK Chang (ed.). DEStech, Lancaster.

112. Babic D, Johnson L, Rauch W, Baxley J, and Johnson M. 2005. Electrochemical power scavenging technology. In *Structural Health Monitoring*, FK Chang (ed.). DEStech, Lancaster.

113. Suster M, Ko WH, and Young DJ. 2004. An optically powered wireless telemetry module for high-temperature MEMS sensing and communication. *J Microelectromech Syst*, 13(3), 536–541.

114. DeHennis AD and Wise KD. 2005. A wireless microsystem for the remote sensing of pressure, temperature, and relative humidity. *J Microelectromecha Syst*, 14(1), 12–22.

115. Esser B, Pelczarski N, Huston D, and Arms S. 2000. Wireless inductive robotic inspection of structures. *Proc IASTED, RA*, Honolulu, HI.

116. Harpster TJ, Hauvespre S, Dokmeci MR, and Najafi K. 2002. A passive humidity monitoring system for *in situ* remote wireless testing of micropackages. *J Microelectromech Syst*, 11(1), 61–67.

117. Spillman Jr WB and Durkee SR. 1995. Contactless interrogation of sensors for smart structures. US Patent 5,433,115.

118. Shinoda H and Oasa H. 2000. Wireless tactile sensing element using stress-sensitive resonator. *IEEE/ASME Trans Mechatronics*, 5(3), 258–265.

119. Sohn H, Greve DW, Oppenheim IJ, and Boscha AK. 2006. Non-contact generation and reception of guided waves using near-field inductive coupling. *Proc 4th China–Japan–US Symp on Structural Control and Monitoring*, Hangzhou.

120. Joshi A, Marble S, and Sadeghi F. 2001. Bearing cage temperature measurement using radio telemetry. *Proc Inst Mech Eng, Part J, J Eng Tribol*, 215, 471–481.

121. Jia Y, Henao-Sepulveda J, and Toledo-Quinones M. 2004. Wireless temperature sensor for bearing health monitoring. *Sensors and Smart Structures Technologies for Civil, Mechanical and Aerospace Systems*, SPIE 5391.

122. French ML, Iftekharunddin KM, Leeper DR, Samy RP, and Hwang WR. 2003. Bearing with wireless self-powered sensor unit. US Patent 6,535,135.

123. Renwen C, Baoqi T, Yong C, and Ke X. 1999. Health monitoring of rotorcraft using a novel non-contact signal transmission system. *Proc 2nd Intl Workshop on Structural Health Monitoring,* Stanford. Technomic, Lancaster.

124. Song KD, Yi WJ, Chu SH, and Choi SH. 2003. Microwave-driven thunder materials. *Microw Opt Technol Lett,* 36(5), 331–333.

125. Song KD, Choi SH, Golemblewski WT, Henderson K, and King G. 2004. Rectenna performance under a 200 W amplifier microwave. *Smart Electronic, MEMS, BioMEMS, and Nanotechnology,* SPIE 5389.

126. Brown WC, George RH, Heenan NI, and Wonson RC. 1969. Microwave to DC converter. US Patent 3,434,678.

127. Landis GA. 2005. Charging of devices by microwave power beaming. US Patent 6,967,462.

128. Zhao X, Qian T, Mei G, Kwan C, Zane R, Walsh C, Paing T, and Popovic Z. 2007. Active health monitoring of an aircraft wing with an embedded piezoelectric sensor/actuator network: II.wireless approaches. *Smart Mater Struct,* 16, 1218–1225.

129. Kim J, Yang SY, Song KD, Jones S, Elliot, and Choi SH. 2006. Microwave power transmission using a flexible rectenna for microwave-powered aerial vehicles. *Smart Mater Struct,* 15, 1243–1248.

130. Chuang J, Thomson DJ, and Bridges G. 2004. Wireless strain sensor based on resonant RF cavities. *Smart Electronic, MEMS, BioMEMS, and Nanotechnology,* SPIE 5389.

131. Spillman WB and Andresen RP. 1990. Optically powered sensor system with improved signal conditioning. US Patent 4,963,729.

132. Feng MQ. 1998. Optically powered electrical accelerometer and its field testing. *J Eng Mech,* 124(5), 513–519.

133. Finkenzeller K. 2003. *RFID Handbook,* 2nd Ed. Wiley, Chichester.

134. Dickerson NP, Simonen JT, Andringa MM, Wood SL, and Neikirk DP. 2005. Wireless low-cost corrosion sensors for reinforced concrete structures. *Sensors and Smart Structures Technologies for Civil, Mechanical, and Aerospace Systems,* SPIE 5765.

135. Bowles A, Gore J, and Tomka G. 2005. An amorphous alloy stress sensor for wireless, battery-free applications. *Sensors and Smart Structures Technologies for Civil, Mechanical, and Aerospace Systems,* SPIE 5765.

136. Andringa MM, Neikirk DP, and Wood SL. 2004. Unpowered wireless analog resistance sensor. *Sensors and Smart Structures Technologies for Civil, Mechanical and Aerospace Systems,* SPIE 5391.

137. Matsuzaki R and Todoroki A. 2004. Passive wireless strain monitoring of tire using capacitance change. *Sensors and Smart Structures Technologies for Civil, Mechanical and Aerospace Systems,* SPIE 5391.

138. Simonen JT, Andringa MM, Grizzle KM, Wood SL, and Neikirk DP. 2004. Wireless sensors for monitoring corrosion in reinforced concrete members. *Sensors and Smart Structures Technologies for Civil, Mechanical and Aerospace Systems,* SPIE 5391.

139. Watters DG, Jayaweera P, Bahr AJ, Huestis DL, Priyantha N, Meline R, Reis R, and Parks D. 2003. Smart Pebble™: Wireless sensors for structural health monitoring of bridge decks. *Smart Systems and Nondestructive Evaluation for Civil Infrastructures,* SPIE 5057.

140. Fletcher RR and Gershenfeld N. 2000. A remotely interrogated temperature sensor based on magnetic materials. *IEEE Trans Magn,* 36(5), 2794–2795.

141. Satou T, Kaneda Y, Nagaoka S, and Ogawa S. 2006. Development of wireless sensors using RFID techniques for concrete structures. *NDE Conf on Civil Engineering,* American Society of Nondestructive Testings, St. Louis.

142. NASA. 2006. *Wireless Measurement of Contact and Motion Between Contact Surfaces,* NASA Technical Briefs, LAR-16849-1.

143. Starr AF, Nemat-Nasser S, Smith DR, and Plaisted TA. 2004. Integrated sensing networks in composite materials. *Sensors and Smart Structures Technologies for Civil, Mechanical and Aerospace Systems*, SPIE 5391.

144. Huston D. 2000. Wireless techniques for structural sensing. *2nd US–Japan Cooperative Research Program on Autoadaptive Media*, Honolulu.

145. Wang ML, Satpathi D, and Ren JY. 1998. Development of wireless structural health monitoring systems. T124-3, *Structural Engineering Worldwide*, Elsevier.

146. Ehrlich J, Zerrouki A, Galisson A, and Demassieux N. 1996. A generic model for smart sensors based data acquisition system. *Smart Sensing, Processing and Instrumentation*, SPIE 2718.

147. Hartsfield N and Ringel G. 1994. *Pearls in Graph Theory A Comprehensive Introduction*. Dover Publications, Mineola, NY.

148. Blundell A. 1982. *Bond Graphs for Modeling Engineering Systems*. Ellis Horwood, Chichester.

149. Newman M, Barabasi AL, and Watts DJ. 2006. *The Structure and Dynamics of Networks*. Princeton University Press, Princeton.

150. Erdös P and Rényi A. 1960. On the evolution of random graphs. *Publicationes Mathematicae*, 6, 290–297.

151. Albert R and Barabasi AL. 2002. Statistical mechanics of complex networks. *Rev Mod Phys*, 74, 47.

152. Pastor-Satorras R, Rubi M, and Diaz-Guilera A. 2003. *Statistical Mechanics of Complex Networks*. Springer, New York.

153. Tsai LW. 2001. *Mechanism Design Enumeration of Kinematic Structures According to Function*. CRC Press, Boca Raton, FL.

154. Bollobas B. 1978. *Extremal Graph Theory*. Dover Publications, Mineola, NY.

155. Krishnamachari B. 2005. *Networking Wireless Sensors*. Cambridge University Press, Cambridge.

156. Casciati F, Casciati S, Faravelli L, and Rossi R. 2004. Hybrid wireless sensor network. *Sensors and Smart Structures Technologies for Civil, Mechanical and Aerospace Systems*, SPIE 5391.

157. Hedman R, Siljander A, and Tikka J. 2003. Embedded microcontroller based networked measurement system with strain gages tailored to fatigue crack detection. In *Structural Health Monitoring*, FK Chang (ed.). DEStech, Lancaster.

158. Jin MH, Wu WJ, Chen CK, Chen YF, Wen CM, Kao CY, Yu SA, Lin YH, Huang JG, Rao H, Hsu CH, and Lee CK. 2004. Hierarchical sensor network architecture for stationary smart node supervision. *Smart Electronic, MEMS, BioMEMS, and Nanotechnology*, SPIE 5389.

159. Civera P, Cravero M, Franco G, and Iabichino G. 2003. Development of an inclinometer chain for wireless data acquisition. In *Structural Health Monitoring*, FK Chang (ed.). DEStech, Lancaster.

160. Ghosal A, Prosser WH, Kirikera G, Schulz MJ, Hughes DJ, and Orisamolli W. 2003. Concepts and development of bio-inspired distributed embedded wired/wireless sensor array architectures of acoustic wave sensing in integrated aerospace vehicles. In *Structural Health Monitoring*, FK Chang (ed.). DEStech, Lancaster.

161. Sharma D, Zadorozhny V, and Chrysanthis PK. 2005. Structural health monitoring with whirlpool. In *Structural Health Monitoring*, FK Chang (ed.). DEStech, Lancaster.

162. Rye P, Schaff K, Huang Y, Ghezzo F, and Nemat-Nasser S. 2007. Embedded distributed sensing network: Integration considerations and findings. In *Structural Health Monitoring*, FK Chang (ed.). DEStech, Lancaster.

163. Deix S and Ralbovsky M. 2007. Intelligent sensor networks—The future of SHM. In *Structural Health Monitoring*, FK Chang (ed.). DEStech, Lancaster.

164. Han L, Newhook JP, and Mufti AA. 2004. Centralized remote structural monitoring and management of real-time data. *Nondestructive Evaluation and Health Monitoring of Aerospace Materials and Composites III*, SPIE 5393.

165. Bakker JD, Postema FJ, and Forster U. 2004. Web-based, fully automated remote monitoring of structures. *Proc IABMAS'04 Bridge Maintenance, Safety, Management and Cost*, Kyoto. Taylor & Francis, London.

166. Chen R and Gong R. 2003. TCP/IP based remote measurement system for structural health monitoring of civil infrastructures. In *Structural Health Monitoring*, FK Chang (ed.). DEStech, Lancaster.

167. Zhu Y, Fu Y, Chen W, Huang S, and Bennett KD. 2003. Health monitoring system for Dafosi cable-stayed bridge. *Smart Systems and Nondestructive Evaluation for Civil Infrastructures*, SPIE 5057.

168. Trego A. 2003. Installation of the autonomous structural integrity monitoring system. In *Structural Health Monitoring*, FK Chang (ed.). DEStech, Lancaster.

169. Beard S, Qing PX, Hamilton M, and Zhang DC. 2004. Multifunctional software suite for structural health monitoring using SMART technology. *Proc 2nd European Workshop on Structural Health Monitoring*. DEStech, Munich.

170. Zhang Y, Li J, and Lu LW. 2006. Sensitivity study of a hysteresis-based damage detection method to Lossy sensor data compression. *Proc 4th China–Japan–US Symp on Structural Control and Monitoring*, Hangzhou.

171. Zhang Y and Li J. 2007. Linear predictor-based lossless compression of vibration sensor data: Systems approach. *J Eng Mech*, 133(4).

172. Milos FS and Karunaratne KS. 2003. Active wireless temperature sensors for aerospace thermal protection systems. *Smart Nondestructive Evaluation and Health Monitoring of Structural and Biological Systems II*, SPIE 5047.

173. Su J, Zhang J, He Y, and Liu W. 2006. Design a wireless sensor network with ZigBee standard for structural health monitoring of bridge. *Proc 4th China–Japan–US Symp on Structural Control and Monitoring*, Hangzhou.

174. Beckerman M. 1997. *Adaptive Cooperative Systems*. Wiley, New York.

175. Clare LP, Pottie GJ, and Agre JR. 1999. Self-organizing distributed sensor networks. *Unattended Ground Sensor Technologies and Applications*, SPIE 3713.

176. Agre JR, Clare LP, Pottie GJ, and Romanov NP. 1999. Development platform for self-organizing wireless sensor network. *Unattended Ground Sensor Technologies and Applications*, SPIE 3713, pp. 257–268.

177. Basheer MR, Derriso M, and Rao VS. 2003. Self organizing wireless sensor networks for structural health monitoring. In *Structural Health Monitoring*, FK Chang (ed.). DEStech, Lancaster.

178. Glaser SD. 2004. Some real-world applications of wireless sensor nodes. *Sensors and Smart Structures Technologies for Civil, Mechanical and Aerospace Systems*, SPIE 5391.

179. Kling R, Adler R, Huang J, Hummel V, and Nachman L. 2005. Intel mote-based sensor networks. *Struct Control Health Monit*, 12, 469–479.

180. Mastroleon L, Kiremidjian AS, Carryer E, and Law KH. 2004. Design of a new power-efficient wireless sensor system for structural health monitoring. *Nondestructive Detection and Measurement for Homeland Security II*, SPIE 5395.

181. Ou J, Li H, and Yu Y. 2004. Development and performance of wireless sensor network for structural health monitoring. *Sensors and Smart Structures Technologies for Civil, Mechanical and Aerospace Systems*, SPIE 5391.

182. Chen Y, Wu W, Chen C, Wen C, Jin M, Gau C, Chang C, and Lee C. 2004. Design and implementation of smart sensor nodes for wireless disaster monitoring systems. *Sensors and Smart Structures Technologies for Civil, Mechanical and Aerospace Systems*, SPIE 5391.

183. Nitta Y, Nagayama T, Spencer Jr. B, and Nishitani A. 2005. Rapid damage assessment for the structures utilizing smart sensor MICA2 MOTE. In *Structural Health Monitoring*, FK Chang (ed.). DEStech, Lancaster.

184. Hashimoto Y, Masuda A, and Sone A. 2005. Prototype of sensor network with embedded local data processing. *Sensors and Smart Structures Technologies for Civil, Mechanical, and Aerospace Systems*, SPIE 5765.

185. Shirgur VL and Rao VS. 2003. Distributed intelligence in wireless sensor networks. *Smart Electronics, MEMS, BioMEMS, and Nanotechnology*, SPIE 5055.

186. Chintalapudi K, Fu TS, Johnson EA, and Govindan R. 2005. Structural damage detection and localization using wireless sensor networks with low power consumption. In *Structural Health Monitoring*, FK Chang (ed.). DEStech, Lancaster.

187. Zimmerman AT and Lynch JP. 2007. Automated damage estimation in wireless sensing networks using parallelized model updating. In *Structural Health Monitoring*, FK Chang (ed.). DEStech, Lancaster.

188. Wolfram S. 2002. *A New Kind of Science*. Wolfram Media.

189. Prabhugord M, Pearson J, Peters K, and Zikry M. 2004. Demonstration of failure identification methodology incorporating sensor degradation. *Sensors and Smart Structures Technologies for Civil, Mechanical and Aerospace Systems*, SPIE 5391.

190. Wang P, Valencia P, Prokopenko M, Price D, and Poulton G. 2003. Self-reconfigurable sensor networks in ageless aerospace vehicles. *Proc 11th Intl Conf on Advanced Robotics*, Portugal.

191. Price DC, Scott DA, Edwards GC, Batten A, Farmer AJ, Hedley M, Johnson ME, Leis CJ, Poulton GT, Prokopenko M, Velencia P, and Wang P. 2003. An integrated health monitoring system for an ageless aerospace vehicle. In *Structural Health Monitoring*, FK Chang (ed.). DEStech, Lancaster.

192. Foreman M, Prokopenko M, and Wang P. 2003. Phase transitions in self-organising sensor networks. *Proc 7th European Conference on Artificial Life*, Germany.

193. Rupp A and Dornbusch T. 2005. Powerful data recording systems for diagnosis and fatigue life monitoring applied at automobiles, commercial vehicles, heavy machinery and wind turbines. In *Structural Health Monitoring*, FK Chang (ed.). DEStech, Lancaster.

5

Data Processing

Primary uses of statistical methods in SHM include the following: (1) *remove randomness or noise* from measurements in an effort to uncover and estimate the values of underlying properties; (2) *condense* large data sets into less voluminous forms with higher information densities; (3) *organize the data* so as to test various hypotheses or to classify the state of the system or process; and (4) *recognize underlying patterns or structures* in the data. The first two of these—removal of randomness and data condensation—are the subject of this chapter. The second two—classification and pattern recognition—are primarily left to the next chapter.

5.1 Probability Distributions

Randomness and uncertainty pervade all physical events and measurements. SHM must deal with uncertain structural loads, properties, and measurements [1]. A rational method of assessing and managing uncertainty describes events in terms of formalized notions of random events and the use of tools from statistics and probability theory. Most random events, including SHM data, are not entirely random. Instead, the data contain underlying structures that, when recognized, can be exploited to advantage. A centerpiece of statistical analysis is the notion of *repeated random events*, which represents the idealized situation where all of the conditions that give rise to an event are constant, except for the random component [2–4].

Random variables are the outcomes of random events. Random variables may be continuous, discrete, or mixed. The data ranges can be finite, semi-infinite, or infinite. The values can be ordered or unordered, as with colors. The data may be one-dimensional or multidimensional, including a heterogeneous mix of data types. An example of a mixed heterogeneous data arises in crack detection. The probability of detecting cracks falls into discrete categories of "Detected" and "Not Detected" and the mixed discrete–continuous probability of a given crack size. A nontrivial SHM issue is to determine the range of possible events a priori. It is worthwhile to note that many more important (and interesting) events are rogues that occur outside of expected ranges.

Probability distribution functions or equivalently *probability density functions* (PDFs) are idealized mathematical representations of random outcomes from repeated events and can often serve as a good approximation to the randomness encountered in nature. The probability of a given outcome is assumed to be always greater than or equal to zero. A common feature of all PDFs is that the measure (sum and/or integral) over the range of possible outcomes equals one. For random variables with N possible discrete outcomes, each with probability p_i,

$$p_i \geq 0 \quad \forall i. \tag{5.1}$$

The sum over possible outcomes is

$$\sum_{i=1}^{N} p_i = 1. \tag{5.2}$$

Random variables with continuous outcomes have continuous probability distributions, $p(x)$. If x varies from x_{min} to x_{max}, then

$$\int_{x_{min}}^{x_{max}} p(x)\,dx = 1. \tag{5.3}$$

$P(a, b)$ is the probability of an event occurring in an interval (a, b), where

$$P(a, b) = \int_{a}^{b} p(x)\,dx. \tag{5.4}$$

Similar relations hold for multidimensional, and mixed continuous and discrete event probability distributions.

Aside from the positivity and unit total probability constraints of Equations 5.1 through 5.3, PDFs can take on many forms and shapes. A PDF encountered in practice can be well-known idealized forms, such as Gaussian, binomial, Rayleigh, and so on, or it can be a unique form associated with the particular circumstances. In many cases, generic descriptors provide a significant amount of information about the nature of a PDF without resorting to a completely detailed description, especially for event outcomes that are numerical and ordered. Typical generic descriptors are the *expected value (mean)* μ, *variance* σ^2, *hinge points*, and *higher-order moments*. The expected value of a discrete random variable is

$$\mu = \sum_{i=1}^{N} x_i p_i. \tag{5.5}$$

The expected value of a continuous random variable is

$$\mu = \int_{x_{min}}^{x_{max}} x p(x)\,dx. \tag{5.6}$$

The variance of a discrete random variable is

$$\sigma^2 = \sum_{i=1}^{N} (x_i - \mu)^2 p_i. \tag{5.7}$$

The variance of a continuous random variable is

$$\sigma^2 = \int_{x_{min}}^{x_{max}} (x - \mu)^2 p(x)\,dx. \tag{5.8}$$

The hinge points split the distribution so that one quarter of the probability lies on one side and three quarters lie on the other.

The mean and variance are examples of a more general calculation known as the *moment*. For a discrete random variable, the kth moment ($k \geq 2$) is

$$\mu_k = \sum_{i=1}^{N} (x_i - \mu)^k p_i, \tag{5.9}$$

and the kth moment for a continuous random variable is

$$\mu_k = \int_{x_{\min}}^{x_{\max}} (x - \mu)^k p(x) \, dx. \tag{5.10}$$

The mean is the first moment and the variance is the second moment. The third moment (normalized by the variance) is the *skewness*. The fourth moment is the *flatness*. The skewness of a symmetric distribution, such as a Gaussian, is zero. The flatness (fourth moment) normalized by the variance minus three is the *kurtosis* (or *curtosis*), κ, where

$$\kappa = \frac{E(x^4)}{[E(x^2)]^2} - 3. \tag{5.11}$$

The kurtosis of a Gaussian distribution is zero. Distributions with a positive kurtosis are *super-Gaussian*. Distributions with a negative kurtosis are *sub-Gaussian* [5].

Comparing the probability moments of measured distributions with those calculated from idealized distributions provides a check on the validity of the assumptions that underlie the system models. For example, a linear system with Gaussian inputs has Gaussian outputs. If the probability moments of the system outputs differ from Gaussian values, then it may be reasonable to question the underlying assumptions, such as system linearity [6].

5.1.1 Joint and Conditional Probability, and Bayes' Theorem

Many situations arise when the outcomes have more than one random variable. The outcome random variables may be independent, or they may depend on one another. Consider the case of two possible outcomes x and y. If x and y are such that when x occurs, y does not occur, then the events are *mutually exclusive* (Figure 5.1). If all possible outcomes are either x or y, then x is the *complement* of y and may be denoted $\bar{y}$ (Figure 5.2). Complementary outcomes satisfy the relation

$$1 = p(y) + p(\bar{y}). \tag{5.12}$$

The *joint PDF* $p(x,y)$ is the probability of two random events, x and y, occurring simultaneously (Figure 5.3). If x and y are *independent*,

$$p(x,y) = p(x)p(y). \tag{5.13}$$

The *conditional probability*, $p(y \mid x)$, is the probability that y will occur given that x has occurred. When x and y are independent, then

$$p(y \mid x) = p(y). \tag{5.14}$$

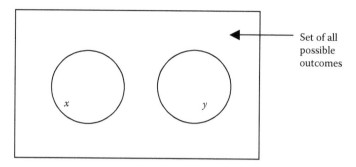

FIGURE 5.1   Venn diagram of two mutually exclusive, but not exhaustive, events x and y.

The region of overlap in the space possible outcomes, that is, $x \cap y$, indicates the probability of events x and y occurring simultaneously (Figure 5.3). The possibility of x and y occurring simultaneously in terms of conditional probabilities is

$$p(x \cap y) = p(x)p(y \mid x) = p(y)p(x \mid y). \tag{5.15}$$

When y and $\bar{y}$ exhaustively cover the space of possible events,

$$p(x) = p(x \cap y) + p(x \cap \bar{y}), \tag{5.16}$$

$$p(x) = p(y)p(x \mid y) + p(\bar{y})p(x \mid \bar{y}). \tag{5.17}$$

Combining Equations 5.15 and 5.17 produces *Bayes' theorem* for two events:

$$p(y \mid x) = \frac{p(y)p(x \mid y)}{p(y)p(x \mid y) + p(\bar{y})p(x \mid \bar{y})}. \tag{5.18}$$

A more general form of Bayes' theorem arises when N mutually exclusive events y_1, y_2, ..., y_N span the set of possible outcomes:

$$p(y_i \mid x) = \frac{p(y_i)p(x \mid y_i)}{\sum_{j=1}^{N} p(y_j)p(x \mid y_j)}. \tag{5.19}$$

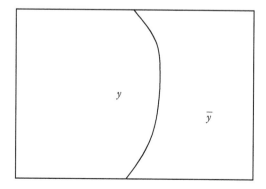

FIGURE 5.2   Venn diagram of two mutually exclusive and exhaustive outcomes.

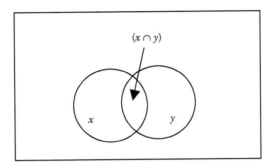

FIGURE 5.3 Venn diagram of two overlapping events.

Bayes' theorem can be a powerful tool in decision making since it has the quantitative ability to incorporate past knowledge or prior belief. Usage of the term "Bayesian" often extends beyond that of Bayes' theorem in Equation 5.19 to describe the analysis of more complicated cases where every possible outcome and the associated probability are known. Chapter 6 discusses some of these issues in more detail.

5.1.2 Functions of PDFs

Functions of single or multiple random variables are also random variables. An example is forming a single random variable y as a function of the random variables $x_1, x_2, \ldots, x_n$:

$$y = f(x_1, x_2, \ldots, x_n). \tag{5.20}$$

Significant simplification arises if y is a linear combination, L, of the x_i:

$$y = L = c_1 x_1 + c_2 x_2 + \cdots + c_n x_n. \tag{5.21}$$

Assuming that the random variables $x_1, x_2, \ldots, x_n$ have means $\mu_1, \mu_2, \ldots, \mu_n$ and variances $\sigma_1, \sigma_2, \ldots, \sigma_n$, then the mean, μ_L, is

$$\mu_L = c_1 \mu_1 + c_2 \mu_2 + \cdots + c_n \mu_n \tag{5.22}$$

and the variance, σ_L^2, is

$$\sigma_L^2 = \{c\}^{\mathrm{T}} [\Sigma] \{c\}, \tag{5.23}$$

where

$$\{c\}^{\mathrm{T}} = [c_1 \quad c_2 \quad \cdots \quad c_n] \tag{5.24}$$

and

$$[\Sigma] = \begin{bmatrix} \sigma_1^2 & \sigma_{12} & \cdots & \sigma_{1n} \\ \sigma_{21} & \sigma_2^2 & & \\ \vdots & & \ddots & \\ \sigma_{n1} & & & \sigma_n^2 \end{bmatrix}. \tag{5.25}$$

The cross-covariances σ_{ij} in Equation 5.25 are

$$\sigma_{ij} = E(x_i x_j) = \int (x_i - \mu_i)(x_j - \mu_j) p(x_i, x_j) \, dx_i \, dx_j. \tag{5.26}$$

When the random variables x_i and x_j are *independent*, then the cross-covariances vanish, that is,

$$\sigma_{ij} = 0, \quad i \neq j \tag{5.27}$$

and

$$\sigma_L^2 = c_1^2 \sigma_1^2 + c_2^2 \sigma_2^2 + \cdots + c_n^2 \sigma_n^2. \tag{5.28}$$

A primary motivation for taking repeated measurements is the reduction of PDF spreads. In the case where the random variables in L are statistical repeats, then

$$\mu_1 = \mu_2 = \cdots = \mu_N = \mu \tag{5.29}$$

and

$$\sigma_1 = \sigma_2 = \cdots = \sigma_N = \sigma. \tag{5.30}$$

If

$$a_1 = a_2 = \cdots = a_N = \frac{1}{N}, \tag{5.31}$$

then

$$\mu_L = \mu \tag{5.32}$$

and

$$\sigma_L = \frac{1}{\sqrt{N}} \sigma. \tag{5.33}$$

Thus the addition of data from repeated experiments reduces the spread of mean value statistics. These results readily generalize to other statistical measures that are linear or relatively smooth nonlinear combinations of random variables.

5.1.3 Common PDFs

The probability and statistics literature describes and tabulates a multitude of PDFs. The following lists some of the common SHM-pertinent PDFs.

The *binomial* PDF $B(n, p)$ applies to the case of a finite number, x, of discrete binary outcomes with probability p in n independent trials [7]:

$$p(x) = B(n, p) = \binom{n}{x} p^x (1 - p)^{n-x}, \quad x = 0, 1, 2, \ldots, n. \tag{5.34}$$

The *Poisson* PDF is an approximation of the binomial PDF for the case where n is large and p is small and $\lambda = np$ [7]:

$$p(x) = B(\lambda) = e^{-\lambda} \lambda^x / x!, \quad x = 0, 1, 2, \ldots. \tag{5.35}$$

The *Gaussian* or *normal* PDF has a range of $-\infty$ to ∞. If $\mu = 0$ and $\sigma^2 = 1$, the PDF takes the standard normal form

$$p(x) = \varphi(x) = \frac{1}{\sqrt{2\pi}} e^{-x^2/2}. \tag{5.36}$$

The skewness and kurtosis of the normal Gaussian PDF both equal zero (fourth moment equals three) [4,6].

Shifting and stretching the Gaussian PDF along the x-axis, so that $\mu \neq 0$ and $\sigma^2 \neq 1$, transforms the distribution into

$$p(z) = \varphi\left(\frac{x-\mu}{\sigma}\right). \tag{5.37}$$

A linear combination of Gaussian random variables is also Gaussian with the mean and variance given by Equations 5.22 and 5.23 [4]. A linear combination of independent non-Gaussian random variables tends to be Gaussian by virtue of the Central Limit Theorem [2,4]. Owing to the wide applicability of the Central Limit Theorem, it is often (although sometimes erroneously) assumed that the distribution of a random variable is Gaussian, unless there is reason to believe otherwise. Deviations from Gaussian behavior can reveal important underlying structures in a system, such as nonlinearities [6,8].

The *multiple variable Gaussian* distribution is

$$p(x_1, x_2, \ldots, x_n) = \frac{1}{(2\pi)^{n/2}(\det \Sigma)^{1/2}} \exp\left(-\frac{1}{2}\{x-\mu\}^T[\Sigma]\{x-\mu\}\right), \tag{5.38}$$

where Σ is the covariance matrix defined in Equation 5.25 and $\{x - \mu\}$ is a column vector of the x_i [7].

The *two-parameter lognormal* distribution is

$$p(x) = \varphi\left[\frac{\ln(x/c)}{\zeta}\right]. \tag{5.39}$$

with $c > 0$ and $\zeta > 0$ as adjustable parameters. This distribution is useful in applications such as constructing fragility curves for reliability analysis [9].

The *beta* family of distributions are 1-D and continuous over a finite range, a to b. Adjusting the two parameters c and d alters the shape of the beta distribution into a variety of useful forms [10],

$$p(x) = \left(\frac{1}{b-a}\right) \frac{\Gamma(a+b)}{\Gamma(a)\Gamma(b)} \left(\frac{x-a}{b-a}\right)^{c-1} \left[1 - \left(\frac{x-a}{b-a}\right)\right]^{d-1} \begin{cases} a \leq x \leq b \\ 0 \leq a < b \\ c, d > 0 \end{cases}. \tag{5.40}$$

$\Gamma(a)$ is the *gamma* function defined as

$$\Gamma(a) = \int_0^\infty x^{a-1} e^{-y} \, dy. \tag{5.41}$$

When a is a positive integer,

$$\Gamma(a) = (a-1)! \tag{5.42}$$

The mean, μ_β, and variance, σ_β, of the beta distribution are

$$\mu_\beta = a + (b - a)\left(\frac{c}{c+d}\right), \tag{5.43}$$

$$\sigma_\beta^2 = (b - a)^2 \left[\frac{cd}{(c+d)^2 (c+d+1)}\right]. \tag{5.44}$$

A characteristic property of the beta distribution is that it takes on a variety of different forms by varying the parameters a, b, c, and d. The special case of $c = 0$ and $d = 1$ causes the beta distribution to become the *uniform* distribution, $U(x)$, with an equal probability over the interval $[a, b]$:

$$p(x) = \left(\frac{1}{a - b}\right) \tag{5.45}$$

with mean, μ_U, and variance, σ_U,

$$\mu_U = \frac{a + b}{2}, \tag{5.46}$$

$$\sigma_U^2 = \frac{(b - a)^2}{12}. \tag{5.47}$$

A random variable formed as the square root of the sum of the squares of two Gaussian random variables with zero means and equal variance has a *Rayleigh* distribution.

$$p(x) = \frac{x}{\sigma^2} e^{-x^2/2\sigma^2}. \tag{5.48}$$

The range of the Rayleigh distribution is 0 to ∞. A common use of the Rayleigh distribution is in the analysis of stationary random time histories with a Gaussian amplitude distribution. The envelope function of the time history is a random variable with a Rayleigh distribution. *Rician* statistics describe the envelopes of Gaussian signals with the addition of a coherent (sinusoidal) component [11]. As an example application of envelope statistics, Hedl et al. compared the Rayleigh, Rice, K, Nakagami, Generalized Nakagami, Compound, and Weibull distributions for ultrasound-based corrosion detection [12]. The Weibull distribution performed best in this particular application. The *Maxwell* PDF is the square root of the sum of the squares of three Gaussian random variables. The Maxwell PDF arises in statistical mechanics with the description of velocities of atoms in a perfect gas.

The *chi-squared* PDF with r degrees of freedom is the sum of squares of a set of r Gaussian independent variables, each with the same variance [7]:

$$p(x) = X_r^2(x) = \frac{x^{(x/2)-1} e^{-x/2}}{2^{r/2} \Gamma(r/2)}. \tag{5.49}$$

The mean, μ_X, and variance, σ_X^2, are

$$\mu_X = r, \tag{5.50}$$

$$\sigma_X^2 = 2r. \tag{5.51}$$

The chi-squared distribution can determine whether observed data are consistent with the assumptions of a sum of squares Gaussian structure [4].

The F distribution is the ratio of two chi-squared variables (X_a^2 and X_b^2) weighted by the respective degree of freedom counts a and b.

$$p(x) = F_{a,b}(x) = \frac{bX_a^2}{aX_b^2}. \tag{5.52}$$

The mean, μ_F, and variance, σ_F^2, are

$$\mu_F = \frac{b}{b-2}, \quad b > 2, \tag{5.53}$$

$$\sigma_F^2 = \frac{[2b^2(a+b-2)]}{[a(b-2)^2(b-4)]}. \tag{5.54}$$

The F distribution can compare groups of data to determine whether they come from different or similar distributions.

The F_{r1} PDF is the *Student's t* distribution.

$$p(x) = \frac{\Gamma((r+1)/2)(1 + (x^2/r))^{-((r+1)/2)}}{\sqrt{r\pi}\,\Gamma(r/2)}. \tag{5.55}$$

The mean, μ_t, and variance, σ_t^2, are

$$\mu_t = 0, \tag{5.56}$$

$$\sigma_t^2 = \frac{r}{r-2}. \tag{5.57}$$

This PDF can compare samples from small data sets where the variance is not known a priori.

5.1.4 Rare Event Distributions

Extreme events are those that occur in the outer ranges of PDFs, that is, the tails. *Rare event distributions* are specialized PDFs that represent the behavior of extreme events [13]. The *Gumbel* or *Fisher–Tippett Type I* is a common rare event PDF.

$$p(x) = a \exp\left[-a(x-b) - e^{-a(x-b)}\right] \begin{cases} -\infty \le x \le \infty \\ a > 0 \\ b < \infty \end{cases}. \tag{5.58}$$

The mean, μ_G, and variance, σ_G^2, are

$$\mu_G \cong b + \left(\frac{0.577}{a}\right), \tag{5.59}$$

$$\sigma_G^2 = \frac{\pi^2}{6a^2}. \tag{5.60}$$

The *Fréchet* or *Fisher–Tippett Type II* is a second common rare event PDF [10].

$$p(x) = \left(\frac{a}{b}\right)\left(\frac{b}{x}\right)^{a+1} \exp\left[-\left(\frac{b}{x}\right)^a\right] \begin{cases} x \geq 0 \\ a, b > 0 \end{cases} \tag{5.61}$$

The mean, μ_{II}, and variance, σ_{II}^2, are

$$\mu_{II} = b\Gamma\left[1 - \left(\frac{1}{a}\right)\right], \tag{5.62}$$

$$\sigma_{II}^2 = b^2\left\{\Gamma\left[1 - \left(\frac{2}{a}\right)\right] - \Gamma^2\left[1 - \left(\frac{1}{a}\right)\right]\right\}. \tag{5.63}$$

The Weibull PDF is [7]

$$p(x) = \lambda^\beta \beta x^{\beta-1} \exp[-(\lambda x)^\beta], \quad x \geq 0. \tag{5.64}$$

The mean μ_W, and variance, σ_W^2, are

$$\mu_W = \frac{1}{\lambda}\Gamma\left(1 + \frac{1}{\beta}\right), \tag{5.65}$$

$$\sigma_W^2 = \frac{1}{\lambda^2}\left[\Gamma\left(1 + \frac{2}{\beta}\right) - \Gamma^2\left(1 + \frac{1}{\beta}\right)\right]. \tag{5.66}$$

It should be noted that the above three distributions have a similar appearance and are actually subsets of a generalized version of a family of extreme event distributions.

The generalized *Pareto* distribution is [14]

$$p(x) = 1 - \left(1 + \frac{\xi x}{\tilde{\sigma}}\right)^{-1/\xi} \tag{5.67}$$

with $x > 0$ and $(1 + \xi x/\tilde{\sigma}) > 0$.

5.2 Statistics on Unordered Data

Similar to random variables, data come in many forms. Data can be quantitative, with specific numbers attached to each datum; or can be nonquantitative, such as labels, subjective evaluations, or beliefs. Quantitative data may be integer or noninteger, with discrete or continuous, and finite or infinite ranges. The data can be unordered and unorganized, or the data can be organized into ordered series, images, or more complex formats, such as databases. Heterogeneous sets and organizations of these different data types are routinely encountered. Bridge management systems collect and assemble heterogeneous image, structural, and inspection data for sets of bridges [15].

An initial concern in processing random data is to understand the random nature of the data. This usually involves estimating the characteristics of the PDFs. This may include calculations of the *histogram* and the statistical moments of distribution, that is, mean, variance, skewness, kurtosis, and so on. A histogram is a set of counts of how many data lie within individual nonoverlapping ranges. These ranges, also known as bins, nominally span the space of possible outcomes. If the data distributions are not well known a priori, then selecting appropriate bin sizes and locations requires a certain amount of experimentation [8]. The *mean*, $\overline{X}$, *variance*, s^2, *skewness* γ_1, and *kurtosis* γ_2 of a set of univariate quantitative data samples are

$$\overline{X} = \frac{1}{N} \sum_{i=1}^{N} x_i, \tag{5.68}$$

$$s^2 = \frac{1}{N-1} \sum_{i=1}^{N} (x_i - \overline{X})^2 = \frac{1}{N-1} \sum_{i=1}^{N} (x_i^2 - \overline{X}^2), \tag{5.69}$$

$$\gamma_1 = \frac{\sum_{i=1}^{N} (x_i - \overline{X})^3}{s^3(N-1)}, \tag{5.70}$$

$$\gamma_2 = \frac{\sum_{i=1}^{N} (x_i - \overline{X})^4}{s^4(N-1)} - 3. \tag{5.71}$$

The *covariance* of a set of N pairs of samples (x_i and y_i) of two random variables, x and y, is [16]

$$R_{xy} = \frac{1}{N-1} \sum_{i=1}^{N} (x_i - \overline{X})(y_i - \overline{Y}). \tag{5.72}$$

The covariance normalized by the variances of the individual data sets, s_x and s_y, is the *correlation*, ρ_{xy}:

$$\rho_{xy} = \frac{R_{xy}}{(s_x^2 s_y^2)^{1/2}}. \tag{5.73}$$

The correlation of two variables always lies between -1 and 1. Highly correlated data have $|\rho_{xy}| \approx 1$. If $|\rho_{xy}| \approx 0$, the data are uncorrelated. Situations with the correlation in an intermediate range between 0 and 1 generally require more sophisticated methods to establish whether two variables are related [17]. The correlation concept also applies to the collection of multiple simultaneous samples of random variables. In this case, multiple pairs of correlations between the several random variables readily assemble into a matrix of correlations with a form similar to the variance–covariance matrix in Equation 5.25. It should be noted that the moments and correlations calculated from data samples generally differ from the PDF moments. Random components that reflect the sampling process cause the difference.

5.3 Statistics on Ordered Data

Many data sets are ordered. A common case is the sampling of data as a function of time or position. Ordered data set information generally has a higher degree of structure than that from unordered sets. The associated data processing techniques must account for this added structure. Initial processing steps on ordered data alter the format to be more suitable for analysis and/or presentation. Preprocessing includes the removal of offsets, drifts, trends, noise, data in particular frequency bands, and outliers. The following is a discussion of common statistics on ordered data. For consistency, it is assumed that the data are temporally ordered. Spatially ordered data can be treated similarly.

Fitting and subtracting a line from the data with the following formulae will remove an offset and linear drift:

$$b_1 = \frac{12 \sum_{i=1}^{N} i x_i - 6(N+1) \sum_{i=1}^{N} x_i}{N(N^2 - 1)}, \tag{5.74}$$

$$b_0 = \frac{1}{N} \sum_{i=1}^{N} x_i - b_1 \frac{(n+1)}{2}, \tag{5.75}$$

$$\hat{x}_i = x_i - b_0 - b_i i. \tag{5.76}$$

Figure 5.4 shows a sample time history before and after trend and offset removal. It should be noted that the process of demeaning and detrending requires adding a long sequence of numbers. Avoiding round-off errors with the sums in Equations 5.74 and 5.75 often requires double precision arithmetic.

Situations where the data contain an unwanted low-frequency drift may warrant using more sophisticated drift removal techniques, such as with a digital high-pass filter (Figure 5.5) [18,19].

A *smoothing* or *low-pass* filter removes high-frequency content in a signal, including unwanted noise. The *moving-average* (MA) filter is a sample low-pass filter that averages

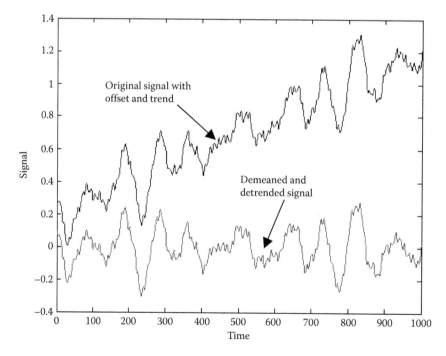

FIGURE 5.4 Demeaned and detrended data.

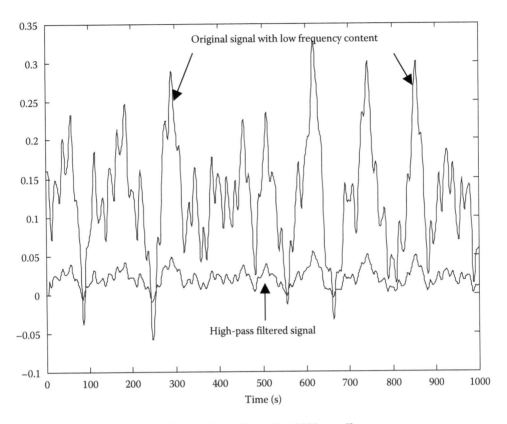

FIGURE 5.5 High-pass filtered data (10 pole Bessel filter with 0.25 Hz cutoff).

each datum with its neighbors, that is,

$$\hat{x}_i = \sum_{i=-a}^{a} w_i x_i,$$ (5.77)

where w_i are weighting factors and n is a small integer. Multiple passes of smoothing filters can apply additional smoothing as needed. Figure 5.6 shows a raw signal and a version smoothed with a 25-point equal-weight MA filter.

Data sets often contain outliers. These outliers may be due to events of minor significance, or may indicate the occurrence of an important event. A *moving-median* filter is effective at removing isolated outliers [8,20]. As an example, Figure 5.7 shows a time history of torsional strain data from a rotating graphite–fiber composite drive shaft. A wireless link transmitted the strain data. Telemetry errors in the wireless link produced data dropouts at random times. Figure 5.8 shows the same data set with the outliers being removed by a three-point moving median filter. Li et al. used a similar moving three-point median technique to precondition AE data [21]. Applying outlier removal techniques requires a bit of caution when the source of the outliers is unknown. As an example, Shenton et al. describe a case of bridge strain data with large spike-like outliers [22]. From an initial analysis, it was not possible to determine whether actual strain spikes caused the data spikes (a potentially troublesome situation) or whether instrumentation anomalies were the source. A series of retests with different transducers, including peak strain detect sensors,

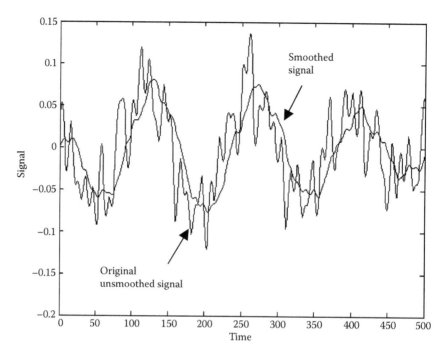

FIGURE 5.6 Original raw signal and the signal after smoothing by the application of a 25-point equal-weight MA filter.

led to the conclusion that the strain data spikes were likely the result of radio interference affecting the electronic instrumentation. If there are gaps in the data, it may be possible to fill in the gaps by making more comprehensive assumptions regarding the nature of the data. An example is an assumption of long-term periodicity [23].

A variety of band-pass and band-reject filters can selectively filter signal components in specific bands. Specific details regarding the design of these filters can be fairly recondite [18,19]. However, modern software packages can design high-performance digital filters with minimal low-level input by the user. Phase distortion by the filter is a concern. One

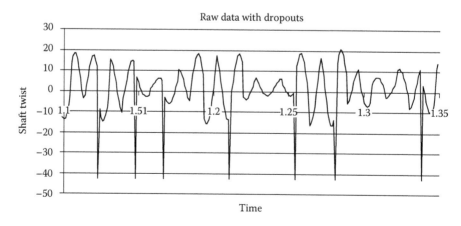

FIGURE 5.7 Torsional shaft data with dropouts due to telemetry errors. (Data courtesy of G. Spencer.)

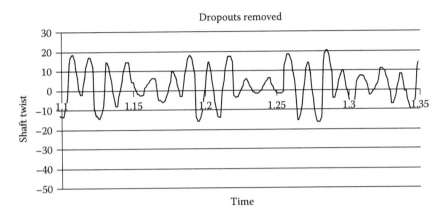

FIGURE 5.8 Torsional shaft data with telemetry-error dropouts corrected by a three-point moving median filter.

method of avoiding phase distortion is to filter the time history twice—once in a forward direction and once in a reverse direction. The intent is to balance and minimize the phase distortions.

The *discrete Fourier transform* (DFT) is an alternative method of removing signal components in specified frequency bands. The first step transforms the data into the frequency domain. The next step replaces the Fourier components in selected frequency bands with zeroes. The final step reconstructs the modified signal in the time domain. Fourier domain filtering is straightforward to apply and can produce very effective frequency domain performance. However, the process can introduce phase shifts and imaginary components into the time domain version of the reconstructed signal.

5.3.1 Auto- and Cross-Correlation Functions

The *autocorrelation* function of a single ordered data set (time history), with zero mean, is the expected value of the product of two points in the data set with a time lag [24,25]:

$$R_{xx}(\tau) = R_{xx}(m\Delta t) = E[x_i x_{i+m}]. \tag{5.78}$$

When the lag interval equals zero, the autocorrelation equals the variance:

$$R_{xx}(0) = \sigma_x^2. \tag{5.79}$$

Using the variance to normalize the autocorrelation

$$\rho_{xx}(\tau) = \frac{R_{xx}(\tau)}{\sigma_x^2} \tag{5.80}$$

produces a function with the properties

$$\rho_{xx}(\tau) = \rho_{xx}(-\tau), \tag{5.81}$$

$$\rho_{xx}(0) = 1, \tag{5.82}$$

$$|\rho_{xx}(\tau)| \le 1. \tag{5.83}$$

A common autocorrelation estimator for a finite-length data sample is

$$R_{xx}(\tau_m) = R_{xx}(m\Delta t)\frac{1}{N-|m|}\sum_{n=1}^{N-m} x_i x_{i+m}, \quad 0 \le m < N. \tag{5.84}$$

Δt is the time interval between samples. A recommendation is to calculate the auto-correlation for lags that extend only up to 5% or 10% of the total time history [26].

The *cross-correlation* between two simultaneously sampled time histories is

$$R_{xy}(\tau) = R_{xy}(m\Delta t) = E[x_i y_{i+m}]. \tag{5.85}$$

The variance-normalized cross-correlation is

$$\rho_{xy}(\tau) = \frac{R_{xy}(\tau)}{(\sigma_x^2 \sigma_y^2)^{1/2}} \tag{5.86}$$

with the property

$$\rho_{xy}(\tau) = \rho_{yx}(-\tau). \tag{5.87}$$

A common cross-correlation estimator is

$$R_{xx}(\tau) = R_{xx}(m\Delta t) = \frac{1}{N-|m|}\sum_{n=1}^{N-m} x_i y_{i+m}, \quad 0 \le m < N. \tag{5.88}$$

An example use of cross-correlation functions is the location of leaks in pipes from measurements of acoustic/vibration signals at two points by Gao et al. [27]. A simplified operating explanation is that a leak source generates acoustic/vibration waves. The waves travel in opposite directions down the pipe. Acoustic/vibration wave energy detected at two points will be similar, but will have a time (phase) lag due to an elastic wave time of travel difference. The time lag of maximum cross-correlation multiplied by the wave speed indicates the relative distance to the leak.

5.3.2 Regression Analysis and System Identification

Regression analysis is a family of statistical procedures that fit generic mathematical models with adjustable parameters to data. *System identification* is regression analysis to determine the model parameters from input and output measurements [20,28]. An example is the mathematical modeling of live load deflections of structures with finite element methods. The model requires material and element-level properties, such as local stiffnesses values. System identification obtains a more accurate estimate of the local stiffnesses from the measured deflection data [29]. The updated material properties aid in the load or safety rating of the bridge.

Regression models with the mathematical form of linear functions of the unknown parameters are particularly convenient and allow for a straightforward identification of the unknown parameters from input–output data with linear algebra. If the data consist of a single input channel, x, and a single output channel, y, then a model formed as a linear combination of unknown parameters b_j and prespecified functions, $f_j(x)$, of the input variables is

$$y = b_0 + b_1 f_1(x) + b_2 f_2(x) + \cdots + b_N f_N(x). \tag{5.89}$$

The prespecified functions $f_j(x)$ can be nonlinear, but should be single-valued.

Determining the unknown parameters b_i usually involves a selection that minimizes modeling errors. Mathematical convenience favors a linear additive error model, with a sum of the squares of errors as the error criterion to be minimized. For the case of N data pairs of input and output measurements, X_i and Y_i, a linear difference representation of the ith measurement error, ε_i, is

$$Y_i = y_i + \varepsilon_i \tag{5.90}$$

and

$$Y_i = b_0 + b_1 f_1(X_i) + b_2 f_2(X_i) + \cdots + b_M f_M(X_i) + \varepsilon_i, \quad i = 1, 2, \ldots, N. \tag{5.91}$$

Condensing into a matrix format,

$$\underset{N\times 1}{[Y]} = \underset{N\times M}{[f]} \; \underset{M\times 1}{[b]} + \underset{N\times 1}{[\varepsilon]} . \tag{5.92}$$

The *least-squares* estimation method selects the model parameters, b_j, to minimize the weighted (or unweighted) squared error

$$\underset{1\times 1}{\varepsilon^2} = \underset{1\times N}{[\varepsilon]^T} \; \underset{N\times N}{[W]} \; \underset{N\times 1}{[\varepsilon]} = \underset{1\times N}{\left\{[Y] - [f][b]\right\}^T} \underset{N\times N}{[W]} \underset{N\times 1}{\left\{[Y] - [f][b]\right\}}, \tag{5.93}$$

where $[W]$ is a weighting matrix. The choice of $[W]$ is usually the identity matrix, unless guidance is given to the contrary.

$$\varepsilon^2 = [\varepsilon]^T [I][\varepsilon] = [\varepsilon]^T [\varepsilon]. \tag{5.94}$$

One method of determining the parameters b_j that minimize the total error is to take the partial derivative of ε^2 with respect to each b_j and set the result equal to zero. This leads to a system of linear equations in the b_j. A simpler and largely equivalent approach is to use the pseudoinverse [16,30,31].

$$\underset{M\times 1}{[b]} = \underset{\underset{M\times M}{\left(\underset{M\times N}{[f]^T} \; \underset{N\times M}{[f]}\right)^{-1}}}{} \; \underset{M\times N}{[f]^T} \; \underset{N\times 1}{[Y]} . \tag{5.95}$$

Occasionally the $M \times M$ matrix $[f]^T[f]$ is ill-conditioned and the successful solution requires specialized techniques, changing the model or collecting new data with independent information.

Least-squares system identification usually works well if

1. The system model (Equation 5.89) is a reasonable approximation of the behavior of the system.
2. The errors, ε_i, are independent and randomly distributed, with zero mean.
3. The number of measurements, N, is greater than or equal to the number of unknowns, M.
4. The mathematical structure of the model represents the unknown parameters as independent degrees of freedom.

5. The experimental design produces data that correspond to varying system degrees of freedom independently. This helps to ensure that the matrix inversion in Equation 5.95 exists and is well-conditioned [32].

If the assumption regarding independence of the errors is not valid, it may still be possible to perform regression analysis system identification by using other techniques, such as the medians of the residuals [20]. When the model does not allow a linear model of the form of Equation 5.89, then a nonlinear model may be more appropriate. Performing regression to nonlinear models requires replacing the matrix manipulations of Equation 5.95 with more involved nonlinear programming. As an example, Fukunaga and Hu identified impact force locations on a plate from an array of piezoelectric sensors with nonlinear programming [33].

5.3.3 Fourier and Spectral Methods

Fourier or *spectral* analysis decomposes functions into a sum (or integral) of component functions [34,35]. One of the main rationales for using Fourier methods is that a system will often process an individual component of an input into components in the output in a well-characterized manner. Simultaneously decomposing the input and output functions of a system can often provide insight as to the behavior of the system. The choice of component functions is not unique. If the system properties do not vary with space or time, then sines and cosines are usually appropriate component functions. If the system properties vary with space or time (as with radial or spherical geometries), then other functions (e.g., Bessel functions or LeGendre polynomials) are more natural in a decomposition. Another advantage of Fourier methods is that they readily subsume to statistical analyses [36]. For example, it is often possible to distinguish random noise from a primary coherent signal by statistical Fourier analysis [37].

Fourier analysis proceeds by noting that a function will decompose into Fourier components if it is piecewise continuous and periodic (with period T) so that

$$f(t+T) = f(t),\tag{5.96}$$

or if the function is defined only on a finite length interval. The decomposition is into an infinite set of discrete components

$$f(t) = \frac{a_0}{2} + \sum_{n=1}^{\infty} a_n \cos(n\omega t) + b_n \sin(n\omega t),\tag{5.97}$$

where

$$a_n = \frac{2}{T} \int_a^{a+T} f(t) \cos(n\omega t)\, dt, \quad n \geq 0,\tag{5.98}$$

$$b_n = \frac{2}{T} \int_a^{a+T} f(t) \sin(n\omega t)\, dt, \quad n \geq 1,\tag{5.99}$$

$$\omega = 2\pi f = \frac{2\pi}{T}.\tag{5.100}$$

The Fourier transform is an appropriate decomposer for a function, $f(t)$, that spans from $-\infty$ to ∞. The decomposition is into a continuous set (function) of Fourier components $F(f)$:

$$F(f) = \int_{-\infty}^{\infty} f(t)e^{-j2\pi ft} \, dt. \tag{5.101}$$

The inverse Fourier transform reconstructs the original function $f(t)$:

$$f(t) = \int_{-\infty}^{\infty} F(f)e^{j2\pi ft} \, df. \tag{5.102}$$

A sufficient, but not necessary, condition for convergence of Equations 5.101 and 5.102 is

$$\int_{-\infty}^{\infty} |f(t)| \, dt < \infty. \tag{5.103}$$

A *stationary* time history may be loosely defined as one in which the statistical properties do not change with time. Many (most) stationary time histories do not satisfy the absolute value integrability requirement of Equation 5.103 and do not have Fourier transforms that converge. Nonetheless, it is often practical to calculate the Fourier transform of a stationary time history from a finite-duration subset of the infinite duration signal, which automatically satisfies Equation 5.103. This is the *finite Fourier transform* [24]

$$X(f, T) = \int_{0}^{T} x(t) \, e^{-j2\pi ft} \, dt. \tag{5.104}$$

Temporally ordered experimental data are often a finite number of discrete samples, N, with a sampling time step interval Δt. Fourier decomposition of discrete data requires discrete Fourier methods [24,34]. The DFT and an inverse with a frequency resolution, $\Delta f = 1/(N\Delta t)$, are

$$X_k = X(k\Delta f) = \Delta t \sum_{n=1}^{N} x_n \exp\left[-\frac{2\pi jkn}{N}\right], \quad k = 1, 2, \ldots, N, \tag{5.105}$$

$$x_n = x(n\Delta t) = \Delta f \sum_{k=1}^{N} X_k \exp\left[\frac{2\pi jkn}{N}\right], \quad n = 1, 2, \ldots, N. \tag{5.106}$$

The *magnitudes*, c_k, of the Fourier coefficients are

$$c_k = [\text{Re}(X_k)^2 + \text{Im}(X_k)^2]^{1/2}. \tag{5.107}$$

The *phase angles*, φ_k, are

$$\varphi_k = a \tan 2[\text{Re}(X_k), \text{Im}(X_k)], \tag{5.108}$$

where $a \tan 2(x, y)$ is the two-argument inverse tangent function that avoids phase ambiguities.

The *fast Fourier transform* (FFT) is an efficient DFT algorithm. The FFT takes advantage of the distributive law of multiplication and the repetitive nature of sinusoidal functions [38]. The FFT is most efficient when applied to data sets with lengths that are powers of 2 [34]. A detail of note is that computer implementations of the FFT take advantage of data symmetries. The raw outputs from the FFT may appear in a compact and cryptic format that requires a bit of tedious deciphering by the end user.

5.3.4 Spectral Estimates

Multiple statistical spectral estimators are available [34]. One calculates the spectrum as the Fourier transform of the autocorrelation by the Wiener–Khinchin relation [24]

$$S_{xx}(f) = 2 \int_0^\infty R_{xx}(\tau) e^{-j2\pi f \tau} \, d\tau. \tag{5.109}$$

An entirely equivalent approach defines the spectrum in terms of the finite Fourier transform as

$$S_{xx}(f) = \lim_{T \to \infty} \frac{1}{T} E[X^*(f,T)X(f,T)]. \tag{5.110}$$

Applying Equation 5.110 directly to a single sample time history with the DFT produces a raw *periodogram* spectral estimate [34,36]:

$$S_{xx}(f_k) = S_{xx}(k\Delta f) = \frac{1}{N\Delta t} X_k^* X_k. \tag{5.111}$$

Calculating statistical estimates of the spectral properties of stationary random time histories typically involves processing multiple data sets. The source of the data sets may be samples of separate time histories, or samples of different portions of the same time history. The magnitudes, c_k, of the DFTs of different data samples from a stationary time history tend to have magnitudes, c_k, of the Fourier coefficients that tend to cluster around mean values, while the phase angles vary uniformly over 0 to 2π. In these cases, the *power spectral density* (also known as the *spectrum*), $S(f)$, quantifies the amount of power, $P(f_1, f_2)$, in a given frequency interval $[f_1, f_2]$ with the integral

$$P(f_1, f_2) = \int_{f_1}^{f_2} S(f) \, df. \tag{5.112}$$

An important point in using power spectra is to note that they can be one-sided and defined over a positive range, or two-sided and defined over both negative and positive ranges. Both definitions are equally valid, but they differ by a factor of 2 [24]. The one-sided spectrum is more convenient for visual display and interpretation. The two-sided spectrum has a degree of mathematical appeal due to time-axis symmetry, and some algorithmic convenience.

A raw periodogram often has a rough or hashy appearance. It is often desired to reduce the variance and to smooth the appearance of the periodogram power spectral density (PSD) estimate. A naïve approach is to sample and process a longer time history in the periodogram. Unfortunately, this produces the curious result of not reducing the variance from frequency to frequency. Instead it creates more finely spaced bins with the same hashy variance. An explanation for the hash is that Equation 5.111 uses only one sample of power per frequency bin. These samples behave as ordinary statistical samples where the individual samples vary about a mean value. In portions of the spectrum that are smooth with respect to frequency, samples from neighboring points have similar variances about an expected value, but are uncorrelated with one another. The variability of the individual neighboring samples causes the hashy appearance. A viable approach to smoothing the periodogram is to start with the same longer time history, but to use point-by-point averaging of periodograms calculated from shorter period time histories. This technique, known as *Welch averaging*, produces a smoother PSD estimate, at the expense of a coarser frequency domain resolution. The decrease in the frequency resolution Δf provides multiple samples of spectral amplitude per frequency bin. Averaging the multiple samples for each frequency bin reduces the variance of the spectral estimate.

Another spectral smoother multiplies the time history by a window function, $W(t)$, before calculating the periodogram:

$$x_w(t) = W(t)x(t). \qquad (5.113)$$

The choice of weighting function affects the performance of the spectral estimator [26,39]. Most windows have values near unity in the middle of the interval and taper to zero at the ends of the window interval. The *Hanning window*, $W_H(t)$, is commonly used to aid with analyzing stationary time histories [24],

$$W_H(t) = \begin{cases} \dfrac{1}{2}\left[1 - \cos\left(\dfrac{2\pi t}{T}\right)\right], & 0 \le t \le T, \\ 0 & \text{otherwise.} \end{cases} \qquad (5.114)$$

The *flat top window* is good for cases where sinusoids of fixed frequency and phase dominate the signal. If the time history is a decaying sinusoid, as occurs with impact-hammer vibration testing of structures, then an exponentially decaying window may be appropriate [39]. Spectral windows alter the amplitude of the time histories and reduce (bias) the amount of power appearing in the spectral estimate. A multiplicative scaling factor can correct for spectral window scaling bias. For example, the correction factor for the Hanning window is $[8/3]^{1/2}$.

One explanation of the spectral smoothing action of windows is that the tapered ends reduce the effects of Fourier mismatches, that is, Gibbs effects, at the ends of the time interval for spectral components that lie between the discrete sampled f_k. In the frequency domain, the action of the Hanning window is equivalent to a three-point MA filter applied to the spectral amplitudes. More aggressive frequency domain MA filters can provide even more spectral smoothing as needed [34].

It is often of interest to fit curves directly to spectral estimates for system identification and characterization. Vibration modal analyses, especially *ambient vibration survey* (AVS) techniques, routinely use spectral curve fitting for model identification. As an example, Hammad and Issa determined the fractal dimension of cracked concrete surfaces by fitting a straight line through a log–log plot of the power spectrum in wavenumber space of the measured crack surface height [40].

5.3.5 Cross-Spectra

The *cross power spectrum* describes the frequency domain distribution of power between two simultaneously sampled random time histories. The Wiener–Khintchine relation gives the cross power spectrum, $S_{xy}(f)$, of two signals, $x(t)$ and $y(t)$, as the Fourier transform of the cross-correlation function, $R_{xy}(\tau)$,

$$S_{xy}(f) = \int_{-\infty}^{\infty} R_{xy}(\tau)\, e^{-j2\pi f \tau}\, d\tau. \qquad (5.115)$$

The expected value of the finite Fourier transform is an alternative but equivalent definition

$$S_{xy}(f) = \lim_{T \to \infty} \frac{1}{T} E\left[X^*(f,T)Y(f,T)\right]. \qquad (5.116)$$

The cross-spectrum has the symmetry properties [24]

$$S_{xy}(-f) = S_{xy}^*(f) = S_{yx}(f).$$
(5.117)

The simultaneous sampling of multiple time histories allows for the calculation of multiple combinations of spectra and cross-spectra. These spectra naturally assemble into a matrix $[S(f)]$:

$$[S(f)] = \begin{bmatrix} S_{11}(f) & S_{12}(f) & \cdots & S_{1n}(f) \\ S_{21}(f) & S_{22}(f) & & \\ \vdots & & \ddots & \\ S_{n1}(f) & & & S_{nn}(f) \end{bmatrix}.$$
(5.118)

5.3.6 Coherence Functions

Two signals measured simultaneously in different parts of a system may, or may not, be physically related. The correlation function (Equation 5.73) provides a coarse estimate of the degree of dependency between the signals. The *coherence function* $\gamma_{xy}^2(f)$ is a finer scale frequency domain measure of two-channel dependence:

$$\gamma_{xy}^2(f) = \frac{|S_{xy}(f)|^2}{S_{xx}(f)S_{yy}(f)},$$
(5.119)

where

$$0 \le \gamma_{xy}^2(f) \le 1.$$
(5.120)

Values of $\gamma_{xy}^2(f)$ that are close to 1 indicate a high level of correlation between two signals at the frequency f. Values that are close to 0 indicate a low level of correlation. Data corresponding to frequency ranges with low coherence should be treated with suspicion and possibly avoided. Similarly, data with coherences precisely equal to 1 should also be treated with suspicion. A common application of coherence functions is to check the quality of FRF estimates used in vibration modal testing.

Statistical estimates of the coherence function require using statistical spectrum estimators. For Welch averaging, the coherence estimate is

$$\gamma_{xy}^2(f) = \frac{\left| \sum_{n=1}^{N} Y_n^*(f) X_n(f) \right|^2}{\left[\sum_{n=1}^{N} X_n^*(f) X_n(f) \right] \left[\sum_{n=1}^{N} Y_n^*(f) Y_n(f) \right]}.$$
(5.121)

The number of averages must be greater than one for the coherence function estimator to be of use. Otherwise $\gamma_{xy}^2(f)$ equals one automatically. When $x(t)$ and $y(t)$ are linearly dependent, the coherence function will also equal one, aside from the effects of noise contamination [41].

The simultaneous acquisition of multiple data channels allows for the use of more general coherence functions. An example is the case of collecting two vectors of simultaneous time histories, $\{X(t)\}^T = [x_1(t), x_2(t), \ldots, x_M(t)]$ and $\{Y(t)\}^T = [y_1(t), y_2(t), \ldots, y_N(t)]$. One method forms an $M \times N$ matrix of ordinary coherences using Equation 5.119 or 5.121 to calculate each element of the matrix. Another approach uses weighted linear combinations of elements of the two sets of vectors, $x_w(t)$ and $y_w(t)$, that is,

$$x_w(t) = a_1 x_1(t) + a_2 x_2(t) + \cdots + a_M x_M(t) = \{a\}^T \{x(t)\}, \tag{5.122}$$

$$y_w(t) = b_1 y_1(t) + b_2 y_2(t) + \cdots + b_N y_N(t) = \{y(t)\}^T \{b\}. \tag{5.123}$$

The *weighted multiple coherence* between $x_w(t)$ and $y_w(t)$ is

$$\gamma_{xyw}(f) = \frac{\{a\}^T [S_{xy}(f)]\{b\}}{\left\{ \{a\}^T [S_{xx}(f)]\{a\} \right\}^{1/2} \left\{ \{b\}^T [S_{yy}(f)]\{b\} \right\}^{1/2}}, \tag{5.124}$$

where $[S_{xx}]$, $[S_{xy}]$, and $[S_{yy}]$ are multichannel spectral density matrices and $\{a\}$ and $\{b\}$ are weighting vectors. Multiple coherences can indicate the frequency ranges over which an output vector appears as a linear combination of components of the input vector. An example from modal vibration testing uses mode shapes as weighting factors in Equation 5.124 [41]. When both sets of vectors, $\{x(t)\}$ and $\{y(t)\}$, are of equal length, and are proportional to one another, then the mutual multiple coherence automatically equals one. This can serve as a check on the dimensionality and independence of the data streams in the inputs and outputs.

5.3.7 FRF Estimators

FRFs are frequency domain representations of the dynamics of linear systems. Figure 5.9 shows a single-input–single-output linear system. In the time domain, the input, output, and system impulse response functions are $f(t)$, $x(t)$, and $h(t)$, respectively. The corresponding Fourier transform frequency domain representations are $F(f)$, $X(f)$, and the FRF $H(f)$. The FRF is also called the *transfer* function in some circles, but such terminology is more commonly reserved for LaPlace transform representations of system with the LaPlace variable $s = j\omega$. The convolution theorem gives a description of the time domain dynamics of the linear system in terms of a convolution integral.

$$x(t) = \int_{-\infty}^{t} h(t - \tau) f(\tau) \, d\tau. \tag{5.125}$$

Application of the Fourier transform and convolution theorem to Equation 5.125 replaces the convolution integral with an algebraic frequency domain expression

$$X(f) = H(f)F(f) \tag{5.126}$$

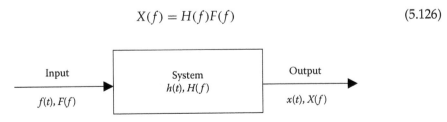

FIGURE 5.9 Linear single-input–single-output system with both time domain and frequency domain components indicated.

or

$$H(f) = \frac{X(f)}{F(f)}.$$ (5.127)

A typical input–output experiment determines the dynamics of a system by applying a known input (force) to the structure and measures the response (acceleration or displacement). A frequency domain analysis calculates the FRF using Equation 5.127. However, noise and other measurement errors inevitably contaminate the measurements. Repeated measurements and averaging generally improves the FRF estimate. The presence of additive noise in the ith repeat modifies Equation 5.126 to

$$\hat{X}_i(f) - \eta_i(f) = H_i(f)^*[\hat{F}_i(f) - v_i(f)],$$ (5.128)

where $\hat{X}_i(f)$ and $\hat{F}_i(f)$ are the Fourier transforms of the noise-free outputs and inputs, respectively, and $\eta_i(f)$ and $v_i(f)$ are the Fourier transforms of the noise. When the input measurement is noise-free, that is, $v_i(f) = 0$, then the $H_1(f)$ least-squares estimate of $H(f)$ is

$$H_1(f) = \frac{\sum_{i=1}^{N} X_i(f)F_i^*(f)}{\sum_{i=1}^{N} F_i(f)F_i^*(f)}.$$ (5.129)

When there is no noise on the output, that is, $\eta_i(f) = 0$, then the $H_2(f)$ least-squares estimate of $H(f)$ is

$$H_2(f) = \frac{\sum_{i=1}^{N} X_i(f)X_i^*(f)}{\sum_{i=1}^{N} F(f)_i X_i^*(f)}.$$ (5.130)

More sophisticated least-squares and multichannel matrix FRF estimates can combine and weight the noise on both the input and output as necessary [41]. Most of these estimates consistently converge to the same values when the measurement noise reduces to zero and the assumption of linearity holds. A case where these differences are important arises when estimating damping from modal FRF resonance peaks. For vibrating structures, the $H_1(f)$ estimate tends to underestimate peak heights at resonances and overestimate damping when using the spectral width method. The $H_2(f)$ estimate tends to overestimate peak height at resonance and underestimate damping.

5.3.8 Time Series Modeling

Time series analysis models the input–output behavior of systems from discrete time domain data series [19,25,42]. Figure 5.10 shows a generic time series model of a system. The system operates on the inputs, w_n, and produces outputs, y_n. Items outside the dashed box are directly measurable. Items inside the dashed box are not directly measurable. The *exogenous* input x_n is an observable and possibly controllable input to the system. z_n is a measurable output. ε_n and η_n are random input and output components, respectively.

A standard assumption is that the random components are not measurable. Instead, the random components add to the measurable input and output signals. It is also usually assumed that ε_n and η_n are zero mean processes with known variances. Common additional assumptions are the random behaviors of independence, whiteness, stationarity, and Gaussian PDFs. The addition of noise to the input is of the form

$$w_n = x_n + \varepsilon_n.$$ (5.131)

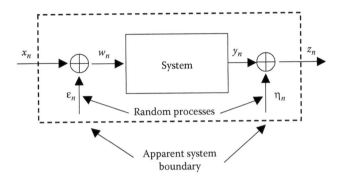

FIGURE 5.10 Time series representation of the behavior of a system with discrete inputs and outputs at time step n.

The measurable output z_n combines with a random component η_n and becomes

$$z_n = y_n + \eta_n. \tag{5.132}$$

x_n, y_n, w_n, z_n, ε_n, and η_n may be time series of scalars or vectors. Common additional assumptions are that the system is linear and time-invariant. It is possible to relax these restrictions and modify the analysis as needed.

Useful tools that aid in developing mathematical models of time series systems are the backward shift operator, B, the forward shift operator, F, and the backward difference operator, ∇. Following the notation of Box and Jenkins, definitions of these operators are [25]

$$Bx_n = x_{n-1}, \tag{5.133}$$

$$B^m x_n = x_{n-m}, \tag{5.134}$$

$$Fx_n = B^{-1}x_n = x_{n+1}, \tag{5.135}$$

$$\nabla x_n = x_n - x_{n-1} = (1 - B)x_n. \tag{5.136}$$

An examination of Equations 5.133 through 5.136 indicates that these operators are linear. Linear combinations of the B, F, and ∇ operators form higher-order linear operators. For example, a linear combination of backward shift operators acting on the past outputs, $\varphi(B)$, is an AR operator of order p with the form

$$\varphi(B) = 1 - \varphi_1 B - \varphi_2 B^2 - \cdots - \varphi_p B^p, \tag{5.137}$$

where φ_n are weighting coefficients. *Vector autoregressive* (ARV) operators act on vectors of time series data. If the exogenous input x_n is zero, the AR output, y_n, is

$$\varphi(B)y_n = w_n, \tag{5.138}$$

$$y_n = \varphi_1 y_{n-1} + \varphi_2 y_{n-2} + \cdots \varphi_p y_{n-p} + w_n. \tag{5.139}$$

A *Markov process* is a random process that depends only on the previous state. MA operators of order q that act on the input signal w_n with a linear combination of backward shift operators, $\theta(B)$, are of the form

$$\theta(B) = 1 - \theta_1 B - \theta_2 B^2 - \cdots - \theta_q B^q, \qquad (5.140)$$

$$y_n = \theta(B) w_n, \qquad (5.141)$$

where q is the order of the MA operator and θ_n are coefficients. A combination of the AR and MA operators forms an *autoregressive moving-average* (ARMA) model

$$\varphi(B) y_n = \theta(B) w_n. \qquad (5.142)$$

Autoregressive moving average with exogenous inputs (ARMAX) models include deterministic inputs, x_n, with additive noise on the input, ε_n, and a separate independent additive noise component, η_n, on the output. The corresponding system model is

$$\varphi(B) y_n + \eta_n = \theta(B)(\varepsilon_n + x_n). \qquad (5.143)$$

Applying a stationary and random input to an ARMA model produces a stationary and random output. A consequence is that ARMA models are particularly well suited for characterizing systems when the input and output are stationary and random. Modeling the dynamics of nonstationary processes with time series techniques requires modifying the ARMA models. Adding integration or summation terms to the output time series can produce a nonstationary time series. Modifying the input signal operator $\varphi(B)$ with a backward shift operator is an equivalent approach. When $\phi(B)$ is an AR operator and d is an integer that represents the order of the integration operator (usually 0, 1, or 2), the modified input operator is

$$\varphi(B) = \phi(B)[1 - B]^d. \qquad (5.144)$$

ARMA models that combine an input operator with a backward difference as in Equation 5.144 are known as *autoregressive integrated moving-average* (ARIMA) models.

Determining estimates of the AR, MA, and ARMA coefficients $\varphi(B)$ and $\theta(B)$ from finite samples of input and output data are staples of time series analysis. The process of estimating model coefficients is a compromise between a best match of the observed data, robustness with respect to modeling new data, simplicity, and computational effort. A host of algorithms that estimate time series coefficients are readily available. Most of these algorithms adjust the time series coefficients to match input and output data with minimal estimation errors, usually with second-order statistics, that is, correlation functions. The *Burg algorithm* efficiently calculates AR coefficients [42,43], and is often called the *maximum entropy spectrum* estimator because it maximizes the entropy of the spectral estimate when using exact values (not estimates) of the output correlations [44].

Choosing the number of coefficients used in a particular time series model is an important consideration. Using a large number of coefficients produces a highly accurate time series model at the expense of increased computational complexity and burden. Using a large number of coefficients runs the risk of overfitting the data and producing a nonrobust model of the system. Box and Jenkins recommend following the principle of *parsimony*. The goal is to use the minimum number of coefficients in a model that enables the model to work properly [25]. Frequency domain and transfer functions representations of AR, MA, and ARMA models are reciprocal polynomials, polynomials, and rational functions of frequency, respectively. Since rational functions tend to fit arbitrary functions with fewer coefficients than reciprocal polynomials or polynomials, ARMA models can be superior to AR and MA models from a perspective of parsimony. Determining ARMA models coefficients tends to be more computationally intensive than determining AR or MA coefficients. For AR models, the *Akaike information criterion* (AIC) guides selecting the number of coefficients by balancing between the goodness of fit and the number of parameters [45,46].

Frequency domain representations of time series models can provide insight into the underlying structure of both the model and the system. The first step in frequency domain conversion is to take the *z-transform* of the discrete time series [44]

$$X(z) = \sum_{n=-\infty}^{\infty} x_n z^{-n} = Z(x_n). \tag{5.145}$$

z is a complex number that lies within a region of convergence of the transformed time series. The z-transform shares many properties with the classical integral transforms (Fourier, LaPlace, etc.). These include linearity, scalability, and convolution. The z-transform of the backward shift operator B is

$$Z[B(x_n)] = Z(x_{n-1}) = z^{-1}Z(x_n). \tag{5.146}$$

The ARMA model in the z-domain becomes

$$Z(y_n) = \left(\frac{1 - \theta_1 z^{-1} - \theta_2 z^{-2} \cdots - \theta_q z^{-q}}{1 - \varphi_1 z^{-1} - \varphi_2 z^{-2} \cdots \varphi_p z^{-p}} \right) Z(w_n) = H(z), \tag{5.147}$$

where $H(z)$ is the system transfer function. Rewriting $H(z)$ in terms of q zeros, β_i, and p poles, α_i, produces the form [47]

$$H(z) = \frac{1 - \theta_1 z^{-1} - \theta_2 z^{-2} \cdots - \theta_q z^{-q}}{1 - \varphi_1 z^{-1} - \varphi_2 z^{-2} \cdots \varphi_p z^{-p}} = \frac{\prod_{i=1}^{q} (1 - \beta_i z^{-1})}{\prod_{i=1}^{p} (1 - \alpha_i z^{-1})}. \tag{5.148}$$

Placing z on the unit circle in the complex domain converts the representation to the frequency domain with the substitution

$$z = e^{j2\pi f \Delta t}. \tag{5.149}$$

Frequency domain representations of time series models provide a means of estimating power spectra from a discrete time series. These methods generally rely on the notion that a white noise input into a linear system (filter) produces an output signal with a power spectrum equal to the squared magnitude of the system FRF (Figure 5.11). Fitting an ARMA model using system outputs based on an assumption of white noise input provides the necessary information for a spectral estimate. Perhaps the most common of these techniques is with an AR system model, that is, to use the Burg or maximum entropy spectral estimation algorithm. ARMA and AR algorithms are useful in estimating spectra that undergo large changes in short frequency ranges. An example is the case of using the half-width method for damping estimates in lightly damped structures that vibrate near to resonance [48–50]. Proakis and Manolakis give further refinements to these parametric spectral estimation techniques [44].

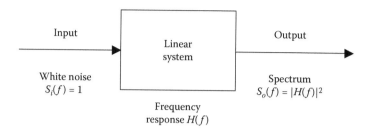

FIGURE 5.11   Concept of an ARMA parametric spectral estimator.

5.3.9   Time–Frequency Analysis

SHM data signals can contain time-varying frequency, amplitude, and phase components. Traditional Fourier and spectral analysis methods may be inadequate for analyzing such transient and non-stationary signals. The difficulty is that Fourier and spectral techniques decompose time histories into sinusoidal components. Since sinusoids extend from negative to positive infinity without change, it is not possible to readily ascertain the temporal location of frequency components without reassembling the time domain signal with an inverse Fourier transform.

Time–frequency analysis techniques decompose data signals into forms that separate the time-varying amplitude, frequency, and phase content. Several of these techniques have potential SHM utility. The *Hilbert transform* aids in calculating the time-dependent amplitude of an oscillating signal, such as the signal received from a gated sine wave ultrasound source [51]. The *short-term Fourier transform, Hilbert–Huang transform* (HHT), and *Wigner–Ville distribution* (WVD) are all useful for decomposing the time versus frequency content of transient signals. The *wavelet* technique can be useful when the frequency content spans multiple scales. Most of the SHM applications of time–frequency methods to date have been to provide descriptive insights. The various techniques discriminate different types of behavior, such as the dynamics of a structure before and after the occurrence of damage. There may be considerable potential for the evolution of these time–frequency analysis techniques into more quantitative approaches.

An uncertainty relation between time and frequency is

$$\sigma_x \sigma_f \geq \frac{1}{4\pi}, \tag{5.150}$$

where σ_x^2 is the variance of $x(t)$ and σ_f^2 is the variance of the Fourier transform of $X(f)$ [38,52].

5.3.10   Short-time Fourier Transform

The short-time Fourier transform (STFT) is a relatively simple time–frequency decomposer. The STFT calculates spectral statistics, such as the periodogram, from a short time interval [53]. Typical time periods range from 64 to 256 data points. The center point of the time interval locates the time base of the frequency data. Moving the center point of the analysis window creates a new time location. The new interval may, or may not, overlap with previous analysis windows. The choice of time period and overlap coverage behavior between intervals is a data and user-preference-driven decision. Figure 5.12 shows nonoverlapping

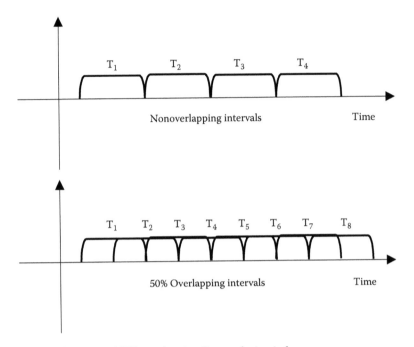

FIGURE 5.12 Nonoverlapping and 50% overlapping time analysis windows.

and 50% overlapping time analysis windows. Since time and frequency are conjugate variables, it is not possible to measure simultaneously with precision the time and frequency content of an event.

The ability of the STFT to separate frequency content and timing makes it attractive for processing highly transient oscillating signals. As examples, Papy et al. and Bang et al. used the STFT to identify AE signals from distributed fiber optic sensors embedded in carbon fiber-reinforced composites, and Pinto et al. used the STFT to locate damage in plates by the propagation of Lamb waves [54–56]. Frarey describes an interesting case of the utility of the STFT [57]. A large rotating machine (650 MW turbine generator) experienced intermittent transient high-amplitude vibrations. While the operators of the machine certainly realized that there was a problem, the standard practice of Welch averaging multiple time histories to form a spectrum estimate washed out the effect of a single transient event. In an effort to understand better the observed behavior, the next step was to calculate high-resolution spectra from long time histories. Again the effect was to wash out the effect of the short-term broad-banded transient. A superior approach to capturing and analyzing the transient was to use an STFT centered on the transient event.

5.3.11 Hilbert Transform

The *Hilbert transform*, $X_H(t)$, of a signal, $x(t)$, is the integral transform (using the Cauchy principal value) [52]

$$X_H(t) = H[x(t)] = \frac{1}{\pi} \int_{-\infty}^{\infty} \frac{x(\tau)}{\tau - t} \, d\tau. \tag{5.151}$$

A standard Hilbert transform application calculates the *envelope* of an oscillating signal with a slowly varying amplitude, phase, and frequency (Figure 5.13). Examples of such

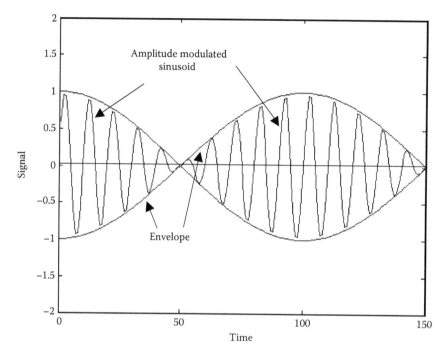

FIGURE 5.13 Amplitude-modulated sine wave with envelope.

calculations arise in SHM when using amplitude-modulated, or gated, sinusoids to form signals for nondestructive evaluation, such as with EM or ultrasonic waves. The envelope can help to estimate the arrival time and strength of reflected and transmitted waves. Each pixel in the image is a gray-scale-encoded amplitude of an A-scan from a particular location and point in time. The effect is that the B-scan replaces time with depth into the surface and produces a spatially coherent 2-D image of subsurface features. Halabe and Pyakurel used the Hilbert transform to help with imaging subsurface metal object defects in wooden logs [58].

One method of determining the envelope considers the oscillating signal, $x(t)$, as being the real part of a complex function, $z(t)$, known as the *analytic signal*:

$$z(t) = x(t) + jy(t) = A(t)\exp[j\varphi(t)]. \tag{5.152}$$

The imaginary component, $y(t)$, is an amplitude-modulated sinusoid with the same frequency and envelope as $x(t)$, but with a 90° phase shift. The envelope (or amplitude) function, $A(t)$, is

$$A(t) = |z(t)| = [x^2(t) + y^2(t)]^{1/2}. \tag{5.153}$$

The instantaneous phase, $\varphi(t)$, and instantaneous frequency, $\omega(t)$, are

$$\varphi(t) = \tan^{-1}\left(\frac{y(t)}{x(t)}\right), \tag{5.154}$$

$$\omega(t) = \frac{d\varphi(t)}{dt}. \tag{5.155}$$

The imaginary part, $y(t)$, of the analytic signal is often a theoretical construct and is not measured directly. In many cases it is possible to construct $y(t)$ from the measured signal with the Hilbert transform

$$y(t) = X_H(t) = \frac{1}{\pi} \int_{-\infty}^{\infty} \frac{x(\tau)}{\tau - t}\, d\tau. \tag{5.156}$$

A rationale for using Equation 5.156 in Equations 5.152 through 5.155 is the assumption that $y(t)$ is the same signal as $x(t)$, but with a 90° phase shift. The Hilbert transform of a pure sine or pure cosine wave automatically produces a 90° phase shift [7], that is,

$$H[\sin \omega t] = \frac{1}{\pi} \int_{-\infty}^{\infty} \frac{\sin \omega \tau}{\tau - t}\, d\tau = \cos \omega \tau, \quad \omega > 0, \tag{5.157}$$

$$H[\cos \omega t] = \frac{1}{\pi} \int_{-\infty}^{\infty} \frac{\cos \omega \tau}{\tau - t}\, d\tau = -\sin \omega \tau, \quad \omega > 0. \tag{5.158}$$

Since almost any well-behaved function, $x(t)$, decomposes as a sum (integral) of sines and cosines and since the Hilbert transform is a linear operator, the Hilbert transform shifts each Fourier sine and cosine component of $x(t)$ by 90°. Envelope calculation with the Hilbert transform proceeds by multiplying the complex DFT of $x(t)$ by j to produce a 90° phase shift of each Fourier component (Figure 5.14). The inverse Fourier transform of the phase-shifted function, $y(t)$, corresponds to the Hilbert transform. The final step calculates the envelope

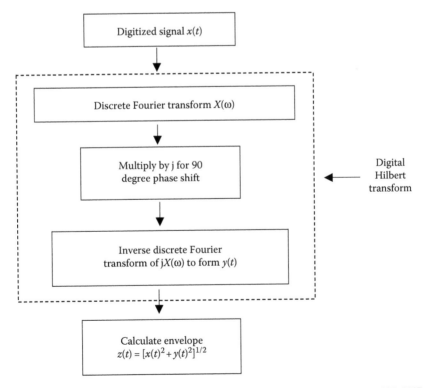

FIGURE 5.14 Envelope calculation with digital Hilbert transform. (Adapted from Lyon RH. 1987. *Machinery Noise and Diagnostics*, Butterworths, Boston.)

amplitude with Equation 5.153 [59]. Francoeur et al. measured elastic wave amplitudes in the 10–50 kHz range with the Hilbert transform envelope so as to monitor the health of lap joints [60].

When a system is linear and causal, that is, responses only occur after the application of inputs, the real and imaginary components of the system FRF form a Hilbert transform pair [38]. Systems with real and imaginary FRF components that do not form a Hilbert transform pair are noncausal and/or nonlinear. This result can limit certain mathematical models of systemic or structural behavior. An example is the so-called structural or hysteretic damping model. This model uses a complex stiffness to represent dissipative forces instead of velocity-dependent viscous forces. While complex stiffness damping has the advantage of providing frequency-independent damping in response to harmonic loading, it produces noncausal responses to transient loads [61]. Worden and Manson confirmed the validity of using the Hilbert transform to assess nonlinearity by examining the frequency domain behavior of a nonlinear-stiffness Duffing with a Volterra series expansion [62]. Loh and Li checked the amount of nonlinearity in a system with the Hilbert transform [63].

5.3.12 HHT and Empirical Mode Decomposition

The HHT extends the Hilbert transform to the analysis of signals that are the sum of distinct oscillating waveforms. *Empirical mode decomposition* (EMD) of signals forms components with distinct oscillating waveforms, possibly with time-varying amplitude, frequency, and phase. Such decompositions can be useful in SHM applications where many signals are composed of multiple oscillatory components. Changes in the state of the structure, for example, damage, can cause changes to the oscillatory behavior. The HHT has the potential to identify the existence and timing of these changes.

The first EMD step fits an envelope to the time history and uses the envelope to remove the low-frequency components. The next step repeats the process and removes the next lowest frequency oscillatory component from the remainder. Repeated application of low-frequency component removal yields a decomposition of the form [64,65]

$$x(t) = \sum_{n=1}^{N} c_n(t) + r_N(t),$$ (5.159)

where $c_n(t)$ are the oscillating component functions and $r_N(t)$ is a residual function that ideally is either the mean or mean plus trend of the signal. Based on these assumptions, an approximation of the original time history is a sum of oscillating components

$$x(t) = Re\left\{\sum_{n=1}^{N} a_n(t) \exp\left[j \int \omega_n(t)\, dt\right]\right\}.$$ (5.160)

The HHT, $H(\omega, t)$, is defined as

$$H(\omega, t) = \sum_{n=1}^{N} \tilde{H}_n(\omega, t) = \sum_{n=1}^{N} a_n(t).$$ (5.161)

EMD–HHT SHM applications appeared quickly after the publication of the original algorithm. Zhang et al. used the HHT to analyze earthquake ground motion signals [64]. Salvino et al. used phase information from EMD–HHT analysis of signals to detect damage in underwater composite plates [66]. Pines and Salvino studied the sensitivity of EMD–HHT techniques for detecting damage in helicopter blades [67]. The phase component proved to be more sensitive to damage than the magnitude component, especially at lower frequencies. Yang et al. identified natural frequencies and damping in tall buildings from an HHT analysis of ambient vibration data [68]. Li et al. examined the degradation of reinforced concrete columns subjected to earthquake loading with the HHT [69]. Wang et al. detected the debonding of piezoelectric transducers from composite beams by EMD–HHT techniques [70]. Vincent et al. compared EMD–HHT methods for SHM with wavelet techniques and concluded that EMD–HHT methods have an intrinsic advantage of being data adaptive, but that the wavelet methods are very good at detecting singularities in dynamic data caused by sudden damage [71]. Chen and Xu, Xu and Chen, and Jha et al. used EMD to detect incipient damage in a laboratory model three-story building from property changes [72–74]. Algernon and Wiggenhauser found that the HHT was effective for analyzing concrete IE waveforms [75]. The reason for the success of the HHT seems to be that IE waveforms are composed of a few decaying sinusoids, each of which usually has a distinct associated elastic wavemode.

5.3.13 Wigner–Ville Distribution

The WVD, $\varphi_{xx}(\omega, t)$, is a time–frequency signal decomposition with an appearance that is similar to the Fourier transform of a short-time autocorrelation [53]:

$$\Phi_{xx}(\omega, t) = \frac{1}{2\pi} \int_{-\infty}^{\infty} x^* \left(\frac{t - \tau}{2} \right) x \left(\frac{t + \tau}{2} \right) e^{-j\omega\tau} \, d\tau. \tag{5.162}$$

The WVD and its more general form, the Wigner distribution, were some of the earliest time–frequency decomposition methods to appear [76]. They are subsets of the even more general Cohen's class of time–frequency distribution functions [77]. Computational complexity is a disadvantage of the Wigner–Ville method. Nonetheless, the Wigner–Ville approach continues to find utility. As an example, Chen and Wissawapaisal used the time–frequency discrimination capability of the WVD to measure ultrasound wave dispersion in multistrand prestressing cables, and then correlated the dispersion with tension [78]. Williams et al. found that the WVD was useful for machinery monitoring applications where the transient consisted of a chirp-type signal, but was not as useful in cases of mixed signals [79]. In a ca. 1993 study, Brancaleoni et al. found the Wigner distribution to have advantages over the Hilbert transform when attempting to analyze nonstationary structural dynamic signals for damage assessments [80].

5.3.14 Wavelets

Wavelet decompositions are similar to Fourier methods in that integrating the original function against a set of basis functions determines the component amplitudes. A primary distinction lies in the nature of the basis functions. Wavelet basis functions are limited in range in the time domain, whereas sinusoidal Fourier basis functions extend to ± infinity [81]. Limitation on the temporal support allows for time localization by sliding the center of the wavelet basis functions along the time axis (Figure 5.15). Another distinguishing feature of wavelets is the self-similar time domain scaling of the wavelet basis functions

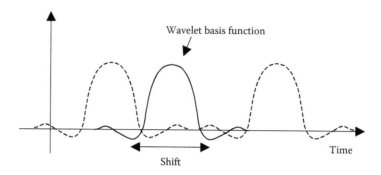

FIGURE 5.15   Horizontal shifting of wavelet basis function for time localization.

(Figure 5.16). This is in contrast with Fourier basis functions that scale linearly with frequency. Octave band analysis techniques used in acoustics and electronics are examples of self-similar scaling. Fourier basis functions scale linearly in frequency. An advantage of self-similar scaling is the efficient representation of information that spans broad frequency ranges. The combination of a sliding center and variable width scaling of the wavelet basis functions enables the dual processes of temporal location and width scaling (decomposition with textural fluctuations).

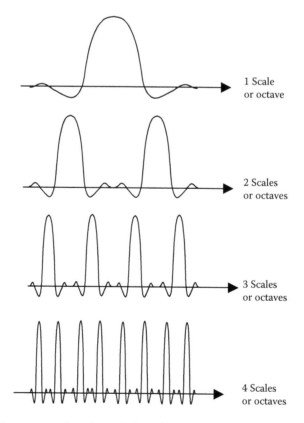

FIGURE 5.16   Self-similar octave scaling of wavelet basis functions.

An equivalent view of wavelet decomposition is the passing of a signal through a bank of band-pass filters with differing center frequencies and bandwidths [38,82,83]. Each filter has a *scale-invariant* bandwidth, that is, the same bandwidth normalized by the center frequency. The arrangement of the center frequencies and bandwidths of the filters spans the entire frequency range of the input signal. $x(t)$ decomposes as

$$x(t) = x_1(t) + x_2(t) + \cdots + x_n(t). \tag{5.163}$$

The historical roots of using filter banks as a means of time–frequency decomposition traces back at least as early as the Renaissance with the appearance of musical notes on octave scales. In the twentieth century, engineers concerned with describing signals that span multiple scales of frequency (as occurs in acoustics or radio engineering) frequently used analog filter banks to decompose the signals (Figure 5.17).

The *continuous wavelet transform* (CWT), $W(u, s)$, of a function, $x(t)$, is

$$W(a, t) = \int_{-\infty}^{\infty} x(\tau) \frac{1}{\sqrt{a}} g^* \left(\frac{t - \tau}{a} \right) d\tau. \tag{5.164}$$

$g(t)$ is a generic wavelet kernel known as the *mother wavelet* [84–87]. The transform variable t corresponds to time location. The transform variable, a, is a scale variable.

The choice of mother wavelet affects the nature of the decomposition and is not unique. Some forms have properties that are useful in specific SHM applications. The Morlet (or Gabor) wavelet is a Gaussian kernel multiplied by an oscillating Fourier component:

$$g(t) = e^{j\omega_0 t} e^{-t^2/2} = e^{-t^2/2} [\cos(\omega_0 t) + j \sin(\omega_0 t)]. \tag{5.165}$$

An attractive feature of the Morlet wavelet is that it minimizes the spread, or uncertainty, in the time and frequency domains [87,88]. Closely related is the *S-transform*, which allows

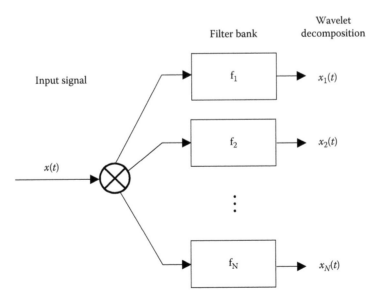

FIGURE 5.17 Filter bank interpretation of wavelet decomposition. (Adapted from Strang G and Nguyen T. 1996. *Wavelets and Filter Banks*, Wellesley-Cambridge Press, Wellesley.)

the user to adjust the degree of time–frequency resolution [89]. The favorable time–frequency resolution properties of the S-transform enabled Wang and Tansel to produce statistical features for a genetic algorithm (GA) analysis in Lamb wave damage detection [90].

Since time and frequency are conjugate variables with respect to the Fourier transform, they are not independent. The mean square dispersion resolutions, Δt and Δf, are bound by an uncertainty relation [52]. For generic wavelets with a scale factor stretch a, the time and frequency resolutions become Δt_g and Δf_g. The time and frequency resolutions for the Morlet wavelet are

$$\Delta t = a\Delta t_g = \frac{a}{\sqrt{2}}, \tag{5.166}$$

$$\Delta f = \frac{\Delta f_g}{a} = \frac{1}{a(2\pi)\sqrt{2}}. \tag{5.167}$$

The Gaussian component of the Morlet wavelet helps provide an optimal balance of resolution between time and frequency. The combined time and frequency resolution forms a *Heisenberg box* in a 2-D plot of time versus frequency (Figure 5.18) [84]. Resolution measures such as these quantify the performance of a statistical signal processing effort, such as resolving two modes with closely spaced frequencies.

A primary application of wavelet analysis in SHM decomposes signals into time–frequency components for use in further analysis and interpretation through methods such as statistical pattern recognition. Al-Khalidy and Dragomir-Daescu used wavelet filter banks to decompose ball bearing acceleration signals for early fault detection [91]. Jiang et al. detected prefailure fault conditions in silicon wafer saws using wavelets [92]. Kim and Ewins concluded that the directional harmonic wavelet transform was superior to the STFT for distinguishing subharmonic resonances in rotating machines that arose due to cracks in the rotor [93]. Jiang et al. applied Bayesian wavelet denoising techniques for structural system identification [94]. Taghvaei et al. used wavelets as a filter to denoise pipe leak

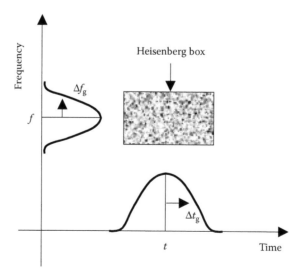

FIGURE 5.18 Frequency resolution and Heisenberg box for Morlet wavelet. (Adapted from Kijewski T and Kareem A. 2003. *Comput Aided Civ Infrastruct Eng*, 18, 339–355.)

acousto-elastic signals before proceeding onto a cepstral analysis [95]. Ip et al. identified different Lamb wave modes with wavelet decomposition [96].

A second application of wavelet transforms identifies abrupt changes or singularities in the higher-order derivatives of a signal, such as may occur in the passage of a wave group or in the curvature of a structural member with a defect [97,98]. Ahmad and Kundu used Gabor wavelets to detect the arrival times of ultrasound signals in pipe inspections [99]. Sohn et al. detected arrival times and calculated damage-sensitive features from Lamb waves in plates with wavelets [100]. Hou et al. detected changes in the modal parameters of vibrating structures with wavelets [101]. Lihua improved damage localization estimates with ETDR signals through the use of wavelets [102]. Zhao and Yuan implemented wavelet transforms in hardware with digital signal processors [103]. Wavelet analysis techniques directly extend to 2-D and 3-D data sets for image compression and feature extraction [104].

The *Lipschitz exponent* (also known as the *Holder exponent*), α, quantifies and locates abrupt changes in functions. The assumption is that an approximation to the function in question is a set of piecewise continuous polynomial functions. The functions have singularities at the joints between the pieces, when the derivatives exceed the degree, m, of the polynomial, $p_m(t)$. The Lipschitz exponent, α, quantifies the strength of a singularity at a point τ. m is the largest integer satisfying

$$\left| f(t) - p_m(t) \right| \leq K \left| t - \tau \right|^{\alpha}. \tag{5.168}$$

Larger levels of localized structural damage have stronger singularities in curvature at the damage site and smaller Lipschitz exponents.

Wavelet transforms can estimate the Lipschitz exponent with mother wavelets that fluctuate, such as the higher-order derivatives of the Gaussian function [105]. If the order of the fluctuation in the signal matches that of the smooth fluctuation in the signal, then the transform will have larger values than if there is a mismatch. Dawood et al. used this approach to identify low levels of damage in beams from the curvature of vibration mode shapes [97]. Xiaorong et al. estimated Lipschitz exponents with wavelet transforms and then used the exponents to identify delaminations in CFRP composites [106]. Robertson et al. detected part loosening from a Holder exponent wavelet analysis of dynamic response data [107]. An alternative method of identifying singularities uses *kriging interpolation* as developed by Iwasaki et al. for locating delaminations in CFRP composites with electrical potential measurements [108].

5.3.15 Vibration Modal Analysis

Modal analysis is a set of regression methods that determine vibration mode shapes, frequencies, and damping ratios of structures from experimental data [39,109]. Under certain circumstances, further processing can yield estimates of the mass, damping, and stiffness ($[M]$, $[C]$, and $[K]$) matrices as well as assessments of the performance and health of the structure [110]. Most modal analysis methods use input–output identification approaches that correlate measurements of the forces that act on the structure with measured vibration responses. When it is impractical to apply and measure forces, AVS and *random decrement* procedures can be useful alternatives.

Linearized equations of motion for a structure modeled with N degrees of freedom are [111,112]

$$[M]\{\ddot{r}\} + [C]\{\dot{r}\} + [K]\{r\} = \{F(t)\}, \tag{5.169}$$

where $[M]$ is an $N \times N$ mass matrix, $[C]$ is an $N \times N$ damping matrix, $[K]$ is an $N \times N$ stiffness matrix, $\{r\}$ is the $N \times 1$ coordinate displacement vector, and $\{F\}$ is an $N \times 1$ force vector matrix. Modal analysis uses the forces, $\{F\}$, as the inputs and the motions, $\{r\}$ or $\{\ddot{r}\}$, as the outputs. At the heart of modal analysis is the assumption that a set of independent modes, each with an associated natural frequency and damping ratio, superpose to form the observed vibrations. In terms of modal motions, the displacement, $\vec{r}_i$, at a point i on the structure becomes

$$\vec{r}_i(x, y, z, t) = \sum_{n=1}^{N} q_n(t) \vec{\varphi}_n(x, y, z), \tag{5.170}$$

where $q_i(t)$ are the modal amplitudes (generalized coordinates) and $\vec{\varphi}_n(x, y, z)$ are the mode shapes.

If the same coordinate transform that diagonalizes $[M]$ and $[K]$ also diagonalizes $[C]$, then the $q_n(t)$ and $\vec{\varphi}_n$ are real. A sufficient condition for diagonalization is for $[C]$ to be proportional to $[M]$ and $[K]$, that is,

$$[C] = \alpha[M] + \beta[K]. \tag{5.171}$$

When the damping matrix does not permit diagonalization (as is often the case), the resulting generalized coordinates, eigenvalues (frequencies), and mode shapes are complex. The complex frequencies are $s_n = \sigma_n + j\omega_n$. The real part, σ_n, corresponds to the damping (with a negative sign). The imaginary part, ω_n, corresponds to the natural frequency. Although complex coordinates and modes pose a bit of a visualization challenge for humans, computers readily calculate and manipulate them as needed.

Experimental modal analysis is a multistep procedure. The main steps are as follows: (1) *Collect input and output modal test data.* Accelerometers are probably the most common vibration motion sensor. Displacement meters, laser scanners, strain gages, and FBGs are also possibilities [113]. (2) *Estimate natural frequencies and damping ratios.* This typically involves an analysis of the peaks on the FRFs. Time domain algorithms are also possible. (3) *Calculate the mode shapes.* Use curve fitting in the time and/or frequency domains. (4) *Assess the quality of the estimates.* (5) *Adjust and recalculate the results as necessary.* Discard spurious mode shapes.

Identifying the modal properties of a structure with N independent mode shapes requires a minimum of N spatially independent measurements. A minimal test system uses a single applied force (often an instrumented hammer or shaker), a single transducer (often an accelerometer), and an assumption that the structural properties do not change between repeat tests. A typical test keeps one transducer (force or motion) fixed at a single location for all the repeat tests. The other transducer roams across the structure to create a set of combined input–output locations that span the structure. The use of multiple transducers for simultaneous multipoint data averaging can reduce the test time. Avoiding difficulties with spatial and temporal aliasing requires sampling with sufficient spatial and temporal resolution. Resolving modes with closely spaced frequencies may require specialized techniques with high-frequency resolutions. Possibilities include the use of a shaker with harmonic excitation that sweeps very slowly through the resonant frequencies, and high-resolution spectral estimators such as the maximum entropy method. Resonant frequencies that shift slightly due to parametric changes or nonlinearities can be especially difficult to resolve with precision and can confound damping estimates.

The processing of the raw input–output data generally proceeds by forming a set of time domain *impulse response functions* in terms of *residues A_{pqn}* [109]

$$h_{pq}(t) = \sum_{n=1}^{N} (A_{pqn}\, e^{\lambda_n t} + A_{pqn}^*\, e^{\lambda_n^* t}) \tag{5.172}$$

or frequency domain FRFs

$$H_{pq}(\omega) = \sum_{n=1}^{N} \left(\frac{A_{pqn}}{j\omega - \lambda_n} + \frac{A_{pqn}^*}{j\omega - \lambda_n^*} \right). \tag{5.173}$$

The subscripts p and q are the input and output locations. n is the mode number. The λ_n are complex eigenvalues, with the real component being the negative of the damping ratio and the imaginary component being the natural frequency. When using accelerometers to collect the data, it is usually necessary to convert the raw acceleration data into equivalent displacements.

An example is the modal analysis of a taut string stretched between two fixed points, with tension T, length L, and mass/unit length ρ. The natural frequencies f_n and mode shapes φ_n are

$$f_n = \frac{n}{2L} \left[\frac{T}{\rho} \right]^{1/2}, \quad n = 1, 2, \ldots, \tag{5.174}$$

$$\varphi_n = \sin\left(\frac{n\pi x}{L} \right). \tag{5.175}$$

The first three modes appear in Figure 5.19. The mass-normalized residues for input point 1 are

$$\begin{aligned}
A_{111} &= \frac{1}{2}, & A_{121} &= \frac{\sqrt{2}}{2}, & A_{131} &= \frac{1}{2}, \\
A_{112} &= 1, & A_{122} &= 0, & A_{132} &= -1, \\
A_{113} &= \frac{1}{2}, & A_{123} &= \frac{-\sqrt{2}}{2}, & A_{133} &= \frac{1}{2}.
\end{aligned} \tag{5.176}$$

If it is assumed that the modes are real with a damping ratio of 2%, then the magnitude and phase of the FRF correspond to the responses of points 1 to 3 that appear in Figures 5.20 and 5.21. The FRFs show large peaks at the *resonances* and *antiresonances* at interlacing locations. The amplitudes of the resonant peaks depend on the modal amplitude. The width of the peak indicates the modal damping. The phase angle indicates the sign of the modal amplitude. Rapid changes in the phase angle correspond to the presence of a resonance.

Modal analysis techniques can be quite effective when the vibrations are linear. Linearity has many implications. One is reciprocity, that is, the motion at point i due to a force applied at point j equals the motion at point j due to the same force applied at point i. Another is that the FRF should be independent of the amplitude of the motions. When the structure behaves in a nonlinear manner, the principle of superposition breaks down, but the concept of modal vibrations often retains utility.

All structures exhibit some level of nonlinear amplitude-dependent behavior. When testing large structures, amplitude-dependent effects at small amplitudes can be of particular

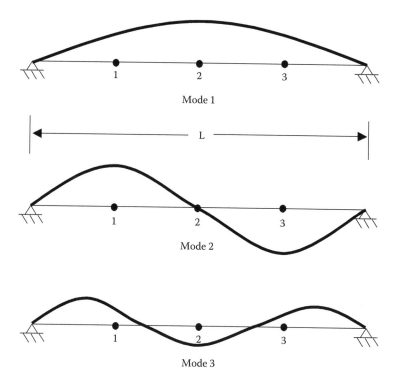

FIGURE 5.19 First three mode shapes of a vibrating string.

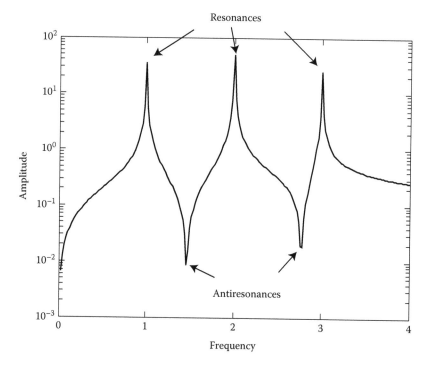

FIGURE 5.20 FRF amplitude for applied force and measurement at point 1.

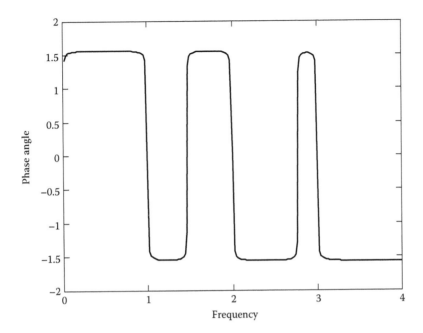

FIGURE 5.21 FRF phase angle for applied force and measurement at point 1.

concern. At small-amplitude motions many jointed structures exhibit amplitude-dependent effects, such as stiction and/or backlash. Although the modeling of stiction and backlash effects is difficult, developments in the use of rational semi-empirical microfriction models render some of these problems tractable [114]. Many precision instruments avoid stiction by replacing bearing joints with elastic flexures [115]. Nonlinearities also arise from sources such as material properties, cracks, delaminations, geometric stiffening, and bolted joints with backlash. Changes in structural states can cause changes in the level and type of nonlinear behavior.

All sets of linear vibration mode shapes have common collective mathematical properties, including orthogonality when weighted with respect to mass and linear independence. The nature of these properties is such that relatively simple statistical analyses can check the quality of the measured or calculated mode shapes and/or check on the underlying assumption of linearity. Two common modal quality assessment statistics are the *modal assurance criterion* (MAC) and the *complementary modal assurance criterion* (COMAC).

The MAC measures the similarity between two mode shapes. The MAC is the normalized dot product between two mode shape vectors $\{\varphi_i\}$ and $\{\varphi_j\}$:

$$\text{MAC}(i,j) = \frac{\left| \{\varphi_i\}^\text{T} \{\varphi_j^*\} \right|^2}{(\{\varphi_i\}^\text{T} \{\varphi_i^*\})(\{\varphi_j\}^\text{T} \{\varphi_j^*\})}. \tag{5.177}$$

Similar modes have a MAC close to 1. Dissimilar modes have a MAC close to 0. The MAC calculation approximates a modal orthogonality dot product, but lacks mass matrix weighting and is not strictly an orthogonality check. The rationale for the success of the MAC is that the typical distribution of nodal coordinates is fairly homogeneous with respect to the mass of the structure and the corresponding mass matrix is approximately proportional to the identity matrix.

The COMAC is a node point estimate of the modal similarity of two sets of modes, φ_{ik} and ψ_{ik}. The COMAC for point k is

$$\text{COMAC}(k) = \frac{\left| \sum_{i=1}^{N} \varphi_{ik} \psi_{ik}^* \right|^2}{\left[\sum_{i=1}^{N} \varphi_{ik} \varphi_{ik}^* \right] \left[\sum_{i=1}^{N} \psi_{ik} \psi_{ik}^* \right]}. \tag{5.178}$$

COMAC values near 1 indicate that the mode shape sets are similar.

A common modal analysis technique uses windowed DFTs to calculate FRFs, and then extracts the modal parameters by frequency domain curve fitting. Since vibration modal analysis assumes linearity, superposition procedures allow using a variety of other algorithms with specific advantages for certain testing situations, including time domain techniques that avoid Fourier transforms [39,109]. For example, Yan et al. compared the wavelet transform and HHT as tools for modal parameter estimation [116]. A conclusion was that both tools were adequate, but that the wavelet transform was more robust and had a stronger theoretical underpinning.

It may be possible to bypass modal decompositions and to identify structural parameters directly. This can be the case if the structure has a known layout with predictable vibration properties. Takewaki and Nakamura developed a technique for identifying stiffness and damping properties in shear buildings by spectral curve fitting with a noise-tolerant method that fits at low frequencies [117].

5.3.16   AVSs: Random Decrement

Controlling and/or measuring structural forces may be impractical. Examples are when a structure is too large, such as a suspension bridge, or when the ambient forces are so large that they overwhelm any potentially available test force, as in an operating rocket motor. In these circumstances it may be possible to use AVS techniques to identify modal properties. An AVS requires only measurements of the outputs and assumptions regarding the nature of the inputs. AVS techniques are particularly amenable to structural and modal situations where ambient loading is random and broad-banded.

The first step in an AVS records motions due to vibrations at several points on the structure, preferably simultaneously. The next step calculates spectra and cross-spectra from the motion data. Curve fitting to the spectra provides an estimate of structural properties. When the structure is lightly damped and the ambient loading is broad-banded, the spectra have sharp peaks at frequencies that correspond to the resonances of the structure. The precise shapes of these spectra depend on the shape and magnitude of the spectrum of the applied force. If the spectra of the applied forces are effectively white or fairly smooth over the narrow width of the spectral peaks, then the spectral peaks have the same shape as results from subjecting the system to pure white noise. Correlating the heights of the spectral peaks measured at multiple points produces the mode shape amplitudes. Since the magnitudes of the applied ambient forces are not known, it is possible to determine the FRFs only to within an arbitrary multiplicative constant. Cross-spectra and cross-correlations of motions measured simultaneously at multiple points indicate the relative phases (real and complex) of the amplitudes [118]. Relative phases are sensitive to structural changes and have potential for use in damage detection and localization [119].

The *random decrement* technique is another AVS method [41,120]. The concept is that the motion at a point, $x(t)$, on a linear vibrating structure is the superposition of an initial condition response (homogeneous solution), $x_h(t)$, and a forced response (particular solution), $x_p(t)$:

$$x(t) = x_h(t) + x_p(t). \tag{5.179}$$

When a set of collected motion time histories, $\{x_i(t)\ i = 1, 2, \ldots, n\}$, all have the same initial conditions but have different subsequent applied external forces, then the initial condition terms, $x_h(t)$, are equal and the forced response terms, $x_{pi}(t)$, are random and different, that is,

$$x_i(t) = x_h(t) + x_{pi}(t). \tag{5.180}$$

A time domain average of the preselected time histories eliminates the random forced responses, $x_{pi}(t)$, and leaves the initial condition response, $x_h(t)$,

$$\langle x_i(t) \rangle = \langle x_h(t) + x_{pi}(t) \rangle = x_h(t) + \langle x_{pi}(t) \rangle = x_h(t). \tag{5.181}$$

One advantage of the random decrement method is the ability to set the initial condition threshold at sufficiently high amplitudes so as to capture the behavior of the structure only when the motions are large enough to break out of Stribeck and stiction effects. Vector and modal-based multipoint extensions of the technique are possible [121]. Abe et al. obtained sufficient information for convergence from a stack of 5000 random decrement time histories to conduct a dynamic analysis of the Hakucho Bridge (Muroran Gulf, Hokkaido, Japan) [122]. Cremona et al. used the random decrement approach to determine modal properties of railway bridges with high-speed trains as the excitation source [123]. Li et al. demonstrated the effectiveness of random decrement analysis for damage detection in fiber-reinforced composites [124]. Liang and Lee proposed using a lag method that makes more use of the time history for those situations where the amplitude of the initial starting point is not important [125].

Multiple bridges have been the subject of AVSs [126–130]. Katsuchi et al. found that AVS measurements of damping on the Akashi Kaikyo Bridge typically had large degrees of variability [131]. A wavelet screening approach that identified particular modal components in the observed motions before using an eigensystem realization algorithm considerably reduced the variability. Nayeri et al. used an AVS to track structural changes in a multi-story building as it underwent a major seismic retrofit [132]. Rubin conducted an AVS of an offshore oil platform [133]. Onboard machinery produced large narrowband inputs at consistent frequencies. Following the removal of the narrowband components, an analysis of the AVS data successfully measured vibration modes and frequency shifts with different sea loading conditions.

5.3.17 Kalman Filter State Estimation

The *Kalman filter* is a recursive least-squares technique that provides up-to-date estimates of the state of a system based on analyzing a sequence of measured system outputs. Useful properties of the Kalman filter are as follows: (1) The Kalman filter can estimate more state variables than the number of measured outputs. (2) A fairly simple mathematical criterion can establish if the system states are measurable (observable). (3) Due to flexibility at the detail level of the recursive algorithm, the experienced user can tailor the performance of the Kalman filter to increase the estimation performance for a particular situation.

Following the notation used by Stengel, a first-order matrix differential equation, that is, the *state space* form, describes the behavior of many linear dynamical systems [31]:

$$\{ \underset{n\times1}{\dot{x}} \} = [\underset{n\times n}{F}(t)]\{ \underset{n\times1}{x} \} + [\underset{n\times m}{G}(t)]\{ \underset{m\times1}{u} \} + [\underset{n\times n}{L}(t)]\{ \underset{n\times1}{w} \}, \tag{5.182}$$

$$\{ \underset{p\times1}{z} \} = [H(t)]\{ \underset{n\times1}{x} \} + \{ \underset{p\times1}{v} \}. \tag{5.183}$$

$\{x\}$ is an $n \times 1$ column vector of system state variables. Position, velocity, and acceleration, as well as more complicated quantities, such as nonlinear friction forces, can be state variables. $[F]$ is an $n \times n$ system dynamics matrix. $[G]$ is an $n \times m$ feedback gain matrix. $\{u\}$ is an $m \times 1$ feedback vector. $[L]$ is an $n \times n$ noise transition matrix. $\{w\}$ is an $n \times 1$ vector of system noise components. $\{z\}$ is a $p \times 1$ observation vector. $[H]$ is a $p \times n$ system observation matrix. $\{v\}$ is a $p \times 1$ observation noise vector. The Kalman filter technique combines the observations, $\{z\}$, with system information, $[F]$, $[G]$, and $[H]$, to form an estimate of the state, $\{x\}$, while accounting for the noise terms, $\{w\}$ and $\{v\}$.

As an example, a second-order ordinary differential equation can usually model the dynamics of a linear single degree of freedom mechanical oscillator

$$\ddot{y} + 2\zeta\omega_0\dot{y} + \omega_0^2 y = K_0 u(t), \tag{5.184}$$

where $y(t)$ is the displacement, ζ is the damping ratio, ω_0 is the natural frequency, K_0 is the feedback gain (which is set equal to one in the case of an external force), and $u(t)$ is the applied force. A state variable form of Equation 5.184 is

$$\begin{bmatrix} \dot{x}_1 \\ \dot{x}_2 \end{bmatrix} = \begin{bmatrix} 0 & 1 \\ -\omega_0^2 & -2\zeta\omega_0 \end{bmatrix} \begin{bmatrix} x_1 \\ x_2 \end{bmatrix} + \begin{bmatrix} 0 \\ K_0 \end{bmatrix} u(t) + \begin{bmatrix} w_1 \\ w_2 \end{bmatrix}, \tag{5.185}$$

$$z(t) = \begin{bmatrix} 1 & 0 \end{bmatrix} \begin{bmatrix} x_1 \\ x_2 \end{bmatrix} + v(t) = y(t) + v(t). \tag{5.186}$$

The state vector, $\{x\}$, is

$$\{x\} = \begin{bmatrix} x_1 \\ x_2 \end{bmatrix} = \begin{bmatrix} y \\ \dot{y} \end{bmatrix}. \tag{5.187}$$

The vector $\{w\}$ and scalar v represent the noise components in the observations and applied forces.

A system is *observable* when the observed measurands, $\{z\}$, contain sufficient information to enable estimating values for all of the system states, $\{x\}$ [31,134,135]. Conversely, a system is *controllable* if the appropriate applied force, $\{u\}$, can produce specified values for all of the n states, $\{x\}$. Based on results derived from the Cayley–Hamilton theorem, an *observability matrix* $[O]$ of rank n guarantees observability, where

$$[O] = \begin{bmatrix} H^T & F^T H^T & (F^T)^2 H^T & \ldots (F^T)^{n-1} H^T \end{bmatrix}. \tag{5.188}$$

A controllability matrix $[C]$ with a rank of n guarantees controllability, where

$$[C] = \begin{bmatrix} G & FG & F^2G & \ldots & F^{n-1} C \end{bmatrix}. \tag{5.189}$$

Practical considerations of sensitivity and robustness to perturbations and noise, along with quantization difficulties and nonlinearities, detract from the strict guarantees of observability and controllability given by Equations 5.188 and 5.189.

The discrete-time equivalent state space representation of a linear system with an equal interval sampling period Δt is

$$\{x_k\} = [\Phi_{k-1}]\{x_{k-1}\} + [\Gamma_{k-1}]\{u_{k-1}\} + [\Lambda_{k-1}]\{w_{k-1}\}, \tag{5.190}$$

$$\{z_k\} = [H_k]\{x_k\} + \{n_k\}, \tag{5.191}$$

where

$$[\Phi_k] = e^{[F_k]\Delta t}, \tag{5.192}$$

$$[\Gamma_k] = [\Phi_k - I][F_k]^{-1}[G_k], \tag{5.193}$$

$$[\Lambda_k] = [\Phi_k - I][F_k]^{-1}[L_k]. \tag{5.194}$$

It is assumed that the error terms $\{w_k\}$ and $\{v_k\}$ are random, zero mean, and possibly nonstationary white noise vector processes. The covariance matrices, $[Q'_k]$ and $[R_k]$, of the error term vectors are

$$[Q'_k] = E[\{w_k\}\{w_k\}^T], \tag{5.195}$$

$$[R_k] = E[\{n_k\}\{n_k\}^T]. \tag{5.196}$$

A similar error term is the difference between the estimated states $\{\hat{x}\}$ and the true states $\{x\}$. $[P_k]$ is the covariance matrix of these estimation errors:

$$[P_k] = E[(\{x_k\} - \{\hat{x}_k\})(\{x_k\} - \{\hat{x}_k\})^T]. \tag{5.197}$$

The Kalman filter is an iterative time stepping procedure. Past state estimates, $\{\hat{x}_{k-1}(+)\}$, predict the present states, $\{\hat{x}_k(-)\}$, with the *state estimate extrapolation*

$$\{\hat{x}_k(-)\} = [\Phi_{k-1}]\{\hat{x}_{k-1}(+)\} + [\Gamma_{k-1}]\{u_{k-1}\}. \tag{5.198}$$

The *extrapolation estimate covariance matrix*, $[P_k(-)]$, is

$$[P_k(-)] = [\Phi_{k-1}][P_{k-1}(+)][\Phi_{k-1}]^T + [\Lambda_{k-1}][Q'_{k-1}][\Lambda_{k-1}]^T. \tag{5.199}$$

The *Kalman filter gain*, $[K_k]$, is

$$[K_k] = [P_k(-)][H_k]^T[[H_k][P_k(-)][H_k]^T + [R_k]]^{-1}. \tag{5.200}$$

The Kalman filter gain combines with the present measurands to form an improved *state estimate update*

$$\{\hat{x}_k(+)\} = \{\hat{x}_k(-)\} - [K_k][\{z_k\} - [H_k]\{\hat{x}_k(-)\}]. \tag{5.201}$$

The *covariance estimate update* is

$$[P_k(+)] = [[P_k(-)]^{-1} + [H_k]^T[R_k]^{-1}[H_k]]^{-1}. \tag{5.202}$$

The choice of initial values $\hat{x}_0(+)$ and $P_0(+)$ can have a large effect on the convergence rate of the Kalman filter.

At first glance, the Kalman filter may not appear to be very useful for SHM since it estimates the states based on known system properties. Many SHM analyses attempt to do the inverse, that is, estimate system (structural) properties from measurements of inputs and outputs. Nonetheless, many SHM applications can benefit from Kalman filter techniques. For example, information about structural states can identify changes in the structural properties [136]. Fritzen and Mengelkamp used changes in output residual errors as a damage indicator [137]. Jones et al. identified states from AVSs with Kalman filters [128]. Shi et al. extended this approach to the identification of structural parameters and input forces from an observation of output motions [138]. Kalman filters can help to combine data from heterogeneous sensor arrays, such as strain gages, accelerometers, and/or GPS transducers, to form accurate displacement estimates [139,140]. Sophisticated data acquisition triggering and event timing applications, such as AE, can make use of Kalman filter state estimates. An interesting and useful property of the Kalman filter is that it readily accommodates multichannel data that are collected at different, and even intermittent, rates. This is useful in navigation where intermittent information, such as that from a GPS transducer, updates high data rate information from an inertial accelerometer.

It is also possible to add system properties as state variables. The resulting equations may be nonlinear. A Taylor series linearization known as the *extended Kalman–Bucy filter* (EKBF) can identify system properties as state variables [141,142]. Yang et al. used the EKBF to identify state changes and damage in vibrating structures [143]. Burnett et al. found that the EKBF was more computationally efficient than state space identification methods based on Gauss–Newton gradient search techniques [144]. Ghanem and Ferro used an ensemble Kalman filter to account for nonlinear and non-Gaussian effects in structural identification, including identifying parameters related to the nonlinear hysteretic Bouc–Wen model [145]. Wu and Smyth used an *unscented Kalman filter* also to identify parameters related to the nonlinear hysteretic Bouc–Wen model [146]. Using the full nonlinear behavior instead of a truncated Taylor series is one of the rationales for using the unscented KBF.

5.4 Image Processing

Continued advances in image collection and processing technologies create many economical and useful imaging options for the structural engineer [147]. Statistical image processing serves at least five purposes: (1) *Image enhancement*: The human eye has very specific requirements in terms of size, intensity, color, and variability. Image enhancement techniques can transform and optimize image data formats for human viewing. (2) *Registration*: Two images of the same structure invariably do not align. Differences in imaging systems, formats, camera angles, distortions, and so on are unavoidable. Registration geometrically transforms one image to match the geometry of another [148,149]. (3) *Quantitative statistical image processing*: It is often desired to calculate specific statistical quantities from an image or an region of interest (ROI). (4) *Image compression*: Image data files are inherently large. (5) *Feature identification and location*: Individual features in an image correlate with structural features such as cracks, struts, aggregate, potholes, and so on.

Most image data formats have common features. The image presented to the viewer is an array of pixels (or voxels) that cover a small area (volume). The value assigned to each pixel is typically a 1-, 8-, 16-, or 24-bit binary integer. The pixel data may represent either a black–gray–white intensity value or a color by mixing red, green, and blue palettes. More data bits generally correspond to more information in the pixels. 1-bit encoding is a simple black and white representation. Eight bits generally encode gray-scale image pixels. Full color often uses 8 bits for each of the three primary colors for a total of 24 bits per colored pixel. Arranging the pixels in the proper locations forms the image. The specific details of the pixels placement, that is, in rectangular or interlaced layouts, can have a large effect on perceived image quality.

An uncompressed image data storage format stores each pixel as an individual binary integer datum in an array. This format, known as a *bitmap*, is inefficient and results in excessively large arrays. Data compression reduces the size of the image files. An important characteristic of image data compression and reconstruction formats is whether any information is lost in the compression–storage–transmit–decompression process. Some algorithms will not lose any information in the compression process. An example is a fax machine that scans a line of pixels horizontally across the page. If an ROI has all the same color or pixel values, such as large amounts of white space in a text document, then the pixel values along that portion of the scanned line are identical. Instead of storing the data corresponding to each pixel, it is more efficient to count the number of contiguous pixels with the same value and then to transmit the pixel value and the number of contiguous pixels. Fax machine compression is relatively easy to implement, but the technique is suboptimal. It does not take advantage of pattern-based or 2-D compression opportunities. The *Joint Photographic Experts Group* (jpeg or jpg) algorithms are standardized, efficient 2-D compression techniques that use a 2-D cosine transform over a small region and truncate the spatial high-frequency components [150]. Varying the amount of discarded spatial high-frequency information adjusts the amount of compression and lost information.

Image enhancement is often necessary for effective viewing and analysis. The purpose of image enhancement is not to change the information content, but instead to convert the image into a more readily usable form. One method is histogram equalization. The process alters pixel intensity values so that a histogram of the intensities is nearly uniform [150]. Figure 5.22 shows the underside of a bridge and a prototype beam-traversing inspection robot. The underside structure of the bridge is difficult to examine because of bright sunlit

FIGURE 5.22 Uncorrected view of the underside of a bridge and a prototype inspection robot. (Adapted from Huston DR, Miller J, and Esser B. 2004. *Smart Structures and Materials 2004: Sensors and Smart Structures Technologies for Civil, Mechanical, and Aerospace Systems*, Shih-Chi Liu (ed.). SPIE 5391.)

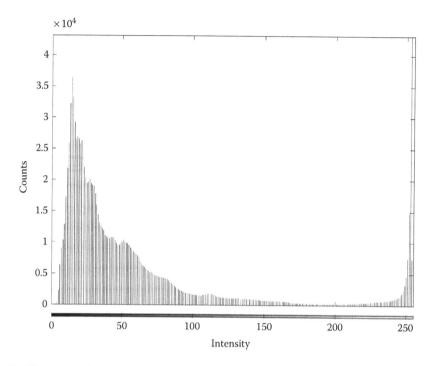

FIGURE 5.23 Histogram of intensity of the uncorrected picture in Figure 5.22.

features in the background. Figure 5.23 is a histogram of the image intensity. Figure 5.24 is the same image but it has been corrected by balancing the histogram. The underside girders are clearly visible. Figure 5.25 is the histogram of the corresponding balanced image. Abdel-Ghaffar et al. used histogram equalization to improve the contrast of x-ray tomography reconstruction of the core features in concrete test cylinders [151].

FIGURE 5.24 Histogram-balanced view of the underside of a bridge and a prototype inspection robot. (Adapted from Huston DR, Miller J, and Esser B. 2004. *Smart Structures and Materials 2004: Sensors and Smart Structures Technologies for Civil, Mechanical, and Aerospace Systems*, Shih-Chi Liu (ed.). SPIE 5391.)

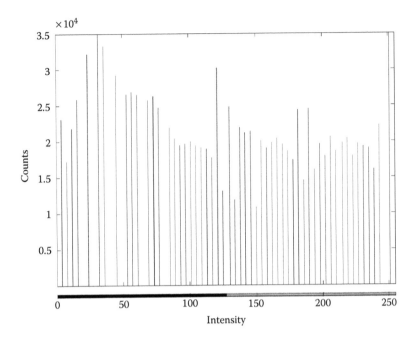

FIGURE 5.25   Histogram of intensity of the corrected picture in Figure 5.24.

Edge detection and enhancement algorithms can aid in determining the positions of objects and cracks [150]. Many implementations of edge detection algorithms are essentially trial and error processes that tweak algorithm-operating parameters until a visually satisfactory result is achieved. Figure 5.26 shows the results of edge detection by applying the Sobel algorithm to the underside of the bridge shown in Figure 5.24. Interpreting the

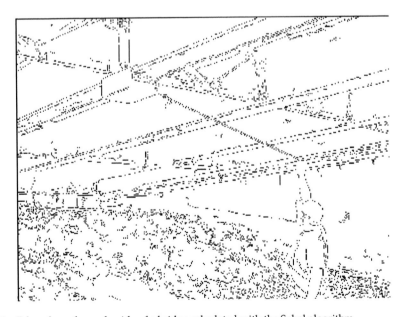

FIGURE 5.26   Edges from the underside of a bridge calculated with the Sobel algorithm.

significance of machine-detected crack patterns remains a matter of research. For example, Cao et al. developed a methodology based on assessing the fractal dimension of crack patterns in concrete that correlates with the overall level of damage in the concrete [152].

Edges are just one of a myriad of possible features extractable from an image. Fu and Moosa used subpixel interpolation techniques to obtain high-resolution estimates of structural displacements from digital images [153]. Al-Nuaimy et al. applied some of these methods to GPR image processing [154]. Mansouri et al. used image correlation methods with satellite-based synthetic aperture radar images to detect earthquake damage in structures [155]. Multicamera systems can provide 3-D profiles of objects. An example is the measurement of railroad rail and wheel profiles [156]. Dai et al. used edge and feature identification techniques to identify the size and location of aggregate gravel in asphalt samples for use in subsequent micromechanical finite element analysis [157]. Lee and Chang detected bridge steel corrosion neural network processing of features extracted from digital images [158]. Yao et al. detected and classified concrete cracks with digital imaging techniques [159].

References

1. Pan Q, Prater JB, and Aktan E. 2007. Identification of civil engineering structures and uncertainty. In *Structural Health Monitoring*, FK Chang (ed.). DEStech, Lancaster.
2. Feller W. 1968. *An Introduction to Probability Theory and Its Applications*, Vol. 1, 3rd Ed. Wiley, New York.
3. Box GP, Hunter WG, and Hunter JS. 1978. *Statistics for Experimenters*. Wiley, New York.
4. Guttman I, Wilks SS, and Hunter JS. 1982. *Introductory Engineering Statistics*, 3rd Ed. Wiley, New York.
5. Hyvarinen A, Karhunen J, and Oja E. 2001. *Independent Component Analysis*. Wiley, New York.
6. Piersol JS. 1990. *Nonlinear System Analysis and Identification from Random Data*. Wiley, New York.
7. Rade L and Westergren B. 1995. *Mathematics Handbook for Science and Engineering*. Birkhauser, Boston.
8. Tukey JW. 1977. *Exploratory Data Analysis*. Addison-Wesley, Reading.
9. Shinozuka M, Feng MQ, Lee J, and Naganuma T. 2002. Statistical analysis of fragility curves. *J Eng Mech*, 126(12), 1224–1231.
10. Hart GC. 1982. *Uncertainty Analysis, Loads, and Safety in Structural Engineering*. Prentice-Hall, Englewood Cliffs, NJ.
11. Rice SO. 1954. Mathematical analysis of random noise. *Bell System Technical J*, Vol. 23, 1944 and Vol. 24, 1945, reprinted in *Selected Papers on Noise and Stochastic Processes*, N Wax (ed.). Dover Publications, Mineola, New York.
12. Hedl R, Hamza R, and Gordon GA. 2007. Automated corrosion detection using ultrasound lamb waves. In *Structural Health Monitoring*, FK Chang (ed.). DEStech, Lancaster.
13. Gumbel EJ. 1958. *Statistics of Extremes*. Dover Publications, Mineola, New York.
14. Coles S and Simiu E. 2003. Estimating uncertainty in the extreme value analysis of data generated by a hurricane simulation model. *J Eng Mech*, 129(11), 1288–1294.
15. Becker SM, McDaniel WT, and Baeverstad E. 2006. Development and implementation of the highway structures information system for the Wisconsin Department of Transportation. Paper no. 06-1480, 86th Annual Meeting, TRB, Washington, DC.
16. Mandel J. 1964. *The Statistical Analysis of Experimental Data*. Dover Publications, Mineola, New York.
17. Shewhart WA. 1986. *Statistical Method from the Viewpoint of Quality Control*. Dover Publications, Mineola, New York.
18. Hamming RW. 1998. *Digital Filters*. Dover Publications, Mineola, New York.

19. Oppenheim AV and Schafer RW. 1975. *Digital Signal Processing*. Prentice-Hall, Englewood Cliffs, NJ.
20. Mosteller F and Tukey JW. 1977. *Data Analysis and Regression*. Addison-Wesley, Reading.
21. Li Z, Li F, Li XS, and Yang W. 2000. P-wave arrival determination and AE characterization of concrete. *J Eng Mech*, 126(2), 194–200.
22. Shenton III HW, Chajes MJ, Finch WW, and Sivakumar B. 2002. Long-term monitoring of the Newburgh-Beacon Bridge. *Proc Structural Materials Technology V, an NDT Conference*, Cincinnati.
23. Lanata F. 2004. Processing of incomplete time-series from continuous static monitoring. *Proc 2nd European Workshop on Structural Health Monitoring*, DEStech, Munich.
24. Bendat JS and Piersol AG. 1980. *Engineering Applications of Correlation and Spectral Analysis*. Wiley-Interscience, New York.
25. Box GE and Jenkins GM. 1976. *Time Series Analysis Forecasting and Control*. Holden-Day, San Francisco.
26. Blackman RB and Tukey JW. 1958. *The Measurement of Power Spectra*. Dover Publications, Mineola, New York.
27. Gao Y, Brennan MJ, Joseph PF, Muggleton JM, and Hunaidi O. 2005. On the selection of acoustic/vibration sensors for leak detection in plastic water pipes. *J Sound Vib*, 283, 927–941.
28. Draper NR and Smith H. 1998. *Applied Regression Analysis*, 3rd Ed. Wiley-Interscience, New York.
29. Lohr M and Dinkler D. 2004. Damage detection in structures using a parameter identification method. *Proc 2nd European Workshop on Structural Health Monitoring*, DEStech, Munich.
30. Campbell SL and Meyer CD. 1979. *Generalized Inverses of Linear Transformations*. Dover Publications, Mineola, New York.
31. Stengel RF. 1994. *Optimal Control and Estimation*. Dover Publications, Mineola, New York.
32. Hildebrand FB. 1987. *Introduction to Numerical Methods*. Dover Publications, Mineola, New York.
33. Fukunaga H and Hu N. 2004. Health monitoring of composite structures based on impact force identification. *Proc 2nd European Workshop on Structural Health Monitoring*, DEStech, Munich.
34. Bloomfield P. 1976. *Fourier Analysis of Time Series: An Introduction*. Wiley, New York.
35. Hildebrand FB. 1976. *Advanced Calculus for Applications*. Prentice-Hall, Englewood Cliffs, NJ.
36. Bendat JS and Piersol AG. 2000. *Random Data: Analysis and Measurement Procedures*, 3rd Ed. Wiley-Interscience, New York.
37. Rus G, Lee SY, Gallego R, Lee RC, and Wooh SC. 2004. Impact testing of a perforated plate using complete frequency information. *Proc 2nd European Workshop on Structural Health Monitoring*, DEStech, Munich.
38. Bracewell RN. 2000. *The Fourier Integral and Its Applications*, 3rd Ed. McGraw-Hill, New York.
39. Ewins DJ. 2001. *Modal Testing: Theory and Practice*, 2nd Ed. Research Studies Press Ltd., Baldock, England.
40. Hammad AM and Issa MA. 1994. A new nondestructive technique to investigate surface topography. *New Experimental Techniques for Evaluating Concrete Material and Structural Performance*, ACI SP-143-11.
41. Allemang RJ, Brown DL, and Rost RW. 1987. Measurement techniques for experimental modal analysis. *Experimental Modal Analysis and Dynamic Component Synthesis*, Vol. II, AFWAL-TR-87-3069, Air Force Wright Aeronautical Laboratories.
42. Robinson EA. 1978. *Multichannel Time Series Analysis with Digital Computer Programs*, 2nd Ed. Holden-Day, San Francisco.
43. Burg JP. 1972. The relationship between maximum entropy spectra and maximum likelihood spectra. *Geophysics*, 37, 375–376.
44. Proakis JG and Manolakis DG. 1988. *Introduction to Digital Signal Processing*. Macmillan, New York.
45. Akaike H. 1974. A new look at the statistical model identification. *IEEE Trans Autom Control*, AC-19(6), 716–723.

46. Allen DW, Inmann DJ, and Farrar CR. 2003. Optimization of time domain models applied to structural health monitoring. In *Structural Health Monitoring*, FK Chang (ed.). DEStech, Lancaster.

47. Martin RJ. 2000. A metric for ARMA processes. *IEEE Trans Signal Process*, 48(4), 1164–1170.

48. Vandiver JK and Campbell JK. 1979. An evaluation of the natural frequencies and damping ratios of three similar offshore platforms using maximum entropy spectral analysis. *ASCE Spring Convention*, Boston, MA.

49. Huston D. 1986. The effects of upstream gusting on the aeroelastic behavior of long suspended-span bridges. PhD dissertation, Civil Engineering, Princeton University.

50. Okabayashi T and Okumatsu T. 2004. High accurate estimation of structural frequency with AR model and shaking-table test. *Proc IABMAS'04 Bridge Maintenance, Safety, Management and Cost*, Kyoto. Taylor & Francis, London.

51. Xie JH. 2006. Application of Hilbert transform and LS-SVM to active damage monitoring for composite materials. *Proc 4th China–Japan–US Symposium on Structural Control and Monitoring*, Hangzhou.

52. Papoulis A. 1962. *The Fourier Integral and Its Applications*. McGraw-Hill, New York.

53. Newland DE. 1993. *An Introduction to Random Vibrations, Spectral, and Wavelet Analysis*, 3rd Ed. Longman House, London.

54. Papy JM, Van Huffel S, Rippert L, and Wevers M. 2003. On-line detection method for transient waves applied to continuous health monitoring of carbon fiber reinforced polymer composites with embedded optical fibers. *Modeling, Signal Processing and Control*, SPIE 4049.

55. Bang HJ, Kim DH, Kang HK, Hong CS, and Kim CG. 2003. Optical fiber sensor systems for simultaneous monitoring of strain and damage in smart composites. *Smart Sensor Technology and Measurement Systems*, SPIE 5050.

56. Pinto P, Hurlebaus S, Jacobs LJ, and Gaul L. 2003. Localization and sizing of discontinuities using lamb-waves. In *Structural Health Monitoring*, FK Chang (ed.). DEStech, Lancaster.

57. Frarey JL. 2002. How wrong signal processing can lead to problems. *Proc 26th Annual Meeting*, Vibration Institute.

58. Halabe UB and Pyakurel S. 2006. 3D GPR imaging for detection of internal defects in wooden logs. NDE *Conf on Civil Eng*, Amer Soc Nondes Test, St. Louis.

59. Lyon RH. 1987. *Machinery Noise and Diagnostics*. Butterworths, Boston.

60. Francoeur D, Pasco Y, Micheau P, and Masson P. 2007. A reflectivity damage detection approach for lap joint structures in the medium frequency range. In *Structural Health Monitoring*, FK Chang (ed.). DEStech, Lancaster.

61. Scanlan RH and Mendelson A. 1963. Structural damping. *AIAA J*, 1, 938–939.

62. Worden K and Manson G. 1998. Random vibrations of a duffing oscillator using the volterra series. *J Sound Vib*, 217(4), 781–789.

63. Loh CH and Li TY. 2003. A signal-based approach for the identification of nonlinear structural dynamic system. In *Structural Health Monitoring*, FK Chang (ed.). DEStech, Lancaster.

64. Zhang RR, Ma S, Safak E, and Hartzell S. 2003. Hilbert–Huang transform analysis of dynamic and earthquake motion recordings. *J Eng Mech*, 129(8), 861–875.

65. Huang NE. 2002. Empirical mode decomposition apparatus, method and article of manufacture for analyzing biological signals and performing curve fitting. US Patent 6, 381, 559.

66. Salvino LW, Rasmussen EA, and Pines DJ. 2004. Detecting structural damage using adaptive feature extraction from transient signals. *Health Monitoring and Smart Nondestructive Evaluation of Structural and Biological Systems III*, SPIE 5394.

67. Pines D and Salvino L. 2003. Sensitivity of Hilbert magnitude and phase to structural damage. *Smart Structures and Integrated Systems*, SPIE 5056.

68. Yang JN, Lei Y, Lin S, and Huang N. 2004. Identification of natural frequencies and dampings of in situ tall buildings using ambient wind vibration data. *J Eng Mech*, 130(5), 85–95.

69. Li YF, Chang SY, Tzeng WC, and Huang K. 2004. The seismic response of RC bridge column analyzed by using the Hilbert–Huang transform. *Proc IABMAS'04 Bridge Maintenance, Safety, Management and Cost*, Kyoto. Taylor & Francis, London.

70. Wang B, Jiang Y, Li Y, Wang X, and Xiong K. 2003. Experimental research on the response characteristic of piezoelectric transducers debonding from composite beam. In *Structural Health Monitoring*, FK Chang (ed.). DEStech, Lancaster.

71. Vincent HT, Hu SL, and Hou Z. 1999. Damage detection using empirical mode decomposition method and a comparison with wavelet analysis. *Proc 2nd Intl Workshop on Structural Health Monitoring*, Stanford, Technomic, Lancaster.

72. Chen J and Xu YL. 2004. Application of EMD for structural damage detection: Experimental investigation. In *Advanced Smart Materials and Smart Structures Technology*, FK Chang, CB Yun, and BF Spencer, Jr. (eds.). DEStech, Lancaster.

73. Xu YL and Chen J. 2004. Structural damage detection using empirical mode decomposition: Experimental investigation. *J Eng Mech*, 130(11), 1279–1288.

74. Jha R, Xu S, and Ahmadi G. 2005. Health monitoring of a multi-level structure based on empirical mode decomposition and Hilbert spectral analysis. In *Structural Health Monitoring*, FK Chang (ed.). DEStech, Lancaster.

75. Algernon D and Wiggenhauser H. 2006. Impact-echo signal processing. *NDE Conf on Civil Eng*, Amer Soc Nondes Test, St. Louis.

76. Xu L, Guo JJ, and Jiang JJ. 2002. Time-frequency analysis of a suspension bridge based on GPS. *J Sound Vib*, 254(1), 105–116.

77. Cohen L. 1966. Generalized phase-space distribution functions. *J Math Phys*, 7, 781–786.

78. Chen HL and Wissawapaisal K. 2002. Application of Wigner–Ville transform to evaluate tensile forces in seven-wire prestressing strands. *J Eng Mech*, 128(11), 1206–1214.

79. Williams WJ, Zalubas EJ, and Rohde MM. 2001. Machine failure prediction using reduced interference distributions. *Proc 25th Ann Mtg Vibration Institute*, Cincinnati.

80. Brancaleoni F, Spina D, and Valente C. 1993. Damage assessment from the dynamic response of deteriorating structures. In *Safety Evaluation Based on Identification Approaches*, HG Natke, GR Tomlinson, and JTP Yao (eds.). Vieweg, Braunschweig.

81. Hernandez E and Weiss G. 1996. *A First Course on Wavelets*. CRC Press, Boca Raton.

82. Strang G and Nguyen T. 1996. *Wavelets and Filter Banks*. Wellesley-Cambridge Press, Wellesley.

83. Randall RB. 1977. *Frequency Analysis*. Bruel and Kjaer, Naerum.

84. Kijewski T and Kareem A. 2003. Wavelet transforms for system identification in civil engineering. *Comput Aided Civ Infrastruct Eng*, 18, 339–355.

85. Grossman A, Kronland-Martinet R, and Morlet J. 1989. Reading and understanding continuous wavelet transforms. In *Wavelets Time-Frequency Methods and Phase Space*, JM Combes, A Grossmann, and P Tchamitchian (eds.), 2nd Ed. Springer, Berlin.

86. Daubechies I. 1989. Orthonormal bases of wavelets with finite support—connection with discrete filters. In *Wavelets Time-Frequency Methods and Phase Space*, JM Combes, A Grossmann, and P Tchamitchian (eds.), 2nd Ed. Springer, Berlin.

87. Chui CK. 1992. *An Introduction to Wavelets*. Academic Press, Boston.

88. McNamarra J, Palmer M, and Lanza di Scalea F. 2003. Health monitoring of railroad tracks by wavelet analysis. *Smart Systems and Nondestructive Evaluation for Civil Infrastructures*, SPIE 5057.

89. Stockwell RG, Mansinha L, and Lowe RP. 1996. Localization of the complex spectrum: The S transform. *IEEE Trans Signal Process*, 44(4), 998–1001.

90. Wang X and Tansel IN. 2007. Modeling the propagation of Lamb waves using a genetic algorithm and S-transformation. *Struct Control Health Monit*, 6(1), 25–37.

91. Al-Khalidy A and Dragomir-Daescu D. 2003. Ball bearing early fault detection using wavelet analysis. In *Structural Health Monitoring*, FK Chang (ed.). DEStech, Lancaster.

92. Jiang Z, Chonan S, Kawashima K, Muto K, and Ichihara W. 1997. Condition monitoring of silicon-wafer slicer cutting crystal ingots. In *Structural Health Monitoring Current Status and Perspectives*, F-K Chang (ed.). Technomic Publishing, Lancaster.

93. Kim J and Ewins DJ. 1999. Monitoring transient vibrations in rotating machinery using wavelet analysis. *Proc 2nd Intl Workshop on Structural Health Monitoring*, Stanford, Technomic, Lancaster.

94. Jiang X, Mahadevan S, and Adeli H. 2007. Bayesian wavelet packet denoising for structural system identification. *Struct Control Health Monit*, 14, 333–356.
95. Taghvaei M, Beck SM, and Staszewski WJ. 2007. Leak detection in pipeline networks using low-profile piezoceramic transducers. *Struct Control Health Monit*, 14, 1063–1082.
96. Ip KH, Tse PW, and Tam HY. 2004. Extraction of patch-induced lamb waves using a wavelet transform. *Smart Mater Struct*, 13, 861–872.
97. Dawood TA, Shenoi RA, Veres SM, and Sahin M. 2004. Low level damage characterization in FRP sandwich beams using the Lipschitz exponent. *Proc 2nd European Workshop on Structural Health Monitoring*, DEStech, Munich.
98. Spanos PD, Failla G, Santini A, and Pappatico M. 2006. Damage detection in Euler–Bernoulli beams via spatial wavelet analysis. *Struct Control Health Monit*, 13, 472–487.
99. Ahmad R and Kundu T. 2004. Guided wave technique to detect defects in pipes using wavelet transform. *Proc 2nd European Workshop on Structural Health Monitoring*, DEStech, Munich.
100. Sohn H, Park G, Wait JR, and Limback NP. 2003. Wavelet based analysis for detecting delamination in composite plates. In *Structural Health Monitoring*, FK Chang (ed.). DEStech, Lancaster.
101. Hou Z, Hera A, Liu W, and Hendrickson D. 2003. Identification of instantaneous modal parameters of time-varying systems using wavelet approach. In *Structural Health Monitoring*, FK Chang (ed.). DEStech, Lancaster.
102. Lihua S. 1999. Structural damage detection by ETDR method and wavelet analysis of the measured signals. *Proc 2nd Intl Workshop on Structural Health Monitoring*, Stanford, Technomic, Lancaster.
103. Zhao X and Yuan S. 2003. Wavelet analysis module design for on-line composite health monitoring. In *Structural Health Monitoring*, FK Chang (ed.). DEStech, Lancaster.
104. Mallat S. 1989. Multifrequency channel decompositions of images and wavelet models. *IEEE Trans Acous Speech Signal Process*, 37, 2091–2110.
105. Hong JC, Kim YY, Lee HC, and Lee YW. 2002. Damage detection using the Lipschitz exponent estimated by the wavelet transform: Application to vibration modes of a beam. *Intl J Solid Struct*, 39, 1803–1816.
106. Xiaorong Z, Baoqi T, and Shenfang Y. 1999. Study on delamination detection in CFRP using wavelet singularity analyses. *Proc 2nd Intl Workshop on Structural Health Monitoring*, Stanford, Technomic, Lancaster.
107. Robertson AN, Farrar CR, and Sohn H. 2003. Singularity detection for structural health monitoring using Holder exponents. *Mech Syst Signal Process*, 17(6), 1163–1184.
108. Iwasaki A, Todoroki A, and Sakai S. 2004. Delamination identification of CFRP Laminates using Kriging interpolation method. *Proc 2nd European Workshop on Structural Health Monitoring*, DEStech, Munich.
109. Allemang RJ and Brown DL. 1987. *Experimental Modal Analysis and Dynamic Component Synthesis*, AFWAL-TR-87-3069, Air Force Wright Aeronautical Laboratories, December.
110. Avitabile P. 2003. Twenty years of structural dynamic modification—a review. *Sound Vib*, (January), 14–25.
111. Craig Jr RR. 1981. *Structural Dynamics*. Wiley, New York.
112. Clough RW and Penzien J. 1993. *Dynamics of Structures*, 2nd Ed. McGraw-Hill, New York.
113. Cusano A, Cappoluongo P, Campopiano S, Cutolo A, Giordano M, Felli F, Paolozzi A, and Caponero M. 2004. Measuring modal parameter variations on composite structures by fiber Bragg grating sensors. *Proc 2nd European Workshop on Structural Health Monitoring*, DEStech, Munich.
114. Armstrong-Helouvry B, Dupont P, and Canudas de Wit C. 1994. A survey of models, analysis tools and compensation methods for the control of machines with friction. *Automatica*, 30(7), 1083–1138.
115. Lobontu N. 2002. *Compliant Mechanisms: Design of Flexure Hinges*. CRC Press, Boca Raton.

116. Yan BF, Miyamoto A, and Shao XD. 2004. Comparison of time-frequency analysis based methods for modal parameter extraction of bridge structure. *Proc IABMAS'04 Bridge Maintenance, Safety, Management and Cost*, Kyoto. Taylor & Francis, London.

117. Takewaki I and Nakamura M. 2005. Stiffness-damping simultaneous identification under limited observation. *J Eng Mech*, 131(10), 1027–1035.

118. Farrar CR and James III GH. 1997. System identification from ambient vibration measurements on a bridge. *J Sound Vib*, 205(1), 1–18.

119. Beskhyroun S, Mikami S, Yamazaki T, and Oshima T. 2004. Structural damage detection and localization using cross spectral density. In *Advanced Smart Materials and Smart Structures Technology*, FK Chang, CB Yun, and BF Spencer Jr (eds.). DEStech, Lancaster.

120. Fasana A, Garibaldi L, Giorcelli E, Marchesiello S, and Ruzzene M. 1999. Comparison between three identification techniques of a road bridge dynamic test. *Nondestructive Evaluation of Bridges and Highways III*, SPIE 3587.

121. Asmussen JC, Brincker R, and Ibrahim SR. 1999. Statistical theory of the vector random decrement technique. *J Sound Vib*, 226(2), 329–344.

122. Abe M, Fujino Y, Kajimura T, Yanagihara M, and Sato M. 1999. Monitoring of a long span suspension bridge by ambient vibration measurement. *Proc 2nd Intl Workshop on Structural Health Monitoring*, Stanford, Technomic, Lancaster.

123. Cremona C, Hallak P, Alvandi A, Ducret D, Inchauspe MH, and Dieleman L. 2004. Dynamic monitoring of a high speed railway bridge. *Proc IABMAS'04 Bridge Maintenance, Safety, Management and Cost*, Kyoto. Taylor & Francis, London.

124. Li HC, Weis M, Herszberg I, and Mouritz AP. 2004. Damage detection in a fibre reinforced composite beam using random decrement signatures. *Proc 2nd European Workshop on Structural Health Monitoring*, DEStech, Munich.

125. Liang Z and Lee GC. 2003. Lag superposition method for structural analysis through ambient vibration. *Smart Nondestructive Evaluation and Health Monitoring of Structural and Biological Systems II*, SPIE 5047.

126. Abdel-Ghaffar AM and Scanlan RH. 1985. Ambient vibration studies of Golden Gate bridge: I. Suspended structure. *J Eng Mech*, 111(4), 463–482.

127. Abdel-Ghaffar AM and Scanlan RH. 1985. Ambient vibration studies of Golden Gate bridge: II. Pier-tower structure. *J Eng Mech*, 111(4), 483–499.

128. Jones NP, Shi T, Ellis JH, and Scanlan RH. 1995. System identification procedure for system and input parameters in ambient vibration surveys. *J Wind Eng Ind Aero*, 54/55, 91–99.

129. Gentile C. 2004. Dynamic-based assessment of a cable-stayed bridge. *Proc IABMAS'04 Bridge Maintenance, Safety, Management and Cost*, Kyoto. Taylor & Francis, London.

130. Dusseau RA, Jones JR, Ramachandran RP, Sun CC, and Arr G. 2006. Field-measured natural frequencies of the Delaware Memorial Bridge. Paper 06-0186, 86th Annual Meeting, TRB, Washington, DC.

131. Katsuchi H, Yamada H, and Kusuhara S. 2006. Modal-damping identification of Akashi Kaikyo Bridge by wavelet screening. *Proc 4th China–Japan–US Symposium on Structural Control and Monitoring*, Hangzhou.

132. Nayeri RD, Masri SF, and Chassiakos AG. 2007. Application of structural health monitoring techniques to track structural changes in a retrofitted building based on ambient vibration. *J Eng Mech*, 133(12).

133. Rubin S. 1980. Ambient vibration survey of offshore platform. *J Eng Mech*, 106(EM3), 425–441.

134. Grewal MS and Andrews AP. 2001. *Kalman Filtering*. Wiley, New York.

135. Kailath T. 1979. *Linear Systems*. Prentice-Hall, Englewood Cliffs, NJ.

136. Omenzetter P and Brownjohn JM. 2004. Application of time series and Kalman filtering for structural health monitoring of a bridge. *Proc 2nd European Workshop on Structural Health Monitoring*, DEStech, Munich.

137. Fritzen CP and Mengelkamp G. 2003. A Kalman filter approach to the detection of structural damage. In *Structural Health Monitoring*, FK Chang (ed.). DEStech, Lancaster.

138. Shi T, Jones NP, and Ellis JH. 2000. Simultaneous estimation of system and input parameters from output measurements. *J Eng Mech*, 126(7), 746–753.

139. Reich GW and Sanders B. 2003. Structural shape sensing for morphing aircraft. *Smart Structures and Integrated Systems*, SPIE 5056.

140. Smyth AW. 2004. The potential of GPS and other displacement sensing for enhancing acceleration sensor monitoring array data by solving low frequency integration problems. *Proc IABMAS'04 Bridge Maintenance, Safety, Management and Cost*, Kyoto. Taylor & Francis, London.

141. Ray LR, Ramasubramanian A, and Townsend J. 2001. Adaptive friction compensation using extended Kalman–Bucy filter friction estimation. *Control Eng Pract*, 9, 169–179.

142. Pan J and Wang R. 2005. Nonlinear observability in the structural dynamic identification. *Sensors and Smart Structures Technologies for Civil, Mechanical, and Aerospace Systems*, SPIE 5765

143. Yang JN, Lin S, and Zhou L. 2004. Identification of parametric changes for civil engineering structures using an adaptive Kalman filter. *Sensors and Smart Structures Technologies for Civil, Mechanical and Aerospace Systems*, SPIE 5391

144. Burnett GC, McCallen DB, and Noble CR. 2000. Updating FEM using state-space based signal processing techniques. *Nondestructive Evaluation of Highways, Utilities, and Pipelines IV*, SPIE 3995.

145. Ghanem R and Ferro G. 2006. Health monitoring for strongly non-linear systems using the ensemble Kalman filter. *Struct Control Health Monit*, 13, 245–259.

146. Wu M and Smyth AW. 2007. Application of the unscented Kalman filter for real-time nonlinear structural system identification. *Struct Control Health Monit*, 14, 971–990.

147. Benamoun M and Mamic GJ. 2002. *Object Recognition Fundamentals and Case Studies. Springer*, London.

148. Hajnal JV, Hill DL, and Hawkes DJ. 2001. *Medical Image Registration*. CRC Press, Boca Raton.

149. Hill DL, Batchelor PG, Holden M, and Hawkes DJ. 2001. Medical image registration. *Phys Med Biol*, 46, R1–R45.

150. Russ JC. 1999. *The Image Processing Handbook*, 3rd Ed. CRC Press, Boca Raton.

151. Abdel-Ghaffar AM, Leahy RM, Masri SF, and Synolakis CE. 1992. A feasibility study for a concrete core tomographer. In *Nondestructive Testing of Concrete Elements and Structures*, F Ansari and S Sture (eds.). ASCE, New York.

152. Cao M, Ren Q, and Qiao P. 2006. Nondestructive assessment of reinforced concrete structures based on fractal damage characteristic factors. *J Eng Mech*, 132(9), 924–931.

153. Fu G and Moosa AG. 2002. An optical approach to structural measurement and its application. *J Eng Mech*, 128(5), 511–520.

154. Al-Nuaimy W, Huang Y, and Shihab S. 2002. Unsupervised segmentation of subsurface radar images. *Proc 9th Intl Conf on GPR*, Santa Barbara.

155. Mansouri B, Houshmand B, and Shinozuka M. 4330. Building change/damage detection in Seyman-Turkey using ERS SAR data. *Smart Systems for Bridges, Structures, and Highways*, SPIE 4330

156. Bachinsky GS. 1995. The electronic BAR gauge (a customized optical rail profile measurement system for rail grinding applications). *Nondestructive Evaluation of Aging Railroads*, SPIE 2458.

157. Dai Q, Sadd MH, Parameswaran V, and Shukla A. 2005. Prediction of damage behaviors in asphalt materials using a micromechanical finite-element model and image analysis. *J Eng Mech*, 131(7), 668–677.

158. Lee S and Chang LM. 2006. Digital image processing methods for bridge coating management and their limitations. Paper 06-1372, 86th Annual Meeting, TRB, Washington, DC.

159. Yao B, Zhao Q, Li F, and Chen H. 2007. Design and application of bridge non-contact damage detection system. *Intl Conf Health Monit Struct Mater Environ*, Nanjing.

160. Huston DR, Miller J, and Esser B. 2004. Adaptive, robotic, and mobile sensor systems for structural assessment. *Smart Structures and Materials 2004: Sensors and Smart Structures Technologies for Civil, Mechanical, and Aerospace Systems*, Shih-Chi Liu (ed.). SPIE 5391.

6

Data Interpretation

6.1 Introduction

Effective application of statistical tools empowers the SHM engineer [1–4]. The availability of a wide variety of powerful statistical tools poses a challenge to the SHM engineer. How does one choose the best statistical tool for a particular application? There is often not a clear answer. This chapter describes common and potentially valuable SHM statistical methods, with an emphasis on higher-level tools that may be useful in interpretation and prediction.

The data that serve as inputs to SHM statistical analyses come in a variety of formats, including the following: *Numbers*: Such data may correspond to the measurement of a physical quantity. *Integer counts*: An example might be the number of times the midspan deflection of a bridge exceeds a certain level. *Ordered grades*: These include semiquantitative measurements, such as the rating of a bridge on a 1–10 scale. *Nonquantifiable information*: This includes items such as the names of materials suppliers or types of structures. *Predigested data*: These items are often called *statistical features*.

One organizing scheme for statistical methods classifies according to the amount of belief, information, and certainty inserted into and extracted from the analysis (Figure 6.1). In general, the use of strong assumptions regarding the underlying nature of observed phenomena facilitates drawing definitive conclusions from the statistical analysis. Weaker assumptions give more flexibility, but the conclusions are inherently less definitive. *Hypothesis testing* is a strong assumption technique. Hypothesis testing determines the probability of occurrence of the observed data being consistent with strong assumptions, that is, hypotheses. The probability of an occurrence gives an indication of the validity, or lack thereof, of the hypotheses and/or underlying assumptions. *Exploratory data analysis* (EDA) comprises a set of weak assumption techniques. EDA organizes, processes, and presents the data in a form that allows for quickly spotting patterns and trends. EDA methods find use in forming plans of action for remediation, repair, and subsequent more detailed monitoring activities. *Confirmatory data analysis* (CDA) techniques lie in between hypothesis testing and EDA. CDA techniques generally require some level of human intervention, but can produce quantitative results for decision-making. These intermediate techniques include regression analysis and more sophisticated (and less controllable at a detailed level) fitting techniques. *Neural networks* (NNs), *support vector machines* (SVMs), *genetic algorithms* (GAs), *cellular automata* (CA), *statistical pattern recognition* (SPR), and *data mining* are examples. There are also methods that combine information and degrees of belief from past experience with statistics and sensors, that is, the Bayesian methods.

It is possible to evaluate and quantify the performance of statistical measures with standard metrics. The *expected value* is perhaps the most common performance metric. When the expected value of a statistic equals the "true" value of the parameter under consideration, then the statistic is *unbiased*. If the expected value of the statistic deviates from the true value, then the statistic is *biased*. *Validity* is the range of confidence of a measure. *Consistency*

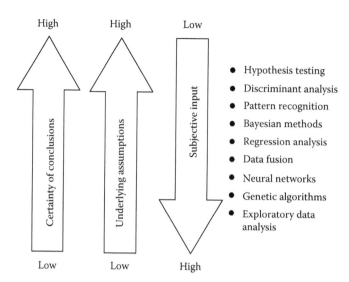

FIGURE 6.1   Classification of SHM statistical analysis techniques. (Adapted from Hall DL. 1992. *Mathematical Techniques in Multisensor Data Fusion.* Artech House, Norwood.)

is the ability of a statistical estimator to converge to the true value as the number of samples increases. Estimators can be biased, yet consistent. *Efficiency* is the ratio of the variance of the optimum statistical measure of a particular quantity to the variance of the statistic being used. A statistical measure is *resistant* if it is relatively unaffected by changes in a small portion of the data. *Robustness* is similar to resistance, but has a broader interpretation [5,6].

It should be noted that the appropriate choice of statistical technique for a given situation is contextual and depends highly on the specific nature of the data. The *No Free Lunch* and *Ugly Duckling theorems* clarify these notions by stating that if no a priori knowledge or assumptions regarding the data exist, then there are no context-independent measures of classifier performance and that no sets of statistical features are guaranteed to be superior to others [7].

6.2 Standardization of Measurement Uncertainty

Prior to conducting detailed data analysis, it is advisable to assess the *accuracy, precision, uncertainty,* and *reliability* of the data. All measurements contain uncertainty. Measurements are at best inaccurate estimates of the "true" values of physical phenomena. Improved measurement techniques can reduce—but cannot eliminate—uncertainty.

Practical considerations for the effective exchange of measurement quality information require using standardized terminology. A *measurand* is the physical quantity being measured, that is, length, temperature, light intensity, and so on. *Measurement error* is the difference between the true value of a measurand and that reported by the measurement process. *Uncertainty* reflects the level of knowledge or confidence in a measurement. A smaller uncertainty implies a larger degree of confidence in the measurement. The *accuracy of measurement* is the closeness of agreement between the result of the measurement and the true value of the measurement. Due to random effects, a measurement may have a high degree of uncertainty, but by chance can be quite close to the true value. A recommendation is to avoid using the term *inaccuracy*, as it is not a term with a standardized definition. *Precision* is distinguished from accuracy. Precision has multiple definitions. One definition of precision is the repeatability of a measurement. Another definition is the smallest gradation on the measurement instrument. A third definition uses precision synonymously with accuracy. Shewhart indicates that assessing measurement accuracy, uncertainty, and reliability is a circular process in which no absolute level of certainty can be attained, that is, "Careful writers in the theory of errors, of course, have always insisted that accuracy involves in some way or another, the difference between what is true, whereas precision involves the reproducibility of what is observed" [9]. Some authors use the term *conventional true value* to indicate what is assumed to be the correct representation of physical reality [10]. Mandel provides statistics-based definitions of accuracy and precision [11].

Measurement errors result from both *systematic* and *random effects*. Systematic effects are inherent in a particular measurement system or process. An example is the use of a ruler to measure the length of objects. If the ruler changes length following manufacture and calibration, then measurements with the ruler contain systematic errors. A suitable data transformation, that is, a *correction*, can remove many systematic effects. The coefficients used in correction formulae are known as *correction factors*.

The *repeatability* of measurements is the closeness of agreement between measurements of the same measurand carried out under changed conditions of measurement, often with a different instrument or technique. *Coverage* is the portion of a measurement range that a confidence interval covers within a given percentage. The symbol $\pm$ usually indicates 95% confidence interval coverage.

Quantifying uncertainty may use either a *Type A* evaluation that statistically analyzes the data, or a *Type B* evaluation based on past experience, scientific evidence, and manufacturer's specifications. The *standard uncertainty*, u_i, corresponds to one standard deviation of error.

A common data processing technique combines multiple measurements, x_i, with a functional relation to create a single measurand, y,

$$y = f(x_1, x_2, \ldots, x_N). \tag{6.1}$$

The uncertainties in the individual x_i combine to create uncertainty, u_c, in y. An estimate of the combined uncertainty, u_c, is

$$u_c^2 = \sum_{i=1}^{N} \left(\frac{\partial f}{\partial x_i} \right)^2 u_i^2 + 2 \sum_{i=1}^{N-1} \sum_{j=i+1}^{N} \frac{\partial f}{\partial x_i} \frac{\partial f}{\partial x_j} u_{ij} \tag{6.2}$$

where u_i is the standard uncertainty of x_i and u_{ij} is the standard cross-uncertainty of x_i and x_j. When the individual measurands, x_i, are independent, the cross-uncertainties vanish and Equation 6.2 reduces to

$$u_c^2 = \sum_{i=1}^{N} \left(\frac{\partial f}{\partial x_i} \right)^2 u_i^2. \tag{6.3}$$

When $f(x_1, x_2, \ldots, x_N)$ is linear with coefficients, a_i, that is,

$$y = f(x_1, x_2, \ldots, x_N) = a_1 x_1 + a_2 x_2 + \cdots + a_N x_N, \tag{6.4}$$

and the individual measurands, x_i, are independent, Equation 6.3 simplifies to

$$u_c^2 = a_1^2 u_1^2 + a_2^2 u_2^2 + \cdots + a_N^2 u_N^2. \tag{6.5}$$

Combining multiple uncertainties affects the estimation of the coverage interval for y. When the x_i values have Gaussian PDFs, then the Student-t distribution can estimate the coverage factors. For linear functional relations, the number of degrees of freedom, v_{eff}, for use in the Student-t distribution is

$$v_{\text{eff}} = \frac{u_c^4(y)}{\sum_{i=1}^{N} a_i^4 u_i^4 / v_i}. \tag{6.6}$$

v_i is the number of degrees of freedom for each measurement x_i [12].

Multivariate uncertainty analyses naturally arise when multiple factors affect the output of a sensor. As an example, Liu et al. describe a multifactor uncertainty analysis in luminescent pressure-sensitive paints that measure pressures in wind tunnel testing. The uncertainty analysis included consideration of coating thickness, temperature, and oxygen diffusion [13]. Similarly, Strauss et al. provide a multifactor probabilistic sensitivity analysis for resistive strain gages [14].

Uncertainties in both the model and the data complicate extending estimates of measurement uncertainties to modeling and prediction uncertainties. Especially problematic is the extension of predictions to regions that extend beyond that of the range of the measured data. The combination of simplified metamodels, model validation, and uncertainty quantification can provide some relief [15].

An additional issue of uncertainty arises when the measurands are functions of independent variables, for example, signal amplitude as a function of time. When the signal tapers to zero at $\pm\infty$, it has a quantifiable location and spread about the location. The first moment normalized by the zeroth moment is the center of gravity, μ_t, which locates a central point of the signal on the time axis. The second moment about the center of gravity, σ_t^2, describes the spread, or uncertainty, of the function.

$$\mu_t = \frac{\int_{-\infty}^{\infty} tf(t)\,dt}{\int_{-\infty}^{\infty} f(t)\,dt}, \tag{6.7}$$

$$\sigma_t^2 = \int_{-\infty}^{\infty} t^2 f(t - \mu_t)\,dt. \tag{6.8}$$

It is common to measure or interpret signal behavior in the frequency domain with a Fourier transform

$$X(\omega) = \int_{-\infty}^{\infty} f(t)e^{-j\omega t}\,dt. \tag{6.9}$$

The uncertainty, σ_f, in the spread, or localization, of the transformed function is

$$\sigma_f^2 = \int_{-\infty}^{\infty} \omega^2 X(\omega)\,d\omega. \tag{6.10}$$

The uncertainties in the spread of the original and transformed functions derive from the same signal and are mutually dependent. The *uncertainty relation* is a general result based on the Schwartz inequality that constrains time and frequency domain uncertainty. In this context, the uncertainty relation takes the form

$$\sigma_t^2 \sigma_x^2 \geq \frac{1}{4\pi^2}. \tag{6.11}$$

Equality occurs in Equation 6.11 when $x(t)$ (and hence $X(\omega)$) are Gaussian functions [16].

6.3 Exploratory Data Analysis

Most statistical analyses require some knowledge or assumptions about the PDFs underlying the observed random processes. When little is known about the PDFs, it is usually advisable to examine the data to identify underlying structures before applying more quantitative techniques, such as hypothesis testing and classification. Such processes combine physical reasoning, data fitting, hypothesis testing, and judicious approximations. EDA can aid this effort [17,18]. The concern of EDA is primarily with describing the overall appearance of the data, and not with the confirmation of hypotheses or fitting data to identify model parameters. EDA techniques include determining the location and amount of spread of the data, and smoothing and transforming the data into forms that are more tractable for further viewing and analysis. Placing the data into graphical formats for quick analysis with direct plotting, forming histograms, and simple data transformations are useful first steps [19]. Higher levels of analysis include selectively removing the contributions of various inputs (treatments) so as to identify how the other inputs affect the data. The exploratory nature of EDA may be sufficient for many SHM applications.

Mean or median value statistics can quickly locate the bulk of the data. The median may be a superior locator, because it is less sensitive than the mean to outliers in the tails. Calculating the variance or locating the *quartile medians* (also known as the hinges) are similarly standard methods of determining the data spread. Quartile medians can be superior to the variance as spread estimators because of the insensitivity to outliers. Mosteller and Tukey give spread estimators that are even more robust than the quartile means, but are more computationally complicated [6].

The various methods used in the graphical representation of data sets are quite diverse [19]. These include *histograms, stem and leaf plots*, and *box and whisker plots* (a.k.a. *box plots*). Histograms represent a sorting and counting of data into bins or ranges. Properly choosing the width of the bins usually requires a bit of trial and error. When the bin spacing is too fine, the counts are low and the histogram has a rough appearance. When the bin spacing is too coarse, the histogram can smear readily apparent features due to a lack of resolution. More sophisticated methods, such as the *maximum entropy* algorithm, can also provide PDF estimates [20]. A stem and leaf plot is essentially a histogram with added information. In the output, values are plotted in the vertical axis (as the stem) and variable (rightmost) digits in the data are listed in the rows (as leafs). The box and whisker plots show the medians, quartile medians (hinges), and extremes in graphical format.

An example histogram appears in Figure 6.2. The data are torsional strains of a FRP composite drive shaft recorded after passing through a wireless telemetry link. The histogram

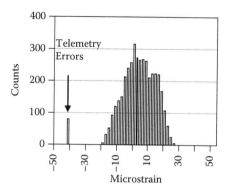

FIGURE 6.2 Histogram of wireless torsion strain data in fiber reinforced composite shaft with intermittent telemetry errors appearing as an isolated bin on the left.

indicates that the strain data are reasonably Gaussian, with the exception of a highly populated bin on the left side of the plot. Subsequent analysis attributed the left-side bin outlier data to dropout errors in the wireless telemetry system.

Transforming data can improve tractability and balance for review. The choice of transform is often a matter of trial and error and/or experience. Simple and useful transforms include raising the data to a power (negative, positive, integral, or fraction) and/or taking the logarithm. When the data are counts of occurrences in specified intervals, that is, histograms, Tukey recommends using the square root of the counts as a transform [17]. One motivation for the use of square roots is that the PDFs of Gaussian data are exponentials of squares. Identifiable physical circumstances that affect the data, such as periodicity, are also sources for suitable data transforms. Many ordered data sets can benefit from smoothing transforms to remove some of the rougher high-frequency components from the data and leave the lower-frequency components largely undisturbed.

Moving away from EDA towards more quantitative confirmatory statistical methods usually requires some knowledge of the PDFs underlying the data. Gaussian PDFs occur when circumstances cause the central limit theorem to apply, or when data are outputs from a linear system with Gaussian inputs. Gaussian PDFs have many attractive properties, including being completely describable by two parameters (the mean and the variance), the preservation of the Gaussian character under linear transformation, and a wide variety of tabulated mathematical manipulations and associated statistical tests [11,21,22]. A linear combination of Gaussian random variables is also a Gaussian random variable. Unknown Gaussian PDFs generally come in three forms. The first is when the mean is unknown and the variance is known. An example arises in the use of scientific instruments for measurement. The measurements collected by many instruments have a known and repeatable variance (or uncertainty), but have stable drifts in the mean from one set of measurements to another. Repeated measurements could produce statistics with Gaussian PDFs with unknown means, but with known variances. When the variance is known and the mean is unknown, then either average or median statistics can estimate the mean. The second case is when the mean and variance are both unknown. This requires using the measured data to estimate simultaneously the mean and variance. The Student-*t* and Fisher F PDFs appear in the subsequent analysis. When using random variables with non-Gaussian PDFs, classification techniques can still be straightforward to apply, but the specific details are not as widely tabulated [5]. The third case is when the mean is known and the variance is unknown. A common occurrence of this case is the measurement of mean square power in AC-coupled zero-mean signals.

In spite of the convenience and mathematical simplicity of Gaussian PDFs, many (if not most) PDFs encountered in practice are non-Gaussian. One case is where the data reside on only one side of the number line due to physical considerations, such as measurements of ultimate tensile strength. Another is where a system or a statistical process performs nonlinear transformations of the data. An example statistic is the square root of the sum of the squares of two equal-variance zero-mean Gaussian random variables. This statistic arises in the analysis of the amplitude and envelopes of time-varying signals. The resulting PDF is a Rayleigh distribution.

A particularly important case of non-Gaussian behavior appears in applications that require accurate descriptions of the outside tails of the distribution. Information regarding the shape of the tails needs to be accurate in order to model the behavior of rare and extreme events. Gaussian PDFs are typically inadequate for describing extreme events because the tails drop off too rapidly. When the primary interest is estimating the rate and size of rare or extreme events, then it is important to use distributions with non-Gaussian PDFs that properly account for the shape of the tails [23–25].

6.4 Bayes' Theorem

Many modern uses of the term "Bayesian" describe a wide class of techniques that include cases where the known outcomes are mutually exclusive and span the space of possible events. Bayes' theorem generally concerns estimating the probability of the occurrence of events based on the availability of supplemental information. Use of Bayes' theorem requires that the outcome of events, y_i, be mutually exclusive, span the universe of possible events, and have known or assumed probability measures. The supplemental information, x_j, may come from a variety of sources, including data from present experiments, prior experiments, and subjective beliefs. For discrete event outcomes, x_i and y_i, Bayes' theorem is [22]

$$p(y_i \mid x_j) = \frac{p(y_i)p(x_j \mid y_i)}{\sum_{k=1}^{N} p(y_k)p(x_j \mid y_k)}. \tag{6.12}$$

The outcome from using Bayes' theorem is a posterior PDF for y_i that incorporates the supplemental information. The form of Equation 6.12 is a straightforward noncontroversial application of probability theory. Extensions of Bayes' theorem can incorporate subjective beliefs and can automate learning processes. Such methods can combine prior information, such as historical data, personal beliefs and hunches, and the certainty of this prior information, with newly acquired information to produce a new and improved model of reality. The method converts the prior information into likelihood functions and then into PDFs for use in Equation 6.12. The spread of the prior conditional PDF $p(x_j \mid y_i)$ is an indication of the strength of the prior information. A small spread indicates that the prior information is strong. A large spread corresponds to prior information being relatively weak. A uniform distribution indicates a minimum of prior knowledge.

When all of the PDFs on the right-hand side of Equation 6.12 are known, then it is in principle a relatively straightforward calculation to estimate the desired conditional probability. However, it is rare for all of these PDFs to be known. A common didactic situation is the use of a collection of a series of measurements $(x_1, x_2, \ldots, x_n)$ to estimate a statistical parameter, θ, that describes the PDF of a system. Instrument calibration is a possible example. A well-built instrument may produce measurements that have a Gaussian distribution with a known and stable variance about an unknown mean, θ. A conditional PDF expresses the joint PDF of θ and the measured data, x_i. Following the notation by Guttman et al. [22],

$$p(\theta, x_1, x_2, \ldots, x_n) = p(\theta)p(x_1, x_2, \ldots, x_n \mid \theta), \tag{6.13}$$

where $p(\theta, x_1, x_2, \ldots, x_n)$ is the joint PDF, $p(\theta)$ is the prior PDF, and $p(x_1, x_2, \ldots, x_n \mid \theta)$ is the conditional PDF of the particular data values x_i being measured given a value of θ. When the measurements, x_i, are independent, then

$$p(x_1, x_2, \ldots, x_n \mid \theta) = \prod_{i=1}^{n} f(x_i \mid \theta) = l(\theta \mid x_1, x_2, \ldots, x_n), \tag{6.14}$$

where $f(x_i \mid \theta)$ is the conditional PDF of x_i given that θ has occurred and $l(\theta \mid x_1, x_2, \ldots, x_n)$ is the *likelihood function*, that is, the likelihood of θ occurring based on the measurements $x_1, x_2, \ldots, x_n$. The new (posterior) PDF of θ given the combination of the prior PDF, the measured data, and an assumed functional form for the conditional PDFs, $f(x_i \mid \theta)$, is

$$p(\theta \mid x_1, x_2, \ldots, x_n) = Kp(\theta)l(\theta \mid x_1, x_2, \ldots, x_n). \tag{6.15}$$

K is a normalization constant that sets the total probability measure equal to 1, that is,

$$\frac{1}{K} = \int_{\text{All } \theta} p(\theta)l(\theta \mid x_1, x_2, \ldots, x_n) \, d\theta. \tag{6.16}$$

In principle, Equation 6.15 incorporates all known information concerning the distribution of θ, that is, qualitative information known prior to the tests, the measured data, and the functional forms of the PDFs. A weakness in this approach is that it assumes the functional forms of the PDFs and may give a false impression of precision [26].

The basis of the simplification in Equation 6.14 is the assumption of independent conditional PDFs. This simplification, known as the *naïve Bayes* approach, considerably reduces the amount of calculation and information required for estimation [27]. The above formulae readily extend to the case of multiple unknown parameters θ_i. A practical issue with using Bayes' theorem is that the integrals are often difficult to calculate, especially in the presence of multiple parameter combinations. Approximate methods, such as the *Markov chain Monte Carlo, Gibbs sampling*, and *Metropolis–Hastings* algorithms, provide some relief [28,29]. For example, Glaser et al. used the Metropolis–Hastings algorithm to estimate damage in beams from static measurements and dual-scale coarse-fine finite-element meshes [30].

One application of Bayes' theorem is to produce improved PDFs by combining information from recent measurements with prior estimates of PDFs. Kayser and Nowak used flange thickness measurements of corroded beams to predict the shear strength of steel beams [31]. In this case, y_i is the actual shear strength of the beam and x_j is the strength predicted from flange thickness measurements. $P(y_i)$ is the probability of a beam having the strength y_i. $P(y_i \mid x_j)$ is the probability of the actual strength having a value y_i given that measurements of flange thicknesses predict a strength of x_j. $P(x_j \mid y_i)$ is the probability of flange thickness measurements predicting a strength x_j given that the actual strength is y_i.

The Bayesian methods also readily apply to the problem of classification. The addition of information from a small training data set updates the prior probabilities of class properties.

Example applications include updating estimates of structural property estimates to deduce levels of damage and classification [32–34]. Shalaby and Reggin trained an automated pavement analyzer with manual, rutting, and roughness indicators to classify pavements as new, good, fair, or poor using a Bayesian maximum likelihood approach [35].

6.5 Hypothesis Testing and Two-Way Classification

Hypothesis testing quantifies the validity of hypotheses. The technique assumes that a hypothesis is either true or false. Hypothesis testing does not prove or disprove the validity of a hypothesis. Instead, the analysis indicates the probability of the measured data occurring, given that the hypothesis is true or false.

Classification is similar to hypothesis testing. Observed data provide the impetus for classifying a system as being in one state or another. In many respects, hypothesis testing is a case of classification that classifies the validity of a hypothesis. Figure 6.3 shows the overall flow of hypothesis testing and two-way classification algorithms. Since the practice of SHM engineering tends to be more of an art than a science, classification is more common than hypothesis testing as an SHM data interpretation mode.

The first steps in hypothesis testing and classification require elucidating and understanding all of the underlying assumptions, such as information about the system being

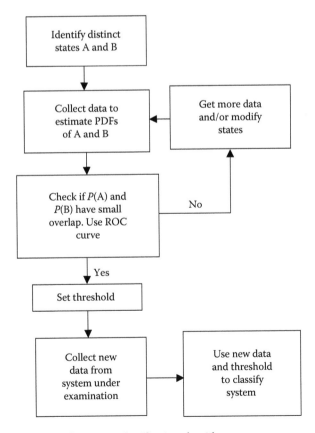

FIGURE 6.3 Hypothesis testing and two-way classification algorithm.

examined, and the PDFs that describe variations in the data. The next step is to calculate the probabilities of the measured data occurring in the context of the assumptions. When the probability exceeds a preset threshold, the hypotheses or assigned classifications are accepted as probably being correct. A 95% probability is the most common criterion for hypothesis acceptance (or rejection).

Two-way classification assumes that the system is in one of two states (A or B) and attempts to decide the state based on an analysis of measured data. When the data consist of a single scalar measurement (or single scalar statistical feature), single-valued hard thresholds can discriminate between states. Data that lie on one side of a prescribed threshold value, x_T, indicate that the system belongs in one state (A) and data that lie on the other side of the threshold indicate that the system belongs in another state (B). The discrimination rule is *Case I*: if $x_i < x_T$, the system is in state A; or *Case II*: if $x_i > x_T$, the system is in state B.

Hard-threshold, two-way classification can be straightforward to implement. Once the threshold is set, all that is needed is to check whether the value is above or below the threshold. However, uncertainty in the data inherently leads to uncertainty in the classification and the need for probabilistic methods to control errors. The discriminating power of threshold-based classification depends on the nature of the PDFs that relate system states to the measurements. An example is assessing whether a bridge girder can carry a particular load (state B) or is unable to carry the load safely (state A). The data might be a set of thickness measurements using calipers. Variability and errors due to uncertainty in caliper accuracy, the measurement technique, and data handling can all introduce uncertainty into the classification scheme.

A key issue in using hard thresholds for classification is determining where to set the threshold. Figure 6.4 shows the case of two states with distinctly separate PDFs. In this case, it is fairly easy to set the discrimination threshold, x_T. When there is more overlap of the PDFs, as in Figure 6.5, then discrimination is more difficult, and the choice of a suitable threshold is not inherently obvious.

Five cases arise with hard-threshold, two-way classification: *Case 1*: *Correct identification* of a system as being in state A, using the criterion $x < x_T$. *Case 2*: *Incorrect identification* of a system being in state B, using the criterion $x > x_T$. The system is actually in A. This is a Type I error. *Case 3*: *Incorrect identification* of a system being in state A, using the criterion $x < x_T$. The system is actually in B. This is a Type II error. *Case 4*: *Correct identification* of an

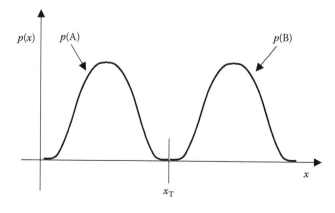

FIGURE 6.4   PDFs for A and B with minimal overlap and unambiguous threshold location.

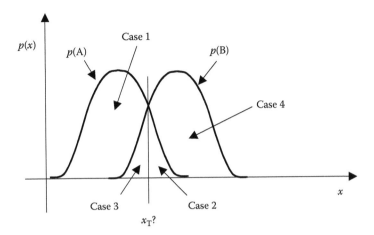

FIGURE 6.5 PDFs for A and B with substantial overlap and no obvious position for threshold cutoff.

object as being in state B using the criterion $x > x_T$. *Case 5: The situation is not well understood and the system is not in state A or B.* One method of eliminating the possibility of Case 5 is to reformulate the classification into a hypothesis that forces the system to be in either of the states A and not-A, where not-A corresponds to state B. This approach has the potential of introducing misleading interpretations.

When the behavior of the system is sufficiently well understood to eliminate the possibility of Case 5 errors, the application of a threshold criterion for classification produces a result that lies in one of the first four cases (Table 6.1). Denoting P_1, P_2, P_3, and P_4 as the probability of occurrence of the four cases give values where

$$P_A + P_B = 1, \tag{6.17}$$

$$P_1 = P_A - P_2 = \int_{-\infty}^{x_T} p_A(x)\,dx = \int_{-\infty}^{x_T} p(x \mid A)\,dx, \tag{6.18}$$

$$P_3 = P_B - P_4 = \int_{-\infty}^{x_T} p_B(x)\,dx = \int_{-\infty}^{x_T} p(x \mid B)\,dx. \tag{6.19}$$

The relative rate of occurrence of the four cases depends on the rates of occurrence of A and B, the amount of overlap of the PDFs of A and B, and the choice of threshold value [36].

TABLE 6.1

Four Possible Classifications and Associated Probabilities

	True Value	
Assigned Value	**A**	**B**
A	Correct, P_1	Type II error, P_3
B	Type I error, P_2	Correct, P_4

Shifting x_T to the left in Figure 6.5 causes the selection rates of Case 1 and Case 3 to decrease, and the selection rates of Case 2 and Case 4 to increase. Conversely, shifting the threshold value to the right increases the selection rates of Case 1 and Case 2, while decreasing the selection rates of Case 3 and Case 4.

When there is significant overlap in the PDFs, as in Figure 6.5, errors are unavoidable. Reducing the overlap reduces the error rate. Methods for reducing the amount of overlap include increasing the relative PDF location separation distance and reducing the PDF spreads by methods such as collecting more data. Reducing the PDF spread is often easier than relocating the center. An alternative approach to increasing the discriminating power is the addition of data from different forms of measurements. Many data fusion and other SPR techniques exploit these approaches. Nonetheless, situations inevitably arise where PDF overlap continues to persist. Choosing a suitable threshold value choice can be crucial to the success of the classification and the overall SHM effort. Possible methods for threshold setting range from arbitrary ad hoc methods to more quantitative methods, such as the Bayes and the Neyman–Pearson approaches, that optimize the relative rates of Type I and Type II errors (Case 2 and Case 3) [8,36,37].

The *Fisher discriminant ratio* (FDR) or *linear Fisher discriminant* is a variance-normalized measure of the discriminating power of a statistic [7]. For the cases of discriminating between A and B in Figures 6.4 and 6.5, the FDR is

$$\text{FDR} = \frac{|\mu_A - \mu_B|^2}{[\sigma_A^2 + \sigma_B^2]}. \tag{6.20}$$

The FDR can evaluate the strength of a particular statistical feature. As examples, Michaels et al. used the FDR to evaluate the ability of various statistical features to distinguish damaged from undamaged conditions in ultrasonic wave measurements, and Aldrin and Knopp distinguished crack types with eddy current inspection of metal fasteners [38,39].

The *receiver operating characteristic* (ROC, pronounced as "rock") curve characterizes the discriminating strength of a particular measurement. The ROC curve plots the probability of Case 1 (P_1) versus Case 3 (P_3) as the threshold parameter x_T is varied (Figure 6.6). When the datum x_i is a strong discriminator of A and B, the curve rises quickly to a value near one and then bends to the right. When x is a weak discriminator, the ROC tracks close to the line running at 45° from (0, 0) to (1, 1).

The Bayes method selects the classification threshold, x_T, by minimizing a *decision cost function*, C, that is a linear combination of the decision probabilities:

$$C = c_1 P_1 + c_2 P_2 + c_3 P_3 + c_4 P_4. \tag{6.21}$$

When the costs of a correct decision are negligible, that is, c_1 and $c_4 \approx 0$, the decision cost reduces to

$$C = c_2 P_2 + c_3 P_3. \tag{6.22}$$

Setting c_2 to a high value by moving x_T to the right corresponds to the cost of a Type II (Case 2) error being high. Conversely, setting c_3 to a high value by moving x_T to the left corresponds to the cost of a Type I (Case 3) error being high. An example is the case of load rating classification of a bridge that estimates whether a bridge can carry a particular load (state B) or is not able to carry the load (state A). Case 1 corresponds to the bridge being

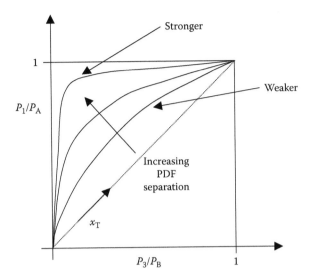

FIGURE 6.6 ROC curves showing discrimination strength versus threshold setting and separation of PDFs.

correctly declared as capable of carrying the load without collapsing. Case 2 corresponds to the bridge being erroneously declared as capable of carrying the load, but is unable to do so and will collapse. In this case, the error (Type I) has large human and economic costs. Case 3 corresponds to the bridge being erroneously declared as unfit for carrying the load, but is actually able to do so. This is a Type II error. The economic costs are traffic disruption and wasted resources. Case 4 is when the bridge is correctly declared as being able to carry the load. All of the four cases above have nonzero decision costs. A plausible relative cost coefficient ranking is

$$c_4 < c_1 \leq c_3 < c_2. \tag{6.23}$$

Setting a threshold is relatively straightforward, if the PDFs of states A and B are known. The first step of the *Neyman–Pearson* technique selects a desired error rate for a particular error type (usually the Type I false-alarm error). The next is to set the threshold at a value that is sufficiently high to produce the selected error rate. Ideally, a large historical database provides sufficient information to estimate the PDFs. It is possible to use smaller data sets at the expense of reduced statistical power.

The lack of sufficient prior data sets can prevent following the classification algorithm of Figure 6.3. Nonetheless, statistics-based classifications are possible. An example is determining whether two data sets were drawn from the same or different distributions. An SHM application is to determine if a structure has changed due to damage, repair, or aging. *Analysis of variance* (ANOVA) and related statistical techniques are powerful tools for examining these issues that extend to cases of multiple possible outcomes by using a larger classification matrix of the type in Table 6.1. Such matrices are known as *confusion matrices* [40].

ANOVA separates data into groups or, more formally, tests hypotheses that the variations in the data are due to an assumed classification scheme. A mathematically convenient and often justifiable assumption is that each datum is the sum of a treatment (or controlled) effect component, and a random variation (or noise) component. When drawing data, Y_{1i}

and Y_{2j}, from two separate sets, this assumption appears as

$$Y_{1i} = X_i + \varepsilon_{1i}, \quad i = 1, 2, \ldots, p, \tag{6.24}$$

$$Y_{2j} = X_j + \varepsilon_{2j}, \quad j = 1, 2, \ldots, q. \tag{6.25}$$

Y_{1i} and Y_{2j} are the measurements. X_i and X_j are the expected or true values of a statistic derived from a measurand. ε_{1i} and ε_{2j} are the measurement errors, or noise. A standard assumption is that the errors have Gaussian PDFs with zero means and equal variances. X_i and X_j may be simple linear statistics, such as the mean, or more complicated variants, such as coefficients fit by multiterm regression analysis. Using an F distribution to compare the size of the variance terms ε_{1i} and ε_{2j} produces a check as to whether the two data sets come from the same or different underlying distributions. When the two data sets come from the same population and the error terms are zero mean Gaussian variables, then the degree of freedom (DOF)-weighted ratio of the summed variables is an F distribution

$$F_{pq} = \frac{q \sum_{i=1}^{p} \varepsilon_{1i}^2}{p \sum_{j=1}^{q} \varepsilon_{2j}^2}. \tag{6.26}$$

Calculating and removing X_i and X_j from the data set by least squares or similar processes causes the sums of the variances in Equation 6.26 to lose DOF, that is, the integers p and q reduce accordingly [11,21]. When X_i is the mean, the PDF of the average (mean) statistic condenses to the F distribution with $p - 1$ and 1 DOF. This is equivalent to Student's t distribution with $p - 1$ DOF. Usually the size of the right-hand tail serves as a check for the validity of the hypothesis. Occasionally, the left-hand tail appears in a test, particularly as a check to see if the data fit too well, perhaps due to equipment malfunction or scientific fraud. More general variants of the F-test are possible [11,21]. A weakness of the F-test is that it is very sensitive to outliers and non-Gaussian data.

A nontrivial question of particular interest to SHM is "Has a structure or system changed?" Changes in the data may be the result of routine statistical variations or may be due to an underlying change in the system. A Chi-squared test can assess if observed variations are due to system changes, for example, damage, or are merely routine variations. For example, Fritzen and Mengelkamp applied a Chi-squared test to the analysis of the residuals of Kalman filter outputs to decide whether a dynamical system has changed [41]. Poulimenos and Fassos used the F-test for on-line fault detection in nonstationary structural systems [42]. Iwasaki et al. used an F-test to assess damage to jet fan mounts [43,44].

Shewhart examined the problem of detecting changes in the context of statistical quality control for manufacturing [45]. Many of these techniques readily extend to SHM by examining whether the health of a structure is in stasis or is changing. A central concept of the Shewhart approach is whether the health (quality) of a system is in a state of *statistical control*. The existence of a numerically quantifiable *structural health metric*, X, allows for an assessment as to whether a system is in statistical control. The probability, dy, that a structure will have a health metric in the range X to $X + dX$ is a function of X and m parameters λ_i, that is,

$$dy = f(X, \lambda_1, \lambda_2, \ldots, \lambda_m) \, dX. \tag{6.27}$$

When the PDF of the health metric is stable over time, the health of the system is in statistical control. When one (or more) of the λ_i varies in a known manner, such as due to seasonal trends, it is possible to identify and remove the variation, usually by direct subtraction.

Shewhart's procedure analyzes sequential data by first parsing the data into subgroups. If no rationale for selecting subgroup membership and size arises, then the recommendation is to use subgroups of 3–6 data points. When the underlying data are Gaussian, then the statistical behavior of both the means and variances of the subgroups is well understood. When the underlying data are not Gaussian, the subgroup means still tend to be Gaussian due to the action of the central limit theorem. This may not be the case for subgroup variances. A *Shewhart control chart* sequentially plots the subgroup means and subgroup variances. About 99.7% of the subgroup means should fall within 3 sigma limits. Deviations from the 3 sigma limits should prompt an examination to determine if the system has changed, or if this is an ordinary rare event (Figure 6.7).

There are numerous published studies of attempts at using statistical analyses to detect important structural changes [46]. For example, Allen et al. used sequential probability test ratios to identify joint damage in models of buildings [47]. Wang and Ong detected progressive damage in concrete frames using Hotteling's T^2 control chart technique [48]. Fu et al. found subtle changes in surface geometry that may indicate damage with the statistical processing of correlations from a laser range finder [49]. Lloyd et al. implemented an automated bridge monitoring system on a concrete box girder bridge across the Kishwaukee River in Rockford, IL (USA) [50]. Numerous shear cracks plague the bridge. LVDT rosettes provided date for crack monitoring. Additional measurements included the continual monitoring of the modal frequencies. Omenzetter et al. used two multichannel approaches to identify outliers [51]. The first was a multichannel discrete wavelet transform. An assumption was that outlier behavior in the structure caused significant changes in the wavelet coefficients. The second approach fitted an ARMA model to a training set of multichannel data, and then ran new data through the ARMA model. Statistics based on the *Mahalanobis* distance (see the next section for a definition) quantified deviations between the predicted and measured data. Exceeding a preset deviation threshold prompted the detection of outliers. Mattson and Pandit detected and located damage in structures by examining the outliers from an ARV analysis of dynamic data with Grubbs Hypothesis and Lilliefors Hypothesis tests [52].

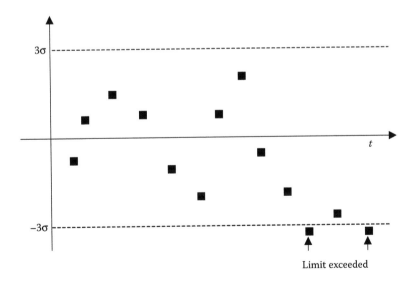

FIGURE 6.7   Control chart.

A key issue in aiding the detection of outlier events that indicate systemic changes is the removal of known environmental effects that distort the data and are independent of outlier phenomena. Ambient temperature shifts can affect both the cracks and modal frequencies. Regression-based temperature effect removal leaves a time series that is more amenable to an outlier analysis. Ko et al. developed an NN approach to identify temperature-induced modal vibration shifts based on one year of data taken at the Ting Kau Bridge [53]. Sohn et al. used auto-associative NNs to account for ambient temperature shifts that alter the dynamic behavior of a computer hard disk drive [54]. The auto-associative NN matches the input variables to the output variables in a manner that causes environmental variables, such as temperature, to appear as hidden variables. Schwabacher used machine learning algorithms to detect outliers in 90-channel rocket engine monitoring data [55]. The algorithms operated in an unsupervised mode, but nonetheless detected anomalies that were well known to the test engineers.

6.6 Logarithmic Measures and Information Theory

Information theory is a broad class of techniques that use logarithmic measures of uncertainty [56,57]. One motivation for the use of logarithms is to compress amplitude scales that vary over orders of magnitude. A second (and often related) motivation is to use logarithms to convert expressions involving multiplication into sums. At least three results from information theory find applications in SHM. The first is the use of logarithmic measures, often in the form of Fisher discriminants or Fisher information matrices, to quantify the discriminating power of a particular statistical measure. The second quantifies uncertainties in communication channels with a logarithmic measure known as *entropy*. The third (and less developed) application is to use logarithmic measures to describe the state of health and uncertainty of the prognosis of a structural system.

An example of logarithmic information measures of information in statistical discrimination is the case of two-state classification (A or B) based on a measurement x as shown in Figures 6.4 and 6.5. The ratio of the probability of the system being in state A versus B given the value of measurement x using Bayes' theorem (Equation 6.12) is

$$\frac{p(A \mid x)}{p(B \mid x)} = \frac{p(A)p(x \mid A)}{p(B)p(x \mid B)}.$$ (6.28)

Taking the logarithm and rearranging produces an additive measure of the difference of the odds in favor of the system being in state A or B based on the value of the measurement of x information in terms of the datum x versus the odds before the measurement [57].

$$\ln\left[\frac{p(x \mid A)}{p(x \mid B)}\right] = \ln\left[\frac{p(A \mid x)}{p(B \mid x)}\right] - \ln\left[\frac{p(A)}{p(B)}\right].$$ (6.29)

Jiang and Mahadevan used this approach to quantify the health of a structure where state A corresponds to the hypothesis that the structure is healthy and state B corresponds to the hypothesis that the structure is damaged [58].

The *mutual information*, $I(a_i, b_j)$, between two measurements a_i and b_i is [59]

$$I(a_i, b_j) = \ln\left[\frac{P_{AB}(a_i, b_j)}{P_A(a_i)P_B(b_j)}\right].$$ (6.30)

When the measurements a_i and b_j are independent,

$$P_{AB}(a_i, b_j) = P_A(a_i)P_B(b_j) \tag{6.31}$$

and $I(a_i, b_j) = 0$. The *average mutual information* between a set of measurements A and B is

$$I_{AB} = \sum_{a_i, b_j} P(a_i, b_j) \ln \left[\frac{P_{AB}(a_i, b_j)}{P_A(a_i)P_B(b_j)} \right]. \tag{6.32}$$

The *Fisher information matrix* is an $n \times n$ matrix that depends on the variable x and n parameters $\theta_1, \theta_2, \ldots, \theta_n$ [57]:

$$F_{ij} = \int p(x, \underset{\sim}{\theta}) \left\{ \frac{\partial \ln \left[p\left(x, \underset{\sim}{\theta}\right) \right]}{\partial \theta_i} \right\} \left\{ \frac{\partial \ln \left[p\left(x, \underset{\sim}{\theta}\right) \right]}{\partial \theta_j} \right\} dx. \tag{6.33}$$

One of the more common logarithmic information measure applications is Shannon's concept of entropy that represents uncertainty in transmitted and received signals [60]. Shannon's approach begins with the assumptions that the entropy $H(p_1, p_2, \ldots, p_n)$ for a signal with n possible outcomes and associated probabilities $p_1, p_2, \ldots, p_n$ has the following properties: (1) H is continuous with changes in the p_i. (2) An increase in the number of possible outcomes corresponds to an increase in entropy. If all of the p_i are equal, that is, $p_i = 1/n$, then H should increase monotonically with n. (3) Entropy adds in sequential choices when weighted with the probability of the outcomes. Figure 6.8 illustrates the concept of additive sequential entropy where one choice should equal the entropy of two equivalent successive choices, that is,

$$H\left(\frac{1}{2}, \frac{1}{3}, \frac{1}{6} \right) = H\left(\frac{1}{2}, \frac{1}{2} \right) + \frac{1}{2} H\left(\frac{2}{3}, \frac{1}{3} \right). \tag{6.34}$$

The multiplicative weighting with probability in the sequential choices naturally leads to a logarithmic measure of uncertainty, that is, the entropy H:

$$H = -K \sum_{i=1}^{n} p_i \ln(p_i), \tag{6.35}$$

where K is a constant.

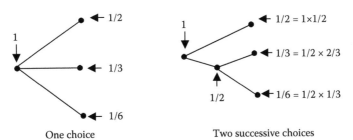

One choice   Two successive choices

FIGURE 6.8 Probabilities associated with one choice and an equivalent set of two successive choices. (Adapted from Shannon CE and Weaver W. 1949. *The Mathematical Theory of Communication.* University of Illinois Press, Urban and Chicago.)

Properties of the entropy are as follows: (1) $H = 0$ if all of the $p_i = 0$, except one of them. This corresponds to the case of complete certainty in the outcome. (2) H is a maximum when $p_i = 1/n$ for all i. (3) For two events x and y with combined probability $p(i,j)$, the *mutual entropy* $H(x,y)$ is

$$H(x,y) = -\sum_{i,j} p(i,j) \ln[p(i,j)]. \qquad (6.36)$$

The *marginal entropies* $H(x)$ and $H(y)$ are

$$H(x) = -\sum_{i,j} p(i,j) \ln \left[\sum_{j} p(i,j) \right],$$

$$H(y) = -\sum_{i,j} p(i,j) \ln \left[\sum_{i} p(i,j) \right]. \qquad (6.37)$$

Schrödinger in a popular press book that discusses the thermodynamics of living creatures introduced a normalized version of entropy called the *negentropy*, $J(x)$ [61]:

$$J(x) = H(x_{\text{Gauss}}) - H(x), \qquad (6.38)$$

where $H(x_{\text{Gauss}})$ is the entropy of the Gaussian variable with the same mean and variance as x. Interesting properties of negentropy are the following: (1) It is always nonnegative. (2) Gaussian distributions have zero negentropy (for infinite domains). (3) Negentropy is invariant under a linear transformation of coordinates. Multidimensional negentropy formulations are possible [5].

The *cepstrum* is a frequency domain logarithmic statistic. The cepstrum is the inverse Fourier transform of the logarithm of the power spectrum of a signal. The logarithm is useful in cases where the power spectrum appears as a product of subcomponents, as in ARMA models (Chapter 5). The logarithm in the cepstrum calculation converts the products into sums that are then conveniently manipulated with an inverse Fourier transform into the time domain for interpretation. The complex cepstrum of a signal modeled as the output from white noise input into an ARMA process containing p poles, α_i, q zeros, β_i, and variance σ^2 is [62,63]

$$C_n = \begin{cases} \dfrac{1}{n} \left(\sum_{i=1}^{p} \alpha_i^n - \sum_{i=1}^{q} \beta_i^n \right) & n > 0, \\[2ex] \ln(\sigma^2) & n = 0, \\[2ex] \dfrac{1}{n} \left(\sum_{i=1}^{p} \bar{\alpha}_i^{-n} - \sum_{i=1}^{q} \bar{\beta}_i^{-n} \right) & n < 0. \end{cases} \qquad (6.39)$$

Taghvaei et al. used the complex cepstrum on acousto–elastic signals collected with PZT transducers to identify the magnitude of leaks in piping networks [64].

6.7 Multichannel and Heterogeneous Data Sets

Many SHM systems use arrays of different types of sensors to collect multichannel heterogeneous data sets. Heterogeneous sensor data often contain more types of information than homogeneous sensor data. Heterogeneous data sets can challenge the investigator by being large and difficult to comprehend without the aid of automated techniques. The automated processing of heterogeneous data sets can make use of techniques such as data fusion, SPR, *data mining, metadata techniques, principal component analysis* (PCA), and *independent component analysis* (ICA) [65,66].

Decision-making with multisensor data requires rational methods of comparing data sets, such as with *generalized distances* [27]. Distances are scalars and intuitively easy to understand. When two data sets are similar, they have a small separation distance. Dissimilar data sets have large separation distances. There are a variety of generalized distance measures with specific capabilities for different phenomena. All valid distance metrics, $d(x_i, x_j)$, between the data sets x_i and x_j, with individual elements x_{ik} and x_{jk}, share the following common traits:

1. $d(x_i, x_j) \geq 0$ for all x_i and x_j and $d(x_i, x_j) = 0$ if and only if $x_i = x_j$.
2. $d(x_i, x_j) = d(x_j, x_i)$ for all x_i and x_j.
3. $d(x_i, x_j) \leq d(x_i, x_k) + d(x_k, x_j)$ for all x_i, x_j, and x_k.

The *Euclidean* distance metric, d_E, is

$$d_E = \left\{ \sum_{k=1}^{n} \left[x_{ik} - x_{jk} \right]^2 \right\}^{1/2}. \tag{6.40}$$

d_E is useful in cases of comparing homogeneous data sets, that is, the set elements have the same format and the same units. Structural deflections measured in the same units at multiple points under different loading conditions are an example of data sets that might be appropriate for comparison with d_E. More abstract situations are also amenable to the use of Euclidean metrics, especially when aided by coordinate transformations and weighting factors. For example, Martin proposed using a weighted Euclidean metric for comparing two ARMA models (M and M') with the metric applied to complex cepstrum coefficients (c_n and c'_n) [67]:

$$d(M, M') = \left(\sum_{n=1}^{\infty} n \left| c_n - c'_n \right|^2 \right)^{1/2}. \tag{6.41}$$

Due to the logarithmic nature of the cepstrum, the weighted Euclidean metric for stable AR models with orders p and p', and corresponding poles α_i and α'_i, converts the sum into a product

$$d(M, M') = \left[\ln \frac{\prod_{i=1}^{p} \prod_{j=1}^{p'} \left(1 - \alpha_i \alpha_j'^* \right) \prod_{i=1}^{p'} \prod_{j=1}^{p} \left(1 - \alpha_i' \alpha_j^* \right)}{\prod_{i=1}^{p} \prod_{j=1}^{p} \left(1 - \alpha_i \alpha_j^* \right) \prod_{i=1}^{p'} \prod_{j=1}^{p'} \left(1 - \alpha_i' \alpha_j'^* \right)} \right]^{1/2}. \tag{6.42}$$

Zheng and Mita used this metric to examine the health state of a building model before and after damage [63].

Replacing the square exponent in Equation 6.40 with other values generalizes the Euclidean distance to the *Minkowski* distance, d_p

$$d_{\mathrm{E}} = \left\{ \sum_{k=1}^{n} [x_{ik} - x_{jk}]^p \right\}^{1/p}. \qquad (6.43)$$

If $p = 1$, the Minkowski distance becomes the *city block* or *Manhattan metric* and measures distances along the coordinate axes. In the limiting case of $p = \infty$, the Minkowski distance is the maximum absolute value of the individual differences in the data sets being compared. The *maximum* metric can compare extreme values in data sets, such as the maximum strain measured in an array of strain gages on a structure when exposed to different loading conditions. Values of p between 1 and 2 can be useful when suppressing the effects of data outliers is a concern [7].

The choice of physical units in heterogeneous data sets can bias Euclidean measures of distance. Suitable scaling factors can remove the bias. The reciprocal of the variance of each component measurement element is a possible scaling factor. A similar, but more comprehensive approach uses the *Mahalanobis* distance, d_{M}, to remove second-order bias effects, where

$$d_{\mathrm{M}}^2(x_i, x_j) = (x_i - x_j)^{\mathrm{T}} \Sigma^{-1} (x_i - x_j). \qquad (6.44)$$

$(x_i - x_j)$ is a column vector of individual differences in the data sets and Σ is the multichannel variance–covariance matrix of the data sets.

Calculating statistical features from raw data is a common source of heterogeneous multichannel data sets. For example, Fillipenko et al. used kurtosis, crest factor, RMS, and other statistics calculated from a single bearing vibration reading to prognosticate the health of a bearing on a rotating machine [68]. Fan et al. used the kurtosis of wavelet-decomposed signals to identify salient features in weak ultrasound signals [69]. Haritos and Owen calculated statistical features for an NN pattern recognition approach to damage detection with a Mahalanobis distance measure of modal vibration test results [70]. Gul and Catbas detected changes in the dynamic properties of a steel frame in the laboratory by calculating the Mahalanobis distance on time series statistical features [71].

Classification with heterogeneous and multivariable data sets is similar in principle to two-way classification, but more complicated. When the data are multidimensional, hard-threshold classification involves determining if the datum lies within or without a bounded volume in a multidimensional space [72]. There are many algorithms available for determining suitable boundaries. Most of these use an original subset of data with known classifications to paint the boundary surface. Overfitting the data is a concern, especially with small training data sets.

Logistic regression analysis can often capture the salient features of multichannel data in terms of discrete outcomes. A standard application is discriminating between two classes of objects, c_1 and c_2, by analyzing multivariate data vectors $\mathbf{x}$ [27]. If it is valid to assume that the probability of an object being in class c_1 given the data point $\mathbf{x} = \{x_1, x_2, \ldots, x_N\}^{\mathrm{T}}$ is given in terms of a logistic function using the coefficient vector $\boldsymbol{\beta} = \{\beta_1, \beta_2, \ldots, \beta_N\}$, then

$$p(c_1 \mid \mathbf{x}) = \frac{1}{1 + \exp(\boldsymbol{\beta}^{\mathrm{T}} \mathbf{x})}. \qquad (6.45)$$

The probability of an object being in class c_2 given the data point $\mathbf{x}$ is

$$p(c_2 \mid \mathbf{x}) = \frac{\exp(\boldsymbol{\beta}^T \mathbf{x})}{1 + \exp(\boldsymbol{\beta}^T \mathbf{x})}. \tag{6.46}$$

An attractive feature is that the logarithm of the odds ratio of c_2 and c_1, given $\mathbf{x}$, simplifies to a linear relation

$$\ln\left[\frac{p(c_2 \mid \mathbf{x})}{p(c_1 \mid \mathbf{x})}\right] = \boldsymbol{\beta}^T \mathbf{x} = \beta_1 x_1 + \beta_2 x_2 + \cdots + \beta_N x_N. \tag{6.47}$$

The discrimination rule usually hinges on the sign of $\boldsymbol{\beta}^T \mathbf{x}$, that is, if $\boldsymbol{\beta}^T \mathbf{x} > 0$ then the object is in c_2, and if $\boldsymbol{\beta}^T \mathbf{x} < 0$ then the object is in c_1. Determining the $\boldsymbol{\beta}$ coefficients may generally require using an iterative maximum likelihood algorithm. For example, Yan et al. used logistic regression as the basis for programming the actions of a watchdog prognostics agent for autonomous machine condition monitoring [73]. The *boosting* technique is an additive logistic regression method for combining multiple classifier methods that may avoid overfitting problems [74].

6.8 Principal Components Analysis

PCA extracts features from multichannel data with a transform that produces a reduced-size and more information-dense data representation. PCA operates on second-order statistics to create a linear rectangular projection transform that condenses the data

$$\{\xi_j\} = [A_R]^T \{x_j\}. \tag{6.48}$$
$$\underset{M \times 1}{} \quad \underset{M \times N}{} \underset{N \times 1}{}$$

$\{x_j\}$ is an $N \times 1$ vector representing the original multichannel data at time j. $\{\xi_j\}$ is an $M \times 1$ reduced-order representation, that is, a principal components representation, of the data at time j where $M < N$. $[A_R]$ is an $M \times N$ rectangular coordinate transformation and projection matrix. A properly chosen projection matrix will retain the bulk of the original data second-order variability in the reduced size data.

PCA uses the following assumptions and algebraic results in choosing the coordinate transformation and projection matrix $[A_R]$: (1) Correlated data channels collectively contain less information than uncorrelated data channels. (2) A reduced-order projection of the data onto fewer channels captures much of the information contained in correlated data channels. (3) Second-order variance–covariance statistics quantify much of the information overlap between data channels. (4) The $N \times N$ correlation matrix, Σ, describes the inter-channel correlations. (5) Any square and invertible linear transformation of the original data alters the data but does not lose information (assuming no round-off or other calculation errors). (6) Using the eigenvectors of the correlation matrix as vectors in the transformation matrix produces new uncorrelated channels. (7) Determining the eigenvectors of the correlation matrix is a straightforward solution of

$$[\Sigma]\{a_i\} = \lambda_i \{a_i\}. \tag{6.49}$$

(8) Ordering the eigenvalues in descending rank from largest to smallest as a diagonal matrix $[\Lambda]$ and then assembling corresponding normalized eigenvectors to form the matrix $[A]$ produces the relations

$$[A] = [\{a_1\} \quad \{a_2\} \quad \ldots \quad \{a_n\}], \tag{6.50}$$

$$[\Lambda] = [A]^T [\Sigma][A] = \begin{bmatrix} \lambda_1 & 0 & \ldots & 0 \\ 0 & \lambda_2 & & \\ \vdots & & \ddots & \vdots \\ 0 & & \ldots & \lambda_n \end{bmatrix} \tag{6.51}$$

with the inverse transform

$$[\Sigma] = [A][\Lambda][A]^T \tag{6.52}$$

and the normalization condition

$$[A]^T[A] = [I] = \begin{bmatrix} 1 & 0 & \ldots & 0 \\ 0 & 1 & & \\ \vdots & & \ddots & \vdots \\ 0 & & \ldots & 1 \end{bmatrix}. \tag{6.53}$$

(9) The variance–covariance matrix decomposes into a singular value form

$$[\Sigma] = \lambda_1[S_1] + \lambda_2[S_2] + \cdots + \lambda_n[S_n], \tag{6.54}$$

where

$$[S_i] = \{a_i\}\{a_i\}^T. \tag{6.55}$$

$[S_i]$ is the singular matrix corresponding to the ith eigenvalue and vector.

The contribution of each term on the right-hand side of Equation 6.54 usually diminishes as i increases. This is due to both the decreasing size of the eigenvalues and the increasing amount of wiggling in the eigenvectors. A rank order plot of the eigenvalues can indicate the distribution of information (as can be captured with second-order statistics) and the potential strength of a reduced-size PCA representation. When the curve falls off quickly, the lower eigenvalues contain the bulk of the information (Figure 6.9). When the rate of fall-off is slower, as in Figure 6.10, a good representation requires more eigenvalues.

Multiplication of the original multichannel data vectors by the $n \times m$ projection matrix $[A_R]$ creates a reduced-size representation in terms of new variables $\xi_1, \xi_2, \ldots, \xi_M$, where $M \leq N$

$$[A_R] = [\{a_1\} \quad \{a_2\} \quad \ldots \quad \{a_M\}], \tag{6.56}$$

$$\{\xi_R\} = [A_R]^T\{x\}. \tag{6.57}$$

An attractive property of PCA is that the computations are routine matrix manipulations [5]. Additionally, PCA requires only minimal assumptions regarding the PDFs underlying the data variations. PCA requires only a knowledge of the variance–covariance matrix and an assumption that this matrix describes the bulk of the random variations of the data. When data variations are Gaussian, the variance–covariance matrix describes all of the statistical

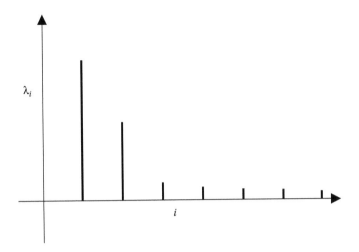

FIGURE 6.9 Rapid falloff of principal components.

variations in the data, and the PCA technique can be very effective. When the components in new data sets corresponding to the lower eigenvalues contain the bulk of the variation in the original data, then the need for fewer principal components eases the separation and classification of data with methods such as NNs [75]. When the data variations are non-Gaussian, the variance–covariance matrix provides an incomplete statistical description and the PCA technique loses power.

There are multiple examples of PCA in the SHM literature. Shinozuka and Rejaie compared and assessed the level of damage in urban areas following major earthquakes by processing satellite gathered images with PCA [76]. De Boe et al. detected damage in substructures with dynamic measurements with PCA [77]. Nedushan and Chouinard used PCA to form reduced-order models of the deformation of concrete dams [78]. Hu and Han identified temperature-induced confounding effects in vibration-based damage detection with a nonlinear variant of PCA [79]. Teng et al. used a spatially weighted PCA technique,

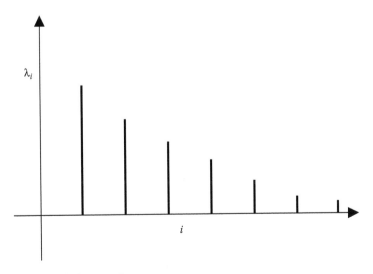

FIGURE 6.10 Gradual falloff of principal components.

known as a proper orthogonal decomposition, to determine reduced-order representations of wind loads on buildings from an array of pressure measurements [80]. Kerschen et al. used PCA to assess and validate the integrity of sensor arrays [81].

6.9 Independent Component Analysis

ICA and PCA are similar techniques. ICA identifies independent components in multichannel data [5]. ICA techniques also go by the names *blind signal separation* (BSS) or *blind source separation*. Both ICA and PCA are examples of statistical techniques known as unsupervised learning, or cluster analysis. These techniques segregate and cluster data based on selected features, but are largely unsupervised [7]. ICA applies a coordinate transformation to multichannel data to produce data channels with statistical independence up through fourth-order moments. PCA uses a coordinate transformation to produce new data channels that are uncorrelated, that is, independent up through second-order moments. Statistical independence up through fourth-order moments is a stronger requirement than the second-order requirement of being uncorrelated. ICA and PCA are virtually equivalent for Gaussian variables where second-order statistics provide a complete description of the random variations in the data. ICA and PCA differ in the handling of non-Gaussian data through the use of higher-order statistics. A disadvantage of higher-order statistics is the high degree of sensitivity to outliers in the data.

A goal of ICA is to find m independent components $\{y_j\}$ in a set of measured n-channel data vectors $\{x_j\}$. ICA forms independent components with the data transformation $A(\{x_j\})$:

$$\{y_j\} = A\left(\{x_j\}\right),\tag{6.58}$$

$$\{x_j\} = \begin{bmatrix} x_{1j} \\ x_{2j} \\ \vdots \\ x_{nj} \end{bmatrix},\quad \text{and}\quad \{y_j\} = \begin{bmatrix} y_{1j} \\ y_{2j} \\ \vdots \\ y_{mj} \end{bmatrix},\quad m \leq n.\tag{6.59}$$

Most ICA formulations use a linear coordinate transformation $[A]$ with an equal length of input and output vectors.

$$\{y_{kj}\} = [A_{ki}]\{x_{ij}\}.\tag{6.60}$$

The matrix $[A]$ is not unique. $[A]$ belongs to a family of transforms that produces relatively similar (usually the same to within a sign or phase angle change) outcomes. The *fast ICA transform* is an efficient method of calculating $[A]$ from measured data sets. Alternative BSS algorithms use NNs [82].

ICA has yet to find extensive application to SHM data processing. One combines ICA with maximum information statistical analysis to detect impending bearing failures in rotating machines [83]. Zang et al. used ICA to preprocess structural vibration data to produce statistical features for an NN damage detection scheme [84]. In an NDE image processing application, Abujarad and Omar demonstrated that fast ICA can help to identify subsurface features [85]. Fan and Zheng identified fault conditions in gearboxes from the vibration signals with BSS [86]. An example from electrophysiology is the removal of heartbeat signals from multichannel electromyography (EMG) data. Figure 6.11 shows multichannel EMG data collected from an array of surface electrodes attached to the skin of the lower back of

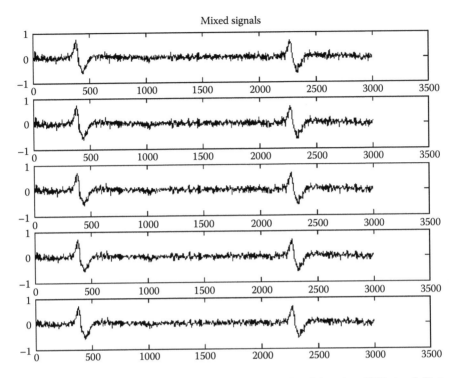

FIGURE 6.11 Multichannel lumbar muscle EMG data contaminated with heartbeat EKG signal. (Data courtesy of Z. He.)

an upright human. The collection of these data is part of ongoing effort to identify structural and neuromuscular differences between people with and without low back pain—a very challenging task [87]. An electrical heartbeat signal contaminates the data in each channel with a synchronized wave. Figure 6.12 shows the same data set after decomposition into independent components by the fast ICA algorithm. The ICA method isolated the heartbeat signal for removal by straightforward algebraic subtraction.

6.10 Fuzzy Interpretations

Fuzzy sets and logic are generalizations of classical set and logic theory. Proponents claim that fuzzy methods mimic some aspects of human reasoning. As such, using fuzzy methods may be advantageous in a variety of SHM applications.

A primary distinction between fuzzy and classical sets is the nature of the membership. Classical set theory uses a crisp representation of set membership. An element is exclusively either a member of a set or not a member. Fuzzy set theory allows for graded or partial membership of elements in multiple sets. Figure 6.13 shows a classical crisp assignment of membership based on a measurement, x. Figure 6.14 shows a fuzzy assignment of membership based on the same measurement. A potential disadvantage of crisp assignment is that small changes in the measurand x can cause a change in the set assignment. Fuzzy sets allow for partial assignment into different classes, such as a rating that is a combination of "good" and "fair." A strength of fuzzy set theory is that it quantifies subjective concepts

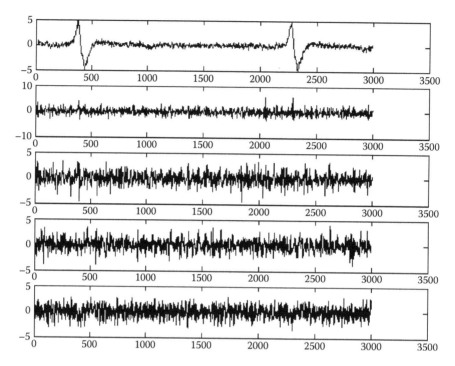

FIGURE 6.12 Multichannel lumbar muscle EMG data decomposed into independent components with FAST ICA algorithm. (Data courtesy of Z. He.)

that are easily understood by humans [88]. In a sense this allows for quantifiable means of subjective classification and reasoning. Figure 6.15 shows the primary steps in fuzzy set theory and reasoning.

Quantitative operations with both classical and fuzzy sets require quantitative allocations of set membership. A typical approach is to use a variable associated with the individual set elements. Crisp sets determine membership as being either inside or not inside a set. For example, in the crisp set representation of Figure 6.13, a system with measured value x_1 lies solely in set B. The fuzzy set representation of Figure 6.14 shows the fuzzy partial

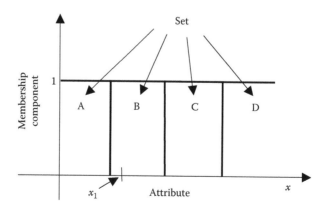

FIGURE 6.13 Crisp representation of set membership versus a single attribute in classical set theory.

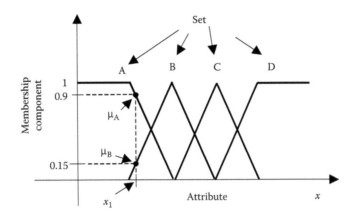

FIGURE 6.14 Fuzzy representation of membership component value per class versus single attribute measurement using normalized convex functions.

membership of x_1 in both sets A and B, with values $\mu_A = 0.9$ and $\mu_B = 0.15$, respectively. For reasons primarily due to mathematical expedience, most fuzzy set membership functions have a maximum value of 1 and a convex shape.

Classification and reasoning with fuzzy sets is a matter of combining the fuzzy information concerning inputs to produce crisp outputs. The Mamdani method may be the most established fuzzy combination technique. The method uses logical "if ... then ..." type statements combined with preassigned weightings. An intermediate output is a set of fuzzy outcomes. Forming a crisp output requires *defuzzification* techniques that combine the fuzzy outcomes. A centroid weighting of the fuzzy outputs is a standard and convenient approach (Figure 6.15).

Using fuzzy set theory can fuse multiple measurand data sets for SHM. The condition rating and management of buried pipelines is an example. It is difficult and expensive to measure the condition of buried pipes. Condition management practices invariably rely on subjective expert assessments. Kleiner et al. developed a fuzzy methodology for combining expert assessments of buried water pipelines with Markov process modeling to produce an overall fuzzy condition assessment [89,90]. Calibrating the fuzzy assessment against probability of failure is a key step. Reda Taha et al. detected damage on a bridge using wavelet features from simulated truck passage dynamic data with a fuzzy fusion approach

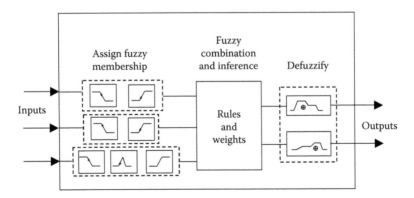

FIGURE 6.15 Fuzzy approach to inference and classification.

[91]. Meegoda et al. classified image damage features collected during the machine inspection of culverts with fuzzy sets [92]. Luo and Chou used fuzzy sets combined with cluster regression analysis to predict pavement performance and reliability [93]. Sun et al. noted that crack width and size are difficult to measure with high precision and are natural ingredients for a fuzzy analysis of the reliability of RC bridges [94]. Li et al. assessed the condition of concrete bridge decks with a fuzzy combination of load data and field conditions [95]. Wen et al. combined fuzzy reasoning with an unsupervised NN to detect and locate damage from structural dynamic data [96].

6.11 Data and Sensor Fusion

Data fusion merges data from multiple sensors to enhance the assessment of a system (or structure). Parameter estimation and classification for decision-making, and reducing the cost of an overall sensor system both may benefit from the application of data fusion techniques.

An assumption underlying many statistical analyses is that collecting more data provides more information. When using a single sensor, the collection of additional data initially tends to provide more useful information. There often reaches a point where gathering and analyzing more data from the same sensor yields diminishing returns. It may be more expedient to add another sensor of the same type and to collect data at a new location. Adding even more sensors of the same type may provide additional information, but again will lead to diminishing returns. A subsequent improved strategy may be to add a different sensor type that collects different types of data. An example is using a strain gage to detect damage in a solid. A single strain gage measures strain at a point. Deviations in the strain from nominally undamaged levels at that point can point to the occurrence of damage. Adding and then collecting data from more strain gages can indicate strain gradients and possibly the location of damage. Adding new varieties of sensors, such as a dye penetrant combined with a machine vision system, should augment detecting the presence and size of surface cracks.

A textbook example of data fusion is the combination of IR imaging and radar signal data for flight vehicle location [8]. IR detectors generally have very good lateral resolution, but are weak in downrange resolution. Conversely, radars tend to have weak lateral resolution, but have a better downrange resolution. Fusing data from both sensors may lead to better feature identification and location. A structural example is concrete bridge deck monitoring [97,98]. GPR can locate the depth and possibly the thickness of delaminations, but lacks the same level of lateral resolution. An IR camera can locate the position of the delaminations (Figures 6.16 through 6.18). As an example, Zheng and Ng used GPR locate the position of post-tension cable ducts in large concrete bridge U-girders followed by IE measurement of concrete depth [99]. Adding a chain drag test or other NDE measurements, such as IE, ultrasonic pulse velocity, or FWD, to the experimental protocol may further improve the quality of the assessments [100–105]. Similar results apply to airport runways and to systems that combine X-ray and ultrasound imaging directly for data fusion [106,107].

Ad hoc data fusion methods overlay and present data from different sensors to humans for interpretation. *Ad hoc* methods are effective when the data sets are relatively simple and the interface is well designed [108,109]. For example, Huang et al. used *ad hoc* fusion to compare fiber optic, IR, ultrasonic, AE, and SMART layer sensing on a CFRP panel as it was loaded and damaged [110]. Similarly, the X-33 reusable launch vehicle is an example

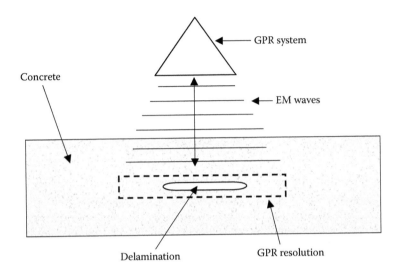

FIGURE 6.16 GPR location of a delamination in concrete with good depth resolution and weak lateral resolution.

of the application of heterogeneous sensing with multiple distributed fiber optic sensors [111]. An array of FBGs for strain measurements, palladium hydride FBG hydrogen sensors, distributed Raman temperature sensors, and AE sensors monitored the cryogenic fuel tanks on this vehicle. Shepard and Hollos showed that combining AE and ultrasonic Lamb wave sensors gives improved estimates of crack detection and location in plates [112]. Liu et al. fused laser ranging surface contour measurements with multifrequency eddy current measurements to diagnose the extent of corrosion in aircraft lap joints [113]. The fusion process included a subtraction of the visual laser and eddy current images. Figure 6.19 shows a graphical comparison of gas permeability and shear wave velocity images taken

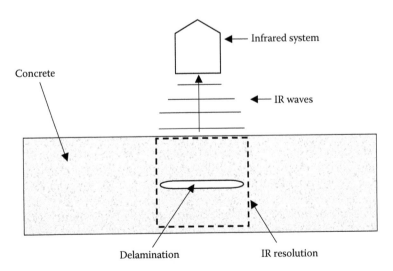

FIGURE 6.17 IR imaging location of a delamination in concrete with weak depth and high-quality lateral resolution.

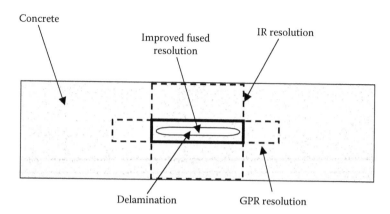

FIGURE 6.18 Fused GPR and infrared location of delamination in concrete with improved resolution.

on a concrete beam following freeze–thaw (F/T) cycling. The correlation of permeability versus shear wave velocity indicates the presence of damage.

Heuristic rule-based methods can also be effective in fusion-based classification. A simple heuristic approach for decision-making uses a threshold value of a single measured scalar, as in Figure 6.4. A heuristic rule-based approach is relatively easy to implement, once the decision rules are set. The challenge is to assign suitable decision rules that are robust under changing and varied circumstances. One approach gives each sensor an effective weighted vote with the *Borda voting algorithm* [114]. Other similar approaches combine thresholds with logical operations. Figure 6.20 shows a simple one-layer fusion system. More complicated topologies are possible, such as the hierarchical variant shown in Figure 6.21 [8,36,37,115,116].

Automated fusion methods formulated on a rational basis including prior knowledge of individual sensor performance, variance, and sensitivity are an alternative to ad hoc methods [8,36,37]. A fusion system may produce either a hard or a soft classification. Hard classification makes a decision as to whether or not an object or event falls into a particular class. Soft classification passes along information that contains attributes about the system that indicate probabilities as to whether an object falls into a particular class. An example is classifying whether a system is in one of two classes, A_1 or A_2. A hard classification system assigns the system as being exclusively in class A_1 or A_2. A soft classification system gives one of three possible responses: (1) There is enough information for the structure to be classified as belonging to A_1. (2) There is enough information for the structure to be classified as belonging to A_2. (3) There is insufficient information for classification.

The *Dempster–Shafer* (DS) approach attempts to mimic human thought processing of potentially ambiguous evidence in a two-step process. The first step quantifies information (evidence) with different degrees of reliability with *belief functions, m-values, and plausibility functions*. This includes the case of the information being sufficiently inadequate to draw a conclusion. The second step combines the evidence from two or more sources to produce fused belief functions, *m*-values, and plausibility functions by *Dempster's rule of combination* or other suitable combination rules [117].

Belief functions extend traditional notions of probabilities to consider what is believable based on the available evidence. An example is the case of using sensors to determine if a structural element has a crack. If sensor A is deemed to be 90% reliable and A indicates the detection of a crack, then the degree of belief in the existence of a crack is 0.9. The degree of

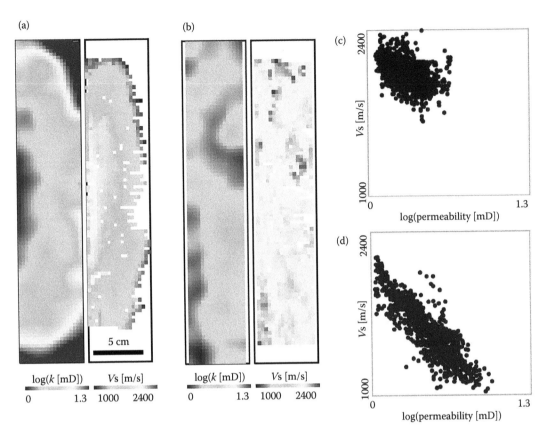

FIGURE 6.19 Upscaled permeability (k^*) and shear-wave velocity (Vs) for visibly damaged concrete beam after 270 F/T cycles (A) and incipiently damaged beam after 5 F/T cycles (B). In the case of the beam having undergone 270 F/T cycles, there is a clear inverse correlation between velocity structure and permeability (D). The correlation between permeability and velocity is poor for the sample having experienced only 5 F/T cycles (C). (Courtesy of Bussod G., New England Research, Inc.)

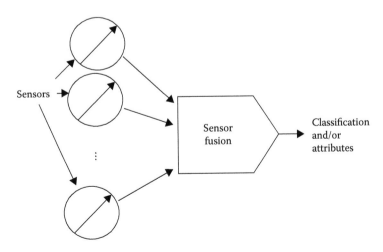

FIGURE 6.20 Single-level fusion topology.

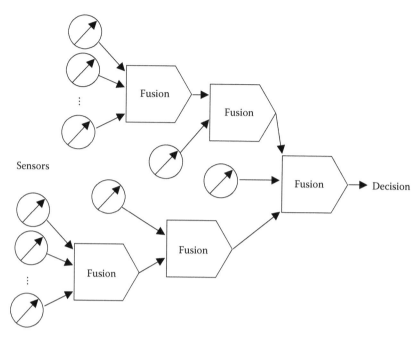

FIGURE 6.21   Possible hierarchical topology sensor fusion system.

belief in there not being a crack is 0. The measurement gives evidence about the presence of a crack, but gives no evidence about the presence of no cracks.

The mathematical representation of evidence and belief functions uses multivalued mappings of probabilities [118,119]. A *frame* (also known as a *frame of discernment*), Θ, is an exhaustive set of nonoverlapping outcomes, $a_1, a_2, \ldots, a_n$. An example of a frame is the states of a structural element being cracked, a_1, or uncracked, a_2. m-values are uncertainties or probability masses, $m(B)$, assigned to subsets, B, of the frame with the normalization condition

$$\sum_{B \subseteq \Theta} m(B) = 1. \tag{6.61}$$

For the case of two possible outcomes, a_1 and a_2, for example, cracked or not cracked, and ignoring the null set, there are three subsets of outcomes: $A_1 = \{a_1\}$, cracked; $A_2 = \{a_2\}$, uncracked; and $A_3 = \{a_1, a_2\}$, cracked and uncracked, that is, there is insufficient information to draw a firm conclusion. The m-values associated with the subsets are $m_1 = m(A_1)$, $m_2 = m(A_2)$, and $m_3 = m(A_3)$. The proper subsets are $B_1 = A_1$ and $B_2 = A_2$. The belief that an event has occurred is the sum of the m-values of all proper subsets, B_i, of Θ that contain the event.

$$\text{Bel}(A) = \sum_{B \subset A} m(B). \tag{6.62}$$

The only proper subset that contains $A = \{a_1\}$ is $B_1 = A_1$. The belief that a_1 has occurred is

$$\text{Bel}(A) = m(B_1) = m_1. \tag{6.63}$$

The plausibility, $\mathrm{Pl}(A)$, of an event, A, is the sum of all possible occurrences that are consistent with the evidence, where

$$\mathrm{Pl}(A) = \sum_{A \cap B \neq \varnothing} m(B) \tag{6.64}$$

or

$$\mathrm{Pl}(A) = 1 - \mathrm{Bel}(\sim A). \tag{6.65}$$

For the case of two possible outcomes, the plausibility that $A = \{a_1\}$ is

$$\mathrm{Pl}(A) = m(B_1) + m(B_3) = m_1 + m_3 = 1 - m_2. \tag{6.66}$$

In many respects, belief and plausibility provide upper and lower bounds of the knowledge concerning the occurrence of state a_1.

It is possible to have another set of independently derived belief functions for the same frame. In SHM, such a situation would arise with the use of an additional sensor. This second set of belief functions may agree or disagree with the first set. Dempster's rule of combination is a method of combining the two sets of belief functions to form a new set [120]. If for a given frame there is a set of outcomes, A, and associated m-values, $m_1(A)$, and a second set of outcomes, B, and associated m-values, $m_2(B)$, then Dempster's rule combines A and B to form a new set of outcomes, C, and m-values, $m(C)$, according to

$$m(C) = \frac{\sum m_1(A) m_2(B) \mid A \subset \Theta; B \subset \Theta; A \cap B = C}{\sum m_1(A) m_2(B) \mid A \subset \Theta; B \subset \Theta; A \cap B \neq \varnothing}. \tag{6.67}$$

The implications for SHM of the Dempster–Shafer approach are that an additional sensor has the potential of increasing or decreasing the degree of belief in an event. If sensor A has a reliability p_A and sensor B has a reliability p_B, then the belief in the occurrence of an event based on the signals, that is, evidence, from A and B depends on both the reliability of the sensors and whether the sensors agree or disagree. The possibilities are given below [121]:

1. If A and B agree, then the degree of belief is $1 - (1 - p_A)(1 - p_B)$.

2. If A and B give different but consistent readings, the degree of belief in both assertions being true is $p_A p_B$.

3. If A and B make assertions that contradict, then the degree of belief in A is $p_A(1 - p_B)/(1 - p_A p_B)$ and the degree of belief in B is $p_B(1 - p_A)/(1 - p_A p_B)$.

If the classifications are exhaustive and there are no overlapping classifications, then the DS method reverts back to hard classification schemes that use ordinary probabilities. A Bayesian extension of DS is the *generalized evidence processing* technique [8,36,37].

The development of many early data fusion applications concerned target recognition. Due to the ubiquity of difficult and ambiguous sensing scenarios, data fusion subsequently has appeared in a wide variety of sensing applications [8,122]. Multiple SHM data fusion applications make use of multichannel and heterogeneous sensor data. For example, Fu et al. describe the design of a bridge SHM system where the choice of multiple sensor types increased the performance of the system when the design was constrained to a modest number of sensor channels [123]. Another example arises in the use of satellite imagery to assess community-wide damage following catastrophic events. The VIEWS system, developed by Adams et al., combines pre- and postevent satellite images of a neighborhood with ground truth data consisting of images and visual observations [124]. A GPS system automatically registers the ground truth data to corresponding locations on the images. Smyth, and Wagner and Oertel fused GPS signals and double-integrated accelerometer signals to produce a superior structural displacement estimator [125,126]. Hundhausen et al. fused on-line in-flight SHM vibration data with pre- and postflight NDE data to assess the health of aerospace TPSs [127]. Tsutsui et al. fused fiber optic sensor data in the form of changes in transmitted light intensity, Bragg grating strain, and timing of Bragg grating signals for triangulation to detect, quantify, and locate damage in composites [128]. Camerino and Peters localized damage by fusing fiber optic sensor data via a structural flexibility approach [129]. A machinery application combines vibration monitoring with oil analysis to assess the health of large rotating machinery and gearboxes [130]. Goebel et al. used the DS approach as a fuzzy reasoner in machinery prognostics [131]. The technique successfully fused multichannel heterogeneous sensor data with an approximation that considered only the kth nearest neighbor to reduce the computational burden.

Several similar techniques compete with the DS approach to data fusion [132]. Liu et al. concluded that the Bayesian approach is more practical than DS in metal eddy current crack detection applications because of issues regarding the ambiguity of selecting a suitable probability mass [133]. Guo compared Bayesian, DS, and fuzzy methods for fusing frequency and modal vibration data to detect damage [134]. The conclusion was that all three methods were superior to using a multiple damage location assurance criterion and a frequency change damage detection method. Osegueda et al. compared several data fusion methods to combine information from eddy current, pulse echo, and magnetic resonance methods to identify and locate faults in plates [135]. After considering union, DS, and modified DS methods, they developed an approach that also included information about the measurement location proximity near the edge of the plate. A threshold value compared information in the form of a sum of the squares of variance-normalized feature quantifiers. The net effect was a classification scheme that retained the same number of false negatives, but reduced the number of false positives by 50%. Liang et al. found the DS technique to be effective in making decisions about the conditions of bridges from site data and NN analysis [136]. Bao et al. used DS analysis in the first stage of a two-stage vibration modal damage detection scheme [137]. The first stage used the DS technique to fuse local and global information to identify plausible damage sites. The second stage used a GA to refine the estimates.

6.12 Statistical Pattern Recognition and Data Mining

Many aspects of SHM are matters of SPR [138]. Structural damage forms patterns that are sometimes easy to recognize, and at other times more subtle and difficult to recognize. Humans are very good at SPR. An untrained human can easily detect gross damage and faults. Highly skilled humans can identify and recognize degrees of damage and dangerous situations that are not always obvious to the novice. In the past couple of decades, SPR has emerged as a standalone field of statistical analysis [37]. Duda indicates that the design of most SPR systems follows the steps of data collection, feature choice, model choice, training, and evaluation [7]. Hierarchical and layered approaches are common (Figure 6.22). An example due to Emamian et al. processes multichannel AE signals by first conditioning

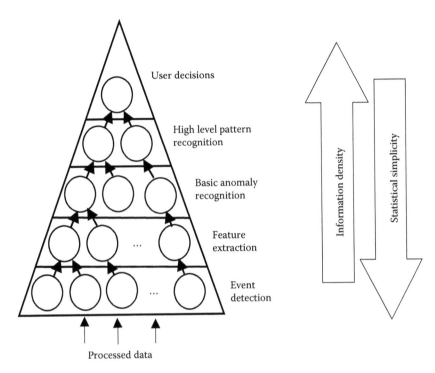

FIGURE 6.22 Hierarchical approach to statistical pattern recognition. (Adapted from Havira RM and Chen J. 1995. *Nondestructive Evaluation of Aging Railroads,* SPIE 2458.)

the data, applying a short-term FTT, using PCA to condense the date channels, and then feeding the data into a self-organizing NN [139].

Collecting high-quality data can significantly improve the power of an SPR effort. Issues of data quality extend beyond the collection of data with a minimum of errors to include issues of information density. Often the first step following data collection selects and uses low-level statistical features. Good features should increase information density. Typical statistical features of single-channel data are means, variances, higher-order moments, spectral peaks, and wavelet components. Features for multichannel data processing include PCA components, ICA components, and mode shapes. Image processing is ripe with possible choices of features, including the following: *intensity measures*: mean intensity, standard deviation of intensity distribution, higher-order moments of intensity, and other characteristics of intensity histograms including median, range, and 10–90 percentile range; *contrast measures*: edge content, gradient magnitude, intensity measures that measure spread; spatial moments of intensity: second-order moments (2-D), displacement of intensity centroid, moment of inertia of intensity field, and higher-order moments of intensity; *entropy measures*; *characteristics of bright areas*: size, maximum intensity, shape descriptors; texture descriptors: Fourier transform content and various wavelet transforms; and *multiscale measures*. The particular choice of feature is not unique, but can have a large impact on the ease and performance of the SPR process. For example, feature statistics based on surface curvatures or displacement variances may be superior to ordinary displacements for the identification of localized damage, such as cracks [141–143]. It is possible to use more quantitative methods to optimize feature selection [144]. Hilson et al. found GAs to be useful when feature selection choice is a matter of combinatorial optimization [145].

SPR algorithm setup and tuning requires access to existing knowledge bases to determine the relations between statistical features and the significant data patterns. Traditionally, the human brain has been the only tool available for sorting through complex databases to produce SPR classifiers. The human brain is a powerful tool, but is limited in the amount and complexity of data that it can assimilate. The human brain often makes effective use of heuristics as a simplifying strategy. For situations with a relatively simple system, it is possible to set up and tune the classifier based solely on physical principles. An example is identifying imbalances in rotating machinery. The mechanics of imbalances are well understood and the data signals (bearing forces) are distinctive (synchronous with shaft angle and proportional to the square of shaft speed). If the system is more complicated, then alternative approaches, such as examining a database of the past performance of similar systems, may be necessary.

SHM applications of SPR techniques are reaching a high degree of sophistication [146]. Allen et al. developed a modular software program that accommodates multiple SPR methods [147]. Haddad assessed damage levels in composite structures with an SPR processing of acousto–ultrasonic signals [148]. Mikumo et al. classified cracks in concrete slabs with SPR techniques applied to image data [149]. The first step conditioned the data and then transformed the geometry to align the spatial registration. Automated algorithms extracted intensity histogram and crack topology features. A pattern classifier determined the type and extent of the cracking pattern from the feature data. Havira and Chen used hierarchical methods to detect automatically known classes of defects on railroad rails [140]. Banavas et al. developed a system that matches strain time histories in a power plant steam pipe with similar historical transients [150]. Such an approach has the potential to automate the analysis of large structural data sets.

Computer-automated feature detection systems are possible. Ideal situations for automation are those where the algorithm can be tuned in an initial setup phase and then operated with minimal maintenance. If the system behavior is not well understood, or if the operational parameters change with time, it is advantageous to use an adaptive algorithm that modifies the features and classification parameters based on new data and circumstances (Figure 6.23).

Data mining analyzes large databases with the goal of recognizing and locating data patterns corresponding to events of interest. Data mining is analogous to geological mining operations that extract commodity-rich ores from the earth and leave commodity-poor deposits behind. Many aspects of data mining are simply the application of SPR to large databases. Data mining techniques dissect through the complexity underlying the structure of large databases. This includes distinguishing local patterns versus global variations in the data. An example SHM application occurs in the maintenance of fleets of vehicles or structures. Typical operations accumulate information regarding inspection and maintenance activity for each vehicle or structure. When an unexpected failure occurs, data mining techniques look for patterns that match those of the unexpected failure. The pattern matches provide the basis for deciding how to manage and prevent similar failures in the rest of the fleet. For example, Cruz et al. predicted the ultimate shear strength of plate girders with data mining techniques [151].

Hand et al. list six principal steps in data mining [27]:

1. *EDA*: This takes a first-cut look at the data to organize and attempt to determine some of the overall structures and patterns.
2. *Descriptional modeling and cluster analysis*: Determining trends in the data, groupings, and using tools such as BSS.

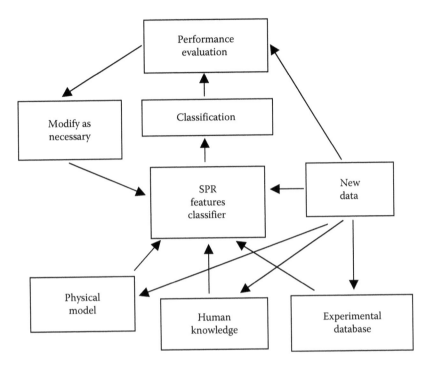

FIGURE 6.23   Information flow in adaptive SPR system.

3. *Predictive modeling, classification, and regression*: Empirical descriptions of the data, regression analysis.

4. *Discovery of patterns and rules*: Combining empirical results with an understanding of underlying principles of operation.

5. *Retrieval by content*: Using the tools established above to analyze rapidly existing databases to determine the content and location of specific patterns.

6. *Data management*: The storage of data in structures and formats that ease the recovery of important data.

Three of the primary data mining algorithms are the following [27]:

1. *Classification and regression trees* (CART): This uses data attributes to place items into specific quantitative and qualitative bins. Logical operations applied in a sequential branching tree-like fashion are a possibility. An example is the case of determining if an airplane is prone to corrosion. The subset of airplanes that land on carriers at sea may likely be prone to more aggressive corrosion problems than land-based aircraft. The analysis of aircraft maintenance records for patterns may benefit from the logical distinction of land-based and naval planes.

2. *Backpropagation*: These are the various forms of NNs that use multilayer perceptrons for data mining. The backpropagation methods are essentially regression analysis approaches to data classification. Nonlinear multilayer operators distinguish the technique from traditional linear regression techniques.

3. *A priori*: The *a priori* algorithm searches through a database for groupings attributes that produce a specified level of probability of occurrence. Logical operations can

often provide the attributes; for example, if A occurs and if B occurs, then C occurs with probability p. The algorithm searches through all possible combinations of attributes to find those with a probability exceeding a given threshold. This can be a computationally intensive task for even a modest number of attributes.

The inherently large nature of many databases causes the application of many of these steps and algorithms to be nontrivial. One approach quantifies the notion of the *interestingness* of patterns in the data [27]. If θ and φ are sets of attributes of particular members of a database, the *J-measure* is a robust quantification of how much the occurrence of θ indicates that φ may occur. The *J*-measure is essentially a cross-entropy calculation of the form

$$J(\theta \Rightarrow \varphi) = p(\theta) \left\{ p(\varphi \mid \theta) \ln \left[\frac{p(\varphi \mid \theta)}{p(\varphi)} \right] + [1 - p(\varphi \mid \theta)] \ln \left[\frac{1 - p(\varphi \mid \theta)}{1 - p(\varphi)} \right] \right\}. \quad (6.68)$$

6.13   Data-Driven Signal Processing

Adapting SHM statistical analysis procedures to the specific nature of the observed data can improve the analysis and strengthen the confidence in the conclusions. This is especially true in emergent and unexpected situations. Figure 6.24 shows a generic nonlinear adaptive multichannel signal processor. The internal operations of the data processor change to accommodate and exploit structures in the data (Figure 6.25).

Classification is a typical SHM signal processing application that commonly uses boundaries in a parameter space to segregate and classify the data. Figure 6.26 shows some of the issues involved with an example that considers the classification of three types of objects using a two-parameter data space. Concerns are nonlinear and overlapping boundaries between object type and the need for developing best-fit compromise-type classifiers that do not overfit training data.

EDA is a first-cut data-driven adaptive data processing technique that combines relatively simple statistical processing with subjective human input and guidance for adaptation. In the latter half of the twentieth century, a new wave of unstructured data-adaptive data processing procedures appeared, that is, NNs, GAs, SVMs, and CAs. The best uses and applications for these methods remain a matter of research and debate. It appears that NNs and SVMs work well when a substantial amount of training data is available and the

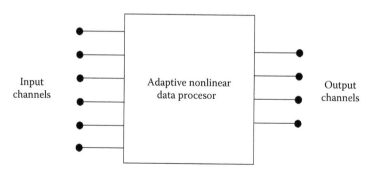

FIGURE 6.24   Adaptive multichannel nonlinear data processor.

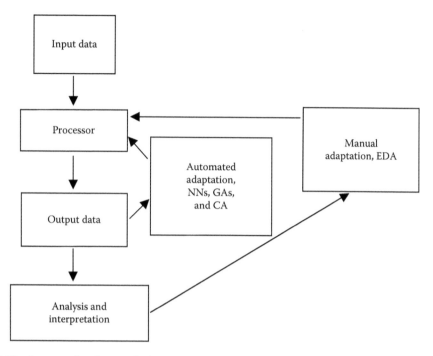

FIGURE 6.25 Automated and manual adaptive signal processing.

situation being modeled is similar to that in the training data set. GAs are strong at solving problems of combinatorial optimization. CAs work well when the organizing dynamics of the automata are similar to the dynamics of the systems being analyzed.

6.13.1 Neural Networks

Artificial NNs comprise a wide class of adaptive data processing algorithms. NNs process input data with a routinely nonlinear algorithm to produce outputs by adjusting the internal data processing parameters to improve the quality of the outputs as needed. In many respects, such adjustments are similar to the learning processes employed by biological systems. Haykin gives the following definition of an NN [152]:

> A NN is a massively-parallel distributed processor made up of simple processing units, which has a natural propensity for storing experimental knowledge and making it available for use. It resembles the brain in two respects: (1) The network acquires knowledge from its environment through a learning process. (2) Interneuron connection strengths, known as synaptic weights, store the acquired knowledge.

An NN has local processors, known as *neurons*, and weighted network-style interconnections between the neurons, known as *synaptic weights*. Figure 6.27 is a schematic of the operation of a generic artificial neuron. The neuron converts a set of input signals, x_i, into an output signal, y. The process weighs and sums the inputs with values w_i, includes an offset term as an input as needed, and runs the summed signal, v, through an activation function, φ. Adjusting the synaptic weights is a primary means of altering the behavior of the neuron and implementing learning. The *activation function* is a primary source of the nonlinear data

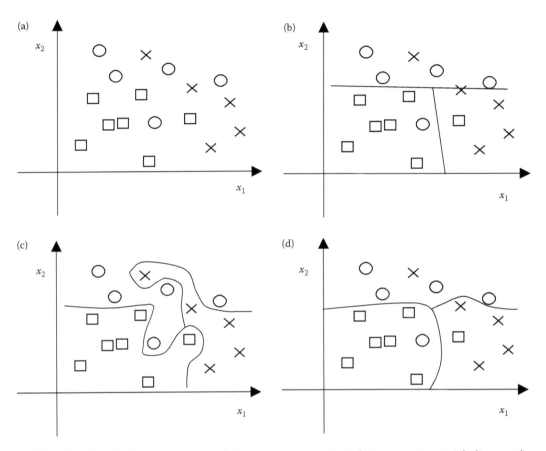

FIGURE 6.26 Classification of three types of objects in a two-parameter data space using straight-line, overfit, and curvilinear boundaries: (a) original data; (b) straight line; (c) overfit; and (d) curvilinear.

processing in an NN. Typical activation functions vary from 0 to 1. The *logistic function* with the adjustable slope parameter a is a common activation function.

$$\varphi(v) = \frac{1}{1 + \exp(-av)}. \tag{6.69}$$

An infinite variety of possible NN architectures is possible. Directed graphs can represent many of these architectures. Once the NN achieves a certain level of architectural complexity, it has the ability to fit a wide variety of nonlinear functions to the input and output data. One of the more common variants is the *multilayer perceptron* (Figure 6.28).

The behavior of a trained and operating NN is essentially one of nonlinear approximation. NNs can be good at separating data that result from different situations, that is, classes, differently. Citing early work by Anderson, Haykin lists four primary rules of NN operation [152]:

1. Similar input data from the same classes should produce similar representations in the NN and similar outputs.
2. Data from different classes should be given different representations in the network.

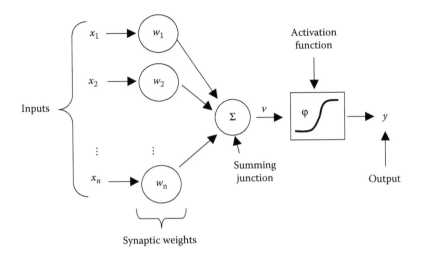

FIGURE 6.27 Schematic of operation of artificial neuron. (Adapted from Haykin S. 1999. *Neural Networks a Comprehensive Foundation*. 2nd Ed. Prentice-Hall, Upper Saddle River, NJ.)

3. Important features in the data should be represented by more neurons than unimportant features.
4. Prior information as to the behavior of the data should be built directly into the NN so as to accelerate the learning process.

Giurgiutiu and Kropas-Hughes describe five different NN architecture variants with potential utility in SHM [153]: (1) *Feedforward*: Compute the outputs directly from inputs without any feedback. (2) *Autoassociative/heteroassociative*: Feedforward NNs that use multi-layer perceptrons. (3) *Recurrent*: Use feedback from the output to the input. (4) *Competitive*:

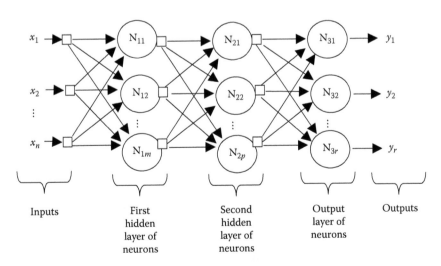

FIGURE 6.28 Multilayer perceptron with two hidden layers. (Adapted from Haykin S. 1999. *Neural Networks a Comprehensive Foundation*. 2nd Ed. Prentice-Hall, Upper Saddle River, NJ.)

Calculate the degree of match between stored patterns and input patterns, usually with a generalized distance measure. (5) *Probabilistic*: Use a hybrid approach with a radial basis layer, a feedforward layer, and a competitive layer. For the particular application considered, the probabilistic network approach proved to be effective in recognizing patterns in impedance spectra due to damage in plates.

Training with input–output data is a first task following the establishment of NN architecture. Training adjusts the neuron synaptic weights to values that optimize the ability of the NN to operate as a nonlinear approximation. The training starts with random or only partially preassigned synaptic weights. The next steps feed the input data into the NN, calculate the output, compute the error between the input and output, and adjust the synaptic weights accordingly. *Backpropagation* is a common NN training algorithm. Many other variants are also possible, such as the probabilistic NN [154]. A second task uses the trained NN to classify and analyze data. Retraining with new data as needed is possible.

A key to the success of NNs is the quality and size of the training data set [155]. If the training data set is large enough and contains data that are similar to the new situations being analyzed, then NNs can be quite effective [156]. An appropriate choice of statistical features as inputs can significantly enhance the performance of an NN. Trial and error is a predominant feature selection technique. One method of quantifying the strength of individual features is to examine the weighting values formed by the NN for each feature. The more important features should have bigger weighting factors [157].

The ability to process complicated data sets is a strength of NNs. In many situations the requisite processing is just too complex for the human mind to comprehend [158]. The complexity of SHM data sets and requisite interpretations often favor the use of NNs [159]. An example is using a limited number of sensors to detect damage that may occur over multiple scales [160]. Kudva et al. made the following conclusions when using NNs for damage detection: (1) The training sets have to be carefully chosen. (2) It is easier to determine damage location than damage magnitude. (3) A hierarchical NN scheme appears to be more efficient than one big network [161].

NNs find use in many SHM applications. A common theme is that complicated, but fairly stable, sensing situations are good candidates for NN signal processing [162]. Worden et al. used a multilayer perceptron with backpropagation training to identify damage stages in a planar truss structure [163]. Liang et al. detected damage in a water distribution system from a minimal set of pressure readings using NNs [164]. Lemoine et al. used NNs to process electrical potential measurements taken on polyethylene plates impregnated with carbon black to detected damage [165]. Spillman et al. processed and classified damage in a beam with data from a distributed fiber optic sensor with NNs [159]. Masri et al. detected changes in nonlinear structural systems with NNs [166]. Duran et al. used NNs to classify damage inside pipes by means of a robot-mounted laser vision system [167]. Zheng et al. detected and located delaminations in composites with piezoelectric patches using a computer-derived training set [168]. Hyland and Suresh et al. used a hierarchical and modular interconnection scheme for anomaly and damage detection [169,170]. Hatem et al. compared five different NN classifiers (generalized regression, linear, backpropagation with and without regularization, and radial basis) and a GA to determine which method is best capable of identifying damage in composite structures from vibration data [171]. The generalized regression network performed the best, with results confirming that detecting the presence of damage was easier than locating the damage. Tse and Atherton found that recurrent NNs were superior to feedforward NNs for prognostic predictions of machine failures from vibration feature analysis [172]. Tang and Sato developed an NN based on H_∞ filters that account for hysteretic structural behavior and are robust against model and exogenous input uncertainties [173]. Pan et al. used probabilistic NNs to analyze multichannel data from the Tsingma Bridge near Hong Kong [174]. Yun et al. demonstrated a committee of NN architecture for damage detection [175]. The individual NNs are trained independently, and vote as a committee to determine if damage is present in the structure. Govindaraju and Sikorski used NNs to predict the durability of thermal barrier coatings (TBC) in turbine blades [176]. The method fuses data from several NDE instruments, coating application data, and turbine operational data to predict the durability.

A disadvantage of NNs occurs when new data are substantially different from the training data. This can be a problem for SHM applications. Most structures are unique. The key items of interest, that is, damages that are sufficiently severe to warrant attention, are rare. Finding a suitable training data set that contains the appropriate damage patterns and features in the midst of patterns corresponding to healthy structures can be difficult. Additional disadvantages include issues related to the black box nature of NNs that limit human control and the availability of multiple variants of architectures without clear guidance as to which one to use.

6.13.2 Support Vector Machines

SVMs are data adaptive classification methods that segregate classes of objects with hyperplanes in a high-dimension attribute space [7,152,177]. A nonlinear transformation converts the original attribute space into the higher-dimension space. A *support vector* is a datum derived from a representative subset of the training data. The support vectors aid in determining the optimal hyperplane for class segregation. Advantages of the SVM approach are that it readily generalizes to include a wide variety of statistical learning techniques, and that it provides a means of naturally controlling the complexity of statistical models with the *Vapnik–Chervonenkis dimension*.

Due to the relative novelty of the technique, SVM classifiers have yet to appear much as SHM classifiers. Worden and Lane used an SVM to classify damage from ball bearing and truss vibration signals [178]. Mita and Hagiwara distinguished damaged from undamaged cases in building modal vibration frequencies with an SVM analysis [179]. In a similar study using FBG strain gages and accelerometers, Hayano and Mita found that combining strain and acceleration measurements strengthened the ability of an SVM to identify damage in a vibrating model building [180]. Yun et al. used a two-stage SVM to first identify damage and then to locate damage with PZT active sensors on railroad tracks [181]. Bulut et al. chose using an SVM over an NN for the real-time analysis of large channel count dynamic data in the form of wavelet features from large structures [182]. The perceived advantage of using an SVM was the speed of calculation and convergence on a classification. He and Yan used wavelets as features for an SVM-based method of detecting damage from dynamic structural data [183]. Zhang et al. extended the SVM technique to regression analysis for the identification of AR and ARMA models of structural dynamics [184].

6.13.3 Genetic Algorithms

GAs are adaptive data processing methods that mimic aspects of biological evolution and are well suited to solving problems of combinatorial optimization [185]. A key ingredient of GAs is the use of software entities known as *agents*. Each agent has a set of attributes that affects individual behavior and performance. The behavior of an individual agent depends on the combined action of a set of attributes. The attribute vector, {P}, for a particular agent with N parameters with values P_i is

$$\{P\} = \begin{bmatrix} P_1 \\ P_2 \\ \vdots \\ P_N \end{bmatrix}. \tag{6.70}$$

The GA attempts to find an optimal set of attribute parameters by combinatorial processes that mimic evolution. In terms of the biological analogy, the vector of attributes $\{P\}$ is analogous to a chromosome. The components, P_i, are analogous to genes. If n_i is the number of possible variations of the ith parameter, the total number of possible agent configurations, C, is

$$C = \prod_{i=1}^{N} n_i. \tag{6.71}$$

For even modest values of N, the number of possible permutations can be huge. Most of the agent variations are worthless, but a few are of high value, that is, have high fitness.

Implementing a GA is a recursive multistep process (Figure 6.29). A typical first step sets up the structure of the agents, their interactions with the environment, and then measures of fitness for a particular application. The second step places the agents in an environment and observes the performance of individual agents. During this step, some agents perform better than others, that is, some are more fit. The third step modifies the population of agents so as to increase the overall fitness by retaining those agents that perform the desired tasks well, that is, the fit agents, and culling from the population those agents that do not perform so well, that is, the unfit. The remaining population contains superior permutations of attributes. These fit agents then interchange information by mixing the attributes and produce new agents in a process analogous to biological breeding. This forms a new population of agents. The subsequent steps subject the agents to repeated cycles of fitness evaluation, culling, and interbreeding until the process converges upon highly fit agents. Practical implementations often use additional techniques to accelerate the convergence. These include the identification of elite agents that are very fit and pass onto the next generation without modification, and the introduction of new agents with random properties [186]. The final step terminates the algorithm based on a convergence criterion. In many cases, highly optimized populations of classifiers emerge from the data after only a couple of generations. Some keys to the successful use of GAs are (1) good training set; (2) selection of parameter space that has sufficient flexibility to converge, but not too much; and (3) choice of optimum evolutionary laws.

One SHM GA application optimizes statistical signal processing and classification. For complicated multiparameter classifiers or signal processors, selecting the ideal attributes may be tedious and time consuming, especially when using conventional optimization algorithms. The data processing fitness can be a measure of combined optimization of the sensitivity and selectivity of the statistical classifier. The agent population is a set of statistical signal processing algorithms. Each individual algorithm has an attribute vector of adjustable parameters. The environment is a set of individual statistics, such as SHM data features combined with outcomes in the form of specific damage types.

Multiple example applications of GAs appear in the SHM literature. Raich and Liszkai used GAs to optimize sensor layout for damage detection from vibration data [188]. Spillman and Huston determined an optimal layout of a distributed fiber optic sensor in a 2-D

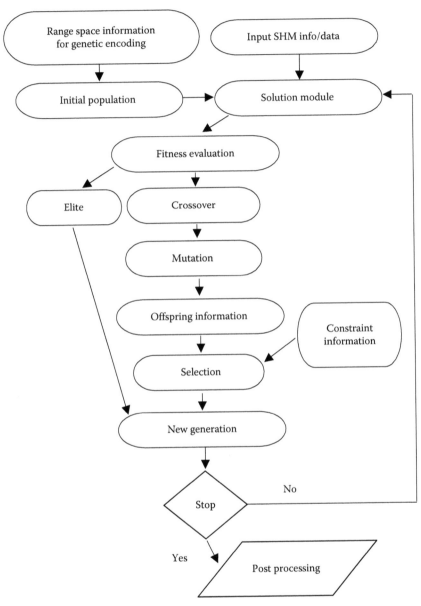

FIGURE 6.29 Genetic algorithm for SHM. (Adapted from Nag A et al. 2002. *Smart Mater Struct*, 11, 899–908.)

pattern with a GA [189]. Li et al. extended these techniques to design the layout of distributed optical fiber sensors in beds [190]. Au et al. demonstrated the effectiveness of GAs to detect damage in single and three span bridges using pre- and postdamage modal data [191]. Franco et al. showed that GAs can identify structural parameters from dynamic data [192]. Kong et al. and Ananda Rao et al. developed GA techniques for determining damage in structures from changes in modal vibration properties [193,194]. The method uses local stiffness changes as parameters to be determined from an optimization based on a residual force vector metric. The method was successful in identifying damage locations, even in the presence of noise. Nag et al. used GAs to identify delaminations in composite beams with spectra of elastic wave motions as the inputs [187]. Yan and Yuan used GAs to identify impact loads in composites [195]. Perera et al. developed flexibility-based damage indices based on the dynamic testing of structures using GAs [196]. Ying et al. detected damage in truss bridge structures from strain data with a GA [197]. Thanh et al. estimated dynamic traffic loads on bridges with GAs [198]. Vishnuvardhan et al. used GAs to estimate material properties for the individual layers in multilayer composite panels [199]. Ma et al. determined Bouc–Wen hysteresis model parameters with an evolutionary approach [200].

6.13.4 Cellular Automata, Agents, and Particle Swarms

A CA is an algorithmic agent that interacts with other agents in a user-defined environment. Some of the possible advantages of the agent-based approach are flexibility, adaptability, and the ability to use legacy systems. A simple set of rules usually governs the nature of the inter-agent interactions, in the form of growth, duplication, interbreeding, and interactions of the automata. Since the collective action of a population of automata can be exceedingly complex and difficult to simulate *a priori*, CAs have the potential for exploring solution spaces in new and unique situations in unpredictable manners. The operation of CA often is simply a matter of setting up the environment, turning the agents loose, and letting them self-organize into patterns that reflect the environment and rules of organization [201].

CAs have yet to be used extensively in SHM data processing. An example application is the processing of image data to identify regions of damage or anomalous behavior, such as damage in a bridge deck [202]. Figure 6.30 shows a survey of concrete bridge deck variations. The rules for the CAs favored those that consumed image intensity as "food." Figure 6.31 shows the principal details of the algorithm. Liu et al. combined GAs with CA for automated feature extraction from images [203]. The self-organization of ad hoc WSNs largely corresponds to the action of CA. Esterline et al. proposed an agent-based framework for vehicle SHM [204].

Particle swarm optimization (PSO) is a set of techniques with many similarities to CAs. PSO methods can solve a wide variety of complicated optimization problems with swarms of particles (agents) that move through solution spaces [205]. The particles communicate with one another both locally and globally to assess the optimality of the solution-space positions of the individual particles. This communication produces a collective intelligence that guides the swarm of particles to an optimal solution. PSO is a relatively novel technique and has yet to receive much attention in the SHM literature. Tang et al. used PSO in the system identification of dynamic structural behavior [206].

6.14 Decision-Making, Trees, and Networks

Decision networks can extend and generalize many classification methods previously described for the treatment of more complicated situations. Directed graphs describe the organization and information flow in many of these networks. The graph nodes represent decision processes. The directed edges (arrows) indicate the flow of information following the making of decisions. Decision networks can be particularly useful when the situation is complicated and the various classifying situations do not allow for quantitative interpretations. For example, it is difficult to compare quantitatively corrosion and cracking damage. Figure 6.31 shows a rudimentary decision network for classifying damage from a combination of image and vibration analysis.

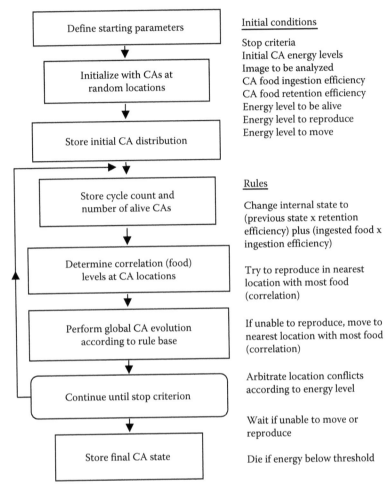

FIGURE 6.30 Flowchart for cellular automaton image analysis algorithm. (Adapted from Spillman Jr WB and Huston DR. 1998. *Opt Eng*, 37(3), 898–903.)

Belief networks are a subset of decision networks. A belief network incorporates probabilistic or belief-type aspects into the decision-making process. A *Bayesian belief network* (BBN) is a directed acyclic graph that represents the flow of information in the form of conditional probabilities through a sequence of possible events [207,208]. Each node in the BBN represents a local application of Bayes rule for calculating conditional probabilities. Figure 6.32 shows a simple BBN. Each node represents a state of possible outcomes. For example, node A can take on values a, where $a = a_1, a_2, \ldots, a_n$.

BBNs can be effective if there is sufficient knowledge about the behavior of a given system to construct a network and the behavior of the system is such that the network connections are stable. The BBN can "learn" by the updating of conditional probabilities. An advantage of the BBN approach is that it takes advantage of known conditions that are independent of one another. This considerably reduces the number of conditional probability distributions required and the subsequent computational complexity. An example is the BBN developed by Lebeau and Wadia-Fascetti that determines the load rating of prestressed concrete beams and backcalculates probable causes for low rating (Figure 6.33) [209,210]. Melcher et al. used

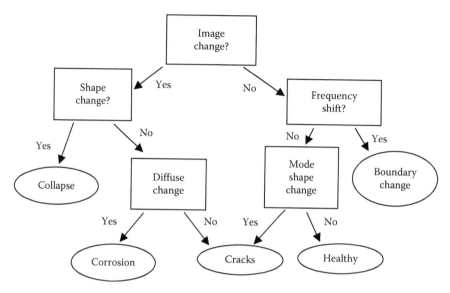

FIGURE 6.31   Decision network with nonquantitative outcomes for combination of image and vibration damage diagnostics.

BBNs to validate sensor and fault detection processors in complex rocket engine health monitoring systems [211]. Neil et al. found that BBNs were particularly adept at updating overly optimistic prognostic predictions in the maintenance of a fleet of vehicles [212].

The Bayes approach is one of several possible methods for updating system representations and algorithms based on the acquisition of additional data. NNs and GAs are possible alternatives. Loizos and Karlaftis compared artificial NNs, hierarchical tree-based regression, and multivariate adaptive regression techniques to determine the ability of each method to predict pavement damage from several other variables, such as construction processes [213]. Mizuno et al. used a decision tree to assess the vulnerability of bridges to corrosion [214]. Ulku et al. respective mined bridge condition rating databases with the aid of decision trees [215].

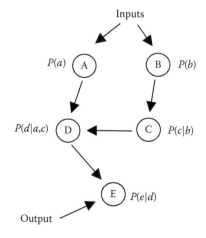

FIGURE 6.32   Simple BBN. (Adapted from Duda RO, Hart PE, and Stork DG. 2004. *Pattern Classification*, 2nd Ed. Wiley, New York, NY.)

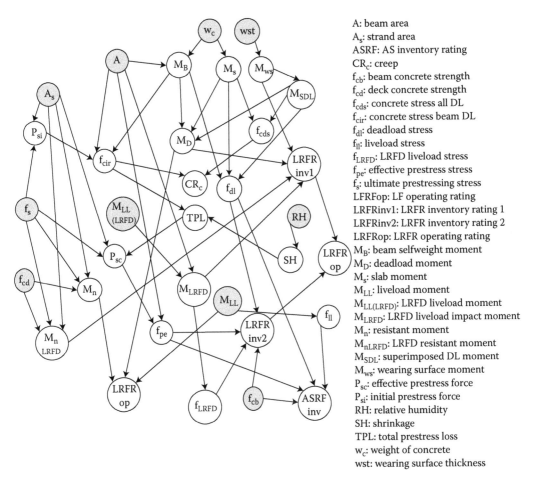

A: beam area
A_s: strand area
ASRF: AS inventory rating
CR_c: creep
f_{cb}: beam concrete strength
f_{cd}: deck concrete strength
f_{cds}: concrete stress all DL
f_{cir}: concrete stress beam DL
f_{dl}: deadload stress
f_{ll}: liveload stress
f_{LRFD}: LRFD liveload stress
f_{pe}: effective prestress stress
f_s: ultimate prestressing stress
LFRFop: LF operating rating
LRFRinv1: LRFR inventory rating 1
LRFRinv2: LRFR inventory rating 2
LRFRop: LRFR operating rating
M_B: beam selfweight moment
M_D: deadload moment
M_s: slab moment
M_{LL}: liveload moment
$M_{LL(LRFD)}$: LRFD liveload moment
M_{LRFD}: LRFD liveload impact moment
M_n: resistant moment
M_{nLRFD}: LRFD resistant moment
M_{SDL}: superimposed DL moment
M_{ws}: wearing surface moment
P_{sc}: effective prestress force
P_{si}: initial prestress force
RH: relative humidity
SH: shrinkage
TPL: total prestress loss
w_c: weight of concrete
wst: wearing surface thickness

FIGURE 6.33 Bayes network used to conduct load rating and low rating diagnosis on prestressed concrete beam. (Adapted from LeBeau K and Wadia-Fascetti S. 2007. *J Bridge Eng*, 12(6), 785–793.)

6.15 Displacement Estimation and Static Analysis

A common SHM task is inferring the global deformation of a structure from local measurements. Many of the calculations are similar to those of navigation. An integration of local position increments determines a global position. An example is to use an array of strain gages to determine global deformation patterns [216].

When structural mechanics are largely linear, then a linear combination of local sensor measurements in the form of an $M \times 1$ vector $\{z\}$ may provide a good estimate of a global $N \times 1$ displacement vector $\{u\}$ [217,218]. The corresponding matrix equation is

$$\{u\}_{N \times 1} = [A]_{N \times M} \{z\}_{M \times 1}, \tag{6.72}$$

where $M \geq N$. A combination of structural analyses and/or calibrations can determine the structural matrix $[A]$. A typical approach proceeds by assuming that the displacement $\{u\}$

is a linear combination of deflection modes:

$$\{u\}_{N\times1} = [\Phi]_{N\times P} \{\eta\}_{P\times1}, \tag{6.73}$$

where $[\Phi]$ is an $N \times P$ matrix of structural mode shape column vectors and $\{\eta\}$ is a $P \times 1$ vector of generalized displacement coordinates. The choice of the mode shapes used in $[\Phi]$ is not unique. Vibration mode shapes, finite element analyses, or other more ad hoc techniques are possible sources for mode shapes. Sensor data, such as localized strains, are similarly expressible as a linear combination of the generalized coordinates.

$$\{z\}_{M\times1} = \{\varepsilon\}_{M\times1} = [\Phi_\varepsilon]_{M\times P} \{\eta\}_{P\times1}, \tag{6.74}$$

where $\{\varepsilon\}$ is an $M \times 1$ vector of strain measurements and $[\Phi_\varepsilon]$ is an $M \times P$ strain influence matrix comprised of strain mode shapes.

$$\{\eta\}_{P\times1} = [\Phi_\varepsilon]^+_{P\times M} \{\varepsilon\}_{M\times1} = [\Phi_\varepsilon]^+_{P\times M} \{z\}_{M\times1} . \tag{6.75}$$

Using the strain to calculate the displacements requires taking the inverse of $[\Phi_\varepsilon]$ (or generalized inverse $[\Phi_\varepsilon]^+$, if $[\Phi_\varepsilon]$ is not square).

$$\{u\}_{N\times1} = [\Phi]_{N\times P} [\Phi_\varepsilon]^+_{P\times M} \{\varepsilon\}_{M\times1} = [A]_{N\times M} \{\varepsilon\}_{M\times1} = [A]_{N\times M} \{z\}_{M\times1} . \tag{6.76}$$

As examples, Jones et al. used least squares to estimate cantilever plate deflections from an array of multiplexed FBG strain sensors and Vazquez et al. used an inverse finite element method to predict displacements from arrays of strain readings [219,220]. Baldwin et al. extended this approach with NNs to estimate the shape of cantilevered plates from an array of multiplexed FBG grating strain sensors [221].

In the case of acceleration measurements, it is possible to estimate displacements by a double integration of accelerations. Straightforward attempts at double integration are prone to biases and initial condition errors. Park et al. developed a double integration algorithm that reduces initial condition errors [222]. Inertial navigation methods use sophisticated algorithms, such as Kalman filters, that may also be applicable to integrative structural displacement estimation. A Kalman filter can incorporate uncertainties in sensor measurements, and measurements that occur at uneven rates to update the location estimate [223]. Reich and Sanders used Kalman filters with the influence matrix approach of Equation 6.76 to estimate displacements in a plate from an array of fiber-optic strain sensors [217]. Smyth combined acceleration integration with GPS updating to form displacement estimates on large structures [125].

Static shape measurement may provide valuable information regarding the effective stiffness of internal elements and possibly damage in a structure. For example, Sanayei et al. used statically measured shapes to estimate underlying damage in a structure [224]. Di Paola et al. suggest that it is possible to use an integral equation method to assess damage in beams from static deflection measurements [225]. Sain and Kishen identified cracks in beams with static measurements of curvature [226]. Sumant and Maiti detected cracks in beams during quasi-static testing with a pair of PZT patches located on opposite sides of the beam [227]. Difficulties with ill-posedness of structural influence matrices often arise. Under certain circumstances, statistical regularization techniques can alleviate these difficulties [228,229]. Harursampath et al. found that the coupling stiffness of pretwisted composite anisotropic beams is highly sensitive to delaminations [230]. Murphy et al. found that an opening and closing delaminated beam exhibited a predictable non-linear load–displacement behavior [231]. Buda and Caddemi developed a procedure for identifying isolate damage in Bernoulli–Euler beams based on measurements of deflected shapes [232].

6.16 Vibration and Modal Analysis Techniques

Structural vibration testing has a rich history that dates back to the earliest days of mechanical analysis. Over the past two or three decades, a powerful set of tools has emerged that enables relatively easy modal vibration testing. This includes specialized transducers, data acquisition systems, and sophisticated prepackaged algorithms for extracting modal properties from test data [233]. There is considerable interest in using modal vibration testing for SHM. Some of the main reasons are that it is relatively easy to measure vibration modal properties and that the modal properties reflect underlying structural properties in a manner that may correspond to the location and extent of damage. It is common to assume that damage reduces structural stiffnesses. Damage can also affect damping, prestress, inertia, connectivity, and linearity.

There are at least four uses of vibration testing for SHM: (1) Identify and resolve a vibration problem that affects the performance or endurance of the structure. (2) Assess the vibration characteristics of a structure to verify design code provisions and mathematical models of structural behavior. (3) Use the vibration data as a tool to assess the health of a structure [234–236]. (4) Use vibrations as a means of testing the damage tolerance of a structure and to guide the design of more robust variants.

Many variants of vibration testing and analysis are available. The choice and success of the particular variant depends on whether the underlying mechanics of the structure are linear or nonlinear, and the availability of predamage vibration data (Figure 6.34). Methods based on principles of linear structural behavior tend to collect linear vibration data in the form of resonant frequencies, damping ratios, and mode shapes, and then process the data to identify patterns that correspond to damage levels and locations. Methods based on principles of nonlinear structural behavior usually require an assumption that damage corresponds to measurable nonlinear mechanical effects. The detection of deviations from linear behavior can signify the presence of damage. When predamage data are available, then damage detection may be simply a matter of detecting changes in vibration behavior (linear or nonlinear) (Figure 6.35) [237]. When no predamage data are available, then the analysis requires assumptions concerning the nature of damage on the structural behavior.

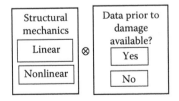

FIGURE 6.34 Coarse combinatorial taxonomy of vibration testing for SHM.

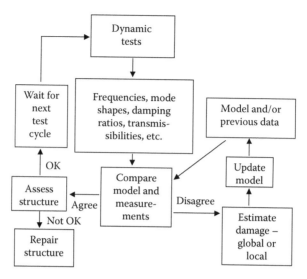

FIGURE 6.35 Process for assessing damage with vibration tests. (Adapted from Hamamoto T. 2003. *Smart Systems and Nondestructive Evaluation for Civil Infrastructures,* SPIE 5057.)

Farrar et al. describe three general approaches to system identification for damage detection [238]: (1) *Change in model form*: Fit a linear model to dynamic data before and after an event. When the structural properties of the structure change due to damage, the parameters of the linear model may change in a detectable manner. (2) *Model error detection*: This method fits a mathematical model to the data to create a predictive error criterion. As the structure changes, such as due to damage, the level of errors changes (usually increases). (3) *Direct identification of nonlinearities*: Damage is usually a nonlinear process that can exhibit nonlinear dynamical behavior. A complication is that damage may occur at multiple sites or may occur in a manner that is undetectable by the sensor system [239].

A strong motivator for using vibration testing in SHM is the success of vibration testing in the health management of rotating machinery [240–243]. Some of the reasons underlying the successes of vibration testing are mentioned below:

1. Rotating machines undergo similar, if not identical, loading during each revolution. The dynamical responses during each revolution also tend to be identical. This simplifies the signal processing and enables the use of phase angle-specific averaging.

2. Rotating machines tend to be made in lots of nominally identical units with similar failure modes and operational damage symptoms.

3. The failure modes of rotating machines are usually relatively few in number and well defined. Correspondences between the influence of the individual failure modes and the vibration behavior of the operating machine can be well understood. When the failure mode of a machine is not anticipated a priori, then vibration analysis may not be successful.

Spijker and DeMatteo report an interesting case of a rogue failure mode confounding diagnostic efforts in a rotating machine [244]. Vibration problems in an electric generator boiler feed pump disrupted operations at a commercial facility. Due to the large economic losses caused by the disruption, it was decided to apply a slew of modern vibration tests to the pump in an attempt to diagnose the problem. The vibration tests were unsuccessful. Ultimately, upon disassembly, it was determined that the source of the vibrations was a metal file being accidentally dropped into the pump during previous maintenance activities. Since none of the diagnostic instruments were trained to recognize the vibration symptoms caused by a piece of foreign matter inside the pump, the vibration tests failed to provide a correct diagnosis. It is conceivable, however, that such a test could be devised.

In spite of the overall relatively high level of success with using vibrations for diagnostics and prognostics on rotating machinery, there remain questions as to whether vibration testing is a viable and reliable method of structural assessment. Some of the arguments that favor the use of vibrations as a means for SHM are the following:

1. Significant damage often affects the mechanical properties of a structure.
2. The mechanical properties of the structure dictate the vibration properties.
3. Changes in a structure due to damage can cause changes in the vibration properties of a structure.
4. Vibration properties are relatively easy to measure.
5. Damage is a nonlinear process and may be detectable using nonlinear vibration analyses.
6. A variety of specific structures, such as pottery, baseball bats, or watermelons, can be quickly identified as healthy or defective, based on the vibration response.

There are several counter arguments to the viability of using vibration-based diagnostics and prognostics for SHM:

1. Often, predamage vibration data are unavailable for postdamage comparisons.
2. Much significant structural damage occurs in a form that does not affect global vibrations. Localized damage, such as corrosion, cracking, or damage to connection details, may give little indication of damage.
3. Local damage in a critical load path may leave a structure in a very precarious state, but may not cause significant shifts in the vibration properties of a structure.
4. Benign changes to the structure can confound SHM assessments. An example is changes in the support fixity due to thermal expansions. These changes may cause large changes in the vibration properties of a structure while being of minor consequence in terms of structural health.

6.16.1 Assessment with Methods Based on Linear Vibration

Investigators have devoted considerable effort to developing methods of using modal vibration measurements for SHM. Four groups of linear vibration-based SHM methods are (1) frequency domain analysis, (2) time domain analysis, (3) mode shape methods, and (4) pattern recognition methods. Linear vibration-based diagnostic efforts can be successful if there is a direct link between the failure and damage modes of a structure, and the measurable linear vibration properties; and if there is a minimum of confounding effects. Linear models are attractive due to relative mathematical simplicity, modal decomposition, and the validity of superposition methods. A ready supply of ever more capable vibration test systems with specialized hardware and software can measure linear vibration modal properties with relative ease. Industry-wide adoption of standard data formats, such as the *universal file format*, aids in the easy exchange of vibration data [245].

6.16.1.1 Frequency Domain Methods

Several reasons favor the use of frequency domain methods in vibration-based SHM. Frequency domain vibration behavior is intimately connected with the mechanical properties of a structure. The measurements usually only require a transducer, an excitation source, and a microprocessor. Frequency domain vibration statistics such as PSDs or FRFs are relatively easy to measure and calculate.

When a structure vibrates as a set of independent modes, the individual modal resonant frequencies, f_{ni}, of structures are

$$f_{ni} = \frac{1}{2\pi}\left[\frac{k_{ei}}{m_{ei}}\right]^{1/2}, \tag{6.77}$$

where f_{ni}, k_{ei}, and m_{ei} are the resonant frequency, effective stiffness, and effective mass of the *i*th mode, respectively. The stiffness and mass, k_{ei} and m_{ei}, are global quantities that reflect the potential and kinetic energy associated with the *i*th mode. Changes in the effective modal mass or stiffness shift the resonant frequencies. Since the effective stiffness and mass are global properties, the resonant frequencies tend to be sensitive to global changes to the mechanical properties. Resonant frequencies are significantly less sensitive to local changes in stiffness and mass. It should also be noted that the square root operation in Equation 6.77 blunts the sensitivity of resonant frequency shifts. Resonant frequency measurements can be effective in evaluating material properties of laboratory specimens [246] (Figure 6.36).

A successful SHM application of resonant frequency shift tracking is the measurement of tension in cables and stays [247]. Success lies in the straightforward connection between the observable states and the parameters of interest, with the presence of few confounding variables. Geometric nonlinearities cause cables and stays to stiffen against lateral deflection with increased tension. Similarly, the resonant frequency of post-tensioned concrete beams depends on the amount of axial compression. Under controlled laboratory conditions, Kim et al. demonstrated that resonant frequency shifting in pretensioned concrete (PC) beams due to a loss of negative geometric stiffening correlates strongly with a loss of cable tension [248].

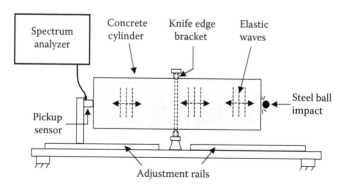

FIGURE 6.36 Apparatus for measuring material properties from resonant frequency according to ASTM C215 and ASTM C666.

Under certain circumstances, high compression loads in beams and skins can cause the resonant frequencies to shift, with the fundamental mode shifting to a higher-frequency [249]. This is a curious and counterintuitive effect (see Chapter 11).

Damage to structures often reduces the stiffness of the damaged elements. Detecting damage-induced loss of stiffness by resonant frequency shifting measurements is practical when the structural arrangement is such that the damaged element is a significant dynamic participant in a mode with a measurable resonant frequency. In one of the early reports, Cawley and Adams used modal frequency shifts to detect and locate damage [250]. Their method addressed the problem of detecting damage in 2-D plate-like structures with changes in the modal frequencies. A key assumption was that the stiffness matrix, mass matrix, and vibration properties were all known prior to the occurrence of damage, and that the mechanical effect of the damage was to reduce the stiffness of the structure. Kato and Shimada measured modal frequencies on a PC bridge with an AVS [251]. The first modal frequency declined significantly as the damage levels approached structural failure, but the damping level remained relatively unchanged. Elongated structures with fairly simple forms, such as pipes, are also good candidates for such techniques. One approach compares measurements of multiple modal frequencies with mathematical models of structural behavior [252]. Morassi and Rovere used a statistic that minimized a generalized distance, $F(\alpha_c)$, between the first M experimental modal frequencies, f_{je}, and the corresponding frequencies from an analytic calculation, f_{ja},

$$F(\alpha_c) = \sum_{j=1}^{M} \left[\frac{f_{je}^2 - f_{ja}^2(\alpha_c)^2}{f_{je}^2} \right]^2 . \qquad (6.78)$$

The parameters α_c represent changes in stiffness parameters due to damage. Varying α_c values can minimize $F(\alpha_c)$. The technique works under restricted conditions, but may give indeterminate estimates otherwise. It is a bit less common, but similar approaches apply to modal changes induced by changes in the mass, such as due to ice buildup on the leading edges of aeronautical structures [253].

There are numerous examples of frequency-shifting methods in the SHM literature. Mazurek and DeWolf correlated resonant frequency shifts with damage in bridge structures [254]. Morita and Teshigawara demonstrated with shaker table tests how the modal frequencies of a building model degrade in a repeatable manner with earthquake-type loading [255]. Melhem and Iyer used a FWD to measure frequency shifts and damage in reinforced concrete beams for bridges [256]. Capozucca and Cerri identified corrosion damage in reinforced concrete beams by vibration frequency shifts [257]. Vestroni and Capecchi detected localized damage in simply supported beams from natural frequency measurements [258]. Gheorghiu et al. correlated fundamental frequency shifts as high as 29% with fatigue damage in CFRP composite-layer and steel-reinforced concrete beams [259]. Stubbs experimentally detected damage in a cantilever beam by analyzing the changes in modal frequencies [260]. Rytter et al. correlated natural frequency shifting with crack size in vibrating cantilevers [261]. Li et al. developed a methodology based on finite element formulations of structural mechanics with improved crack modeling capabilities to identify and locate cracks in beams [262]. Wang used NNs to detect damage in tied arch bridges from an analysis of modal frequency shifts [263]. Qian and Mita found that NNs combined with Parzen window estimations of MDOF vector PDFs produced a frequency domain damage identification scheme for multistory buildings with a small training set [264]. Ackers et al.

identified the presence of cracks in heavy truck wheel end spindles by detecting shifts in the resonant frequency of the spindle with the wheel removed [265].

Power spectra or FRFs often contain significant useful information in addition to resonant frequencies. Damping ratios, antiresonances, and frequency band power can all indicate the presence of structural changes, and possibly the presence of damage. A simple, but useful, statistic is the area under a power spectrum. Park and Zhou found that the area under the spectrum of received ultrasonic waves dropped off and correlated strongly with damage in FRP plates [266]. Yoder and Adams determined that both peak spectral responses and the broadband spectral area corresponded to increased bead area damage in tires [267]. Hearn and Testa report laboratory tests of controlled damage causing measurable decreases in the frequencies of vibration, and modest increases in the damping ratios of welded steel frames and wire ropes [268]. Liberatore and Carman suggest a method for measuring frequency-shifting damage by the RMS energy in a spectral band centered at an undamaged resonance peak [269]. Shifts in resonance due to damage cause measurable shifts in the RMS energy in the preselected undamaged resonance bands. Fanning and Carden combined FRF single-input–single-output curve fitting with mechanical structural models by an optimization approach to detect and locate damage in structures [270]. Cade et al. detected faults in rotor/magnetic bearing systems by examining vibration signals that pass through digital wavelet filter banks [271]. Manson et al. used vibration transmissibility to locate damage in airplane wing structures [272]. Blodgett showed that delaminated rebars in concrete exhibit high-frequency vibrations that correspond to the increased mobility of the rebar caused by the delamination [273]. Schulz et al. concluded that if a structure is of a fairly simple form, then analyzing detailed measurements of FRFs can prove useful in identifying damage [274]. Fang and Tang found that PCA was a useful method of analyzing FRFs over broad frequency ranges for damage detection [275]. White et al. used broadband frequency response techniques to assess the quality of adhesively bonded composite repairs [276].

Wavelets and other time-frequency signal processing methods can measure power in spectral bands, when a simple power spectrum is insufficient. Multisensor wavelet analysis can tease out the presence of spatial coherence in spectral bands, which can correlate with specific damage modes [277]. Melhem and Kim used wavelet and modal methods to assess sequentially changing damage in a concrete beam [278]. The damage caused a distinctive increase in signal energy at higher wavelet numbers and a decrease in the first three natural frequencies by an amount up to 26%. Bland and Kapania detected small amounts of incipient damage in cantilever beams by using the energy in particular wavelet bands as features in an NN analysis [279].

Confounding effects can complicate using vibration frequency data in SHM [280]. Carrion et al. describe measurements of prestressed concrete beams subjected to reinforcing bar corrosion and cracking [281]. Confounding effects prevented an easy detection of damage by tracking resonant frequency shifts. Rebar corrosion caused the prestress to relax, reduced the negative geometric stiffening of the prestress, and caused the flexural frequencies to increase. Corrosion cracking in the concrete caused a reduction in effective section modulus, which caused the frequencies to decrease. The competing effects of prestress and section loss effectively prevented the use of simple flexural vibration frequency measurements. Gyekenyesi et al. found that the first spectral moments in acousto-ultrasonic tests were highly sensitive to both damage and mounting support details, which effectively negated the utility of using the spectral moment [282]. The diffuse field decay rate (a form of bulk material damping) was found to also be sensitive to damage, but not as sensitive to support conditions. Dharmaraju and Sinha report that tracking antiresonances was not a good indicator of cracks in modal tests of free–free beams [283]. Johnson et al.

determined that daily and seasonal temperature fluctuations and water table level correlated with significant changes in the fundamental natural frequencies of a soil–structure interaction testbed [284].

When the damage correlates well with modal frequency shifts and there are minimal confounding effects, frequency shift measures and MAC assessments of modal changes can be effective for damage detection [285]. Johnson et al. used *embedded sensitivity functions* to detect damage in vibrating structures by measuring the FRFs of a structure before and after damage has occurred [286]. Changes in the FRFs can be correlated with localized, that is, embedded, FRF sensitivities to determine the presence of localized damage, that is, change in stiffness. Johnson and Adams report similar results using vibration and acoustic transmissibility to detect structural damage [287]. As another example, shifts in vibration modal property can characterize the failure and damage modes of ceramic candles in coal burners. Since these failure modes of the burners are well understood and are the principal cause of modal property shifts, the frequency shifts are good indicators of damage [288]. Marwala used pseudomodal energies (defined as the integral of the area under the real and imaginary parts of an FRF near a resonance peak) as inputs to a modal vibration-based damage detection scheme [289]. The rationale is that pseudomodal energies are more information dense with respect to structural damage than raw FRFs.

Most of the above discussion of frequency domain techniques concentrated on the amplitudes of the frequency domain behavior. Occasionally phase angle behavior can be of interest. An example is the *narrow band electromechanical impedance* (NBEMI) method developed by Stepinski and Engholm [290]. The NBEMI technique uses narrow frequency band excitation and response analysis at selected frequencies. Examining changes in the phase angle of the response can provide indications of structural health. The HHT technique is another method of quantifying the phase angle of the response. Soh and Bhalla, and Lim et al. used EMI testing with PZT patches to assess the state of cure and damage in concrete and trusses [291,292].

6.16.1.2 *Time Domain Methods*

Direct processing of vibration signals in the time domain to produce statistical features is an alternative method of extracting useful SHM information. Many time domain methods are recursive time stepping algorithms including AR(X) methods, Shewhart charts, Kalman filters, and related techniques. Lei et al., Adams and Farrar, Nair et al., and Lu and Gao used autoregressive with exogenous input (ARX), ARMA, ARMAX modeling before and after potential damage as a damage identification and location technique [293–297]. An advantage of these methods and other similar input–output analysis methods is the possibility of conducting rapid real-time assessments of structural conditions and the preprocessing of data in distributed sensor networks [298]. Torkamani and Hart used time domain optimization methods to estimate changes in the stiffness of buildings from dynamic data [299]. Ge and Soong used a regularization method from optimization theory to process experimental time domain data for damage detection [300,301]. An advantage is that the method does not require an *a priori* assumption of the type of damage. Peelamedu et al. determined impact locations on plates from a time domain analysis of piezoelectric patch signals [302].

6.16.1.3 *Mode Shape Methods*

Analyzing spatial multipoint distributions of frequency domain behavior can provide a more comprehensive picture of structural behavior than that provided by single-point methods. A standard method processes multipoint frequency domain data to produce statistical features in the form of vibration mode shapes [233]. Mode shapes are relatively easy to measure, and reflect underlying structural properties. There is considerable interest in using mode shape information in SHM efforts.

Some of the reasons that favor using vibration modal information in SHM analysis are the following:

1. Mode shapes provide a spatial description of mechanical properties.
2. Damage to the structure may affect mechanical properties and mode shapes.
3. When predamage modal data are available, comparisons with the postdamage data can help to detect changes in mechanical properties.
4. When modal vibrations cause damage, then the distribution pattern of the damage often reflects the shapes and associated participation factors of the offending modes. Such a distribution of damage may be sensitive to modal damage identification methods [303].
5. Macro-scale damage to the structure, such as cracks or delaminations, can produce damage-specific distinctive mode shape patterns. This can alleviate the need for predamage modal data.

Some of the arguments against the utility of using modal analysis in SHM are the following:

1. Mode shapes are global quantities that reflect the integration of local mechanical properties over the whole structure. Local property changes due to localized damage contribute to mode shapes along with the rest of the undamaged structure. The result is the smearing of localized property changes with the undamaged properties. It is difficult to detect the local property changes by a cursory analysis of the mode shapes.
2. Reconstructing localized changes in structural properties from modal information requires using many higher-order modes. It is often difficult to measure a sufficient number of modes to achieve adequate spatial resolution.

While the outcome of the pro and con arguments above is not conclusive, one result is that it is generally not easy to detect damage by a direct examination of mode shapes. It is more common to use mode shapes as statistical features that feed into more definitive statistical analyses. Pandey and Biswas recognized that a modal representation of a structural flexibility matrix is more sensitive to damage than the inverse (stiffness matrix), especially when using a truncated set of modes [304]. This method successfully detected damage in experiments with beams undergoing flexural vibrations. The rationale for using the flexibility technique begins with the notion that the vibration modes of an undamped (or classically damped) structure are orthogonal when weighted with respect to the mass matrix. A normalized version of mass-weighted modal orthogonality is

$$[\Phi]^T[M][\Phi] = [I], \tag{6.79}$$

where $[\Phi]$ is an $N \times N$ matrix with the individual vibration mode shapes, φ_i, as column vectors,

$$[\Phi] = [\varphi_1 \mid \varphi_2 \mid \cdots \mid \varphi_N], \tag{6.80}$$

and $[M]$ is an $N \times N$ mass matrix. Expanding the stiffness and flexibility matrices in terms of outer products of the individual modes, that is, in a singular value decomposition, produces representations that can quantify the contributions of the individual modes. The $N \times N$ stiffness matrix $[K]$ decomposes as

$$[K] = [M][\Phi][\Omega][\Phi]^{\mathrm{T}}[M] = [M]\left(\sum_{i=1}^{N} \omega_i^2 \varphi_i \varphi_i^{\mathrm{T}}\right)[M], \tag{6.81}$$

where $[\Omega]$ is an $n \times n$ diagonal matrix of the squares of the modal frequencies, ω_i. The $N \times N$ flexibility matrix $[F]$ decomposes as

$$[F] = [\Phi][\Omega]^{-1}[\Phi]^{\mathrm{T}} = \sum_{i=1}^{N} \frac{1}{\omega_i^2} \varphi_i \varphi_i^{\mathrm{T}}. \tag{6.82}$$

An examination of the stiffness decomposition Equation 6.81 indicates the presence of frequency-squared coefficients for each term. The increase in value of these coefficients with mode number complicates efforts at forming an approximate representation of the stiffness matrix by using a truncated series that includes only the lower modes. Conversely, an examination of the flexibility decomposition Equation 6.82 indicates the presence of coefficients that vary as the inverse of the frequency squared for each term. A truncated expansion using the lower-order modes can approximate the flexibility matrix. Since it is usually more convenient to measure and to calculate the lower-order modes, it is more convenient to use flexibility matrices than stiffness matrices for limited information reconstructions of structural properties. This advantage plays out when attempting to detect damage from changes in modal properties.

A variety of damage detection, location, and severity estimation activities have successfully used mode shape data as statistical features. Many of these have used specialized algorithms that are fairly complicated in implementation, but generally follow the philosophy of expanding the dynamical equations of motions in terms of unknown perturbations. A fitting of the perturbed equations of motion to pre- and postdamage data produces an estimate of the damage.

Stubbs and Osegueda developed a perturbation method for extracting changes in structural properties from changes in measured modal properties [305]. This technique forms the basis for many more sophisticated subsequent developments in modal vibration damage assessment. An assumption is that the structure has N DOF and linear dynamics, that is,

$$[M]\{\ddot{x}\} + [C]\{\dot{x}\} + [K]\{x\} = \{F(t)\}, \tag{6.83}$$

where $[M]$, $[C]$, and $[K]$ are $N \times N$ matrices representing the mass, damping, and stiffness of the structure. $\{x\}$ and $\{F\}$ are $N \times 1$ matrices (column vectors) of displacement and force, respectively, for each DOF. The method assumes that the mass, damping, and stiffness matrices depend on damage sensitive parameters (m_i, c_i, and k_i):

$$[M] = [M(m_1, m_2, \ldots, m_N)], \tag{6.84}$$

$$[C] = [C(c_1, c_2, \ldots, c_P)], \tag{6.85}$$

$$[K] = [K(k_1, k_2, \ldots, k_B)]. \tag{6.86}$$

The number of parameters (N, P, and B) for each matrix need not be equal. A lightly damped structure with viscous or proportional damping has mechanics that allow for determining the mode shapes and frequencies from the undamped eigenproblem

$$[[K] - \lambda[M]]\{\varphi\} = \{0\}. \tag{6.87}$$

An N DOF system has N modes, $\{\varphi_i\}$, eigenvalues, λ_i, and modal natural frequencies of vibration, f_i, such that

$$\lambda_i = \frac{\{\varphi_i\}^T[K]\{\varphi_i\}}{\{\varphi_i\}^T[M]\{\varphi_i\}} = \frac{K_i}{M_i}, \quad i = 1, 2, \ldots, N, \tag{6.88}$$

$$\lambda_i = (2\pi f_i)^2, \tag{6.89}$$

$$C_i = \{\varphi_i\}^T[C]\{\varphi_i\}. \tag{6.90}$$

The fractional changes in the eigenvalues are $\{z\} = z_i = \delta\lambda_i/\lambda_i$. Similar fractional changes are $\{\alpha\} = \alpha_i = \delta k_i/k_i$, $\{\beta\} = \beta_i = \delta m_i/m_i$, and $\{\gamma\} = \gamma_i = \delta c_i/c_i$ for the stiffness, mass, and damping, respectively. Expressions for the modal sensitivities in terms of first-order variations are

$$[F] = F_{ij} = \frac{k_j}{K_i}\left(\frac{\partial K_i}{\partial k_j}\right), \tag{6.91}$$

$$[G] = G_{ij} = \frac{m_j}{M_i}\left(\frac{\partial M_i}{\partial m_j}\right), \tag{6.92}$$

$$[D] = D_{ij} = \frac{c_j}{C_i}\left(\frac{\partial C_i}{\partial c_j}\right). \tag{6.93}$$

Typical methods for calculating the stiffness sensitivity $[F]$ and mass sensitivity $[G]$ derive from numerical structural analysis or direct analytical procedures. In principle, numerical analysis can also yield a damping sensitivity $[D]$, but experimental approaches may be more effective. An expansion of the modal frequency sensitivity in terms of the structural property sensitivities that neglects higher-order terms is

$$\{z\} = [F]\{\alpha\} - [G]\{\beta\} - [D]\{\gamma\}. \tag{6.94}$$

A typical unknown is the change in the fractional stiffnesses $\{\alpha\}$. A solution of Equation 6.94 for $\{\alpha\}$ with a generalized inverse of $[F]$ is

$$\{\alpha\} = [F]^+\left(\{z\} + [G]\{\beta\} + [D]\{\gamma\}\right). \tag{6.95}$$

When all of the mode shapes and modal sensitivities are known, Equation 6.95 can estimate damage in the form of a localized stiffness reduction. Difficulties arise with this technique when only a limited number of mode shapes and modal sensitivities are known, and those that are known have uncertainties. Calculating damage with incomplete modal information tends to be an ill-posed problem. Specialized techniques, such as those using penalty-based linearizing regularizations, can provide some relief [306].

    The SHM literature contains numerous citations of modal-based damage detection efforts. Yu et al. concluded from numerical studies that when the structural mechanics are of a sufficiently simple form, such as having a tridiagonal stiffness matrix, then analyzing a reduced-order model of the dynamics taken with a limited amount of measurements can identify and locate stiffness-based damage [307]. Khiem and Lien used mode shapes to identify multiple cracks in beams [308]. A minimum of twice as many modes shapes as cracks to be located is required. Cha and Tuck-Lee developed a similar approach to model updating based on frequency response data using systems with diagonal mass matrices [309]. Amani et al. identified damage in laboratory building models using nonproportional viscous damping and an AVS [310]. Xue et al. applied Grey control system models to detected structural damage from dynamic data [311].

Since vibration measurements can only be taken at a limited number of points, it is difficult to capture the dynamics of all the degrees of freedom. However, it is sometimes possible to extrapolate the behavior from a limited number of measurements to determine the behavior of the entire structure and possibly to identify damage [312]. Yoshimoto et al. developed a MIMO testing strategy for shear type buildings subjected to seismic damage [313]. Koh et al. showed that this it is possible to combine Kalman filter state estimation with a GA technique in the analysis of simple multistory buildings [314]. Pothisiri and Hjelmstad detected damage with numerical simulations in more complicated structures by an adaptive element subgrouping approach that started with the sensors optimally prepositioned [315]. Avitabile describes dynamic modification methods for manipulating the equations of motion of a dynamic structural system to estimate the effect of changes to structural subsystems [316]. Lim et al. detected damage in trusses with a real-time modal parameter updating technique [317]. Zhu and Xu found that modal sensitivity analysis was a good indicator of damage in periodic structures [318]. Zhao and DeWolf conducted a series of modal analyses on highway bridges under traffic loading and concluded that changes in the bridge structure appeared as changes in the modal properties [319]. Catbas et al. identified deck support bearing damage in an overpass bridge with a detailed MIMO modal analysis that found a new mode when the structure was damaged with a large displacement amplitude at the damaged bearing [320]. High-accuracy modal measurements aided in this effort. Catbas and Aktan indicate that measured mode shapes can construct a virtual deflection of the structure under a virtual load. Localized large virtual deflections can indicate localized damage and incorporate into a damage index [321]. Zhou et al. found that random excitation modal testing tended to produce more variability in the results for use in damage detection than harmonic methods [322]. Ching and Beck used Bayesian modal updating techniques to detect damage in a large model building as part of a benchmark study [33]. Damage in the form of loosened diagonal braces caused significant changes in the modal properties and could be readily detected. Damage to joints did not cause much of a change in the measured modal properties and was difficult to detect. Shi et al. proposed a method of damage detection based on modal strain energy change and an improved version that includes analytical estimates of modal properties to reduce modal truncation errors [323,324]. Numerical simulations indicate the potential for success. Onyemelukwe et al. demonstrated that boundary element methods are useful for numerical studies of the effect of cracking on vibration behavior due to the ability of the procedure to localize the crack mechanics in the analysis [325]. Chen and Chen, and Song et al. localized damage from incomplete modal data by combining the results with changes in finite element models [326,327]. The method is effective on arch dams. Materazzi and Breccolotti conducted a series of modal tests, including the calculation of MACs and COMACs, on concrete beams subjected to severe damage [328]. A conclusion was that the classic modal methods were poor indicators of damage in reinforced concrete beams. Swann et al. found that it was difficult to detect small delaminations in composites with modal damage indices

[329]. Franchetti et al. were slightly more successful with polymer reinforced concrete (PRC) precast concrete beams in that modal analysis methods damage could detect and localize damage when the beam was severely damaged [330]. It was not possible to detect and localize with certainty for more modest levels of damage. Park et al. located fairly severe levels of damage with substructure flexibility analysis of modal information. Jang et al. combined modal system identification tests to successfully identify damage in a structure in the lab [331]. Shi et al. used a two-step approach where the mode shapes are examined first and then frequency data updated the damage estimates [332]. Wong et al. developed a method that identifies damage at multiple sites through an iterative multiparameter perturbation method that operates on linear vibration modal data [333]. Porfiri et al. used mode shape analysis to identify the electromechanical properties of structures laden with piezoelectric transducers [334].

Combining pre- and postdamage modal information with detailed numerical models, that is, finite element models, opens up the possibility for additional damage detection algorithms. Lee and Liang suggest that modal damping and energy transfer may be sensitive indicators of structural damage [335]. Fritzen and Bohle used a modal energy criterion to identify and localize relatively severe bridge pier undercutting damage with a formulation that assumed that the mass of the structure did not change [336]. Li et al. developed a modal strain energy decomposition method that localizes damage in 3-D frame structures using only a few modes [337]. A key feature is that it decomposes the strain energy in the elements of the frame structure into axial and transverse coordinates. The method is not as effective at determining the level of damage. Liu et al. developed a two-stage procedure for localizing and evaluating the severity of damage [338]. The first step locates the damage by a residual modal force method. The second step evaluates the severity by a combination of modal energy and element energy analysis. Numerical simulations on a truss structure confirmed the validity of the technique. If the damage locations are known, but not the level of damage, then it may be possible to use the *cross-modal strain energy method* (CMSEM) developed by Hu et al., which pre- and postmultiplies the mass and stiffness matrices by the vibration mode shapes collected before and after damage to form cross-modal strain energy estimates [339]. A relatively straightforward algebraic manipulation yields the change in stiffness due to damage. An advantage of CMSEM is that it readily accommodates different numbers of modes in the pre- and postdamage modal vector sets. Jaishi and Ren combined modal test data with a finite element model of a structure to update the model properties and estimate damage [340]. The method successfully identified damage in simply supported beams. Gray et al. developed a multisite damage localization technique using low-rank matrix SVD updates [341].

Bernal developed the damage locating vector (DLV) method based on estimating the structural flexibilities from modal properties [342]. Ji et al. and Gao et al. verified this approach with experiments on 2-D and 3-D truss structures [343,344]. The results were the identification of damaged elements along with false positives. Duan et al. extended the DLV approach to ambient vibration measurements [345].

The development of advanced measurement technologies continues to enhance vibration-based SHM efforts. Of particular interest are full-field scanning systems that measure complete mode shapes with a high degree of resolution. Detailed modal measurements enable localized detection of underlying structural changes. Modal curvature techniques can detect damage that affects the bending stiffness of structural sections in beams, plates, and shells. Ratcliffe et al. detected local changes in deformation patterns and possibly damage in a bridge structure [346]. Gauthier et al. experimentally verified the validity of higher-order modal derivatives to detect damage in beams [347]. Kim et al. combined damage index methods with modal curvature flexibility estimates to detect damage in plates [348]. Reynders et al. determined curvatures from modal vibration data to assess the health of the Tilff prestressed concrete box girder bridge with an array of SOFO fiber optic strain gages [349]. Pai and Jin, Sharma et al., and Hoffmann et al. used laser vibrometry to measure beam and plate modal vibrations with sufficient spatial resolution to detect damage and to compare different identification algorithms [350–352]. Otieno et al. used a laser vibrometer and modal updating to locate damage in laboratory models [353]. Vikhagan et al. used full-field vibrations of composite sandwich panels with optical shearography to indicate the presence of damage [354].

It is possible to tune modal testing techniques to enhance damage detection. Piezoelectric patches can excite local vibration modes and frequency shifts in a manner that indicates damage. Likewise, conducting modal tests with variable stroke and frequency shakers allows for tuning the shaker excitation to be sensitive to structural changes and damage [355].

Random variations and measurement errors can confound using pre- and postmeasured modal vibration properties as a means of damage detection and location. If possible, it is prudent to understand the nature of the variations in pre- and postdamage data to resolve some of these difficulties. Vanik et al. suggest using a Bayesian probabilistic approach that first conducts a series of modal tests on the healthy structure to determine the PDFs [356]. It is possible to quantify and compensate for the confounding effects of benign temperature changes on modal vibration properties [357–359]. For example, Sikorsky et al. measured the temperature and modal property variations of a concrete bridge over a 24-month period [360]. The temperatures during the modal tests ranged from 10.5°C to 40.5°C, with corresponding frequency shifts in the first and second mode of less than 0.4%. The conclusion was that the temperature-induced changes did not adversely affect the damage localization results, but did affect the level of confidence for a given damage location. Kullaa eliminated environmental and operational effects with a factors analysis approach [361]. Yan et al. used both ARX and PCA modeling to compensate for temperature shifts in a controlled test where pier settlement damage affected the dynamics of a bridge [362]. Peeters and De Roeck found that the eigenvalues of the Z24 concrete box girder shifted with temperature in a nonlinear, but repeatable, manner [363]. Ciloglu et al. found, in a series of laboratory tests of bridge models, that the modal properties tended to be more sensitive to changes in the boundary conditions than to damage to the structure [364].

6.16.1.4 *Pattern Recognition Methods*

Damage-induced alterations in dynamic structural behavior can produce data signal changes that are too subtle and/or complicated for interpretation with simple analyses of spectral peaks and/or modal properties. Automated pattern recognition methods may enhance damage detection and location. Doebling et al. describe a custom software toolbox for the recognition of patterns in dynamic data that correspond to structural damage [365]. The custom software toolbox has four main types of vibration pattern identification algorithms:

1. Methods that estimate the strain energy in local elements before and after damage

2. Methods that estimate the flexibility matrix before and after damage

3. Finite element correlation methods

4. Nonlinear damage identification methods

Fasteners are in many respects excellent candidates for using pattern recognition techniques on vibration data gathered in the vicinity of the fastener. Fastener vibration properties can change dramatically with loosening. Park et al. conducted a series of tests where frequency domain impedance shifts indicated cracks and bolt loosening in members for civil structures [366]. Ritdumrongkul et al. and Mascarenas et al. demonstrated the detectability of changes to the vibratory response of bolted joints with bolt tension [367,368]. The detection method examined the change in EMI of piezoceramic elements attached near the joint. Another example is a bolt-looseness-detection washer that detects shifts in the modal vibration properties of the piezoelectric washer [369]. Shimada et al. used vibration data collected with FBGs to detect missing fasteners [370]. Johnson et al. demonstrated a method where the shape of spectra measured at a riveting hammer indicated the quality of aircraft riveting operations [371]. The amount of spectral power in the 65–75 Hz and 88–118 Hz ranges distinguished good from substandard riveting operations, with good riveting operations consistently showing large amounts of power in both these ranges, as opposed to the defective rivets, which did not. Olson et al. correlated modal vibration pattern shifts with missing and loose mounting bolts on a square aluminum plate [372]. Missing bolts were easy to identify; loose bolts were not as readily obvious. Yang and Chang detected bolt loosening on TPS panel mounts based on an analysis of statistical features derived from measurements of vibration transmission through the joints [373,374].

The complexity of extracting information from vibration signals naturally leads to the use of computer-based pattern recognition methods, such as NNs [375]. Such methods have been successful when used with controlled conditions where signals from the damaged structure can train the NN [376]. Feng et al. used baseline forced and ambient vibration measurements of highway bridges for damage detection with the aid of NN analysis [377]. Reddy and Ganguli used radial basis functions with NNs to model stiffness-based damage detection in rotating helicopter blades [378]. The NN initially used 10 modes for damage detection. Upon training the NN and then conducting a parametric study of modal sensitivity, a reduced set of five modes proved to be effective for damage detection. Michaels et al. used a Fisher discriminant approach to compare multiple statistical features from ultrasonic testing of plates [379]. An example of the trend towards the application of increasingly complicated linear vibration analysis techniques is the NN committee approach by Marwala [380]. The technique uses three independent NNs to analyze FRF, modal property, and wavelet transform data. The three NNs then use a weighted committee-voting scheme to classify the state of health of the structure (Figure 6.37). If the statistical features of FRF, modal properties, and wavelet transforms are largely independent, then the use of a committee structure can increase the strength of the overall classification scheme. Supplemental dependent features, such as modal properties that depend on FRFs, tend to not increase the strength of the classification.

Novelty detection attempts to detect changes in system behavior by detecting changes, that is, novel events, in statistical features. SHM applications of novelty detection typically arise in the continuous periodic monitoring of structures where damage causes detectable changes in statistical features [381,382]. While the methods of novelty detection may apply to most SHM statistical features, the bulk of the applications to date use features derived from vibrations or dynamic testing and the application of statistical methods such as NNs and Mahalanobois distance. Examples of successful novelty detection methods were on an airplane wing panel and on a small airplane in the laboratory [383]. Sundararaman et al. detected simulated dislocation buildup damage in cryogenic tank Li–Al mounts by monitoring the frequency characteristics of propagating elastic waves [384]. Difference histograms of DFTs and discrete wavelet transforms of the transforms of the transmitted waves

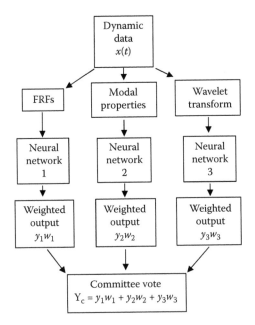

FIGURE 6.37 Committee of NNs votes on multiple dynamic data features. (Adapted from Marwala T. 2000. *J Eng Mech*, 126(1).)

indicated the presence of damage. Ma et al. demonstrated, via computer simulations, that a control theory concept of time domain monitoring of deviations can detect and locate damage in a vibrating structure [385]. Chan et al. used NN novelty detection to identify cable tension shifts in the Tsing Ma bridge [386].

Wavelets are a relatively novel signal-processing tool. Establishing the realms of the utility of wavelets for SHM analysis remains a matter of research. Identifying transient wave signal arrival times is perhaps one of the more promising applications. Tian et al. used wavelets to detect and locate damage in beams by analyzing and decomposing the signals in terms of flexure waves [387]. Marzani et al. used wavelets to determine elastic wave arrival times for pavement monitoring applications [388]. Numerical simulations by Chang and Sun showed that a wavelet packet signature method may be sufficiently sensitive to flexural rigidity changes to identify and locate damage at multiple sites [389]. Peng et al., Gu et al., and Gubbala and Rao located damage in plates by using wavelets to determine the arrival times of Lamb waves [390–392]. Hong and Kim developed methods of optimizing the mother wavelet for damage localization [393]. Dawood et al. used the CWT to determine a Lipschitz exponent as an identifier of damage in composite sandwich beams [394]. Amaravadi et al. detected mode shape changes due to damage and repair patches with 2-D wavelets [395].

Wavelet analysis appears in a variety of other SHM dynamic signal processing algorithms. Hou et al. used wavelet analysis to tease out damage-induced structural changes in real time [396,397]. Wimmer and DeGiorgi applied the CWT to the problem of damage detection in plates [398]. The method met with varied degrees of success. Certain edge damage modes were easy to detect; others were more difficult. Liew and Wang decomposed signals from vibrating beams with wavelets to identify the location of a nonpropagating crack. The higher-order wavelets, that is, those with a smaller spatial resolution, showed greater sensitivity to the emergence of cracks than that which was readily apparent from a traditional vibration modal analysis [399]. These results came from numerical studies that show the potential for methods where the data acquisition system has a fine spatial resolution, such as with a scanning optical vibrometer. Sun and Chang reached similar conclusions with the wavelet packet transform [400]. Masuda et al. demonstrated the effectiveness of wavelet analysis for damage detection using ambient vibrations in beams [401]. Similarly, Li et al. used wavelets to extract features of dynamic structural motions for use in NN damage detection. Rizzo and Lanza di Scalea used discrete wavelet methods to analyze guided waves in tendons and cables [402]. Law et al. identified damage in concrete beams with wavelets [403]. Kumar and Rao combined vibration testing with finite element analysis to infer that particular spans of aged reinforced concrete bridges have lost deck and diaphragm stiffness [404].

Since wavelet analysis produces statistics from dynamic data, it is possible to use wavelet statistics in pattern recognition algorithms. For example, Yan and Gao used harmonic wavelet analysis methods to produce features for use in a Fisher discriminant analysis of rotating machinery dynamic data [405].

A summary of the above-described efforts at using linear vibrations and structural dynamic methods to detect, locate, and assess structural damage is one of partial success. The most effective methods seem to be those where (1) there is a clearly identified conceptual link between the structural damage and vibration properties, (2) only minimal or well-understood confounding effects affect the vibration properties, and (3) the vibrations are relatively easy to measure.

6.16.2 Nonlinear Systems Approaches

The arguments for using nonlinear mechanics to test structures for SHM are in many respects very similar to those that favor using linear mechanics. Damage may initiate and/or alter nonlinear structural dynamics. A combination of mechanical models with carefully chosen experimental methods and pattern recognition techniques may effectively identify, locate, and quantify the extent of the damage. The behavior and mathematical descriptions of nonlinear structural interactions are generally more complicated than those of linear variants [406]. The increased level of complexity may provide enhanced opportunities for sensing and prognosis, but likely require increased mathematical sophistication.

Many structural nonlinearities are fairly weak. Such nonlinearities may alter quantitative aspects of the dynamic response of a structure, but not the overall qualitative nature. Equivalent linearization and harmonic balance methods are often effective with the analysis of weak nonlinearities [407]. Under certain circumstances, nonlinear stiffness and damping induce qualitative changes in behavior that do not appear in linear systems. *Attractors*, *limit cycles*, and *chaotic* behavior are examples. *Phase plane* representations, *bifurcation* analysis, and the *center manifold theorem* can help to explain these phenomena. SPR methods can also identify damage-induced nonlinear mechanical behavior.

A generic model of nonlinear structural dynamics for an SDOF system is

$$m\ddot{x}(t) + r[x(t), \dot{x}(t), r(t)] = u(t), \tag{6.96}$$

where m is the mass, $x(t)$ is the displacement, $r(t)$ is a nonlinear structural restoring force that depends on structural displacement, velocity, and the past restoring force, and $u(t)$ is an applied external force. The *Bouc–Wen* model (also spelled *Bonc–Wen*) is a popular method of modeling dynamic hysteretic nonlinearities in structures [408].

$$\dot{r} = \left(\frac{1}{\eta}\right)\left[A\dot{x} - \nu\left(\beta\,|\dot{x}|\,|r|^{n-1}\,r - \gamma\dot{x}\,|r|^{n}\right)\right]. \tag{6.97}$$

Varying the five parameters η, A, ν, β, and n produces a wide variety of hysteretic behaviors that encompass many observed structural hysteresis effects.

System-identification techniques can be quite effective in determining nonlinear system parameters that best fit the data. These techniques generally require that the nonlinear system model be qualitatively correct and the size of the measured motions be of sufficient amplitude to drive the system into nonlinear behavior [409]. Other hysteresis models are available, such as that due to Ashrafiz et al. that model hysteretic action as the collective behavior of parallel springs with the same stiffness and different yield points [410]. Nonlinear curve fits to the model were particularly successful when the assumed distribution of the yield point behavior (Rayleigh or Weibull) reflected that of the underlying distribution.

Damage to a structure can alter the mechanics so that nonlinear effects become more pronounced. An example is a crack that opens and closes as a system vibrates. The structure stiffens as the crack closes and softens as the crack opens. Such a nonlinearity can cause behaviors, such as superharmonic vibrations, that do not appear in uncracked structures with linear mechanics, [411]. Bolted joints are attractive testbeds for damage identification procedures based on nonlinear mechanics. Bolt tension can cause dramatic changes in the mechanics. As an example, Todd et al. used *chaotic attractor mapping* to detect bolt preload loss across a lap joint [412].

6.16.2.1 *Harmonic Analysis Approaches*

Harmonic analysis is a standard tool for analyzing nonlinear systems, particularly those with isolated nonlinear elements. The methods analyze the outputs from harmonic inputs. When a system is perfectly linear, the response to a single-frequency harmonic input is a same-frequency single harmonic output with an amplitude change and a phase shift. The frequency of the harmonic output will be the same as the input. The amplitude and phase are usually different. When the system has nonlinear elements, the response to a single-frequency harmonic input is a periodic multiple-frequency output. The output frequencies are often integer multiples, reciprocal multiples, and algebraic combinations of the multiples of the primary input frequency. An example is the single-frequency input response of a system with a nonlinear spring that has a polynomial force–displacement relation

$$F = k_0 + k_1 x + k_2 x^2 + k_3 x^3 + \cdots . \tag{6.98}$$

When the displacement, x, is a harmonic oscillating at circular frequency ω in complex form,

$$x = X e^{j\omega t}, \tag{6.99}$$

then the force, F, contains harmonic components

$$F = k_0 + k_1 X e^{j\omega t} + k_2 X e^{j2\omega t} + k_3 X e^{j3\omega t} + \cdots . \tag{6.100}$$

If the input contains multiple frequencies, then the output becomes more complicated, but the analysis can often still be tractable.

Opening and closing cracks are repeatable and fairly understandable examples of damage inducing nonlinear stiffness. An example occurs in rotating shafts and the interaction of the resonant frequency and rotating frequency with the nonlinear bending stiffness due to a breathing crack. Quinn et al. showed that such a machine is sensitive to forced excitation at the frequency Ω_F where [413]

$$\Omega_F = |\Omega_R - \Omega_C|. \tag{6.101}$$

Ω_R is the operational rotational frequency and Ω_C is the critical shaft speed. Experiments using active magnetic bearings as the excitation source confirmed theoretical predictions along these lines. Testa et al. found that the nonlinear action of an opening and closing crack increased the effective damping of vibrations [414]. Bovsunovsky and Surace developed a mathematical model of superharmonic vibrations in cantilever beams with opening and closing cracks [415]. Experiments on four different metal specimens generally confirmed the predicted second-harmonic distortion ratios. The results of such an approach can be highly sensitive to the level of damping in the structure. Hurlebaus et al. developed numerical methods for imaging harmonic distortions due to damage, such as cracks [416]. Das et al. identified the effect of delamination damage in composites by a harmonic analysis technique using the matching pursuit decomposition signal processing technique based on the hypothesis that the opening and closing of a delamination causes harmonic distortions to elastic waves [417]. *Vibro-acousto-ultrasonics* is a method for detecting fatigue cracks with a combination of acoustic and ultrasonic guided waves [418]. The operating principle is that the low-frequency vibrations and acoustic waves cause cracks to open and close with the passage of the wave crests. Crack opening causes a nonlinear change in the scattering and propagation of high-frequency ultrasound elastic waves. If there are cracks present, the low-frequency waves modulate the high-frequency waves. Without cracks, the level of mixing between low- and high-frequency waves reduces. Similar effects occur when delaminations in a composite open and close [419,420]. The mode shapes will change and the nonlinear effect of crack closing causes vibrations to occur at harmonic multiples of the primary harmonic. Stauffer et al. found that early stage concrete damage generated detectable second- and third-order harmonics in ultrasonic waves with a fundamental frequency in the range of 50–75 kHz [421].

6.16.2.2 Volterra Series

A Volterra series (also known as a Volterra–Weiner series) is a nonlinear extension to convolution integral representations of system behavior [422]. Figure 6.38 shows a generic system with input $x(t)$ and output $y(t)$. A linear model of the behavior of a causal system gives the output in terms of a convolution integral of the input with a kernel function $I_1(\tau)$:

$$y(t) = \int_{-\infty}^{t} I_1(\tau)x(t - \tau)\,d\tau. \tag{6.102}$$

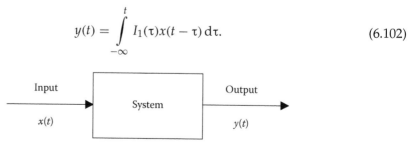

FIGURE 6.38   Generic system with input and output.

A Volterra series uses additional higher-order convolutions to capture nonlinear behavior.

$$y(t) = \int\limits_{-\infty}^{t} I_1(\tau)x(t-\tau)\,d\tau + \int\limits_{-\infty}^{t}\int\limits_{-\infty}^{t} I_2(\tau,\tau')x(t-\tau)x(t-\tau')\,d\tau\,d\tau' + \cdots. \tag{6.103}$$

The Volterra series can be useful in situations where the system is weakly nonlinear and the series converges quickly.

Smyth et al. developed adaptive nonlinear curve fitting procedures to identify the Bouc–Wen model parameters in Equation 6.97 [423]. As an alternative they used a technique that combined Volterra–Weiner series with NNs. Luong used Volterra series to analyze data from impulse testing of electric power transmission pylons with nonlinear mechanics [424]. Zhang et al. used Volterra series as part of a frequency domain formulation of the dynamics and subsequent input–output system identification of MDOF dynamical systems with polynomial-type nonlinearities [425]. Worden and Manson developed expressions for the frequency domain behavior of a nonlinear Duffing oscillator with a truncated Volterra series [426].

6.16.2.3 *Attractors, Chaos, and Phase Space Methods*

Differential equations are excellent tools for describing many dynamic structural behaviors. A reasonably general representation uses multiple coupled first-order *ordinary differential equations* (ODEs) with state variables $x = [x_1, x_2, \ldots, x_N]^T$ and control parameters $c = [c_1, c_2, \ldots, c_K]^T$ of the form

$$\dot{\mathbf{x}} = \mathbf{F}(\mathbf{x}, \mathbf{c}). \tag{6.104}$$

An inertial dynamical system with N degrees of freedom usually requires N coupled second-order ODEs in a mathematical model. State variable methods replace the N second-order ODEs with an equivalent set of $2N$ coupled first-order ODEs in the form of Equation 6.104. The state variable approach has the benefit of algebraic simplicity at the expense of more equations.

For the case of a linear and time-invariant system, Equation 6.104 simplifies to

$$\dot{\mathbf{x}} = \mathbf{A}(\mathbf{c})\mathbf{x} + \mathbf{f}(t), \tag{6.105}$$

where $\mathbf{A}(\mathbf{c})$ is an $N \times N$ matrix with values that depend on the K control parameters $\mathbf{c}$. The initial conditions are

$$x_0 = x(0). \tag{6.106}$$

The structure of $\mathbf{A}$ governs the stability of the linear system. Several equivalent methods can establish system stability. These include determining the sign of the real part of the eigenvalue, determining when $\mathbf{A}$ is positive definite, and determining the signs of the principal minors of $\mathbf{A}$ [427]. These stability quantifiers change with variations in the parameters $\mathbf{c}$. Along these lines, Shukla and Frederick examined the applicability of using a distance measure to a cubic bifurcation as an approach to damage prediction [428].

SDOF models can capture the behavior of many nonlinear systems. An inertial dynamic SDOF system has two state variables, usually with $x_1 = x$ and $x_2 = dx/dt$. *Phase plane plots* capture the behavior of nonlinear single DOF systems where x and dx/dt are the abscissa

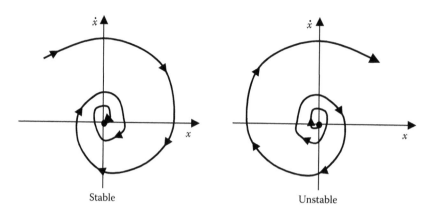

Stable　　　　　　　　　　Unstable

FIGURE 6.39　Phase plane behavior of stable and unstable single point attractors.

and ordinate, respectively, in a 2-D Cartesian space. The velocity and displacement state evolution of the system forms a curve in the phase plane. SDOF systems with zero applied force tend to converge on and diverge from fixed points and curves called *attractors* and *limit cycles*. Figure 6.39 shows the trajectories of an SDOF system with stable and unstable single-point attractors. An example of a stable SDOF system is the decaying free vibrations of a damped oscillator following an initial condition perturbation.

Figure 6.40 shows a system with an unstable attractor at the origin. The motion initially grows outward, but then converges onto a simple closed limit cycle curve. The trajectory moves inward and onto the limit cycle. Limit cycle behavior in SHM applications is relatively rare. Structural examples include slip–stick vibrations and vortex-shedding lock-in fluid–structure interactions.

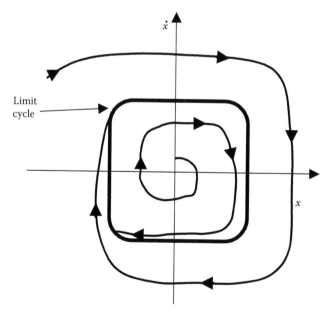

FIGURE 6.40　Convergence on a limit cycle in the phase plane.

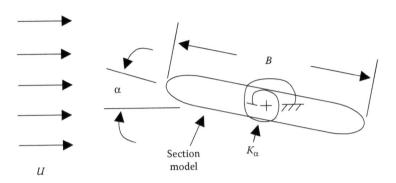

FIGURE 6.41   SDOF torsion aeroelastic bridge deck section model.

Since limit cycles act as attractors, the presence of limit-cycle behavior can confound measurements that assume a fixed-point attractor. An example arises in the wind tunnel section model testing of long-span bridge decks for aeroelastic stability, where SDOF torsional behavior is a dominant aeroelastic issue (Figure 6.41). The flutter derivative model is a common linear representation of aeroelastic behavior [429].

$$\ddot{\alpha} + 2\zeta_\alpha \omega_\alpha \dot{\alpha} + \omega_\alpha^2 \alpha = \frac{\rho U B^3}{2I} K A_2^*(K)\dot{\alpha}, \tag{6.107}$$

where α is the section twist angle, ζ_α is the torsional damping, ω_α is the torsional circular natural frequency, ρ is the density of air, U is the mean wind speed, B is the chord or width of the section, and A_2^* is the flutter derivative corresponding to negative torsional aerodynamic damping. Stable systems have large aeroelastic damping and a corresponding largely negative value of A_2^*. Aeroelastic instabilities arise when A_2^* is positive. K is the dimensionless real component of the aeroelastic frequency ω, that is,

$$K = \frac{B\omega}{U}. \tag{6.108}$$

The typical wind tunnel measurement of A_2^* is a free-vibration test with an elastically mounted section model placed in a steady wind to determine the effective aeroelastic damping and stiffness. Figure 6.42 shows the nature of the torsional time histories that correspond to stable, borderline stable, and unstable conditions. Figure 6.43 shows the A_2^* flutter derivative corresponding to a deck section with torsional SDOF flutter proclivities.

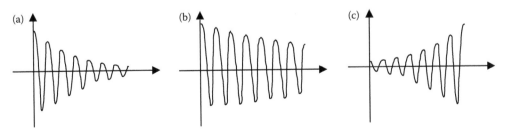

FIGURE 6.42   Stable, borderline stable, and unstable SDOF aeroelastic section model response following an initial perturbation. (a) Stable with attractor at origin; (b) borderline stable; and (c) unstable with attractor at infinity.

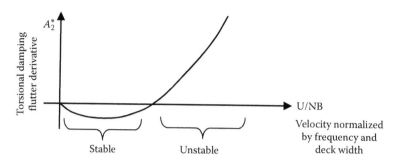

FIGURE 6.43   Torsional aeroelastic flutter derivative for an unstable bridge deck section.

Vortex-shedding lock-in is a type of aeroelastic behavior that is distinct from the single DOF flutter that plagues long-span bridges. Many bluff bodies subjected to steady flow produce highly separated wakes with alternating von Karman vortices. The shedding frequency, N, is proportional to flow velocity, U, and inversely proportional to cross-section dimension, D. The dimensionless Strouhal number, S, quantifies this relation:

$$S = \frac{ND}{U}. \qquad (6.109)$$

S is approximately 0.2 for many structures. As the wind speed increases, so does the shedding frequency. When the shedding frequency equals a structural resonant frequency, the structure also vibrates. The interaction of structural and vortex shedding vibrations is more complicated than that of forced resonance vibrations. The frequency of vibration of both vortex shedding and structural vibration coalesce, that is, lock-in, over a frequency band surrounding the natural frequency of vibration. Figure 6.44 qualitatively shows this behavior.

During lock-in, the structural motions converge onto a limit cycle. Vortex-shedding lock-in can confound flutter derivative testing, which assumes that the system converges onto a fixed-point attractor. Figure 6.45 shows the free-vibration response of a section model combined with both small and large limit cycles. When the initial displacement for the free vibration test exceeds that of the limit cycle, the attraction of the limit cycle can give flutter estimates that correspond to a stable situation (Case A), even though the bridge continually moves. When the initial displacement is less than the limit cycle amplitude, the attraction of the limit cycle can give flutter derivative estimates that correspond to an

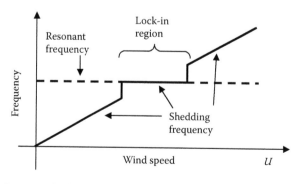

FIGURE 6.44   Aeroelastic vortex-shedding lock-in of an elongated bluff body.

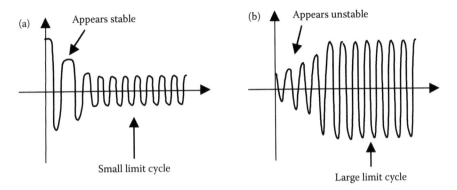

FIGURE 6.45 Confounding of flutter derivative measurements due to the limit-cycle attraction of vortex-shedding lock-in.

unstable situation (Case B). The concept of a Hopf bifurcation qualitatively explains some of this behavior (Figure 6.46) [430]. Vortex shedding can also confound and distort wind tunnel flutter derivative measurements. Figure 6.47 shows flutter derivatives measured on the Deer Isle Bridge (Deer Isle, ME, USA). Vortex shedding causes anomalous negative $A_2^*(K)$ readings for reduced velocity values of 1.7, 3.4, 5.1, and 6.6. It is interesting to note that these values are approximately harmonic multiples.

Bolted connections are excellent candidates for the study of nonlinear mechanics and SHM. The nonlinearities present in bolted connections can be fairly strong and often show repeatable dependence on bolt tightness parameters. Tightly fastened bolted connections that do not slip may behave as a linear elastic solid. Loosened connections can slip, rattle, and move with minimal resistive force in gaps and act elastically at the end points. The mechanics of loose bolted connections can be highly nonlinear. As an example, Shin used nonlinear system identification to detect bolt loosening [431]. The model assumed that the incremental equation

$$[M]\{\Delta \ddot{u}_j\} + \left[C(y_j)\right]\{\Delta \dot{u}_j\} + \left[K(x_j)\right]\{\Delta u_j\} = \{\Delta f_j\} \tag{6.110}$$

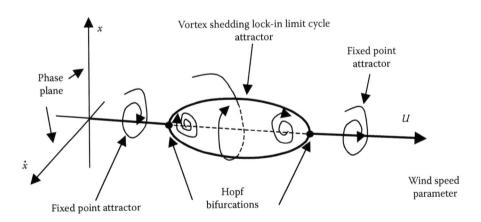

FIGURE 6.46 Hopf bifurcation interpretation of vortex-shedding lock-in.

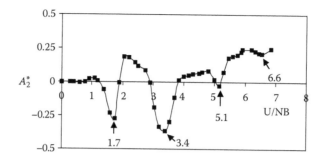

FIGURE 6.47 Distortion of flutter derivatives on H-shaped section model due to vortex-shedding lock-in. (Data collected at FHWA G.S. Vincent Wind Tunnel, McLean, VA.)

can describe a structure that has nonlinear time-varying mechanics. j is a time-step increment, $[M]$ is a constant mass matrix, $[C(y_j)]$ is a damping matrix with time-varying parameters y_j, $[K(x_j)]$ is a stiffness matrix with time-varying parameters x_j, $\{\Delta u_j\}$ is the incremental displacement vector, and $\{\Delta f_j\}$ is the incremental force vector. Allen et al. demonstrated the sensitivity of vibrations of bolted joints to looseness as a test case in the use of SPR methods for damage detection [147]. Similarly, Mita and Taniguchi demonstrated the effectiveness of using SPR methods for identifying damage to bolts in a multiple bolt connection [432]. The method used piezoelectric transducers to induce elastic waves that interacted with the bolts. Features for the SPR algorithm included a variety of spectral measures of the received signals, such as spectral peak half-width.

Phase plane plots provide good insights into the behavior of systems that move in response to initial conditions, for example, free vibrations, but are less useful for describing systems under the influence of external forces. Modified phase plane plots, known as *Poincare maps*, are usually better tools for describing the behavior of dynamical systems with periodic external forces or variations in system parameters [430]. A Poincare map is an input-synchronized sampling of the phase plane behavior of the system. For a selected phase angle of the input, the Poincare map plots the phase space configuration as a single point. Each repeated input cycle at the same phase angle plots another point in the phase plane. Since the responses of linear systems to harmonic forcing are also harmonic at the same frequency, the Poincare map of a linear system appears as a single point. Nonlinear systems with sub- and superharmonc response behavior produce Poincare maps with multiple points. The Poincare maps of chaotic systems have complicated infinite-point distribution patterns. Such patterns may be sensitive to damage that alters the nonlinear dynamics of the system [433].

The notion that trajectories in a phase plane describe the mechanics of SDOF systems readily extends into higher-dimensional phase spaces for MDOF systems. Cusamano and Chelidze, and Cusumano et al. examined systems with damage that appears as a slowly growing change in the overall system properties [434–436]. The method takes advantage of separate time scales: a slow time scale that covers the change in system properties, and a fast time scale that is needed to describe system dynamics. The different time scales of the evolution of the system properties allow for writing the overall system properties in a hierarchical format

$$\dot{\mathbf{x}} = \mathbf{F}\,[\mathbf{x}; u(\varphi), t]\,, \tag{6.111}$$

$$\dot{\varphi} = \varepsilon g(\mathbf{x}, \varphi, t), \tag{6.112}$$

where **x** is a vector of state variables, F is a nonlinear multivariable state transition function, and u is a function of the system parameters. Damage alters the system parameters and causes the damage to warp the phase space trajectories in a measurable manner. An assumption of damage growth that follows a power law leads to a straightforward prognosis calculation of time of failure. Epureanu et al. used a somewhat similar approach to detect damage in plates subjected to supersonic flow [437]. The combination of nonlinear structural forces due to in-plane stresses with piston-theory supersonic aerodynamics formed a nonlinear aeroelastic system [438]. Damage to the structure in the form of a reduced stiffness changed the attractors and chaotic behavior in the phase plane and in Poincare maps. Todd et al. used nonlinear attractor mapping to detect bolt preload loss [412].

Nonlinear systems often exhibit more complicated and exotic behaviors than that of simple attractors and bifurcations. Chaotic motion under the influence of a strange attractor is an example [430,439,440]. The nonlinear mechanics of chaotic systems are such that certain regions of phase space have trajectories with the interesting property that neighboring points diverge from one another while other regions have nonlinearities that tend to cause the system to return to the diverging region. This combination of divergence and return causes similar but nonrepeating behaviors. The behavior at first glance appears to be quite random, but instead has a significant underlying structure. Mechanical devices associated with games of chance follow this pattern. An example is the shaking of a pair of dice in a cup in order to randomize the state prior to tossing. The impact of the dice on the walls of the cup causes rotations that are highly sensitive to the configuration (position in state space) prior to impact. The combination of the walls and the shaking tends to return the dice to similar regions of phase space. Slight deviations in initial conditions can cause vastly different outcomes. Truly identical initial conditions and shaking actions should cause the same outcome. (Random effects due to quantum mechanics and thermal fluctuations prevent true repeats from ever occurring.)

The *Lyapunov exponent* measures the divergence of closely matched points in phase space [441]. The concept is that locally linearized homogeneous differential equations of the form of Equation 6.105 represent the local evolution of a dynamical system in phase space, especially at nonsingular points. Defining $\delta x(t)$ as the separation between two neighboring points in an N-dimensional phase space and λ_i as the eigenvalues of the state transition matrix $[A]$, the separation evolves in the near term as an exponential equation in a singular value decomposition form with constant vectors k_i, that is,

$$\delta x(t) \approx \sum_{i=1}^{N} \delta x(0) k_i^T e^{\lambda_i t}. \tag{6.113}$$

In Hamiltonian systems with phase space volume preserving flows, the λ_i sum to zero. Typically it is the maximum Lyapunov exponent, λ_{max}, that dominates the separation rate. Positive values of λ_{max} combined with a negative sum of λ_i can correspond to chaotic systems.

The SHM literature contains a modest number of citations regarding the identification of chaotic behavior and the potential usage in SHM activities. These generally fall into two categories. The first covers cases of systems with chaotic mechanical behavior. Livingston and Jin identified the Lyapunov exponents for a cable-stayed bridge that suffered from ambient traffic and wind–rain vibrations [442]. Some of the higher-order modes exhibited positive exponents while the sum over all of the exponents remained negative—a clear characteristic of chaotic motions. A crack that opens and closes as a structure vibrates causes a nonlinear change in the stiffness. The response of a structure with such a nonlinearity to harmonic forcing can under the appropriate circumstances be chaotic. Cracked rotors with consistent crack opening and closing combined with harmonic input may be prime candidates for chaotic vibration crack identification [443]. The second category uses chaotic dynamics as a tool to probe the health and state of a structural system. Since the Lyapunov exponent describes some of the fundamental qualitative aspects of the behavior of dynamical systems, Todd et al. used the Lyapunov exponent as a statistical feature to detect the dimensionality of the dynamic attractor under chaotic excitation [444]. Jin et al. tracked the Lyapunov exponent as a means of detecting damage [445]. Seaver et al. confirmed some of these predictions by measuring the response of a plate with weld defects to forced chaotic Lorenz oscillator vibration and then estimating the Lyapunov dimension from the correlation dimension [446]. This use of higher-order statistics in the analysis is similar to ICA approaches. Similarly, Moniz et al. also used a Lorenz oscillator to test for damage in a composite beam [447]. Nichols et al. used a chaotic signal as the excitation for system identification [448]. The concept is that chaotic signals are broad banded, yet deterministic and controllable. Using a chaotic signal that can maximally excite the modal properties of the undamaged structure forms a test signal that may be highly sensitive to modal property-affecting damage. Han et al. also found that chaotic signals were favorable for damage detection in sonic IR imaging applications [449].

Nonlinear dynamics and vibrations of structures can be quite complicated. Nonetheless, the dynamics may exhibit patterns where system parameters (such as damage or bolt preload) affect the nonlinear behavior in repeatable and predictable manners. Parameter changes usually cause a smooth change in the behavior of the system. Evolving parameter changes cause singularities to appear in the dynamical behavior of the system. In many cases, these singularities are of a lower-order dimension, such as bifurcations, cusps, and limit cycles [450,451].

The *center manifold theorem* is a dynamic version of the notion that most system singularities are lower order, even if the system has a large number of degrees of freedom [452]. Lower-order dynamic system singularities include limit cycle and pitchfork bifurcations. The singularity at the point of emergence of a limit cycle is known as a Hopf bifurcation. The emergence of vortex shedding limit cycle behavior with a variation in wind speed is an example of a Hopf bifurcation (Figure 6.46). Brown and Adams demonstrated the use of the center manifold theorem to describe changes in the vibration modal properties in a structure with evolving damage [453]. Shukla notes that many biological sensors, such as cilia-based sound sensors in the cochlea, use self-tuned Hopf bifurcations [454]. The nonlinearities in these sensors enable the detection of vibratory signals over wide dynamic ranges, in spite of the large viscous forces at play. This may inspire designs for new sensors.

Methods of *catastrophe theory* describe singular phenomena where a potential function governs the mechanics of a system (such as elastic potential energy) [455]. In these situations, the potential energy $V(\mathbf{x})$ of a multiple DOF system with a vector of states $\mathbf{x}$ and vector of control parameters $\mathbf{c}$, and with an equilibrium point at $\mathbf{x} = \mathbf{0}$ expands in a Taylor Series as

$$V(\mathbf{x}, \mathbf{c}) = V_0(\mathbf{c}) + \mathbf{V}_1^{\mathrm{T}}(\mathbf{c})\mathbf{x} + \frac{1}{2}\mathbf{x}^{\mathrm{T}}\mathbf{V}_2(\mathbf{c})\mathbf{x} + \cdots . \tag{6.114}$$

Since potential energy is arbitrary to within an additive constant, it is possible to set the first term, V_0, to zero. At a point of equilibrium, the second term vanishes, that is,

$$\frac{\partial V(\mathbf{x}, \mathbf{c})}{\partial \mathbf{x}} = \mathbf{0} \tag{6.115}$$

and

$$\mathbf{V}_1(\mathbf{c}) = \mathbf{0}. \tag{6.116}$$

The third term is a quadratic form based on the *Hessian* or *tangent stiffness* matrix $\mathbf{V}_2$. This is usually the lowest-order term to survive following a Taylor series expansion about equilibrium. As such, the algebraic nature of $\mathbf{V}_2$ governs much of the qualitative behavior of the system. The Hessian matrix is usually symmetric. This implies that the eigenvalues are real. Similar to the case of a first-order linear dynamical system, if all of the eigenvalues are positive, the system is stable when subjected to small perturbations. If some of the eigenvalues are negative, then the system is unstable with respect to small perturbations. If none of the eigenvalues equal zero, the equilibrium point is called a *Morse critical point*. If at least one equals zero, the equilibrium point is called a *non-Morse critical point*. In these cases higher-order terms in the Taylor series expansion of the potential play a significant role in describing the system behavior. The behavior about a Morse critical point is unaffected by small perturbations in the control parameters. At a non-Morse critical point, small changes in the control parameters can dramatically alter the behavior of the system. Jumps, cusps, and other catastrophes involving only a few coordinates occur. These lower-order singularities tend to dominate the observed behavior of systems, even though the systems may have high numbers of degrees of freedom. Such effects can appear in structural mechanics and may be the basis of SHM efforts. Common examples are buckling and snap-through phenomena, for example, with Mises truss [456,457]. Figure 6.48 shows how a cubic nonlinearity can cause a cusp-like catastrophe in a bolted joint with bolt tightness as a variable parameter [453].

6.16.2.4 Probability Density and Information-Based Methods

The application of concepts from probability, statistics, and information theories can provide added insights into the behavior of nonlinear systems. A variety of circumstances cause

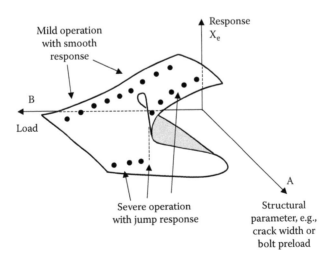

FIGURE 6.48 Cubic nonlinearity with smooth operation under mild loading and cusp bifurcation catastrophe under severe loading corresponding to nonlinear damage mechanics due to cracks or loosening bolts. (Adapted from Brown RL and Adams DE. 2003. *J Sound Vib*, 262(3), 591–611.)

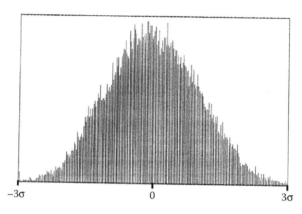

FIGURE 6.49 Histogram of Deer Isle bridge section model wind tunnel aeroelastic vertical displacements when subjected to turbulent buffeting. (Data collected at FHWA G.S. Vincent Wind Tunnel, McLean, VA.)

system signals to have non-Gaussian amplitude histograms even though the inputs are Gaussian. These include amplitude-dependent nonlinearities, fixed point attractors, limit cycles, and chaotic behavior. Figures 6.49 and 6.50 show histograms of linear and nonlinear aeroelastic behavior observed in a wind tunnel [458]. Estimating the PDF from experimental data can give an insight into the underlying nonlinear processes. Fasel et al. demonstrated that analyzing the PDFs (histograms) of structural vibration data extracted by ARX methods with extreme value statistics readily detected the presence of mechanical nonlinearities due to bolt loosening [459]. Park et al. also combined ARX methods with extreme value statistics to detect damage with EMI testing of structural dynamics [460].

Logarithmic measures of probability, that is, entropies, can also quantify nonlinear structural behaviors. Nichols et al. found that the time-delay transfer entropy was an effective measure of nonlinear interactions in two structural dynamics time histories, and was superior to time-delay mutual information statistics [461].

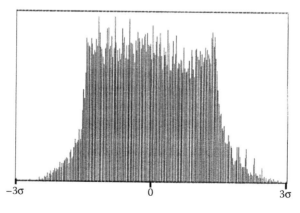

FIGURE 6.50 Histogram of Deer Isle bridge section model wind tunnel aeroelastic vertical displacements when subjected to vortex-shedding. (Data collected at FHWA G.S. Vincent Wind Tunnel, McLean, VA.)

References

1. Sohn H, Farrar CR, Hemez FM, Shunk DD, Stinemates DW, and Nadler BR. 2003. A review of structural health monitoring literature: 1996–2001. *Los Alamos National Laboratory Report,* LA-13976-MS.
2. Mayer D, Herold S, and Azroot H. 2004. Structural health monitoring of smart structures using pattern recognition and adaptive digital filters. *Proc 2nd European Workshop on Structural Health Monitoring.* DEStech, Munich.
3. Brownjohn JM, Moyo P, Omenzetter P, and Chakraoorty S. 2004. Interpreting data from bridge performance and health monitoring systems. *Proc IABMAS'04 Bridge Maintenance, Safety, Management and Cost,* Kyoto. Taylor & Francis, London.
4. McCann D, Jones NP, and Ellis JH. 1998. Toward consideration of the value of information in structural performance assessment. In *Structural Engineering Worldwide.* GL Fenves, NK Srivastava, AH Ang, RG Domer (eds.). Elsevier, Amsterdam, T216-6.
5. Hyvarinen A, Karhunen J, and Oja E. 2001. *Independent Component Analysis.* Wiley, New York.
6. Mosteller F and Tukey JW. 1977. *Data Analysis and Regression.* Addison-Wesley, Reading, MA.
7. Duda RO, Hart PE, and Stork DG. 2004. *Pattern Classification,* 2nd Ed. Wiley, New York.
8. Hall DL. 1992. *Mathematical Techniques in Multisensor Data Fusion.* Artech House, Norwood.
9. Shewhart WA. 1986. *Statistical Method from the Viewpoint of Quality Control.* Dover Publications, Mineola, NY.
10. Habel WR. 2004. Fiber optic sensors for deformation measurements: Criteria and method to put them to the best possible use. *Smart Sensor Technology Measurement Systems,* SPIE 5384.
11. Mandel J. 1964. *The Statistical Analysis of Experimental Data.* Dover Publications, Mineola, NY.
12. Taylor BN and Kuyatt CE. 1994. Guidelines for evaluating and expressing the uncertainty of NIST measurement results. *NIST Technical Note* 1297, Gaithersburg.
13. Liu T, Guille M, and Sullivan JP. 2001. Accuracy of pressure sensitive paint. *AIAA J,* 39(1), 103–112.
14. Strauss A, Pascale G, Bonfiglioli B, Bergmeister K, and Di Muro P. 2004. Probability and sensitivity analysis of strain measurements in FRP. *Struct Control Health Monit,* 11, 55–74.
15. Hemez FM, Robertson AN, and Cundy AL. 2003. Uncertainty quantification and model validation for damage prognosis.In *Structural Health Monitoring,* FK Chang (ed.). DEStech, Lancaster.
16. Papoulis A. 1962. *The Fourier Integral and Its Applications.* McGraw-Hill, New York.
17. Tukey JW. 1977. *Exploratory Data Analysis.* Addison-Wesley, Reading, MA.
18. Hartwig F and Dearing BE. 1979. *Exploratory Data Analysis.* Sage Publications, Newbury Park.
19. Tufte ER. 1983. *The Visual Display of Quantitative Information.* Graphics Press, Cheshire.
20. Livingston RA and Jin S. 2003. Fractal interpolation method for maximum entropy analysis of chaotic structural response data. In *Structural Health Monitoring,* FK Chang (ed.). DEStech, Lancaster.
21. Box GP, Hunter WG, and Hunter JS. 1978. *Statistics for Experimenters.* Wiley, New York.
22. Guttman I, Wilks SS, and Hunter JS. 1982. *Introductory Engineering Statistics,* 3rd Ed. Wiley, New York.
23. Gumbel EJ. 1958. *Statistics of Extremes.* Dover Publications, Mineola, NY.
24. Bolotin VV. 1989. *Prediction of Service Life for Machines and Structures.* ASME Press, New York.
25. Simiu E and Scanlan RH. 1996. *Wind Effects on Structures,* 3rd Ed. Wiley, New York.
26. Kozine I. 2006. Updating probability intervals with the beta-Bernoulli model. *NATO Advanced Research Workshop, Computational Models of Risks to Infrastructure,* Primosten, Croatia.
27. Hand D, Mannila H, and Smyth P. 2001. *Principles of Data Mining.* MIT Press, Cambridge.
28. Woodward P. 2005. BugsXLA: Bayes for the common man. *J Statist Software,* 14(5), 1–18.
29. Tang L, Kacprzynski GJ, Roemer MJ, and Walsh B. 2007. A graphical modeling and analysis tool for Bayesian based structural reliability analysis. In *Structural Health Monitoring,* FK Chang (ed.). DEStech, Lancaster.

30. Glaser RE, Lee CL, Nitao JJ, Hickling TL, and Hanley WG. 2007. Markov chain Monte Carlo-based method for flaw detection in beams. *J Eng Mech*, 133(12), 1258–1267.

31. Kayser JR and Nowak AS. 1987. Evaluation of corroded steel bridges. *Proc ASCE Structures Congress*, Bridges and Transmission Line Structures, L Tall, (ed.). ASCE Publication, 1987, Orlando, FL, pp. 35–46.

32. Papadimitriou C. 2004. Bayesian inference applied to structural damage detection. *Proc 2nd European Workshop on Structural Health Monitoring*. DEStech, Munich.

33. Beck JL and Au SK. 2002. Bayesian updating of structural models and reliability using Markov chain Monte Carlo simulation. *J Eng Mech*, 128(4), 380–391.

34. Ching J and Beck JL. 2004. Bayesian analysis of the phase II IASCASCE structural health monitoring experimental benchmark data. *J Eng Mech*, 130(10), 1233–1244.

35. Shalaby A and Reggin A. 2006. Maximum likelihood classification of pavement surface condition data in urban road networks. Paper no. 06-0185, 86th Annual Meeting, TRB, Washington, DC.

36. Varshney PK. 1997. *Distributed Detection and Data Fusion*. Springer, New York.

37. Webb A. 2002. *Statistical Pattern Recognition*, 2nd Ed. Wiley, West Sussex.

38. Michaels JE, Cobb AC, and Michaels TE. 2004. A comparison of feature-based classifiers for ultrasonic structural health monitoring. *Health Monitoring and Smart Nondestructive Evaluation of Structural and Biological Systems III*, SPIE 5394.

39. Aldrin JC and Knopp JS. 2005. Method for crack characterization with noise invariance for eddy current inspection of fastener sites. *Rev Prog QNDE*, 25, 315–322.

40. Derriso M, Olson S, Braisted W, DeSimio M, Rosenstengel J, and Brown K. 2004. Detection of fastener failure in a thermal protection system. *Smart Electronic, MEMS, BioMEMS, and Nanotechnology*, SPIE 5389.

41. Fritzen CP and Mengelkamp G. 2003. A Kalman filter approach to the detection of structural damage. In *Structural Health Monitoring*, FK Chang (ed.). DEStech, Lancaster.

42. Poulimenos AG and Fassos SD. 2004. Vibration-based on-line fault detection in non-stationary structural systems *via* a statistical model based method. *Proc 2nd European Workshop on Structural Health Monitoring*. DEStech, Munich.

43. Iwasaki A, Kuwahara H, Todoroki A, and Sugiya T. 2004. Automatic damage diagnosis of existing civil structure using statistical method. *Proc 2nd European Workshop on Structural Health Monitoring*. DEStech, Munich.

44. Iwasaki A, Todoroki A, Sugiya T, and Sakai S. 2004. Damage diagnosis for SHM of existing civil structure with statistical diagnostic method. *Health Monitoring and Smart Nondestructive Evaluation of Structural and Biological Systems III*, SPIE 5394.

45. Shewhart WA. 1931. *Economic Control of Quality of Manufactured Product*. van Nostrand, New York.

46. Lee SW and Lee SG. 1998. Instrumented monitoring of prestressed segmental bridge on ambient traffic. In *Structural Engineering Worldwide*. GL Fenves, NK Srivastava, AH Ang, RG Domer (eds.). Elsevier, Amsterdam, T116-2.

47. Allen DW, Sohn H, Worden K, and Farrar CR. 2002. Utilizing the sequential probability ratio test for building joint monitoring. *Nondestructive Evaluation and Health Monitoring of Aerospace Materials and Civil Infrastructures*, SPIE 4704.

48. Wang Z and Ong KC. 2007. Structural health monitoring of reinforced concrete frames for progressive damage using Hotelling's T^2 control. In *Structural Health Monitoring*, FK Chang (ed.). DEStech, Lancaster.

49. Fu G, Moosa AG, and Peng J. 2001. Optical nondestructive evaluation for structures. *Res Nondes Eval*, 13(1), 37–59.

50. Lloyd GM, Wang ML, Wang X, and Love J. 2003. Recommendations for intelligent bridge monitoring systems: Architecture and temperature-compensated bootstrap analysis. *Smart Systems and Nondestructive Evaluation for Civil Infrastructures*, SPIE 5057.

51. Omenzetter P, Brownjohn JM, and Moyo P. 2003. Identification of unusual events in multi-channel bridge monitoring data using wavelet transform and outlier analysis. *Smart Systems and Nondestructive Evaluation for Civil Infrastructures*, SPIE 5057.

52. Mattson SG and Pandit SM. 2006. Damage detection and localization based on outlying residuals. *Smart Mater Struct*, 15, 1801–1810.

53. Ko JM, Chak KK, Wang JY, Ni YQ, and Chan TH. 2003. Formulation of uncertainty model relating modal parameters and environmental factors by using long-term monitoring data. *Smart Systems and Nondestructive Evaluation for Civil Infrastructures*, SPIE 5057.

54. Sohn H, Worden K, and Farrar CR. 2001. Novelty Detection under Changing Environmental Conditions. *Smart Systems for Bridges, Structures, and Highways*, SPIE 4330.

55. Schwabacher M. 2005. Machine learning for rocket propulsion health monitoring. SAE 2005-01-3370, *World Aerospace Congress Transitions*.

56. Adami C. 2004. Information theory in molecular biology. *Phys Life Rev*, 1, 3–22.

57. Kullback S. 1968. *Information Theory and Statistics*. Dover Publications, Mineola, NY.

58. Jiang X and Mahadevan S. 2007. Bayesian wavelet probabilistic method for nonparametric damage detection of structures. In *Structural Health Monitoring*, FK Chang (ed.). DEStech, Lancaster.

59. Trendafilova I, Heylen W, and Van Brussel H. 2001. Measurement point selection in damage detection using the mutual information concept. *Smart Mater Struct*, 10, 528–533.

60. Shannon CE and Weaver W. 1949. *The Mathematical Theory of Communication*. University of Illinois Press, Urbana and Chicago.

61. Schrödinger E. 1944. *What is Life?* Cambridge University Press, Cambridge.

62. Oppenheim AV and Schafer RW. 1975. *Digital Signal Processing*. Prentice-Hall, Englewood Cliffs, NJ.

63. Zheng H and Mita A. 2007. Two-stage damage diagnosis based on the distance between ARMA models and pre-whitening filters. *Smart Mater Struct*, 16, 1829–1836.

64. Taghvaei M, Beck SM, and Staszewski WJ. 2007. Leak detection in pipeline networks using low-profile piezoceramic transducers. *Struct Control Health Monit*, 14, 1063–1082.

65. Thompson PD. 2004. Bridge management systems: Where should we go from here?. *Proc IABMAS'04 Bridge Maintenance, Safety, Management and Cost*, Kyoto. Taylor & Francis, London.

66. Wong JM and Stojadinovic B. 2005. Metadata and network API aspects of a framework for storing and retrieving civil infrastructure monitoring data. *Sensors and Smart Structures Technologies for Civil, Mechanical, and Aerospace Systems*, SPIE 5765.

67. Martin RJ. 2000. A metric for ARMA processes. *IEEE Trans Signal Process*, 48(4), 1164–1170.

68. Fillipenko A, Brown S, and Neal A. 2002. Vibration analysis for predictive maintenance of rotating machines. US Patent 6,370,957.

69. Fan X, Zuo MJ, and Wang X. 2006. Identification of weak ultrasonic signals in testing of metallic materials using wavelet transform. *Smart Mater Struct*, 15, 1531–1539.

70. Haritos N and Owen JS. 2004. The use of vibration data for damage detection in bridges: A comparison of system identification and pattern recognition approaches. *Intl J Struct Health Monit*, 3(2), 141–164.

71. Gul M and Catbas N. 2007. Identification of structural changes by using statistical pattern recognition. In *Structural Health Monitoring*, FK Chang (ed.). DEStech, Lancaster.

72. Sakellariou JS, Petsounis KA, and Fassois SD. 2004. A functional model based method for fault detection and identification in stochastic structural systems. *Proc 2nd European Workshop on Structural Health Monitoring*. DEStech, Munich.

73. Yan J, Lee J, and Pan YC. 2005. Introduction of watchdog prognostics agent and its application to elevator hoistway performance assessment. *J Chin Inst Ind Eng*, 22(1), 56–63.

74. Friedman J, Hastie T, and Tibshirani R. 2000. Additive logistic regression: A statistical view of boosting. *The Ann Statist*, 28(2), 337–407.

75. Le Clerc J, Haywood J, Stazewski W, and Worden K. 2004. Impact detection in an aircraft composite panel—a neural network approach. *Proc 2nd European Workshop on Structural Health Monitoring*. DEStech, Munich.

76. Shinozuka M and Rejaie A. 2001. Damage assessment from remotely sensed images using PCA. *Smart Systems for Bridges, Structures, and Highways*, SPIE 4330-01.

77. De Boe P, Al-Min Y, and Golinval JC. 2003. Substructure damage detection by principal component analysis: Application to environmental vibration testing. In *Structural Health Monitoring*, FK Chang (ed.). DEStech, Lancaster.

78. Nedushan BA and Chouinard LE. 2003. Multivariable statistical analysis of monitoring data for large constructed facilities. In *Structural Health Monitoring*, FK Chang (ed.). DEStech, Lancaster.

79. Hu LP and Han DJ. 2006. Vibration-based damage detection for bridges considering the influence of changing environment using NLPCA. *Proc 4th China–Japan–US Symp on Structural Control and Monitoring*, Hangzhou.

80. Teng J, Yan CW, Ma BT, and Liu HJ. 2006. Study on the identification of fluctuating wind pressure in SHM for large-span structures. *Proc 4th China–Japan–US Symp on Structural Control and Monitoring*, Hangzhou.

81. Kerschen G, De Boe P, Golinval JC, and Worden K. 2005. Sensor validation using principal component analysis. *Smart Mater Struct*, 14, 36–42.

82. Amari SI and Chichocki A. 1998. Adaptive blind signal processing-neural network approaches. *IEEE*, 86(10), 2026–2048.

83. Li X, Du R, and Guan XP. 2004. Utilization of information maximum for condition monitoring with applications in a machining process and a water pump. *IEEE/ASME Trans Mechatronics*, 9(4), 711–714.

84. Zang C, Friswell MI, and Imregun M. 2004. Structural damage detection using independent component analysis. *Struct Health Monit*, 3(1), 69–83.

85. Abujarad F and Omar A. 2006. GPR detection of landmine by FastICA. *Proc 11th Intl Conf on Ground Penetrating Radar*, Columbus.

86. Fan Y and Zheng G. 2007. Blind separation of gearbox vibration signal and it application to fault diagnosis. In *Structural Health Monitoring*, FK Chang (ed.). DEStech, Lancaster.

87. He R, Fox J, and Krag M. 2004. Private communication to the author, University of Vermont.

88. Tanaka K. 1997. *An Introduction to Fuzzy Logic for Practical Applications*. Springer, New York.

89. Kleiner Y, Rajani B, and Sadiq R. 2006. Failure risk management of buried infrastructure using fuzzy-based techniques. *J Water Supply Res Tech: Aqua*, 55(2), 81–94.

90. Kleiner Y, Sadiq R, and Rajani B. 2006. Modelling the deterioration of buried infrastructure as a fuzzy markov process. *J Water Supply Res Tech: Aqua*, 55(2), 67–80.

91. Reda Taha MM, Noureldin A, and Ross TJ. 2004. A fuzzy-aided wavelet damage recognition for intelligent structural health monitoring. *Proc 2nd European Workshop on Structural Health Monitoring*. DEStech, Munich.

92. Meegoda JN, Juliano TM, and Banerjee AA. 2006. Framework for automatic condition assessment of culverts. Paper no. 06-2414, 86th Annual Meeting, TRB, Washington, DC.

93. Luo Z and Chou EY. 2006. Pavement condition prediction using clusterwise regression. Paper no. 06-0637, 86th Annual Meeting, TRB, Washington, DC.

94. Sun XY, Wang HL, and Huang CK. 2007. Reliability analysis of existing reinforced concrete bridges for serviceability limit state of maximum crack width. *Proc Intl Conf on Health Monitoring Structure and Material Environment*, Nanjing.

95. Li J, Song B, Wu L, and Zhou H. 2007. Damage assessment of concrete bridge slab by field monitoring data. *Proc Intl Conf on Health Monitoring Structure and Material Environment*, Nanjing.

96. Wen CM, Hung SL, Huang CS, and Jan JC. 2005. Unsupervised fuzzy neural networks for damage detection of structures. *Struct Control Health Monit*, 14, 144–161.

97. Maser KR, Horschel MG, and Grivas DA. 2002. Integration of ground penetrating radar and infrared thermography for bridge deck assessment. *Proc Structural Materials Technology V, an NDT Conference*, Cincinnati.

98. Weil GJ. 1995. Non-destructive, remote sensing technologies for locating subsurface anomalies on railroad track beds. *Nondestructive Evaluation of Aging Railroads*, SPIE 2458.

99. Zheng YH and Ng KE. 2006. Detection of voids in tendon ducts of post-tensioned U-beams for an existing bridge. *NDE Conf on Civil Engineering*, American Society for Nondestructive Testing, St. Louis.

100. Lorenzi A, Filho LC, Campagnolo J, Streider A J, and Caetano LF. 2004. Using non-destructive testing for monitoring concrete elements of sizable structural members. *Proc Structural Materials Technology VI, an NDT Conference*, Buffalo.

101. Hammons MI Von Quintus H, Geary GM, Wu PY, and Jared DM. 2006. Detection of stripping in hot mix asphalt. Paper no. 06-0658, 86th Annual Meeting, TRB, Washington, DC.

102. Wiggenhauser H, Gardei A, Kohl C, Maierhofer C, Krause M, and Streicher D. 2004. Development and application of echo-methods for NDE of post tensioned concrete bridges. *Proc IABMAS'04 Bridge Maintenance, Safety, Management and Cost*, Kyoto. Taylor & Francis, London.

103. Hertlein BH and Calhoun BL. 2006. Field experience with nondestructive tools for bridge deck condition assessment. *NDE Conf on Civil Engineering*, American Society for Nondestructive Testing, St. Louis.

104. Cao H. 2006. Implementation of NDT techniques in an underground tunnel investigation. *NDE Conf on Civil Engineering*, American Society for Nondestructive Testing, St. Louis.

105. Parrillo R, Sanati G, and Roberts R. 2006. Integrated FWD and GPR pavement thickness data acquisition and storage. *NDE Conf on Civil Engineering*, American Society for Nondestructive Testing, St. Louis.

106. Moropoulou A, Avdelidis NP, Koui M, Aggelopoulos A, and Karmis P. 2002. Infrared thermography and ground penetrating radar for airport pavements assessment. *Nondes Test Eval*, 18(1), 37–42.

107. Lasser RS, Lasser ME, and Gurney JW. 2005. Apparatus for multimodal plane wave ultrasound imaging. US Patent 6,971,991.

108. Davis AG, Olson CA, and Michols KA. 2001. Evaluation of historic reinforced concrete bridges. *Proc ASCE Structures Congress*, Structures—A Structural Engineering Odyssey Proceedings of Structures Congress 2001, Washington, DC, PC Chang (ed.). ASCE.

109. Danay A, Adeghe L, and Hindy A. 1993. Diagnosis of the cause of the progressive concrete deformations at Saunders dam. *Concrete Intl*, 15(9), 25–33.

110. Huang J, Rose J, Gordon J, and Boucher R. 2005. Structural sensor testing for space vehicle applications. *Industrial and Commercial Applications of Smart Structures Technologies*, SPIE 5762.

111. Melvin L, Childers B, Rogowski R, Prosser W, Moore J, Frogatt M, Allison S, Wu MC, Bly J, Aude C, Bouvier C, Zisk E, Enright E, Cassadaban Z, Reightler R, Sirkis J, Tang I, Peng T, Wegreich R, Garbos R, Mouyos W, Aibel D, and Bodan P. 1997. Integrated vehicle health management for aerospace vehicles. In *Structural Health Monitoring Current Status and Perspectives*. F-K Chang, (ed.). Technomic, Lancaster.

112. Shepard D and Hollos A. 2005. Combined sensor signals for structural health monitoring applications. In *Structural Health Monitoring*, FK Chang (ed.). DEStech, Lancaster.

113. Liu Z, Forsyth DS, Safizadeh S, Genest M, Fahr A, and Marincak A. 2005. Fusion of visual and eddy current inspection results for the evaluation of corrosion damage in aircraft lap joints. *Health Monitoring and Smart Nondestructive Evaluation of Structural and Biological Systems IV*, SPIE 5768.

114. Levner E. 2006. Environmental risk ranking: Theory and applications for emergency planning. *Computational Models of Risks to Infrastructure*. NATO Advanced Research Workshop, Primosten, Croatia.

115. Chun X, Weilian Q, and Dongmei T. 2005. An application of data fusion technology in structural health monitoring and damage identification. *Smart Sensor Technology and Measurement Systems*, SPIE 5758.

116. Su Z, Wang X, Chen Z, and Ye L. 2007. A hierarchical data fusion scheme for identifying multi-damage in composite structures with a built-in sensor network. *Smart Mater Struct*, 16, 2067–2079.

117. Sentz K and Ferson S. 2002. Combination of evidence in Dempster–Shafer theory. *Technical Report*, Sandia National Laboratories, SAND 2002-0835.

118. Dempster AP. 1968. A generalization of Bayesian inference. *J Roy Statist Soc*, Ser B, 30(2), 205–247.

119. Sun L, Srivastava RP, and Mock TJ. 2006. An information systems security risk assessment model under Dempster–Shafer theory of belief functions. *J Manage Inf Syst*, 22(4), 109–142.

120. Shafer G. 1981. Constructive probability. *Synthese*, 48, 1–60.

121. Shafer G. 1990. Perspectives on the theory and practice of belief functions. *Intl J Approx Reason*, 4, 323–362.

122. Costlow, T. 2005. Safety drives sensor growth. *Autom Eng Intl*, 113, 65–70.

123. Fu Y, Zhu Y, Chen W, Huang S, and Bennett KD. 2003. Remote state monitoring system for Hongcaofang Crossroad Bridge. *Smart Systems and Nondestructive Evaluation for Civil Infrastructures*, SPIE 5057.

124. Adams BJ, Huyck CK, Mio M, Cho S, Ghosh S, Chung HC, Eguchi RT, Houshmand B, Shinozuka M, and Mansouri B. 2004. The bam (Iran) earthquake of December 26, 2003: Preliminary reconnaissance using remotely sensed data and VIEWS system. *MCEER Earthquake Investigation Report*, Bam12-26-03.

125. Smyth AW. 2004. The potential of GPS and other displacement sensing for enhancing acceleration sensor monitoring array data by solving low frequency integration problems. *Proc IABMAS'04 Bridge Maintenance, Safety, Management and Cost*, Kyoto. Taylor & Francis, London.

126. Wagner JF and Oertel T. 2007. Generalizing integrated navigation systems for structural health monitoring. In *Structural Health Monitoring*, FK Chang (ed.). DEStech, Lancaster.

127. Hundhausen RJ, Adams DE, Derriso M, Kukuchek P, and Alloway R. 2004. Loads, damage identification and NDE/SHM data fusion in standoff thermal protection systems using passive vibration-based methods. *Proc 2nd European Workshop on Structural Health Monitoring*. DEStech, Munich.

128. Tsutsui H, Kawamata A, Kimoto J, Isoe A, Hirose Y, Sanda T, and Takeda N. 2003. Impact damage detection system using small-diameter optical fiber sensors wavily embedded in CFRP laminate structures. *Industrial and Commercial Applications of Smart Structures Technologies*, SPIE 5054.

129. Camerino M and Peters K. 2003. Flexibility approach for damage localization suitable for multi-scale sensor fusion. *Smart Sensor Technology and Measurement Systems*, SPIE 5050.

130. Kumar S. 2004. Vibration and oil analysis detect a gearbox problem. *Vibrations*, 20(3), 7–13.

131. Goebel K, Eklund N, and Bonanni P. 2005. Fusing competing prediction algorithms for prognostics. *Proc IEEE Aerospace Conference*, Big Sky, MT.

132. Koks D and Challa S. 2003. *An Introduction to Bayesian and Dempster–Shafer Data Fusion*, DSTO-TR-1436, Australia Defence Science and Technology Organisation, Edinburgh, Australia.

133. Liu Z, Forsyth DS, Safizadeh MS, Lepine BA, and Fahr A. 2003. Quantitative interpretation of multi-frequency eddy current data by using data fusion approaches. *Nondestructive Evaluation and Health Monitoring of Aerospace Materials and Composites II*, SPIE 5046.

134. Guo HY. 2006. Structural damage detection using information fusion technique. *Mech Syst Signal Process*, 20, 1173–1188.

135. Osegueda RA, Seelam SR, Mulupuru B, and Kreinovich V. 2003. Statistical and Dempster–Shafer techniques in testing structural integrity of aerospace structures. *Smart Nondestructive Evaluation and Health Monitoring of Structural and Biological Systems II*, SPIE 5047.

136. Liang Z, Chen W, Wang D, Fu Y, and Zhu YA. 2005. Technique for damage diagnosis of bridge structure based on information fusion. *Nondestructive Evaluation and Health Monitoring of Aerospace Materials, Composites and Civil Infrastructure IV*, SPIE 5767.

137. Bao YQ, Li H, and Ou JP. 2006. Study of two-stage structural damage detection combining both global and local information. *Proc 4th China–Japan–US Symp on Structural Control and Monitoring*, Hangzhou.

138. Farrar CR, Duffey TA, Doebling SW, and Nix DA. 1999. A statistical pattern recognition paradigm for vibration-based structural health monitoring. *Proc 2nd Intl Workshop on Structural Health Monitoring*, Stanford. Technomic, Lancaster.

139. Emamian V, Kaveh M, Tewfik AH, Shi Z, Jacobs LJ, and Jarzynski J. 2003. Robust clustering of acoustic emission signals using neural networks and signal subspace projections. *EURASIP J Appl Signal Process*, 3, 276–286.

140. Havira RM and Chen J. 1995. High speed rail flaw pattern recognition and classification. *Nondestructive Evaluation of Aging Railroads*, SPIE 2458.

141. Li Q, Shan D, and Zhang Y. 2006. Damage detection of bridge structures based on SOFM neural networks. *Proc 4th China–Japan–US Symp on Structural Control and Monitoring*, Hangzhou.

142. Li ZX, Yang XM, and Ding YA. 2006. Novel method of damage identification for beams. *Proc 4th China–Japan–US Symp on Structural Control and Monitoring*, Hangzhou.

143. Cross K, Jha R, Whelan M, Janoyan K, and Gangone M. 2007. Bridge health monitoring using linear and nonlinear approaches: Experimental validation. In *Structural Health Monitoring*, FK Chang (ed.). DEStech, Lancaster.

144. DeSimio M, Olson S, and Derriso M. 2005. Decision uncertainty in a structural health monitoring system. *Smart Structures and Integrated Systems*, SPIE 5764.

145. Hilson G, Worden K, and Manson G. 2004. Feature selection for a neural damage locator using a genetic algorithm. *Proc 2nd European Workshop on Structural Health Monitoring*. DEStech, Munich.

146. Hickman GA, Gerardi JJ, and Feng Y. 1991. Application of smart structures to aircraft health monitoring. *J Intel Mater Struct Syst Struct*, 2(3), 411–430.

147. Allen DW, Clough JA, Sohn H, and Farrar CR. 2003. A software tool for graphically assembling damage identification algorithms. *Smart Systems and Nondestructive Evaluation for Civil Infrastructures*, SPIE 5057.

148. Haddad YM. 2003. On the nondestructive identification of the mechanical response of engineering materials: A pattern recognition and classification approach. In *Structural Health Monitoring*, FK Chang (ed.). DEStech, Lancaster.

149. Mikumo Y, Hirokane M, Furata H, Kusunose Y, and Yasuda K. 2004. A study of crack classification in concrete slabs by using pattern recognition methods. *Proc IABMAS'04 Bridge Maintenance, Safety, Management and Cost*, Kyoto. Taylor & Francis, London.

150. Banavas G, Burkhart DA, Cohen DA, Levi AF, Soffer B, and Barnoski MK. 2000. An intelligent monitoring and communication system. *Nondestructive Evaluation of Highways, Utilities, and Pipelines IV*, SPIE 3995.

151. Cruz PJ, Quintela H, and Santos MF. 2004. Prediction of ultimate shear resistance of non-prismatic tapered plate girders using data mining techniques. *Proc IABMAS'04 Bridge Maintenance, Safety, Management and Cost*, Kyoto. Taylor & Francis, London.

152. Haykin S. 1999. *Neural Networks a Comprehensive Foundation*. 2nd Ed. Prentice-Hall, Upper Saddle River, NJ.

153. Giurgiutiu V and Kropas-Hughes C. 2003. Comparative study of neural network damage detection from a statistical set of electro-mechanical impedance spectra. *Smart Nondestructive Evaluation and Health Monitoring of Structural and Biological Systems II*, SPIE 5047.

154. Specht DF. 1990. Probabilistic neural networks. *Neural Networks*, 3, 109–118.

155. Cowley PH. 1993. The application of neural networks to condition monitoring of non-linear structures. In *Safety Evaluation Based on Identification Approaches*, HG Natke, GR Tomlinson, and JTP Yao (eds.). Vieweg, Braunschweig.

156. Jenq ST and Lee WD. 1997. Identification of hole defect for GFRP woven laminates using neural network scheme. In *Structural Health Monitoring Current Status and Perspectives*. F-K Chang, (ed.). Technomic, Lancaster.

157. Dellomo MR. 1999. Helicopter gearbox fault detection: A neural network based approach. *J Vib Acoust*, 121, 265–272.

158. Wu X, Ghaboussi J, and Garrett Jr JH. 1992. Use of neural networks in detection of structural damage. *Comput Struct*, 42(4), 649–659.

159. Spillman Jr WB, Huston DR, Fuhr PL, and Lord JR. 1993. Neural network damage detection in a bridge element. *Smart Sensing, Processing and Instrumentation*, SPIE 1918.

160. Studer M and Peters K. 2003. Neural network application in damage identification using multi-scale sensing data. *Modeling, Signal Processing and Control*, SPIE 5049.

161. Kudva JN, Munir N, and Tan PW. 1992. Damage detection in smart structures using neural networks and finite element analyses. *Smart Mater Struct*, 1, 108–112.

162. Kesavan A, Deivasigamani M, John S, and Herszberg I. 2005. Damage criticality assessment in complex geometric structures using static strain response-based signal processing techniques. *Smart Structures and Integrated Systems*, SPIE 5764.

163. Worden K, Ball AD, and Tomlinson GR. 1993. Fault location in a framework structure using neural networks. *Smart Mater Struct*, 2, 189–200.

164. Liang J, Xiao D, Zhao X, and Zhang H. 2004. Online monitoring of seismic damage in water distribution systems. *Sensors and Smart Structures Technologies for Civil, Mechanical and Aerospace Systems*, SPIE 5391.

165. Lemoine GI, Love KW, and Anderson TA. 2003. An electric potential-based structural health monitoring technique using neural networks. In *Structural Health Monitoring*, FK Chang (ed.). DEStech, Lancaster.

166. Masri SF, Smyth AW, Chassiakos AG, Caughey TK, and Hunter NF. 2000. Application of neural networks for detection of changes in nonlinear systems. *J Eng Mech*, 126(7), 666–676.

167. Duran O, Althoefer K, and Seneviratne LD. 2003. Pipe inspection using a laser-based transducer and automated analysis techniques. *IEEE/ASME Trans Mechatronics*, 8(3), 401–409.

168. Zheng S, Wang X, and Liu L. 2004. Damage detection in composite materials based upon the computational mechanics and neural networks. *Proc 2nd European Workshop on Structural Health Monitoring*. DEStech, Munich.

169. Hyland DC. 1997. Connectionist algorithms for structural system identification and anomaly detection. In *Structural Health Monitoring Current Status and Perspectives*, FK Chang, (ed.). Technomic, Lancaster.

170. Suresh S, Omkar SN, Ganguli R, and Mani V. 2004. Identification of crack location and depth in a cantilever beam using a modular neural network approach. *Smart Mater Struct*, 13, 907–915.

171. Hatem TM, Abuelfoutouh MN, and Negm HM. 2004. Applications of genetic algorithms and neural networks to health monitoring of composite structures. *Proc 2nd European Workshop on Structural Health Monitoring*. DEStech, Munich.

172. Tse PW and Atherton DP. 1999. Prediction of machine deterioration using vibration based fault trends and recurrent neural networks. *J Vib Acoust*, 121, 355–362.

173. Tang H and Sato T. 2004. Structural damage detection using neural network and h∞ filter algorithm. *Health Monitoring and Smart Nondestructive Evaluation of Structural and Biological Systems III*, SPIE 5394.

174. Pan J, Guo S, and Wu Z. 2006. The state of art on damage inspection and identification based on probability neural network. *Proc 4th China–Japan–US Symp on Structural Control and Monitoring*, Hangzhou.

175. Yun CB, Lee JW, Kim JD, and Min KW. 2003. Damage estimation using committee of neural networks. *Smart Nondestructive Evaluation and Health Monitoring of Structural and Biological Systems II*, SPIE 5047.

176. Govindaraju MR and Sikorski RL. 2007. Thermal spray TBC durability issues. In *Structural Health Monitoring*, FK Chang (ed.). DEStech, Lancaster.

177. Vapnik VN. 1999. *The Nature of Statistical Learning Theory*, 2nd Ed. Springer, New York.

178. Worden K and Lane AJ. 2001. Damage identification using support vector machines. *Smart Mater Struct*, 10, 540–547.

179. Mita A and Hagiwara H. 2003. Damage diagnosis of a building structure using support vector machine and modal frequency patterns. *Smart Systems and Nondestructive Evaluation for Civil Infrastructures*, SPIE 5057.

180. Hayano H and Mita A. 2005. Structural health monitoring system using FBG sensor for simultaneous detection of acceleration and strain. *Sensors and Smart Structures Technologies for Civil, Mechanical, and Aerospace Systems*, SPIE 5765

181. Yun CB, Park S, and Inman DJ. 2006. Health monitoring of railroad tracks using PZT active sensors associated with support vector machines. *Proc 4th China–Japan–US Symp on Structural Control and Monitoring*, Hangzhou.

182. Bulut A, Singh A K, Shin P, Fountain T, Jasso H, Yan L, and Elgamal A. 2005. Real-time nondestructive structural health monitoring using support vector machines and wavelets. *Advanced Sensor Technologies for Nondestructive Evaluation and Structural Health Monitoring*, SPIE 5770.

183. He HX and Yan WM. 2007. Structural damage detection with wavelet support vector machine: Introduction and applications. *Struct Control Health Monit*, 14, 162–176.

184. Zhang J, Sato T, and Iai S. 2006. Novel support vector regression for structural system identification. *Struct Control Health Monit*, 14, 609–626.

185. Mitchell M. 1996. *An Introduction to Genetic Algorithms*. MIT Press, Cambridge.

186. Coverley PT and Stazewski WJ. 2003. Impact damage location in composite structures using optimized sensor triangulation procedure. *Smart Mater Struct*, 12, 795–203.

187. Nag A, Mahapatra DR, and Gopalalkrishanan S. 2002. Identification of delamination in composite beams using spectral estimation and a genetic algorithm. *Smart Mater Struct*, 11, 899–908.

188. Raich AM and Liszkai TR. 2003. Multi-objective genetic algorithm methodology for optimizing sensor layouts to enhance structural damage identification. In *Structural Health Monitoring*, FK Chang (ed.). DEStech, Lancaster.

189. Spillman Jr W and Huston D. 1997. Detection, location and characterization of point perturbations over a two dimensional area using two spatially weighted distributed fiber optic sensors. *Smart Sensing, Processing, and Instrumentation*, SPIE 3042.

190. Li J, Kapania WB and Spillman Jr WB. 2004. Placement optimization of fiber optic sensors for a smart bed using genetic algorithms. AIAA-2004-4334, *Proc 10th AIAA/ISSMO Multidisciplinary Analysis and Optimization Conference*, Albany.

191. Au FT, Cheng YS, Tham LG, and Bai ZZ. 2003. Structural damage detection based on a micro-genetic algorithm using incomplete and noisy modal test data. *J Sound Vib*, 259(5), 1081–1094.

192. Franco G, Betti R, and Lus H. 2004. Identification of structural systems using an evolutionary strategy. *J Eng Mech*, 130(10), 1125–1139.

193. Kong F, Li SJ, and Suzuki Y. 2006. Structural health monitoring using genetic algorithms. *Proc 4th China–Japan–US Symp on Structural Control and Monitoring*, Hangzhou.

194. Ananda Rao M, Srininvas J, and Murthy BS. 2004. Damage detection in vibrating bodies with genetic algorithms. *Comput Struct*, 82, 963–968.

195. Yan G and Yuan FG. 2007. Identification of impact loads for composites using genetic algorithms. In *Structural Health Monitoring*, FK Chang (ed.). DEStech, Lancaster.

196. Perera R, Ruiz A, and Manzano C. 2007. Structural damage evaluation combining flexibility and a fault location indicator. In *Structural Health Monitoring*, FK Chang (ed.). DEStech, Lancaster.

197. Ying W, Renda Z, and Lewen Z. 2007. Damage identification method of Truss Bridge based on a change of strain using genetic algorithm. In *Structural Health Monitoring*, FK Chang (ed.). DEStech, Lancaster.

198. Thanh TN, Perry MJ, and Koh CG. 2007. Moving force identification: A time domain genetic algorithm approach. In *Structural Health Monitoring*, FK Chang (ed.). DEStech, Lancaster.

199. Vishnuvardhan J, Krishnamurthy CV, and Balasubramaniam K. 2007. Genetic algorithm based reconstruction of the elastic moduli of orthotropic plates using an ultrasonic guided wave single-transmitter-multiple-receiver SHM array. *Smart Mater Struct*, 16, 1639–1650.

200. Ma F, Ng CH, and Ajavakom N. 2006. On system identification and response prediction of degrading structures. *Struct Control Health Monit*, 13, 347–364.

201. Wolfram S. 2002. *A New Kind of Science*. Wolfram Media, Champaign, IL.

202. Spillman Jr WB and Huston DR. 1998. Cellular automata for image analysis of damage in large structures. *Opt Eng*, 37(3), 898–903.

203. Liu J, Tang YY, and Cao YC. 1997. An evolutionary autonomous agents approach to image feature extraction. *IEEE Trans Evol Comput*, 1(2), 141–158.

204. Esterline A, Gandluri B, Sundaresan M, and Sankar J. 2005. Verified models of multiagent systems for vehicle health management. *Modeling, Signal Processing, and Control*, SPIE 5757.

205. Clerc M and Kennedy J. 2002. The particle swarm—explosion, stability, and convergence in a multidimensional complex space. *IEEE Trans Evol Comput*, 6(1), 58–73.

206. Tang H, Fukuda M, and Xue S. 2007. Particle swarm optimization for structural system identification. In *Structural Health Monitoring*, FK Chang (ed.). DEStech, Lancaster.
207. Kurowicka D, Goossens L, and Morales O. 2006. Use of Bayesian belief nets as a technique to compute risks of complex systems. *Computational Models of Risks to Infrastructure*. NATO Advanced Research Workshop, Primosten.
208. Charniak E. 1991. Bayesian networks without tears. *AI Mag*, 12(4), 50–63 (Winter).
209. LeBeau K and Wadia-Fascetti SA. 2006. Bi-directional probabilistic bridge load rating model. Paper no. 06-2507, 86th Annual Meeting, TRB, Washington, DC.
210. LeBeau K and Wadia-Fascetti S. 2007. Comparative probabilistic initial bridge load rating model. *J Bridge Eng*, 12(6), 785–793.
211. Melcher KJ, Sowers TS, and Maul WA. 2005. Meeting the challenges of exploration systems: Health management technologies for aerospace systems with emphasis on propulsion. *NASA* TM-2005-214026.
212. Neil M, Fenton N, Forey S, and Harris R. 2001. Using Bayesian belief networks to predict the reliability of military vehicles. *Comput Control Engrg J*, 12(1), 11–20.
213. Loizos A and Karlaftis MG. 2006. Neural networks and non-parametric statistical models: A comparative analysis in pavement condition assessment. Paper no. 06-1108, 86th Annual Meeting, TRB, Washington, DC.
214. Mizuno Y, Fujino Y, and Abe M. 2004. Proposal of data management architecture for bridge management system. *Proc IABMAS'04 Bridge Maintenance, Safety, Management and Cost*, Kyoto. Taylor & Francis, London.
215. Ulku AE, Attanayake U, and Aktan HM. 2006. Remaining service life estimation of bridge components by knowledge discovery techniques. NDE *Conf on Civil Engineering*, American Society for Nondestructive Testing, St. Louis.
216. Inaudi D, Casanova P, Kronenberg S, Marazzi S, and Vurpillot S. 1997. Embedded and surface mounted fiber optic sensors for civil structural monitoring. *Industrial and Commercial Applications of Smart Structures Technologies*, SPIE 3044.
217. Reich GW and Sanders B. 2003. Structural shape sensing for morphing aircraft. *Smart Structures and Integrated Systems*, SPIE 5056.
218. Duncan RG and Raum MT. 2005. Fiber-optic shape and position sensing. In *Structural Health Monitoring*, FK Chang (ed.). DEStech, Lancaster.
219. Jones RT, Berkoff TA, Bellemore DG, Early DA, Sirkis JS, Putnam MA, Frieble EJ, and Kersey AD. 1996. Cantilever plate deformation monitoring using wavelength division multiplexed fiber Bragg grating sensors. *Smart Sensing, Processing and Instrumentation*, SPIE 2718.
220. Vazquez SL, Tessler A, Quach CC, Cooper EG, Parks J, and Spangler JL. 2005. *Structural Health Monitoring Using High-Density Fiber Optic Strain Sensor and Inverse Finite Element Methods*, NASA/TM-2005-213761.
221. Baldwin CS, Salter TJ, and Kiddy JS. 2004. Static shape measurements using a multiplexed fiber Bragg grating sensor system. *Smart Sensor Technology Measurement Systems*, SPIE 5384.
222. Park KT, Park HS, Kim SH, and Lee KW. 2004. Development of IVE algorithm for monitoring of bridge's vertical displacement. *Proc IABMAS'04 Bridge Maintenance, Safety, Management and Cost*, Kyoto. Taylor & Francis, London.
223. Zarchan P and Musoff H. 2000. *Fundamentals of Kalman Filtering—a Practical Approach*, Vol. 190. Progress in Astronautics and Aeronautics. AIAA, Reston.
224. Sanayei M, Imbaro GR, McClain JA, and Brown LC. 1997. Structural model updating using experimental static measurements. *J Struct Eng*, 123(6), 792–798.
225. Di Paola M and Bilello C. 2004. An integral equation for damage identification of Euler–Bernoulli beams under static loads. *J Eng Mech*, 130(2), 1–10.
226. Sain T and Kishen JM. 2006. Damage assessment in beams using inverse method. *J Eng Mech*, 132(4), 337–344.

227. Sumant PS and Maiti SK. 2006. Crack detection in a beam using PZT sensors. *Smart Mater Struct*, 15, 695–703.

228. Yeo I, Shin S, Lee HS, and Chang SP. 2000. Statistical damage assessment of framed structures from static responses. *J Eng Mech*, 126(4), 414–421.

229. Baumeister J. 1993. Identification of parameters, ill-posedness and adaptive systems. In *Safety Evaluation Based on Identification Approaches*, HG Natke, GR Tomlinson, and JTP Yao (eds.). Vieweg, Braunschweig.

230. Harursampath D, Harish AB, and Kiran S. 2007. Online structural health monitoring of pretwisted anisotropic beams. In *Structural Health Monitoring*, FK Chang (ed.). DEStech, Lancaster.

231. Murphy KD, Nichols JM, and Motley S. 2007. Nonlinear mechanics of delaminated beams. In *Structural Health Monitoring*, FK Chang (ed.). DEStech, Lancaster.

232. Buda G and Caddemi S. 2007. Identification of concentrated damages in Euler–Bernoulli beams under static loads. *J Eng Mech*, 133(8), 225–234.

233. Ewins DJ. 2001. *Modal Testing: Theory and Practice*, 2nd Ed. Research Studies Press Ltd, Baldock.

234. Doebling SW, Farrar CR, and Prime MB. 1998. A summary review of vibration-based damage identification methods. *Shock Vibr Dig*, 30(2), 91–105.

235. Sharma MJ. 1987. Bridge inspection program for the Baltimore metro system. *Proc 4th Ann Intl Bridge Conference*, IBC-87-30, Pittsburgh.

236. Kashangaki TA. 1991. On orbit damage detection and health monitoring of large space trusses status and critical issues. *Proc 32 AIAA SDM Conference*, AIAA-91-1181-CP, Baltimore.

237. Hamamoto T. 2003. Experimental verification for damage detection strategies of multistory buildings based on vibration monitoring. *Smart Systems and Nondestructive Evaluation for Civil Infrastructures*, SPIE 5057.

238. Farrar CR, Sohn H, and Robertson AN. 2004. Applications of nonlinear system identification to structural health monitoring. *Proc 2nd European Workshop on Structural Health Monitoring*. DEStech, Munich.

239. Araki Y and Miyagi Y. 2005. Mixed integer nonlinear least-squares problem for damage detection in truss structures. *J Eng Mech*, 131(7), 659–667.

240. Wowk V. 1992. *Machinery Vibration*. McGraw-Hill, New York.

241. Zhang C, Kurfess TR, Danyluk S, and Liang SY. 1999. Dynamic modeling of vibration signals for bearing condition monitoring. *Proc 2nd Intl Workshop on Structural Health Monitoring*, Stanford. Technomic, Lancaster.

242. Bach H and Markert R. 1997. Determination of the fault position in rotors for the example of a transverse crack. In *Structural Health Monitoring Current Status and Perspectives*. F-K Chang, (ed.). Technomic, Lancaster.

243. Keiner H and Gadala MS. 2002. Crack parameter identification using inverse analysis of on-line vibration measurements. *Proc 26th Annual Meeting*. Vibration Institute, Pittsburgh.

244. Spijker JS and DeMatteo T. 2003. The case of the missing file. *Proc 27th Annual Meeting*. Vibration Institute, New Orleans.

245. Allemang RJ and Brown DL. 1987. Universal file formats. In *Experimental Modal Analysis and Dynamic Component Synthesis*, Vol. V. AFWAL-TR-87-3069, Air Force Wright Aeronautical Laboratories, Wright-Patterson AFB.

246. Nili M and Vahedi F. 2006. The effects assessing of the supplementary cementitious materials on the durability of roller compacted concrete for pavement construction against freezing-thawing cycles by a nondestructive test. *NDE Conf on Civil Engineering*, American Society for Nondestructive Testing, St. Louis.

247. Jeong W and Kim NS. 2004. Estimation of tension forces of assembly stay cables connected with massive anchorage block. *Proc IABMAS'04 Bridge Maintenance, Safety, Management and Cost*, Kyoto. Taylor & Francis, London.

248. Kim JT, Ryu YS, and Yun CB. 2003. Vibration-based method to detect prestress-loss in beam-type bridges. *Smart Systems and Nondestructive Evaluation for Civil Infrastructures*, SPIE 5057.

249. Ali B, Sundaresan MJ, Schulz MJ, and Hughes D. 2005. Early detection of local buckling in structural members. *Smart Structures and Integrated Systems*, SPIE 5764.

250. Cawley P and Adams RD. 1979. The location of defects in structures from measurements of natural frequencies. *J Strain Anal*, 14(2), 49–57.

251. Kato M and Shimada S. 1986. Vibration of PC bridge during failure process. *J Struct Eng*, 112(7), 1692–1703.

252. Morassi A and Rovere N. 1997. Localizing a notch in a steel frame from frequency measurements. *J Eng Mech*, 123(5), 422–432.

253. Hickman GA, Gerardi JJ, and Feng Y. 1991. Application of smart structures to aircraft health monitoring. *J Intel Mater Syst Struct*, 2(3), 411–430.

254. Mazurek DF and DeWolf JT. 1990. Experimental study of bridge monitoring technique. *J Struct Eng*, 116(9), 2532–2549.

255. Morita K and Teshigawara M. 2004. Damage detection and estimation of a steel frame through shaking table test and measurements. *Sensors and Smart Structures Technologies for Civil, Mechanical and Aerospace Systems*, SPIE 5391.

256. Melhem HG and Iyer SK. 1998. Using the falling weight deflectometer for highway bridge rating. In *Structural Engineering Worldwide*. GL Fenves, NK Srivastava, AH Ang, and RG Domer (eds.). Elsevier, Amsterdam, T200-6.

257. Capozucca R and Cerri MN. 2000. Identification of damage in reinforced concrete beams subjected to corrosion. *ACI Struct J*, 97(6).

258. Vestroni F and Capecchi D. 2000. Damage detection in beam structures based on frequency measurements. *J Eng Mech*, 126(7), 761–768.

259. Gheorghiu C, Rhazi JE, and Labossière P. 2005. Impact resonance method for damage detection in RC beams strengthened with composites. *Nondestructive Evaluation and Health Monitoring of Aerospace Materials, Composites and Civil Infrastructure IV*, SPIE 5767.

260. Stubbs N and Osegueda R. 1990. Global damage detection in solids—Experimental verification. *Intl J Anal Exp Modal Anal*, 5(2), 81–97.

261. Rytter A, Krawczuk M, and Kirkegaard PH. 2000. Experimental and numerical study of damaged cantilever. *J Eng Mech*, 126(1), 60–65.

262. Li B, Chen XF, Ma JX, and He Z. 2005. Detection of crack location and size in structures using wavelet finite element methods. *J Sound Vib*, 285, 767–782.

263. Wang GH. 2003. Study on damage identification of tied arch structure based on modal parameters. *Proc 4th China–Japan–US Symp on Structural Control and Monitoring*, Hangzhou.

264. Qian Y and Mita A. 2007. Structural damage identification using parzen-window approach and neural networks. *Struct Control Health Monit*, 14, 576–590.

265. Ackers S, Evans R, and Adams DE. 2007. Crack detection in a wheel end spindle using modal impacts. In *Structural Health Monitoring*, FK Chang (ed.). DEStech, Lancaster.

266. Park SW and Zhou M. 2000. Nondestructive assessment of the post-impact properties of FRP composite laminates. In *Structural Materials Technology IV*, S Allampalli (ed.). Technomic, Lancaster.

267. Yoder NC and Adams DE. 2007. An experimental forced response tire model and its application to the near real-time monitoring of bead area damage in rolling tires. In *Structural Health Monitoring*, FK Chang (ed.). DEStech, Lancaster.

268. Hearn G and Testa RB. 1991. Modal analysis for damage detection in structures. *J Struct Eng*, 117(10), 3042–3063.

269. Liberatore S and Carman GP. 2003. Damage detection of structures based on spectral methods using piezoelectric materials. In *Structural Health Monitoring*, FK Chang (ed.). DEStech, Lancaster.

270. Fanning PJ and Carden EP. 2003. Damage detection based on single-input–single-output measurements. *J Eng Mech*, 129(2), 202–209.

271. Cade IS, Keough PS, and Sahinkaya MN. 2005. Fault identification in rotor/magnetic bearing systems using discrete time wavelet coefficients. *IEEE/ASME Trans Mechatronics*, 10(4), 391–396.

272. Manson G, Worden K, and Allman D. 2003. Experimental validation of a structural health monitoring methodology: Part III. Damage location on an aircraft wing. *J Sound Vib*, 259(2), 365–385.

273. Blodgett DW. 2003. Detection of re-bar delamination using modal analysis. *Smart Systems and Nondestructive Evaluation for Civil Infrastructures*, SPIE 5057.

274. Schulz MJ, Pai PF, Naser AS, Thyagarajan SK, Brannon GR, and Chung J. 1997. Locating structural damage using frequency response reference functions and curvatures. In *Structural Health Monitoring Current Status and Perspectives*. F-K Chang, (ed.). Technomic, Lancaster.

275. Fang X and Tang J. 2005. Damage detection by statistical analysis of vibration signature. *Sensors and Smart Structures Technologies for Civil, Mechanical, and Aerospace Systems*, SPIE 5765.

276. White C, Whittingham B, Li H, Bannister M, and Mouritz A. 2007. Frequency response techniques for SHM of adhesively bonded composite repairs. In *Structural Health Monitoring*, FK Chang (ed.). DEStech, Lancaster.

277. Medda A and deBrunner V. 2007. Wavelet packet sub-band beamforming for SHM. In *Structural Health Monitoring*, FK Chang (ed.). DEStech, Lancaster.

278. Melhem H and Kim H. 2003. Damage detection in concrete by Fourier and wavelet analysis. *J Eng Mech*, 129(5), 571–577.

279. Bland SM and Kapania RK. 2004. Damage identification using wavelet packet analysis and neural networks. In *Advanced Smart Materials and Smart Structures Technology*. FK Chang, CB Yun, and BF Spencer, Jr. (eds.). DEStech, Lancaster.

280. Katafygiotis LS and Lam HF. 1997. A probabilistic approach to structural health monitoring using dynamic data. In *Structural Health Monitoring Current Status and Perspectives*. F-K Chang, (ed.). Technomic, Lancaster.

281. Carrion F, Torres-Acosta A, Quintana JA, Mullins JA, and Sen R. 2003. Corrosion damage evaluation of pre-stressed beams using an inverse wave propagation method. *Smart Systems and Nondestructive Evaluation for Civil Infrastructures*, SPIE 5057.

282. Gyekenyesi AL, Harmon LM, and Kautz HE. 2002. The effect of experimental conditions on acousto-ultrasonic reproducibility. *Nondestructive Evaluation and Health Monitoring of Aerospace Materials and Civil Infrastructures*, SPIE 4704.

283. Dharmaraju N and Sinha JK. 2005. Some comments on use of antiresonance for crack identification in beams. *J Sound Vib*, 286, 669–671.

284. Johnson EA, Asghari A, and Nigbor RL. 2006. Summary of environmental effects on identified parameters of the NEES SFSI test structure EA. *Proc 4th China–Japan–US Symp on Structural Control and Monitoring*, Hangzhou.

285. Jiang Z, Chonan S, Kawashima K, Muto K, and Ichihara W. 1997. Condition monitoring of silicon-wafer slicer cutting crystal ingots. In *Structural Health Monitoring Current Status and Perspectives*. F-K Chang, (ed.). Technomic, Lancaster.

286. Johnson TJ, Yang C, Adams DE, and Ciray S. 2004. Structural damage identification using embedded sensitivity functions. *Smart Structures and Integrated Systems*, SPIE 5390.

287. Johnson TJ and Adams DE. 2002. Transmissibility as a differential indicator of structural damage. *J Vib Acoust*, 124, 634–641.

288. Chen HL and Kiriiakidis AC. 2003. Stiffness evaluation and damage detection of ceramic candle filters. *J Eng Mech*, 126(3), 308–319.

289. Marwala T. 2004. Fault classification using pseudomodal energies and probabilistic neural networks. *J Eng Mech*, 130(11), 1346–1355.

290. Stepinski T and Engholm M. 2007. Structural health monitoring of composite structures for temperature varying applications. In *Structural Health Monitoring*, FK Chang (ed.). DEStech, Lancaster.

291. Soh CK and Bhalla S. 2005. Calibration of piezo-impedance transducers for strength prediction and damage assessment of concrete. *Smart Mater Struct*, 14, 671–684.

292. Lim YY, Bhalla S, and Soh CK. 2006. Structural identification and damage diagnosis using self-sensing piezo-impedance transducers. *Smart Mater Struct*, 15, 987–995.

293. Lei Y, Kiremidjian AS, Nair KK, Lynch JP, and Law JH. 2003. An enhanced statistical damage detection algorithm using time series analysis. In *Structural Health Monitoring*, FK Chang (ed.). DEStech, Lancaster.

294. Adams DE and Farrar CR. 2002. Classifying linear and nonlinear structural damage using frequency domain ARX models. *Struct Control Health Monit*, 1(2), 185–201.

295. Nair KK, Kiremidjian AS, and Law KH. 2006. Time series-based damage detection and localization algorithm with application to the ASCE benchmark structure. *J Sound Vib*, 291(1–2), 349–368.

296. Lu Y and Gao F. 2005. A novel time-domain auto-regressive model for structural damage diagnosis. *J Sound Vib*, 283, 1031–1049.

297. Yang JN, Lin S, and Zhou L. 2004. Identification of parametric changes for civil engineering structures using an adaptive Kalman filter. *Sensors and Smart Structures Technologies for Civil, Mechanical and Aerospace Systems*, SPIE 5391.

298. Koh BH, Dharap P, Nagarajaiah S, and Pham MQ. 2005. Input error function formulation for real-time structural damage monitoring. In *Structural Health Monitoring*, FK Chang (ed.). DEStech, Lancaster.

299. Torkamani MA and Hart GC. 1998. A time domain approach for estimation of structural parameters of buildings. In *Structural Engineering Worldwide*. GL Fenves, NK Srivastava, AH Ang, and RG Domer (eds.). Elsevier, Amsterdam, T216-2.

300. Ge L and Soong TT. 1998. Damage identification through regularization method. I: Theory. *J Eng Mech*, 124(1), 103–108.

301. Ge L and Soong TT. 1998. Damage identification through regularization method. II: Applications. *J Eng Mech*, 124(1), 109–116.

302. Peelamedu SM, Naganathan NG, Dukkipati RV, and Srinivas J. 2005. Impact identification for a metallic plate using distributed smart materials. *Smart Mater Struct*, 14, 449–456.

303. Diez JL, Gordo AG, Pascual RA, Revuelta EC, and Rejado CC. 2004. Tuned condensation method for finite element models guidelines. *Proc 2nd European Workshop on Structural Health Monitoring*. DEStech, Munich.

304. Pandey A K and Biswas M. 1994. Damage detection in structures using changes in flexibilty. *J Sound Vib*, 169(1), 3–17.

305. Stubbs N and Osegueda R. 1990. Global non-destructive damage evaluation in solids. *Intl J Anal Exper Modal Anal*, 5(2), 67–79.

306. Weber B, Paultre P, and Proulx J. 2005. Improved iterative regularization for vibration-based damage detection. *Nondestructive Evaluation and Health Monitoring of Aerospace Materials, Composites and Civil Infrastructure IV*, SPIE 5767.

307. Yu J, De Angelis M, Imbimbo M, and Betti R. 2003. Damage detection in reduced order models of linear structural systems. In *Structural Health Monitoring*, FK Chang (ed.). DEStech, Lancaster.

308. Khiem NT and Lien TV. 2004. Multi-crack detection for beam by the natural frequencies. *J Sound Vib*, 273, 175–184.

309. Cha PD and Tuck-Lee JP. 2000. Updating structural system parameters using frequency response data. *J Eng Mech*, 126(12), 1240–1246.

310. Amani MG, Riera JD, and Curadelli RO. 2007. Identification of changes in the stiffness and damping matrices of linear structures through ambient vibrations. *Struct Control Health Monit*, 14, 1155–1169.

311. Xue ST, Mita A, Qian YY, Xie LY, and Zheng HT. 2005. Application of a grey control system model to structural damage identification. *Smart Mater Struct*, 14, S125–S129.

312. Rahai A, Bakhtiari-Nejad F, and Esfandiari A. 2007. Damage assessment of structure using incomplete measured mode shapes. *Struct Control Health Monit*, 14, 808–829.

313. Yoshimoto R, Mita A, Okada K, Iwaki H, and Shiraishi M. 2003. Damage detection of a structural health monitoring system for a 7-story seismic isolated building. *Smart Systems and Nondestructive Evaluation for Civil Infrastructures*, SPIE 5057.

314. Koh CG, Tee KF, and Quek ST. 2003. Stiffness and damage identification with model reduction technique. In *Structural Health Monitoring*, FK Chang (ed.). DEStech, Lancaster.

315. Pothisiri T and Hjelmstad KD. 2003. Structural damage detection and assessment from modal response. *J Eng Mech*, 129(2), 135–145.

316. Avitabile P. 2003. Twenty years of structural dynamic modification-a review. *Sound Vib*, (January), 14–25.

317. Lim TW, Bosse A, and Fisher S. 1996. Structural damage detection using real-time modal parameter identification algorithm. *AIAA J*, 34(11), 2370–2376.

318. Zhu HP and Xu YL. 2005. Damage detection of mono-coupled periodic structures based on sensitivity analysis of modal parameters. *J Sound Vib*, 285, 365–390.

319. Zhao J and DeWolf JT. 2002. Dynamic monitoring of steel girder highway bridge. *J Bridge Eng*, 7(6), 350–356.

320. Catbas FN, Brown DL, and Aktan AE. 2004. Parameter estimation for multiple-input multiple-output modal analysis of large structures. *J Eng Mech*, 130(8), 921–930.

321. Catbas FN and Aktan AE. 2002. Condition and damage assessment: Issues and some promising indices. *J Struct Eng*, 128(8), 1026–1036.

322. Zhou Z, Sparling BF, and Wegner LD. 2005. Damage detection on a steel-free bridge deck using random vibration. *Nondestructive Evaluation and Health Monitoring of Aerospace Materials, Composites and Civil Infrastructure IV*, SPIE 5767.

323. Shi ZY, Law SS, and Zhang LM. 2000. Structural damage detection from modal strain energy change. *J Eng Mech*, 126(12), 1216–1223.

324. Shi ZY, Law SS, and Zhang LM. 2002. Improved damage quantification from elemental modal strain energy change. *J Eng Mech*, 128(5), 512–529.

325. Onyemelukwe O, Mirmiran A, and Chopra M. 1995. Application of boundary element method to structural damage assessment. *Proc ASCE Structures Congress* XIII, Boston, MA, M Sanayei (ed.). ASCE.

326. Chen XD and Chen HP. 2006. Localisation and quantification of damage in arch dams based on incomplete vibration modal data. *Proc 4th China–Japan–US Symp on Structural Control and Monitoring*, Hangzhou.

327. Song W, Dyke S, and Yun GJ. 2007. FE model updating for structural damage localization and quantification in high-dimension SHM problem. In *Structural Health Monitoring*, FK Chang (ed.). DEStech, Lancaster.

328. Materazzi AL and Breccolotti M. 2004. Reliability of the dynamic methods for evaluating structural concrete bridge integrity. *Proc IABMAS'04 Bridge Maintenance, Safety, Management and Cost*, Kyoto. Taylor & Francis, London.

329. Swann C, Chattopadhyay A, and Ghoshal A. 2003. Characterization of delamination in using damage indices. In *Structural Health Monitoring*, FK Chang (ed.). DEStech, Lancaster.

330. Franchetti P, Pavan C, and Modena C. 2004. Vibration based damage assessment of PRC precast beams. *Proc 2nd European Workshop on Structural Health Monitoring*. DEStech, Munich.

331. Jang JH, Yeo I, Shin S, and Chang SP. 2002. Experimental investigation of system-identification-based damage assessment on structures. *J Struct Eng*, 128(5), 673–682.

332. Shi ZY, Law SS, and Zhang LM. 2000. Damage localization by directly using incomplete mode shapes. *J Eng Mech*, 126(6), 656–660.

333. Wong CN, Zhu WD, and Xu GY. 2004. On an iterative general-order perturbation method for multiple structural damage detection. *J Sound Vib*, 273, 363–386.

334. Porfiri M, Maurini C, and Pouget J. 2007. Identification of electromechanical modal parameters of linear piezoelectric structures. *Smart Mater Struct*, 16, 323–331.

335. Lee GC and Liang Z. 1999. Development of a bridge monitoring system. *Proc 2nd Intl Workshop on Structural Health Monitoring*, Stanford. Technomic, Lancaster.

336. Fritzen CP and Bohle K. 2004. Damage identification using a modal kinetic energy criterion and output-only modal data-application to the Z24 bridge. *Proc 2nd European Workshop on Structural Health Monitoring*. DEStech, Munich.

337. Li H, Yang H, and Hu SL. 2006. Modal strain energy decomposition method for damage localization in 3D frame structures. *J Eng Mech*, 132(9), 941–951.
338. Liu W, Gao WC, and Sun Y. 2006. A two-step damage detection method of large scale space truss. *Proc 4th China–Japan–US Symp on Structural Control and Monitoring*, Hangzhou.
339. Hu SJ, Wang S, and Li H. 2006. Cross-modal strain energy method for estimating damage severity. *J Eng Mech*, 132(4), 429–437.
340. Jaishi B and Ren WX. 2006. Damage detection by finite element model updating using modal flexibility residual. *J Sound Vib*, 290, 369–387.
341. Gray MA, Parker DL, Frazier WG, and Cuevas P. 2007. Multi-site damage localization using least-squares optimization with low rank svd updates. In *Structural Health Monitoring*, FK Chang (ed.). DEStech, Lancaster.
342. Bernal D. 2002. Load vectors for damage localization. *J Eng Mech*, 128(1), 7–14.
343. Ji X, Qian J, and Xu L. 2007. Damage diagnosis of a two-storey spatial steel braced-frame model. *Struct Control Health Monit*, 14, 1083–1100.
344. Gao Y, Spencer Jr BF, and Bernal D. 2007. Experimental verification of the flexibility-based damage locating vector method. *J Eng Mech*, 133(10), 1043–1049.
345. Duan Z, Yan G, Ou J, and Spencer BF. 2005. Damage localization in ambient vibration by constructing proportional flexibility matrix. *J Sound Vib*, 284, 455–466.
346. Ratcliffe CP, Gillespie JW, Heider D, Eckel DA, and Crane RM. 2000. Experimental investigation into the use of vibration data for long term monitoring of an all composite bridge. *Nondestructive Evaluation of Highways, Utilities, and Pipelines IV*, SPIE 3995.
347. Gauthier JF, Whalen TM, and Liu J. 2008. Experimental validation of the higher-order derivative discontinuity method for damage identification. *Struct Control Health Monit*, 15, 143–161.
348. Kim BH, Stubbs N, and Park T. 2005. Flexural damage index equations of a plate. *J Sound Vib*, 283, 341–368.
349. Reynders E, De Roeck G, Gundes Bakir P, and Sauvage C. 2007. Damage identification on the Tilff Bridge by vibration monitoring using optical fiber strain sensors. *J Eng Mech*, 133(2), 185–193.
350. Pai PF and Jin S. 2000. Locating structural damage by detecting boundary effects. *J Sound Vib*, 231(4), 1079–1110.
351. Sharma VK, Ruzzene M, and Hanagud S. 2005. Damage index estimation in beams and plates using laser vibrometry. In *Structural Health Monitoring*, FK Chang (ed.). DEStech, Lancaster.
352. Hoffmann S, Wendner R, and Strauss A. 2007. Comparison of stiffness identification methods for reinforced concrete structures. In *Structural Health Monitoring*, FK Chang (ed.). DEStech, Lancaster.
353. Otieno AW, Liu P, Rao VS, and Koval LR. 2000. Damage detection using modal strain energy and laser vibrometer measurements. *Smart Structures and Integrated Systems*, SPIE 3985.
354. Vikhagan E, Haugland SJ, Vollen MW, Vines L, Wang G, and Jensen AE. 2004. Vibration characterization for damage detection in sandwich structures using shearography techniques. *Proc 2nd European Workshop on Structural Health Monitoring*. DEStech, Munich.
355. Jung H, Park Y, and Park K. 2003. Damage location identification *via* select feedback probing. In *Structural Health Monitoring*, FK Chang (ed.). DEStech, Lancaster.
356. Vanik MW, Beck JL, and Au SK. 2000. Bayesian probabilistic approach to structural health monitoring. *J Eng Mech*, 126(7), 738–745.
357. Kim JT, Park JH, Lee YH, Lee BJ, and Lee JS. 2004. Risk-alarming of structural damage in plate-girder bridges under uncertain temperature condition. *Proc IABMAS'04 Bridge Maintenance, Safety, Management and Cost*, Kyoto. Taylor & Francis, London.
358. Wang X, Wang ML, Zhao Y, Chen H, and Zhou LL. 2004. Smart health monitoring system for a prestressed concrete bridge. *Sensors and Smart Structures Technologies for Civil, Mechanical and Aerospace Systems*, SPIE 5391.
359. Pardo de Vera C and Guemes JA. 1997. Embedded self-sensing piezoelectric for damage detection. In *Structural Health Monitoring Current Status and Perspectives*. F-K Chang, (ed.). Technomic, Lancaster.

360. Sikorsky C, Stubbs N, and Guan F. 2003. The impact of natural frequency variation on damage detection. In *Structural Health Monitoring*, FK Chang (ed.). DEStech, Lancaster.
361. Kullaa J. 2004. Structural health monitoring under variable environmental or operational conditions. *Proc 2nd European Workshop on Structural Health Monitoring*. DEStech, Munich.
362. Yan AM, Golinval JC, Peeters B, and De Roeck G. 2004. A comparative study on damage detection of Z24-bridge: One-year monitoring with varying environmental conditions. *Proc 2nd European Workshop on Structural Health Monitoring*. DEStech, Munich.
363. Peeters B and De Roeck G. 2001. One-year monitoring of the Z24-bridge: Environmental effects versus damage events. *Earthq Eng Struct Dyn*, 30, 149–171.
364. Ciloglu K, Catbas FN, Pervizpour M, Wang A, and Aktan A. 2001. Structural identification of phenomenological physical properties with controlled mechanisms of uncertainty. *Smart Systems for Bridges, Structures, and Highways*, SPIE 4330.
365. Doebling SW, Farrar CR, and Cornwell PJ. 1998. Development of a general purpose code to couple experimental modal analysis and damage identification algorithms. *Proc 1998 Structural Engineers World Congress*, San Francisco.
366. Park SH, Roh Y, Yi JH, Yun CB, Kwak HK, and Lee SH. 2004. Impedance-based damage detection for civil infrastructures. *Sensors and Smart Structures Technologies for Civil, Mechanical and Aerospace Systems*, SPIE 5391.
367. Ritdumrongkul S, Abe M, Fujino Y, and Miyashita T. 2003. Quantitative health monitoring of bolted joints using piezoceramic actuator-sensor. *Smart Systems and Nondestructive Evaluation for Civil Infrastructures*, SPIE 5057.
368. Mascarenas DD, Park G, and Farrar CR. 2005. Monitoring of bolt preload using piezoelectric active devices. *Health Monitoring and Smart Nondestructive Evaluation of Structural and Biological Systems IV*, SPIE 5768.
369. Nakahara T, Yamamoto M, Ohya Y, and Okuma M. 2004. Bolt loosening using vibration characteristics of thin plate with piezoelectric elements. *Sensors and Smart Structures Technologies for Civil, Mechanical and Aerospace Systems*, SPIE 5391.
370. Shimada A, Urabe K, Kikushima Y, Takahashi J, and Kageyama K. 2003. Detection of missing fastener based on vibration mode analysis using fiber Bragg grating (FBG) sensors. *Smart Structures and Integrated Systems*, SPIE 5056.
371. Johnson TJ, Manning R, Adams DE, Sterkenburg R, and Jata K. 2004. Vibration-based structural health monitoring of tool-part interactions during riveting operations on an aircraft fuselage structure. *Proc 2nd European Workshop on Structural Health Monitoring*. DEStech, Munich.
372. Olson SE, DeSimio MP, and Derriso MM. 2006. Fastener damage estimation in a square plate. *Struct Health Monit*, 5(2), 173–183.
373. Yang J and Chang FK. 2006. Detection of bolt loosening in C–C composite thermal protection panels: I. Diagnostic principle. *Smart Mater Struct*, 15, 581–590.
374. Yang J and Chang FK. 2006. Detection of bolt loosening in C–C composite thermal protection panels: II. Experimental verification. *Smart Mater Struct*, 15, 591–599.
375. Xu H and Humar J. 2005. Application of artificial neural networks in vibration based damage detection. *Nondestructive Evaluation and Health Monitoring of Aerospace Materials, Composites and Civil Infrastructure IV*, SPIE 5767.
376. Chen SS and Kim S. 1995. Neural network based sensor signal monitoring of instrumented structures. *Proc ASCE Structures Congress XIII*, Boston, MA, M Sanayei (ed.). ASCE.
377. Feng MQ, Kim DK, Yi JH, and Chen Y. 2004. Baseline models for bridge performance monitoring. *J Eng Mech*, 130(5), 562–569.
378. Reddy RR and Ganguli R. 2003. Structural damage detection in a helicopter rotor blade using radial basis function neural networks. *Smart Mater Struct*, 12, 232–241.
379. Michaels JE, Cobb AC, and Michaels TE. 2004. A comparison of feature-based classifiers for ultrasonic structural health monitoring. *Health Monitoring and Smart Nondestructive Evaluation of Structural and Biological Systems III*, SPIE 5394.
380. Marwala T. 2000. Damage identification using committee of neural networks. *J Eng Mech*, 126(1).

381. Worden K, Manson G, and Allman D. 2003. Experimental validation of a structural health monitoring methodology: Part I. Novelty detection of a laboratory structure. *J Sound Vib*, 259(2), 323–343.

382. Card L and McNeill DK. 2004. Novel event identification for SHM systems using unsupervised neural computation. *Nondestructive Evaluation and Health Monitoring of Aerospace Materials and Composites III*, SPIE 5393.

383. Manson G, Worden K, and Allman D. 2003. Experimental validation of a structural health monitoring methodology: Part II. Novelty detection on a gnat aircraft. *J Sound Vib*, 259(2), 345–363.

384. Sundararaman S, Haroon M, Adams DE, and Jata K. 2004. Incipient damage identification using elastic wave propagation through a friction stir al-li interface for cryogenic tank applications. *Proc 2nd European Workshop on Structural Health Monitoring*. DEStech, Munich.

385. Ma TW, Yang HT, and Chang CC. 2003. Structural health monitoring using time domain residual generator technique. In *Structural Health Monitoring*, FK Chang (ed.). DEStech, Lancaster.

386. Chan TH, Ni YQ, and Ko JM. 1999. Neural network novelty filtering for anomaly detection of Tsing Ma Bridge cables. *Proc 2nd Intl Workshop on Structural Health Monitoring*, Stanford. Technomic, Lancaster.

387. Tian J, Li Z, and Su X. 2003. Crack detection in beams by wavelet analysis of transient flexural waves. *J Sound Vib*, 261(4), 715–727.

388. Marzani A, Rizzo P, Lanza di Scalea F, and Benzoni G. 2004. Mobile acoustic system for the detection of surface-breaking cracks in pavement. *Smart Sensor Technology Measurement Systems*, SPIE 5384.

389. Chang CC and Sun Z. 2003. Locating and quantifying structure damage using spatial wavelet packet signature. *Smart Systems and Nondestructive Evaluation for Civil Infrastructures*, SPIE 5057.

390. Peng G, Yuan SF, and Xu YD. 2004. Damage localization of two-dimensional structure based on wavelet transform and active monitoring technology of lamb wave. *Nondestructive Evaluation and Health Monitoring of Aerospace Materials and Composites III*, SPIE 5393.

391. Gu H, Song G, and Qiao P. 2004. Delamination detection of composite plates using piezoceramic patches and wavelet packet analysis. *Health Monitoring and Smart Nondestructive Evaluation of Structural and Biological Systems III*, SPIE 5394.

392. Gubbala R and Rao VS. 2003. Health monitoring of adhesively bonded composite patch repair of aircraft structures using wavelet transforms of lamb wave signals. *Smart Structures and Integrated Systems*, SPIE 5056.

393. Hong JC and Kim YY. 2003. The optimal selection of mother wavelet shape for the best time-frequency localization of the continuous wavelet transform. *Modeling, Signal Processing and Control*, SPIE 4049.

394. Dawood TA, Shenoi RA, Veres SM, Sahin M, and Gunning MJ. 2003. Damage detection in a sandwich composite beam using wavelet transforms. *Modeling, Signal Processing and Control*, SPIE 4049.

395. Amaravadi K, Rao V, and Derriso MM. 2003. Structural health monitoring of composite patch repairs using two-dimensional wavelet maps. In *Structural Health Monitoring*, FK Chang (ed.). DEStech, Lancaster.

396. Hou Z, Noori M, and St Amand R. 2000. Wavelet-based approach for structural damage detection. *J Eng Mech*, 126(7), 677–683.

397. Hou Z, Hera A, Liu W, and Hendrickson D. 2003. Identification of instantaneous modal parameters of time-varying systems using wavelet approach. In *Structural Health Monitoring*, FK Chang (ed.). DEStech, Lancaster.

398. Wimmer SA and DeGiorgi VG. 2004. Detecting damage through the application of a CWT based algorithm. *Health Monitoring and Smart Nondestructive Evaluation of Structural and Biological Systems III*, SPIE 5394.

399. Liew KM and Wang Q. 1998. Application of wavelet theory for crack identification in structures. *J Eng Mech*, 124(2), 152–157.

400. Sun Z and Chang CC. 2002. Structural damage assessment based on wavelet packet transform. *J Struct Eng*, 128(10), 1354–1361.

401. Masuda A, Sone A, and Yamamoto S. 2004. Health monitoring of beam structures using traveling waves induced by ambient loads. *Sensors and Smart Structures Technologies for Civil, Mechanical and Aerospace Systems*. SPIE 5391.

402. Rizzo P and Lanza di Scalea F. 2004. Discrete wavelet transform to improve guided-wave-based health monitoring of tendons and cables. *Sensors and Smart Structures Technologies for Civil, Mechanical and Aerospace Systems*, SPIE 5391.

403. Law SS, Li XY, and Lu ZR. 2006. Structural damage detection from wavelet coefficient sensitivity with model errors. *J Eng Mech*, 132(10), 1077–1087.

404. Kumar R and Rao MV. 2003. Prestressed concrete bridges in India. *Concrete Intl*, 25(10), 75–81.

405. Yan R and Gao RX. 2004. Harmonic wavelet packet transform for on-line system health diagnosis. *Sensors and Smart Structures Technologies for Civil, Mechanical and Aerospace Systems*, SPIE 5391.

406. Nayfeh AH and Mook DT. 1979. *Nonlinear Oscillations*. Wiley, New York.

407. Jezequel L and Lamarque CH. 1993. Nonlinear mechanical systems, sinusoidal excitation signal and modal analysis: Theory. In *Safety Evaluation Based on Identification Approaches*, HG Natke, GR Tomlinson, JTP Yao (eds.). Vieweg, Braunschweig.

408. Wen YK. 1976. Method for random vibration of hysteretic systems. *J Eng Mech Div* (ASCE) 102(EM2), 249–262.

409. Tan GE and Pellegrino S. 1997. Non-linear dynamic identification: An application to prestressed cable structures. *J Sound Vib*, 208(1), 33–45.

410. Ashrafiz SA, Smyth AW, and Betti R. 2006. A parametric identification scheme for non-deteriorating and deteriorating non-linear hysteretic behaviour. *Struct Control Health Monit*, 13, 108–131.

411. Donskoy D, Ekimov A, Luzzato E, Lottiaux JL, Stoupin S, and Zagrai A. 2003. N-SCAN: New vibro-modulation system for detection and monitoring of cracks and other contact-type defects. *Smart Systems and Nondestructive Evaluation for Civil Infrastructures*, SPIE 5057.

412. Todd M, Chang L, Erikson K, Lee K, and Nichols J. 2004. Nonlinear excitation and attractor mapping for detecting bolt preload loss in an aluminum frame. *Health Monitoring and Smart Nondestructive Evaluation of Structural and Biological Systems III*, SPIE 5394.

413. Quinn DD, Mani G, Kasarda ME, Bash T, Inman DJ, and Kirk RG. 2005. Damage detection of a rotating cracked shaft using an active magnetic bearing as a force actuator-analysis and experimental verification. *IEEE/ASME Trans Mechatronics*, 10(6), 640–647.

414. Testa RB, Zhang W, Smyth AW, and Betti R. 2000. Detection of cracks with closure. In *Structural Materials Technology IV*, S Allampalli (ed.). Technomic, Lancaster.

415. Bovsunovsky AP and Surace C. 2005. Considerations regarding superharmonic vibrations of a cracked beam and the variation in damping caused by the presence of the crack. *J Sound Vib*, 288, 865–886.

416. Hurlebaus S, Kogl M, and Gaul L. 2004. Damage selective imaging by nonlinear response analysis. *Proc 2nd European Workshop on Structural Health Monitoring*. DEStech, Munich.

417. Das A, Papandreou-Suppappola A, Zhou X, and Chattopadhyay A. 2005. On the use of the matching pursuit decomposition signal processing technique for structural health monitoring. *Smart Structures and Integrated Systems*, SPIE 5764.

418. Staszewski WJ and Buderath M. 2004. Vibro-acousto-ultrasonics for fatigue crack detection and monitoring in aircraft components. *Proc 2nd European Workshop on Structural Health Monitoring*. DEStech, Munich.

419. Zak A, Krawczuk M, and Ostachowicz W. 1999. Vibration analysis of composite plate with closing delamination. *Proc 2nd Intl Workshop on Structural Health Monitoring*, Stanford. Technomic, Lancaster.

420. Hanagud S and Luo H. 1997. Damage detection and health monitoring based on structural dynamics. In *Structural Health Monitoring Current Status and Perspectives*. F-K Chang, (ed.). Technomic, Lancaster.

421. Stauffer JD, Woodward CB, and White KR. 2005. Nonlinear ultrasonic testing with resonant and pulse velocity parameters for early damage in concrete. *ACI Mater J*, 102–M14.

422. Casti JL. 1985. *Nonlinear System Theory*. Academic Press, New York.

423. Smyth AW, Masri SF, Kosmatopoulos EB, Chassiakos AG, and Caughey TK. 2002. Development of adaptive modeling techniques for nonlinear hysteretic systems. *Intl J Nonlinear Mech*, 37(8), 1435–1451.

424. Luong MP. 1993. Safety evaluation of electric pylons using impulse testing. In *Safety Evaluation Based on Identification Approaches*, HG Natke, GR Tomlinson, JTP Yao (eds.). Vieweg, Braunschweig.

425. Zhang JH, Natke HG, Qiu Y, and Zhang XN. 1993. The identification of nonlinear systems with statistically equivalent polynomial systems. In *Safety Evaluation Based on Identification Approaches*, HG Natke, GR Tomlinson, JTP Yao (eds.). Vieweg, Braunschweig.

426. Worden K and Manson G. 1998. Random vibrations of a duffing oscillator using the volterra series. *J Sound Vib*, 217(4), 781–789.

427. Hildebrand FB. 1992. *Methods of Applied Mathematics*. Dover Publications, Mineola, NY.

428. Shukla A and Frederick A. 2004. Smart damage prediction: A distance to bifurcation based approach. *Health Monitoring and Smart Nondestructive Evaluation of Structural and Biological Systems III*, SPIE 5394.

429. Scanlan RH and Tomko JJ. 1971. Airfoil and bridge deck flutter derivatives. *J Eng Mech*, 97(6), 1717–1737.

430. Thompson JM and Stewart HB. 1986. *Nonlinear Dynamics and Chaos*. Wiley, Chicester.

431. Shin S, Oh SH, and Jo JY. 2004. Structural damage assessment by nonlinear time-domain SI scheme. *Proc IABMAS'04 Bridge Maintenance, Safety, Management and Cost*, Kyoto. Taylor & Francis, London.

432. Mita A and Taniguchi R. 2004. Active damage detection method using support vector machine and amplitude modulation. *Sensors and Smart Structures Technologies for Civil, Mechanical and Aerospace Systems*, SPIE 5391.

433. Epureanu BI, Yin SH, and Derriso MM. 2005. High-sensitivity damage detection based on enhanced nonlinear dynamics. *Smart Mater Struct*, 14, 321–327.

434. Chelidze D. 2004. Identifying multidimensional damage in a hierarchical dynamical system. *Nonlinear Dyn*, 37(4), 307–322.

435. Cusamano JP and Chelidze D. 2005. Phase space warping-a dynamical approach to diagnostics and prognostics. *IUTAM Symp on Chaotic Dynamics and Control of Systems and Processes in Mechanics, Solid Mechanics and Its Applications*, Vol. 122. Springer, New York.

436. Cusumano J, Chelidze D, and Chatterjee A. 2003. General method for tracking the evolution of hidden damage or other unwanted changes in machinery components and predicting remaining useful life. US Patent 6,567,752.

437. Epureanu BI, Yin SH, and Derriso MM. 2004. Attractor-based damage detection in a plate subjected to supersonic flows. *Health Monitoring and Smart Nondestructive Evaluation of Structural and Biological Systems III*, SPIE 5394.

438. Dowell EH and Ilgamov M. 1988. *Studies in Nonlinear Aeroelasticity*. Springer, New York.

439. Guckenheimer J and Holmes P. 2002. *Nonlinear Oscillations, Dynamical Systems, and Bifurcations of Vector Fields*. Springer, New York.

440. Moon FC. 1987. *Chaotic Vibrations*. Wiley, New York.

441. Eckmann JP and Ruelle D. 1985. Ergodic theory of chaos and strange attractors. *Rev Mod Phys*, 57(3), 617–656.

442. Livingston RA and Jin S. 2006. Nonlinear dynamics simulation of local chaotic behavior in the Bill Emerson cable-stayed Bridge. *NDE Conf on Civil Engineering*, American Society for Nondestructive Testing, St. Louis.

443. Sawicki JT, Wu X, Gyekenyesi AL, and Baaklini GY. 2005. Application of nonlinear dynamics tools for diagnosis of cracked rotor vibration signatures. *Nondestructive Evaluation and Health Monitoring of Aerospace Materials, Composites and Civil Infrastructure IV*, SPIE 5767.

444. Todd MD, Nichols JM, Pecora LM, and Virgin LN. 2001. Vibration-based damage assessment utilizing state space geometry changes: Local attractor variance ratio. *Smart Mater Struct*, 10, 1000–1008.

445. Jin S, Livingston RA, and Marzougui D. 2001. Stochastic system invariant spectrum analysis applied to smart systems in highway bridges. *Smart Systems for Bridges, Structures, and Highways*, SPIE 4330.

446. Seaver M, Nichols JM, Trickey ST, and Todd MD. 2003. Weld line degradation assessment using chaotic attractor property analysis. *Smart Nondestructive Evaluation and Health Monitoring of Structural and Biological Systems II*, SPIE 5047.

447. Moniz L, Nichols JM, Nichols CJ, Seaver M, Trickey ST, Todd MD, Pecora LM, and Virgin LN. 2005. A multivariate, attractor-based approach to structural health monitoring. *J Sound Vib*, 283, 295–310.

448. Nichols JM, Hunter MJ, Seaver M, and Trickey ST. 2003. The role of excitation in attractor-based structural health monitoring. In *Structural Health Monitoring*, FK Chang (ed.). DEStech, Lancaster.

449. Han X, Islam M, Li W, Loggins V, Lu J, Zeng Z, Favro LD, Newaz G, and Thomas RL. 2003. Acoustic chaos in sonic infrared imaging of cracks in aerospace components. In *Structural Health Monitoring*, FK Chang (ed.). DEStech, Lancaster.

450. Arnold VI. 1992. *Catastrophe Theory*. Springer, Berlin.

451. Nataraju M, Adams DE, and Rigas E. 2003. Nonlinear dynamics for modeling and predicting damage accumulation. In *Structural Health Monitoring*, FK Chang (ed.). DEStech, Lancaster.

452. Walgraef D. 1997. *Spatio-Temporal Pattern Formation*. Springer, New York.

453. Brown RL and Adams DE. 2003. Equilibrium point damage prognosis models for structural health monitoring. *J Sound Vib*, 262(3), 591–611.

454. Shukla A. 2005. Bio-inspired design of sensor system for damage prognosis. *Health Monitoring and Smart Nondestructive Evaluation of Structural and Biological Systems IV*, SPIE 5768.

455. Gilmore R. 1993. *Catastrophe Theory for Scientists and Engineers*. Dover, Mineola, NY.

456. Huston DR. 1987. Snap through and bifurcation in a simple structure. *J Eng Mech*, 113(12), 1977.

457. Zhang Y, Wang Y, Li Z, Huang Y, and Li D. 2007. Snap-through and pull-in instabilities of an arch-shaped beam under an electrostatic loading. *J Microelectromech Syst*, 16(3), 684–693.

458. Huston D. 1986. The effects of upstream gusting on the aeroelastic behavior of long suspended-span bridges. PhD dissertation, Princeton University, Civil Engineering.

459. Fasel TR, Sohn H, and Farrar CR. 2003. Application of frequency domain ARX models and extreme value statistics to damage detection. *Smart Systems and Nondestructive Evaluation for Civil Infrastructures*, SPIE 5057.

460. Park G, Rutherford AC, Sohn H, and Farrar CR. 2005. An outlier analysis framework for impedance-based structural health monitoring. *J Sound Vib*, 286, , 229–250.

461. Nichols JM, Seaver M, Trickey ST, Todd MD, Olson C, and Overbey L. 2005. Detecting nonlinearity in structural systems using the transfer entropy. *Phys Rev E*, 72, 046217.

7

Adaptive and Automated Sensing

Unpredictability is common in many aspects of SHM. If structural behaviors were entirely predictable, there would be little need for SHM. This is not always the case. Many aspects of structural loading, response, damage formation, and damage propagation are unpredictable. Designing and implementing an SHM effort without accounting for unpredictability risks inefficiency, and the failure to collect and act upon important data. Choosing the proper measurement location, timing and operational parameters of an SHM effort may be crucial for realizing successful outcomes. It is, often, not always practical to set up SHM systems with sufficient sensing resolution to capture all unpredictable phenomena of interest. Adaptation is a strategy successfully used by many biological and manmade systems to counter the effects and even take advantage of unpredictable stressors and environmental changes [1,2]. Applying such techniques to SHM has the potential of significantly enhancing the performance of many SHM systems [3].

High-value structures, with predictable failure modes, such as dams, routinely use pre-placed hard-wired permanent sensor suites for SHM [4]. When the failure and damage modes are less predictable, the possibility of missing vital data with a fixed preplaced system and the inherent limited structural coverage arises. An example is the space shuttle Columbia. Being the first shuttle, an extra suite of 800 strain, temperature, and other sensors measured and recorded structural conditions and performance, with the primary purpose being for postflight and not real-time analyses. During Columbia's final flight, the system did not provide any inkling that fatal structural damage had occurred prior to the final descent. The postmortem data clearly showed the sequential loss of sensor channels to indicate the rapid progression of damage during descent. The underlying causes were somewhat elusive and ultimately attributed to ice impact onto the TPS during launch several days earlier [5]. As an adaptation following the Columbia accident, it is planned to include additional sensing capabilities in future shuttle missions, such as front panel impact sensors and thorough in-flight visual inspections. Similarly, the International Space Station now uses externally mounted cameras for damage inspection and evaluation [6].

Common features of adaptive SHM sensor systems are: (1) An underlying motivation that drives the adaptation. One possibility is to maximize the information content per measurement. (2) The system must be able to alter its operating behavior in response to the classification of input data. (3) The state alteration process must be stable. A potential problem is switching states too often. Adding hysteresis to state-switching cycles enhances stability. Setting the proper level of hysteresis usually requires a trial and error effort. If the level of hysteresis is set too high, the system will not switch states quickly enough to make an effective data capture. If the hysteresis is too small, the system may switch states too often.

Until recently, humans have been the primary system for adaptive SHM. Trained humans can recognize and focus their attention on critical situations for more detailed examination.

An example of this approach is the assessment of postconstruction leaks in the Boston Central Artery tunnel [7]. A fairly significant leak (300 gpm) occurred in the tunnel soon after opening on September 15, 2004. Following plugging of the leak, a team of inspectors conducted a fairly rapid assessment in December 2004. The team proceeded by first identifying the likely cause of the accident (deficiencies in slurry wall construction), next identifying portions of the tunnel with similar construction details that may have similar deficiencies, and then examining the potential problem areas in more detail. Most of the inspections relied on visual examinations to determine if there were any minor leaks at potentially vulnerable portions of the tunnel. The inspection appeared to make little use of automated instruments for sensing and NDE of subsurface features in the tunnel.

High-risk and hard to reach locations render impractical the use of humans in many applications. Additionally, biases, imperfect memories, and variable performance can degrade human inspection performance [8,9]. Automated and mobile systems are a means of providing adaptability to a machine-based SHM system for situations where using humans is impractical. Patterson-Hine et al. recommend using automated diagnostic systems in the following cases [10]:

1. The automated system can provide valuable information that could not be obtained at all, or quickly enough to be useful, without the automated system.

2. The automated system offers significant improvements in the quality of information over human-performed diagnostic activities, such as increased accuracy or consistency.

3. The automated system can perform the diagnostic function at a lower cost than human-performed diagnosis.

A schematic of a hierarchical adaptive SHM data collection process is shown in Figure 7.1. The adaptive SHM process uses multiple inspection sweeps, each with a different data collection protocol. The sensitivity and selectivity of the first pass are set at levels so as to identify all (or most) possible items of interest, that is, damage and damage precursors. The intention is to avoid missing important items in the first pass while running the risk of high false alarm rates. Next, more expensive protocols and procedures with high sensitivity and high selectively evaluate potential problem areas. The intention is to produce fewer false positive alarms in this second pass. For example, if a routine inspection or operation of the structure identifies a crack, subsequent examinations will take a closer look at the crack. If the close-up examination determines that the crack is potentially significant or dangerous, prudence may require attaching crack-width-monitoring instruments, reading the output at regular intervals and/or tripping an alarm. Figure 7.2 shows an adaptive inspection process for evaluating the postearthquake condition of structures [11].

Many (if not most) SHM systems suffer from sensor and/or data acquisition faults. Detecting, diagnosing, and correcting for sensor system faults is often quite important. In a complicated multichannel heterogeneous sensor system, it may not be practical to use operator-based detection and correction methods. Instead automated methods may be needed. Many of the same techniques used for structural fault detection apply equally well to the detection of SHM system faults. For example, Kerschen et al. used principal component analysis (PCA) analysis of data streams to detect sensor faults [12].

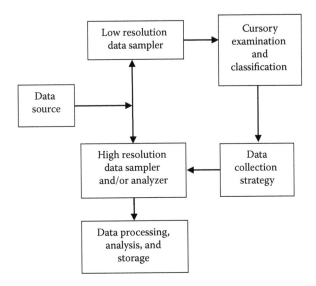

FIGURE 7.1 Two-stage hierarchical adaptive data collection process.

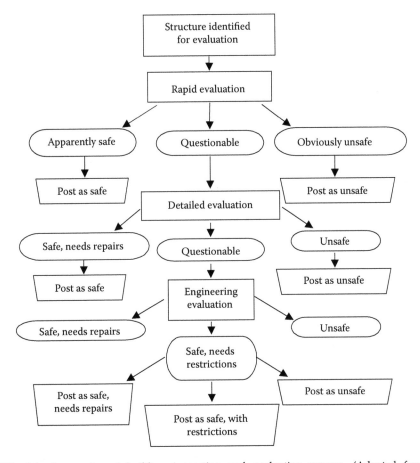

FIGURE 7.2 Adaptive postevent building inspection and evaluation process. (Adapted from Applied Technology Council. 2004. ATC-45 Field Manual: Safety Evaluation of Buildings after Windstorms and Floods, Redwood City.)

7.1 Data-Adaptive Data Processing

Perhaps the easiest adaptive SHM technique to implement is a data-driven modification of the data processing algorithms. Adaptive data processing forms the basis of many of the "modern" signal processing and statistical analysis techniques discussed in Chapters 5 and 6. One approach estimates damage location and extent in a two-step hierarchical process that largely mimics the action of Figure 7.1, without resorting to changes in the data collection protocols and procedures [13–17]. The first step examines statistical features from the data, such as mode shapes or mechanical impedances, to determine the location of possible damage. The next step improves the estimate of the damage location by processing more accurate (and usually more expensive) frequency and stiffness estimates. As an example Sim et al. used novelty detectors to modify a damage classification algorithm based on observed multipoint transmissibilities [18]. Safak proposed a time domain-filter fitting algorithm for damage detection in structures during and after earthquakes by adaptive data processing that operates in a fast mode during earthquakes and in slower modes during quiescent interludes [19]. Relational database techniques that classify and organize data based on the nature of the data are also potential data-adaptive signal processing applications [20].

7.2 Data-Adaptive Data Acquisition

Adapting the operational behavior of sensor and data acquisition systems as circumstances warrant is an effective and often necessary means of improving overall SHM system performance. Many methods of data-adaptive sensing are relatively low cost and easy to implement. These include selecting which data to acquire, store, transmit, and analyze; and tuning the sensor system to match the amplitude and frequency characteristics of the data.

7.2.1 Triggering, Input Buffering, and Autoranging

Triggering, input buffering, and *autoranging* techniques match the operational parameters of the data acquisition system to that of the analog sensor data. A data acquisition trigger selects when to acquire data. A simple trigger strategy activates the system when data signals exceed predetermined thresholds. More complex multichannel rule-based amplitude and time-duration weighted triggers are possible. The practical implementation of triggering is adaptive. The process examines past sets of gathered data to see if they were worth collecting and adjusts the trigger parameters as needed.

Many transient events, such as impacts or earthquake loads, produce data that are difficult to capture with a trigger-activated data acquisition system. Losing important pretrigger data is a risk. A remedy is to use a circular pretrigger data buffer to collect data continuously. The circular buffer contains all of the data from a predetermined past interval. Newly acquired data overwrite the oldest data stored in the buffer. Upon completion of a successful event trigger, the data acquisition system saves the pretrigger data along with the new data. The downside to using pretrigger circular buffers is the high power requirements of operating the data acquisition system continuously.

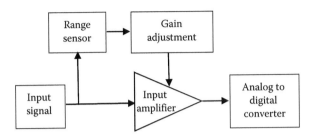

FIGURE 7.3 Autoranging circuit.

Autoranging facilitates acquiring data streams with uncertain and/or large dynamic ranges. Autoranging automatically adjusts the input amplifier gain levels to match the analog data signal amplitudes to the dynamic range of the ADC. Figure 7.3 shows a schematic of a simple autoranging circuit. An interesting alternative to autoranging uses digital data acquisition systems with large dynamic ranges, for example, 24-bit ADCs. Such techniques off-load the need for adaptation onto the data analysis algorithms. However, to be effective, the analog signal processing hardware upstream of the ADCs must also be able to process signals with high fidelity over large dynamic ranges.

7.2.2 Data-Adaptive Scheduling and Reporting

Distributed and networked data acquisition systems readily accommodate complicated adaptive triggering, acquisition, and data transmission strategies. Spatially variable triggers, with considerable configuration and operational flexibility, are possible. Distributed arrays of wireless battery-powered sensors are a prime example of systems that take advantage of spatially and temporally adaptive scheduling and reporting. Implementations of adaptation may be local where each sensor responds autonomously to the local environment, globally where a central controller broadcasts control signals to the network, or regionally where the sensor nodes interact and adapt in complex patterns.

Electric power consumption is a principal constraint on battery-powered sensor systems. Depending on the desired level of activity, a system can operate at different levels of power consumption. Wireless data transmission is often the biggest power hog. A low-power mode of operation is one that takes measurements at scheduled intervals, for example, once every hour, and then intermittently transmits small data sets to a central host computer for storage and further analysis. If circumstances warrant, the system could switch to a medium-power-consumption mode, with an increased level of activity. The sensors then collect and transmit data at higher rates. A high-power-consumption mode of operation is for the local sensor nodes to continuously collect and transmit data in a near to real time mode. Variations include operating different regions of the structure in different modes with a hierarchical organization, and having a minimal set of sensors continuously operating in a sentry alarm mode. Ideally, only those sensors and transmission links that need to be active are active.

7.2.3 Self-Assembling Sensor Networks

Certain manifestations of sensor networks are self-assembling (see Chapter 4). It is conceivable that the self-assembly protocols and implementations operate so as to maximize the density and/or amount of transmitted information. One biomimetic adaptive approach is

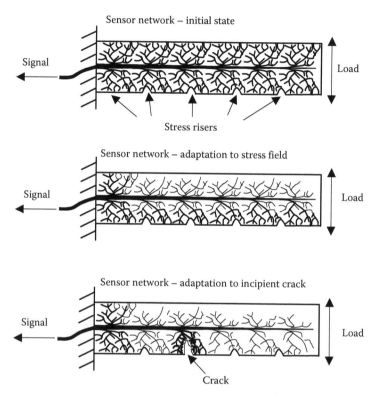

FIGURE 7.4   Biomimetic sensor adaptation based on information content link pruning and strengthening.

to place a redundant high-density set of sensors throughout the structure in a generic non-specific manner. The distribution could be homogeneous or biased with a higher density of sensors in known regions of significance, for example, hands and tongues have high nerve cell densities. Upon operation, those sensors that transmit the most useful information have their network links strengthened, for example, they transmit more often. Those that contain less useful information transmit less often or are disconnected from the system. Figure 7.4 shows how such an adaptation might occur near a stress riser in a structure. Such a process would mimic the development of the optic nerve in vertebrates [2]. Early in a creature's lifetime, those nerve cells that transmit useful information from the eye to the brain are strengthened. Those that do not are pruned. A consequence of this process is that neglecting to stimulate the optical nerve at appropriate early stages of development results in permanent blindness.

7.2.4 Squinting and Zooming

The human eye is very adept at adapting operational parameters to optimize image collection. Eyes can quickly pan wide areas, identify features of interest, autofocus on those features, apply enhanced neural signal processing by focusing items of interest on the part of the retina with the highest density of receptors and concentrate the activities of the brain on that object. Modern digital camera and image processing systems are similarly capable. Cameras can zoom, pan, tilt, and autofocus on command, but presently lack the dynamic range of the human eye.

Adaptive vision and imaging systems have considerable SHM potential. An example is a system that correlates past and present images of the structure. The system detects differences for more detailed examination. Geometric differences and cracking are obvious measurands. Color differences and IR imaging can indicate surface conditions, for example, corrosion, and by inference subsurface conditions. Long-term image storage provides a historical record of the structure for use in future evaluations and for implementing adaptive novelty detection and signal processing. Proper image registration and effective data processing algorithms are nontrivial technical challenges associated with these techniques.

A modest number of applications of adaptive image processing appear in the SHM literature. Liu et al. used adaptive image processing to locate and identify corrosion in aircraft skin rivets [21]. Duran et al. describe a pipe crawling inspection robot with a laser vision system combined with artificial neural networks to detect geometric flaws inside of pipes [22]. Other imaging NDE instruments, such as ultrasonic corrosion detectors, can also use squinting and precision focusing techniques [23]. Steady progress continues with regard to software-based pattern and object recognition capabilities [24]. Ueda et al. developed an electrical-property-change-based crack monitoring system for CFRP composites. The system uses a zooming technique to identify zigzag cracks [25]. Kress developed a Smart Wide area Imaging Sensor System (SWISS) that scans plate-like structures with ultrasound waves for damage detection [26]. Controlled scanning with ultrasound waves examines regions of interest. Williams et al. evaluated adaptive squinting methodology for NDE sampling of large concrete structures [27,28]. The sampling strategy initially is to take NDE data at random locations across the structure. Those locations showing distress prompt an examination of neighboring locations. Random sampling followed by adaptive sampling did not prove to be more effective than systematic first scans.

An example of systematic sampling followed by adaptive sampling appears in Figure 7.5. Satellite images of the Bam citadel in Iran before and after the December 2003 earthquake with zoom insets indicate locations of severe damage [29]. Bennett et al. combined machine image processing with human input to discern crack locations in a robotic pavement crack sealer [30].

Analogs of squinting and zooming appear in adaptive data processing algorithms of non-imaging data. For example, Hjelmstad used an adaptive parameter grouping approach to detect damage from sparse sensor data sets [31]. The method uses a hierarchical approach to identify and localize damage. McGugan et al. used acoustic emission measurements as part of a composite wind turbine blades test program [32]. The acoustic emission measurements provided an early warning of impending failures that warranted a closer follow-up examination by other NDE methods. Wen et al. used an adaptive signal processing technique incorporated with a mobile sensor system to detect leaks in underground pipes with acoustic waves [33].

7.2.5 Active Sensing Systems

Embedded actuators and sensors can combine their collective actions to form an active structural interrogation system. Such systems can actively interrogate structures in an automated and an adaptive manner. Examples include the use of piezoelectric elements as active stimulation sources combined with arrays of sensing elements, such as piezoelectric patches or FBGs as sensors [34]. Awata and Mita developed an active source system to

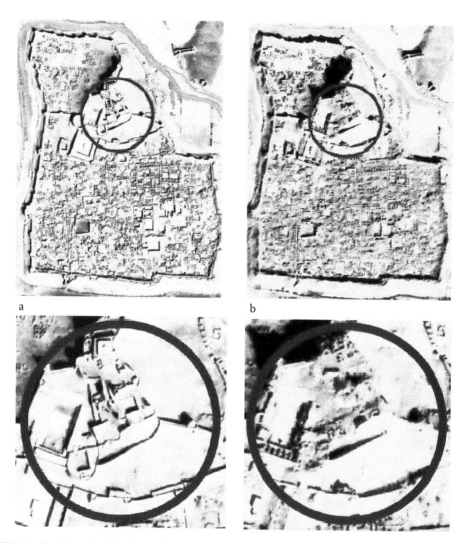

FIGURE 7.5 The Bam citadel in Iran before and after the December 26, 2003 earthquake with close-up zoom. (Picture courtesy of the Digital Globe; Adams BJ et al. 2004. The Bam (Iran) Earthquake of December 26, 2003: Preliminary Reconnaissance Using Remotely Sensed Data and VIEWS System. MCEER Earthquake Investigation Report, Bam 12-26-03.)

generate controlled ultrasonic waves for ball bearings damage detection. The ultrasound waves served as carrier signals for narrow band acoustic signals generated by damage in ball bearings [35]. Sweeping the source carrier frequency produced an enhanced diagnostic capability. Similarly, Jiang et al. tuned piezoelectric patch excitation frequencies to increase damage detection performance [36]. The tuning process used a numerical model of damage scenarios with updating.

It is possible for a sensing system to identify its own state of health and to adjust the operation of the sensing system accordingly. White and Adams demonstrated the use of self-diagnostics on the piezoelectric source–receiver diagnostic system for TPS health monitoring [37]. The system detects the degradation of the transducer mounts by broadband FRF shifts. If the degradation is modest, the system can compensate, that is, self-heal, by adjusting the drive signal amplifier gains.

7.3 Mobile Sensor Systems

Mobility is another enabler for adaptive SHM. Mobile sensor systems may be particularly useful for short-term monitoring, routine inspection, or in emergent situations that require the rapid application of the sensors to the structure, and for situations where it is impractical or dangerous to insert humans for inspection. Mobile sensors can attach onto a structure or can operate in a standoff noncontact imaging mode. Attachable sensors require convenient means of attachment and, if desired, a convenient means of removal. Standoff imaging systems require a vehicle or other means to move about the structure, such as with beamsteering. Handheld units are useful in many applications.

7.3.1 Deployable Sensors

Deployable and redeployable sensors are convenient for adaptive and short-term sensing. Sensor deployment modalities include attaching or inserting into structures, and removal and placement at new locations, as is needed. Adhesives, bolting, clamp-on, or magnetic attachments can provide an adequate attachment grip for short-term sensor deployment. Beeswax and petroleum-based equivalents are common temporary adhesives for small accelerometers. Plaster of Paris can attach larger sensors to horizontal surfaces. Foil strain gages are a bit more of an attachment challenge. Attaching the gages directly to structures can produce high-quality data, but is labor and skill intensive. An alternative is weldable systems that use a strain gage preattached to a metal plate, and the tack welding of the plate onto the structure. Mechanical clamping of the plate is another possibility. Figure 7.6 shows a clamp-on Z-gage that measures strain with strain gages preattached to an isolated shear plate. The strain gages transduce the relative longitudinal motions of the clamped ends of the Z-gage. Schulz and Commander report that clamp-on strain transducers require 3–5 minutes of installation time per transducer versus 30 minutes for the direct bonding of strain gages to the structure [38].

Magnets can attach accelerometers and other sensors to ferromagnetic structures. Simple magnets have an inherently harsh force-displacement attachment action. More complex

FIGURE 7.6   Clamp-on Z-gage. (Adapted from Huston D et al. 2003. *Proc 3rd Intl Workshop on Structural Health Monitoring*, Stanford University.)

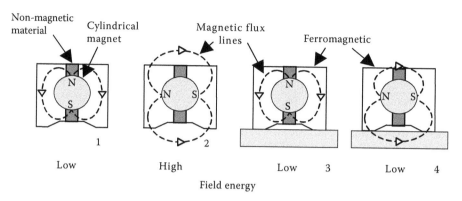

FIGURE 7.7 Four states of the smooth-force magnetic attachment system. (Adapted from Huston DR, Miller J, and Esser B. 2004. *Smart Structures and Materials 2004: Sensors and Smart Structures Technologies for Civil, Mechanical, and Aerospace Systems*, SPIE 5391.)

magnetic mechanisms can provide a smoother attachment process. Figure 7.7 shows a magnetic circuit system with four energy states—three low and one high. Moving in a continuous low-energy state sequence, such as 1-3-4-3-1, provides for a smooth low force and energy attachment and removal.

The SHM literature contains many examples of using deployable sensors. Tsuchimoto et al. moved sensors (accelerometers) to different locations to locate and quantify damage by wave propagation methods [39]. Reed et al. developed PZT disks to attach quickly to plates for ultrasonic monitoring of impact damage defects in plates [40]. Popular press reports describe ballistically deployed sensor modules with possible SHM applicability [41]. Wireless battery-powered systems enable rapid deployment techniques. Rong and Cuffari report that the benefits of wireless deployable strain sensors for bridge fatigue monitoring outweigh the difficulties associated with limited telemetry ranges and the need for battery replacement (Figure 7.8) [42]. Tikka et al. measured the growth of cracks in aircraft structures with strain gages positioned near known cracks [43]. Combining the strain readings with a finite element stress analysis of the structure and crack geometry enabled crack growth monitoring with the strain reading.

Ultrasound transducers typically require direct mechanical contact with a structure in order to take readings. A standard method uses a liquid or gel in the interface as a coupling agent. Dry contact provides good transmission by establishing a solid contact point, that is, Hertzian point contact, preloaded rollers or compliant polymer interfaces [44–46]. Kozlov et al. found that transverse shearing transducer motions were superior to longitudinal motions for dry contact testing due to a better impedance match with the possible elastic waves in the solid [47]. The decoupling of dynamic and static transducer–solid contact forces is another explanation of the dry-contact effectiveness of shear-motion transducers. Pardini et al. discuss in detail the development of a remotely controlled mobile ultrasonic inspection system with knuckle joints in liquid waste storage tanks [48]. Gräfe and Krause, and Cetrangolo and Popovics developed an air-coupled ultrasound test system for concrete [49,50]. A key feature for the success of the latter study was the use of quarter-wavelength Styrofoam impedance-matching layers attached to the transducers.

A distributed communications network preexisting in the structure can assist in mobile and rapid sensor deployment by providing convenient communications and possibly power. Ethernet/Internet communication protocols have the bandwidth to support arrays of point sensors and image data [51].

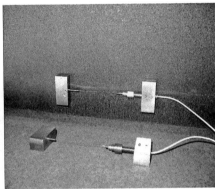

FIGURE 7.8 Deploying a magnetically attachable strain sensor on a bridge girder. (Adapted from Huston DR, Esser B, and Miller J. 2004. *Proc 2nd European Workshop on Structural Health Monitoring*, DEStech, Munich; Sensor provided by Microstrain Inc.)

7.3.2 Handheld Systems

Handheld systems are convenient enablers of adaptive SHM. Handheld systems can carry a variety of NDE instruments, that is, eddy current, ultrasound, optical, and magnaflux. Successful handheld sensor design requires considerations of human–machine interactions that include: minimization of safety hazards, weight minimization, convenient grips, flexible cables, minimal requirements for setup and fiddling in the field, readouts that are visible in sunlight, and easy-to-interpret display fields [52]. It is likely that the continued development of miniaturized NDE instruments will produce more sophisticated and highly capable systems, such as multichannel sensors with imaging capabilities [53] (Figure 7.9).

The SHM literature contains numerous descriptions of the use of handheld SHM/NDE systems. Snell and Duinkherjav evaluated the performance of handheld devices for locating reinforcing bars and cover in reinforced concrete (RC) [54]. Giurgiutiu and Xu developed a portable EMI analyzer [55]. Portable optical shearography systems are now available [56,57]. Komsky used a rolling ultrasonic test head for mobile inspection of aircraft structures [58]. Nazarian et al. developed an integrated handheld system, known as a PSPA, that uses a combination of ultrasonic body waves, ultrasonic surface waves, impact echo, impulse response, and spectral analysis of surface wave techniques for pavement assessments [59].

7.3.3 Endoscopes

Endoscopes are thin articulated imaging systems for insertion into tight spaces for inspection. Endoscopes transmit images by rigid optics, often with long *graded index* (GRIN) lenses, coherent fiber optic bundles, or miniature cameras. The use of endoscopes in structural inspection is fairly limited. However, endoscopes have found use in the inspection of items

FIGURE 7.9 Handheld instrument that combines concrete cover and half-cell measurements. (Photograph courtesy of Proceq SA, Switzerland.)

such as anchorages, tendons, and inaccessible joints [60]. It is possible to use articulated robotic snake arms in high-performance endoscope inspection systems.

7.3.4 Inspection Vehicles

Specialized machines transport and place human inspectors in locations they cannot conveniently climb or reach. Inspection vehicles are also useful for inspecting pavements and roadways, especially when large road areas need to be inspected and/or it is dangerous to stop traffic [61]. Figures 7.10 and 7.11 show a roadway inspection vehicle that is equipped with multiple sensors including GPS and accelerometers [62].

7.3.5 Automated Scanning Systems

Automated systems can potentially transfer many repetitive, tedious, and dangerous measurement tasks from humans to machines. Most of the automated NDE/SHM systems used so far act on structures with a sufficiently simple form to be scanned easily with highly repetitive measurements, for example, pipes. Fully automated systems require mechanical staging to move the sensors, a control system to activate the sensor, and signal processors to condition and interpret the data. Semiautomated systems use humans in the control loop.

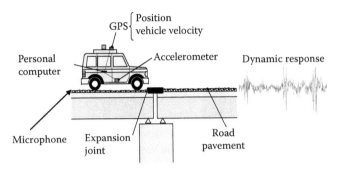

FIGURE 7.10 Concept of the Vehicle Intelligent Monitoring System (VIMS). (From Fujino Y et al. 2005. *Sensors and Smart Structures Technologies for Civil, Mechanical, and Aerospace Systems*, SPIE 5765. With permission.)

Reported uses of automated systems for SHM are now fairly numerous. Mares et al. moved ultrasound transducers across a plate with an automated mechanical staging system [63]. Rezai et al. used an automated ultrasound imaging system mounted on a magnetic base with a three degrees of freedom (DOF) articulated arm to detect defects, such as cracks in steel butt welds, that used propagating ultrasonic Lamb waves [64]. Henderson et al. developed automated chain-drag concrete slab testers to alleviate some of the difficulties with using humans for chain drag testing in high-traffic areas [65]. The systems mimic human chain drag methods with an automated chain dragger and digital spectral analysis of received acoustic signals. Delaminated concrete sections produced acoustic signals with significantly larger amounts of power in the 2–8 kHz range than intact nondelaminated sections. Colla assessed the integrity of post-tensioning strand ducts with an automated impact echo instrument [66]. Wiggenhauser et al. and Streicher et al. combined GPR, ultrasound,

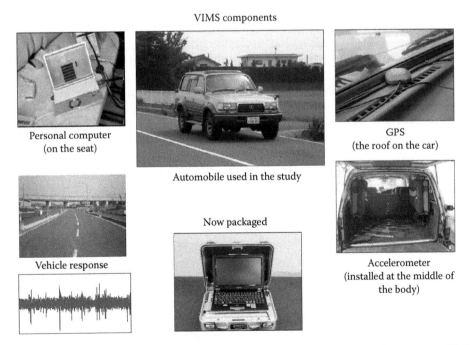

FIGURE 7.11 The VIMS multisensor road and bridge inspection vehicle. (Photographs courtesy of Y. Fujino, Univ. of Tokyo.)

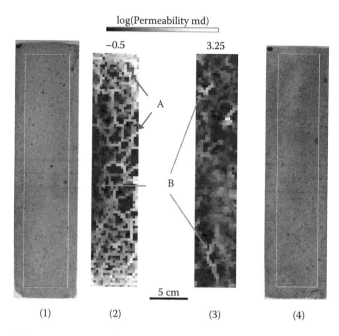

FIGURE 7.12 Permeability measurements of two concrete beams 30 cm × 8 cm × 8 cm subjected to controlled air freeze and water thaw (F/T) tests. (1) and (2) show the photographic image and gas permeability map, respectively, of a sample with 270 F/T cycles and visible honeycomb structure of fracture cells on its surfaces. (3) and (4) show the gas permeability and photographic image, respectively, of a sample after only 5 F/T cycles with incipient organized damage not visually detectable but with detectable permeability changes. (Concrete beams courtesy of C. Korhonen, CRREL; NDE permeability images courtesy of G. Bussod, New England Research, Inc.)

and impact-echo NDE inspection of post-tensioned concrete structures with an integrated automated system [67,68]. Haardt et al. inspected bridges and tunnels with a combination of automated visible light imaging, IR imaging, impact echo, and GPR instruments [69]. The system registered the geometric data of the several scanners with that of the structure for data fusion analyses. Chen et al. suggest that it is possible to automate bridge monitoring by interrogating wireless sensor networks on bridges with inspection vehicles that collect data as they pass over the bridge [70]. Capuruço et al. used automated methods to measure pavement condition and roughness from image data [71]. Hao et al. automated the magnetic fluorescent particle crack detection of railcar wheels with machine vision hardware and crack detection software based on Canny edge detection [72]. Figure 7.12 shows the results of testing concrete beams for freeze-thaw damage with a gas permeability sensor mounted onto a mechanical scanning system. The sensor and scanning system produces measurements with sufficiently fine spatial resolution to identify the formation of spatial damage patterns.

7.3.6 Sensing and Inspection Robots

Some useful SHM tasks for robots are: (1) Sensing structural conditions; (2) Installing and servicing sensors; and (3) Repairing structures. Structures with relatively simple geometries that do not allow for easy and/or human access are prime candidates for robotic SHM. Maintenance activities on pipes, railroad rails, and hazardous storage tanks already routinely use robots for inspection. Other applications on more complicated structures, such as bridges and aircraft, may see expanded future use.

Until recently, using robots for SHM efforts would have been viewed as impractical. The development of robotic and automated inspection systems is at present in a state of rapid growth. The convergence of technologies from fields such as information processing, military endeavors, and hobbyists is such that workable SHM robots are being realized [73]. Many of the design constraints that apply to human-operated machines and inspection systems also apply to robotic inspection systems [74]. Robots require intelligence, mobility, and power. Robots must not create safety hazards, such as falling on, pinching, crushing, or electrocuting people and/or damaging the structure [75].

Sensors on SHM robots can indicate structural conditions and robot operational parameters. Since human-based visual structural inspection can be very effective, it is natural to expect that robot-mounted machine-vision surrogates may also be effective. Digital cameras with tilt, pan, zoom, autofocus, and image enhancement often exceed the capabilities of human eyes. Digital cameras have the ability to store and transmit images with high fidelity. Using machine vision systems to automatically detect and ascertain the level of damage requires developing suitable image processing and statistical pattern recognition algorithms [76]. Figure 7.13 shows an articulated robot arm with an autofocusing digital microscope. Dent detection was possible with a system that used a 2.4 GHz telemetry link to transmit video data back to a host computer. Deutschl developed a machine vision inspection robot for the railroad industry [77]. The system attaches to a rail car device for in place inspection of rail tracks and automatically detects defects such as flakes, cracks, grooves, or break-offs by digital image processing. Coccia et al. developed a noncontacting ultrasound rail inspection system that uses laser light as a thermoelastic source and air-coupled microphones as receivers [78]. Another passive vision system can sense cracks in sheet metal in automotive manufacturing [79]. The system detects the presence of splits and through cracks greater than 0.2 mm.

Robots can potentially carry multiple onboard sensors, including virtually any of the handheld NDE systems used by human inspectors. Sensor actuation requires a bit of sophistication. Ultrasound measurement is a case in point. Most ultrasound transducers need a firm mechanical contact between the transducer and the structure, such as with a coupling agent. Figure 7.14 shows a servo-actuated *Articulated Ultrasound Robot Arm* (AURA) that

FIGURE 7.13 Versacam™ articulated robot arm with autofocusing digital microscope. (Photograph courtesy of the Applied Research Associates, Inc.)

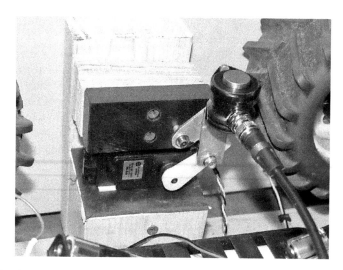

FIGURE 7.14   AURA for steel girder inspection. (Adapted from Huston DR, Esser B, and Miller J. 2004. *Proc 2nd European Workshop on Structural Health Monitoring*, DEStech, Munich.)

measures bridge girder thickness, with typical data appearing in Figure 7.15. The design of the 4-bar linkage used available mechanical advantage with a high-force output range of the stroke corresponding to the anticipated ranges of transducer contact.

RFID technology encompasses a broad class of remotely powered miniaturized embedded electronic systems (see Chapter 4) [80]. Modern RFID systems may include sensors with microprocessors, but still have a limited range. Robots have the ability to provide RFID sensors with close-up telemetry and wireless EM power supply [81]. Through a modest amount of continued development, it is likely that robots will also be able to service and install sensors in remote locations.

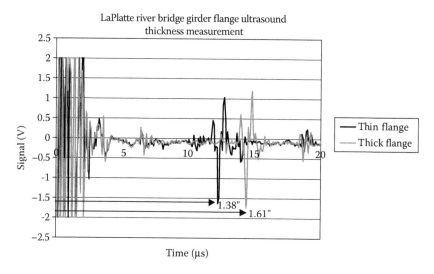

FIGURE 7.15   Bridge girder thickness measurements taken with a robot deployed ultrasound sensor. (Adapted from Huston DR, Esser B, and Miller J. 2004. *Proc 2nd European Workshop on Structural Health Monitoring*, DEStech, Munich.)

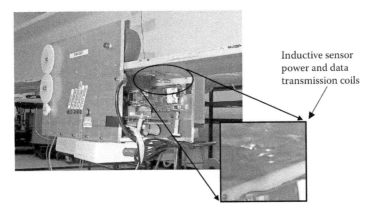

Inductive sensor power and data transmission coils

FIGURE 7.16 Laboratory version of a beam crawling robot with an inductively powered telemetry system. (Telemetry system courtesy Microstrain, Inc.)

Figure 7.16 shows an early laboratory version of an inspection robot that rolls along a beam. The robot locates and inductively powers sensor modules. The modules then transmit data back to a central data logger/processor. Figure 7.17 shows beam flexure strain data collected by a surface attached sensor module and transmitted by a wireless link to a central host as the beam was loaded [81]. The robot supplied power to the sensor module via a wireless short-range low-frequency inductive interconnect.

Some robots have fixed points of attachment. Others are mobile. Fixed robots can potentially access large amounts of power and tend to not be constrained much by weight. The SHM applications of fixed robots are fairly limited. One application uses fixed robots to inspect parts during structural fabrication [82]. In contrast, mobile robots can move over, around, and through structures to gain access for inspection and have potentially greater utility. Weight and power limitations are significant design constraints.

A primary requirement of mobile robotic structural inspection systems is sufficient mobility to move to points of interest. Building a single all-purpose mobile SHM robot at present seems to be impractical due to the complex and varied geometries of many structures. Specialization for specific structural forms is an attractive option. Simple structural forms such

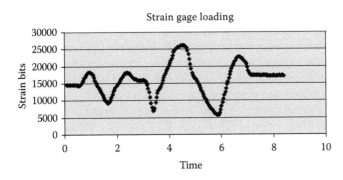

FIGURE 7.17 Strain data collected by a beam crawling robot. (Adapted from Huston DR et al. 2001. *Health Monitoring and Management of Civil Infrastructure Systems*, SPIE 4337-24; Telemetry system courtesy Microstrain, Inc.)

FIGURE 7.18 Beam crawling robot on the LaPlatte River Bridge in Shelburne, VT. (Adapted from Huston DR, Esser B, and Miller J. 2004. *Proc 2nd European Workshop on Structural Health Monitoring*, DEStech, Munich.)

as the inside of a pipe are especially amenable to such approaches [83]. Biological inspiration of robot motion is an attractive design philosophy, but is nontrivial to implement [73,84,85].
   Possible mobility solutions include:

1. *Rolling*: Certain simple structural geometries accommodate rolling motions. Constrained rolling with circumferential grips around a girder, cable, or similar elongated member reduces navigational complexity and the risk of the robot falling or snagging. Figure 7.18 shows a beam crawling robot with a design for traversing a multispan steel-girder-composite concrete-deck bridge (US Rte. 7 across the LaPlatte River in Shelburne, VT). Although this bridge has a simple girder design with readily accessible girder beams, minimal clearance obstructions due to inter-girder diaphragms and variable spanwise flange thicknesses constrain the design. Small wheels that pass through the diaphragm obstructions and a drive train derived from a moderately expensive radio-controlled toy truck formed an effective mobility solution. Remotely switchable autonomous and teleoperation radio control provided direction and control for the movements. Figure 7.19 shows a robot for the inspection of above-ground storage tanks. Figure 7.20 is of a buried power cable inspection robot.

2. *Climbing*: Climbing robots that can maneuver over complex geometries may be useful in structural inspection [86]. Upon reaching the desired location, the arms can reconfigure themselves for manipulation. Reliable traction and grip aid in most crawling applications. When the structure is ferromagnetic, that is, made of steel, *Magnetic On/off Robot Attachment Devices* (MOORADs) based on low energy switching magnetic circuits are possible gripping solutions [87–90]. An early application of switchable magnetic feet for robot inspection systems is the Neptune liquid storage tank inspection system [91]. Figure 7.21 shows a biped inspection robot with MOORAD feet and an articulated onboard camera. Magnetic wheels are also a possibility [92]. Many structures are nonferromagnetic. Another gripping solution is to use van der Waals forces to cling to smooth surfaces, in a manner similar to that of a gecko lizard [93–95]. Such materials can grip nonmagnetic surfaces, but durability and grip engagement remain problematic. Pneumatic sucker feet are another option [96,97].

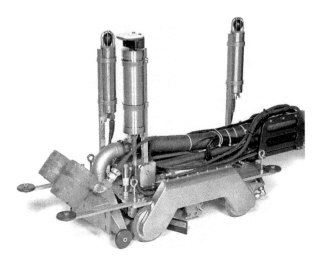

FIGURE 7.19 An above-ground storage tank inspection robot. (Photograph courtesy of the TechCorr USA, LLC.)

3. *Flying*: Flying may overcome many of the difficulties with rolling and climbing. RC airplanes and helicopters may be viable cost-effective solutions. Airplanes may be awkward for structural inspection since they cannot hover. Helicopters can hover, but pose control and endurance challenges. Figure 7.22 shows a camera-equipped RC helicopter with damage detected on a concrete retaining wall. Safety is a big

FIGURE 7.20 A buried power cable inspection robot built at the University of Washington. (Photo courtesy of A. Mamishev; From Jiang B and Mamishev AV. 2004. *IEEE Transactions on Power Delivery*, 19(3), 912–918. With permission.)

FIGURE 7.21 A biped inspection robot with magnetic feet and an articulated camera.

concern with helicopters [98]. The California Department of Transportation developed an aerial platform system for conducting bridge inspections without traffic delays. The system used a shrouded vertical-axis fan to provide the lift [99].

4. *Blimping*: Blimps can hover long term with minimal power consumption. Small hobby and advertising blimps are commercially available. These may readily adapt to SHM applications. Buffeting by strong winds limits blimp utility.

5. *Floating and Swimming*: RC boats may be a good option for inspecting bridge piers and dams. A common application for underwater swimming robots is the deep-water inspection of high-value objects and structures where it is impractical to use human divers. Continued technical development and cost reduction will likely make such robots economical for underwater structural inspectors at shallower depths that typically use human divers. A pipe inspection robot appears in Figure 7.23.

FIGURE 7.22 Radio-controlled helicopter and damage detected in concrete retaining wall by onboard camera via wireless transmission. (Adapted from Huston DR, Esser B, and Miller J. 2004. *Proc 2nd European Workshop on Structural Health Monitoring*, DEStech, Munich.)

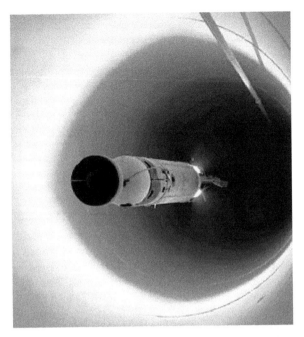

FIGURE 7.23 Tunnel inspection vehicle developed by Woods Hole Oceanographic Institution. (Photograph by Tom Kleindinst, Woods Hole Oceanographic Institution.)

Robotic navigation is a difficult technical challenge. Robot navigation can be *teleoperated*, *semi-autonomous*, and *fully autonomous*. Teleoperated navigation uses a human to exert high- and low-level control over the motion and path selection, often with the aid of visual feedback. Semiautonomous navigation and control systems use humans to give high-level commands and the robot executes low-level control over the motion. Fully autonomous navigation has the robot select the path. Teleoperation can train the robot to move and to become fully or semiautonomous for future inspections of the same components. Software can map the interior of the structure as the robot moves, possibly with the aid of maps generated from CAD drawings and with machine vision input to determine the path. For example, Scott used automated positioning and signal processing of GPR signals for concrete bridge deck inspection with a robot [100]. Qian and Ferrari used entropy measures of information combined with data-fusing Bayesian networks to plan the path of GPR sensing systems for buried mine and ordinance detection [101].

Power supply and consumption are critical design constraints for robotic systems. The constraints are much more demanding for mobile robots than tethered varieties. Untethered robots can only use onboard power harvested from the local environment. Base docking stations can replenish the power supply on mobile robots, download data, and program modified inspection protocols, as necessary. In an example of energy harvesting, Wilcox inspected the interior of a gas pipeline with an untethered robot that extracted power from the gas flows [102].

Most of the commercially successful robots operate on structures with relatively simple geometries and operate in regions that are difficult for human access. Pipe inspection robots are probably the most common application in this category. Schempf et al. inspected cast iron pipes with a robot that used magnetic circuit transducers [103]. Carpenter et al. developed corrosion inspection robots for pipes that use feeler gages, video cameras, and other

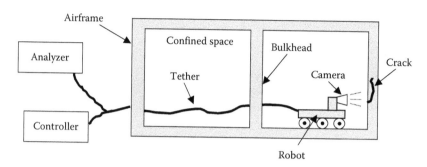

FIGURE 7.24   Concept for robot vision inspection of confined compartments in aircraft.

sensors [104]. Storage tanks are another application with a relatively simple geometry in which robots can be useful SHM inspectors. Schempf and Albrecht inspected the inside of steel liquid storage tanks with a robot that used magnetic tracks [91]. On/off magnetic circuits propel the robot and keep it in contact with the tank surface. Ultrasonic transducers mounted on the robot measure the tank wall thickness. Hudson used a mobile robot platform, known as the Polecat Pro, to inspect steel light poles [105]. Hwang and Kim used a small mobile robot to inspect composite panels by analyzing the sound emanating from mechanical taps [106,107]. NASA has proposed a mini-rover for deploying sensors for lunar or planetary exploration [108].

Confined spaces are an excellent opportunity for using robots for inspection and maintenance activities. An example application is aircraft (Figure 7.24). Present aircraft inspection techniques require a costly and labor-intensive disassembly of the aircraft for human inspection. Disassembly inherently causes damage, such as the scratching of protective coatings, excessively loading elements, enlarging of rivet holes, and substandard reassembly. Replacing some or all of the disassembly inspection by using miniaturized inspection robots that move inside confined aircraft structural compartments can positively impact economic maintenance objectives. Snake geometry fixed and mobile configurations are possibilities [109,110].

7.4   Adaptive Sensor Location and Inspection Timing Strategies

Adapting sensor location to damage scenarios as they emerge may be more effective than keeping the sensors fixed in preset locations. Such techniques require a combination of strategies that determine new and improved locations and a means of moving the sensors to new locations (Figure 7.25). One goal of adaptive sensor location techniques is to place the sensors in positions that maximize the information density and/or value of the collected data. The value of specific data features depends on the details of the failure modes of a given structure, and the value of detecting damage. An example is the detection of cracks. Detecting microcracks may not provide much valuable information. Cracks smaller than a critical threshold may not grow and may not cause trouble. Detecting large cracks may or may not provide valuable information. In certain circumstances, detecting large cracks can be very valuable because detection may prevent a catastrophe. In other circumstances, detection of a large crack may not provide much value because the damage is so severe that it cannot be repaired and it may not be possible to avert a major failure. Detection of intermediate-size cracks may be most valuable because they can be repaired with relative

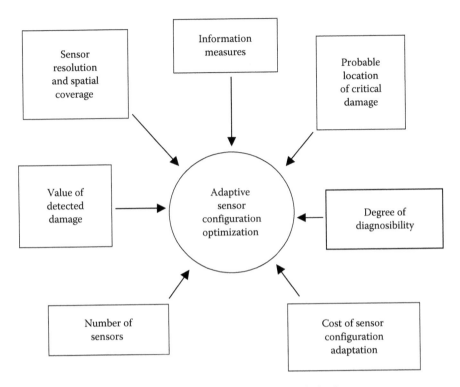

FIGURE 7.25   Issues associated with adaptive sensor configuration optimization.

ease and minimal costs. An example is aluminum skin on airplanes. Holes for rivet fasteners are stress risers that normally produce small fatigue cracks following a series of flight operations. The growth of rivet hole fatigue cracks is a well-understood and expected phenomenon (Figures 7.26 and 7.27). Detecting the presence of such small fatigue cracks may not provide much valuable information. When the fatigue cracks grow faster than expected, or coalesce in a manner that forms a larger crack that grows to span two or more

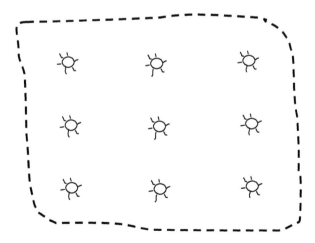

FIGURE 7.26   Airplane aluminum skin section with small predictable fatigue cracks around rivet holes.

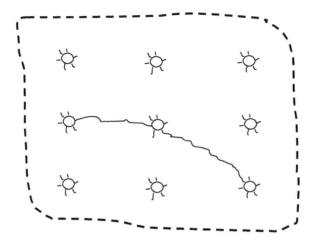

FIGURE 7.27 Airplane aluminum skin section with an unusual and possibly dangerous crack that spans rivet holes.

rivet holes, early detection and subsequent close-up monitoring can provide very valuable information.

Crack location is also important. Some cracks are benign. Others are dangerous. An example is a concrete floor slab. Since flexure cracks are relatively benign, flexure crack detection may not provide much useful information. Shear cracks are very dangerous, but may not be worth detecting because shear crack failures often occur very rapidly. Shrinkage cracks may be very valuable to detect. During the pouring and curing process, shrinkage cracks can occur, especially with high-performance concrete. Detecting shrinkage cracks at early stages of cure creates the opportunity for palliatives, such as wetting the surface of the concrete.

Moving beyond the qualitative discussion above remains a matter of active research. Kouchmeshky et al. developed a genetic algorithm evolutionary method for identifying structural damage from a minimum of physical measurements using an adaptive sensor location technique [111]. Zhang et al. used algebraic manipulations of a *fault signature matrix* (FSM) to optimize and adapt sensor configurations to changing circumstances [112]. An FSM is a rectangular array with elements that take on the binary values of 0 or 1. The rows represent sensor outputs. The columns represent fault conditions. Uniquely identifying a fault condition from an array of sensor outputs requires that the associate column vector be independent of the other column vectors. Table 7.1 shows an example of a relatively simple FSM for a system with four fault conditions (C_1, C_2, C_3, C_4) and five sensor outputs (S_1, S_2, S_3, S_4, S_5). The sensor outputs may be direct On/Off switching from a transducer or may be more complicated, such as multisensor fusion output. Using the output from all five sensors enables a unique identification of the four fault conditions. The associated 5×4 FSM has a rank of 4. Using the output from the reduced set of sensors (S_2, S_3, S_4, S_5) creates a situation where the C_1 cannot be uniquely distinguished from C_4 and the associated FSM is of reduced rank. It is possible to incorporate probabilistic and information-value weightings into the analysis. Particle swarm optimization (PSO) proved to be an effective optimization technique for this problem.

Many of the issues regarding adaptive sensor location strategies also apply to adaptive inspection timing. Setting up an inspection schedule to optimize the amount and value of useful information gathered versus the cost of inspection can benefit from an adaptive

TABLE 7.1

Example of Relatively Simple FSM

Sensor	Condition			
	C_1	C_2	C_3	C_4
S_1	1	0	1	0
S_2	0	1	0	0
S_3	1	0	1	1
S_4	0	1	1	0
S_5	1	1	0	1

approach. Qin and Wang used such a strategy to time the inspection of concrete girders for the inspection of chloride-induced reinforcing and post-tension steel corrosion [113]. Corrosion processes in concrete structures are sufficiently well understood to establish utility functions that describe the expected benefit and cost of inspections.

7.5 General Principles

The practice of SHM requires accommodating complex interactions of mechanics, sensors, information, and people over multiple scales of time and positions. This may be particularly true for systems with mature technologies and moderately long track records of operation. For mature systems, many of the likely failure modes have already occurred and have been corrected by system modification and/or redesign. The predictable failures due to simple interactions have been eliminated and unpredictable failure modes, possibly due to more complex interactions, are more likely to occur. An example is modern highway bridge design practices that proscribe building structures with fracture critical members. While such practices may eliminate the occurrence of fracture-critical collapses, this leaves the situation where the failure modes of nonfracture critical bridges are difficult to predict and hence prevent. There are several strategies for designing and operating SHM systems in such complex circumstances. These include:

1. *Simplification*: Develop a clear understanding of the behavior of the structure. This inevitably requires simplification by ignoring extraneous effects to increase the understanding and predictability of future events. An example would be the use of fixed preinstalled sensor systems in dams to detect known precursor-measurands of damage and failure. A possible explanation for the utility of such ad hoc simplifications is the Center Manifold Theorem [114].

2. *Hierarchical and Compartmental Organization*: Organizing complex systems into a hierarchy of simpler subunits often produces workable simplifications. Examples include hierarchical databases, image compression, wavelet decompositions, star or tree-like sensor topologies, tree-style decision rules for system operation and data processing.

3. *Adaptation*: The SHM system adapts and modifies itself to changing and emerging behaviors. Triggering, autoranging, input buffering, squinting, mobility, using adaptive statistical analyses, and redesign are all potential adaptation modalities. Stochastic adaptive techniques, such as partially observable Markov decision processes, are another possibility [115].

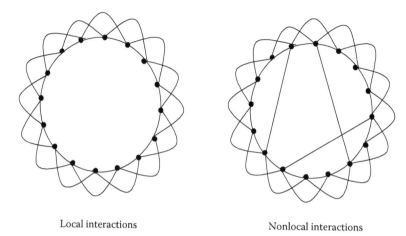

Local interactions   Nonlocal interactions

FIGURE 7.28   Networks with local only and with nonlocal long-range (small world) interactions. (Adapted from Watts DJ and Strogatz SH. 1998. *Nature*, 393(4), 440–442.)

4. *Self-Organizing Cooperative-Adaptive Systems*: A curious property of multiple inter-acting distributed autonomous systems is that groups of individual systems often cooperatively self-organize into global behaviors that are not inherently obvious from an examination of the local properties. Examples include social insects, flock-ing behaviors of birds, schooling behavior in fish, mob behavior in humans, the wave at sporting events, and progressive collapse or global instabilities in struc-tures [116,117]. Self-organizing cooperative behavior is not completely understood, but has been studied extensively [2]. Further complications arise if there are long-range (small world) interactions between entities. Nonlinear behavior in long-range interactions can lead to unexpected and dramatic changes to the whole system in response to seemingly minor events (Figure 7.28). An example is the interactions between rain gutters on the roof of a building. The rain gutters normally interact only slightly with the roofing system by using the roof as a means of structural sup-port. If the gutters become clogged, perhaps by a bird nest, and prevent drainage during a heavy rainstorm, then the gutters and the roof become tightly coupled through the puddling of water and may cause the roof to collapse along with the gutters. Network models can represent many salient aspects of long-range (small world) interactions [118].

In many respects, issues of SHM are similar to those of complex adaptive systems [119]. Figure 7.29 shows a graded taxonomy of simple to complex adaptive systems. While com-plex adaptive systems may defy a precise definition, they do exhibit a variety of common interactions and behaviors [120,121].

1. A complex adaptive system is a collection of individual agents. The individual agent's actions are based on internal rules. The collective actions of groups of agents are not always easily predicted from the internal rules of agents.

2. The combined interactions between agents and the system are adaptive. The interconnections between agents change the local environment and behavior of agents.

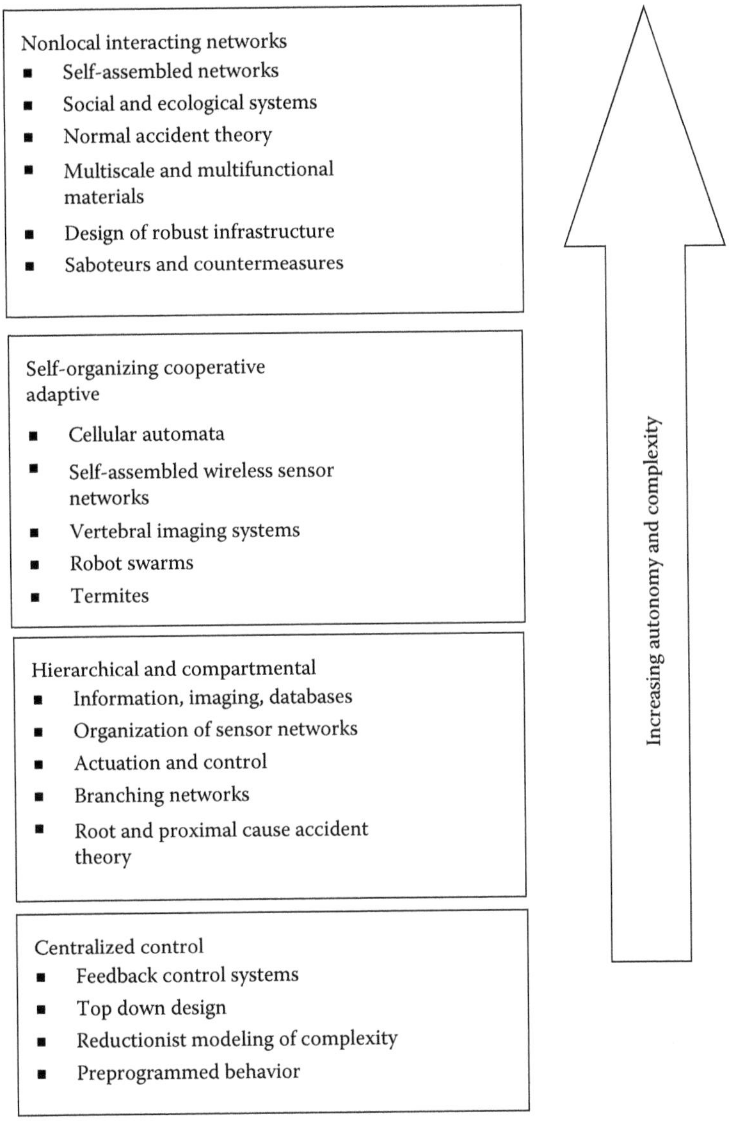

Nonlocal interacting networks
- Self-assembled networks
- Social and ecological systems
- Normal accident theory
- Multiscale and multifunctional materials
- Design of robust infrastructure
- Saboteurs and countermeasures

Self-organizing cooperative adaptive
- Cellular automata
- Self-assembled wireless sensor networks
- Vertebral imaging systems
- Robot swarms
- Termites

Hierarchical and compartmental
- Information, imaging, databases
- Organization of sensor networks
- Actuation and control
- Branching networks
- Root and proximal cause accident theory

Centralized control
- Feedback control systems
- Top down design
- Reductionist modeling of complexity
- Preprogrammed behavior

Increasing autonomy and complexity

FIGURE 7.29 System and adaptation complexity versus autonomy. (Adapted from Huston DR, Esser B, and Miller J. 2004. *Proc 2nd European Workshop on Structural Health Monitoring*, DEStech, Munich.)

3. Systems embed within other systems and coevolve. The inherent self-organization of groups of agents through locally applied rules leads to continually emerging and novel behavior.

4. Complex systems can form stable and self-assembled patterns out of chaotic conditions. The boundaries between different parts of the system can be fuzzy and intertwining.

5. Unpredictability and paradox are natural occurrences in complex systems. An example is safety improvement initiatives where a safety innovation in one part of a complex system causes unexpected accidents in other parts of the system.

6. Networks of systems, especially those with weak long-range interactions, can be a dominant form of complex systems behavior. Long-range links can lead to the strong coupling of previously uncoupled phenomena.

A major research question is how to use adaptive self-organizing cooperative system phenomena to design and build useful engineering systems. Many of the ideas from complex systems science have yet to be fully utilized by the SHM and other engineering practices. One reason may be that many standard approaches in engineering design and engineering science aim at simplifying complex situations so that they are more tractable. Stochastic techniques can model many cases of excessively complex systems [118]. The situation is analogous to turbulence modeling in fluid mechanics. Some models are reductionist, such as the direct numerical simulation methods. Other models are primarily statistical, such as the Kolmogorov spectral approach. Other models occupy a more intermediate regime, such as the large eddy simulation methods. In the midst of these mathematical models lie researchers and engineers that attempt to control and/or design systems to exploit turbulence, eddies and laminar flow for specific applications.

References

1. Stengel RF. 1994. *Optimal Control and Estimation.* Dover Publications, Mineola, NY.
2. Beckerman M. 1997. *Adaptive Cooperative Systems.* Wiley, New York.
3. Huston D. 2001. Adaptive sensors and sensor networks for structural health monitoring. *Symposium on Complex Adaptive Structures,* SPIE 4512-24.
4. US Army Corps of Engineers. 1980. Engineering and design instrumentation for concrete structures, EM 1110-2-4300, Department of the Army, Office of the Chief of Engineers, Washington, DC.
5. NASA. 2003. Columbia Accident Investigation Board Report, Vol. 1, CAIB, Aug.
6. Associated Press. 2003. Orbiting Junk Blamed for Space Station Clang, Nov. 28.
7. Osborn PW, Jakovich G, Nimis R, and Greer M. 2005. FHWA Tunnel Leak Assessment. Boston Central Artery Interim Report, US Dept of Trans, Federal Highway Administration.
8. Moore M, Phares B, Graybeal B, Rolander D, and Washer G. 2000. Reliability of Visual Inspection for Highway Bridges Vol 1 Final Report FHWA-RD-01-120, Federal Highway Administration, McLean, VA.
9. Moore M, Phares B, Graybeal B, Rolander D, and Washer G. 2000. Reliability of Visual Inspection of Highway Bridges Vol II: Appendices FHWA-RD-01-121, US Federal Highway Administration.
10. Patterson-Hine A, Aaseng G, Biswas G, Narasimhan S, and Pattipati K. 2005. A Review of Diagnostic Techniques for ISHM Applications. First Intl Forum on Integrated System Health Engineering and Management in Aerospace, Napa.
11. Applied Technology Council. 2004. ATC-45 Field Manual: Safety Evaluation of Buildings after Windstorms and Floods, Redwood City.
12. Kerschen G, De Boe P, Golinval JC, and Worden K. 2005. Sensor validation using principal component analysis. *Smart Mater Struct,* 14, 36–42.
13. Shi ZY, Law SS, and Zhang LM. 2000. Damage localization by directly using incomplete mode shapes. *J Eng Mech,* 126(6), 656–660.
14. Yang SM and Lee GS. 1999. Effects of modeling error on structural damage diagnosis by two-stage optimization. *Proc 2nd Intl Workshop on Structural Health Monitoring,* Stanford, Technomic, Lancaster.

15. Lopes V, Park G, Cudney HH, and Inman DJ. 1999. Smart structures health monitoring using artificial neural network. *Proc 2nd Intl Workshop on Structural Health Monitoring*, Stanford, Technomic, Lancaster.

16. Pothisiri T and Hjelmstad KD. 2003. Structural damage detection and assessment from modal response. *J Eng Mech*, 129(2), 135–145.

17. Ni YQ, Wang BS, and Ko JM. 2002. Constructing input vectors to neural networks for structural damage identification. *Smart Mater Struct*, 11, 825–833.

18. Sim PL, Worden K, and Manson G. 2004. An adaptive structural health monitoring strategy using novelty detectors. *Proc 2nd European Workshop on Structural Health Monitoring*, Munich, DEStech.

19. Safak E. 2004. Post-earthquake structural condition assessment by real-time vibration monitoring. *Proc 2nd European Workshop on Structural Health Monitoring*, Munich, DEStech.

20. Tiwari DN and Brown TG. 2004. Monitoring for the safety of bridge—Starting with the ice forces. *Proc IABMAS'04 Bridge Maintenance, Safety, Management and Cost*, Kyoto. Taylor & Francis.

21. Liu Z, Forsyth DS, and Marincak A. 2003. Preprocessing of edge of light images: Towards a quantitative evaluation. *Nondestructive Evaluation and Health Monitoring of Aerospace Materials and Composites II*, SPIE 5046.

22. Duran O, Althoefer K, and Seneviratne LD. 2003. Pipe inspection using a laser-based transducer and automated analysis techniques. *IEEE/ASME Trans. on Mechatronics*, 8(September)3.

23. Blackshire JL, Hoffmann J, Kropas-Hughes C, and Tansel I. 2003. Microscopic NDE of hidden corrosion. *Testing, Reliability, and Application of Micro- and Nano-Material Systems*, SPIE 5045.

24. Benamoun M and Mamic GJ. 2002. *Object Recognition Fundamentals and Case Studies*. Springer-Verlag, London.

25. Ueda M, Todoroki A, Shimamura Y, and Kobayashi H. 2002. Novel zooming method for delamination monitoring of CFRP laminates using electrical potential change. *Nondestructive Evaluation and Health Monitoring of Aerospace Materials and Composites III*, SPIE 5393.

26. Kress KP. 2003. Integrated imaging ultrasound SWISS. In *Structural Health Monitoring*, FK Chang (ed.). DEStech, Lancaster.

27. Williams TJ, Nozick LK, Sansalone MJ, and Poston RW. 2006. Sampling techniques for evaluating large concrete structures: Part I. *ACI Structural J*, 103(3), 399–408.

28. Williams TJ, Nozick LK, Sansalone MJ, Poston RW, and Kesner K. 2006. Sampling techniques for evaluating large concrete structures: Part II. *ACI Structural J*, 103(4), 505–512.

29. Adams BJ, Huyck CK, Mio M, Cho S, Ghosh S, Chung HC, Eguchi RT, Houshmand B, Shinozuka M, and Mansouri B. 2004. The Bam (Iran) Earthquake of December 26, 2003: Preliminary Reconnaissance Using Remotely Sensed Data and VIEWS System. MCEER Earthquake Investigation Report, Bam12-26-03.

30. Bennett DA, Feng X, and Velinsky SA. 2003. Robotic machine for highway crack sealing. Transportation Research Record 1827 Highway Pavements and Structures Maintenance and Security, Paper no. 03-3355.

31. Hjelmstad KD. 1998. Damage Detection in Civil Engineering Structures. T118-4, *Structural Engineering Worldwide*, Elsevier.

32. McGugan M, Bech T, Sorensen BF, Jorgensen E, and Kristensen OJ. 2003. Improving the structural testing of wind turbine blades by monitoring acoustic emission. In *Structural Health Monitoring*, FK Chang (ed.). DEStech, Lancaster.

33. Wen Y, Li P, Na X, and Yang Y. 2005. Mobile instrument for intelligent leak detection and location on underground water supply pipelines. *Smart Sensor Technology and Measurement Systems*, SPIE 5758.

34. Ogisu T, Shimanuki M, Kiyoshima S, Okabe Y, and Takeda N. 2004. Development of damage monitoring system for aircraft structure using a PZT actuator/FBG sensor hybrid system. *Industrial and Commercial Applications of Smart Structures Technologies*, SPIE 5388.

35. Awata A and Mita A. 2004. Active diagnostics of ball bearing using amplitude modulation of ultrasonic wave. *Proc 2nd European Workshop on Structural Health Monitoring*, Munich, DEStech.

36. Jiang LJ, Tang J, and Wang KW. 2005. An improved damage identification method using tunable piezoelectric transducer circuitry. *Sensors and Smart Structures Technologies for Civil, Mechanical, and Aerospace Systems*, SPIE 5765.
37. White JR and Adams DE. 2007. Actuator-sensor pair excitation tuning and self diagnostics for damage identification of a sandwich plate. In *Structural Health Monitoring*, FK Chang (ed.). DEStech, Lancaster.
38. Schulz JL and Commander BC. 1995. Efficient field testing for load rating railroad bridges. *Nondestructive Evaluation of Aging Railroads*, SPIE 2458.
39. Tsuchimoto K, Ikeshita T, and Kitagawa Y. 2003. Diagnostic system for structural damage and degradation: Damage identification based on a mode parameter and wave propagation. *Smart Systems and Nondestructive Evaluation for Civil Infrastructures*, SPIE 5057.
40. Reed RD, Osmont D, and Dupont M. 2004. Monitoring of impact damages in stiffened and non-stiffened carbon epoxy plates using bonded PZT disks. *Proc 2nd European Workshop on Structural Health Monitoring*, DEStech, Munich.
41. Lipeles J. 2002. Smart bullet. US Patent 6,422,507.
42. Rong AY and Cuffari MA. 2004. Structural health monitoring of a steel bridge using wireless strain gages. *Proc Structural Materials Technology VI, an NDT Conference*, Buffalo.
43. Tikka J, Hedman R, and Siljander A. 2003. Strain gage capabilities in crack detection. In *Structural Health Monitoring*, FK Chang (ed.). DEStech, Lancaster.
44. Khuri-Yakub BT, Degertekin FL, and Pei J. 1997. Dry contact ultrasonic sensors for structural monitoring. In *Structural Health Monitoring Current Status and Perspectives*, F-K Chang (ed.), Technomic Publishing, Lancaster.
45. Godinez-Azcuaga VF, Gostautas RS, Finlayson RD, Miller R, and Trovillion J. 2004. Nondestructive evaluation of FRP wrapped concrete columns and bridges. *Proc Structural Materials Technology VII, an NDT Conference*, Buffalo.
46. Komsky IN. 2005. Modular dry-coupled ultrasonic probes for field inspections of multilayered aircraft structures. *Health Monitoring and Smart Nondestructive Evaluation of Structural and Biological Systems IV*, SPIE 5768.
47. Kozlov VN, Samokrutov AA, and Shevaldykin VG. 2006. Ultrasonic equipment for evaluation of concrete structures based on transducers with dry point contact. *NDE Conference on Civil Engineering*, American Soc for Nondestructive Testing, St. Louis.
48. Pardini AF, Alzheimer JM, Crawford SL, Diaz AA, Gervais KL, Harris RV, Reichers DM, Samuel TJ, Schuster GJ, Tucker JC, and Roberts RA. 2001. Development of a Remotely Operated NDE System for Inspection of Hanford's Double Shell Waste Tank Knuckle Regions. Annual Report PNNL-13682, US Dept of Energy, Sept. 2001
49. Gräfe B and Krause M. 2006. Basic investigation with air-coupled ultrasonic echo for concrete elements. *NDE Conference on Civil Engineering*, American Soc for Nondestructive Testing, St. Louis.
50. Cetrangolo G and Popovics JS. 2006. Measurement of P-wave velocity through concrete using air coupled transducers. *NDE Conference on Civil Engineering*, American Soc for Nondestructive Testing, St. Louis.
51. Todoroki A, Shimamura Y, and Inada T. 1999. Plug and monitor system via Ethernet with distributed sensors and CCD cameras. *Proc 2nd Intl Workshop on Structural Health Monitoring*, Stanford, Technomic, Lancaster.
52. Yarrington LW. 2000. Ultrasonic evaluation of in place fracture critical bridge elements. In *Structural Materials Technology IV*, S. Allampalli (ed.). Technomic Press, Lancaster.
53. Malas JC, Kropas-Hughes CV, Blackshire JL, Moran T, Peeler D, Frazier WG, and Parker D. 2003. Micro and nano NDE systems for aircraft: Great things in small packages. *Testing, Reliability, and Application of Micro- and Nano-Material Systems*, SPIE 5045.
54. Snell LM and Duinkherjav Y. 2003. Safe coring and drilling areas found quickly, efficiently. Concrete Intl, July.
55. Giurgiutiu V and Xu B. 2004. Development of a field-portable small-size impedance analyzer for structural health monitoring using the electromechanical impedance technique.

Sensors and Smart Structures Technologies for Civil, Mechanical and Aerospace Systems, SPIE 5391.

56. Findeis D and Gryzagoridis J. 2004. A comparison of the capabilities of portable shearography and portable electronic speckle pattern interferometry. *Nondestructive Evaluation and Health Monitoring of Aerospace Materials and Composites III*, SPIE 5393.

57. Livingston RA, Ceesay J, Amden AM, and Newman JW. 2006. Development of a portable laser shearography system for crack detection in bridges. *NDE Conference on Civil Engineering*, American Soc for Nondestructive Testing, St. Louis.

58. Komsky IN. 2004. Rolling dry-coupled transducers for ultrasonic inspections of aircraft structures. *Health Monitoring and Smart Nondestructive Evaluation of Structural and Biological Systems III*, SPIE 5394.

59. Nazarian S, Baker MR, and Crain K. 1997. Movable Seismic Pavement Analyzer. US Patent 5,614,670.

60. Poston RW, Frank KH, and West JS. 2003. Enduring Strength. Civil Engineering, Sept. 2003

61. Marzani A, Rizzo P, Lanza di Scalea F, and Benzoni G. 2004. Mobile acoustic system for the detection of surface-breaking cracks in pavement. *Smart Sensor Technology Measurement Systems*, SPIE 5384.

62. Fujino Y, Kitagawa K, Furukawa T, and Ishii H. 2005. Development of Vehicle Intelligent Monitoring System (VIMS). *Sensors and Smart Structures Technologies for Civil, Mechanical, and Aerospace Systems*, SPIE 5765.

63. Mares JR, Osegueda R, Kaukuri N, and Kreinovich V. 2004. Geometric approach to detecting and locating cracks in thin plates by lamb wave reflection: Case of moving transducer. *Health Monitoring and Smart Nondestructive Evaluation of Structural and Biological Systems III*, SPIE 5394.

64. Rezai A, Washer G, and Moore M. 2004. Laboratory and field evaluation of automated ultrasonic testing for inspection of Butt-Welds during fabrication. *Proc Structural Materials Technology VI, an NDT Conference*, Buffalo.

65. Henderson RE, Dion GN, and Costley RD. 1999. Acoustic inspection of concrete bridge decks. *Nondestructive Evaluation of Bridges and Highways III*, SPIE 3587.

66. Colla C. 2002. Scanning impact-echo NDE of post-tensioning ducts in concrete bridge beam. *Proc Structural Materials Technology V, an NDT Conference*, S Allampalli and G Washer (eds). Cincinnati.

67. Wiggenhauser H, Gardei A, Kohl C, Maierhofer C, Krause M, and Streicher D. 2004. Development and application of echo-methods for NDE of post tensioned concrete bridges. *Proc IABMAS'04 Bridge Maintenance, Safety, Management and Cost*, Kyoto. Taylor & Francis, London.

68. Streicher D, Kohl C, Wiggenhauser H, and Petz J. 2006. Assessment of post tensioned box girder bridges applying automated non-destructive testing methods investigations on two box girder bridges of the highway A23-Süd-Ost-Tangente in Vienna with Radar, Ultrasonic Echo and Impact-Echo. Paper 06-1990, 86th Annual Meeting, TRB, Washington, DC.

69. Haardt P, Krause M, Streicher D, and Krause KJ. 2004. Scanning NDT-methods for the inspection of highway structures. *Proc IABMAS'04 Bridge Maintenance, Safety, Management and Cost*, Kyoto. Taylor & Francis, London.

70. Chen SE, Callahan D, Jones S, Zheng L, Biswas P, Siswobusono PE, El Yamak B, Malpeker AD, and Lokonath R. 2004. Design of a remote monitoring technique of bridge integrity using a wireless drive-by network. *Proc Structural Materials Technology VI, an NDT Conference*, Buffalo.

71. Capuruço RA, Tighe SL, Ningyuan L, and Kazmierowski T. 2006. Performance evaluation of sensor-based and image-based technologies for automated pavement condition surveys. Paper 06-1648, 86th Annual Meeting, TRB, Washington, DC.

72. Hao H, Li L, and Deng LY. 2005. Vision system using linear CCD cameras in fluorescent magnetic particle inspection of axles of railway wheelsets. *Health Monitoring and Smart Nondestructive Evaluation of Structural and Biological Systems IV*, SPIE 5768.

73. Hrynkiw D and Tilden MW. 2002. *Junkbots, Bugbots and Bots on Wheels*. McGraw-Hill, New York.

74. Huston D, Esser B, Miller J, and Wang X. 2003. Robotic and mobile sensor systems for structural health monitoring. *Proc 3rd Intl Workshop on Structural Health Monitoring*, Stanford University.

75. Schraft RD, Graf B, Traub A, and John D. 2001. A mobile robot platform for assistance and entertainment. *Industrial Robot*, 28(1), 29–34.

76. Benmokrane B, Quiriron M, El-Salakawy E, Debaikey A, and Lackey T. 1993. Fabry-Perot sensors for the monitoring of FRP reinforced bridge decks. *Nondestructive Evaluation and Health Monitoring of Aerospace Materials and Composites III*, SPIE 5393.

77. Deutschl E, Gasser C, Niel A, and Werschonig J. 2004. Defect detection on rail surfaces by a vision based system. IEEE Intelligent Vehicles Symposium.

78. Coccia S, Bartoli I, Lanza di Scalea F, Rizzo P, and Fateh M. 2007. Non-contact ultrasonic rail defect detection system: Prototype development and field testing. In *Structural Health Monitoring*, FK Chang (ed.). DEStech, Lancaster.

79. de la Fuente E, Trespaderne FM, and Gayubo F. 2003. Detection of small splits in car-body manufacturing. *Proc IASTED International Conference on Signal, Processing, Pattern Recognition, and Applications*, 354–359, ACTA Press.

80. Finkenzeller K. 2003. *RFID Handbook*, 2nd Ed. Wiley, Chichester.

81. Esser B, Pelczarski N, Huston D, and Arms S. 2000. Wireless inductive robotic inspection of structures. *Proc IASTED*, RA, Honolulu.

82. Georgeson G. 2002. Recent advances in aerospace composite NDE. *Nondestructive Evaluation and Health Monitoring of Aerospace Materials and Civil Infrastructures*, SPIE 4704.

83. Horodinca M, Doroftei I, Mignon E, and Preumont AA. 2002. Simple architecture for pipe inspection robots. *Intl Colloquium on Mobile and Autonomous Systems*, Magdeburd.

84. Taubes G. 2000. Biologists and engineers create a new generation of robots that imitate life. *Science*, 288(5463), 80–83.

85. Bar-Cohen Y. 2002. Biologically inspired robots as artificial inspectors. *E-J of Nondes Test*, 7(1), http://www.ndt.net/article/v07n01/barcohen/barcohen.htm.

86. Clerc JP and Wiens GJ. 2004. Reconfigurable inspection robots with climbing capabilities. *ANS 10th International Conference on Robotics and Remote Systems for Hazardous Environments*, Gainesville, FL.

87. Huston DR, Esser B, and Miller J. 2004. Adaptive and mobile structural health monitoring. *Proc 2nd European Workshop on Structural Health Monitoring*, DEStech, Munich.

88. Moon F. 1984. *Magneto-Solid Mechanics*. Wiley, New York.

89. Levesque GN. 1942. Releasable Permanent Magnet Holding Device. US Patent 2,280,437.

90. Iwasaki H and Kaisha FJ. 1983. Magnet Base for Tool. US Patent 4,393,363.

91. Schempf H and Albrecht B. 1994. Reconfigurable Mobile Vehicle with Magnetic Tracks. US Patent 5,363,935.

92. Kim JH, Lee JC, and Kim JY. 2001. RISYS: An advanced reactor vessel inspection system with underwater mobile robots. *Proc 10th Asia-Pacific Conference on Non-Destructive Testing*, Brisbane, Australia.

93. Geim AK, Dubonis SV, Grigorieva IV, Novoselov KS, Zhukov AA, and Shapoval SY. 2003. Micro fabricated adhesive mimicking gecko foot-hair. *Nat Mater*, 2, 461–463.

94. Berengueres J, Urago M, Saito S, Tadakuma K, and Meguro H. 2006. Gecko inspired electrostatic chuck. *IEEE Intl Conference on Robotics and Biomimetics*, ROBIO '06, Kunming.

95. Ge L, Sethi S, Ci L, Ajayan PM, and Dhinojwala A. 2007. Carbon nanotube-based synthetic gecko tapes. *PNAS*, 104(26), 10792–10795.

96. Pack RT, Iskarous MZ, and Kawamura K. 1996. Climber Robot. US Patent 5,551,525.

97. Bar Cohen Y. 1997. Autonomous rapid inspection of aerospace structures. *E-J of Nondes Test*, 2(11), http://www.ndt.net/article/aero1197/yosi2/yosi2.htm.

98. Associated Press. 2003. Man Killed in Freak Accident with Radio-Controlled Helicopter. Nov 3.

99. Woo D. 1995. Robotics in highway construction and maintenance. *Public Roads*, 58(3), 26–34.

100. Scott ML. 1999. Automated Characterization of Bridge Deck Distress Using Pattern Recognition Analysis of Ground Penetrating Radar Data. PhD Dissertation, Civil and Environmental Engineering, Virginia Polytechnic Institute and State University.

101. Qian M and Ferrari S. 2005. Probabilistic deployment for multiple sensor systems. *Sensors and Smart Structures Technologies for Civil, Mechanical, and Aerospace Systems*, SPIE 5765.

102. Wilcox BH. 2003. Gas pipe explorer. *NASA TSP NAS* 7-918.

103. Schempf H, Mutschler E, Goltsberg V, and Crowley W. 2003. GRISLEE: Gasmain repair and inspection system for live entry environments. *Intl J Robotics Research*, 22 (July–August), 7–8.

104. Carpenter J, Stanton D, and Kumar A. 1991. Corrosion inspection by a robotic crawler. Paper 563, *Proc NACE Annual Conference*.

105. Hudson K. 2000. *LabVIEW-Controlled Robot Climbs and Inspects Highway Lighting Towers*. A National Instruments Corporation Publication.

106. Hwang JS and Kim SJ. 2003. Validation of tapping sound analysis for damage detection of composite structures. *Nondestructive Evaluation and Health Monitoring of Aerospace Materials and Composites II*, SPIE 5046.

107. Hwang JS and Kim SJ. 2004. Development of an automated tapping device for practical application of tapping sound. *Nondestructive Evaluation and Health Monitoring of Aerospace Materials and Composites III*, SPIE 5393.

108. Trebi-Ollenu A and Kennedy BA. 2002. Minirovers as test beds for robotic and sensor web concepts Fido Rover. *NASA Tech Brief*, 26(11); JPL New Technology Report NPO-30342, Nov. 2002

109. Buckingham R and Graham A. 2005. Snaking around in a nuclear jungle. *Industrial Robot: An Intl J*, 32 (2), 120–127.

110. Nilsson M. 1998. Snake robot free climbing. *IEEE Control Systems*, 18(1), 21–26.

111. Kouchmeshky B, Aquino W, Bongard JC, and Lipson H. 2007. Coevolutionary strategy for structural damage identification using minimal physical testing. *Intl J Num Meth Engrg*, 69(5), 1085–1107.

112. Zhang Y, Fu J, Yen IL, Bastani F, Tai AT, Chau S, Vatan F, and Fijany A. 2006. QoS adaptive ISHM systems. ICTAI '06 *18th IEEE Intl Conference on Tools with Artificial Intelligence*, Arlington, VA.

113. Qin Q and Wang JX. 2006. Optimum bridge inspection/repair schedules based on time-dependent reliability and pre-posterior decision making. *Proc 4th China–Japan–US Symposium on Structural Control and Monitoring*, Hangzhou.

114. Gilmore R. 1993. *Catastrophe Theory for Scientists and Engineers*. Dover, New York.

115. Corotis R, Ellis JH, and Jiang M. 2005. Modeling of risk-based inspection, maintenance and life-cycle cost with partially observable Markov decision processes. *Structure and Infrastructure Engineering*, 1(1), 75–84.

116. Barabasi AL. 2002. *Linked*. Perseus Publishing, Cambridge.

117. Huang YL, Yen T, and Chen WF. 1996. A monitoring system for high-clearance scaffold systems during construction. *Proc ASCE Struct Congress XVI*, pp 719–726, Chicago.

118. Albert R and Barabasi AL. 2002. Statistical mechanics of complex networks. *Reviews of Modern Physics* 74, 47.

119. Bar-Yam Y. 1997. *Dynamics of Complex Systems*. Westview Press, Cambridge.

120. Plsek PE and Greenhalgh T. 2001. The challenge of complexity in health care. *BMJ*, 323, 625–628.

121. Plsek PE and Wilson T. 2001. Complexity, leadership, and management in healthcare organizations. *BMJ*, 323, 746–749.

122. Huston DR, Pelczarski N, Esser B, Gaida G, Arms S, and Townsend C. 2001. Wireless inspection of structures aided by robots. *Health Monitoring and Management of Civil Infrastructure Systems*, SPIE 4337-24.

123. Jiang B and Mamishev AV. 2004. Robotic monitoring of power systems, *IEEE Transactions on Power Delivery*, 19(3), 912–918.

124. Huston DR, Miller J, and Esser B. 2004. Adaptive, robotic, and mobile sensor systems for structural assessment. *Smart Structures and Materials 2004: Sensors and Smart Structures Technologies for Civil, Mechanical, and Aerospace Systems*, Shih-Chi Liu (ed.). SPIE 5391.

125. Watts DJ and Strogatz SH. 1998. Collective dynamics of 'small world' networks, *Nature*, 393(4), 440–442.

8

Health Monitoring and Prognosis

8.1 Health Monitoring as a Concept

Health is a term traditionally used to describe biological systems. Health has also gained currency as a descriptor of the state of structural systems. Some of the similarities between health life cycles for humans and structures appear in Figures 8.1 and 8.2.

Some of the correspondences between health assessments in medicine and structural engineering are as follows:

1. *Health is a complicated concept and is difficult to quantify with a simple scalar rating value.* One approach is to use adjectives. Health can be "excellent, good, fair, or poor." Adjectival descriptions lack specificity, yet their meanings are easily understood. Numerical quantifications are more definite, but not much more so. Examples are the 1-10 APGAR ratings for neonates and the AASHTO 0–9 rating for bridges [1,2]. In both cases larger ordinal values represent good health and lower values represent poor health. Both assessment procedures use specific evaluation criteria in an attempt to minimize intertester variability. The APGAR rating is attributed to Dr. Virginia Apgar. The original rating assigned two points for each category of color, heart rate, reflex irritability, muscle tone, and respiratory effort; or as an acronym with *A*ppearance, *P*ulse, *G*rimace, *A*ctivity, and *R*espiration.

2. *There is a dichotomy between uniqueness and commonality of healthy and unhealthy conditions.* Humans are unique. Each individual has different medical characteristics, potential disease conditions, and medical histories. Conversely, many medical conditions are common among all people. Many structures are unique. Each structure has unique problems and histories. Most structures have common problems, such as fracture, fatigue, corrosion, and so on.

3. *The biological interactions in humans are very complex and often poorly understood.* In spite of spectacular advances in modern medical practice, the development of many (perhaps most) effective remedies and treatments is empirical, often with only a minimal scientific basis. Structural interactions are not nearly as complex as those in humans, but are nonetheless sufficiently complex to be only partially understood. Structural design and evaluation have become much more rationally and scientifically based in the past couple of decades, primarily due to increased capabilities of computerized analysis and structural sensing methods. Many of the interpretations, analyses, and design codes remain empirically based.

4. *Diseases, injuries, and damage can be acute, or chronic and degenerative.* Acute disease or injury usually requires a brief one-time course of medical treatment. Chronic and degenerative diseases require long-term management. When chronic diseases are not properly managed, especially with early-stage intervention, then the maladies

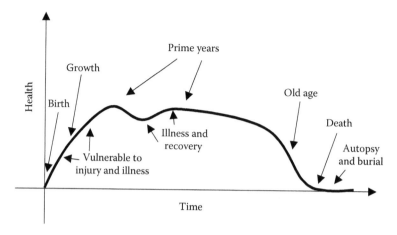

FIGURE 8.1   Possible human health life cycle.

can progress to more severe stages where treatments are much more expensive and relatively ineffective. Structures also suffer from acute and chronic degenerative injuries and degradation. An acute injury might be an impact loading that when properly repaired can largely be forgotten. Corrosion is often a chronic degenerative condition. Early identification and treatment can prevent severe damage due to corrosion [3].

5. *Most people and structures can survive for extended periods of time with little monitoring or treatment.* A modest level of routine medical checks and a healthy lifestyle can often prolong a person's life. Minimal medical checks of vital signs, such as heartbeat, blood pressure, aspiration, and bodily fluid tests, will catch many of the major life-threatening conditions. Predicting when a person will die of natural causes is very difficult. Likewise, most structures will survive for extended periods of usage without health monitoring or treatment. A modest amount of routine inspection and monitoring combined with routine maintenance activities, along with keeping operational loads to within specified limits, can lead to long-standing structures. Predicting the time of failure or collapse of a structure subjected to routine in-service or external loads can be very difficult.

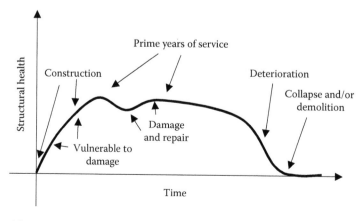

FIGURE 8.2   Possible structural health life cycle.

6. *Disease and structural fault diagnosis are imprecise arts.* Medical diagnosis requires piecing together vague bits of information from reported symptoms and test results. Diagnosing the source of many structural problems often requires piecing together information from a minimal set of measurements. Many structural measurements are only partially indicative of underlying conditions.

7. *Health diagnoses can indicate the need for remedial actions.* People in poor health are subjected to closer scrutiny, follow-up medical examinations, and treatment. As medical care technologies continue to improve, the procedures for assessing health are becoming more complicated and expensive, and require highly skilled operators. Similarly, structures and structural details determined to be in poor health are often repaired and put back into service. Structures in poor health should be inspected and monitored more frequently, especially critical details with damage. Structural inspection techniques are becoming increasingly sophisticated and complicated, and require more highly trained personnel.

8. *The health maintenance of both humans and structures can benefit from holistic considerations.* Eastern (Chinese) medical practices use global qualitative methods of diagnosis and treatment. For humans, diet, lifestyle, and social considerations can be vital. For structures, holistic approaches to design, material selection, fabrication, loading, and maintenance operations may also be useful [4].

9. *The complexity of interactions and the empirical nature of medical practice often create situations where it is difficult for a single physician or engineer to determine the best treatment methods.* Using specialists and specified procedures can improve overall medical outcomes. Issues of complexity also apply to structures where improved communication and the development of standardized test, inspection, and maintenance methods can lead to improved structural health engineering outcomes.

Perhaps the biggest difference between medical practice and structural health engineering is that biological systems can heal themselves. Very few, but some, structural systems possess self-healing capabilities [5–9]. Self-sealing fuel tanks are perhaps the most notable example [10,11].

A primary application of SHM is to enable making informed decisions about where and when to apply maintenance palliatives. A second application is to combine SHM information with estimates of future loads to establish a *prognosis* as to how damage might progress and how long the structure may survive without failing or requiring maintenance. Many of these activities are processes of SPR [12–14].

The multitude of possible structural behaviors and interactions often complicate and confound the utility of simple quantitative health assessments. Easy-to-measure quantities, such as geometry, may provide only limited clues as to the health prognosis of a structure. Other bits of information, such as incipient failure modes, are more valuable, but may be much more difficult to determine. Figure 8.3 shows the conceptual relations between ease of measurement, complexity, information value, and uncertainty.

8.2 Visual Inspections

A visual inspection is generally the first and most common assessment of structural health. A definition of visual inspection is "All unaided inspection/evaluation techniques that use the five senses with only very basic tools (e.g., flashlights, sounding hammers, tape measures,

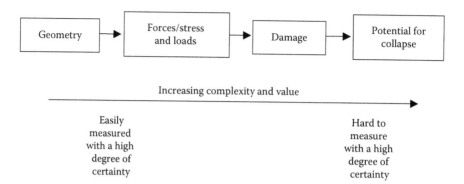

FIGURE 8.3   Complexity and certainty over a range of structural health measurements.

plumb bobs, etc.)" [15]. Gross deficiencies are readily apparent to the untrained human inspector. Trained personnel can identify the severity of less obvious deficiencies, and can produce quantitative evaluations that are useful indicators for health and maintenance engineering activities.

Hearn surveyed and compared bridge inspection practices in multiple countries in North America, Europe, and South Africa [16]. In an effort to quantify and improve the performance of human inspectors, the U.S. Federal Highway Administration (FHWA) conducted an extensive study of the accuracy, repeatability, and factors that affect inspection [15]. The study included experiments where teams of experienced bridge inspectors conducted repeated, randomized, anonymous inspections of seven bridges in northern Virginia and southern Pennsylvania. Surveys of the inspectors assessed self-reports of health, experience, engineering qualifications, job interest, and relationships with management, along with measurements of visual acuity.

The results of the FHWA study are as follows: (1) For routine visual inspections, the standard deviation of the rating on a 10-point scale ranged from 0.58 to 0.94. Particularly large variability appeared in the assessment of concrete bridge deck delaminations. (2) Factors affecting visual inspection included subject-dependent factors, such as visual acuity, age, experience, personality, and so on; physical and environmental factors, such as lighting and access; task factors, such as inspection time, fault mix and conspicuity, complexity; and organizational factors, such as the number of inspectors, training, social factors, and so on. (3) Most of the inspection teams did not use licensed professional engineers. (4) Testing the visual acuity of inspectors was almost nonexistent. (5) There was a substantial variability in time predicted and time required for inspection. (6) Inspectors were hesitant to assign extreme "low" or "high" condition ratings, and instead grouped the ratings near the center of the scale.

Using detailed and agreed-upon terminology for describing damage and defects reduces the variability of visual inspection results [17,18]. Another factor is to use a detailed inspection plan so as to avoid missing important details. For example, Raina lists many of the details that should be inspected in concrete bridges [19]. Shirakura et al. developed a reliability-based checklist for inspecting concrete bridges [20]. Structural Engineering Institute (SEI)/ASCE 11–99 describes many of the key issues related to existing building inspection and structural condition assessment [21]. Veletzos et al. found that internet-based training tools that included visual catalogs of damage types were an effective means of increasing the uniformity and effectiveness of post-earthquake assessments [22].

Electronic aids to visual inspection can increase the efficiency and performance of the inspection process [23–25]. Ward monitored corrosion-prone maritime aircraft with digital image acquisition equipment [26]. Upon visually detecting corrosion, the capture and storage of digital images in a database map all detected regions of corrosion on the same model of aircraft to form an integral part of a fleet-wide corrosion maintenance program. The use of virtual reality techniques may provide further enhancement of post-inspection evaluations [27]. The combination of electronic image acquisition for detailed post-inspection examination is an attractive possibility [28]. Applying automated reasoning methods to the visual inspection data, such as the neuro-fuzzy system used by Kawamura et al., can provide further enhancements [29]. Dan et al. and Rajani et al. suggested fuzzy logic approaches as a rational means of combining qualitative verbal inspection data to produce condition ratings [30,31]. Kaito and Abe used the inherent variability in condition ratings to develop a probabilistic approach to bridge maintenance [32].

Safety for both the public and the inspectors is of paramount concern. Bridge and traffic-sign bridge inspection often requires dangerous traffic pattern alterations that pose hazards to both motorists and inspectors [33]. Falls, enclosed-space suffocation, disease from vermin, poisoning, and electric shock are all hazards for structural inspectors. Underwater inspection and inspection inside structures with noxious chemicals, such as airplane wings, can be particularly hazardous. At a minimum, all inspection activities should conform to government-imposed safety regulations [34].

A principal difficulty with visual inspection is that the inspector usually has access to only the outer surface of structural elements and has difficulty determining the nature and extent of subsurface features. A trained inspector can interpret subtle surface changes as indicators of subsurface conditions. For example, efflorescence may be a strong indicator of underlying problems in building façades [35]. Subsurface corrosion activity can be particularly difficult to assess with only a visual inspection [36]. The result is that structural disassembly is common with aircraft corrosion inspection. A study by Shdid et al. compared visual ratings with postdeconstruction flexural strength of prestressed concrete bridge pier columns [37]. The visual ratings on a scale of 1–4 did not correlate well with the ultimate flexural strength of the columns.

8.3 Condition Assessment

Those structures where failures pose major safety and economic hazards to society are often the subject of codified condition assessment procedures. Since the present condition of a structure depends in large part on the past history of the structure, collecting and using extensive structure-specific records significantly enhances rating procedures. Such records include construction plans, as-built drawings, material test data, maintenance and repair history, accident records, flood data, traffic data, and inspection history.

An example is the federal regulations in the United States that require the inspection of most bridges every two years according to procedures defined by AASHTO [2,38]. AASHTO recognizes five types of inspections: (1) *Initial inspection*: An examination of a structure when it is first put into service. (2) *Routine inspection*: Regularly scheduled inspections that include detailed inspections of interesting items and the recalculation of load ratings as appropriate. (3) *Damage inspection*: An unscheduled inspection that initiates when environmental or human actions may have damaged the structure. The level of damage inspection depends on the amount of perceived damage. (4) *In-depth inspection*: This is a close-up, hands-on inspection of structural members that is more detailed than a routine inspection and may involve using NDE equipment. (5) *Special inspections* Scheduled for perceived potential deficiencies, such as scour or incipient cracking [39].

The results of an AASHTO-type evaluation include a *condition rating* and a *load rating*. The condition rating combines functional capabilities of the structure with the amount of deterioration or damage. Condition ratings are on a scale of 0–9, with larger values indicating a better condition. The use of a combined condition rating leaves open the possibility of physically sound, but functionally deficient, bridges being heavily de-rated. An example of a functionally deficient bridge may be one that is too narrow for existing traffic patterns. The load rating is the estimated live load carrying capacity of an existing bridge. The calculation of a load rating combines information from existing bridge plans supplemented by information gathered from a field inspection.

A primary outcome of inspections is to indicate whether there are critical deficiencies that require a rapid response. The geometric configuration of the structure should be consistent with the plans on record. The inspections should apply an extra degree of scrutiny to fatigue-prone details and fracture critical members, that is, those that are in tension and do not have redundant members that can carry the load temporarily if the member fails. Fractures often initiate as fatigue cracks at stress concentrations near specific details, such as welds and joints [40]. Depending on the magnitude of the loads and the material, fatigue cracks may grow by brittle or ductile cracking. Brittle cracking is particularly dangerous because the cracks can grow very quickly with little warning and can progress to a rapid collapse. All NDE material evaluations should follow published standards, when practical, and should use standardized and ergonomically designed reporting forms.

Defects classification must account for both the present damage severity and the potential for growth into more serious defects. Karbhari et al. suggest the following defect classification scheme for composite structures [41].

1. *Aesthetic*: Defects with a poor aesthetic appearance, but do not affect the strength, stability, or performance of the structure.

2. *Quality assurance/quality control critical*: Defects due to poor quality assurance and control during fabrication.

3. *Structural performance/integrity*: Defects that directly affect structural performance and integrity. The defects are of sufficient severity as to be likely to cause substantial reduction in performance.

4. *Short-term durability*: Defects that are likely to grow to a stage that critically affects stability and performance shortly after exposure to environmental and service loads.

5. *Long-term durability*: Defects that are likely to grow to a stage that critically affects stability and performance after extended exposure (perhaps 5 or more years) to environmental and service loads.

Bridge decks are an example of high-performance components. Virtually all bridge decks degrade with time. Agrawal et al. found that the average degradation in the condition rating of a bridge is fairly predictable across the New York State system of bridges [42]. Such information is useful for overall maintenance planning efforts. However, it is necessary to detect damage at the detail level on individual bridge decks in order to guide maintenance activities that can arrest the progression of early-stage damage. The development of advanced instruments for assessing bridge deck condition continues to be an area of active research [43]. Invisible subsurface delaminations are a principal bridge deck defect that when detected and corrected early in a stage of growth can provide large economic benefits. Techniques for assessing the extent of delaminations in concrete bridge decks include coring, chain drag, elastic wave methods, infrared thermography, half-cell, and GPR [44]. All these methods have limited resolution and specificity. A workaround solution combines the data from multiple instrument types. Huston et al. found that chain drag and PSPA elastic wave measurements tended to be correlated, whereas GPR and half-cell data were also correlated [45]. Rhazi and Omar found similar correlations between indications of distress from half-cell and GPR data [46]. Corrosion of post-tensioning strands is also a big concern [47]. In a study of parking garages, Puri and Moser found that GPR was a good tool for locating post-tensioning strands and for identifying gross failures, but that magnetic scanning provided a better resolution estimate of severed strands [48].

Sensors and NDE instruments aid many inspection processes [49]. Although the use of these instruments can enhance the amount of information gained with an inspection, such a result is not guaranteed. Comstock and Garrison recommended a set of procedures for the ultrasonic testing of hanger pins in bridges [50]. These procedures are sufficiently generic to paraphrase as follows:

1. Inspection personnel must have training specific to both the instrument and structural elements being inspected, along with adequate opportunities to build and maintain these skills.

2. Inspectors must have confidence in their skills based on good training, solid experience, and continuing practice.

3. Inspection personnel must fully understand the capacities and limitations inherent in the process and make recommendations in light of this understanding. A common understanding of the reliability of NDE is essential to turn inspector recommendations into structural management decisions.

4. An independent agency should certify the NDE and overall inspection program on a regular basis to identify deficiencies and generally to keep the program aligned to the larger needs of the owners of structures.

8.4 Damage Indices

Damage indices are simplified and condensed representations of the complicated circumstances surrounding structural damage. Most damage indices are scalars, such as the AASHTO 0–9 rating scale for bridges, where 0 is worst and 9 is best. The condensed nature of the information in damage indices makes them convenient tools for maintenance and planning purposes. For example, Sugimoto et al. used a 1–5 structural member health-rating index to establish system-wide repair and maintenance strategies for bridges with genetic algorithms [51]. Another example is the Pontis Bridge deck evaluation 1–5 index, with 1 being best and 5 being worst, and the Pontis-inspired highway sign bridge rating system in California [52,53]. Not all damage indices provide the same level of performance and reliability. Williams et al. examined eight different damage indices as predictors of seismic resistance in reinforced concrete elements [54]. Displacement-based indices were superior for this specific application. Baladi developed a framework for relating pavement condition indices with expectations of pavement longevity and performance [55].

Many damage indices express the observed behavior as normalized deviations from average behaviors [56]. Defining θ_j as an observed behavior, δ as an offset (usually the mean $\bar{\theta}$ or zero), $\hat{\theta}$ as a normalizing factor (usually the mean, $\bar{\theta}$, or the standard deviation, σ_θ), and k as an adjustable scale factor (typically 1 or 2, but other values appear), to obtain

$$z = \left[\frac{\sum_{i=1}^{N} (\theta_i - \delta)^k}{\hat{\theta}k} \right]^{1/k}.$$ (8.1)

Multiple versions of deviation from normal damage indices appear in the SHM literature, particularly in the interpretation of vibration test results. Stubbs et al. used pre- and postdamage stiffness ratios as estimates for the local element as the damage indicator β_j [57]:

$$\beta_j = \frac{E_j}{E_j^*} = \frac{1}{2} \left[\frac{\left(\int_a^b \{\varphi_i^*\}^2 \, dx \int_0^L \{\varphi_i''^*\}^2 \, dx \right) \int_0^L \{\varphi_i''\}^2 \, dx}{\left(\int_a^b \{\varphi''\}^2 \, dx \int_0^L \{\varphi''\}^2 \, dx \right) \int_0^L \{\varphi_i''^*\}^2 \, dx} + 1 \right]$$ (8.2)

E_j and E_j^* are the pre- and postdamage elastic moduli of beam element j that runs from a to b for a beam that runs from 0 to L. $\phi_i(x)$ and $\phi_i(x)^*$ are the pre- and postdamage ith mode shapes. Shi et al. proposed a similar curvature-based damage index [58]. Hernandez and Fujino used this index to assess the health of RC railway viaducts [59]. A limitation on the utility of Equation 8.2 is the difficulty with measuring the second derivative, that is, curvature, with sufficient precision for practical applications. A possible workaround proposed by Salgado and Cruz uses a similar damage index based on the continuous wavelet transform scale number 3 coefficients to identify damage in beams [60]. Xiang and Wang found that a damage index based on average vibration mode curvature effectively assessed damage in cable-stayed bridges [61]. Masuda et al. defined a damage index based on flexibility, where β_i is a local flexibility index [62]. Guy et al. used an index defined as the sum of deviations over a frequency spectrum to assess damage in composite plates by Lamb wave leakage energy losses [63]. Ma and Pines used a damage index based on the relative phase lag between points on a structure subjected to dynamic motions, such as an earthquake [64]. Damage in the form of loss of stiffness produces a detectable increase in the phase lag. Cheng and Hu detected damage in bridge structures using pre- and postdamage vibration data with a damage index based on the modal-weighted change in energy at the structural element level [65].

Quadratic and other nonlinear extensions to the generic damage index in Equation 8.1 appear as needed. Ihn and Chang used a damage index based on the energy of scattered ultrasound Lamb waves [66,67]:

$$\text{Damage index} = \left[\frac{\text{Scattered energy of wave}}{\text{Baseline energy of wave}} \right]^k.$$ (8.3)

The ultrasound wave is typically an s0 Lamb wave for crack detection and an a0 Lamb wave for delamination detection. The exponent k is an empirically determined gain factor such that $0 < k \leq 1$. A gain factor of 0.5 can produce a damage index that is linear with crack length. Song et al. used a somewhat similar energy-based damage index, I, to evaluate concrete bents by a wavelet analysis of elastic wave energy produced by piezoelectric patches [68]:

$$I = \sqrt{\frac{\sum_{j=1}^{2^n} \left(E_{i,j} - E_{h,j} \right)^2}{\sum_{j=1}^{2^n} E_{h,j}^2}}.$$ (8.4)

A value of $I = 0$ corresponds to a perfectly healthy structure. $E_{i,j}$ is the total wavelet decomposed energy at time i for frequency band j for the damaged structure. $E_{h,j}$ is the corresponding wavelet decomposed energy for the healthy structure. A drawback to using strain energy damage indices is the requirement of high-quality and high-resolution modal data. Ho and Ewins, in a series of numerical simulations, concluded that a strain energy damage index was not effective at assessing the severity of the damage [69]. Tseng found that a similar damage index known as the *root mean square deviation index* (RMSD) for changes in the conductance (real part of admittance) was effective in identifying damage in piezoelectric patch impedance measurements of structural elements [70]:

$$\text{RMSD}(\%) = \sqrt{\frac{\sum_{j=1}^{N} \left(G_j^1 - G_j^0\right)^2}{\sum_{j=1}^{N} \left(G_j^0\right)^2}} \times 100, \tag{8.5}$$

where G_j^0 and G_j^1 are the conductances of the original undamaged and damaged structures, respectively. Naidu and Soh used a somewhat similar index to track frequency shifts in vibrating structures in an effort to combine electromechanical impedance testing with finite element analysis [71]. Taking advantage of imaging data provided by *scanning laser Doppler velocimetry*, Ayers et al. developed a spatially distributed damage index based on the ratio of incident and reflected Lamb waves [72].

A damage index proposed by Newhook for quantifying the possibility of delamination in bonded composite layers from strain measurements is [73]

$$\theta_i = \frac{\varepsilon_{i+1} - \varepsilon_{i-1}}{2h}, \tag{8.6}$$

where θ_i is a strain gradient damage index, ε_i is the strain at point i, and h is the spacing between strain readings. Values of θ_i close to 0 indicate that the composite layer is in pure tension and that a delamination has occurred.

Damage and health indices are often weighted combinations of qualitative and quantitative factors. For example, Sobjano and Tawfiq used an overall composite index of pavement condition known as the *pavement condition index* (PCI) [74]:

$$\text{PCI} = \lambda_1 \text{RCI} + \lambda_2 \text{SDI} + \lambda_3 \text{SRI} + \lambda_4 \text{SCI}, \tag{8.7}$$

where RCI is the riding comfort index, SDI is the surface distress index, SRI is the skid resistance index, SCI is the structural condition index, and λ_i is a relative weighting factor. The indices are calculated on a 0–10 scale (0 = failed, 10 = excellent) based on a combination of qualitative observations and quantitative NDE measurements. Betti et al. used a *wire damage index* (WDI) to describe the state of suspension bridge cables [75]:

$$\text{WDI} = N^b + 0.5N^4 + 0.2N^{32}, \tag{8.8}$$

where N^b is the percent of broken wires, N^4 is the percent of corrosion grade 4 wires, and N^{32} is the percent of corrosion grade 3 and grade 2 wires.

Vibrating structures with measured modal properties create the opportunity for using a variety of vibration-based damage indices [76]. Ma and Asundi proposed a *dynamic damage factor* (DDF) for systems with multiple lightly damped vibration modes [77]. The rationale for the DDF is that damage to a structure causes it to become more flexible and the natural frequencies of vibration shift to lower values. The definition of the multimode DDF is

$$\text{DDF} = \left[\frac{1}{n} \sum_{i=1}^{n} w_i \left(\frac{f_{i,\text{damaged}}}{f_{i,\text{undamaged}}} \right)^2 \right]^{1/2} < 1, \tag{8.9}$$

where n is the number of modal natural frequencies measured, w_i is a modal weighting factor, and $f_{i,\text{damaged}}$ and $f_{i,\text{undamaged}}$ are the damaged and undamaged natural frequencies for the ith mode, respectively. A pristine undamaged structure has an assumed DDF of 1. Structures subject to ballistic impacts often suffer severe localized damage that involves surface penetration, plastic deformation, and material removal. Melchor-Lucero et al. propose using a *single impact damage index* (SIDI) to quantify impact damage [78]:

$$\text{SIDI} = 0.05 \, V_{\text{dis}} + 1.0 \, V_{\text{pen}}, \tag{8.10}$$

where V_{dis} is displaced volume and V_{pen} is penetrated volume.

A *reliability index* or *safety index*, β, can be useful when there is sufficient statistical information available concerning the nature of the structural loads and resistance capabilities [79,80]. When the loads and strengths of a structure are normally distributed, the reliability index is the number of standard deviations of a one-sided Gaussian probability tail corresponding to a probability of failure, P_F. Assuming that the structure fails when the loads exceed the ability to resist the loads, then

$$\beta = \frac{\mu_R - \mu_L}{\sqrt{\sigma_R^2 + \sigma_L^2}}, \tag{8.11}$$

where μ_R and μ_L are the mean of the resistance strengths and loads on the structure, respectively (in the same units). σ_R and σ_L are the standard deviation of the resistance strengths and loads on the structure, respectively. The probability of failure in terms of the reliability index is

$$P_F = \int_{-\infty}^{-\beta} \frac{1}{\sqrt{2\pi}} \exp\left(-\frac{u^2}{2} \right) du. \tag{8.12}$$

The larger the value of β, the more reliable the system. For example, if $\beta = 2.3, P_F = 10^{-2}$; and if $\beta = 6, P_F = 10^{-9}$. The reliability index extends to cases when the probability distributions of loads and resistance strengths are non-Gaussian. The technique calculates the probability of failure based on the non-Gaussian distributions and then converts the probability of failure into a reliability index, β, based on an assumed equivalent Gaussian distribution, as in Equation 8.12.

The advent of modern computers and database management systems makes it practical to use rating indices that contain more information than that of a scalar. Catbas and Aktan described the possibilities of using indices for condition and damage assessment based on modal, strain, and temperature changes [81]. Koyluoglu et al. used hysteretic softening during cyclic earthquake loads as the principal damage indicator [82]. Lazarov and Trendafilova used the plate thickness ratios in damaged and undamaged states as an index for use in vibration identification of damage [83]. Graphical and multimedia techniques can enhance information presentation [84,85]. Wu and Su used a hierarchical structure of sensor and condition ratings to form an overall scalar health index for dams [86].

Replacing scalar-based damage indices with vector-based variants offers the possibility of conveying more information at the expense of increased complexity and ambiguity of interpretation. For example, Yam et al. used a damage index vector formed from components of wavelet decompositions of elastic wave motions as features for a neural network (NN) damage detection effort applied to honeycomb composites [87]. The damage index vector $\mathbf{V}_D$ is

$$\mathbf{V}_D = \left\{ 1 - \frac{U^d_{k,1}}{U^0_{k,1}} \quad 1 - \frac{U^d_{k,2}}{U^0_{k,2}} \quad \cdots \quad 1 - \frac{U^d_{k,2^{k-1}}}{U^0_{k,2^{k-1}}} \right\}^T, \tag{8.13}$$

where $U^d_{k,j}$ and $U^0_{k,j}$ are the energy contents in damaged and undamaged wavelet signal decomposition bands, respectively. Lam et al. used a *damage signature index* (DSI) for vibration-based damage assessments, with the definition

$$\mathrm{DSI}_i = \frac{\{\varphi_{ui}\} - \{\varphi_{di}\}}{f^2_{ur} - f^2_{dr}}, \tag{8.14}$$

where DSI_i is the DSI for the ith mode, $\{\varphi_{ui}\}$ and $\{\varphi_{di}\}$ are measured undamaged and damaged ith mode shape vectors, respectively, and f_{ur} and f_{dr} are the natural frequencies for undamaged and damaged reference modes, respectively. The first mode is a suggested reference mode [88].

8.5 Load Testing

Load testing is the evaluation of structural behavior with the controlled application of loads. The complexities and uncertainties of structural interactions often combine with the imperatives of public safety to make load testing the preferred method for rating structural load-carrying capacity. Load testing for the purpose of load rating presents a bit of a conundrum. Determining with certainty the ultimate load-carrying capacity of a structure requires testing to failure. Loading to failure prior to putting the structure into service is inherently undesirable. Subfailure load testing procedures are necessary for many applications.

AASHTO recognizes three types of load tests [2]: (1) *Load evaluation* is the measurement of in-service loads, such as traffic, thermal stresses, or wind loads. The measured loads can include both static and dynamic components. An example of load evaluation from the operation of railroads is the measurement of compressive thermal stresses in continuously welded railroad rails. When the compressive stresses become too large, lateral buckling and derailments can result [89]. (2) *Diagnostic testing* attempts to establish how a particular load affects individual members or structural systems, possibly to verify the numerical stress analysis of a complicated structure. For example, Kiddy et al. used an array of fiber optic Bragg grating strain gages in a load test to verify finite element model predictions of the stresses in a deep sea submersible vehicle with a complicated structural form [90]. Grace et al. used load tests to evaluate the load distributions in a CFRP-reinforced concrete bridge [91]. (3) *Proof load testing* establishes the maximum safe live load by applying a submaximal load and observing the response. Proof testing is a routine practice for bridges before putting them into service. The traditional proof test places heavily loaded vehicles and other objects on the structure while observing the response and overall behavior. Modern proof load tests can make use of SHM instruments and dynamic testing, such as robotic tacheometers (Total Station optical surveying instruments), laser instruments, mode shape identification, and statistical pattern recognition (SPR) methods, including NNs [92–102]. Information concerning the variability of member strengths and construction details enable incorporating probability into the interpretation of proof load testing results [103]. Aerospace structures undergo similar testing at both the component and full-scale level [104]. The use of extensive SHM instruments in these test regimes can reduce overall program costs considerably by the early identification of deficient designs [105]. Dynamic load testing with vehicles is rather uncommon, but can be a viable technique, especially if the analysis includes a dynamic model of the structure response [106]. Load testing to failure is relatively rare for large structures, but such tests can provide valuable information such as failure modes and load paths [107,108].

Relatively simple measurements of structural behavior during load tests often produce valuable information. One possibility is displacement sensing. More sophisticated measurements with multichannel structural sensors are possible and perhaps advisable.

Standard preparations for a load test combine a condition survey with structural analyses. These preparations reduce the possibility of damage or collapse during testing, and give an indication of the probable load distribution for the design of effective sensor locations. Usually the measurands are mechanical quantities, such as strain, displacement, acceleration, load, and applied pressure. Statically determinate structures permit a relatively straightforward interpretation of the load measurements. Statically indeterminate, nonlinear, and 3-D effects confound load measurement interpretations that rely on simple mechanical models [109]. When possible, it is best to measure the distribution of loads throughout the separate redundant load paths. Additional concerns arise in the interpretation of strain gage measurements to calculate loads. End effects can easily overwhelm the assumptions of St. Venant's principle. Portable, or rapidly deployable, sensors, including optical instruments, are attractive options for load testing [110–112]. The sensors can also be part of a permanently installed monitoring system [113]. Infrared thermography can detect active damage processes during load testing, such as due to microcracking, yielding, or fatigue [114].

Load testing at submaximal loads inherently runs into questions of mechanical nonlinearity and secondary structure coupling. All structural systems are nonlinear to some extent. At low load levels, joint and bearing stiction, that is, Stribeck effects, are active nonlinearities [115]. The structure may behave linearly at higher and perhaps routine service-level loads. The structure may again behave nonlinearly at higher loads as the individual members yield, crack, and redistribute the loads. Figure 8.4 shows some of the qualitative aspects of these mechanics.

An important case of submaximal load testing arises when asking bridges to carry loads that significantly exceed the posted loads, that is, *superloads*. Phares et al. described a procedure for determining whether a bridge can carry a superload [116]. The first step is to measure the response of the bridge with fairly heavy proof loads. The heavy proof load serves to calibrate a detailed numerical model of the bridge mechanics. The calibrated model predicts the behavior of the bridge in response to the superload. If a judgment determines that the bridge can safely carry the superload, then SHM during the subsequent passage of the superload can provide additional information regarding the mechanics of the load distributions and possibly the extent of any damage. Rohrmann et al. reported on an effort in which the load monitoring instruments also serve a dual role as a long-term SHM system [117]. Wood et al. compared calibrated detailed finite element analyses with measurements taken on a composite- concrete-deck–steel-girder bridge undergoing

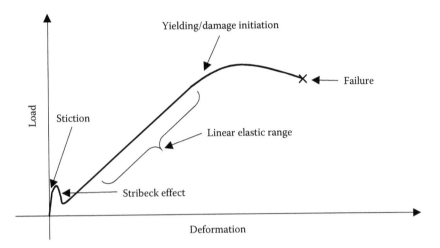

FIGURE 8.4 Typical nonlinear load–displacement relation.

superloads [118]. The analysis accurately predicted the measured load, once it properly modeled potentially nonlinear behavior, such as the degree of composite shear action in the deck.

Structures with poorly understood behaviors or with new designs that lack a long track record of performance may also warrant load testing [119]. There are only a handful of laboratories around the world with full-scale large structure testing capabilities [120]. Hag-Elsafi et al. conducted a series of load tests on an FRP bridge retrofit [121]. The testing consisted of load displacement measurements immediately following installation of the retrofit and repeated measurements three years later. In this case, the FRP retrofit showed minimal degradation. Kosmatka tested a composite bridge by both field load testing, and laboratory cyclic whiffletree load testing under accelerated loading conditions that exceeded the lifetime design parameters for the bridge [122]. For this case, peak strain measurements combined with modal testing indicated that there was minimal degradation in the strength of the bridge. Abdullah et al. measured the friction resistance of bolted FRP joints as a function of surface conditions and bolt tension [123].

An important result from load testing is that the mechanics of even relatively simple bridges are not always well understood. For example, Cai et al. found cases where the rating produced by bridge load tests differed by factors of up to two from that of analytical methods [124]. Saraf and Nowak loaded tested girder bridges with ages of 60 years or more [125]. The bridges showed significantly more stiffness than expected from simple mechanical analyses. An explanation attributed the additional stiffness to the coupling action of secondary structures on the bridges. Similarly, Prader et al. conducted load tests on a series of older undocumented reinforced concrete bridges [126]. The measured stiffness and inferred load-carrying capacity was much larger than that predicted by simple mechanical models.

Dynamic load testing alleviates some of the instrumentation difficulties associated with static testing. A particular advantage is to eliminate the need for measuring displacements relative to a fixed reference frame. Dynamic testing measures accelerations relative to an easily established inertial reference frame. Dynamic testing requires dynamic excitation loads on the structure from ambient effects or devices such as shakers or hammers. Large structures require large loading devices [127]. Quick-release tests are a convenient method of testing fairly large structures for dynamic property evaluation. The typical quick-release test applies a large static load to a structure, such as with a steel cable, and then quickly releases the load, by disconnecting or cutting the cable. Analysis of the ensuing motion produces estimates of the dynamic properties of the structure [128]. Scale model testing of structural systems can provide insight into the effects of complicated interactions that can be computationally intractable. Turbulent wind interaction with large bluff structures is one application that benefits from scale model testing. Concrete failure mechanics is another [129]. Faithful mechanical renditions with scale models require suitable similarity scaling of the several physical effects at play. It is rare that all of the physical effects will scale appropriately and that compromises are necessary. For example, the aeroelastic testing of bridges in a wind tunnel abandons Reynolds number scaling in lieu of a faithful scaling of wind velocity, geometric scale, and structural frequency.

8.6   Load Rating of Structures

The term *load rating* is used as both a noun and a verb. As a noun, a "load rating" is an estimate of the allowable in-service loads that the structure can sustain without damage and the maximum load that the structure can sustain without collapse or failure. As a verb, "load rating" is the determination of allowable in-service loads. Examples are the rating of bridge structures to limit or permit the passage of heavy vehicles and the imposition of flight condition envelopes on airplanes [130]. Due to complexities and uncertainties associated with the mechanics of structural failure, most load-rating techniques use simplified semi-empirical techniques.

The *allowable stress* and *load factor* methods are common procedures for calculating load ratings. Both procedures incorporate experience with similar structures and materials with anticipated failure mechanisms. The allowable stress method calculates the maximum stress in a member under an extreme load, usually with an assumption of linear elastic mechanics, and uses a safety factor that requires strengthening the structure or reducing the loads to safe levels. Maximum stress calculation techniques often derive from first principles, for example, frame or finite element stress analyses. Load test data calibrate and augment the calculations. The *load factor* technique multiplies the loads on the structure by various factors that reflect the degree of uncertainty in the level of the load. For example, AASHTO recommends using the rating formula

$$RF = \frac{C - A_1 D}{A_2 L (1 + I)},$$

(8.15)

where RF is the rating factor for live load-carrying capacity, C is the capacity of the member, A_1 is the dead load factor, D is the dead load effect on the member, A_2 is the live load factor, L is the live load effect on the member, and I is an impact factor for use with the live load. The values of A_1 and A_2 depend on whether the rating is for inventory or operating purposes and whether the method uses the allowable stress or load factor methods. The live load that the structural member can sustain, RT, is the nominal live load on the bridge, W, multiplied by the rating factor, RF,

$$RT = RF(W),$$

(8.16)

where RT is the live load that the member can safely sustain and W is the weight of the live load [2].

McCall et al. developed a load-carrying-capacity rating factor for reinforced concrete slabs [131]. The rating factor combines information from concrete strength measurements, sub-base k-values of stiffness, and slab thickness measurements. Since each source of data individually shows considerable variation, using the data individually in load rating may produce an overly conservative and costly design recommendation. When the outliers for concrete strength, sub-base k-value, and slab thickness all occur independently, a combined rating factor based on Westergaard's equation can give a more cost-effective rating estimate.

Data from load tests and field measurements can provide input to update and improve load-rating procedures. Especially valuable, but relatively rare, data come from tests that simultaneously monitor and load a structure to failure [132,133]. Bayesian methods can effectively implement updates from the data into the load-rating procedures [134]. Estimating and calculating reliability indices can also make use of monitoring data [135].

8.7 Load, Fatigue, and Real-Time Monitoring

Monitoring the specific load environment can provide detailed information as to how hard a structural system works, when the load paths and load distributions in the structure are relatively independent of the amplitude, frequency, and spatial distribution of the applied load. This leads to improved maintenance scheduling, even if the specific components undergoing damage are not monitored [136]. Similarly, gross structural deformation measurements can provide estimates of localized fatigue and damage, even if the elements of concern are not specifically monitored [137]. There are multiple applications of such load and fatigue monitoring in the SHM literature. Vines et al. recorded and estimated the buffeting loading exerted on a composite ship by rough seas with an array of FBG strain sensors [138]. The viability of this approach improves with placing the sensors in locations that indicate the overall loading of the structure and are insensitive to spurious insignificant signals. Careful consideration of the structural loading and details, aided by numerical and mechanical models, can aid in this effort [139]. McColl and Bradfield, and Johnson et al. used a similar approach with airplane fatigue load monitoring [140,141]. The method converts acceleration and strain data from flight operations into gust load factors and combines the information with other mission data, such as the number of landings, to produce a fatigue load estimate. El-Bakry described the design requirements of an operational load monitoring system for commercial aircraft landing gear [142]. A key requirement is to provide useful information for maintenance activities without any disruptions to aircraft operations. Wu et al. measured softening shifts of the Young's modulus in composite structures as they undergo cyclic loading to indicate the progress of fatigue with FBGs [143]. de Leeuw et al. monitored peak strains in continuous welded rails for railroads with autonomous microprocessor-based record and sense systems [144].

Real-time and on-line monitoring systems can help provide rapid SHM assessments. Good candidates are situations where the damage signatures are known and where rapid intervention has potential benefit. Real-time systems require a combination of sensors, data processors, and preselected data processing techniques to all be in place and operational [145–149]. Hard thresholds can detect relatively simple situations. More complicated situations may benefit from the application of sophisticated statistical methods, such as Bayesian networks, and the examination of information-rich statistical features, such as wavelet packages [150–152]. In circumstances that can benefit from real-time intervention, such as dams and rotating machinery, SHM systems that combine measurements with physical models of expected behavior are potentially superior to off-line monitoring systems [153,154].

There are alternative maintenance scheduling methods that do not involve monitoring. These include acting only when damage is sufficiently severe to demand immediate attention, or time of usage. The hypothesis underlying a *time-of-usage maintenance scheduling* approach is that the loads experienced by a structure during a given operational state tend to be sufficiently similar to other nominally similar operational states, and that failure tendencies under these loads are consistent within statistically defined parameters. Time-of-usage maintenance scheduling combines time of use with historical maintenance records. To avoid costly accidents and unexpected service outages, such a strategy inevitably requires the costly removal from service and the maintenance of systems that are healthy and do not require any intervention. This is especially the case for structures with high degrees of variability in damage progression. It may be possible to reduce overall system operational and maintenance costs by using load monitoring in lieu of time-of-usage methods [155].

8.8 Root Cause Analysis and Normal Accident Theory

Understanding the causes and processes of structural failures is an important aspect of SHM. Many failures involve complicated interactions between unexpected forces, material responses, mechanical couplings, societal pressures, and human errors [156–160]. In the past century, primarily due to safety initiatives in the aviation, nuclear power, chemical, and medical industries, two schools of thought have emerged on how to analyze and explain failures or accidents. One is a set of linear or reductionist methods, often known as *root cause analysis* (RCA). RCA attempts to find underlying root causes of accidents or failures, that if corrected will prevent failures. Contrasting with RCA is a set of nonlinear holistic approaches. Perhaps the most well-known nonlinear description is *normal accident theory* (NAT). The premise of NAT is that many situations are complex. The coupling of unexpected interactions inevitably leads to accidents [161]. NAT is a relatively new concept. NAT initially appears as a descriptive taxonomy of accidents or failures. NAT has evolved to a more general approach that also encompasses accidents or failures that are too complex to be decomposed effectively by RCA.

An overriding concept in safety engineering is that measuring, controlling, and reducing the variability of system and subsystem behaviors increases the predictability of overall system behavior. Increased predictability can lead to system operations that simultaneously improve safety, quality, productivity, and cost statistics [162]. To this end, many accident prevention and analysis efforts attempt to take advantage of the *near-miss hypothesis*, and the notion that adverse events are often a convergence of interacting causes and events, each of which would be insufficient to cause the adverse event. The near-miss hypothesis is the notion that many of the same factors and interactions that cause near misses also cause adverse events. Studying the behavior of systems that undergo near misses may elucidate many of the same causes and conditions that give rise to adverse events. Following a near miss, it may be possible to collect data in an unbiased manner, since it may avoid many of the legal entanglements of blame that arise when accidents occur. Unbiased data collection can be more difficult in the event of an accident.

8.8.1 Root Cause Analysis

The premise of RCA methods is that determining the underlying root causes of adverse events facilitates implementing remedial actions to prevent future failures. RCA comprises a broad range of techniques. The common theme is sorting and simplifying complex situations and events. In many respects, the results of RCA-based accident reduction efforts have been quite impressive, especially in the commercial aviation industry.

The analysis of errors, accidents, failures, and other adverse events uses specialized, but not universally accepted, terminology. An example is the definition of a *root cause* [163]. Wilson et al. defined a root cause as the most basic reason for an undesirable condition or problem that, if eliminated or corrected, would have prevented the condition or problem from existing or occurring [164]. Similarly, EQE International, Inc. defines a root cause as the most basic cause that can be identified and modified to prevent recurrence of an event [165]. An *adverse event* is an occurrence or outcome that is less than what was expected. Reason defines an *error* as the failure of a planned sequence of activities to achieve the intended outcome and when the failure cannot be attributed to random events [166]. An *active error* is an error that causes damage, injury, or other similar losses. A *latent error* is an error lurking in the system that has not yet caused damage or injury, but has the potential to do so in the future. A *sentinel event* is where damage has occurred that is above a preset intensity threshold, for example, an engine falls off an airplane during flight. A *near miss* (perhaps better termed as a *near hit*) is an accident that almost occurred but did not, for example, the engine mounting bolts on an airplane in flight loosen to unacceptably large levels, but the engine does not fall off. A *trigger event* is the detection of a near miss of sufficient severity to prompt further investigation, for example, the detection and subsequent retightening of loose engine bolts following a safe landing that prompts an investigation as to why the bolts loosened.

In some circumstances, the root cause of an adverse event may be obvious. In many (if not most) cases, the root cause of an adverse event is not clear and can be the subject of contention. When several events and conditions must converge and occur simultaneously for the occurrence of an adverse event, it can be argued that all these preceding events and causes are root causes. This conflicts with the notion of a single root cause. Requiring a root cause to be an event or condition that is preventable or correctable pares the list of possibilities. If the intention of the RCA is to assign blame or legal responsibility, then the expedience of punitive actions and blame avoidance can strongly influence root cause assignment.

The safety and accident investigation communities typically use several different RCA methods, each of which has advantages and disadvantages. Selecting the best RCA method prior to the analysis is not inherently obvious and can be highly situation dependent. Wilson et al. gave some suggestions, but note that choosing the best RCA method is essentially a trial-and-error process with guidance from heuristics [164]. The following is a list of some of the principal RCA methods.

Change analysis looks at what has changed in the system and environment of the present adverse situation versus previous nonadverse situations. Change analysis is useful for analyzing systems that have a track record of running without adverse events. The technique identifies the changes between the two cases. A critical examination of the changes determines if the change was the cause of the adverse event. It is important to consider that people often resist change and such resistance can inadvertently cause adverse events [164].

The basis of *event and causal factor analysis* is the notion that all events have causes. This is a graphical method of elucidating the causes and conditions required for an event to occur. The method can be particularly helpful in clarifying situations that are somewhat complicated. A criticism is the difficulty in converging upon a consensus for the root cause. Every event and condition has preceding events and conditions.

The *Apollo method* is similar to event and causal analysis methods [163]. The distinguishing feature is a graphical format that tabulates the evidence supporting the conclusions.

The graphical technique facilitates finding solutions to problems, especially the use of brainstorming methods.

Fault tree Analysis organizes and graphically illustrates causal chains of events. Fault tree analysis can be particularly useful in situations where there are many repeated events and circumstances, as in manufacturing, or in distributed infrastructure systems with many similar components [167]. The construction of a fault tree uses a branching top-down approach. EQE International, Inc. gives a generic fault tree with a broad scope and utility in many situations [165]. An abbreviated version appears in Figure 8.5. In many cases, it is possible to formulate and use pre-assembled specialized fault trees for specific applications. An advantage of fault tree analysis is that the hierarchical structure is amenable to computer-based implementations [168].

The first step in constructing a fault tree is to define the *top event*, which may be an error, near miss, bad outcome, or accident. Since there is more than one possible top event, choosing an appropriate top event is not always obvious, but may be essential to a successful outcome of the method. The first layer of underlying events often contains the categories of *personnel, procedures, equipment,* and *other*. Lower levels of causes are filled in, as necessary. The next step identifies the most important causal path to the top event by a process of modifying and pruning the tress. The lowest item on the causal path to the top event is the root cause.

The *Fishbone* method (attributed to Ishikara) has the same logical structure as the fault tree methods, but uses a different graphical format [164]. The layout of causes and effects is

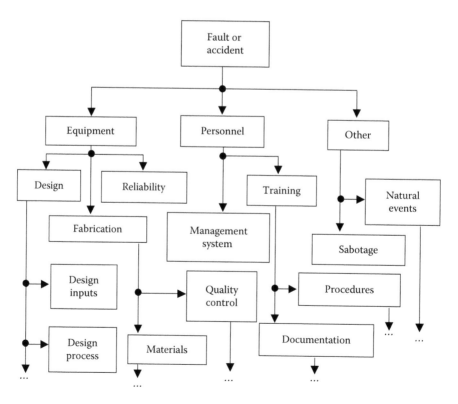

FIGURE 8.5 Abbreviated portion of generic fault tree. (Adapted from EQE International, Inc. 1999. *Root Cause Analysis Handbook,* Government Institutes, Rockville.)

from left to right with diagonal interconnecting lines to a central backbone line. The layout has a fishbone appearance.

Barrier analysis uses a different view of adverse events. Barrier analysis assumes that a host of adverse events could happen. Multiple lines of barriers as defenses prevent most serious adverse events (Figure 8.6). In this context, barriers take on the positive connotation of negating the occurrence of adverse events. For example, preventing an airplane engine from falling off requires that engine mounting bolts be tightened with sufficient torque. However, a prudent design may be to have the bolts play only a secondary fastening role so that the engine stays on, even if the bolts loosen. An additional barrier could be to use cotter pins to prevent nut loosening and vibration and/or bolt tightness sensors to sense and indicate when the engine mounting bolts loosen. Barriers are not always impermeable. Occasionally barriers fail to prevent accidents by a *Swiss cheese*-type alignment of openings (Figure 8.7). Proponents of barrier analysis claim that it is especially useful for facilitating group brainstorming. *Failure mode effects analysis* (FMEA) attempts to elucidate all of the consequences of subsystem failure. *Event trees* are similar to fault trees, except that they describe possible event scenarios following an adverse event. A *bow-tie* graph is a combined representation of fault trees, event trees, adverse events, and barrier analysis (Figure 8.8) [169].

8.8.2 Normal Accident Theory

NAT describes the complex interactions of events and systems that give rise to accidents in seemingly unexpected, unanticipated, and—possibly—unpreventable manners. Perrow led the development of NAT, starting in the 1980s [161].

NAT uses specialized terminology. *Incidents* and *accidents* are unintended events that cause damage to property and/or injury to people. An incident differentiates from an accident by the level of severity, with accidents being more severe than incidents. NAT considers holistic interactions of systems with subsystems. In an effort to organize the complex interactions, NAT uses four levels of systems: *parts, units, subsystems,* and *systems.* Parts are small components, such as rivets. Units are small assemblies of parts, such as a riveted wing skin panel. Subsystems are collections of units, such as an airplane wing. Systems are entire entities, such as an airplane. An accident is a failure of a system or a

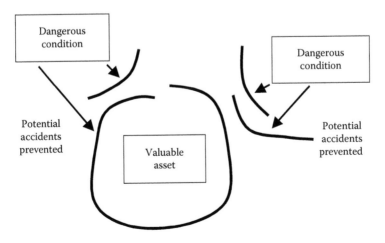

FIGURE 8.6 Safety barriers prevent accidents.

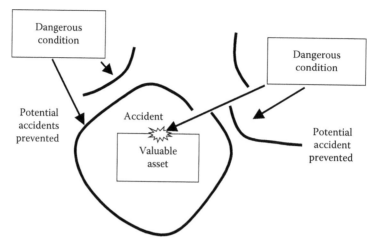

FIGURE 8.7 Safety barriers fail in Swiss cheese mode and an accident occurs.

subsystem and that failure causes severe disruption to ongoing or future uses of the system and people. An incident is the failure of a unit or part that does not cause major disruption.

Accidents and incidents may produce victims. Perrow distinguishes four groups of victims: *First-party victims* are the operators of the system, for example, pilots. First-party victims volunteer to work on the system and may be responsible for the occurrence or prevention of the accident or incident. *Second-party victims* are nonoperators and other personnel, such as passengers on an airplane, or on-site personnel that explicitly or implicitly volunteer to be in proximity to the offending system. *Third-party victims* are people who have

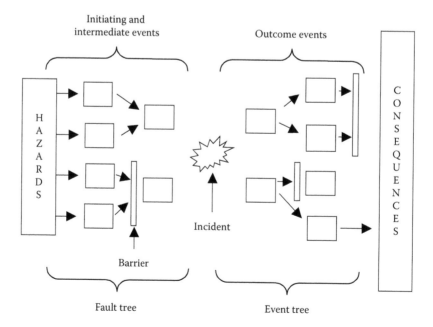

FIGURE 8.8 Bowtie representation of adverse incidents. (Adapted from Kontic B and Kontic D. 2006. *Computational Models of Risks to Infrastructure*, NATO Advanced Research Workshop, Primosten.)

not volunteered to be associated with the system. An example is people on the ground who are hit by a crashing airplane. *Fourth-party victims* are unborn people and future generations.

The principal distinction between NAT and RCA is the complexity of interactions between systems, subsystems, units, and parts. The adjectives *linear* and *complex* distinguish the interaction types. Linear interactions follow a direct path of an observable sequence of events and causes. Figure 8.9 indicates how a set of closely related sequences of interactions may follow a linear path through a space of benign events into adverse events. The premise of RCA is that with a thorough analysis, all adverse events can be traced to root causes by a linear path of causes and conditions. Complex interactions are those that are unexpected and incomprehensible. The incomprehensibility of complex situations makes it difficult to predict and control situations. Doerner indicates that once a situation attains a sufficient level of complexity (this can be as few as a couple of interacting systems), humans lose the ability to predict and correct for these interactions [170]. Heuristics and rigid hierarchical control (as is often attempted in military endeavors and large corporations) may help to reduce complex interactions to simpler and more manageable levels.

Nonlinear interactions exhibit path topologies that are fundamentally different from those of linear interactions. *Cycles* occur when events lead to events that repeat (Figure 8.10). A *limit cycle* is a cyclic process that attracts a nearby process onto a continuously repeating path. Limit cycles are ubiquitous and occur in many circumstances. Notions of cause and effect tend to lose meaning in the context of limit cycles. The interpretations of limit cycles spawning from Hopf bifurcations and other singularities may prove to be more useful than cause and effect approaches. In a path *bifurcation*, the system encounters a fork in the path of a linear sequence of interactions (Figure 8.11). At bifurcation points, the system follows one of two (or more) different paths. The deterministic physical processes that cause the system to select one path versus the other may be so sensitive to fluctuations in the detailed (and often unobservable) state variables that the path selection appears to be random.

Systems and subsystems often interact with one another. *Loosely coupled* systems and subsystems exhibit fairly modest interactions that do not result in significant or unpredictable

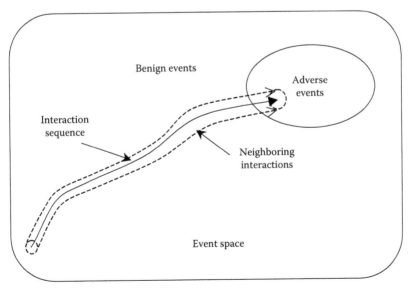

FIGURE 8.9 Linear sequence and neighboring set of interactions that follow a path of benign events to adverse events.

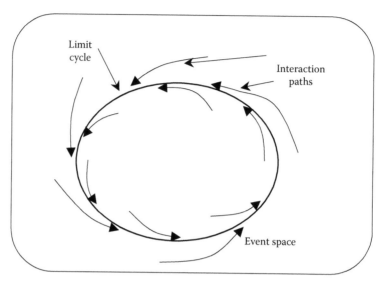

FIGURE 8.10   Convergence of interaction paths on a limit cycle.

changes in the course of behavior (Figure 8.12). *Tightly coupled* systems and subsystems have interactions that are sufficiently strong to cause dramatic and unexpected changes in the course and types of system behavior (Figure 8.13).

Various types of *progressive* and *nonproportional collapse* in structures are examples of loose and tight coupling of subsystems [171]. Structures that do not suffer from progressive collapse have loosely coupled subsystems. For example, if a building has a construction such that the load-bearing columns are loosely coupled, then the removal of a single column will cause a local isolated collapse, but the collapse does not progress laterally or vertically to other parts of the structure [172]. Airplanes are specifically designed to be damage tolerant

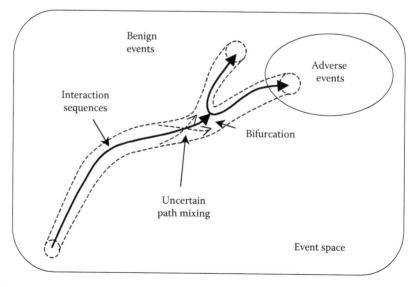

FIGURE 8.11   Interaction paths with uncertain mixing near a bifurcation where one path leads to adverse events and the other does not.

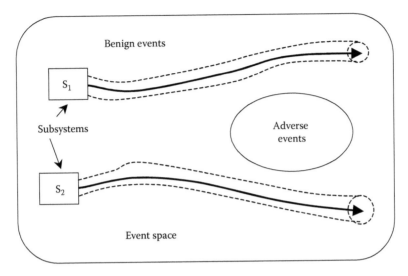

FIGURE 8.12   Loosely coupled subsystems avoid adverse events.

so as to avoid the progressive propagation of damage [173]. If instead, the structural system and subsystems are tightly coupled, then the removal of a single column could cause a progressive and large collapse of the structure. A related concept is that of proportional collapse, that is, the level of damage versus the level of structural insult [174]. Interestingly, a superior structural system is one that is partially coupled. In such a structure, the removal of a single column does not result in a local collapse because the load is redistributed to the remaining columns, possibly due to a catenary action of the floors [175]. These are complex interactions that, if properly understood and implemented, result in a superior structural system [176,177]. An example is highway girder bridges with partial composite-concrete-deck-to-girder interactions that may exhibit a coupling that is hard to predict from first principles, and can result in additional strengthening [178,179]. Quinn et al. described

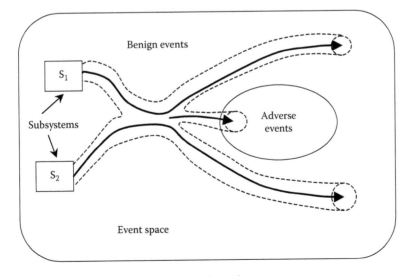

FIGURE 8.13   Tightly coupled subsystems spawn a path to adverse events.

an example of such behavior where the deliberate cutting of several main girders caused severe damage to a highway bridge [180]. In spite of a series of subsequent severe structural insults in terms of overloads, the bridge yielded only slightly, redistributed the loads, and did not collapse. Lee et al. examined the cable loads by natural frequency analysis in the Gi-Lu Bridge after it was damaged in the 1999 Gi-Gi earthquake [181]. The damage caused a redistribution of the loads such that some of the cables were overloaded, but did not fail. Bažant and Verdure recommend monitoring the demolition of structures as a means of better understanding progressive collapse mechanics [182].

High-performance vehicles, such as airplanes, can fail catastrophically due to coupled subsystem failures. Recognizing the potential for catastrophic system coupling can be essential to producing a realistic assessment of the hazards. For example, locating a jet engine near critical components, such as a nexus of redundant hydraulic lines, may enable the failure of a single turbine blade to cut all of the hydraulic lines, defeat all of the redundant systems, and cause a loss of control of the airplane. Fire prevention is an example of the deliberate uncoupling of combustible subsystems and potentially energetic processes.

A coupling versus interaction chart qualitatively describes the relations between linear versus complex interactions, and loose versus tight intersystem coupling (Figure 8.14) [161]. This chart has four sectors. Tightly coupled systems with linear interactions lie in the top left region in Sector 1. An example is a dam. Failure or breaching one part of the dam is tightly coupled to the rest of the dam and may cause systemic failure. The events leading up to dam failures are usually well understood, predictable, and classifiable as linear. Systems that are complex and have tightly coupled interactions lie in the top right side of the interaction chart in Sector 2. A structural example of a complex process is construction. Many construction failures are characterized by the unexpected combination of overloads [183]. Another example is high-performance vehicles. The view of NAT is that systems operating in this regime will inevitably result in large-scale unpredicted and possibly unpreventable accidents. The paradox is that many systems achieve a high boost in performance through the use of complex interactions and the tight coupling of subsystems. Designing high-performance complex and tightly coupled systems that do not suffer from catastrophic, unexpected failures is a difficult engineering challenge. Biological systems routinely solve this problem with a plethora of semichoreographed complex interactions. Linear and loosely coupled systems are those that appear in the lower left side of the chart in Sector 3. These are systems where the events leading up to failure are comprehensible, and the failure of a component does not propagate beyond the local system or region of failure. An example is pavements on a highway. Potholes are troublesome and expensive to repair. However, the causes of pothole formation are fairly well understood, and they do not strongly couple with other systems to cause more damage. Systems that are complex but have loosely coupled interactions operate in Sector 4. A tree is an example. A tree is extremely complex in its interactions between systems and subsystems. However, when the tree is damaged by strong winds, it may shed limbs without sustaining fatal damage. Another example is a university. Interactions among university personnel are exceedingly complex, but many of them are loosely coupled to overall academic mission and are of minor global or long-term consequence.

An interesting attempt at developing an automated system for sorting through complex situations is the *problem knowledge coupler* (PKC) as developed by Weed [184]. PKCs arose in the field of medicine where practitioners have come to realize that the causes and remedies of disease are too complex for comprehension by mere mortals. The PKC forms a network of links between problems and knowledge that are arranged and presented in a useful format for clinicians. Niederleithinger and Helmerich developed a similar system for bridge NDT

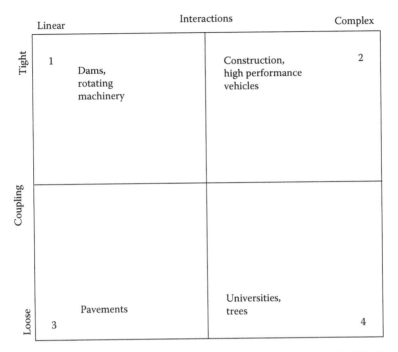

FIGURE 8.14 Coupling versus interactions in systems. (Adapted from Perrow C. 1999. *Normal Accidents.* Princeton University Press, Princeton.)

techniques (Figure 8.15) [185]. Cordier et al. demonstrated that if the knowledge of the system is complete, both fault detection and isolation, and diagnostic approachesconverge on the same result [186].

When sufficient data are available, it may be possible to assess whether accident rates are increasing or decreasing. Hanayasu suggests using sequential probability test ratios with a methodology similar to Shewhart charts to determine if the rate of accidents has changed [187]. Hanayasu applied these methods to examine the rates of structural failure accidents in Japan.

8.8.3   High-Profile Accidents

8.8.3.1   *Space Shuttle Challenger*

Two of the United States Space Shuttles suffered catastrophic failures. The first was Challenger. The shuttle exploded a few minutes after liftoff on January 28, 1986. A blue ribbon panel, known as the Rogers Commission, investigated the causes of the accident. One of the panelists, Richard Feynman, quickly zeroed in on the possible root cause being cold-weather-induced failure of the O-ring seals between segments on the solid rocker boosters. Subsequent testing and detailed analysis indicated that problems with the O-rings were a likely cause of the accident.

A narrow RCA of the accident could be to identify the O-ring as the root cause. The next steps would be to fix the O-ring design, certify the shuttle as being safe, and then move on to more launches. Feynman recognized that the situation was considerably more complex and undertook a more in-depth investigation [188]. Space shuttles are examples of very

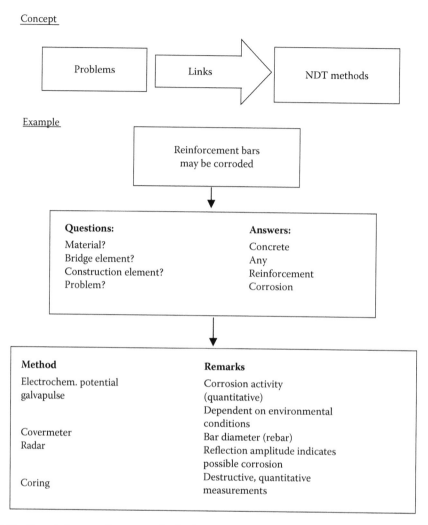

Concept

Example

FIGURE 8.15 Knowledge coupler output for bridge NDT systems. (Adapted from Niederleithinger E and Helmerich R. 2006. *NDE Conf on Civil Eng*, Amer Soc Nondes Test, St Louis.)

high-performance machines with nonlinear and tight coupling between the subsystems and components. The failure of the O-ring in cold weather was only a partially understood phenomenon. The engineers with the pertinent information were unable to make a sufficiently strong case to convince the launch controllers to abort the launch or redesign the system. The close physical proximity of the solid rocket boosters to the liquid oxygen fuel tank caused a tight coupling of the two subsystems and resulted in an unexpected and catastrophic change of events. In his report appended to the Rogers Commission report, R. Feynman stated [189]

> If a reasonable launch schedule is to be maintained, engineering often cannot be done fast enough to keep up with the expectations of originally conservative certification criteria designed to guarantee a very safe vehicle. In these situations, subtly, and often with apparently logical arguments, the criteria are altered so that flights may still be certified

in time. They therefore fly in a relatively unsafe condition, with a chance of failure of the order of a percent (it is difficult to be more accurate) … For a successful technology, reality must take precedence over public relations, for nature cannot be fooled.

8.8.3.2 Space Shuttle Columbia

The next space shuttle accident was with the Columbia, which disintegrated upon reentry on February 1, 2003. The Columbia was the first shuttle launched. A suite of 800 sensors (primarily strain and temperature gages) measured structural behavior (Figure 8.16). A *modular auxiliary data system* (MADS) recorded data from the sensors (Figure 8.17). The MADS survived the accident virtually intact. An examination of the data channels indicated that the temperature rose rapidly upon reentry, starting with the leading portion of the left wing near the wheel well. As the temperatures rose, the MADS stopped recording data from data channels (Figure 8.18) [190]. In spite of a very detailed postmortem analysis of the data, conclusive evidence as to the precise time, nature, and location of the fatal

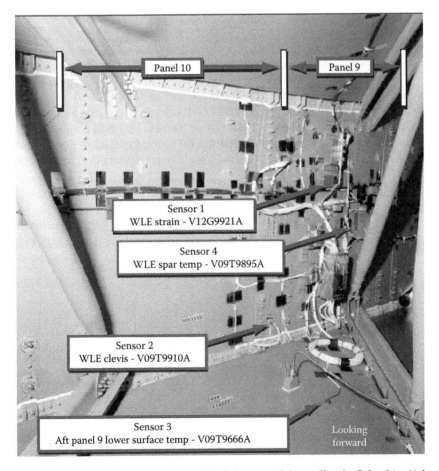

FIGURE 8.16 Sensors attached to the interior of the left wing of Space Shuttle Columbia. (Adapted from NASA. 2003. *Columbia Accident Investigation Board Report*, Vol. 1, CAIB, August, National Aeronautics and Space Administration. http://www.nasa.gov/audience/formedia/features/MP_Photo_Guidelines.html)

FIGURE 8.17   Columbia MADS recorder found virtually intact following the accident. (Adapted from NASA. 2003. *Columbia Accident Investigation Board Report*, Vol. 1, CAIB, August, National Aeronautics and Space Administration. http://www.nasa.gov/audience/formedia/features/MP_Photo_Guidelines.html)

structural damage remains a matter of debate. Ice and insulating foam shedding from the oxygen fuel tank impacting on the heat shield tiles during the initial ascent is the leading consensus candidate for cause of the failure. Contributing to this conclusion is work by Iverson in which *inductive monitoring system* (IMS) algorithms analyzed the sensor data (Figure 8.19) [191]. The IMS software coalesces multichannel data and looks for deviations from nominal behavior.

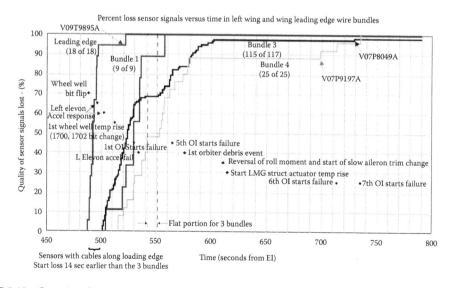

FIGURE 8.18   Quantity of sensor signals lost on Space Shuttle Columbia versus time of reentry. (Adapted from NASA. 2003. *Columbia Accident Investigation Board Report*, Vol. 1, CAIB, August, National Aeronautics and Space Administration. http://www.nasa.gov/audience/formedia/features/MP_Photo_Guidelines.html)

8.8.3.3 Original Tacoma Narrows Bridge

The original Tacoma Narrows Bridge was a suspension bridge with a center span of 853 m (2800 ft) between Tacoma and Gig Harbor in Washington State (US). The bridge opened to traffic on June 1, 1940. Almost immediately after opening, the bridge experienced wind-induced oscillations. Reports of unwanted oscillations prompted the commissioning of Professor F.B. Farquharson at the University of Washington to observe and examine the behavior of the bridge. On November 7, 1940, during a modest wind of 68 kph (42 mph), the bridge experienced uncontrolled large-scale oscillations. Initially the oscillations were vertical. The oscillations subsequently changed to torsional. Professor Farquharson was fortuitously in place and equipped to capture on film the famous footage of the bridge twisting in the wind and ultimately collapsing.

Despite the video documentation of the collapse of the Tacoma Narrows Bridge, confusion lingers as to the root causes of the collapse [192]. Several factors contributed to this confusion: (1) Suspension bridges historically have been vulnerable to wind-induced vibrations, with several spectacular collapses in the nineteenth century. The use of designs based on a heuristic understanding of the wind loads resulted in the design of heavy and stable bridges, for example, the Brooklyn Bridge. By the 1930s, this level of success led to the neglect of dynamic wind loads in the design of bridges. (2) The development of an improved nonlinear mathematical model of the distribution of heavy gravity loads on suspension bridges encouraged designers to build lighter, more efficient structures. The Tacoma Narrows design was a prime example of the new design methods. (3) The Tacoma Narrows Bridge used an H-shaped plate girder for the deck cross section. This shape was

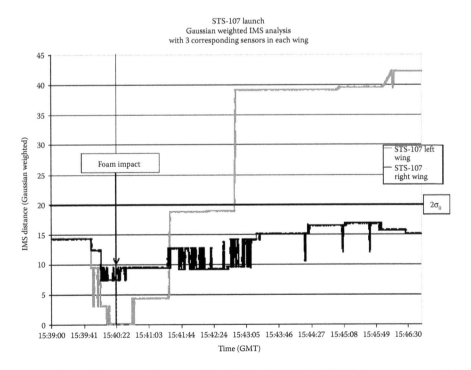

FIGURE 8.19 Inductive Monitoring System analysis of Columbia Shuttle STS-107 wing sensor ascent data showing an anomaly that is attributed to foam impact on the left wing. (Data from Iverson DL. 2005. Inductive system health monitoring with statistical metrics. NASA Ames Research Center, Moffett Field, CA.)

very efficient for carrying vertical traffic loads and static lateral wind loads, but was inherently weak in torsion. Postcollapse wind tunnel testing indicated that the H-shape of the Tacoma Narrows was one of the worst possible shapes for aeroelastic performance [193]. (4) Bluff body aeroelasticity is complicated. Several distinct physical mechanisms can couple aerodynamic and elastic forces to produce vibrations. The Tacoma Narrows Bridge was prone to both vortex-shedding lock-in and single degree of freedom (SDOF) flutter phenomena. Vortex shedding and flutter exhibit similar symptoms, that is, deck vibrations and wake vortices, but have different underlying mechanisms and outcomes. (5) The combination of a light structure with a deck shape that was quite unfavorable for aeroelastic effects was the cause of the collapse. Torsional self-excited SDOF bluff body flutter—not resonant vortex shedding—was the culprit underlying the collapse. However, it is also likely that vortex shedding caused relatively benign oscillations to occur in prior incidents.

8.9 Failure Prognosis

Prognosis is a prediction of the progress of impending system degradation, such as disease or structural damage. Prognosticating future damage requires combining information about material states, structural interactions, and future loading scenarios, (Figure 8.20) [194]. In spite of what often seems to be an overwhelming wealth of available information, predicting the future with certainty is difficult. It is perhaps not surprising that in the famous movie and novel "2001 A Space Odyssey" by Clark and Kubrick, the HAL 9000 computer failure appeared first as a system prognosis error (in an electromechanical antenna pointing system) [195]. Figure 8.2 is a generic plot of health versus lifespan for a structure. When plotting the curves of nominally identical structures on the same graph, the figure may look more like Figure 8.21 with multiple bifurcating life paths. Predicting future outcomes for such structures may be difficult. A similar bifurcating life path picture arises for a single structure with the introduction of possibilities and decisions as to whether to perform corrective maintenance [196]. A primary goal of SHM is to reduce the variability in structural prognostics (Figure 8.22) [197,198].

Robertson and Hemez suggest a four-step process for damage prognosis [199]: (1) determine the failure mode; (2) identify the random variables that affect failure and determine the associated PDFs as a function of accumulated damage; (3) determine the multidimensional limit state beyond which the structure will fail; (4) determine the probability of failure as a function of accumulated damage. This produces a probability of failure for each instant in time, as in Figure 8.23.

There are two main approaches to reducing the variability of structural prognoses. The first uses a reductionist approach with the focus on improved knowledge, often with the aid of statistical and probabilistic methods [200,201]. Better measurements of structural properties and conditions, improved mathematical and statistical models of structural behavior, postmortem analyses of failed structures with similar characteristics, and improved estimates of future loads can all contribute to the building of more accurate (and usually more complicated) models for structural prognosis. Many of these efforts are analogous to the problem of modeling turbulence in fluids, that is, the development of ever more sophisticated and detailed computer models of turbulent fluid mechanics reduces the regions and scales of the models that require statistical treatments. An example of the reductionist approach in SHM is the development of models that predict the progress of chloride penetration attack on reinforced concrete structures. After a considerable amount of

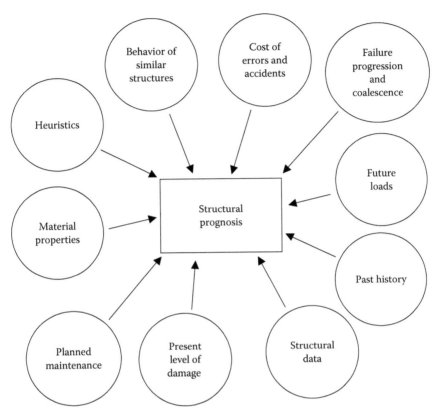

FIGURE 8.20 Some of the factors affecting structural prognosis. (From Iverson DL. 2005. Inductive system health monitoring with statistical metrics. NASA Ames Research Center, Moffett Field, CA.)

research effort, the mechanisms of chloride penetration and attack are now well understood. It is possible to predict the time to failure with some certainty [202–206]. The second method of reducing prognosis variability is to alter the design to facilitate more effective inspection and monitoring. In many cases this involves identifying and altering critical structural

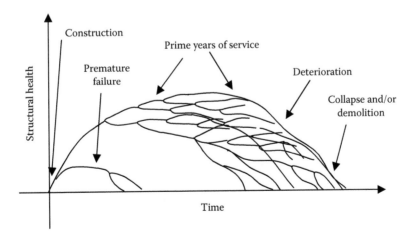

FIGURE 8.21 Tree of possible life paths and health functions for a structure.

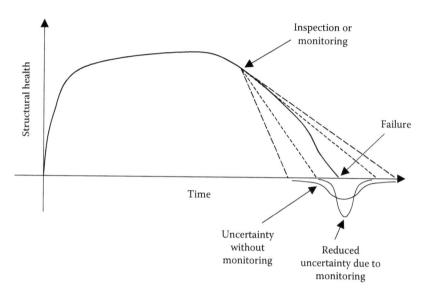

FIGURE 8.22 Monitoring has the potential to reduce variability in prognosis. (Adapted from Bond LJ et al. 2003. Nuclear Energy Research Initiative Project No. 30344, Final Project Report, Pacific Northwest National Laboratory, Richland.)

details so that close-up inspections are more convenient. An example is the practice of not painting wood ladders. Paint obscures defects and creates a safety hazard. Similar arguments recommend against using asphalt overlays and/or steel pans as combined formwork and reinforcing for concrete bridge decks.

Uncoupling potentially deleterious subsystem interactions can improve prognostic efforts. An example is the use of specialized alloys and multilayer component layups in high-pressure pipes and vessels that cause the failure mode to be a leak before a rupture [207]. The presence of leaks provides an early warning of distress. The design of structures

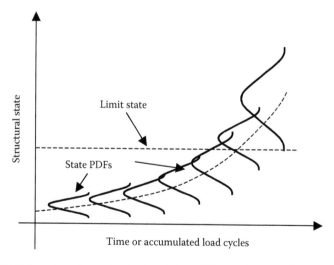

FIGURE 8.23 Prognosis based on change in structural state PDF as a function of time or load accumulation. (Adapted from Robertson AN and Hemez FM. 2004. In *Advanced Smart Materials and Smart Structures Technology*, DEStech, Lancaster.)

with redundant load paths and graceful rather than catastrophic failure modes may also ease prognostication efforts and could improve overall structural performance. In some respects, this is analogous to the practice in fluids engineering of altering turbulent flow patterns so that they become either laminar or fully turbulent, and hence are easier to control and analyze.

A bit of a paradox with both the reductionist and structural redesign approaches is that once the failure modes of a structure become predictable, they are often preventable. This leads to new designs that do not have the same failure modes as the old design. The failure modes of the new designs may again be difficult to predict, especially if they result from the coupling of secondary subsystem interactions. The aviation industry is an exception. The need for minimal weight and high performance leads to the design of parts and systems that fail after a certain amount of usage. The trick is to remove the parts before the predictable failure causes trouble, but not so soon as to waste the remaining useful service lives. *Health and usage monitoring systems* (HUMS) on aircraft, particularly helicopters, aid in this effort. HUMS count the number and size of load cycles, and measure operational data, such as vibrations, to produce predictions of the remaining lifetime of various critical parts that are presumably superior to methods based solely on time-of-flight usage [208–210].

8.9.1 Probabilistic and Age-Based Methods of Prognosis

Owing to the inherent uncertainties in predicting the future, probabilistic methods are rational tools for prognosticating. It is natural to use time as the independent lifetime variable when the loading and/or material degradation effects are fairly constant over time. Alternatively, it is more reasonable to use load as an independent lifetime variable if the loading and/or material degradation effects are dependent on amplitude and/or number of load cycles. In the following discussion, time is predominantly used as the independent variable, but can be easily replaced by load cycling in the proper context. More complicated situations with multiple independent load variables may require the use of similar multivariable analyses.

A relatively simple case of probabilistic failure prediction is the case of operating a system so that when it fails it is either immediately repaired or replaced. In this case, the Poisson distribution

$$Q_k(t) = \frac{(\lambda t)^k}{k!} \exp(-\lambda t), \quad k = 0, 1, \ldots \tag{8.17}$$

describes the probability of k failures in the interval leading up to time t [211,212].

A common prognostic concern is to predict the lifespan, T, of a system or structure. For situations where the structures are not immediately repaired upon the occurrence of damage or failure, it may be expedient to use a scalar health function that varies monotonically with the state of health of a structure. One possibility is to use a scale that runs from 0 to 1, where the health function, $H(t)$, has a value of 1 if the structure is in perfect condition and a value of 0 if the structure is completely failed or sufficiently damaged so as to be unusable. A simplified application of this concept is to classify a structure as either healthy or failed. In these circumstances, the health function undergoes a negative step at the time of failure, T (Figure 8.24).

The *reliability function*, $P(T)$, is the probability of failure-free operation in the interval leading up to time T. It can often be assumed that $P(T)$ is a monotonically decreasing function of T. Since $P(T)$ incorporates the cumulative action of the effects acting in

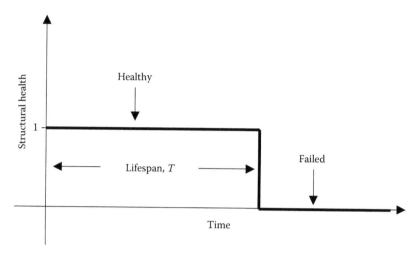

FIGURE 8.24　Health function of a structure exhibiting a simple failure.

the interval leading up to T, it is natural to use a cumulative integral with the density function $P(T)$:

$$P(T) = \int_0^T p(t)\, dt. \tag{8.18}$$

The complementary cumulative probability of a failure occurring in the interval leading up to T is $F_T(T)$, with the corresponding complementary probability density function $p_T(T)$:

$$F_T(T) = 1 - P(T) = 1 - \int_0^T p(t)\, dt = \int_0^T p_T(t)\, dt, \tag{8.19}$$

$$p_T(T) = -\frac{dP(t)}{dt}\Big|_{t=T}. \tag{8.20}$$

The expected value and variance of failure-free operation T are

$$E(T) = \int_0^\infty p_T(T)T\, dT = \int_0^\infty P(T)\, dT, \tag{8.21}$$

$$V(T) = E(T^2) - [E(T)]^2. \tag{8.22}$$

The *failure rate*, $\lambda(t)$, of a collection of identical systems is the rate of failure at time t, normalized by the number that is surviving:

$$\lambda(t) = \frac{[dF_T(t)/dt]}{P(t)} = \frac{[-dP(t)/dt]}{P(t)}. \tag{8.23}$$

If $P(0)$ is the fraction of systems that has already failed when $t = 0$, then

$$P(t) = P(0) \exp\left[-\int_0^t \lambda(\tau)\,dt\right].$$ (8.24)

The failure rate is constant when the probability that a surviving system will fail is independent of age. A constant failure rate simplifies Equation 8.24 to

$$P(t) = P(0)\,e^{-\lambda t}.$$ (8.25)

When all the systems are initially healthy, that is, $P(0) = 1$, then the expected value and variance of failure-free operation for a constant failure rate are

$$\mu_T = \frac{1}{\lambda},$$ (8.26)

$$\sigma_T = \frac{1}{\lambda^2}.$$ (8.27)

Failure rates that vary with time often correspond to aging effects. Power law time variations in the failure rate give rise to failure probabilities with Weibull distributions [79]

$$P(t) = (a\lambda)(t\lambda)^{a-1} \exp\left[-(t\lambda)^a\right] \quad \begin{cases} 0 < t < \infty \\ a, \lambda > 0 \end{cases}.$$ (8.28)

The mean, μ_I, and variance, σ_I^2, are

$$\mu_I = \frac{1}{\lambda}\Gamma\left[1 + \left(\frac{1}{a}\right)\right],$$ (8.29)

$$\sigma_I^2 = \lambda^{-2}\left\{\Gamma\left[1 + \left(\frac{2}{a}\right)\right] - \Gamma^2\left[1 + \left(\frac{1}{a}\right)\right]\right\},$$ (8.30)

where a is the Weibull exponent and λ is a scale factor.

The size of the Weibull exponent indicates the dependence of the failure rate with respect to time. When the Weibull exponent a is greater than one, the failure rate increases with time. Kawakami and Frangopol report finding Weibull parameters that range from 1.47 to 3.97 and scale factors ranging from 0.0106 to 0.0375, based on over a decade of maintenance, inspection, and repair data for steel girders on composite girder highway bridges in Japan [213]. Marotta et al. used Weibull analysis to predict the reliability of rockets [214].

Weibull distribution can also be useful in describing material toughness and resilience to repeated loads. Materials with Weibull exponents greater than one exhibit a toughening behavior. Materials with Weibull exponents less than one are weakening. Determining Weibull exponents from experimental data can be difficult because measurements of the load required to cause failure measurements often produce large variances between samples. The principal explanation is that microscopic and otherwise unobservable damage grows to form significant macroscopic-scale damage.

Large variations in damage accumulation due to unobservable microstructural details cause problems on several fronts. One is that when the scatter is large, ensuring reasonable levels of safety often requires setting sufficiently low allowable load thresholds to prompt unnecessary maintenance and/or removal on healthy systems that do not require such treatment. Another problem is that accurate quantifications of the mean and standard deviations of failure load distributions often require such large numbers of repeat experiments that the testing protocols become prohibitively expensive. Bayesian methods based on previous tests of similar materials, and accelerated and extremal load testing methods, such as HALT, may offer some relief.

Describing the failure probabilities of brittle versus ductile structures is another application of Weibull distribution. Materials with small Weibull exponents tend to be brittle with large failure load variabilities. Materials with large Weibull exponents are ductile and have smaller failure load variabilities. For example, Iarve et al. found that a classical Weibull integral analysis underpredicted the failure strengths of composites with stress concentration damage by about 20–30% [215]. An analysis based on a *critical failure volume* proved to be a better prognosticator.

Material tolerance to damage accumulation falls into two main classes: (1) materials where damage weakens the material and causes an increased susceptibility to damage and (2) materials where damage causes toughening that tends to reduce the accumulation of more damage. The underlying mechanics of both these processes typically span multiple length scales and can be quite complicated. The damage mechanisms range from a micro- and nanoscale accumulation of dislocations to a mesoscale accumulation of cracks along crystal boundaries to a macrostructural accumulation of damage and load transfers between structural elements. The statistical behavior of complex networks may shed some light on these processes [216,217]. For example, Dienes developed a statistical model of crack networks [218].

A *fragility curve* plots the probability of failure as a function of a load parameter, for example, time, earthquake magnitude, g-load of shock, and so on. The two-parameter log-normal PDF is a common representation of fragility curves [219]:

$$F(a; c, \xi) = \Phi\left[\ln(a/c)/\xi\right], \tag{8.31}$$

where a is the applied load, and c and ξ are parameters. Objects made of ductile materials tend to have minimal variations in the loads that cause yielding. The corresponding fragility curves have steep slopes centered on the yield point (Figure 8.25). Brittle materials tend to have large variabilities in the load levels that cause cracking and failure. The corresponding fragility curves are more smoothly sloped. Righiniotis developed a methodology based on log-normal distributions for assessing the probability of fatigue crack growth in bridges with posted maximum weight limits [220]. Polidori et al. gave asymptotic approximation procedures for calculating associated reliability integrals [221]. Saito et al. developed fragility curves for the seismic performance of buildings based on probabilistic assessments of modal vibration changes due to damage [222]. These studies underscored the utility of obtaining probabilistic assessments of the quality of measured data.

Examining subsystem behavior can reveal more complicated, and presumably more accurate, models and predictions of system failures. Subsystem interconnections are principal factors affecting the ability of a structure to survive large and/or small stressors. Qualitatively, designers have intuitively understood and exploited these concepts for millennia. Figure 8.26 shows a system composed of two subsystems. Assuming that the subsystems are either healthy or failed, the total system is in one of four states: *State 1.* $S_1 = [A_H, B_H]$— A is healthy and B is healthy; *State 2.* $S_2 = [A_H, B_F]$—A is healthy and B is failed; *State 3.* $S_3 = [A_F, B_H]$—A is failed and B is healthy; and *State 4.* $S_4 = [A_F, B_F]$—A is failed and B

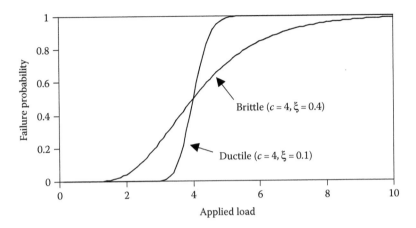

FIGURE 8.25 Fragility curves based on log-normal distribution for brittle and ductile materials.

is failed. Associated with each state is a probability of occurrence. The aggregate system performance may be either functional (healthy) or nonfunctional (failed), depending on the interactions of the subsystems with one another and with the system at large. The probability, $P(S_H)$, that the system is functional is the sum of conditional probabilities of the system being functional for each of the four possible subsystem state configurations:

$$P(S_H) = P(S_H \cap S_1) + P(S_H \cap S_2) + P(S_H \cap S_3) + P(S_H \cap S_4) \tag{8.32}$$

and

$$P(S_H) = P(S_1)P(S_H|S_1) + P(S_2)P(S_H|S_2) + P(S_3)P(S_H|S_3) + P(S_4)P(S_H|S_4). \tag{8.33}$$

Figure 8.27 shows a system with a series configuration of subsystems. In this case, both of the subsystems A and B must be healthy for the system to be functional, that is,

$$P(S_H) = P(S_1) = P(A_H \cap B_H) \tag{8.34}$$

and

$$P(S_H|S_1) = 1, \quad P(S_H|S_2) = 0, \\ P(S_H|S_3) = 0, \quad P(S_H|S_4) = 0. \tag{8.35}$$

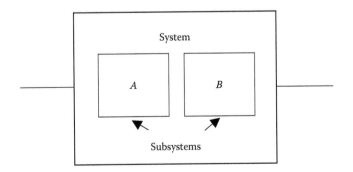

FIGURE 8.26 System with two subsystems.

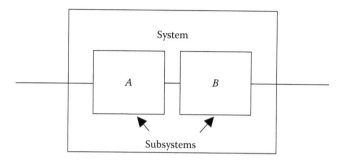

FIGURE 8.27 System with subsystems in series.

When the failure modes of subsystems A and B are independent, then

$$P(S_H) = P(A_H)P(B_H). \tag{8.36}$$

Extending to the case of a system formed with m multiple independent subsystems, A^i, in series produces a multiplicative relation

$$P(S_H) = \prod_{i=1}^{m} P(A_H^i). \tag{8.37}$$

An examination of Equation 8.37 indicates that since all $P(A_H^i)$ are positive but less than one, the addition of more subsystems in series reduces the reliability of the system. If the subsystems all have the same constant failure rate, λ, the probability that the series configuration system will be healthy as a function of time becomes

$$P(t) = P(0)\exp(-m\lambda t). \tag{8.38}$$

Parallel redundant subsystems, as shown in Figure 8.28, are an alternative arrangement. In a parallel system, only one of the parallel subsystems has to be healthy for the system to be functional. The conditional probabilities for redundant subsystems become

$$\begin{array}{ll} P(S_H|S_1) = 1, & P(S_H|S_2) = 1, \\ P(S_H|S_3) = 1, & P(S_H|S_4) = 0, \end{array} \tag{8.39}$$

and

$$P(S_H) = P(A_H \cap B_H) + P(A_H \cap B_F) + P(A_F \cap B_H). \tag{8.40}$$

When the failure probabilities of A and B of the redundant system are independent, and when the subsystems are either healthy or failed, then

$$\begin{aligned} P(S_H) &= P(A_H)P(B_H) + P(A_H)P(B_F) + P(A_F)P(B_H) \\ &= P(A_H) + P(A_F)P(B_H) \\ &= P(B_H) + P(A_H)P(B_F). \end{aligned} \tag{8.41}$$

Since the probability distributions in Equation 8.41 always lie between zero and one, the conclusion is that redundant parallel systems are as reliable as or more reliable than

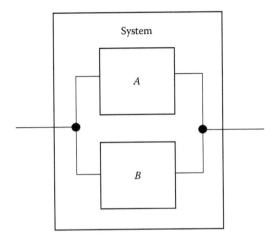

FIGURE 8.28 System with subsystems in parallel.

individual subsystems, which in turn are as reliable as or more reliable than a series of the subsystems. Figure 8.29 shows the possible course of health degradation in a system with two parallel redundant systems.

A more complicated, yet potentially advantageous, situation occurs when the failure of a parallel redundant subsystem redistributes loads onto other subsystems. Some systems are predesigned for a graceful redistribution of loads. An example is the use of parallel strands in the main catenary cables of suspension bridges. The fabrication of parallel strand cables is inherently error prone, with the individual strands having a variation in lengths. During the initial loading, the shortest strands will carry the bulk of the load and often fail. The failure of the short strands causes the load to redistribute to other strands, which then carry the load more uniformly [223]. If the redistribution instead precipitates more failures, then progressive failure may occur.

More complicated subsystem combinations are possible. The system in Figure 8.30 consists of two subsystems (1 and 2) in parallel. Subsystem 1 comprises two sub-subsystems

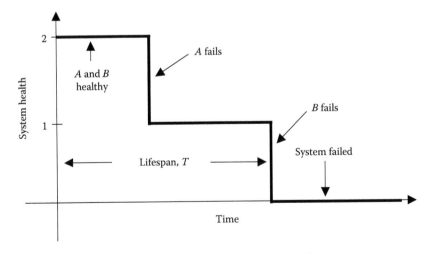

FIGURE 8.29 Failure of redundant system requires multiple subsystem failures.

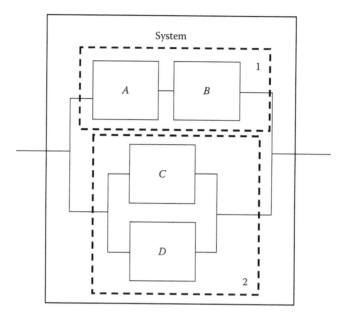

FIGURE 8.30 Mildly complicated system with nested series and parallel subsystems.

(A and B) in a series configuration. Likewise, subsystem 2 comprises two sub-subsystems (C and D) in a parallel configuration. A possible failure sequence appears in Figure 8.31.

These concepts, and substantially more complicated variants, provide a means of analyzing and monitoring the health and reliability of systems and structures. Bazovsky gives a

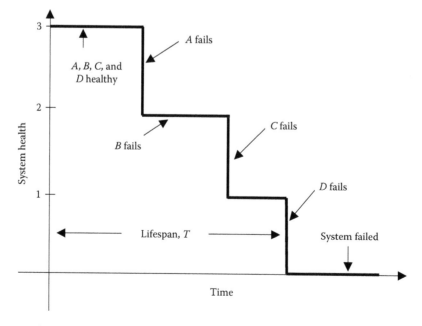

FIGURE 8.31 Failure by sequential failure of subsystems.

comprehensive discussion of many cases of mixed serial and parallel systems with interdependencies [224]. Yun and DeWolf used the concept of series and parallel systems combined with inspection data to assess the reliability of highway truss signs [225]. Dong and Shinozouka used a fragility information analysis to assess the propensity of electric power system transformers to fail in a seismic event, and the potential of failures to subsequently cause systemic power outages [226]. Murachi et al. and Shibata used a similar analysis to assess the robustness of transportation networks and community-wide building infrastructure to earthquakes [219,227].

A principal assumption underlying the conclusion that redundancy increases reliability is that subsystem failures are independent. When subsystems are tightly coupled, the failure of one subsystem influences the failure of other subsystems. A consequence is that a system with tightly coupled redundant subsystems may be considerably less reliable than indicated by an independent redundant systems analysis. A difficulty is that the modes and rates of subsystem coupling may not be obvious.

The analysis of systems with coupled subsystems is generally more complicated than that of systems with uncoupled subsystems. In the redundant-parallel two-subsystem model of Figure 8.32, the failure modes of subsystems A and B are coupled, as indicated by the dashed line. When there is a coupling of failure modes between subsystems, then the terms on the right-hand side of Equation 8.40 become

$$
\begin{aligned}
P(A_H \cap B_H) &= P(A_H)P(B_H|A_H) = P(B_H)P(A_H|B_H), \\
P(A_H \cap B_F) &= P(A_H)P(B_F|A_H) = P(B_F)P(A_H|B_F), \\
P(A_F \cap B_H) &= P(A_F)P(B_H|A_F) = P(B_H)P(A_F|B_H).
\end{aligned}
\tag{8.42}
$$

The details of the coupling interactions between the subsystems determine how individual subsystem failures affect the reliability of other subsystems and the entire system. Usually subsystem coupling reduces reliability, that is, the failure of one subsystem causes other subsystems to fail, possibly in a chain reaction. The propagation of fires throughout a structure is an example. If two subsystems, for example, structural elements, are physically close to one another, a fire in one subsystem may cause the adjacent member to catch fire. Preventing such progressive failures, that is, conflagrations, due to subsystem coupling is the intent of many fire safety code provisions. Under certain circumstances, such as

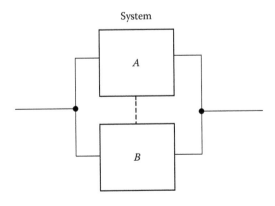

FIGURE 8.32 System with two subsystems in parallel with subsystem coupling indicated by the dashed line.

extreme loads, the failure of one subsystem can protect other subsystems by uncoupling the remaining subsystems from the extreme load. An example is the use of swales to confine flood water runoff damage to a small portion of a roadway. Intelligent systems can sense damage and alter their operational behavior and configuration with strategies such as limping. Bankowski et al. developed a software agent-based system that queries the health states of subsystems and alters the behavior of a high-performance complex vehicle system accordingly [228].

A compromise approach recognizes that the human brain is inherently capable of grasping complex information in a manner that remains difficult to automate with a computer. Alan and Fujino proposed providing web-based SHM data in a virtual reality environment as a means of combining enhanced sensing capabilities with human reasoning [229].

When the subsystem behavior is well understood, then Monte Carlo techniques based on estimates of the remaining life of individual components from load and usage histories can supplement structural prognoses [230]. For example, Garavaglia and Sgambi suggest using a *credibility index* based on error level measures to compare different probability models of structural reliability [231]. A major difficulty with using probabilistic methods for failure prognosis is that the failure probabilities are rarely known with precision. The rise of modern digital database systems has now made it practical to alleviate some of these difficulties by analyzing large databases of damage, deterioration, and failure rates for structures and collections of similar structures [232,233].

Building an accurate computer-based model of a system makes it possible, in principle, to determine failure propensities by examining various loading and state conditions. This may involve analyzing a large number of different damage coupling configurations. Since the number of possible different loadings and material state combinations is huge, simulating and evaluating every possible combination can be prohibitively expensive. A further complication arises if the occurrence of a systematic failure requires a combination of one or more subsystem failures, many of which may be rare events. One approach to reducing the computational load is to use the *importance sampling Markov chain simulation method* developed by Au that calculates the conditional probabilities of the system behavior based on the condition that a subsystem has failed [234].

Developments in the field of complex networks may shed some light on underlying organizing principles of systems for robustness and stability and provide quantitative estimates of performance [216,217,235–237]. Many of the systems studied to date are those that are organized with self-similar properties that scale according to power laws over orders of magnitude. Crucitti used the concept of efficiency as a measure of error and attack tolerance of scale-free networks [238]. A result is that different organizational and assembly architectures of networks produce vastly different levels of robustness and stability. Dodds et al. developed and analyzed the robustness of generalized forms of networks superposed on top of hierarchical frameworks [239]. Graphs are natural tools to represent coarse-scale interactions between systems. The nodes represent subsystems and the edges represent the presence of subsystem coupling. Mathematical tools for describing the properties of graphs are well-established [240,241]. Bond graphs provide a similar, but more detailed representation of system interactions that includes energy flows [242].

Within this context, the question arises as to whether there is a consistent and rational mathematical quantification of health (or damage) [243]. Health is a prediction of the ability of a structure to survive or meet performance specifications in the future. Healthier systems are more likely to resist failure due to external loading or internal degradations, such as aging or corrosion. A suitable health metric might be one that is additive, monotonic, and positive. For systems composed of subsystems operating in series, the reliability of the whole system is the product of the underlying subsystem reliabilities. An additive version of the reliability could be one that uses a logarithmic measure of the probability of subsystem failures. Such an approach may naturally lead to probabilistic health metrics with mathematical form similar to Shannon's entropy. A similar concept is the notion of a redundancy index. Hoshiya and Yamamoto examined the utility of various redundancy indices in the context of the reliability of pipeline systems [244]. A conclusion was that redundancy indices based on Shannon's information entropy formulation seem to be the most useful, with the R_E index being the best of the ones they examined:

$$R_E = \frac{-\sum_{i=1}^{m-1} P_{D_i|D} \log_2(P_{D_i|D}) - P_{D_F|D} \log_2(P_{D_F|D})}{\log_2(m)} \tag{8.43}$$

and

$$P_{D_x|D} = \frac{P(D_x)}{P_D}, \tag{8.44}$$

where D is a damage state; D_i is a damaged, but operational state; and D_F is a damaged, but nonoperational state. It is curious to note that the R_E index has a similar cross-entropy form as the J-measure of interestingness used in data mining (see Chapter 6). Hierarchical damage indices (such as the one due to Wu and Su [86]) may also be amenable to a logarithmic entropy measure.

8.9.2 Damage Accumulation Models

Structures and structural elements fail when the applied loads exceed resistance capabilities. There is an extensive body of literature on the mechanics and methods of modeling damage accumulation in structures. A significant portion addresses the fatigue and fracture of metal [156,211]. Another significant portion addresses damage accumulation in materials, such as concrete and fiber-reinforced composites. Additional studies examine damage due to nonmechanical modes, such as corrosion or fire [245]. *Damage-tolerant* structures are those that can sustain large amounts of cumulative damage without failure, and have gradual failure modes when they do fail. In other situations, load-induced failures occur very quickly, as in brittle or buckling-prone systems. In principle, if the load history is known, combining empirical data with rational mechanical models can provide realistic estimates of damage-level and future load sustainability.

Failures often begin with microscopic cracks and other flaws that grow with time. The growth rate depends on the specific details of load histories and material behavior. Interactions on length scales ranging from those of molecular dimensions to those of the largest structural dimensions can all be important. When a structure sustains damage without repair, then the damage may accumulate to where even the most innocuous of additional loadings leads to failure.

Certain physical processes arrest and/or limit damage accumulation and growth. TRIP steels exhibit a strain-induced phase transition from austenitic to martensitic forms coupled with the appearance of Lüders lines that aggressively arrest the propagation of cracks [246]. Another example is riveted metal construction. Cracks tend not to jump across riveted joints. Certain ductile materials arrest cracks by a plastic redistribution of stress across in front of the crack tip. In other cases, the mechanics do not arrest the damage growth and the damage continues to grow. Other processes can cause the damage growth rate to accelerate. Examples include crack growth causing a stress redistribution that increases crack tip stress, or when cracks coalescence. It is not always clear whether the damage processes are steadily growing, accelerating, or arresting. Such ambiguous cases may warrant repeated inspection or the installation of monitoring instruments, such as crack growth sensors [247,248]. A problem with crack-monitoring instruments is that significant damage may occur prior to the cracks being of a measurable size. Microstructural modeling of material behavior coupled with improved microdamage and subsurface diagnostic tools may be of assistance in improving prognoses in these situations [249,250]. Grundy recommends the following multistep procedure for evaluating cracks: (1) examination of crack surfaces, (2) computation of the stress regimes around cracks, (3) confirmation of the stress regimes by strain measurements under passing trains, (4) estimation of fatigue life in the presence of known cracks, (5) laboratory testing of fatigue behavior in the presence of known cracks, and (6) identifying the causes of cracks [251]. Suzuki et al. followed a similar procedure to determine the causes of cracks on steel bridge piers on the Osaka–Nishinomiya line in southern Japan [252]. The issue was to determine whether the cracks were due to fatigue or brittle fracture. After a detailed study, it was concluded that the Hyogoken–Nanbu earthquake caused brittle fracture. A key factor supporting this conclusion was the result of 24-h operational stress monitoring that indicated that the stress levels were probably too small to cause fatigue cracks.

Similar situations arise in corrosion and wear. Corrosion can be diffuse, look terrible, but cause only minimal structural damage. Corrosion can also be localized with severe damage, such as with pitting, crevice, or stress-crack corrosion, yet appear to be innocuous to the casual observer. The design of many machines uses a break-in period where localized wear of the bearing and other joint surfaces during initial operation cycles causes the parts to fit together. The break-in period ends with the arresting of further wear and with a smooth running machine. In other situations, a small flaw in a bearing may cause vibrations that load the bearing and accelerate the growth of the flaw. Many systems fail during an initial wear-in or burn-in period. During the next stage, the systems operate relatively failure-free with only chance failures. The final stage involves increasing failure rates due to wear-out processes. A *bathtub curve* describes these early failure, break-in, smooth running, and failure phenomena (Figure 8.33) [224].

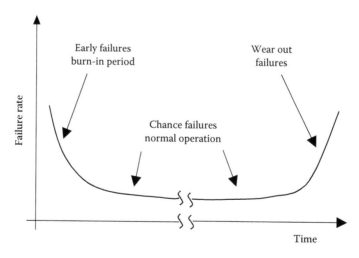

FIGURE 8.33　Bathtub curve showing lifetime failure rates of many typical systems. (Data from Bazovsky I. 2004. *Reliability Theory and Practice.* Dover Publications, Mineola.)

In an attempt to capture and describe some of the complicated behaviors and interactions associated with damage accumulation, Bolotin uses a *damage index* $\psi[t, q(\tau)]$ [211]. t is time. $q(\tau)$ is the set of loadings on the structure for $\tau < t$ and $0 < \psi < 1$. In general, $\psi[t, q(\tau)]$ can be a functional operator of the past loadings $q(\tau)$. ψ may be a scalar, or it may be in a more complicated multidimensional format. When the structure is in pristine condition, $\psi = 0$. When the structure has failed, $\psi = 1$. The vector format has a physical appeal because it treats different damage modes, for example, fatigue and corrosion, separately, at the expense of increased complexity. The loads, $q(\tau)$, may be vector or scalar, depending on the modeling requirements and availability of detailed information.

The functional representation of damage accumulation allows for including the effects of past history, if the damage accumulation rate is independent of history, and instead depends only on the present instantaneous level of damage. A differential rate equation can model the damage accumulation, that is,

$$\frac{d\psi}{dt} = f(\psi, q, t). \tag{8.45}$$

If the structure is not repaired, self-healing, or undergoing a break-in phase, then $f > 0$ and the damage index ψ increases monotonically. In general, rate equations such as Equation 8.45 are simpler to use than functional history-dependent formulations. An example application of such an approach is that of Haroon and Adams, who combined Paris' law of crack growth with load history to form a health prognosis for automotive sway bars (Figure 8.34) [253]:

$$\frac{da}{dN} = E(Y\Delta\sigma\sqrt{\pi a})^m, \tag{8.46}$$

where a is the half crack length, σ is the applied stress, and E, Y, and m are parameters that depend on material properties, specimen geometry, and environmental conditions, respectively. In another example, Testa and Yanev compared the decline of condition ratings for bridge components as functions of time versus maintenance activities [254]. Such assessments can be quite valuable when planning and managing maintenance and replacement decisions.

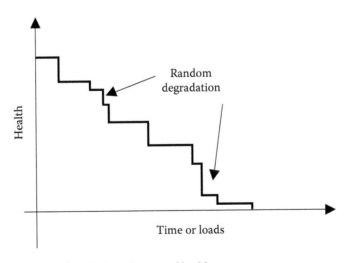

FIGURE 8.34 Gamma process degradation of structural health.

If the damage accumulation rate is history-independent and is independent of the present level of damage, that is, $f(\psi, q, t) = f(q, t)$, then the damage growth rate Equation 8.45 becomes a linear differential equation with the solution

$$\psi(t) = \int_0^t f[q(\tau), \tau]\, d\tau \qquad (8.47)$$

or in a discrete load form

$$\psi_N = \sum_{i=1}^N f(q_i), \qquad (8.48)$$

where Equation 8.48 suggests that damage accumulates as an additive process that depends on the loads in each cycle, and accommodates load histories with varying amplitude cycles. One approach to modeling the behavior of materials susceptible to fatigue cracking, under a sequence of load cycles at stress level σ_i, is to rewrite Equation 8.48 as

$$\psi_K = \sum_{i=1}^K \frac{n_i}{N_i}, \qquad (8.49)$$

where n_i is the number of cycles at σ_i and N_i is the fatigue life at σ_i. Equation 8.49 is the *Palmgren–Miner* accumulative damage rule [156].

The primary advantage of a linear damage accumulation model is mathematical simplicity. Each load causes a specified amount of damage that accumulates. The total damage is a summation of past damage increments. The sum is independent of the order of the loadings. When the loads are all approximately the same size, then order is unimportant and a linear damage accumulation model can be effective in prognosis. Often all that is needed is to measure the load duration or the number of load cycles. This is a standard technique for managing maintenance on airplanes and large rotating machines. When the loads are nonuniform, a linear cumulative rule may lead to erroneous conclusions. For example, consider a structure with a load sequence that consists of repeated small loads and one big load (Figure 8.35). An undamaged structure may be able to tolerate a sequence of many small loads before the application of a large load without causing any damage. Reversing the load sequence by applying the large load first may cause the same sequence of small loads to produce significant damage, even failure.

When the load sequence is uncertain, it may be necessary to overdesign and overmaintain a structure to account for the possibility of worst-case loading scenarios. Including load monitoring into maintenance decisions can alleviate some of the pressure to overdesign and overmaintain. Boller and Biemans used this concept to develop a quantitative estimate of the value of SHM for aircraft [255]. The analysis assumes that if the airplane is operated without SHM, then the aircraft design requires withstanding the worst possible sequence of heavy loads being applied early in the life of an airplane. The result is a potential overdesign. If instead, the airplane operations include SHM, then it may be possible to design a lighter structure and to identify the occurrence of large potentially damaging loads and to repair the structure as necessary. This helps prevent the subsequent accelerated accumulation of damage by small loads that would occur if the damage was not identified and repaired.

More sensitive and more complicated nonlinear accumulation models are possible. Bolotin suggests starting with a self-similarity hypothesis based on the notion that larger

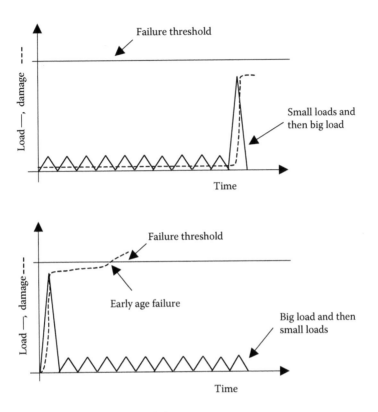

FIGURE 8.35 Load sequence can affect time to failure.

loads cause damage to accumulate more quickly and reduce the time to failure. This leads to building piecewise-linear nonlinear models that scale the effects of time and/or load according to a self-similar hypothesis. Such methods accommodate combined high-cycle and low-cycle fatigue models [156,211,256]. Along these lines, AASHTO recommends using an effective stress range, S_r. This corresponds to a root mean cube weighting of the fraction of cycles in a particular stress range, that is,

$$S_r = \left(\sum_i f_i S_{ri}^3 \right)^{1/3} . \tag{8.50}$$

f_i is the fraction of cycles with a given stress amplitude, S_{ri} [257,258]. Including partial information from load measurements can enhance this technique. For example, Shimizu et al. combined load-test-calibrated member stress measurements with estimates of heavy truck traffic to form a root mean cube fatigue index for individual members on a bridge [259].

Assessing complicated load effects, such as cyclic behavior with variable amplitudes, requires the use of more complicated cycle-counting approaches. The *rainflow algorithm* is one of the more popular of the cycle-counting cumulative damage procedures [260]. The name derive from the graphical implementation of the counting technique in which the tracking of individual half-cycles resembles rainfall dripping off the roof edges of a pagoda. Automated instruments that use the rainflow algorithm can be useful in fatigue and damage monitoring. For example, Brower used an array of fiber optic strain gages to measure fatigue data from risers on deepwater offshore oil production risers and condensed the data with the rainflow algorithm [261]. Leconte et al. estimated the long-term probability of fatigue damage on a bridge structure by first measuring the influence lines by static load tests and combining the results with rainflow histograms of stress levels and historical truckload estimates [262]. Hock et al. combined ultrasound transmission measurements with mechanical models of microcracks to predict the failure of concrete under repeated loading [263]. Fracture toughness depends on many factors, including water-to-cement ratio, and the acoustic ultrasound measurements provide toughness (fracture energy) information for the prognosis models.

An alternative to the rainflow algorithm is to introduce random load sequences directly into damage accumulation models. Sun and Hudson extended the linear Palmgren–Miner law of damage accumulation by modeling the load as a random variable with the specific application of predicting pavement cracking subjected to random traffic loading [264]. Khalak developed a Monte Carlo approach that reduces some of the computational burdens of damage accumulation modeling [265]. When the damage occurs at random times and with random levels, but is independent of the present level of damage and accumulates monotonically, Pandey and van Noortwijk recommend using a gamma process model [266–268]. Bakker and van Noortwijk used gamma process models on inspection data to validate deterioration models in structures [269].

Quasi-continuum damage accumulation models are another approach that used the components of a damage tensor as state variables. This includes such techniques as a detailed multiscale modeling of damage in composites [270]. Experiments with micro-x-ray examination of materials being loaded to failure indicate that using damage as a state variable shows some promise [271].

All of the above methods of assessing fatigue damage presuppose that the propensity of a given structural material to suffer from fatigue is known a priori. This is often not the case, especially with older structures that are made of materials with uncertain compositions and properties. One approach removes samples from the structure for laboratory testing. When removing samples is not practical, *in situ* live load measurements with observations of existing fatigue damage are possible workarounds [272].

Three approaches to ameliorating the effects of large variabilities in damage accumulation are as follows: (1) Use materials that fail in a more predictable manner, for example, use ductile rather than brittle alloys. (2) Construct redundant systems that can tolerate low-load failures in a small number of members. This requires that the failures of the individual members or subsystems be reasonably independent. (3) Measure the condition of the structure as damage accumulates, and act accordingly. In addition to variability in material response, the characterization and control of structural loads is often less than what is desired. Using probabilistic methods coupled with rare event statistics to quantify future and past loadings on structures is often expedient [79,273,274].

8.9.3 Condition-Based Models

A central thrust of SHM is the hypothesis that structural measurements can lead to improved predictions of future structural conditions and can guide superior structural health engineering practices. Quantifying and controlling uncertainties is a key component of the success of these efforts (Figure 8.36).

Presumably, measuring detailed structural information reduces the variability in the prognosis of small damage that can grow to larger damage [275]. The conditional probability of the time of failure of a structure, T, based on the value of a measurand, m, such

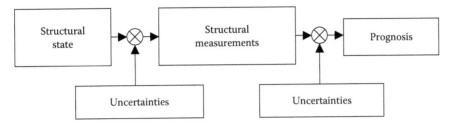

FIGURE 8.36 Structural states, measurements, uncertainties, and prognosis.

as crack length, is

$$p(T|m) = \frac{p(T, m)}{p(m)}. \tag{8.51}$$

The measurement of m produces useful information if the datum corresponds to structural properties that indicate structural health. In the limit where the measurand contains no useful information, that is, $p(T)$ and $p(m)$ are independent,

$$p(T|m) = \frac{p(T)p(m)}{p(m)} = p(T). \tag{8.52}$$

The expected value of T given m is

$$E[T|m] = \mu = \int_0^\infty \frac{Tp(T|m)}{p(m)} \, dT. \tag{8.53}$$

The variance of T given m is

$$V[T|m] = \sigma^2 = \int_0^\infty \frac{(T - \mu)^2 \, p(T|m)}{p(m)} \, dT. \tag{8.54}$$

The ability to detect damage can be highly dependent on the size of the damage and on the characteristics of the measurement instruments. Large levels of damage are usually easier to detect than small levels.

Damage detection in a particular situation with a particular instrument often depends on the sensitivity and selectivity of the sensing instruments. This naturally leads to the use of probabilistic descriptions. A probability distribution attributed to Lewis et al. and Palmberg et al. describes the *probability of detecting* (POD) a crack of length a [276]:

$$\mathrm{POD}(a) = \frac{\alpha a^\beta}{1 + \alpha a^\beta}. \tag{8.55}$$

Figure 8.37 shows the *Lewis–Palmberg detection probabilities* for various values of the parameters α and β. Lopatin and Mahmood gave an example of detecting cracks in solid rocket fuel with a small array of embedded strain gages that combines the readings with a mechanical model and pattern-recognizing statistical analyses [277]. Doctor et al. reported on a comprehensive flaw detection and size distribution study of a nuclear reactor with a cataloging of more than 4,000 flaws [278]. The presence of complex flaws complicated the study.

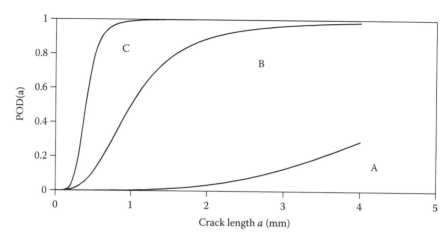

FIGURE 8.37 Lewis–Palmberg probability of crack detection distribution for various values of α and β: A($\alpha =$ 0.0032, $\beta = 3.5$), B($\alpha = 1$, $\beta = 3$), C($\alpha = 100$, $\beta = 5$). (Adapted from Achenbach JD, Moran B, and Zulfigar A. 1997. *Structural Health Monitoring Current Status and Perspectives.* Technomic Publishing, Lancaster.)

The size-dependent value of damage detection depends on the context of the particular application. For example, detecting large levels of damage that may lead to imminent collapse or failure can be quite beneficial. There are many cases where detecting small levels of damage may be more valuable because it is possible to remedy the situation and to prevent the small damage from growing into bigger damage. Small repairs tend to be easier and less expensive than large repairs. Figure 8.38 shows the combination of probability of detection, crack size, crack growth, and value of detection.

Level of safety (LOS) analysis techniques combine probabilistic notions of critical crack size with crack detectability. The approach of Lin et al. assumes that the detected damage is a discrete variable, d, with the states $d_1 =$ "Damage is Detected" and $d_2 =$ "Damage is Not Detected"; and damage size is a continuous positive variable, a [279]. A key assumption is that there exists a critical damage size, a_c. Four possible states exist: (1) *Safe*—damage size is less than the critical size and is detected ($a < a_c$ and $d = d_1$). (2) *Safe*—damage size is less than the critical size and is not detected ($a < a_c$ and $d = d_2$). (3) *Safe*—damage size is greater than the critical size and is detected ($a > a_c$ and $d = d_1$). (4) *Unsafe*—damage size

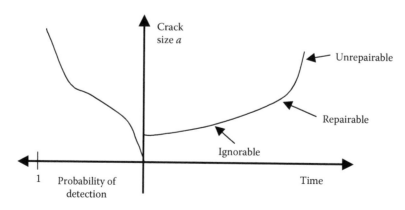

FIGURE 8.38 Probability of detection versus crack size, growth, and value of detection. (Adapted from Shiao M. 2003. *Structural Health Monitoring*, DEStech, Lancaster.)

is greater than the critical size and is not detected ($a > a_c$ and $d = d_1$). This condition is unsafe. Denoting P_i ($i = 1, 2, 3$, or 4) as the probability of state i occurring, the definition of the LOS is

$$\text{LOS} = 1 - P_4. \tag{8.56}$$

When P_4 is small, the LOS approaches 1 and the system is deemed to be safe. Calculating P_4 uses the probability density function of detected damage size, $p_0(a) = p(a|d_1)$, and the probability of damage of size a being detected, $P_D(a) = p(d_1|a)$. The LOS is

$$\text{LOS} = 1 - \frac{\int_{a_c}^{\infty} p_0(a)/P_D(a) \ [1 - P_D(a)] \ da}{\int_0^{\infty} p_0(a)/P_D(a) \ da}. \tag{8.57}$$

When the structural form and material composition is such that detectable damage grows in a relatively slow and predictable manner, then it is often possible to combine damage measurements with statistical estimates of damage growth to improve prognostics. Chelidze and Cusumano et al. monitored slowly growing damage in dynamical systems by tracking changes of the system dynamical properties in phase space [280,281]. An advantage of phase space representations is that they can accommodate nonlinear and even chaotic systems. Kalman filters are a natural method of estimating system states and phase space representations from limited channel count measurements.

Monitoring the behavior of a fleet of similar vehicles provides data for fleet-wide statistical condition-based prognoses. An approach patented by Talbott for machinery monitoring combines periodic measurements of specific symptoms, such as level of vibration and temperature, with postmortem failure analyses to form a condition history matrix $[H_k]$ for each machine [282]:

$$[H_k] = \left[S_i^T | C_i^T | r_i \right], \quad i = 1, \ldots, n, \tag{8.58}$$

where $[S_i]$ = symptom matrix and $[C_i]$ = environmental condition matrix. A prognostic knowledge base $[H]$ combines the condition history information for the entire fleet:

$$[H] = [H_1 | H_2 | \ldots | H_N]. \tag{8.59}$$

Since different systems fail at different rates, $[H]$ is the size of the largest matrix $[H_k]$ and the remainder is filled in with zeros. The system uses the data to form an estimated conditional lifetime PDF based on measured symptoms $E\{R|S^*\}$. Shiao developed a Bayesian update methodology for improved structural prognoses [283].

Another approach to prognosis is to calculate a single scalar statistic from a vector of multiple statistical features and to use the combined information as an indicator. Fillipenko et al. [284] proposed fusing kurtosis, crest factor, root mean square vibration, and other features into a single quantity for predicting bearing failures.

Condition-based prognoses have difficulty if the measurements contain little information about the damage or condition of the structure. For example, Sarhan et al. describe a series of studies on the load-carrying capacity of drilled concrete piles with flaws that are too small to detect positively with traditional NDE methods, but are big enough to be potentially troublesome [285]. Another point of view from statistical information theory is that the structure is attempting to communicate information [286]. Adami argues that information is relative, that is, the observer needs a basis for comparison [287]. Hunt and Hebden indicate that validation and calibration can significantly improve the prognostic capabilities of aircraft SHM systems [288].

8.10 Corrosion and Chemical Damage

Corrosion can attack virtually all metallic structures. Metal loss and the corresponding strength loss are primary manifestations of corrosion-induced damage. Corrosion by-products, such as the swelling of iron oxide in steel-reinforced concrete, can also cause damage by causing the concrete to crack. The crack enables the ingress of more electrolytes and accelerates the corrosion. Chloride ions from deicing salts or ocean spray, and the use of chloride-laden aggregates are especially pernicious in this regard.

Since the mechanics and electrochemistry of corrosion are fairly well-understood, it is possible to formulate rational mathematical models of the prognosis of corrosion. Models are available that correlate chloride exposure, seasonal effects, and measured corrosion currents to predict the time to corrosion [289]. Tarighat used an NN to predict damage due to diffusion of chemical reagents into concrete, especially those that cause carbonation [290]. Schmidt-Dohl et al. developed a monitoring system that fuses data from electrochemical and permeation sensors to form an adaptive prognosis system for predicting damage due to chemical attack on concrete [291]. However, the tendency of certain corrosion processes to become very aggressive in local regions often confounds the predictive capability of many models. Stress corrosion cracking, aluminum lap joints in aircraft, copper water pipes, and steel reinforcing bars in concrete can all be subject to the effect of accelerating local corrosion attack.

Rock anchors are susceptible to localized subsurface corrosion. Determining whether a rock anchor has a corrosion problem is nontrivial. Withiam et al. and Lane and Fishman recommend an approach that combines assessments of corrosion risk with NDT measurements [292,293]. The risk assessment includes considerations of rock chemistry, presence of water, and water chemistry (especially pH) to produce a classification of the corrosion environment as being *normal*, *aggressive*, or *very aggressive*. A power law attributed to Romanoff predicts the amount of corrosion as

$$X = Kt^r, \tag{8.60}$$

where X is the amount of corrosion loss in μm, K is a constant in μm, t is time in years, and r is a constant. Recommended values for K and r appear in Table 8.1. Recommended NDT techniques for identifying rock anchor corrosion include impact echo, ultrasound, and half-cell.

TABLE 8.1

Recommended Factors for Power Law Prediction of Rock Anchor Corrosion

Parameter	Normal	Aggressive	Very Aggressive
K (μm)	35	50	340
r	1.0	1.0	1.0

Source: Withiam JL et al. 2002. Recommended practice for evaluation of metaltensioned systems in geotechnical applications. NCHRP Report 477, National Academy Press, Washington, DC.

8.11 Long-Term Monitoring

Prognosis and maintenance engineering efforts on structures with lengthy expected life-times may benefit from long-term monitoring efforts. This requires instruments that measure significant structure parameters with a minimum of maintenance and/or active intervention by the user. An example application is the long-term SHM of highway bridges using a multichannel suite of instrumentation, including tiltmeters and accelerometers in Connecticut by DeWolf et al. [294]. The instrumentation detects gross changes in the structural behavior. Zilch et al. developed both probabilistic and long-term monitoring to assess temperature-induced fatigue loading in post-tensioned concrete bridge piers [295].

Long-term SHM requires a long-term commitment to collecting and analyzing data on a timely basis. The efforts may be most effective by coupling and comparing the data with appropriate mechanical models of long-term structural behavior. A case in point is the long-term deflection behavior of concrete box-girder bridges [296]. The deflection in the first couple of years is often small, which may lead to incorrect conclusions regarding the size of longer-term deflections that can accelerate in subsequent years due to differential creep and shrinkage.

8.12 Postevent Assessment

It is not uncommon for structures to experience extreme loads that exceed specified maximum operational loads. The structure may be undamaged, damaged but still functional, damaged but repairable, or damaged and beyond repair. Establishing the postevent state of a structure often requires quick assessments. Vibration testing may be an effective protocol. Chen et al. examined the cable-stayed Chi–Lu Bridge in Taiwan following a major earthquake in 1999 [297]. A key measurement was using an AVS to assess the tension in the individual cable stays. Achter et al. measured significant shifts in the vibration properties of a reinforced concrete bridge bent following substantial earthquake-type damage [298]. Similarly, Robinson et al. measured significant reductions in the modal vibration frequencies of a six-span reinforced concrete bridge subjected to major levels of damage prior to deconstruction [299].

References

1. Apgar V. 1953. A proposal for a new method of evaluation of the newborn infant. *Curr Res Anesth Analg*, (July–August), 32, 260–267.
2. AASHTO. 2003. *AASHTO Manual for Condition Evaluation of Bridges*, 1994, 2nd Ed., with revisions through 2002. Amer Assoc State Highway and Trans Officials, Washington, DC.
3. Wolters K and Soffker D. 2004. Improving systems availability by combining reliability and control engineering techniques. *Proc 2nd European Workshop on Structural Health Monitoring*, DEStech, Munich.
4. Chang CC and Yang HT. 2004. From human health monitoring to structural health monitoring. *Proc 2nd European Workshop on Structural Health Monitoring*, DEStech, Munich.
5. Dry C. 2003. Self repairing of composites. *Smart Electronics, MEMS, BioMEMS, and Nanotechnology*, SPIE 5055.
6. Du C. 2005. A review of magnesium oxide in concrete. *Concrete Intl*, (December), 27(12), 45–50.

7. Kim B and Roque R. 2006. Evaluation of healing property of asphalt mixture. Paper No. 06-0596, 86th Annual Meeting of the Transportation Research Board, Washington, DC.

8. White SR, Sottos NR, Geubelle PH, Moore JS, Kessler MR, Sriram SR, Brown EN, and Viswanathan S. 2001. Autonomic healing of polymer composites. *Nature*, 409(February), 15.

9. Esser B, Huston D, Spencer G, Burns D, and Kahn E. 2005. Self-healing cables and composites. *Smart Structures and Materials: Industrial and Commercial Applications of Smart Structures Technologies*, San Diego, SPIE 5762-02.

10. Eckelmeyer EH Jr. 1946. The story of the self-sealing tank. *US Naval Inst Proc*, (722), 205–219.

11. Monk RA, Ohnstad TS, and Henry JJ. 2007. Projectile barrier method for sealing liquid container. US Patent 7,229,673.

12. Sohn H, Farrar CR, Hemez FM, Shunk DD, Stinemates DW, and Nadler BR. 2003. A review of structural health monitoring literature: 1996–2001. *Los Alamos National Laboratory Report*, LA-13976-MS.

13. Farrar CR, Duffey TA, Doebling SW, and Nix DA. 1999. A statistical pattern recognition paradigm for vibration-based structural health monitoring. *Proc 2nd Intl Workshop on Structural Health Monitoring*, Stanford, Technomic, Lancaster.

14. Worden K, Ball AD, and Tomlinson GR. 1993. Fault location in a framework structure using neural networks. *Smart Mater Struct*, 2, 189–200.

15. Moore M, Phares B, Graybeal B, Rolander D, and Washer G. 2000. *Reliability of Visual Inspection for Highway Bridges*, Vol. 1. Final Report FHWA-RD-01-120, Federal Highway Administration, McLean.

16. Hearn G. 2007. *Bridge Inspection Practices—a Synthesis of Highway Practice*, NCHRP Synthesis 375, Transportation Research Board, Washington, DC.

17. ACI. 1984. *Guide for Making a Condition Survey of Concrete in Service*, 201.1R-68, K. Lauer (ed.). Amer Conc Inst., Farmington Hills, MI.

18. Moore M, Phares B, Graybeal B, Rolander D, Washer G. 2000. Reliability of Visual Inspection of Highway Bridges Vol II: Appendices FHWA-RD-01-121, US Federal Highway Administration.

19. Raina VK. 1996. *Concrete Bridges Inspection, Repair, Strengthening, Testing, Load Capacity Evaluation*. McGraw-Hill, New York.

20. Shirakura A, Kawashima Y, Yonezawa Y, and Morikawa H. 2004. Condition rating method for performance evaluation of existing RC bridges. *Proc IABMAS'04 Bridge Maintenance, Safety, Management and Cost*, Kyoto. Taylor & Francis, London.

21. ASCE. 2000. *Guideline for Structural Condition Assessment of Existing Buildings*, SEI/ASCE 11-99, Reston.

22. Veletzos MJ, Panagiotou M, Van Den Einde Y, Restrepo JI, and Sahs S. 2008. Post-earthquake assessment of bridge columns. *Concrete Intl*, (March), 30(3), 61–65.

23. Grosse CU, Kurz JH, Beutel R, Reinhardt HW, Krueger M, Saugk O, Marron PJ, Rothermel K, Meyer J, Bischoff R, and Feltrin G. 2005. Combined inspection and monitoring techniques for SHM of bridges. In *Structural Health Monitoring*, FK Chang (ed.). DEStech, Lancaster.

24. Hahn-Keith K, Stump JL, Charters D, and Andrews L. 2005. Bridge inspections using handheld data collectors. *Trans Res Rec*, Spec Vol CD 11-S, 279–286.

25. Leung A. 1996. Perfecting bridge inspecting. *Civil Eng*, (March), 66(3), 59–61.

26. Ward MD. 2004. An overview of United States coast guard corrosion control through the use of online corrosion mapping. *Proc 2nd European Workshop on Structural Health Monitoring*, DEStech, Munich.

27. Baker DV, Chen SE, and Leontopoulos A. 2000. Visual inspection enhancement via virtual reality. In *Structural Materials Technology IV*, S Allampalli (ed.). Technomic Press, Lancaster.

28. Le Diouron T. 2004. Infrastructure management database system. *Proc IABMAS'04 Bridge Maintenance, Safety, Management and Cost*, Kyoto. Taylor & Francis, London.

29. Kawamura K, Frangopol DM, and Miyamoto A. 2004. Evaluation system for reinforced concrete bridges based on two performance indicators. *Proc IABMAS'04 Bridge Maintenance, Safety, Management and Cost*, Kyoto. Taylor & Francis, London.

30. Dan D, Sun L, and Cheng W. 2004. Online bridge linguistic monitoring methods based on Mamdani fuzzy inference systems. *Proc IABMAS'04 Bridge Maintenance, Safety, Management and Cost*, Kyoto. Taylor & Francis, London.

31. Rajani B, Kleiner Y, and Sadiq R. 2006. Translation of pipe inspection results into condition ratings using the fuzzy synthetic evaluation technique. *J Water Supply Res Tech Aqua*, 55(1), 11–24.

32. Kaito K and Abe M. 2004. Prediction of structural deterioration from inspection ratings. *Proc IABMAS'04 Bridge Maintenance, Safety, Management and Cost*, Kyoto. Taylor & Francis, London.

33. Thorkildsen E. 2002. Challenges to "Hands-On" inspection of overhead sign structures. In *Proc Structural Materials Technology V, an NDT Conference*, S Allampalli and G Washer, (eds.). Cincinnati.

34. OSHA. 1999. *Occupational Safety and Health Standards for the Construction Industry*, 29 CFR Part 1926, United States Department of Labor, Washington, DC.

35. Jergenson P. 2004. The deception of WYSIWIG in façade inspections. *Structures*, (May).

36. Gurjar S and Morgan DR. 1998. Corrosion rate surveys, a tool for maintenance expenditure prioritization and service life evaluation of concrete bridges. *Proc Intl Conf on Corrosion and Rehabilitation of Reinforced Concrete Structures*, Orlando.

37. Shdid CA, Ansley MH, and Hamilton III HR. 2006. Visual rating and strength testing of 40-year old precast, prestressed concrete bridge piling. Paper 06-0212, 86th Annual Meeting, TRB, Washington, DC.

38. McCaffrey B. 2006. Bridge load rating practices in New York State. *NDE Conf on Civil Eng*, Amer Soc Nondes Test, St. Louis.

39. Avent RR, Alawady M, and Heymsfield E. 2001. Inspecting concrete bridge substructures Part 1: Scour. *Concrete Intl*, (January), 23(1), 41–45.

40. Harland JW, Purvis RL, Graber DR, Albrect P, and Flournoy TS. 1986. *Inspection of Fracture Critical Bridge Members*, FHWA-IP-86-26, U.S. Federal Highway Administration.

41. Karbhari VM, Kaiser H, Navada R, Ghosh K, and Lee L. 2005. *Methods for Detecting Defects in Composite Rehabilitated Structures*, SPR 336, Final Report, Oregon Department of Transportation.

42. Agrawal AK, Kawaguchi A, Qian G, and Lagace S. 2004. Deterioration rates of bridge elements in New York State. *Proc IABMAS'04 Bridge Maintenance, Safety, Management and Cost*, Kyoto. Taylor & Francis, London.

43. Chase SB. 2004. A long-term bridge performance monitoring program. *Nondestructive Detection and Measurement for Homeland Security II*, SPIE 5395.

44. Khan MS. 2003. Detecting corrosion-induced delaminations. *Concrete Intl*, (July), 25(7), 73–78.

45. Huston DR, Gucunski N, Maher A, Cui J, Burns D, and Jalinoos F. 2007. Bridge deck condition assessment with electromagnetic, acoustic and automated methods. In *Structural Health Monitoring*, FK Chang (ed.). DEStech, Lancaster.

46. Rhazi J and Omar D. 2006. Condition assessment of existing concrete bridge decks: The half-cell potential versus GPR. *NDE Conf on Civil Eng*, Amer Soc Nondes Test, St. Louis.

47. Poston RW, Frank KH, and West JS. 2003. Enduring strength. *Civil Eng*, (September), 73(9), 58–63.

48. Puri S and Moser DE. 2007. Condition assessment of unbonded post-tensioning strands. *Concrete Intl*, (April), 29(4), 60–64.

49. Aguilar R and Kaslan E. 2002. Use of NDE in the California Bridge Inspection Program. In *Proc Structural Materials Technology V, an NDT Conference*, S Allampalli and G Washer, (eds.). Cincinnati.

50. Comstock G and Garrison L. 2004. Improving the value of UT inspection of in-service bridge pins and shafts: Report from Washington State. *Proc Structural Materials Technology VI, an NDT Conference*, S Allampalli and GA Washer (eds.). Buffalo.

51. Sugimoto H, Watanabe T, and Akadomari K. 2004. Repair cost model and optimum management system for bridges in Hokkaido. *Proc IABMAS'04 Bridge Maintenance, Safety, Management and Cost*, Kyoto. Taylor & Francis, London.

52. Babaei K and Fouladgar A. 1998. Prediction of condition of concrete bridge decks at network level for Pontis Bridge Management System. *Proc Intl Conf on Corrosion and Rehabilitation of Reinforced Concrete Structures*, Orlando.

53. Rowe JB. 2006. Taking overhead sign structure inspection to the next level element-based overhead sign structure inspections. Paper 06-2803, 86th Annual Meeting, TRB, Washington, DC.

54. Williams MS, Villemure I, and Sexsmith RG. 1997. Evaluation of seismic damage indices for concrete elements loaded in combined shear and flexure. *ACI Struct J*, 94(May–June), 3, S29.

55. Baladi GY. 2006. Distress points, distress index, and remaining service life versus asset values. Paper 06-1663, 86th Annual Meeting, TRB, Washington, DC.

56. Barroso LR and Rodriguez R. 2004. Damage detection utilizing the damage index method to a benchmark structure. *J Eng Mech*, 130(2), 142–151.

57. Stubbs N, Kim JT, and Farrar C. 1995. Field verification of a nondestructive damage localization and severity estimation algorithm. *Proc 13th Intl Modal Analysis Conference*, Nashville.

58. Shi ZY, Law SS, and Zhang LM. 2002. Improved damage quantification from elemental modal strain energy change. *J Eng Mech*, 128(5), 512–529.

59. Hernandez JY Jr and Fujino Y. 2007. Image detection in pile foundations using changes in identified properties of an RC railway viaduct. In *Structural Health Monitoring*, FK Chang (ed.). DEStech, Lancaster.

60. Salgado R and Cruz P. 2005. Detecting damage in structures using wavelet analysis. In *Structural Health Monitoring*, FK Chang (ed.). DEStech, Lancaster.

61. Xiang Y and Wang J. 2006. Advance in health monitoring and assessment theory of long span concrete bridge and application. *Proc 4th China–Japan–US Symposium on Structural Control and Monitoring*, Hangzhou.

62. Masuda A, Sone A, and Morita S. 2004. Continuous damage monitoring of civil structures using vibratory gyroscopes. *Sensors and Smart Structures Technologies for Civil, Mechanical and Aerospace Systems*, SPIE 5391.

63. Guy P, Jayet Y, and Goujon L. 2003. Guided waves interaction with complex delaminations application to damage detection in composite structures. *Smart Nondestructive Evaluation and Health Monitoring of Structural and Biological Systems II*, SPIE 5047.

64. Ma J and Pines DJ. 2001. Detecting damage in building structure model under seismic excitation. *Smart Systems for Bridges, Structures, and Highways*, SPIE 4330.

65. Cheng GQ and Hu LH. 2006. Study on damage identification of bridge structure based on the method of element modal strain energy. *Proc 4th China–Japan–US Symposium on Structural Control and Monitoring*, Hangzhou.

66. Ihn JB and Chang FK. 2004. Hot spot monitoring for aircraft structures. In *Advanced Smart Materials and Smart Structures Technology*. FK Chang, CB Yun, and BF Spencer, Jr (eds.). DEStech, Lancaster.

67. Ihn JB and Chang FK. 2004. Detection and monitoring of hidden fatigue crack growth using a built-in piezoelectric sensor/actuator network I. Diagnostics. *Smart Mater Struct*, 13, 609–620.

68. Song G, Gu H, Mo YL, Hsu T, Dhonde H, and Zhu RR. 2005. Health monitoring of a concrete structure using piezoceramic materials. *Sensors and Smart Structures Technologies for Civil, Mechanical, and Aerospace Systems*, SPIE 5765.

69. Ho YK and Ewins DJ. 1999. Numerical evaluation of the damage index. *Proc 2nd Intl Workshop on Structural Health Monitoring*, Stanford, Technomic, Lancaster.

70. Tseng KK. 2006. Smart piezoelectric sensors for structural health monitoring in civil, mechanical, and aerospace systems. *Proc 4th China–Japan–US Symposium on Structural Control and Monitoring*, Hangzhou.

71. Naidu AS and Soh CK. 2004. Damage severity and propagation characterization with admittance signatures of piezo transducers. *Smart Mater Struct*, 13, 393–403.

72. Ayers J, Ruzzene M, and Apetre N. 2007. A wave filtering displacement-based damage measure. In *Structural Health Monitoring*, FK Chang (ed.). DEStech, Lancaster.

73. Newhook JP. 2005. Developing an SHM system for FRP-strengthened beams. *Nondestructive Evaluation and Health Monitoring of Aerospace Materials, Composites and Civil Infrastructure IV*, SPIE 5767.

74. Sobjano JO and Tawfiq KS. 1999. Framework for incorporating nondestructive evaluation (NDE) into pavement and bridge management systems. *Nondestructive Evaluation of Bridges and Highways III*, SPIE 3587.

75. Testa RB, Zhang W, Smyth AW, and Betti R. 2000. Detection of cracks with closure. In *Structural Materials Technology IV*, S Allampalli (ed.). Technomic Press, Lancaster.

76. Maia NM, Silva JM, Almas EA, and Sampaio RP. 2003. Damage detection in structures: From mode shape to frequency response function methods. *Mech Syst Signal Process*, 17(3), 489–498.

77. Asundi AK and Ma J. 2001. Structural health monitoring using a fiber optic polarimetric sensor and a fiber optic curvature sensor—static and dynamic test. *Smart Mater Struct*, 10(2), 181–188.

78. Melchor-Lucero O, Carrasco CJ, Espino L, Fernandez A, and Osegueda RA. 2007. Computer modeling for a generalized approach to measure impact damage. *J Eng Mech*, 133(3), 299–307.

79. Hart GC. 1982. *Uncertainty Analysis, Loads, and Safety in Structural Engineering*, Prentice-Hall, Englewood Cliffs, NJ.

80. Carrión F, López JA, and Balankin A. 2005. Probabilistic model for bridge structural evaluation using nondestructive inspection data. *Nondestructive Evaluation and Health Monitoring of Aerospace Materials, Composites and Civil Infrastructure IV*, SPIE 5767.

81. Catbas FN and Aktan AE. 2002. Condition and damage assessment: Issues and some promising indices. *J Struct Eng*, 128(8), 1026–1036.

82. Koyluoglu HU, Nielsen SR, Abbott J, and Çakmak AS. 1998. Local and modal damage indicators for RC frames subject to earthquakes, *J Eng Mech*, 124, 12.

83. Lazarov B and Trendafilova I. 2004. An investigation on vibration-based damage diagnosis in thin plates. *Proc 2nd European Workshop on Structural Health Monitoring*, DEStech, Munich.

84. Jurado JA, Hernandez S, and Mosquera A. 1999. Safety analysis and multimedia-based software for monitoring of cooling towers. *ACI Struct J*, 96(May–June), 3, S47.

85. Yang S and Lee H. 2006. Enhancing pavement management information system by integrating digital image processing and management features. Paper 06-2154, 86th Annual Meeting, TRB, Washington, DC.

86. Wu Z and Su H. 2005. Dam health diagnosis and evaluation. *Smart Mater Struct*, 14, S130–S136.

87. Yam LH, Yan YJ, Cheng L, and Jiang JS. 2003. Identification of complex crack damage for honeycomb sandwich plate using wavelet analysis and neural networks. *Smart Mater Struct*, 12, 661–671.

88. Lam HF, Ko JM, and Wong CW. 1998. Localization of damaged structural connections based on experimental modal and sensitivity analysis. *J Sound Vib*, 210(1), 91–115.

89. Harrison HD and McCanney TO. 1995. Wayside system for measuring rail longitudinal force due to thermal expansion of continuous welded rail. *Nondestructive Evaluation of Aging Railroads*, SPIE 2458.

90. Kiddy JS, Baldwin CS, and Salter TJ. 2004. Certification of a submarine design using fiber Bragg grating sensors. *Industrial and Commercial Applications of Smart Structures Technologies*, SPIE 5388.

91. Grace NF, Roller JJ, Navarre FC, Nacey RB, and Bonus W. 2004. Load testing a CFRP-reinforced bridge. *Concrete Intl*, (July), 26(7), 51–57.

92. Hernandez ES, Galati N, Myers J, and Nanni A. 2006. Long term monitoring of bridges using a robotic tacheometry system. *NDE Conf on Civil Eng*, Amer Soc Nondes Test, St. Louis.

93. Roufa GJ. 2006. Use of rotating laser systems in load testing of bridges. *NDE Conf on Civil Eng*, Amer Soc Nondes Test, St. Louis.

94. Issa MA and Issa MA. 2004. Dynamic and static tests of prestressed concrete girder bridges in Florida. *Proc IABMAS'04 Bridge Maintenance, Safety, Management and Cost*, Kyoto. Taylor & Francis, London.

95. Ferreira J and Branco FA. 2004. The importance of bridge load tests. *Proc IABMAS'04 Bridge Maintenance, Safety, Management and Cost*, Kyoto. Taylor & Francis, London.

96. Santos LO, Rodrigues J, Min X, and Fernandes JA. 2004. Static and dynamic tests of Salgueiro Maia cable-stayed bridge. *Proc IABMAS'04 Bridge Maintenance, Safety, Management and Cost*, Kyoto. Taylor & Francis, London.

97. Rodriguez J, Santos LO, and Min X. 2004. Static and dynamic tests of Arade Bridge in A2 highway. *Proc 2nd European Workshop on Structural Health Monitoring*, DEStech, Munich.

98. Felix C, Pimentel M, Costa B, Faria R, and Figueiras J. 2004. Monitoring of Oporto Luiz I Steel Bridge during a loading test. *Proc 2nd European Workshop on Structural Health Monitoring*, DEStech, Munich.

99. Alampalli S, Aref A, Schongar G, and Greenberg H. 2003. In-service performance and analytical investigations of FRP superstructure. In *Structural Health Monitoring*, FK Chang (ed.). DEStech, Lancaster.

100. Zoghi M, Schulz J, Plews P, and Foster D. 2003. Nondestructive load testing and evaluation to identify structural health of prestressed concrete bridges. In *Structural Health Monitoring*, FK Chang (ed.). DEStech, Lancaster.

101. Liu PL and Sun SC. 1997. The application of artificial neural networks on the health monitoring of bridges. In *Structural Health Monitoring Current Status and Perspectives*, F.-K. Chang (ed.). Technomic Publishing, Lancaster.

102. Yannotti AP, Alampalli S, O'Connor J, Schongar G, Greenberg H, and Norfolk M. 2000. Proof load testing and monitoring of an FRP composite bridge. In *Structural Materials Technology IV*, S Allampalli (ed.). Technomic Press, Lancaster.

103. Wisniewski D, Cruz PJ, and Henriques AA. 2004. Probability based assessment of acceptable precast girder deflections for proof load testing. *Proc IABMAS'04 Bridge Maintenance, Safety, Management and Cost*, Kyoto. Taylor & Francis, London.

104. Dorrell L and Roach D. 1999. Development of test facility for structural evaluation of composite rotor hubs. *Proc 2nd Intl Workshop on Structural Health Monitoring*, Stanford, Technomic, Lancaster.

105. Goggin P, Huang J, White E, and Haugse E. 2003. Challenges for SHM transition to future aerospace systems. In *Structural Health Monitoring*, FK Chang (ed.). DEStech, Lancaster.

106. Bu JQ, Law SS, and Zhu XQ. 2006. Innovative bridge condition assessment from dynamic response of a passing vehicle. *J Eng Mech*, 132(12), 1372–1379.

107. Scanlon A, Nanni A, and Ragan S. 1994. Experimental techniques for large-scale impact testing of reinforced concrete. *New Experimental Techniques for Evaluating Concrete Material and Structural Performance*, ACI SP-143-16.

108. Miller RA, Aktan AE, and Shahrooz BM. 1994. Destructive testing of decommissioned concrete slab bridge. *J Struct Eng*, 120(7), 2176–2198.

109. Grimmelsman KA and Aktan AE. 2004. Uncertainty and similitude in field research. *Proc IABMAS'04 Bridge Maintenance, Safety, Management and Cost*, Kyoto. Taylor & Francis, London.

110. Kumar R and Rao MV. 2003. Prestressed concrete bridges in India. *Concrete Intl*, (October), 25(10), 75–81.

111. Fuchs PA, Washer GA, Chase SB, and Moore M. 2004. Laser-based instrumentation for bridge load rating. *J Perform Constructed Facilities*, 18(4), 541.

112. Merkle WJ and Myers JJ. 2004. Use of the total station for load testing of retrofitted bridges with limited access. *Sensors and Smart Structures Technologies for Civil, Mechanical and Aerospace Systems*, SPIE 5391.

113. Cruz PJ, Valente I, and Wisniewski DF. 2004. Long-term monitoring of a highway bridge in Portugal. *Proc IABMAS'04 Bridge Maintenance, Safety, Management and Cost*, Kyoto. Taylor & Francis, London.

114. Bremond P and Potet P. 2001. Lock-in thermography: A tool to analyse and locate thermomechanical mechanisms in materials and structures. *Thermosense XXIII*, SPIE 4360.

115. Armstrong-Helouvry B. 1991. *Control of Machines with Friction*. Kluwer Academic, Boston.

116. Phares BM, Wipf TJ, Klaiber FW, and Wood DL. 2004. Bridge load rating for superloads using testing. *Proc IABMAS'04 Bridge Maintenance, Safety, Management and Cost*, Kyoto. Taylor & Francis, London.

117. Rohrmann RG, Said S, and Schmid W. 2004. Health monitoring of a steel bridge under heavy traffic load. *Proc IABMAS'04 Bridge Maintenance, Safety, Management and Cost*, Kyoto. Taylor & Francis, London.

118. Wood SM, Akinci NO, Liu J, and Bowman MD. 2006. Analysis and instrumentation of a steel bridge for investigating the effects of superloads. Paper 06-1903, 86th Annual Meeting, TRB, Washington, DC.

119. Weinmann TL and Lewis AE. 2004. Structural health monitoring and testing for the bridge street bridge deployment project. *Proc Structural Materials Technology VI, an NDT Conference*, Buffalo.

120. Molina FJ, Magonette G, and Renda V. 2003. Health monitoring research activities at the ELSA laboratory. In *Structural Health Monitoring*, FK Chang (ed.). DEStech, Lancaster.

121. Hag-Elsafi O, Alampalli S, and Kunin J. 2002. In-service evaluation of an FRP bridge retrofit. *Proc Structural Materials Technology V, an NDT Conference*, Cincinnati.

122. Kosmatka J. 2003. Evaluating the health of composite military bridges. In *Structural Health Monitoring*, FK Chang (ed.). DEStech, Lancaster.

123. Abdullah B, Minata O, Muranaka A, and Katsuno H. 2004. Strength characteristics of friction type bolted joints for fiber-reinforced composite structural members. *Proc IABMAS'04 Bridge Maintenance, Safety, Management and Cost*, Kyoto. Taylor & Francis, London.

124. Cai CS, Shahawy M, and El-Saad A. 1999. Comparison to bridge load rating based on analytical and field testing methods. *Nondestructive Evaluation of Bridges and Highways III*, SPIE 3587.

125. Saraf V and Nowak AS. 1998. Proof loading of deteriorated steel girder bridges. *J Bridge Eng*, 3(2).

126. Prader J, Grimmelsman K, Jalinoos F, Ghasemi H, Burrows S, Taylor J, Liss F, Moon F, and Aktan E. 2006. Load testing and rating of undocumented reinforced concrete bridges. *NDE Conf on Civil Eng*, Amer Soc Nondes Test, St. Louis.

127. Chang KC, Tsai MH, Hwang JS, and Wei SS. 2003. Field testing of a seismically isolated concrete bridge. *Struct Eng Mech*, 16(3), 241–257.

128. Chen Q, Douglas BM, Maragakis EA, and Buckle IG. 2002. Extraction of hysteretic properties of seismically isolated bridges from quick-release field tests. *Earthquake Eng Struct Dyn*, 31, 333–351.

129. Fan YF, Zhou J, Hu ZQ, and Zhu T. 2007. Study on mechanical response of an old reinforced concrete arch bridge. *Struct Control Health Monit*, 14, 876–894.

130. Legace S. 2004. Practical use of load testing to eliminate permit vehicle restrictions. *Proc Structural Materials Technology VI, an NDT Conference*, Buffalo.

131. McCall C, Phillips D, Basham K, Kelly M, and Suprenant BA. 2004. Quality assurance for industrial slabs. *Concrete Intl*, (March), 26(3), 44–47.

132. Miller RA, Aktan AE, and Shahrooz BM. 1992. Nondestructive and destructive testing of a three span skewed R.C. Slab Bridge. In *Nondestructive Testing of Concrete Elements and Structures*, F Ansari and S Sture (eds.). ASCE, New York.

133. von Rosenvinge T, Acheampong K, and Kidd J. 2004. Capacity testing for deep foundations, the new paradigm. *Structure*, (August) pp. 14–16.

134. Val DV and Stewart MG. 2002. Safety factors for assessment of existing structures. *J Struct Eng*, 128(2).

135. Strauss A and Bergmeister K. 2004. Target reliability index regarding monitoring systems. *Proc IABMAS'04 Bridge Maintenance, Safety, Management and Cost*, Kyoto. Taylor & Francis, London.

136. Carlin A L and Saunders SA. 2004. P-3/S-3 individual aircraft tracking program. *Proc 2nd European Workshop on Structural Health Monitoring*, DEStech, Munich.

137. Lund R and Alampalli S. 2004. Estimating remaining fatigue life of bridge components: A case study. *Proc Structural Materials Technology VI, an NDT Conference*, Buffalo.

138. Vines L, Torkildsen HE, Wang G, Pran K, and Sagvolden G. 2004. Development of fiber optic ship hull health monitoring system for operation of ships within design limits. *Proc 2nd European Workshop on Structural Health Monitoring*, DEStech, Munich.

139. Jensen AE and Wang G. 2004. Finite element analysis of ship hulls for structural health monitoring applications using fiber optic sensor. *Proc 2nd European Workshop on Structural Health Monitoring*, DEStech, Munich.

140. McColl C and Bradfield S. 2004. The influence of extreme weather environments on structural health and fatigue life of WP-3D aircraft. *Proc 2nd European Workshop on Structural Health Monitoring*, DEStech, Munich.

141. Johnson PS, Smith DL, and Hauge ED. 1999. Characterizing aircraft high-cycle fatigue environments using the damage dosimeter. *Proc 2nd Intl Workshop on Structural Health Monitoring*, Stanford, Technomic, Lancaster.

142. El-Bakry. 2005. Commercial aircraft landing gears operational loads monitoring (OLM) systems engineering requirements. In *Structural Health Monitoring*, FK Chang (ed.). DEStech, Lancaster.

143. Wu Z, Zhang B, Wan L, and Wang D. 2004. Fatigue sensor based on fiber Bragg gratings for advanced composite materials. *Proc 2nd European Workshop on Structural Health Monitoring*, DEStech, Munich.

144. de Leeuw B, Brennan FP, and Roffey T. 2003. Remote stress monitoring and reliability analysis of continuously welded rails. In *Structural Health Monitoring*, FK Chang (ed.). DEStech, Lancaster.

145. de Callafon RA. 1999. On-line damage identification using modal based orthonormal functions. *Proc 2nd Intl Workshop on Structural Health Monitoring*, Stanford, Technomic, Lancaster.

146. Nigbor RL and Diehl JG. 1997. Two years' experience using OASIS real-time remote condition monitoring system on two large bridges. In *Structural Health Monitoring Current Status and Perspectives*, F.-K. Chang, (ed.). Technomic Publishing, Lancaster.

147. Lee WH and Shin TC. 1997. Real time seismic monitoring of buildings and bridges in Taiwan. In *Structural Health Monitoring Current Status and Perspectives*, F.-K. Chang, (ed.). Technomic Publishing, Lancaster.

148. Kambli S, Helmicki AJ, and Hunt VJ. 2002. Dedicated real time monitoring of a steel-stringer highway bridge (HAM-126-0881L). *Proc Structural Materials Technology V, an NDT Conference*, S Allampalli and G Washer, (eds). Cincinnati.

149. Chi-Chang Lin CC, Wang CE, Wu HW, and Wang JF. 2005. On-line building damage assessment based on earthquake records. *Smart Mater Struct*, 14, S137–S153.

150. Vanik MW and Beck JL. 1997. A Bayesian approach to structural health monitoring. In *Structural Health Monitoring Current Status and Perspectives*, F.-K. Chang, (ed.). Technomic Publishing, Lancaster.

151. Li A, Ding Y, and Miao C. 2006. Design and realization of structural damage alarming system for the Runyang Yangtse River Bridge. *Proc 4th China–Japan–US Symposium on Structural Control and Monitoring*, Hangzhou.

152. Demetriou MA and Hou Z. 2003. On-line fault/damage detection schemes for mechanical and structural systems. *J Struct Control*, 10, 1–23.

153. Lee SR, Noh JH, Jeong KY, and Shin CG. 2004. Estimation of Rockfill Dam core damage area by data monitoring and numerical analysis. In *Advanced Smart Materials and Smart Structures Technology*, FK Chang, CB Yun, and BF Spencer, Jr (eds.). DEStech, Lancaster.

154. Haase WC and Drumm MJ. 2002. Detection, discrimination and real-time tracking of cracks in rotating disks. *Nondestructive Evaluation and Health Monitoring of Aerospace Materials and Civil Infrastructures*, SPIE 4704.

155. Hochmann D and Duke D. 2005. Aircraft structural health monitoring, a vision. In *Structural Health Monitoring*, FK Chang (ed.). DEStech, Lancaster.

156. Hertzberg RW. 1989. *Deformation and Fracture Mechanics of Engineering Materials*, 3rd (ed.). Wiley, New York.

157. Petroski H. 1992. *To Engineer is Human*. Vintage Books, New York.

158. Levy M and Salvadori M. 1992. *Why Buildings Fall Down*. Norton, New York.

159. Bogner MS. 1994. *Human Error in Medicine*. Lawrence Erlbaum, Hillsdale.

160. Nicastro DH. 1997. *Failure Mechanisms in Building Construction*. ASCE Press, New York.

161. Perrow C. 1999. *Normal Accidents*. Princeton University Press, Princeton.

162. Heinrich HW and Granniss ER. 1959. *Industrial Accident Prevention*. McGraw-Hill, New York.

163. Gano DL. 1999. *Apollo Root Cause Analysis*. Apollian Publications, Yakima.

164. Wilson PF, Dell LD, and Anderson GF. 1993. *Root Cause Analysis*. ASQ Quality Press, Milwaukee.
165. EQE International, Inc. 1999. *Root Cause Analysis Handbook*. Government Institutes, Rockville.
166. Reason J. 1990. *Human Error*. Cambridge University Press, Cambridge.
167. Hwang HH. 1998. Seismic evaluation of electric substation using event tree/fault tree technique. In *Structural Engineering Worldwide*, GL Fenves, NK Srivastava, AH Ang, RG Domer (eds.). Elsevier, Amsterdam, T123-5.
168. NASA. 2002. *Software for Analyzing Root Causes of Process Anomalies*, TSP KSC-12142, US National Aeronautics and Space Administration.
169. Kontic B and Kontic D. 2006. Weaknesses in risk assessment scenario development aimed for spatial planning. *Computational Models of Risks to Infrastructure*, NATO Advanced Research Workshop, Primosten.
170. Doerner D. 1989. *The Logic of Failure*. Perseus Books, Cambridge.
171. Ellingwood BR, Smilowitz R, Dusenberry DO, Duthinh D, Lew HS, and Carino NJ. 2007. *Best Practices for Reducing the Potential for Progressive Collapse in Buildings*, NISTIR 7396, US National Institute of Standards and Technology.
172. Sasani M, Bazan M, and Sagiroglu S. 2007. Experimental and analytical progressive collapse evaluation of actual reinforced concrete structure. *ACI Struct J*, 104(Nov-Dec), S69.
173. O'Brien E. 2004. Structural health monitoring of damage tolerant civil aircraft structures. *Proc 2nd European Workshop on Structural Health Monitoring*, DEStech, Munich.
174. Nair RS. 2004. Progressive collapse basics. *Mod Steel Constr*, (March), 44(3), 37–42.
175. Shenton III HW and Zhao L. 2003. Dead load redistribution due to damage in a large moment frame. In *Structural Health Monitoring*, FK Chang (ed.). DEStech, Lancaster.
176. Shipe JA and Carter CJ. 2003. Defensive design. *Mod Steel Constr*, (November), 43(11), 23–26.
177. Ettouney M, Smilowitz R, and Rittenhouse T. 1996. Blast resistant design of commercial buildings. *ASCE Pract Periodical Struct Des Constr*, 1(1), 31–39.
178. Halling MW, Barr PJ, Womack KC, and Bott SP. 2004. Diagnostic load testing of a continuous, curved steel girder highway bridge. *Proc IABMAS'04 Bridge Maintenance, Safety, Management and Cost*, Kyoto. Taylor & Francis, London.
179. Lai LL and Ressler PR. 2000. NDT and NDE on an I-95 Viaduct. In *Structural Materials Technology IV*, S Allampalli (ed.). Technomic Press, Lancaster.
180. Quinn P, Chajes M, Mertz D, Zoli T, and Volk J. 2004. Inelastic response of a steel girder bridge. *Proc IABMAS'04 Bridge Maintenance, Safety, Management and Cost*, Kyoto. Taylor & Francis, London.
181. Lee ZK, Chang KC, Loh CH, and Chen CC. 2004. Cable force analysis of Gi–Lu cable-stayed bridge after Gi–Gi earthquake. *Proc IABMAS'04 Bridge Maintenance, Safety, Management and Cost*, Kyoto. Taylor & Francis, London.
182. Bažant ZP and Verdure M. 2007. Mechanics of progressive collapse: Learning from World Trade Center and building demolitions. *J Eng Mech*, 133(3), 308–319.
183. Hurd MK. 1985. Formwork safety goes beyond codes and standards. *Concrete Intl*, (April), 7(4), 24–27.
184. Weed LL. 1991. *Knowledge Coupling*. Springer, New York.
185. Niederleithinger E and Helmerich R. 2006. Sustainable bridges—NDT as a tool to assist railway bridge management. NDE *Conf on Civil Eng*, Amer Soc Nondes Test, St. Louis.
186. Cordier MO, Dague P, Lévy F, Montmain J, Staroswiecki M, and Travé-Massuyès L. 2004. Conflicts versus analytical redundancy relations: A comparative analysis of the model based diagnosis approach from the artificial intelligence and automatic control perspectives. *IEEE Trans Syst Man Cybern Part B*, 34(5), 2163–2177.
187. Hanayasu S. 2004. Statistical accident analysis by sequential probability ratio tests in operating systems. *Proc IABMAS'04 Bridge Maintenance, Safety, Management and Cost*, Kyoto. Taylor & Francis, London.
188. Feynman R and Leighton R. 2001. *What Do You Care What Other People Think?: Further Adventures of a Curious Character*. Norton, New York.

189. Feynman R. 1986. Personal observations on reliability of shuttle. *Report of the Presidential Commission on the Space Shuttle Challenger Accident*, NASA, Appendix F.
190. NASA. 2003. *Columbia Accident Investigation Board Report*, Vol. 1. CAIB, August, National Aeronautics and Space Administration.
191. Iverson DL. 2005. Inductive system health monitoring with statistical metrics. NASA Ames Research Center, Moffett Field, CA.
192. Billah KY and Scanlan RH. 1991. Resonance, Tacoma Narrows Bridge failure, and undergraduate physics textbooks. *Am J Phys*, 59(2), 118–124.
193. Scanlan RH and Tomko JJ. 1971. Airfoil and bridge deck flutter derivatives. *J Eng Mech*, 97(6), 1717–1737.
194. Farrar CR, Hemez FM, Park G, Robertson AN, Sohn H, and Williams TO. 2004. An introduction to the Los Alamos Damage Prognosis Project. In *Advanced Smart Materials and Smart Structures Technology*, FK Chang, CB Yun, and BF Spencer, Jr (eds). DEStech, Lancaster.
195. Clark AC. 1968. *2001 A Space Odyssey*. Signet, New York.
196. Johnsen TH, Geiker MR, and Faber MH. 2003. Quantifying condition indicators for concrete structures. *Concrete Intl*, (December).
197. Bond LJ, Jarrell DB, Koehler TM, Meador RJ, Sisk DR, Hatley DD, Watkins KS Jr, Chai J, and Kim W. 2003. On-line intelligent self-diagnostic monitoring system for next generation nuclear power plants. Nuclear Energy Research Initiative Project No. 30344, Final Project Report, Pacific Northwest National Laboratory, Richland.
198. Roemer MJ, Byington CS, Kacprzynski GJ, and Vachtsevanos G. 2005. An overview of selected prognostic technologies with reference to an integrated PHM architecture. *First Intl Forum on Integrated System Health Engineering and Management in Aerospace*, Napa.
199. Robertson AN and Hemez FM. 2004. Reliability analysis of impact-damaged composites. In *Advanced Smart Materials and Smart Structures Technology*, FK Chang, CB Yun, and BF Spencer, Jr (eds). DEStech, Lancaster.
200. Yao JT and Wong FS. 1999. Symptom based reliability and structural health monitoring. *Proc 2nd Intl Workshop on Structural Health Monitoring*, Stanford, Technomic, Lancaster.
201. Mohanty S, Teale R, Chattopadhyay A, Peralta P, and Willhauck C. 2007. Multivariate statistical analysis technique for predictive structural health monitoring. In *Structural Health Monitoring*, FK Chang (ed.). DEStech, Lancaster.
202. Amey SL, Johnson DA, Miltenberger MA, and Farzam H. 1998. Predicting the service life of concrete marine structures: An environmental methodology. *ACI Struct J*, 92(March–April), S20.
203. Li CQ. 2002. Initiation of chloride-induced reinforcement corrosion in concrete structural members—prediction. *ACI Struct J*, 99(March–April), 2.
204. Rafiq MI, Chryssanthopoulos MK, and Onoufriou T. 2004. Sensitivity analysis of chloride induced deterioration in concrete bridges. *Proc IABMAS'04 Bridge Maintenance, Safety, Management and Cost*, Kyoto. Taylor & Francis, London.
205. Xue PF, Xiang YQ, and Zhang YG. 2005. Endurance life prediction of concrete structures exposed to chloride ion penetration environment. *Proc 4th China–Japan–US Symposium on Structural Control and Monitoring*, Hangzhou.
206. Henriksen CF and Stoltzner E. 1993. Chloride corrosion in Danish bridge columns. *Concrete Intl*, 15(8), 55–60.
207. Zheng JY, Sun GY, and Wang LQ. 1998. A review of development in layered vessels using flat-ribbon-wound cylindrical shells. *Intl J Pressure Vessels Piping*, 75(9), 653–659.
208. Anderson R, Kornecki A, and Rajnicek R. 2007. Life limiting rationale for a level D HUMS utilized for maintenance credits. In *Structural Health Monitoring*, FK Chang (ed.). DEStech, Lancaster.
209. Agnello M. 2007. HUMS system design issues for usage monitoring on older aircraft. In *Structural Health Monitoring*, FK Chang (ed.). DEStech, Lancaster.
210. Le DD. 2007. Federal aviation administration perspectives for rotorcraft structural monitoring usage credits. In *Structural Health Monitoring*, FK Chang (ed.). DEStech, Lancaster.

211. Bolotin VV. 1989. *Prediction of Service Life for Machines and Structures*. ASME Press, New York.

212. Guttman I, Wilks SS, and Hunter JS. 1982. *Introductory Engineering Statistics*, 3rd Ed. Wiley, New York.

213. Kawakami Y and Frangopol DM. 2004. Life prediction of damaged structures of Hanshin expressway. *Proc IABMAS'04 Bridge Maintenance, Safety, Management and Cost*, Kyoto. Taylor & Francis, London.

214. Marotta SA, Kudiya A, Ooi TK, Toutanji HA, and Gilbert JA. 2005. Predictive reliability of tactical missiles using health monitoring data and probabilistic engineering analysis. *First Intl Forum on Integrated System Health Engineering and Management in Aerospace*, Napa.

215. Iarve EV, Mollenhauer D, Whitney TJ, and Kim R. 2006. Strength prediction in composites with stress concentrations: Classical Weibull and critical failure volume methods with micromechanical considerations. *J Mater Sci*, 41, 6610–6621.

216. Pastor-Satorras R, Rubi M, and Diaz-Guilera A. 2003. *Statistical Mechanics of Complex Networks*. Springer, New York.

217. Albert R and Barabasi AL. 2002. Statistical mechanics of complex networks. *Rev Mod Phys*, 74, 47.

218. Dienes JK. 2006. On the mean cluster size of a network of cracks. *Struct Control Health Monit*, 13, 169–189.

219. Murachi Y, Orikowski MJ, Dong X, and Shinozuka M. 2003. Fragility analysis of transportation networks. *Smart Systems and Nondestructive Evaluation for Civil Infrastructures*, SPIE 5057.

220. Righiniotis TD. 2004. Fatigue reliability assessment of bridge details based on maximum load specification. *Proc IABMAS'04 Bridge Maintenance, Safety, Management and Cost*, Kyoto. Taylor & Francis, London.

221. Polidori DC, Beck JL, and Papadimitriou C. 1999. New approximations for reliability integrals. *J Eng Mech*, 125(4), 466–475.

222. Saito T, Mase S, and Morita K. 2005. A probabilistic approach to structural damage estimation. *Struct Control Health Monit*, 12, 283–299.

223. Mayrbaural RM and Camo S. 2004. Guidelines for inspection and strength evaluation of suspension bridge parallel wire cables. NCHRP Report 534, TRB, Washington, DC.

224. Bazovsky I. 2004. *Reliability Theory and Practice*. Dover Publications, Mineola.

225. Yun J and DeWolf JT. 2002. Reliability assessment of highway truss sign supports. *J Struc Eng*, 128(11), 1429–1438.

226. Dong XJ and Shinozuka M. 2003. Performance analysis and visualization of electric power systems. *Smart Systems and Nondestructive Evaluation for Civil Infrastructures*, SPIE 5057.

227. Shibata A. 2006. Estimation of earthquake damage to urban systems. *Struct Control Health Monit*, 13, 454–471.

228. Bankowski E, Miles C, and Saboe MS. 2003. Health monitoring and diagnostics of ground combat vehicles. *Modeling, Signal Processing and Control*, SPIE 4049.

229. Alan A and Fujino Y. 2004. Developing a web-based simulation system for bridge health monitoring and maintenance. *Proc IABMAS'04 Bridge Maintenance, Safety, Management and Cost*, Kyoto. Taylor & Francis, London.

230. Kiddy JS. 2003. Remaining useful life prediction based on known usage data. *Nondestructive Evaluation and Health Monitoring of Aerospace Materials and Composites II*, SPIE 5046.

231. Garavaglia E and Sgambi L. 2004. The credibility of lifetime assessment based on few experimental measurements. *Proc IABMAS'04 Bridge Maintenance, Safety, Management and Cost*, Kyoto. Taylor & Francis, London.

232. Ueda K and Hirose T. 2004. Development of a periodic inspection system. *Proc IABMAS'04 Bridge Maintenance, Safety, Management and Cost*, Kyoto. Taylor & Francis, London.

233. Wang X, Foliente G, Nasseri S, and Ye L. 2004. Lowe bridge condition deterioration modeling based on data from level two inspection. *Proc IABMAS'04 Bridge Maintenance, Safety, Management and Cost*, Kyoto. Taylor & Francis, London.

234. Au SK. 2004. Probabilistic failure analysis by importance sampling Markov chain simulation. *J Eng Mech*, 130(March), 3.

235. Barabasi AL. 2002. *Linked*. Perseus Publishing, Cambridge.
236. Buchanan M. 2002. *Nexus Small Worlds and the Groundbreaking Science of Networks*. Norton, New York.
237. Newman M, Barabasi AL, and Watts DJ. 2006. *The Structure and Dynamics of Networks*. Princeton University Press, Princeton.
238. Crucitti P, Latora V, Marchiori M, and Rapisarda A. 2003. Efficiency of scale-free networks: Error and attack tolerance. *Physica* A, 320, 622–642.
239. Dodds PS, Watts DJ, and Sabel CF. 2003. Information exchange and the robustness of organizational networks. *PNAS*, 100, 12516–12521.
240. Hartsfield N and Ringel G. 1994. *Pearls in Graph Theory: A Comprehensive Introduction*. Dover Publications, Mineola.
241. Bollobas B. 1978. *Extremal Graph Theory*. Dover Publications, Mineola.
242. Blundell A. 1982. *Bond Graphs for Modeling Engineering Systems*. Ellis Horwood, Chichester.
243. Cempel C and Natke HG. 1993. Damage evolution and diagnostics in operating systems. In *Safety Evaluation Based on Identification Approaches*, HG Natke, GR Tomlinson, and JTP Yao (eds). Vieweg, Braunschweig.
244. Hoshiya M and Yamamoto K. 2002. Redundancy index of lifeline systems. *J Eng Mech*, 128(9), 961–968.
245. Proverbio E and Epasto G. 2004. Evaluation of fire damage on a prestressed concrete railway bridge. *Proc IABMAS'04 Bridge Maintenance, Safety, Management and Cost*, Kyoto. Taylor & Francis, London.
246. Parker ER and Zackay VF. 1973. Enhancement of fracture toughness in high strength steel by microstructure control. *Eng Fract Mech*, 5, 147–165.
247. Cook SJ, Till RD, and Pearson L. 2000. Fatigue cracking of horizontal gusset plates at arm-to-pole connection of cantilever sign structures. In *Structural Materials Technology IV*, S Allampalli (ed.). Technomic Press, Lancaster.
248. Cluni F, Gioffre M, and Gusella V. 2004. Damage assessment of the Ponte Delle Torri in Spoleto. *Proc IABMAS'04 Bridge Maintenance, Safety, Management and Cost*, Kyoto. Taylor & Francis, London.
249. Nasser L and Tryon R. 2003. Prognostic system for microstructural-based reliability. In *Structural Health Monitoring*, FK Chang (ed.). DEStech, Lancaster.
250. Zilberstein V, Washabaugh A, Grundy D, and Goldfine N. 2003. Fatigue monitoring and adaptive damage tolerance using embedded eddy-current sensor arrays. In *Structural Health Monitoring*, FK Chang (ed.). DEStech, Lancaster.
251. Grundy P. 2004. Reassessment of a wrought iron railway bridge. *Proc IABMAS'04 Bridge Maintenance, Safety, Management and Cost*, Kyoto. Taylor & Francis, London.
252. Suzuki T, Wakatsuki K, Uehira S, Katou H, Sakano M, and Horikawa K. 2004. An investigation on cracking in the rigid steel pier P88 on Osaka–Nishinomiya line. *Proc IABMAS'04 Bridge Maintenance, Safety, Management and Cost*, Kyoto. Taylor & Francis, London.
253. Haroon M and Adams DE. 2007. Damage evolution regression models for prognosis in an automotive sway bar link. In *Structural Health Monitoring*, FK Chang (ed.). DEStech, Lancaster.
254. Testa RB and Yanev BS. 2002. Bridge maintenance level assessment. *Comput Aided Civ Infrastruct Eng*, 17(5), 358–367.
255. Boller C and Biemans C. 1997. Structural health monitoring in aircraft—state-of-the-art, perspectives and benefits. In *Structural Health Monitoring Current Status and Perspectives*, F.-K. Chang, (ed.). Technomic Publishing, Lancaster.
256. Das AK. 1997. *Metallurgy of Failure Analysis*. McGraw-Hill, New York.
257. AASHTO. 1990. *Guide Specifications for Fatigue Evaluation of Existing Steel Bridges*.
258. Halstead JP, O'Connor JS, and Szustak PW. 1999. Fatigue evaluation through field testing. *Proc ASCE Structures Congress*, RR Avent and M Alawady (eds.). New Orleans.
259. Shimizu H, Isoda A, Yamagami T, and Kawakami Y. 2004. Fatigue evaluation and reinforcement for cracks of box girder bridge with steel deck based on actual live loads. *Proc IABMAS'04 Bridge Maintenance, Safety, Management and Cost*, Kyoto. Taylor & Francis, London.

260. Endo T, Mitsunaga K, Takahashi K, Kobayashi K, and Matsuishi M. 1974. Damage evaluation of metals for random or varying loading. *Proc 1974 Symposium on Mechanical Behavior of Materials*, Kyoto.

261. Brower DV. 2003. Real-time fatigue monitoring of deepwater drilling and oil production risers using fiber-optic sensors. In *Structural Health Monitoring*, FK Chang (ed.). DEStech, Lancaster.

262. Leconte R, Goepfer F, Piccardi J, and Cremona C. 2004. Fatigue damage assessment of a welded joint from a composite bridge. Example of the Saint-Vallier Bridge. *Proc IABMAS'04 Bridge Maintenance, Safety, Management and Cost*, Kyoto. Taylor & Francis, London.

263. Hock V, McInerney M, Morefield S, Majumdar A, and Carlyle J. 2005. Investigating fundamental mechanics of microcracks in concrete using acoustic signature modeling. In *Structural Health Monitoring*, FK Chang (ed.). DEStech, Lancaster.

264. Sun L and Hudson WR. 2005. Probabilistic approaches for pavement fatigue cracking prediction based on cumulative damage using Miner's law. *J Eng Mech*, 131(May), 5.

265. Khalak A. 2007. Rapid prediction of remaining life probability distribution based on uncertain environmental conditions under a damage accumulation rule. In *Structural Health Monitoring*, FK Chang (ed.). DEStech, Lancaster.

266. Pandey MD and van Noortwijk JM. 2004. Gamma process model for time-dependent structural reliability analysis. *Proc IABMAS'04 Bridge Maintenance, Safety, Management and Cost*, Kyoto. Taylor & Francis, London.

267. Dufrense F, Gerber HU, and Shiu ES. 1991. Risk theory with the gamma process. *Astin Bull*, 21, 2.

268. Abdel-Hameed M. 1975. A gamma wear process. *IEEE Trans Reliab*, 24(2), 152–153.

269. Bakker JD and van Noortwijk JM. 2004. Inspection validation model for life-cycle analysis. *Proc IABMAS'04 Bridge Maintenance, Safety, Management and Cost*, Kyoto. Taylor & Francis, London.

270. Williams TO, Beyerlein IJ, and Tippetts TB. 2003. Physics based damage modeling of laminated composite plates. In *Structural Health Monitoring*, FK Chang (ed.). DEStech, Lancaster.

271. Venson AR and Voyiadjis GZ. 2001. Damage quantification in metal matrix composites. *J Eng Mech*, 127(3), 291–298.

272. He X, Chen Z, and Huang F. 2004. Fatigue damage reliability analysis for the Nanjing Yangtze River Bridge with the structural health monitoring data. *Proc 2nd European Workshop on Structural Health Monitoring*, DEStech, Munich.

273. Lin YK. 1976. *Probabilistic Theory of Structural Dynamics*. McGraw-Hill, New York.

274. Newland DE. 1993. *An Introduction to Random Vibrations, Spectral, and Wavelet Analysis*, 3rd Ed. Longman House, London.

275. Abdul-Aziz A, Baaklini G, and Bhatt R. 2002. Non destructive evaluation of ceramic matrix composites coupled with finite element analyses. *Nondestructive Evaluation and Health Monitoring of Aerospace Materials and Civil Infrastructures*, SPIE 4704.

276. Achenbach JD, Moran B, and Zulfigar A. 1997. Techniques and instrumentation for structural diagnostics. In *Structural Health Monitoring Current Status and Perspectives*, F.-K. Chang, (ed.). Technomic Publishing, Lancaster.

277. Lopatin C and Mahmood S. 2005. Use of neural nets and embedded sensors for health monitoring of solid rocket motor propellant. In *Structural Health Monitoring*, FK Chang (ed.). DEStech, Lancaster.

278. Doctor SR, Schuster GJ, and Simonen FA. 2000. Density and distribution of fabrication flaws in the shoreham reactor vessel. *Nondestructive Evaluation of Highways, Utilities, and Pipelines IV*, SPIE 3995.

279. Lin KY, Du J, and Rusk D. 2000. *Structural Design Methodology Based on Concepts of Uncertainty*, NASA/CR-2000-209847.

280. Chelidze D and Cusumano JP. 2004. A dynamical systems approach to failure prognosis. *J Vib Acous*, 126(January), 1.

281. Cusumano J, Chelidze D, and Chatterjee A. 2003. General method for tracking the evolution of hidden damage or other unwanted changes in machinery components and predicting remaining useful life. US Patent 6,567,752.

282. Talbott CM. 2002. Condition-based prognosis for machinery. US Patent 6,411,908.
283. Shiao M. 2003. Risk forecasting and updating for damage accumulation processes with inspections and maintenance. In *Structural Health Monitoring*, FK Chang (ed.). DEStech, Lancaster.
284. Fillipenko A, Brown S, and Neal A. 2002. Vibration analysis for predictive maintenance of rotating machines. US Patent 6,370,957.
285. Sarhan HA, O'Neill MW, and Hassan KM. 2002. Flexural performance of drilled shafts with minor flaws in stiff clay. *J Geotech Geoenviron Eng*, 128(12), 974–985.
286. Shannon CE and Weaver W. 1949. *The Mathematical Theory of Communication*. University of Illinois Press, Urbana and Chicago.
287. Adami C. 2004. Information theory in molecular biology. *Phys Life Rev*, 1, 3–22.
288. Hunt SR and Hebden IG. 2001. Validation of the Eurofighter Typhoon structural health and usage monitoring system. *Smart Mater Struct*, 10, 497–503.
289. Feliu S, Gonzalez JA, and Andrade A. 1996. Electrochemical methods for on-site determinations of corrosion rates of rebars. *Techniques to Assess the Corrosion Activity of Steel Reinforced Concrete Structures*, ASTM STP 1276.
290. Tarighat A. 2004. Application of neural network and Monte Carlo method for concrete durability analysis. *Proc IABMAS'04 Bridge Maintenance, Safety, Management and Cost*, Kyoto. Taylor & Francis, London.
291. Schmidt-Dohl F, Bruder S, and Budelmann H. 2004. Monitoring and prognosis of concrete durability under chemical attack. *Proc 2nd European Workshop on Structural Health Monitoring*, DEStech, Munich.
292. Withiam JL, Fishman KL, and Gaus MP. 2002. Recommended practice for evaluation of metal-tensioned systems in geotechnical applications. NCHRP Report 477, National Academy Press, Washington, DC.
293. Lane R and Fishman KL. 2006. Condition assessment and service life estimates for rock reinforcement: Case study. Paper 06-0878, 86th Annual Meeting, TRB, Washington, DC.
294. DeWolf J, Mao J, and Virkler C. 2002. Non-destructive monitoring of bridges in Connecticut. *Proc Structural Materials Technology V, an NDT Conference*, Cincinnati.
295. Zilch K, Penka E, and Buba R. 2004. Fatigue resistance evaluations for existing post-tensioned concrete bridges. *Proc IABMAS'04 Bridge Maintenance, Safety, Management and Cost*, Kyoto. Taylor & Francis, London.
296. Kristek V, Bazant ZP, Zich M, and Kohoutkova A. 2006. Box girder bridge deflections: Why is the initial trend deceptive?. *Concrete Intl*, (January).
297. Chen CC, Chou CC, Chang KC, and Wu WH. 2004. The determination of cable force of Chi-Lu cable-stayed bridge. *Proc IABMAS'04 Bridge Maintenance, Safety, Management and Cost*, Kyoto. Taylor & Francis, London.
298. Achter JL, Halling MW, and Womack KC. 2000. Full-scale reinforced concrete bridge bent condition assessment using forced-vibration testing. *Nondestructive Evaluation of Highways, Utilities, and Pipelines IV*, SPIE 3995.
299. Robinson MJ, Halling MW, and Womack KC. 2000. Condition assessment of a six span full-scale using forced vibration. *Nondestructive Evaluation of Highways, Utilities, and Pipelines IV*, SPIE 3995.

9

Structural Performance Assessment

Structural performance is the ability of a structure to behave in a specified manner under defined loading conditions. Structural performance assessments can provide useful information to owners, end-users, and designers of structures [1,2]. Examples of performance assessments include measurements of the ability to retain a particular geometric configuration to within tolerances, and sustaining specified levels of serviceability, durability, and reliability under normal, extreme, or specified operating conditions [3]. The assessment of structural performance due to normal operational loads may be merely a matter of analyzing routine measurements of normal operations. If the specified loads in the performance criteria occur only rarely, such as during extreme events, then performance assessments may use a combination of low-load experiments, numerical modeling, rating, and other techniques to estimate extreme event performance capabilities.

9.1 Geometric Configuration

Changes in geometry over time and under load are important structural performance quantifiers. Many structural performance specifications pertain to geometric configuration, that is, tolerances placed on dimensions, angles and surface shapes, and connectivity. Geometric assessments can use simple instruments such as liquid levels, plumb bobs, rulers, and tape measures, or they can use more sophisticated devices including electronic surveying instruments, laser scanners, wireless rangefinders, and tiltmeters.

Many types of structures have standardized geometric performance specifications [4]. An example is ACI 117, which specifies concrete geometric tolerances [5]. Measuring long-term trends in geometric stability requires recording and registering the geometric configurations with sufficient precision to be useful. Most structures undergo shape changes due to operational effects, seasonal temperature changes, and aging. The normal operational deformations must be identified so that they can be differentiated from abnormal deformations and may indicate underlying problems [6]. For example, Shahrooz et al. detected abnormal movements of FRP bridge deck slabs due to deficiencies in the underlying deck support system [7]. Since rail geometry is such a crucial parameter in the safe operation of railroads, automated rail geometry assessment methods are well developed [8,9]. These instruments can detect excessive wear on the rails, bulk damage, and gross geometric defects that can lead to derailments. Krawinkler indicates that measuring interstory drift in buildings during earthquakes could provide a significant boost to *performance-based earthquake engineering* assessments [10]. At present there are few effective sensing solutions for interstory drift measurements.

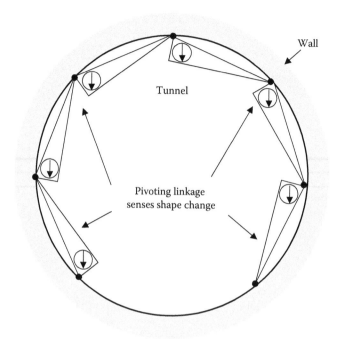

FIGURE 9.1 Bassett convergence system for tunnel shape monitoring with linked array tiltmeter array. (Photograph courtesy of the Durham Geo Slope Indicator, Inc.)

Examples of automated structural geometry assessment instruments abound. These include long gage white light interferometers, GPS instruments, scanning laser instruments, and tiltmeters [11,12]. Figure 9.1 shows a tunnel cross section shape monitoring system, known as the *Bassett Convergence System*. An array of tiltmeter sensors located on the perimeter and connected through a series of pivoting mechanical links provides the data necessary to form an estimate of the cross-sectional shape. Similar measurements with an array of FBGs or other deformation sensors are possible [13,14]. Concrete wall thickness measurements are mandated on all new tunnels constructed in Germany [15]. Kim et al. measured creep column shortening effects in the Petronas Tower in Malaysia with vibrating wire strain gages [16].

9.2 Construction Sensing Systems

Construction is probably the most important life phase of a structure in terms of potential opportunities to contain costs, affect safety, and improve reliability and long-term performance. Errors in construction can degrade structural performance, pose safety hazards to construction workers, and cause premature failures. The construction phase also presents a prime opportunity for installing sensors and sensor networks for health and performance monitoring.

As with completed structures, the most common structural measurements during construction verify geometric configuration. Measuring displacement and position during construction routinely makes use of surveying instruments, levels, plumb bobs, and more sophisticated electronic and optical devices, such as laser levels and EM positioners [17]. An example application of such techniques is a study of dam deformations during construction by Vanadit-Ellis et al. [18]. The measured loads and deformation behavior compared favorably with calculated values, and lent confidence to the integrity of the overall construction process.

Structural fabrication with materials that change from compliant fluids into solid structural forms, for example, RC- or fiber-reinforced composites, is a process that when properly implemented can produce high-performance structures at relatively low cost. Conversely, improper techniques are difficult and expensive to use for correcting defects. These fabrication processes may benefit greatly from structural monitoring activities.

Building RC structures typically relies on temporary formwork. Economical construction requires removing formwork as quickly as possible, without compromising the geometric or safety integrity of the structure. An important consideration is that concrete mixes have temperature-dependent cure rates that vary from batch to batch. Onsite test procedures normally establish the state of cure by using the traditional *test cylinder* method that casts cylinders of the same concrete as it is placed, subjects the cylinders to the same temperature environment as the curing structure, and then tests the strength of the cylinders at specified ages, for example, 7 days. While the test cylinder method is a long-standing staple of construction practice, it may be overly conservative with the result being an inefficient, but fairly safe, construction sequence.

The *concrete maturity method* (CMM) is an instrument-based alternative to test cylinders [19,20]. This procedure pretests a set of cylinders to develop a model of the temperature versus time-rate-of-cure of a particular concrete mix. Combining onsite temperature measurements with pretest cylinder data provides an estimate of the state of cure in the concrete on a real-time basis. This may allow for a more rapid—yet still stable and reliable— construction sequence. CMM enables the timely removal of the concrete formwork without compromising on safety and more rapid whitetopping repaving operations [21–23]. Using electric penetrometers to control the quality of shotcrete placement is another application of real-time measurements facilitating concrete construction [24].

Excessive heat generation due to hydration during the construction of large concrete structures can adversely affect long-term performance. Measurements of internal temperatures generally require multipoint sensing in order to capture the gradients. As an example, Jalinoos and Haramy monitored the temperature of large drilled concrete bridge pier as the concrete cured [25]. The measurements used thermistors. Moving the thermistors along a water-filled shaft provided the distributed temperature measurements.

Fabricating high-performance reinforced polymer composite structural components is a sophisticated process involving changes in the physical states of the polymers at the nano- and microscales, and of the entire composite at the meso- and macroscales. Many opportunities exist to make effective use of monitoring instruments. Measuring the state of cure is a principal consideration. Ultrasound and capacitive cure measurements rely upon changes in the mechanical and electrical properties of the polymer as it cross links. The mechanical properties transition from a viscous liquid state to a gel and then into a solid. The result is an increase in the ultrasound transmissibility and a decrease in the attenuation. The electrical properties also change due to cure. Polymer cross-linking reduces the mobility of dipole molecules. This causes a measurable reduction in bulk capacitance [26]. AE measurements can form an integral part of the initial proof testing of the structure, both following cure and before placing into service. When the number of AE events exceeds the preset threshold levels during a proof test, then it is likely that the strength is deficient.

Measuring critical loads during construction is fairly commonplace. Examples include placing load cells in the load path of rock anchors and posttensioning strands. Temporary structures, such as shores or stay cables, are also excellent opportunities for the effective use of load measurements [27–29]. Tanaka et al. developed a methodology to adjust cable tensions during cable-stayed bridge construction based on real-time stay tension and overall bridge geometry measurements [30]. Lagoda and Olaszek used strain, displacement, and photoelastic techniques to evaluate stress levels during a novel bridge construction process with prefabricated steel plate girders with composite concrete decks [31]. Turer and Ozerkan measured strains that reach 60% of the allowable capacity during the construction of a cable-stayed pedestrian bridge [32]. Gómez et al. measured strain, temperature, wind speed, and displacement during the erection of the Chiapas cable-stayed bridge in Mexico [33]. A detailed numerical analysis of the anticipated construction sequence loads and deflections guided the design of the sensing system so as to be likely to measure the critical values. The numerically predicted behavior agreed quite well with the measured values. Sandford et al. measured stresses in H-piles during the construction of bridges with integrated abutment piers [34]. A solar powered array of strain gages, inclinometers, extensometers, earth pressure cells, and survey points provided the necessary data.

Failing to detect construction defects before embarking on subsequent steps can amplify the defect costs significantly. Concrete foundations are an example. Lew et al. evaluated several assessment techniques for RC foundations during construction that included gamma–gamma logging, cross-hole sonic logging, sonic–echo testing, and impulse response [35]. Cross-hole sonic logging is the most effective method for assessing pile integrity. This method requires the preplacing of sensor access tubes (PVC pipes) into the piling formwork before placing the concrete [36]. Gucunski et al. used a similar impact echo cross-hole forced vibration test procedure for evaluating drilled shaft bridge pier foundations [37]. Helmicki et al. measured the stresses during and immediately following the construction of a steel stringer bridge [38]. The measured stresses were higher than those expected from design calculations but remained well within safety limits. It is unclear at this point whether the members subjected to large construction stresses will be more prone to future damage. This will require long-term monitoring for a definitive answer.

Measuring soil behavior during geotechnical construction can produce a better-finished product at a reduced cost. An example is the monitoring of soil elastic moduli during compaction operations [39]. Measuring structural behavior before, during, and after rehabilitation can provide useful information regarding the overall structural behavior, the behavior of specific troublesome conditions, and the performance of the repair [40]. Monitoring stress waves as they propagate down the piles can indicate the strength and properties of the surrounding soil [41]. This technique has been codified in ASTM D 4945-96. Interpreting the data requires knowledge of whether the soil is sandy or made of clay. Pore water pressure due to soil conditions significantly affects the long-term mechanical behavior of piles [42].

9.3 Operational Load Performance Assessment

Structural performance criteria limit the ranges of allowable responses to external and internal loads. Assessing operational load structural performance usually requires onsite measurements and direct comparisons of applied loads and responses with performance criteria such as the comparison of dead and live loads in concrete box-girder bridges calculated and measured thermally by Thompson et al., and in prestressed concrete bridge girders by Onyemelukwe et al., and soil pressures on integrated bridge abutments by Kerokoski and Laaksonen [43–47].

The amount of data gathered and the duration of operational load performance tests are always important parameters to consider in the planning and design performance assessments. A typical measurement is a short-term process that attempts to extrapolate, often without justification, to longer-term performance assessments. A statistically rigorous approach collects pilot measurements during the initial phase to assess the variability of the data, and then plans and conducts more thorough tests based on the measured level of variability. A complication is that normal operating conditions tend to vary with time (day, month, season, or year). Ideally, both the pilot and subsequent more thorough measurements should span the variation of normal operating conditions. Such an approach is often impractical due to economic pressures and other constraints. Nonetheless, longer-term measurements are particularly useful when important load events occur only rarely, or when the damage accumulates slowly over long periods of time. For example, Yamamoto et al. report on a 5-year monitoring of a building equipped with active mass dampers to resist motions during strong winds and earthquakes [48]. Since this building may have a useful lifetime of decades, or even centuries, the quite substantial 5-year monitoring effort is actually only a moderately sized sample of the early age operational behavior of the building. Rainieri et al. assessed the long-term vibration performance assessment of a building using a continuous modal measurement and analysis approach [49].

9.3.1 Acoustics

Structural acoustic performance is of particular concern to occupants. Excessive levels of noise can damage hearing, cause discomfort, distract, and otherwise annoy the inhabitants. Structures are generally expected to limit the amount of noise transmitted to and from the environment [50]. Certain specialized structures, such as concert halls, should transmit high-fidelity sound. The performance criteria of most other structures are to limit sound transmission and acoustic noise levels. Assessing both the noise abatement and the acoustic projection performance of structures requires specialized sound measurement techniques [51]. Calibrated microphones measure sound at a point, often with directional and frequency dependence. Specialized frequency-domain weighting filters correct the raw acoustic signals to account for human sensitivity [52–54]. The A-band contour filter is a band pass filter with a center frequency of approximately 2300 Hz. The A-band filter design of this is to match the sensitivity of the human ear. The standard notation for A-band filter sound power readings is decibels A-band (dBA).

9.3.2 Corrosion and Chemical Attack

Corrosion is a comprehensive term that encompasses a collection of often-complicated diffusion and metal-oxidizing processes [55]. As such, it is inherently difficult to assess corrosion levels and corrosion resistance with a single measurement. Many situations favor the combination of multiple measurement methods of corrosion assessment. Possibilities include half-cells, chemical assays, visual inspections, and remote-machine visual inspection. An example is the notion that RC corrosion resistance performance of concrete often depends directly on the resistance to penetration by chloride ions due to deicing salts or ocean spray [56]. In a postmortem materials analysis, Kassner et al. found that the decreased permeability of lightweight concrete performed very well on the Woodrow Wilson Bridge deck [57]. Chamberlin and Weyers measured the long-term corrosion performance of bridge decks by a combination of several techniques [58]. Guthrie et al. evaluated the corrosion performance of reinforcing bars in concrete bridge decks with stay-in-place formworks with half-cell potential measurements [59]. Additional performance measurements included Schmidt hammer rebound, chain drag, and visual inspection for distress and chloride concentration. El Maaddawy et al. evaluated the performance of CFRP repairs on RC beams in aggressively corrosive environments [60]. Hayashi et al. evaluated the ability of a set of bridges constructed of weathering steel to resist salt attacks by a combination of detailed estimates of the salt loads and an assessment of the levels of corrosion [61]. In a series of laboratory tests, Hristova showed that the ultimate strength of RC concrete beams degraded linearly with the diameter loss due to corrosion [62].

Porous materials, such as concrete, are especially vulnerable to the diffusion of chemical reagents [63]. Alkali–silica reactions in concrete are an example. Alkali interacts with silica aggregates and leads to excessive expansions and often to severe damages in concrete structures. Mukhopadhyay et al. suggest a performance-based prediction of concrete for *alkali silica reactivity* (ASR) in the power law form [64]:

$$\varepsilon = \lambda \varepsilon_0 \exp\left[-\left(\frac{\alpha}{t_e}\right)^{\beta}\right], \tag{9.1}$$

where

$$t_e = \gamma_H \sum_0^t \Delta t \exp\left[-\left(\frac{E_a}{R}\right)\left(\frac{1}{273 + T} - \frac{1}{273 + T_R}\right)\right], \tag{9.2}$$

where ε is the expansion of the concrete; ε_0 is the ultimate ASR expansion; α, β, λ, γ_H, and E_a are constants; R is the universal gas constant, $8.314\,J/mol^\circ K$; t_e is an equivalent time; and T and T_R are temperature and reference temperature in $^\circ$C, respectively. Hossain used standard concrete permeability tests to evaluate the high-temperature performance of pumice concrete [65]. Air entrainment improves the freeze–thaw capability of concrete. An *air void analyzer* can measure the amount of air entrained in fresh concrete [66].

9.3.3   Moisture

Moisture emissions from concrete slabs affect the performance of flooring overlays, such as carpeting. A standard measure is the *moisture–vapor emission rate* (MVER). Many applications use a performance specification that requires keeping the MVER below 170 or at 280 $\mu g/(s\text{-}m^2)$ [67]. ASTM F 1869 specifies an MVER measurement procedure based on weight changes of calcium chloride samples due to the moisture–vapor absorption of the emissions from a floor slab. Qiu and Yang developed reversible electrical-resistance-based humidity sensors that used ZnO nanometer sized tetrapods as the sensing material [68].

Moisture affects the long-term performance of pavements. Rainwater et al. evaluated the ability of several instruments and laboratory techniques to measure moisture in pavements over a 5-year period [69]. The test instrumentation included a weather station, ETDR, pavement and subgrade temperature sensors, free-drainage pan lysimeters, and resistivity probes. The ETDR probes consistently found higher moisture levels than those determined by laboratory-oven-dried water content measurements. Careful ETDR calibrations based on soil and temperature effect compensations did not entirely eliminate the moisture measurement discrepancy. A conclusion was that confidence in the measurements improves with the use of multiple instrument types. Janoo et al., Kotdawala et al., Look et al., and van Schelt et al. report similar ETDR studies of highway and airport pavements [70–73]. Kassem et al. developed a relative humidity suction technique for measuring moisture in samples of drying asphalt with thermocouples [74]. Healy and van Doorn demonstrated SAR imaging with UWB radar to determine and locate moisture levels in buildings [75]. Pervious concretes are a relatively new paving material that have a large permeability. A principal application is to mitigate rainwater runoff. A series of studies are being undertaken to establish the long-term durability and performance of these materials [76].

9.3.4 Floor Vibrations

Floor vibrations adversely affect people and delicate machines in buildings and vehicles. Maintaining suitably low levels of floor vibrations can be an important performance parameter for many buildings and vehicular structures. Controlling floor vibrations typically requires combining measurements with performance specifications and the application of vibration mitigating designs [77].

Accelerometers or other similar instruments can readily measure floor vibrations. Transducer sensitivity is often an issue. Many floor vibrations of interest are low frequency ($\sim$0–10 Hz) and low amplitude ($\lesssim$0.001 g). Piezoelectric accelerometers, when properly selected, are often adequate. A more expensive, but presently more sensitive, option is to use servo-electric accelerometers with large inertial masses.

The nonstationary nature of floor vibration signals requires the use of specialized signal processing techniques. A standard statistical approach is to use short-term Fourier analysis methods with sliding start times. Advantages of the short-term Fourier transform (STFT) approach are the ease of implementing the fast Fourier transform (FFT) and outcome statistics in standard Fourier formats. An alternative specialized approach calculates a single scalar as the fourth root of the mean fourth power of an acceleration time history that has passed through a frequency-domain filter that matches the vibration sensitivity of humans. The effect is to weigh the larger-amplitude shock-type vibrations more heavily than would be the case with a square root mean square (RMS) statistic. The rationale of such an approach arises in the common experience of hitting potholes while riding in a vehicle. An occasional large pothole is more annoying and damaging to the vehicle than riding on a road with more stationary roughness, that is, rumble, even though both have the same RMS acceleration and power spectra.

The perception, comfort level, and injury response of humans to floor and whole body vibrations are all frequency, amplitude, and duration dependent. Seated humans are particularly susceptible to vertical vibrations in the 4–8 Hz band and to horizontal motions in the 0–2 Hz band [78,79]. ISO 2631 is a standard for measuring and reporting human whole body vibrations in a format that can aid in predicting injury, fatigue, tolerance, and comfort levels [80]. ISO 2631 specifies calculating 1/3 octave acceleration spectra for intervals of stationary vibration power and then compares these spectra with specified limits relative to perceived levels of comfort. A complicated weighting procedure can accommodate most nonstationary signals, with the exception of shock-like motions that have intermittent large accelerations between quiescent periods. The *crest factor* is the ratio of the maximum acceleration in an interval to the RMS of the acceleration. It is common to treat vibration environments with crest factors less than seven or nine as being stationary. Appropriate statistics for assessing human response to mechanical shock remain a matter of research [81].

Floor vibrations also affect the operation of delicate and high-precision machines. Passive vibration isolation systems, such as air tables, attenuate many of the bothersome vibrations at frequencies greater than 10 Hz. This leaves sub-10 Hz vibrations as being of most concern. Potential sources for building vibrations include traffic, underground trains, seismic activity, nearby construction, building mechanical systems (HVAC, plumbing), human movement, air handling, and machinery [82–85]. Pyl et al. developed and validated a source–receiver model for traffic-induced building vibrations [86,87].

The *BBN criteria* (named after the consulting firm Bolt, Berenek, and Neuman) are standard floor vibration assessment criteria for high-precision machinery [88]. These criteria place limits on 1/3 octave spectra of the vibrations (Figure 9.2). One-third (1/3) octave statistics condense a wide variety of vibration waveforms, that is, periodic, random, single frequency, multiple frequency, stationary, and mildly nonstationary, into a reduced set of cumulative spectral values. Highly nonstationary signals, such as impact or blast loads, require different analysis methods. The BBN criteria give five levels (A–E) and suggestions for the use of various machineries: (A) optical microscopes 400×; (B) optical microscopes 1000×; (C) 1 μ feature lithography; (D) TEM's, SEM's, and E-beam writers; and (E) difficult to achieve. If a particular frequency band is of concern, such as 5–10 Hz, then the signals may be filtered through an appropriate band pass filter before tabulating the short-term statistics. A simple frequency-dependent criterion in use at a major semiconductor manufacturing facility is to keep the amplitude of vibrations that occur in the 5–10 Hz range to less than 2.54 μm (100 μ in).

Histograms provide useful estimates of vibration amplitude PDFs. When the vibration histograms are Gaussian, then a single Gaussian source or multiple uncorrelated random sources are likely to be the cause of the vibrations. When the amplitude histograms are non-Gaussian, it may be possible to identify a single coherent source of the vibrations [89,90].

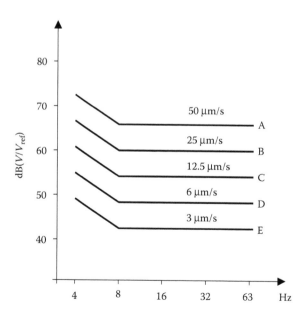

FIGURE 9.2 BBN velocity limits on floor vibrations based on 1/3 octave spectra with $V_{ref} = 1\,\mu$ in/s (25.4 nm/s). (Adapted from Amick H et al. 1991. *Vibration Control in Microelectronics, Optics and Metrology*, SPIE 1619.)

9.3.5 Bridges and Roadways

Fatigue-inducing stresses, displacement levels, and the ability to carry excessive or posted loads are primary bridge performance parameters [91]. Performance assessments of bridges and roadways inevitably require measurement under normal operational loads. Traffic loads generally contain both static and dynamic components. Usually performance criteria for static (or quasi-static) loads dominate decisions regarding the design, rating, and maintenance. In a ca. 2000 study, the California Department of Transportation compared five different methods of measuring the static load performance of pavements [92]. The comparison included variants of the *Benkelman Beam* and the falling weight deflectometer (FWD). FWD testing is often the method of choice due to the convenience of the testing technique.

Short-term strain response assessments can use temporary or quick-attach transducers, such as clamp-on strain gages. Peak strain and fatigue fuse sensors are also possibilities. *Weigh-in-motion* (WIM) technologies measure the weight of traffic and assess the performance of bridges [93–95]. Yamada and Ojio demonstrated that with a judicious placement of strain gages near to the pier and bearing supports, and with a suitable calibration, a WIM system can identify the weight of trucks on the bridge [96]. Yamaguchi et al. showed that it is possible to use WIM systems on bridges with complicated mechanics, such as skewed-approach steel-plate-girder bridges with continuous spans, as long as the load path details are properly considered [97]. Chowdhury found that accelerometers can be as effective as strain gages in determining dynamic load amplification factors for bridges [98]. Strain gage data measured on cracked concrete beams were less reliable, especially those with short gage lengths.

Occasionally small amplitude vibrations are a concern. An important issue is bridges in urban areas. A combination of stopped traffic and stress-constrained designs with high-strength steels creates a situation where safe, but occasionally alarming, vibrations affect motorists [99]. Deflection measurements can combine with other measurements, such as dynamic responses, to produce useful information [100]. For example, Bustos et al. combined FWDs with height differential measuring faultmeters for pavement performance assessments (Figure 9.3) [101].

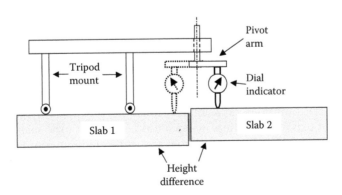

FIGURE 9.3 Faultmeter for measuring height differentials across cracks and faults in pavements. (Adapted from Bustos MG et al. 2006. Evaluation of performance of PCC pavements in mid-Western Argentina, and comparison with LTPP data using HDM-4 distress models. Paper no. 06-2847, 86th Annual Meeting, TRB, Washington, DC.)

9.3.6   Railroads

Maintaining the gage width of railroad rails during the passage of heavy trains is essential to preventing gage-widening derailments. Visual inspection or automated means of determining rail geometry are often inadequate in this regard. Gage widening occurs while the rails are loaded. Specialized rail cars with split axles that provide a lateral load-displacement testing capability can provide a more realistic assessment of the lateral strength performance of railroads under actual load conditions [102,103].

9.3.7   Wind

Wind imposes the principal live load on many structures, including tall buildings and long-span bridges [104]. Localized wind load effects can severely damage structural details, such as windows. More globally distributed wind load effects can cause excessive motions, aeroelastic instabilities, and failures. Bluff structures, that is, bridges, buildings, cables, towers, and so on, may be prone to a variety of aeroelastic behaviors including *buffeting, flutter, galloping, vortex-shedding,* and *wind-rain vibration* phenomena [105–107]. The correct diagnosis of the underlying aeroelastic mechanism directly aids in correcting and retrofitting structures with aeroelastic problems. An example is the analysis of the Vincent Thomas Bridge by He et al. [108]. The analysis developed an aeroelastic model of the dynamics of the bridge with sufficient detail for use in damage identification.

Buffeting is the random pressure loading on a structure due to wind turbulence. Of particular concern to tall buildings is the appearance of localized pressure gusts that can damage windows. Measuring and predicting the behavior of pressure gusts remains a technical challenge. Pressure measurements require careful consideration of the reference pressure in differential transducers. Absolute pressure transducers generally lack the dynamic range necessary both to measure fluctuating wind pressures and to track large barometric pressure shifts during windstorms.

Buffeting-induced swaying in tall buildings can be annoying, long before it reaches levels that cause structural damage [79,80]. Since swaying is typically a low-frequency phenomenon, low-frequency transducers, such as accelerometers, tiltmeters, optical transducers, and GPS instruments, are suitable for measurements. For example, Koike et al. used a GPS to measure the performance of a hybrid damper system in a 51-story building during windstorms [109]. Bosch and Miklofsky report on a long-term study of the buffeting loads on a suspension bridge. The measured buffeting motions correlated with motions predicted from a modal analysis [110]. Xiao and Li measured the motions of the 79-story Di-Wang Tower, located in Shenzhen, during Typhoon Chanzhu in May 2006 [111]. A modified random decrement analysis estimated amplitude-dependent structural damping values that ranged from less than 0.2% for small motions to more than 0.5% for accelerations up to 0.001 g.

Wind loads on tall buildings occasionally cause very expensive glass breakage. The transient and non-Gaussian nature of wind loads complicates the measurement of wind pressures, particularly the peak pressures that are implicated in many cases of glass breakage. Accurate measurements of probability distributions generally require long-duration wind load recordings. When the general nature of the wind load PDF is known, shorter-term measurements combined with parameter fitting may be adequate [112].

Bluff body flutter occurs when the aerodynamic damping on an elongated structure becomes more negative than the inherent structural damping. Some long-span bridges have poor shapes for aerodynamic performance and are particularly prone to flutter. Scale model wind tunnel tests can reliably determine normalized measures of aerodynamic damping for prototypes in the form of flutter derivatives [113].

Field measurements of structural and aerodynamic damping can be technically and financially challenging. Nonetheless, field measurements can provide information that is difficult to obtain otherwise. Since most important civil structures are large, and are generally not easily amenable to forced and/or displacement testing, exceptions include the few structures with built-in active tuned mass dampers for vibration control and the uses of large shakers [114,115]. The damper provides a controlled vibration excitation. Nigbor et al. found that it is possible to use the motions of elevators as the excitation source for buildings [116]. For the more routine case where controlled excitation is impractical, damping measurements may require indirect methods using ambient loads and responses. Possibilities include the analysis of ambient vibration survey (AVS) signals with frequency-domain curve fits, random decrement analysis, and Kalman filter state estimation. A potential problem with AVS is that since the amplitudes of the vibrations tend to be smaller than those encountered in large loading situations, the estimated values of damping may not reflect those that act in a structure subjected to large loads. Uhl and Bogacz developed a real-time method for identifying the damping in airplane wings as a precursor to flutter with an ARMA identification method [117]. The system updated damping value estimates at a rate of a couple times per second.

Galloping appears as large-amplitude and low-frequency oscillations of elongated structures that have an asymmetric cross section [104]. Examples include ice buildup on electricity transmission cables or highway sign bridges. Field measurements and detection of galloping may require SHM operations during ice storms.

Vortex-shedding-induced oscillations occur when the wake downstream from bluff structures forms a series of alternating vortices with a shedding frequency that coincides with a natural frequency of the structure. The shedding frequency, N, is proportional to the wind speed, U, and inversely proportional to the cross-section dimension, D. The *Strouhal number*, S, is the dimensionless proportionality constant defined as

$$S = \frac{ND}{U}. \tag{9.3}$$

Typical values of S are 0.1–0.2.

The *Scruton* number, Sc, is a useful dimensionless number for assessing the level of structural damping and the normalized equivalent for wind-induced oscillations:

$$Sc = \frac{\zeta m}{\rho D^2}, \tag{9.4}$$

where ζ is the damping coefficient, m is the mass (or modal equivalent), ρ is the air density, and D is a cross-sectional dimension [104]. Damping information regarding most structures is fairly scant. An exception is the existence of a comprehensive database of damping measures of buildings in Japan [118]. The bulk of these measured damping ratios lie between 0.5% and 5%.

Large inclined cables with smooth exteriors that are exposed to the elements can suffer from *wind-rain* vibrations [119]. During rainstorms, water beads into rivulets and flows down the cables. The combined mechanics of water rivulets, cable motions, and wind induces large cable vibrations in a manner that is not yet well understood. Visual observations, cable-mounted accelerometers, and video cameras are effective means of identifying and recording the presence of these vibrations. A bit of care should be taken in interpreting the results of the accelerometer measurements, since the higher-frequency modes tend to be amplified as the square of the frequency [120].

Wooden structures have many unique performance issues. These include resistance to moisture ingress, fungal rot, creep, and fire. Long-term multiple measurand assessments may be a superior performance assessment approach [121]. Inappropriate levels of moisture can cause damage and cause fungal growth. Harpster developed a variety of devices to measure humidity, such as a passive wireless capacitance-based humidity sensor [122].

9.4 Extreme Event Performance

Measuring structural performance during extreme events, such as blasts, earthquakes, fire, floods, impacts, heavy traffic, and wind, can provide valuable information. Such measurements are often difficult because extreme events tend to be rare, unanticipated, and of relatively short duration; access is dangerous; line electric power is unreliable; and the dynamic ranges of the measurand and subsequent transducer signals are difficult to predict in advance.

Four strategies for assessing structural performance during extreme events are as follows: (1) direct performance assessment during an extreme event; (2) performance assessment of similar structures or substructures during extreme events; (3) assessment of structural behavior during a subextreme event and extrapolation to extreme event ranges; and (4) postevent damage assessment.

Aggressive test procedures that simulate harsh conditions in a short period of time can be useful for both system development and certification [123]. The goal of many of these methods is to determine the ranges of loads that cause damage and destruction of the object under stress (Figure 9.4). A key aspect is the simultaneous combination of multiple axes and types of loads. The intent is to provoke second-order failure-inducing interactions [124]. Included in these methods are *highly accelerated life testing* (HALT) and *highly accelerated stress screening* (HASS). An example is the combination of six DOF shaking and rapid thermal cycling. Such testing commonly uses non-Gaussian vibration signals with large kurtosis to produce a high level of shock-type loading. The laboratory testing does not faithfully reproduce field conditions. Instead, it reproduces conditions with similar higher-order statistical parameters, such as acceleration Kurtosis [125]. HALT methods attempt to cause component failures that are not anticipated in the design process. HASS is similar to HALT, but it has a different intent. HASS is a screening process for manufactured systems before they are released for usage [126]. HASS is similar in many respects to the burn-in of electronics manufacturing that identifies those parts and systems that are prone to suffering an early life failure. Sen and Shahawy describe a similar accelerated test protocol using temperature, humidity, and load to test FRP concrete bridge decks [127]. Iwata et al. demonstrated the use of the *electric discharging shock force method* of testing RC slabs to determine the crushing resistance of the concrete [128].

Situations arise when it is desired to test with low crest factors as opposed to the large values typical of HALT and HASS. The *Schroeder phased signal generation technique* can produce test signals that match a prescribed power spectrum, yet have low crest factors [129,130]. The technique exploits the low crest factor of individual sinusoids. The Schroeder signal is a sequence of sinusoids with different frequencies that are assembled with phases and amplitudes to give a smooth signal with small peak amplitudes,

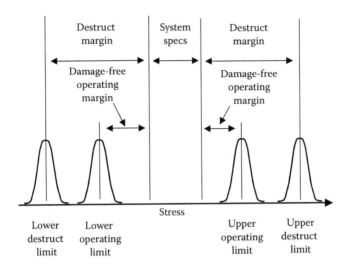

FIGURE 9.4 Probabilistic interpretation of operating without damage and destructive loadings on an object under test. (Adapted from Doertenbach N. 2006. Highly accelerated life testing–testing with a different purpose. *Sound Vib*, 18–23.)

that is, low crest factors. The reduced peak excursions in the time domain minimize the potential for damaging the test structure or driving it into amplitude-dependent nonlinear behavior.

An alternative to testing under extreme loads is to test at subextreme levels and to extrapolate to extreme load performance. While subextreme loading can provide valuable insights, it has difficulty in identifying crucial subsystem and secondary system interactions that may only arise during an extreme event. Examples of subextreme loading include *in situ* static load tests with custom test rigs to assess the blast resistance of windowsills and frames by Ratay, and the use of *in situ* pneumatic four-point bending in strength testers for assessing glass load performance *in situ* on the sides of buildings by Zarghamee et al. [131,132]. Wipf et al. combined subextreme field testing with load-to-failure laboratory testing of precast concrete channel bridges [133]. All of the specimens in the laboratory test exhibited the same failure mode of deck crushing followed by excessive displacement load levels exceeding those predicted with nominal concrete strengths and steel elongations. This predictable failure mode led to increased confidence in the existing rating and evaluation procedures for this particular type of bridge. Kudo et al. measured the dynamic behavior of a passively damped medium-height building in Tokyo using multiple input multiple output (MIMO) system identification techniques [134]. The intent was to evaluate the performance of the building in the potentially much more dynamic circumstances of a strong earthquake. Skolnik et al. conducted similar studies of two buildings in Los Angeles [135]. Of particular interest is the four-story RC Four Seasons Building that suffered significant prior damage in the 1994 Northridge earthquake. For both of these buildings it was found that simple frame models of mass and stiffness were inadequate for providing a detailed prediction of the measured performance.

Laboratory testing of concrete structural elements for earthquake performance is relatively routine due to the appearance of laboratories with the requisite large servo-actuators and test frames. *Pseudodynamic testing* with controlled displacements and measured forces can simulate large earthquake forces without resorting to the use of large shake tables

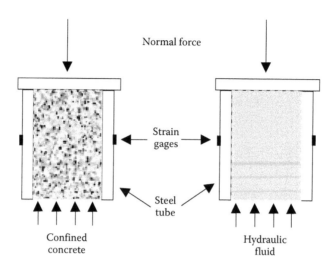

FIGURE 9.5   Hydraulic calibration testing of confined concrete steel tubes. (Adapted from Miao RY and Yang WH. *New Experimental Techniques for Evaluating Concrete Material and Structural Performance*, American Concrete Institute, ACI SP-143-9.)

[136]. As an example, Chung et al. tested the earthquake performance of RC bridge piers built with two different design codes by laboratory-based pseudodynamic techniques [137]. A specialized case arises in the testing of concrete-filled steel tubes. The confining action of the steel tube increases the strength of the concrete under severe loads. Measuring the confining pressure directly is difficult whereas measuring the strain in the steel tube is relatively easy. The strain represents the steel tube deformation along with the elastic–plastic behavior of the concrete. An experimental method of separating the effects is to apply a hydraulic load directly to an equivalent steel tube and to measure the resulting deformation (Figure 9.5) [138].

Earthquake measurements require collecting data, such as strong motion accelerograms, during unpredicted intense, short-duration events. A worldwide effort has been underway to place SHM instruments into buildings and other structures in earthquake regions for examining the performance during and assessing the health after major earthquakes. As an example, Iwaki et al. placed an array of FBG strain sensors in a building fitted with passive dampers for earthquake protection [139]. The sensing system routes onto the Internet for rapid and remote data analysis. The system successfully collected and transmitted building data during a nonmajor earthquake on June 14, 2002.

Column–beam connections in moment resisting buildings were the subject of unexpected brittle fractures in the 1994 Northridge, CA, earthquake [140]. Subsequent studies indicated that this was a common, but undetected, problem in earlier earthquakes. Developing effective methods for detecting this damage and assessing the extent of existing damage remains a technical challenge. A primary difficulty is that architectural details cover most of these connections from sight. Detecting such damage with embedded sensors is an example of one of the basic conundrums of SHM. One approach is to fit the weak joints with damage sensors prior to completing the construction that covers the steel columns and beams. However, if it was known that the columns and beams were so vulnerable, it may have been better to redesign and strengthen the details. Cost-to-benefit ratios less than unity require inexpensive sensors and situations with uncertain, but important and correctable, structural performance.

9.5 Fatigue Performance

Structures subjected to repeated loads that cause stresses to rise above a particular level may suffer from fatigue. Due to the complexities and uncertainties in load paths and levels, it is often not practical to determine the in-service stress loads at fatigue-prone structural details. One approach measures the strain induced in the structural details during a sequence of representative loads. Combining measured loads and lifetime cycle counts forms an improved fatigue estimate [141]. Once it appears that a specific detail is prone to fatigue cracks, a fleet-wide or system-wide examination of similar details is usually warranted. For example, Tokumasu et al. studied a set of fatigue crack-prone beam-to-column connections in a set of traffic hackneyed bridges along the Hanshin Expressway near Osaka [142]. The performance evaluation combined visual inspections with a detailed study of each crack to produce a rational approach to retrofit and management.

9.6 Special and Monumental Structures

The behaviors of novel structural systems are often not well understood. An item of particular concern is the behavior of failure-causing subsystem interactions. As such, novel structural systems inherently have performance uncertainties and are prime candidates for long-term and detailed performance assessments. An example is the use of fiber reinforced composite construction for large civil structures. The structural performance has the potential to be superior to conventional systems, but this needs to be verified [143]. Another example is the use and performance testing of novel bridge deck designs, such as steel-free variants [144]. Guan et al. described a study that monitored a composite highway bridge with multiple accelerometers and other transducers to produce estimates of structural flexibility [145]. The results of the monitoring to date are that the composite bridge has lost some strength, that is, the estimated flexibility has increased. In another example, Nassif et al. examined the in-service performance of new designs for concrete bridge approach slabs [146]. Data from a preconstruction embedded array of load, strain, and temperature sensors determined the amount of shrinking and curling of the slab during concrete cure and early stages of use.

Antique monuments are another class of structures with uncertain performance capabilities that may benefit from SHM. The performance testing of antique monuments is in many respects quite similar to the testing of ordinary structures. An example is the testing of the historic wood arch Kintai Bridge by Yoda that employed a host of modern assessment techniques [147]. Some distinguishing features are that the structural plans and design calculations, along with material properties and specifications, are often nonexistent. There can be severe restrictions on testing. Even slight damage, as in the taking of material samples for further analysis or the hard mounting of transducers, can be severely restricted. Aesthetic considerations related to monitoring instruments can also be severe. It is often necessary to develop specialized structure-specific testing procedures [148]. Bati et al. developed a method where static load-deflection tests can be used to determine the tension in bars that support monumental domes [149].

References

1. Goranson UG. 1997. Jet transport structures performance monitoring. In *Structural Health Monitoring Current Status and Perspectives*, F.-K. Chang, (ed.). Technomic, Lancaster.
2. Feng MQ, Kim DK, Yi JH, and Chen Y. 2004. Baseline models for bridge performance monitoring. *J Eng Mech*, 130(5), 562–569.
3. Sikorsky C, Stubbs N, Bolton R, Choi S, Karbhari V, and Seible F. 2001. Measuring bridge performance using a structural health monitoring system. *Smart Systems for Bridges, Structures, and Highways*, SPIE 4330.
4. Solanki H. 1996. Deflection and minimum thickness requirements of the CEB and ACI Codes. ACI SP 161-2, In *Recent Developments in Deflection Evaluation of Concrete*, EG Nawy (ed.). American Concrete Institute, Farmington Hills, MI.
5. American Concrete Institute. 1990. *Standard Tolerances for Concrete Construction and Materials*, ACI 117–90.
6. Fling RS. 1996. Deflection of a concrete beam and roof slab. ACI SP 161–1, In *Recent Developments in Deflection Evaluation of Concrete*, EG Nawy (ed.). American Concrete Institute, Farmington Hills, MI.
7. Shahrooz BM, Reising RM, Hunt VJ, Helmicki AJ, and Neumann AR. 2002. Testing and monitoring of a five-span bridge with fiber reinforced polymer deck systems. In *Proc Structural Materials Technology V, an NDT Conference*, S Allampalli and G Washer, (eds). Cincinnati.
8. McCarthy WT. 1995. Track geometry measurement on Burlington Northern Railroad. *Nondestructive Evaluation of Aging Railroads*, SPIE 2458.
9. Trosino M, Yang TL, and Kelshaw R. 1995. Developments and applications in the amtrack track geometry measurement car. *Nondestructive Evaluation of Aging Railroads*, SPIE 2458.
10. Krawinkler H. 2006. The role of performance-based earthquake engineering in structural health monitoring and control. *Proc 4th China–Japan–US Symp on Structural Control and Monitoring*, Hangzhou.
11. Inaudi D, Vurpillot S, Glisic B, Kronenberg P, and Lloret S. 1999. Long-term monitoring of a concrete bridge with 100+ fiber optic long-gauge sensors. *Nondestructive Evaluation of Bridges and Highways III*, SPIE 3587.
12. Breuer P, Chmielewski T, Gorski P, and Konopka E. 2002. Application of GPS technology to measurements of displacements of high-rise structures due to weak winds. *J Wind Eng Indus Aero*, 90(3), 223–230.
13. Metje N, Chapman DN, Rogers CD, Kukereka SN, Miao P, and Henderson PJ. 2004. Structural monitoring using optical fiber technology. *Proc Structural Materials Technology VI, an NDT Conference*, Buffalo.
14. Miao P, Kukureka SN, Metje N, Chapman N, Rogers CD, and Henderson P. 2007. Mechanical reliability of optical fibre sensors and SmartRods for tunnel displacement monitoring. *Smart Mater Struct*, 16, 382–390.
15. Kroggel O, Jansohn R, and Pierson R. 2006. Tunnel inspection. *NDE Conf on Civil Engineering*, I Al Qadi and G Washer (eds.). American Society for Nondestructive Testing, St. Louis.
16. Kim W, Oh J, Cho H, and Yom K. 1998. Column shortening of Petronas Towers in Malaysia. In *Structural Engineering Worldwide*, GL Fenves, NK Srivastava, AH Ang, and RG Domer (eds.). Elsevier, Amsterdam, T108-7.
17. Beliveau YJ, Williams JM, King MG, and Niles AR. 1995. Real-time position measurement integrated with CAD: Technologies and their protocols. *J Constr Eng Manage*, 121, 4.
18. Vanadit-Ellis W, Hall RL, and Graham PW. 1989. Analysis of data from instrumentation program: Lock and dam number Red River Waterway, Louisiana. Army Eng Waterways Exp Sta, Vicksburg, MS, *Report*, WES/MP-GL-89-24.
19. Rasmussen RO, Cable JK, and Turner DJ. 2003. Strength measurements using maturity for Portland Cement Concrete Pavement Construction at Airfields. *US Federal Aviation Administration Report*, DOT/FAA-01-G-002-4.

20. Ahmad I, Ali H, Azhar S, Markert L, Vanwhervin D, and Escobar L. 2006. Utilization of maturity method for concrete quality assurance in bridge substructure in Florida. Paper no. 06-2344, 86th Annual Meeting, TRB, Washington, DC.

21. Fox M. 2003. Maturity Method Speeds Form Cycling. *Concrete Intl*, 25(10), 93–96.

22. American Concrete Institute. 2003. *In-place Methods to Estimate Concrete Strength*, ACI 228.1R-03.

23. Jung JS and Cho YH. 2006. The application of maturity method on whitetopping construction. Paper no. 06-2851, 86th Annual Meeting, TRB, Washington, DC.

24. Jolin M, Beaupre D, and Mindess S. 2002. Quality control of d-mix shotcrete during construction. *Concrete Intl*, (October).

25. Jalinoos F and Haramy KY. 2006. Temperature monitoring of drilled shaft foundations using wireline logging. *NDE Conf on Civil Engineering*, I Al Qadi and G Washer (eds.). American Society for Nondestructive Testing, St. Louis.

26. Pelczarski N and Huston D. 2000. Cure monitoring of composite laminates used in the manufacture of snowboards. *Nondestructive Evaluation of Aging Materials and Composites IV*, SPIE 3993.

27. Kothekar AV, Rosowsky DV, and Huston DR. 1998. Investigating the adequacy of vertical design loads for shoring. *J Perform Constr Facil*, 12(1), 41–47.

28. Brayley A, Commander B, and Schulz J. 1994. Monitoring stay cable forces during construction of the Natchez Trace Parkway Arch Bridge. *Proc ASCE Structures Congress XII*, Atlanta.

29. Fang DP, Zhu HY, Geng CD, and Liu XL. 2001. On-site measurement of load distribution in reinforced concrete buildings during construction. *ACI Struct J*, 98(2), 157–163.

30. Tanaka H, Kamei M, and Kaneyoshi M. 1987. Cable adjustment by structural system identification. *Proc Intl Conf on Cable-Stayed Bridges*, Bangkok.

31. Lagoda M and Olaszek P. 2004. Monitoring of the new bridge assembling technology. *Proc IABMAS'04 Bridge Maintenance, Safety, Management and Cost*, Kyoto. Taylor & Francis, London.

32. Turer A and Ozerkan T. 2004. Strain and vibration measurements of a cable-stayed pedestrian bridge. *Proc IABMAS'04 Bridge Maintenance, Safety, Management and Cost*, Kyoto. Taylor & Francis, London.

33. Gómez R, Murìa D, Escobar JA, Sanchez R, Muñoz D, and Vera R. 2004. Modelling and monitoring the launching of the chiapas bridge superstructure. *Proc IABMAS'04 Bridge Maintenance, Safety, Management and Cost*, Kyoto. Taylor & Francis, London.

34. Sandford TC, Davids WG, Hartt SL, and DeLano JG. 2006. Construction-induced stresses in H-piles supporting an integral abutment bridge. Paper no. 06-1725, 86th Annual Meeting, TRB, Washington, DC.

35. Lew M, Zadoorian CJ, and Carpenter LD. 2002. Integrity testing of drilled piles for tall buildings. *Structure*, 9(8), 14–17.

36. Hertlein B and Davis A. 2006. *Nondestructive Testing of Deep Foundations*. Wiley, Chichester.

37. Gucunski N, Balic M, and Nassif HH. 2004. Dynamics and field testing of Doremus Avenue Bridge substructure. *Proc IABMAS'04 Bridge Maintenance, Safety, Management and Cost*, Kyoto. Taylor & Francis, London.

38. Helmicki A, Hunt V, Shell M, Lenett M, Turer A, Dalal V, and Aktan A. 1999. Multidimensional performance monitoring of a recently constructed steel-stringer bridge. *Proc 2nd Intl Workshop on Structural Health Monitoring*, Stanford. Technomic, Lancaster.

39. Petersen DL, Siekmeier J, Nelson CR, and Peterson RL. 2006. Intelligent soil compaction technology, results and a roadmap toward widespread use. Paper no. 06-2914, 86th Annual Meeting, TRB, Washington, DC.

40. Massicotte B and Picard A. 1994. Instrumentation of the grand-mere bridge during strengthening. *New Experimental Techniques for Evaluating Concrete Material and Structural Performance*, American Concrete Institute, ACI SP-143-4.

41. Rausch F, Moses F, and Goble GG. 1972. Soil resistance predictions from pile dynamics. *ASCE J Soil Mech Found Div*, 98(9), 917–937.

42. Morgano CM and White B. 2004. Identifying soil relaxation from dynamic testing. *Proc 7th Intl Conf on the Application of Stresswave Theory to Piles*, Petaling Jaya, Selangor, Malaysia.

43. Shekar V, Aluri S, Laosiriphong K, Petro S, and GangaRao HS. 2004. Field monitoring of teo fiber reinforced bridges. *Proc Structural Materials Technology VI, an NDT Conference,* Buffalo.

44. Olund J, Cardini AJ, D'Attilio P, Feldblum E, and DeWolf JT. 2006. Connecticut's bridge monitoring systems. *NDE Conf on Civil Engineering,* I Al Qadi and G Washer (eds.). American Society for Nondestructive Testing, St. Louis.

45. Thompson MK, Davis RT, and Breen JE. 1998. Indications from instrumentations on a balanced cantilever bridge. In *Structural Engineering Worldwide,* GL Fenves, NK Srivastava, AH Ang, and RG Domer (eds.). Elsevier, Amsterdam, T213-5.

46. Onyemelukwe OU, Issa M, and Mills CJ. 2003. Field measured pre-stress concrete losses versus design code estimates. *Exp Mech,* 43(2), 201–215.

47. Kerokoski O and Laaksonen A. 2006. Monitoring of haavistonjoki bridge abutment performance. Paper no. 06-2005, 86th Annual Meeting, TRB, Washington, DC.

48. Yamamoto M, Higashino M, Toyama K, and Aizawa S. 1998. Five years of wind and earthquake observation results from a building with active mass dampers. In *Structural Engineering Worldwide,* GL Fenves, NK Srivastava, AH Ang, and RG Domer (eds.). Elsevier, Amsterdam, T198-2.

49. Rainieri C, Fabrocino G, and Cosenza E. 2007. Continuous monitoring for performance evaluation of the dynamic response of the school of engineering main building at University of Naples Federico II. In *Structural Health Monitoring,* FK Chang (ed.). DEStech, Lancaster.

50. Ehrlich GE and Gurovich YA. 2004. Typical case study of school sound insulation dBa measurements of sound in an elementary school before and after sound reduction treatment. *Sound Vib,* June, 16–19.

51. Heckl M. 1993. Advantages and disadvantages of intensity measurements. In *Safety Evaluation Based on Identification Approaches,* HG Natke, GR Tomlinson, and JTP Yao (eds). Vieweg, Braunschweig.

52. Reynolds DD. 1981. *Engineering Principles of Acoustics Noise and Vibration Control.* Allyn and Bacon, Boston.

53. Beranek LL. 1988. *Noise and Vibration Control.* Institute of Noise Control Engineers, Washington, DC.

54. Harris CM. 1998. *Handbook of Acoustical Measurements and Noise Control,* 3rd Ed. Acoustical Society of America, Woodbury.

55. Liang MT, Lin LH, and Liang CH. 2002. Service life prediction of existing reinforced concrete bridges exposed to chloride environment. *J Infrastruct Syst,* 8(3), 76–85.

56. Stanish KD, Hooton RD, and Thomas MD. 2000. *Testing the Chloride Penetration Resistance of Concrete: A Literature Review,* FHWA DTFH61-97-R-00022, US Federal Highway Administration.

57. Kassner BL, Brown MC, and Schokker AJ. 2007. Material investigation of the full-depth, precast concrete deck panels of the Old Woodrow Wilson Bridge. Virginia Transportation Research Council, *Final Report,* FHWA/VTRC 08-R2.

58. Chamberlin WP and Weyers RE. 1994. Field performance of latex-modified and low-slump dense concrete bridge deck overlays in the United States, American Concrete Institute, ACI SP 151-1.

59. Guthrie WS, Frost SL, Birdsall AW, Linford ET, Ross LA, Crane RA, and Eggett DL. 2006. Effect of stay-in-place metal forms on performance of concrete bridge decks. Paper no. 06-3082, 86th Annual Meeting, TRB, Washington, DC.

60. El Maaddawy T, Soudkl K, and Topper T. 2007. Performance evaluation of carbon fiber-reinforced polymer-repaired beams under corrosive environmental conditions. *ACI Struct J,* 104(S01), 3–11.

61. Hayashi K, Sakaida M, Furusawa E, Nara S, and Murakami S. 2004. Performance of uncoated weathering steel bridges against deicing salts in Gifu prefecture. *Proc IABMAS'04 Bridge Maintenance, Safety, Management and Cost,* Kyoto. Taylor & Francis, London.

62. Hristova EH, O'Flaherty FJ, Mangat PS, and Lambert P. 2004. Impact of main steel diameter on the flexural capacity of deteriorated reinforced concrete beams. *Proc IABMAS'04 Bridge Maintenance, Safety, Management and Cost*, Kyoto. Taylor & Francis, London.

63. Tarighat A. 2004. Application of neural network and monte carlo method for concrete durability analysis. *Proc IABMAS'04 Bridge Maintenance, Safety, Management and Cost*, Kyoto. Taylor & Francis, London.

64. Mukhopadhyay A K, Shon CS, and Zollinger DG. 2006. Determination of activation energy of alkali silica reaction for minerals and aggregates using dilatometer method. Paper no. 06-2560, 86th Annual Meeting, TRB, Washington, DC.

65. Hossain KM. 2006. Macro- and microstructural investigations on strength and durability of pumice concrete at high temperature. *J Mater Civil Eng*, 18(4), 527–536.

66. Agura DD. 1996. *Air Void Analyzer Evaluation*, FHWA-SA-96-062, US Federal Highway Administration.

67. Suprenant BA. 2003. Use and misuse of moisture-vapor emission tests on concrete slabs. *Concrete Intl*, (Decmber).

68. Qiu Y and Yang S. 2007. ZnO nanotetrapods: controlled vapor-phase synthesis and application for humidity sensing. *Adv Funct Mater*, 17, 1345–1352.

69. Rainwater NR, Drumm EC, Wright WC, Zuo G, and Yoder RE. 2006. Evaluation of instrumentation for monitoring seasonal variations in pavement subgrade water content. Paper no. 06-1889, 86th Annual Meeting, TRB, Washington, DC.

70. Janoo V, Berg RL, Simonsen E, and Harrison A. 1994. Seasonal changes in moisture content in Airport and highway pavements. *Time Domain Reflectometry in Environmental, Infrastructure, and Mining Applications*, US Bureau of Mines, SP 19-94, September.

71. Kotdawala SJ, Hossain M, and Gisi AJ. 1994. Monitoring of moisture changes in pavement subgrades using time domain reflectometry (TDR). *Time Domain Reflectometry in Environmental, Infrastructure, and Mining Applications*, US Bureau of Mines, SP 19-94, September.

72. Look BG, Reeves IN, and Williams DJ. 1994. Field experiences using time domain reflectometry for monitoring moisture changes in road embankments and pavements. *Time Domain Reflectometry in Environmental, Infrastructure, and Mining Applications*, US Bureau of Mines, SP 19-94, September.

73. van Schelt W, de Jong-Hänninen MK, van der Aa JP, and van Loon WK. 1994. Field and laboratory experiments with time domain reflectometry as a moisture monitoring system in road structures. *Time Domain Reflectometry in Environmental, Infrastructure, and Mining Applications*, US Bureau of Mines, SP 19-94, September.

74. Kassem E, Masad E, Bulut R, and Lytton R. 2006. Measurements of moisture suction and diffusion coefficient in hot mix asphalt and their relationships to moisture damage. Paper no. 06-2179, 86th Annual Meeting, TRB, Washington, DC.

75. Healy WM and van Doorn E. 2004. A preliminary investigation on the use of ultra-wideband radar for moisture detection in building envelopes. *ASHRAE Trans*, 110, 2.

76. Hussein FK and Delatte N. 2007. Freeze-thaw durability and nondestructive testing (NDT) of pervious concrete (PC). *Proc NASA/Ohio Space Grant Consortium 2006–2007 Annual Student Research Symposium XV*.

77. Murray TM, Allen DE, and Ungar EE. 1997. *Floor Vibrations Due to Human Activity*, AISC Design Guide 11, American Institute of Steel Construction, Chicago.

78. Dupuis H and Zerlett G. 1986. *The Effects of Whole-Body Vibration*. Springer, Berlin.

79. Griffin M. 1990. *Handbook of Human Vibration*. Academic Press, San Diego.

80. International Standards Organization. 1997. *Mechanical Vibration and Shock—Evaluation of Human Exposure to Whole-Body Vibration*, ISO 2631-1:1997(E).

81. Huston DR, Johnson CC, and Zhao X. 2000. Whole body shock and vibration: frequency and amplitude dependence of comfort. *J Sound Vib*, 230(4), 964–970.

82. Aiello V, Boiero D, D'Apuzzo M, Socco LV, and Silvestri F. 2008. Experimental and numerical analysis of vibrations induced by underground trains in an urban environment. *Struct Control Health Monit*, 15, 315–348.

83. Ungar EE, Zapfe JA, and Kemo JD. 2004. Predicting footfall-induced vibrations of floors. *Sound Vib*, November, 16–22.

84. Richart Jr FE, Hall Jr JR, and Woods RD. 1970. *Vibrations of Soils and Foundations*. Prentice-Hall, Englewood Cliffs.

85. Ohmi T and Yasuda M. 1989. Evaluating passive and active microvibration control technologies. *Microcontamination*, 7(9), 23–29.

86. Pyl L, Degrande G, Lombaert G, and Haegeman W. 2004. Validation of a source-receiver model for road traffic-induced vibrations in buildings. I: Source model. *J Eng Mech*, 130(December), 12.

87. Pyl L, Degrande G, and Clouteau D. 2004. Validation of a source-receiver model for road traffic-induced vibrations in buildings. II: Receiver model. *J Eng Mech*, 130(December), 12.

88. Amick H, Hardash S, Gillet P, and Reaveley RJ. 1991. Design of stiff low-cost floor structures. *Vibration Control in Microelectronics, Optics, and Metrology*, SPIE 1619.

89. Lyon RH. 1987. *Machinery Noise and Diagnostics*. Butterworths, Boston.

90. Piersol JS. 1990. *Nonlinear System Analysis and Identification from Random Data*. Wiley, New York.

91. Shimizu H, Isoda A, Yamagami T, and Kawakami Y. 2004. Fatigue evaluation and reinforcement for cracks of box girder bridge with steel deck based on actual live loads. *Proc IABMAS'04 Bridge Maintenance, Safety, Management and Cost*, Kyoto. Taylor & Francis, London.

92. California Department of Transportation. 2000. Methods of test to determine flexible pavement rehabilitation requirements by pavement deflection measurements. *California Test 356*, Engineering Service Center, Trans Lab.

93. Loshbough RC and Hall DL. 1991. Vehicle weighing in motion apparatus and method. US Patent 5,002,141.

94. Guo L, Chen X, Yu J, Tang Y, Liu R, Rogers R, Leidy J, and Claros G. 2005. Pavement deflection vehicle weighing method with embedded piezoelectric sensor. *Smart Sensor Technology and Measurement Systems*, SPIE 5758.

95. Wang CY, Wang HL, and Chen MH. 2005. Application of FBG sensors on bridge health monitoring & diagnosis. In *Structural Health Monitoring*, FK Chang (ed.). DEStech, Lancaster.

96. Yamada K and Ojio T. 2004. Monitoring service loading by BWIM. *Proc IABMAS'04 Bridge Maintenance, Safety, Management and Cost*, Kyoto. Taylor & Francis, London.

97. Yamaguchi E, Matsuo K, Kawamura S, Kobayashi Y, Mori M, Momota K, and Nishinohara T. 2004. Weigh-in-motion by continuous skew steel-plate-girder bridge. *Proc IABMAS'04 Bridge Maintenance, Safety, Management and Cost*, Kyoto. Taylor & Francis, London.

98. Chowdhury MR. 2000. A comparison of bridge load capacity using accelerometer and strain gage data. In *Structural Materials Technology IV*, S Allampalli (ed.). Technomic, Lancaster.

99. Barker MG and Barth KE. 2006. Field testing serviceability performance of Missouri's first HPS bridge. Paper no. 06-0489, 86th Annual Meeting, TRB, Washington, DC.

100. Huang D. 2006. Field performance of curved steel box girder bridges. Paper no. 06-0151, 86th Annual Meeting, TRB, Washington, DC.

101. Bustos MG, Marcet JE, Cordo OV, Pablo GM, Pereyra MO, and Altamira AL. 2006. Evaluation of performance of PCC pavements in mid-Western Argentina, and comparison with LTPP data using HDM-4 distress models. Paper no. 06-2847, 86th Annual Meeting, TRB, Washington, DC.

102. Carr GA and Stuart C. 1995. Performance based tie/fastener inspection technique using the gage restraint measurement system. *Nondestructive Evaluation of Aging Railroads*, SPIE 2458.

103. Kalay S. 1995. Measurement of track strength using non-destructive evaluation techniques. *Nondestructive Evaluation of Aging Railroads*, SPIE 2458.

104. Simiu E and Scanlan RH. 1996. *Wind Effects on Structures*, 3rd Ed. Wiley, New York.

105. Blevins RD. 1991. *Flow Induced Vibrations*, 2nd Ed. Van Nostrand Reinhold, New York.

106. Irvine HM. 1981. *Cable Structures*. MIT Press, Cambridge.

107. Dowell EH, Crawley EF, Curtiss Jr HC, Peters DA, Scanlan RH, and Sisto FA. 1995. *Modern Course in Aeroelasticity*, 3rd Ed. Kluwer, Dordrecht.

108. He X, Moaveni B, Conte JP, Elgamal A, and Masri S. 2004. System identification of Vincent Thomas Bridge using simulated wing response data. *Proc IABMAS'04 Bridge Maintenance, Safety, Management and Cost*, Kyoto. Taylor & Francis, London.

109. Koike Y, Tanida K, Mataguchi M, Murata T, Imazeki M, Yamada T, Kurokawa Y, Ohrui S, and Suzuki Y. 1998. Application of V-shaped hybrid mass damper to high-rise buildings and verification of damper performance. In *Structural Engineering Worldwide*, GL Fenves, NK Srivastava, AH Ang, and RG Domer (eds.). Elsevier, Amsterdam, T198-4.

110. Bosch HR and Miklofsky HA. 1993. Monitoring the aerodynamic performance of a suspension bridge. *Proc 7th US National Conf on Wind Engineering*, Los Angeles.

111. Xiao YQ and Li QS. 2006. Synchronal monitoring of wind characteristics and wind effects on super tall building during Typhoon Chanzhu. *Proc 4th China–Japan–US Symp on Structural Control and Monitoring*, Hangzhou.

112. Sadek F, Diniz S, Kasperski M, Gioffre M, and Simiu E. 2004. Sampling errors in the estimation of peak wind-induced internal forces in low-rise structures. *J Eng Mech*, 130(February), 2.

113. Scanlan RH and Tomko JJ. 1971. Airfoil and bridge deck flutter derivatives. *J Eng Mech*, 97(6), 1717–1737.

114. Cunha A and Caetano E. 2006. Experimental modal analysis of civil engineering structures. *Sound Vib*, June, 12–20.

115. Zhang D and Johnson EA. 2007. Structural control system design for parameter identification of shear structures. In *Structural Health Monitoring*, FK Chang (ed.). DEStech, Lancaster.

116. Nigbor RL, Hansen M, Tileylioglu S, and Baek JH. 2007. Elevators as repeatable excitation source for SHM in buildings. In *Structural Health Monitoring*, FK Chang (ed.). DEStech, Lancaster.

117. Uhl T and Bogacz M. 2004. Real time modal damping identification—airplane case study. *Proc 2nd European Workshop on Structural Health Monitoring*. DEStech, Munich.

118. Tamura Y. 1998. Damping in buildings in Japan. In *Structural Engineering Worldwide*, GL Fenves, NK Srivastava, AH Ang, and RG Domer (eds.). Elsevier, Amsterdam, T131-2.

119. Hikami Y and Shiraishi N. 1998. Rain-wind induced vibrations of cables in cable stayed bridges. *J Wind Eng Indus Aerodyn*, 29, 2.

120. Jones NP and Porterfield M. 1997. Measurement of stay-cable vibrations. *Proc ASCE Structures Congress XV*, L Kempner Jr and CB Brown (eds.). Portland, OR, ASCE.

121. Wacker JP and Calil Jr C. 2005. Pennsylvania hardwood timber bridges: Field performance after 10 years. *Proc Structural Materials Technology VI, an NDT Conference*, Buffalo.

122. Harpster TJ, Hauvespre S, Dokmeci MR, and Najafi K. 2002. A passive humidity monitoring system for in situ remote wireless testing of micropackages. *J Microelectromech Syst*, 11(February), 1.

123. Laplace PN, Sanders DH, Saiidi MS, Douglas BM, and El-Azazy S. 2005. Performance of concrete bridge columns under shaketable excitation. *ACI Struct J*, (May–June).

124. Doertenbach N. 2001. Highly accelerated life testing—testing with a different purpose. *Sound Vib*, 18–23.

125. van Baren J. 2005. Kurtosis—the missing dashboard knob. *Test Eng Manag*, (October/November), 14–16.

126. Hobbs GK. 2000. *Accelerated Reliability Engineering HALT and HASS*. Wiley, Chichester.

127. Sen R and Shahawy M. 1994. Accelerated bond and durability testing of FRPs for bridge applications. *New Experimental Techniques for Evaluating Concrete Material and Structural Performance*, American Concrete Institute, ACI SP-143-16.

128. Iwata S, Maehata H, and Arai H. 2004. Damage simulation to an RC slab with electric discharging shock force method. *Proc IABMAS'04 Bridge Maintenance, Safety, Management and Cost*, Kyoto. Taylor & Francis, London.

129. Schroeder M. 1970. Synthesis of low-peak-factor signals and binary sequences with low autocorrelation. *IEEE Trans Inf Theory*, 16(1), 85–89.

130. Liu P, Sana S, and Rao VS. 1999. Structural damage identification using time-domain parameter estimation techniques. *Proc 2nd Intl Workshop on Structural Health Monitoring*, Stanford. Technomic, Lancaster.

131. Ratay R. 1995. Field load testing of a drilled-in anchor system. *Proc ASCE Structures Congress XIII*, M Sanayei (ed.). Boston, MA, ASCE.

132. Zarghamee MS, Schwartz TA, and Kan FW. 1997. Evaluation of reliability of building envelopes subjected to wind effects. *Proc ASCE Structures Congress XV*, L Kempner Jr and CB Brown (eds.). Portland, OR, ASCE.

133. Wipf TJ, Klaiber FW, Ingersoll JS, and Wood DL. 2006. Field and laboratory testing of precast concrete channel bridges. Paper no. 06-2291, 86th Annual Meeting, TRB, Washington, DC.

134. Kudo R, Nakamura Y, Mita A, and Harada H. 2007. Performance assessment of a building with passive dampers using MIMO system identification. In *Structural Health Monitoring*, FK Chang (ed.). DEStech, Lancaster.

135. Skolnik D, Taciroglu E, and Wallace JW. 2007. System identification and health monitoring studies on two buildings in Los Angeles. In *Structural Health Monitoring*, FK Chang (ed.). DEStech, Lancaster.

136. Donea J and Jones P. 1991. *Experimental and Numerical Methods in Earthquake Engineering*. Kluwer Academic Publishers, Dordrecht.

137. Chung YS, Park CK, and Meyer C. 2008. Residual seismic performance of reinforced concrete bridge piers after moderate earthquakes. *ACI Struct J*, 105(1), 87–95.

138. Miao RY and Yang WH. 1994. A technique for measuring the interface confining stress in concrete filled steel tubes: Hydraulic analogy. *New Experimental Techniques for Evaluating Concrete Material and Structural Performance*, American Concrete Institute, ACI SP-143-9.

139. Iwaki H, Shiba K, and Takeda N. 2003. Structural health monitoring system using FBG-based sensors for a damage tolerant building. *Smart Systems and Nondestructive Evaluation for Civil Infrastructures*, SPIE 5057.

140. Mahin SA, Malley JO, Hamburger RO, and Mahoney M. 2003. Overview of the U.S. program for reduction of earthquake hazards in steel moment-frame structures. *Earthq Spectra*, 19(2), 237–254.

141. DeWolf JT, Lindsay TR, and Culmo MP. 1997. Fatigue evaluations in steel bridges using field monitoring equipment. *Proc ASCE Structures Congress XV*, L Kempner Jr and CB Brown (eds.). Portland, OR, ASCE.

142. Tokumasu K, Nakamura I, Sakano M, and Yoshihara S. 2004. Inspection and cause of failure of beam-to-column connection in steel bridge piers on Hanshin expressways. *Proc IABMAS'04 Bridge Maintenance, Safety, Management and Cost*, Kyoto. Taylor & Francis, London.

143. Arduini M, Nanni A, and Ramagnolo M. 2004. Performance of one-way reinforced concrete slabs with externally bonded fiber-reinforced polymer strengthening. *ACI Struct J*, 101(2), 193–201.

144. Mufti AA. 1998. Field performance of reinforcing steel-free FRC deck slab for the Salmon River Bridge. In *Structural Engineering Worldwide*, GL Fenves, NK Srivastava, AH Ang, and RG Domer (eds.). Elsevier, Amsterdam, T138-4.

145. Guan H, Karbhari VM, and Sikorksky C. 2004. Health monitoring of a FRP composite bridge augmented by use of web based and wireless technologies. *Proc IABMAS'04 Bridge Maintenance, Safety, Management and Cost*, Kyoto. Taylor & Francis, London.

146. Nassif HH, Suksawang N, and Malhas F. 2004. Health monitoring of new concrete bridge approach slabs. *Proc Structural Materials Technology VI, an NDT Conference*, Buffalo.

147. Yoda T. 2004. Inspection and assessment of the Japanese historical timber bridge: Kintai Bridge. *Proc IABMAS'04 Bridge Maintenance, Safety, Management and Cost*, Kyoto. Taylor & Francis, London.

148. Rossi AP. 1996. Methods and systems for structural behavior assessment of monumental structures, before and after execution of restoration activities. In *Proc Intl Conf on Retrofitting Structures*, R Betti, C Meyer, and B Yanev (eds.). Columbia University, New York.

149. Bati SB and Tonietti U. 2001. Experimental methods for estimating in situ tensile force in tie rods. *J Eng Mech*, 127(12), 1275–1283.

10

Sensor and Health Monitoring System Design

The successful design of SHM systems and components requires accommodating and accounting for a multitude of physical effects, physical constraints, sensing capabilities, human interfaces, economics, and the needs of society at large. Interactions between these various entities can be complicated, and even complex with emergent and unexpected behaviors. Many authors believe that an effective approach is to design and use SHM systems as part of an overall integrated structural or systemic health management effort [1–32]. Despite the apparent complexity of the task, design principles and procedures common to other engineering practices also apply to SHM system design [33]. An overriding principle is that effort expended in developing a good design usually provides high-value payback in terms of reduced fabrication and operational costs, along with increased levels of performance.

10.1 Design Procedures

Original, *adaptive*, and *variant* are three broad categories of engineering design processes. Original designs and inventions involve the creation of new concepts, configurations, and principles. Truly original designs and inventions are relatively rare. The engineering and market viability of original designs are even more rare. Occasionally, original designs produce revolutionary effects on products, markets, and engineering practice. Adaptive design is the reconfiguration of original inventions from other applications to a new problem. Adaptive design is much more common than original design. Altshuller developed a well-known theory of invention, based on adaptive principles, known as Teoriya Resheniya Izobretatelskikh Zadatch (TRIZ) [34]. Variant design is the modification of existing designs to meet new requirements, often with the aid of optimization techniques. The present state of SHM technology is that it is undergoing rapid change and producing original, adaptive, and variant designs. As the technology continues to mature, new SHM system designs will likely be more of the adaptive and variant varieties and less so of the original versions.

Designers work with both *ad hoc* and *systematic* procedures. *Ad hoc* techniques use the insight and intuition of the designer to arrive at a solution. *Ad hoc* techniques are quite effective, especially in the hands of experienced or highly talented designers. A disadvantage of the *ad hoc* techniques is that the design solution can be difficult to justify. Opposed to the *ad hoc* techniques are the systematic design procedures that put engineering design into a rational framework. The steps are well understood, justifiable, and documented. Systematic design generally follows four steps: (1) *task definition*, (2) *conceptual design*, (3) *embodiment design*, and (4) *detailed design* [34–36]. Systematic techniques compel designers to consider a wide swath of concepts and to evaluate the merits of each variant consistently and without bias. Systematic methods, while initially appearing more tedious than the *ad hoc* approaches, offer many advantages, such as the ability to design systems without having a large experience base and the production of a well-documented set of design decisions that are justifiable and transferable to management and other designers. As examples, Habel gives a comprehensive discussion of fiber optic deformation sensor system design, with concepts that are applicable to other sensing technologies, and Hutcheson and Tumer describe the systematic functional design of a health management system for a vented liquid oxygen tank for spacecraft [37,38].

Similar to the systematic approaches are the *Axiomatic Design Methods* promulgated by Suh [39]. Axiomatic design emphasizes the advantage of uncoupling the functional requirements of the system. This leads to simpler optimization methods. Uncoupling can also prevent certain unexpected catastrophic, that is, normal, accidents [40]. Uncoupling, however, may come at the cost of reduced performance. Suh measures the size of a design problem by the number of high-level functional requirements that a system must satisfy during its lifetime [39]. Braha and Maimon extended the axiomatic approach to develop measures of design complexity [41]. Aiding in this effort is the development of software systems with the capability of simulating the behaviors of SHM systems for design purposes [42]. Fang et al. suggest designing an SHM system so that the logical operation of each component is rigorously justified, that is, each separate module should have a relevant purpose and should be capable of independent operation [43].

10.1.1 Task Definition

Proper task definition is critical to SHM system design. Task definition includes specifications as to what is expected of the system and what constraints apply to the operations. If the requirements are not well understood or specified, then large resources may be wasted on a system that collects insufficient data or amasses mountains of useless data. Formulating the task definition in a solution-neutral form prevents biasing the selection of particular systems. Expressing the requirements in a format that can be documented and communicated to other interested parties prevents misunderstandings.

Some of the requirements and specifications that may appear in a list of tasks for an SHM system are as follows:

1. *Active structures and active sensing*: Some structures are equipped with active structural control systems, such as tuned mass dampers or actuators. It may be possible to use the active control system as an excitation source for SHM and system identification activities [44]. Control of normal operational states of the structural system, for example, flight testing, is another source of controlled stimulation.

2. *Aesthetics*: The use of SHM systems can raise aesthetic concerns. The sensors should have minimal and/or positive visual impact.

3. *Calibration*: Does the system need to be self-calibrating? Can it be recalibrated in the field? Is a preinstallation calibration stable and sufficient?

4. *Certification, verification, and documentation*: Many engineered systems that affect public safety and/or large capital assets require certification that they will operate as anticipated [45–49]. What procedures are necessary to certify an SHM system? What design considerations will aid in the certification process? How will the system be documented?

5. *Cost requirements*: Cost requirement analyses should consider the anticipated installation, operational, and maintenance costs incurred throughout the intended

lifetime of a system. Systems that require high levels of maintenance usually have sufficiently large databases of maintenance and cost records to enable a rational analysis of the cost–benefit levels of SHM [50]. Typical costs for sensor and data processing hardware are only a small fraction of the cost of an overall SHM effort. Sensor installation can be expensive and labor intensive. Rybak estimates that it costs about $130/m (USD ca. 2006) to run two channels of instrument cable in an industrial environment. The cabling accounts for approximately two thirds of the hardware and installation costs of a machinery health monitoring system [51]. Bakker et al. indicate that the cost of remote monitoring of 16 highway bridges in the Netherlands equals the cost of one replacement bridge [52]. The economic costs of traffic disruption due to maintenance and/or monitoring activities can be substantial. Who will pay for the long-term system operation, data archives, and data analysis?

6. *Data collection and archiving*: Data collection requirements govern many aspects of SHM system design:

   a. When should the system collect data? Continuously, periodically, after a trigger?

   b. What are the resolution and dynamic ranges of the data acquisition system?

   c. What are the data sampling rates?

   d. What are the data archiving requirements? Does all of the raw data need to be stored? Is it only necessary to store interesting and summary statistics?

   e. How should the data be processed? What information is expected to be extracted from the data?

7. *Duration and survivability*: How long will the SHM system be operated? Some SHM measurements require only short-duration testing. Others, such as corrosion, fatigue, and seasonal thermoelastic effects in large structures and extreme events, may require multiyear measurements [53]. Many critical structural parts sustain severe loading, thermal stresses, and so on. Monitoring the condition of these parts requires sensors that can also survive such severe conditions [54–56]. The miniaturization of sensors requires specialized micro- and nano-engineering and diagnostic techniques to assess and to assure reliability [57].

8. *Installation*: Is this an insertion into a newly built structure, or is it a retrofit installation? Inserting an SHM system into a structure during fabrication offers considerable flexibility as to system design and installation. This requires careful planning of the installation process relative to the construction sequence. Many construction and manufacturing circumstances are harsh, unpredictable, and may damage the SHM system. Effective and damage-free installation requires coordination between multiple workgroups [58]. Retrofit installations have fewer options. Supplementing the data with computer simulations can be very useful in retrofit situations [59]. The monitoring of ancient and historical structures poses additional concerns of minimizing permanent changes to the structure [60].

9. *Maintenance and repair*: Does the SHM system require maintenance? Are embedded sensors accessible, repairable, or expendable? Is redundancy a requirement? Is it worthwhile to use more expensive and reliable sensors? Are cheaper throwaways a better choice? Some authors, such as Denton, suggest that the long-term savings

of more expensive and reliable sensors can reduce lifetime predictive maintenance cost reductions by factors of five or more [61].

10. *Measurand to be measured*: Measurand selection intimately affects the performance of an SHM system. Factors to consider are the ability to measure the desired measurand, the cost of measurement, and the ability of the measurand to indicate structural conditions of interest. SHM systems can measure single or multiple measurands. A heterogeneous mix of measurands offers opportunities for information redundancy and for using data fusion techniques at the expense of additional design complications due to the differing nature of sensor operational requirements, the format of the sensor signals, and the associated time scales [62,63]. Many situations arise, such as in a regulatory requirement that preselect a particular measurand. A recommendation is to avoid taking interesting but nonessential measurements [64].

11. *Purpose of the SHM system*:

    a. Provide early warning of collapse or catastrophic failure

    b. Provide information for scheduling maintenance activities on a demand, rather than on a usage or postevent basis

    c. Determine if the structure meets specified performance criteria

    d. Identify the presence of fault conditions

    e. Provide structural condition assessment information for use in maintenance, operational, and rehabilitation activities

    f. Evaluate the condition of the structure upon discovery of a problem, such as incipient cracks

    g. Assess the integrity of the structure following catastrophic events

    h. Provide an estimate of the remaining lifetime of the structure

    i. Provide information that aids in the development of new design codes and procedures

    j. Assist in construction processes by increasing safety and/or productivity

    k. Establish the viability and performance of novel sensor or SHM systems

    l. Certify the safety of prototype and production designs

    m. Provide a maintenance history and state assessment as part of an economic assessment of a structure

12. *Security*: SHM system design poses many vexing security issues. Vandals, protection of equipment and data, and the prevention of pertinent structural information falling into unwanted hands pose technical challenges that need to be considered in the design of virtually all SHM systems.

13. *Sensor fault detection*: Detecting and diagnosing sensor faults can be useful [65].

14. *Transducer influence*: Sensors often distort and interfere with the behavior of the objects that they measure. Usually, it is desired to minimize such influences. Embedded sensors are of particular concern. Access is difficult and the penetration of lead wires through the outer layers of the structural element can be troublesome [66]. Stiffness matching and stress risers are also matters of concern [67–69]. It is possible that sensing elements, such as distributed fiber sensors, can contribute instead of detracting from structural performance.

TABLE 10.1

Design Requirements of NDE versus SHM Systems

Requirement	NDE system	SHM system
Measurement interval	Periodic	Periodic or continuous
Transducer location	Coupling or noncontact	Structure integrated
Measurement location and modality	Scanning or imaging	Local (hot spots) or averaging (fibers, plate waves)
Performance versus cost	High performance	Low cost
Signal processing	Signal preprocessing on board	Intelligent (processor + interface)
Replacement versus reliability	Easy to replace	Extremely high reliability (10–30 years)
Energy supply and consumption	Energy not critical	Stand-alone energy management

Source: Data from Kroening M, Berthold A, and Meyendorf N. 2005. *Advanced Sensor Technologies for Nondestructive Evaluation and Structural Health Monitoring*, SPIE 5770.

15. *Type of conditions to be monitored*:
    a. Improper or excessive displacements
    b. Degradation of material properties
    c. Cracking
    d. Scour
    e. Fatigue
    f. Corrosion
    g. Moisture penetration
    h. Thermal containment
    i. Load conditions on the structure
    j. Concrete strength—see ACI 228.1R-03 for a detailed discussion of the options [70].

Kroening et al. note that the requirements for the design of SHM systems are different from the requirements of NDE systems [71]. Table 10.1 lists some of the principal design requirements and the differences.

10.1.2 Concept Design

Concept design is a high-level process that considers a broad range of solutions and concepts [35]. A four-step systematic methodology for concept design is as follows:

1. *Requirements specification*: Convert the required and desired tasks into a set of specifications that do not predispose a particular design solution. At this point, it is often preferable to transform quantitative requirements into qualitative statements that represent the form of bare essential requirements. This first step should eliminate requirements with no direct bearing on the function and essential constraints of the system.

2. *System function description*: Examine the logical relations of the required function structure to determine dependence and independence of the various functions. Example independent requirements for an SHM system are that the data processing system should implement a particular algorithm, and that the sensors should operate in a specified temperature range. Example dependent functional requirements are the timing of sensor installation, sensor location, and the need for sensor maintenance.

3. *Concept variant generation*: Generate viable concept variations that satisfy the functional requirements. The goal is to identify every possible workable variation. A recommendation is to use a broad set of thought processes, such as considerations of conventional solutions, examining other designs in detail, Internet and catalog searches, and brainstorming. *Maieutics* is the study of these thought processes [72].

4. *Concept variant evaluation*: Assembling various combinations of subfunction solution principles forms proposed design solutions. Independent functions combine to multiply the number of possible variations. Dependent functions produce a smaller set of variations. Often considering the whole gamut of various design combinations is advantageous. Unfortunately, this may be a bit cumbersome, and even impossible for complicated situations with a large number of possible variants.

The final stage of concept design selects one or a few concept variants for further analysis and evaluation against technical and economic criteria. Selection methods include weighting techniques, uncertainty evaluations, and weak spot identification. If there is no clear-cut winner, different weighting schemes can produce different best-design variants. A major advantage of weighted-value selection systems is the documentation of the selection criteria and processes. Other designers and interested parties can readily evaluate and examine the design decisions.

There is a tension between physics-based SHM systems that can derive a direct relation between damage and observed behavior, and empirical reasoning-based systems that look for signature patterns of damage in the data [73]. Knowledge of potential failure and damage modes for a given structure simplifies many aspects of SHM concept design. For example, Celebi et al. designed a system for the real-time seismic monitoring of steel frame buildings during earthquakes [74]. The integrity of welded frame connections of the steel girders is the primary concern for many of these buildings. Connections can yield or even crack during an earthquake. The hidden location of most connections under walls and flooring makes visual and portable NDE inspection difficult. Instead, the proposed method of inspection is to monitor the interstory motions with accelerometers and to infer interstory displacements. If the interstory displacements are excessive, a likely cause is damaged connections.

10.1.3 Embodiment Design

Embodiment design fleshes out the leading concept variant into a form that is suitable for detailed design [35]. It is at this stage where the implications from the various design constraints and optimization criteria come into play. Some of the items that should be considered are as follows:

1. *Calibration*: The ability to calibrate sensors and to maintain calibration is an important component of most SHM systems. Precalibrating the sensor before installation and packaging so that the calibration remains reliably stable is one

approach. Precalibration works well for certain sensors, such as hermetically sealed piezoelectric accelerometers, vibrating wire gages, and Whittemore displacement gages. In cases where the initial calibration is not sufficiently stable, provisions should be made for in-field calibration. Some in-field calibration strategies include the following: (1) Access the sensor in the field. Apply a traceably measured excitation. Measure the sensor output. (2) Apply known loads to the structure. Measure the sensor output. An example is to drive a truck of known weight across a bridge. (3) Use built-in calibration circuitry in the sensor. An example is the shunt calibration of strain gage circuits. The appearance of MEMS technology and smart sensor techniques opens the door for the development of clever self-calibrating sensor systems.

2. *Clarity*: The purpose of each design step should be clear.

3. *Coverage*: Does the sensor configuration cover the structure so as to measure all items of interest? One approach is to examine the topology of intersecting domains of sensor coverage [75].

4. *Cost and economic benefit*: Ideally, increased public safety combined with reduced maintenance, construction, operational, and overall lifetime costs justifies the expense of an SHM system. Cost estimates should include design, hardware, installation, maintenance, calibration, power, data processing, and data interpretation. Owing to the relative novelty of the technology, there is no large database of published SHM system cost data. A good case for SHM arises with structures where failure poses an immediate threat to public safety, and monitoring can provide sufficient advanced warning so as to prevent or mitigate disasters. Dams are a good example. In 1993, the U.S. Army Corps of Engineers used 38,000 monitoring instruments on 580 dams for an annual operating cost of $15,600,000 (USD) [76]. The high benefit of preventing an unexpected failure or reducing maintenance costs, combined with a clear understanding of which measurements to take and how to interpret the data, often causes the cost–benefit analysis to weigh in favor of using SHM. In other cases, the convergence and crossing of the cost and benefit curves remain to be established. Structures that are suitable candidates for aggressive inspection and monitoring efforts include those that have excessive failure rates, high costs associated with failure, and the potential for failure amelioration by early detection of failure precursors. Lauridsen indicates that the use of such a program produced significant cost savings for the inspection of bridges in Denmark [77].

A rational approach to quantifying the costs of an SHM system examines the overall maintenance and life extension costs of a structure. Unmonitored and unmaintained structures will likely fail before similar monitored or maintained structures. The cost of maintenance and monitoring could be less than replacement. A formula for estimating the overall *life cycle cost* (LCC) is

$$\text{LCC} = C_I + \sum_{i=1}^{m} C_{\text{MBM}}^i + \sum_{j=1}^{n} C_{\text{REP}}^j + C_D, \tag{10.1}$$

where C_I is the initial cost, C_{MBM}^i is the monitoring-based maintenance cost, C_{REP}^j is the repair cost, and C_D is the demolition cost [78–80]. Monitoring for aircraft corrosion instead of disassembly for inspection is another possible cost–benefit opportunity [81].

5. *Durability*: Different sensing systems have different durability requirements. Field installations often expose SHM systems to harsh conditions that quickly render inoperable systems that work well in the laboratory. Some of the issues are moisture, vandalism, lightning strikes, temperature extremes, adverse chemistry, construction activities, ultraviolet radiation in sunlight, rodents, and freeze–thaw actions. One design approach uses ruggedized components, such as those built to military specifications. For example, Sanders et al. used chemical vapor deposition (CVD) diamond as a rugged semiconductor sensing material [82]. Another approach is to pretest the equipment in aggressively harsh conditions, such as with HALT methods [83].

Using the electronic equipment outside on large structures requires lightning protection. There are several methods of providing lightning protection. First, the structures should have well-defined multiple electrical grounds for dissipating the massive currents and voltages of a lightning strike. Adhering to standard electrical construction codes, when applicable, is a necessary step. Next, the equipment should be isolated by a redundant array of fuses that prevent the flow of excess currents and by automated short circuit shunt devices, such as metal oxide varistors, that allow the excess current to bypass vulnerable equipment. An alternative is to use fiber optic links and sensors in lightning critical areas. The resilience of stand-alone wireless sensor modules to lightning damage remains to be established.

6. *Electromagnetic compatibility*: Electronic sensing systems emit and absorb EM radiation. Both EM emission and absorption must be considered in most stages of system and instrument design. Government agencies regulate the levels and frequency bands of allowable EM radiation in a location-specific patchwork set of rules. The absorption of EM radiation can degrade the performance of electronic systems in manners that are often subtle and intermittent [84]. Related issues are system resilience with respect to *electrostatic discharge* (ESD) and *electromagnetic pulses* (EMP). Detailed designs of enclosures typically resolve most ordinary ESD issues. EMP resilient design has traditionally been within the realm of military applications. The potential advent of ubiquitous UWB consumer products raises a new set of possible EMP design issues.

7. *Flexibility and adaptability*: This can significantly enhance SHM system performance. Methods for making the system adaptable are (1) reprogrammable processors to control data acquisition and signal processing functions; (2) scalable sensing system that readily allows for the addition or concentration of sensors at points of interest; (3) mobile and/or rapidly deployable sensors; (4) imaging systems with squinting and focusing capabilities; (5) networks of sensors with reconfigurable topologies and variable operational modalities; and (6) using sensor systems that provide a measurement that is relatively invariant with respect to complicated structural changes and adaptations, as may occur in systems that change structural configuration in different operational environments [85].

8. *Maintainability*: Maintenance-friendly design features include providing easy physical mechanical access to sensor locations, using smart self-diagnosing sensors, having the sensor system configured to quickly identify fault conditions and to set alarms as needed, using modular and off-the-shelf components, and documenting the system design and installation properly.

9. *Modularity*: This uses standardized system components that enable easy configuration assembly and adaptation to different situations. The use of modular components can greatly reduce the costs of a system. Modularity applies to virtually

every component of an SHM system. Instrumented fasteners may be particularly attractive in this regard [86–89].

10. *Open architecture*: Modern SHM systems can be quite complex with multiple interactions between components, systems, data, information, and people. Developing SHM systems from scratch can be a huge undertaking. Instead, the design process is often one of integrating components and subsystems. The use of open-architecture information systems and component interfaces can significantly enhance this process [90]. As an example of the utility of such an approach, the machine monitoring industry has adopted an open architecture for the passage and processing of data known as the *machinery information management open systems alliance* (MIMOSA) standard [91].

11. *Performance*: In the context of an SHM system, this includes frequency response, hysteresis, power consumption, range, reliability, repeatability, sensitivity, speed, stability, and so on.

12. *Power requirements*: SHM systems require power to operate. Many structures have electrical power at the site that is readily available. Since line power often fails during extreme events, provisions for backup power are necessary for taking measurements during an extreme event. Wireless and other stand-alone SHM systems often pose severe design constraints on the power system design. Digital signal processing and data encoding can often provide a power efficiency advantage over analog equivalents [84]. Bond graphs can help to track the energy flow in a system and to analyze design alternatives [92].

13. *Redundancy*: This uses multiple components and multiple-component configurations to create a more robust and fault-tolerant system. If one of the components fails, then the other components take over and carry out the required task. A key feature is that the redundant components of a system are independently capable of performing the desired task. True redundancy requires that components be designed and manufactured independently so that the possibility of a common weakness causing the failure of multiple similar systems is reduced. A potential pitfall is poorly formulated specifications that lead to common faults in independent redundant designs. Sensing systems offer many interesting possibilities for redundancy.

One approach to designing redundant SHM systems uses sensors that are inexpensive and easy to install. This enables using multiple sensors with overlapping ranges of spatial and temporal sensitivity to provide redundant sensing. Deshmukh et al. developed a redundant leading edge panel integrity sensing system for aerospace structures with a schematic appearing in Figure 10.1 [93]. The system design takes advantage of the ability to interleave distributed 1-D sensors. It may be possible to use redundant systems with automated data acquisition system component switching [94]. An alternative and complementary approach uses different sensor types to make the same measurement. An example would be structural temperature measurements that use a heterogeneous array of thermistors, thermocouples, infrared cameras, and/or temperature-sensing paint. Mixing the several transducers on separate data networks further extends the redundancy. Another example is the use of a redundant position sensing system during the construction of the new Tacoma Narrows Bridge [95]. Redundant GPS and manual electronic distance meters monitored caisson position during placement. The importance of using real-time data during caisson touchdown motivated the requirement of sensor system redundancy. Redundancy is also an important component in data

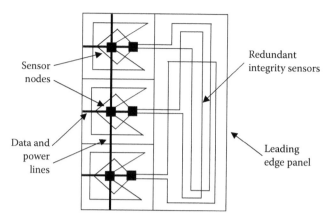

FIGURE 10.1   Redundant sensor layout for critical leading edge integrity sensing in aerospace thermal protection system. (From Deshmukh A, Pepyne D, Hanlon AB, and Hyers RW. 2005. *First Intl Forum on Integrated System Health Engineering and Management in Aerospace*, Napa, CA.)

transmission and storage. Encoding the signals with redundant parity and check-sum bits is routine, as is the practice of storing important data at multiple off-site locations.

14. *Robustness*: The sensing system should be able to tolerate the failure of multiple individual components without significantly degrading system performance [96].

15. *Safety*: This should be a concern in every design step. Safety with regards to SHM system design has many interesting aspects. One issue is that the operation and use of SHM systems should enhance the safety of the overall structural system. A second issue is that the SHM system affects the safety of system installers, inspectors, and the general public. An ideal SHM system safety design would be one that provides accident-preventing safety warnings and alerts, reduces the risk to inspectors, and poses no additional safety hazards to humans or the structure, while reducing the costs and increasing the performance of the overall structural system.

    Safety design can use *direct*, *indirect*, or *warning* methods. Direct methods are usually best, followed by indirect and then warning techniques.

    A direct design does not require action by a human or protective device to ensure safety. An example is to power a sensor system with voltage and amperage levels that are sufficiently low so as to be inherently safe, even in the event of an electrical fault. Direct safety design includes fail-safe systems, that is, those that are safe even in the event of a failure. An example is the use of positive pneumatic pressure to disengage brakes in heavy vehicles. If the air system fails, the brakes engage and the vehicle cannot move. This contrasts with hydraulic brake systems in light vehicles. If the hydraulic system fails, the brakes cannot engage. In complicated situations, *fault-tree* and FMEAs can establish and assess the degree of fail-safe design. A fault-tree diagram indicates the logical flow of consequences due to a fault in the system. FMEA attempts to consider the various possible failure scenarios for a system, especially in the case of cascading subsystem failures. Complex failure modes are notoriously difficult to predict [40]. One method of reducing complexity is to reduce the level of coupling between subsystems.

    Indirect safety methods are physical barriers that prevent access to dangerous conditions. Indirect methods are less desirable than direct methods, since they allow for dangerous conditions to exist. A major disadvantage is that it is often possible to defeat and/or override safety mechanisms. This is especially problematic when the safety barriers compromise system performance.

Warning methods attempt to influence humans to follow safe practices. Safety training requires constant reinforcement of safety principles and enforcement of safety rules. Unfortunately, many warning labels are ignored, hard to read, require literacy, or are not informative. Additionally, humans have difficulty in understanding the risks and consequences associated with unsafe practices.

An example of safety design is fall protection of inspection workers. A direct method of preventing falls is to install an SHM system that allows the inspector to stay on the ground and avoid being placed at dangerous heights during an inspection. An indirect method of fall prevention is to use fall arresting harnesses and railings. Warnings about how to work safely at elevated heights are a necessary, but often ineffective, last line of defense. In spite of aggressive safety programs, fall injuries remain the most common cause of the death of workers in the U.S. construction industry with over 1000 deaths annually [97].

Placing a sensor system in a position where it could cause accidents and injuries, if it falls, raises the sensor mounting design details to a high level of importance. Embedded sensors pose another set of hazards. Embedded sensors may cause stress risers weaken the structure.

16. *Scalability*: This is the ability of a generic system design to work on systems with sizes that vary over large ranges. The sizing of most engineering products varies as a power of size, even for product lines that use discrete sizing [35]. It should be anticipated that sensing on a large structure requires more sensors than on a small structure. The number of required sensors might also scale according to a power of length. An example is using strain sensors to identify crack formation from anomalous strain readings. Strain gages will work for this application if the gages are placed near to the crack [98]. The number of strain sensors required to identify cracks may be proportional to surface area, that is, the length squared. It may be impractical to sustain the appropriate strain gage packing density for crack detection on larger structures [99]. If, instead, the damage detection scheme relies on shifts in behavior of the lowest-order vibration modes, the number of sensors (accelerometers) may be invariant with respect to length.

Designing a scalable SHM system requires a certain amount of care. The use of more sensors increases the bandwidth and signal processing requirements of the sensing system. Possible solutions include using sensor identification numbers with 16-, 32-, or even 64-bit labels. Hierarchical organization of the system offers an interesting possibility for adapting to problems of scale, but may give rise to communication bottlenecks. Mesh network architectures may alleviate some of the bottleneck issues and the expense of a more complicated design. Certain distributed sensors, such as TDR, show considerable potential for scalability.

17. *Sensor fault detection*: SHM systems with a modest amount of sophistication should be able to detect, diagnose, and possibly recover from sensor faults.

18. *Simplicity*: If possible, the design should be simple. One approach is to combine mutual function requirements into a single piece of hardware. Integrated modules that combine sensing, signal conditioning, data processing, and wireless data transmission into a single unit can potentially simplify the design of an overall SHM system. This approach tends to localize complexity and can take advantage of economies of scale.

10.2   Detail Design

Detail design typically appears in the latter stages of an overall design effort. Some SHM design details to consider are as follows:

1. *Cabling*: The detailed design of the cabling system is often critical. For example, the cables used to monitor dams have codified design details concerning length, conduit requirements, terminal locations, and the avoidance of field splices [100].

2. *Component and system durability*: What is the anticipated environment for the sensor system? Are the components rated for an anticipated harsh environment?

3. *Component and system power requirements*: What are the power requirements? How will power be supplied? Are energy efficient or energy harvesting designs required?

4. *Cost*: Are there cost constraints on the components and overall system? Will less expensive and (perhaps) less capable sensors and components be capable of performing the desired tasks?

5. *Data processing*: How will the data be processed? Who is responsible for reviewing and analyzing the data? How much of the data will be archived? Who is responsible for archiving the data?

6. *Embedment*: Does embedding sensors cause unintended consequences, such as stress risers [101]?

7. *Installation*: Will the system be installed during the fabrication of the structure? Will it be a retrofit installation? Who will install the system? What are the safety considerations? How will a high-quality installation be assured?

8. *Modularity*: Will standardized parts and protocols be used? Is this a custom design?

9. *Onboard intelligence*: It is now practical to embed intelligent electronics into many sensor systems [102]. Architecture considerations include the following:

    a. *Application-specific integrated circuits* (ASICs): These have the advantage of high performance at high cost per design. Multiple uses increase the economies of scale of ASICs.

    b. *Field programmable gate arrays* (FPGAs): These have moderate levels of performance at a lower cost per design than ASICs.

    c. *Microprocessors*: These have lower levels of performance than ASICs or FPGAs, but cost less and have the most flexibility in terms of reprogrammability.

    d. *Packaging*: How will the system be packaged? Are aesthetics or ergonomics important issues? Are there any special environmental considerations?

    e. *Sensor calibration*: How is the calibration established? Is the calibration reliable? Can the sensor self-calibrate?

    f. *Sensor dimension*: Point sensor or distributed sensor?

    g. *Sensor networking and data transmission*: What is the physical mechanism of data transmission? How are the data encoded? Does the network architecture, physical transmission, and data protocol follow an industry standard? Is it a custom design?

    h. *Sensor position and mounting*: Are the sensors embedded, surface attached, or standoff? What are the details of the location? Are mounting brackets or other hardware required? Will the structure be altered? Does the sensor cause stress risers or architectural defects?

i. *Sensor selectivity*: Is the sensor signal only dependent on the desired measurand? Do other factors affect and confound the reading? What levels of channel cross-talk are tolerable? For load measurements, are the load paths redundant or statically determinate? Are simple 1-D structural mechanics models valid? Is it necessary to consider 3-D effects?

j. *Sensor sensitivity*: What is the anticipated range of measurement? Does the sensor sensitivity match the anticipated range of data? Is autoranging necessary?

k. *Size*: What are the size constraints on the system and sensors?

l. *Triggering*: Does the system need a data collection trigger? Is a simple threshold adequate? Are complicated multiparameter triggers needed? How will the trigger levels be set?

m. *Weight*: Are there any weight limitations?

10.3 Optimization Techniques and Sensor Placement Strategies

Optimizing SHM system designs requires considering many options. A principal consideration is the type, number, and location of sensors [103]. A structure normally has several, if not many, multiple possible sensor location configurations. Selecting which configuration to use can be a nontrivial exercise. If an SHM system uses N_S distinct sensors spread over N_L possible locations, the number of different sensor configurations, C, is

$$C = \frac{N_L!}{(N_L - N_S)!}. \tag{10.2}$$

If, instead, the N_S sensors are identical, then the number of possible combinations reduces to [104]

$$C = \binom{N_L}{N_S} = \frac{N_L!}{N_S!(N_L - N_S)!}. \tag{10.3}$$

Similar more complicated formulae arise when considering heterogeneous mixes of multiple sensor types [105].

Ad hoc approaches to optimizing sensor system design make use of probabilistic historical data concerning the failure rates of particular structures and subsystems. This includes the identification and dense instrumentation of hot spots. At the detail level, it can be effective to place the sensors in areas that are potentially regions of high stress, distortion, or otherwise of importance, that is, *hot spots* [106–111]. Determining hot spots location may be simply a matter of locating stress risers through a suitable stress analysis. More sophisticated analyses may use probabilistic reliability techniques [112,113]. Kim et al. combined failure modes, failure probabilities, and damage sensitivity with an NN to identify hot spots for sensor placement [114]. Hot spots for accelerometers in vibration measurements are points predicted to have high modal kinetic energies [115].

Traditional rational approaches to optimal design use scalar *cost functions* of system characteristics. Optimization proceeds by varying the design parameters until the minimum cost is achieved. Success, in large part, depends on selecting a suitable cost function. An example is the *structural damping index* (SDI) used by Kang et al. to optimize the locations of piezoelectric sensors and actuators in an attempt to control structural dynamics [116]. The SDI is a weighted sum of modal damping ratios, that is,

$$\text{SDI} = \sum_r 2\varsigma_r \omega_r c_r, \tag{10.4}$$

where ς_r, ω_r, and c_r are the damping ratio, natural frequency, and modal weighting (participation) factor for the rth mode. When the cost function is relatively simple, optimization algorithms such as gradient searches, and linear and nonlinear programming, are straightforward to apply and are often successful [117–121]. Papadoppulis and Garica compared 21 different optimization criteria for sensor placement in the dynamic testing of a cantilever beam and frame truss [122].

The best cost function for optimization is not always inherently obvious. In addition to routine considerations of size, weight, cost, power, and so on, a cost function should include variables that quantify the quality and amount of obtainable information from a particular sensor configuration [104]. Applying such methods usually benefits from some sort of a priori information about expected structural behavior and what outputs are expected from the SHM effort. Including random system and sensor properties, if quantifiable, can be useful [123]. The possible outputs run the gamut of SHM measurement possibilities, for example, corrosion, mode shape, damage, and so on. Restricting the set of possible outputs can simplify cost function formulation and optimization at the expense of system functionality. Zhang proposes using a *quantifiable directed graph* (QDG) to provide a comprehensive assessment of the ability of a sensing network to detect damage [124]. In addition to connectedness, the QDG includes information concerning sensor SNR, signal propagation gains and delays, and the value of detecting damage.

An example case is the selection of optimal sensor location for vibration mode shape determination. If prior information regarding the mode shapes is available (perhaps from a finite element analysis), optimization criteria, such as the minimization of the determinant of a Fisher information matrix or degree of operability, can be useful in selecting the sensor location [125–128]. Meo and Zumpano used the Fisher information matrix to find optimal locations of motion sensors for modal vibration monitoring of a bridge with a method known as the effective independence (EI) method [129]. Heo et al. used a similar approach with kinetic energy instead of modal independence to optimize sensor placement on long span bridges [130]. Kinetic energy criteria may be superior to the modal independence criteria in situations where signal to noise in the measurements is an issue. Song et al. compared the EI, modal assurance criterion (MAC), and *average driving point residue* (ADPR) methods for optimizing the placement of accelerometers for modal identification and ultimately damage detection [131]. The EI method produced a Fisher information matrix with a better condition number. The MAC method produced MAC matrices with better condition numbers. The addition of more sensors improved (decreased) the matrix condition numbers. In a set of numerical studies, Gaitanaros et al. found that optimizing the information entropy as the cost function produced a strain gage array configuration that could identify cracks in plates [132]. Gao et al. optimized sensor placement for rotating machinery SHM based on a principle of maximizing the signal-to-noise ratio combined with rational models and experimental confirmation [133]. For input–output modal vibration testing, Gawronski suggests matching the Hankel singular value vectors of the input actuators to the output sensor with those structural transfer functions [134].

An assumption of linear structural dynamics can help to guide the selection of optimal sensor locations for vibration-based damage detection [135]. Raich and Liszkai suggest optimized sensor placement with a damage information criterion based on the change in the measured FRFs [136].

$$t_{km} = \sum_{j=1}^{n_e} \left(\int_{\omega_0}^{\omega_1} \left| \frac{\partial H_{km}(\omega)}{\partial x_j^s} \right| \right)^2 \tag{10.5}$$

where t_{km} is the information contained in the FRF $H_{km}(\omega)$ due to a damage vector x_j^s. The sums are over the elements in a finite element model. Trendafilova et al. used a mutual information criterion to guide sensor placement (see Chapter 5 for definition) [137]. The procedure places sensors at a distance equal to the first zero of the mutual information function as determined by the cross-correlation between the signals from the two channels. The motivation for this approach is that placing the sensors any closer causes the two sensors to measure redundant information. Placing the sensors farther apart runs the risk of losing information. An interesting case arises when using modal curvature to detect sampling. Increasing the number of sampling points improves the damage estimate, until a state of oversampling occurs in which an excess number of points can cause errors in the curvature and damage estimates [138]. The primary source of the difficulty is that the variance of the error of each sampling point is independent of the number of samples. Increasing the number of samples does not reduce the variance. The independence of the sampling at adjacent points has the effect of distorting the smoothness of the mode shapes and the associated curvatures. (A similar issue arises with the smoothness of periodogram spectral estimates.) Worden and Burrows compared genetic algorithms (GAs), simulated annealing, and a brute force sensor deletion and insertion approaches for detecting damage in a cantilever plate from a modal vibration analysis [139]. All the methods showed a high degree of agreement with regard to the optimal configurations per particular number of sensors. Azarbayejani et al. used a probabilistic approach that considered both the wide range of possible combinations of damage and sensor placements [140]. The method used NNs to sort through the vast range of possibilities.

Systems with heterogeneous sensors and multiple failure modes pose additional sensor system design challenges. Fijany and Vartan suggest using a quantitative representation of the *diagnosability degree* based on an analysis of the system fault signature matrix [141]. Mishalani and Gong developed optimal infrastructure, for example, pavement, sampling procedures based on lifetime costs, and uncertainties using Markov chains [142].

Miyamoto and Matsukawa developed a corrosion and chemical attack (hydrogen sulfide) monitoring system for a sewer system with concrete pipes [143]. Sensor placement selection and overall system design used algorithms that combined existing measurements with expectations of damage progression. Using an array of distributed sensors is another approach of capturing the hot spot behavior. Yang et al. found that an array of at least eight long-gage FBGs could effectively detect cracks in reinforced concrete beams [144].

The design of an entire SHM system may be too complicated for optimization by simply minimizing a cost function. One issue is that SHM systems tend to be a heterogeneous mix of discretely and continuously variable parameters. A relatively simple case is where static testing requires a different sensor configuration than needed for dynamic testing [145]. The result is a cost function with multiple local minima that confound steepest descent and other similar techniques. Nondeterministic optimization algorithms such as GAs, simulated annealing, and cellular automata may be more useful in these more complicated design situations [146–149]. GAs can be effective in solving certain sensor location design problems [150,151]. Reich and Sanders, and Li et al. used GAs to determine optimal sensor placement layouts [152,153]. Spillman and Huston used a GA to select an optimal dual fiber sensor layout for distributed point load sensing on sandwiched plates [154]. Worden et al. developed a strategy in which NNs and GAs decide where to place sensors so as to optimally detect structural damage and faults [155]. Tumer and Stone developed a methodical fault-tolerant system design procedure with potential SHM applicability [156].

The design optimization of a wireless sensing system is a somewhat severe example of the issues [157,158]. There are many possible configurations and variables to consider: (1) network architecture; (2) number of sensors; (3) sensor location; (4) sampling rates of the individual sensors; (5) extent of local data processing and analysis; (6) amount of data to be sent; (7) local power consumption; (8) bandwidth and transmission protocol; and (9) cost. Optimizing the sensor configurations under such complicated circumstances may be nontrivial and may require resort to either *ad hoc* approaches or to sophisticated numerical techniques. All these methods make use of hypotheses and other assumptions regarding the failure and damage modes of the structure. Wu et al. recommend considering the use of a heterogeneous mix of sensing and master nodes [159]. The sensing nodes would have enhanced analog signal processing capability. The master nodes would have enhanced computational powers.

Adaptive sensing strategies use data from measurements to guide selecting those areas that are more critical for sensing. Stress gradients, information, and novelty detection are all measurement strategies. Shi et al. proposed an adaptive placement approach using modal data from a vibrating structure [160]. The technique first calculates eigenvector sensitivities to damage from the modal data. The sensor replacement strategies maximize the measurement of the eigenvector sensitivities. Eckhoff et al. described a data acquisition system that adapts by having both self-healing and self-calibrating capabilities [161]. Beard in an early (ca. 1971) development lays on some fundamental limitations and capabilities of linear systems to recognize and recover from sensor failures using concepts based on state space observabilities [162].

10.4   Design for Inspection

Inspectabilty is an important design consideration for new structures. Design for inspection and maintenance is common in the aviation industry. It is somewhat more intermittent in automotive and civil structural design. Some general principles of design for inspection are (1) early identification in the design process of critical components that require inspection and (2) design of details around critical components so as to give ready access. Examples include catwalks and access ports.

Some details are notoriously difficult to inspect. These include (1) pre- and post-tensioning strands in reinforced concrete members, (2) hanger pins, (3) rock and concrete anchors for cables and ties, (4) scour around bridge piers, (5) concrete bridge decks with asphalt overlays and steel piers, and (6) lap joints with hidden corrosion in airplanes. An example of difficulty in inspectability is the steel arch girder Hoan Bridge in Milwaukee, Wisconsin, USA. Major incipient damage caused the closure of the bridge to traffic on December 13, 2000 [163,164]. The existence of uninspectable fatigue-prone connection details between main girders and lateral bracing received the bulk of the blame for the failure. In many respects, the Hoan Bridge failure also represents a major failure of an SHM effort to provide actionable information for the owners. Although the previous routine inspections and identified cracking prompted the installation of monitoring instrumentation, the information gained from the inspection and monitoring efforts was insufficient to mandate corrective actions to prevent the failure.

10.5 Experimental Design

The proper design of an experimental protocol can greatly enhance the value of the data collected and can reduce the overall testing costs [165]. An example is dynamic bridge testing. Farrar and Doebling, and Felber recommend the following procedures that have sufficient generality to be applicable to many other testing situations [166,167].

1. *Perform a thorough pretest visual inspection*: Unexpected damage or structural details can confuse the data interpretation.
2. *Perform a linearity check*: Many dynamic diagnostic procedures assume linearity and/or explicit deviations from linearity. Linearity is fairly easy to check with reciprocity and other tests.
3. *Perform as many environmental and testing procedure sensitivity tests as possible*: Many benign effects unrelated to structural damage alter structural dynamics. Temperature shifts and gradients that change boundary condition fixities are a common example. Anticipating and correcting for such confounding issues prior to testing and as part of the post-test analysis can significantly improve the overall results.
4. *Perform false-positive studies.*
5. *Perform statistical analysis*: Dynamic bridge testing is essentially similar to all other experimental methods in that errors and uncertainties are inevitable. Quantifying the level of uncertainty can be useful. It should be noted, however, that the expense and impracticality of performing independent repeat tests often precludes a strict adherence to rigorous standards of statistically valid experimental designs.
6. *Design test sequence so as to minimize the movement distances for sensors.*
7. *Design test sequence with preplanned alternative data collection sequences in the event that a given sensor is difficult to position properly, or fails.*
8. *Manage cabling*: Plan, design, and preassemble cabling prior to site visit. Avoid tangling cables. Use color-coded and labeled cables and connectors.

10.6 Extreme Events: Wind, Blast, Flood, and Earthquake

The use of systems to measure the performance and safety of structures in extreme events, such as hurricanes, earthquakes, impacts, fires, and explosions, poses severe technical challenges. Extreme events place severe performance demands on both structural and sensing systems. Issues include survivability, proper triggering, and large dynamic ranges of the data. Extreme event recorders, in general, need their own power supply and must be able to operate autonomously. Safety-critical data acquisition systems should preserve the collected data, in spite of structural failures and catastrophes. In many respects, the performance of extreme event monitoring systems depends heavily on a commitment to long-term maintenance of the system to ensure that it operates when needed [168].

Measurements during extreme events often require specialized transducers. Blast measurements are an example. Many blasts appear as highly transient pressure overloads. Measurements of such transients require specialized techniques that account for the compressibility of the air in the vicinity immediately surrounding the pressure transducer as well as the high-frequency fidelity of the cabling and associated electronics [169]. When selecting accelerometers for mechanical shock measurements, Walter suggests using an accelerometer with a natural period, $T_n = 1/f_n$, less than 20% of the pulse duration, T [170], that is,

$$Tf_n = \frac{T}{T_n} > 5. \tag{10.6}$$

10.7 Real-Time Systems

SHM systems that provide real-time data analysis and assessments can be very useful in situations that require rapid responses and decisions. Examples of possible uses include decisions about allowing infrastructure systems to remain operational immediately following an earthquake or to close, inspect, and repair as necessary; and the prescheduling of maintenance on ships and aircraft as they are returning to the port with the intent of executing a limping or reconfiguration strategy in the event of damage. In many respects, the design and implementation of real-time systems is simply a matter of performing routine SHM activities in a quick and automated fashion. However, this generally requires a robust and well-understood relation between collected data and structural conditions. Choi et al. and Wahbeh et al. describe real-time SHM strategies and design issues for large structures, such as bridges [171,172]. Gorinevsky et al. outline important integration design issues for real-time SHM of commercial aircraft [173]. Yun et al. demonstrated the utility of real-time systems for evaluating the integrity of the Vincent Thomas Suspension Bridge (Los Angeles, CA) with a system based largely on monitoring of the structural dynamics [174]. Mijarez et al. developed a wireless system to provide real-time warning of the flooding of compartments in off-shore structures [175]. Real-time systems that incorporate decision making quickly become complicated. Figure 10.2 shows the structure of a moderately complicated SHM data collection and decision support system.

10.8 Biomimetic and Ecomimetic Design

Biomimetic design is the process of designing systems and devices based on inspiration from biological systems. Biological systems present elegant solutions to difficult issues of optimization in the face of uncertain circumstances. Virtually all biological systems sense and adapt to their environments. Biomimetic design is a topic that has received considerable attention from both serious researchers and dilettantes, but remains in its infancy. *Ecomimetic* design attempts to follow some of the organization principles of ecological systems to the design and organization of engineered systems. Some of the features of biomimetic sensor systems are as follows:

1. *Adaptability*: Virtually all biological systems are adaptive. They respond to environmental conditions by modification of operational parameters and configuration. An

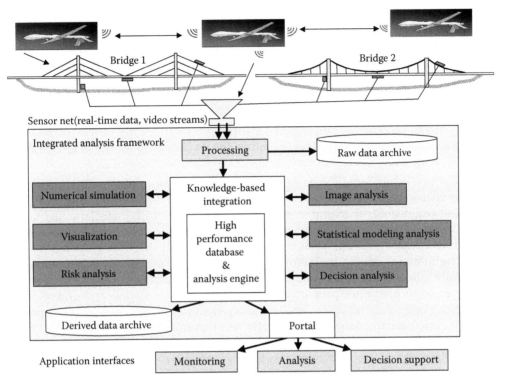

FIGURE 10.2 SHM data collection, processing, and decision support system. (Figure courtesy of A. Elgamal and J.-P. Conte, University of California San Diego, and T. Fountain of San Diego State College.)

interesting biological adaptation is in the growth and development of nervous systems [176]. Most young creatures have an excess of nerves and connections. Nervous system development entails eliminating a substantial number of nerves and altering the connections as needed. The optic nerve is an example. Understimulation of the optic nerve at certain early stages in the development of an animal prevents the removal of certain nervous pathways. Blindness is the result. Implementing such a complex, yet highly effective, process has yet to be fully realized in SHM practice.

2. *Exploitation of nonlinearity*: Many engineering design processes use linear mechanics and systems approaches. The advantages of linearity include mathematical tractability and the relative stability of solution types within a given parameter space. Biological systems are not constrained by issues of mathematical tractability and often use phenomenon linked to nonlinear systems to advantage. An example is the limit cycle response of a neuron to a stimulus that results in information being encoded in voltage pulse trains [177]. The pulse trains are essentially asynchronous binary signals. It is possible, and perhaps advantageous, to fabricate neuronal limit cycle circuits that produce pulse trains as part of microscopic transducer amplifiers [178].

3. *Hierarchical and fractal organization*: Many biological systems have hierarchical and fractal topologies. This includes the nervous systems of chordates. Hierarchical systems exploit and optimize geometric scaling effects. The structure of a typical nervous system enables a broad sensory coverage of tissue without excessively taxing the attention of the central nervous system [179]. The use of materials with

hierarchical inherent self-sensing capabilities is an interesting possibility for SHM systems [180]. As an example, Kirikera et al. describe a sensory architecture based on a network of artificial neurons using piezoelectric ribbon elements [181]. The ribbons detect AE signals and other indicators of damage. Hierarchical organization of SHM systems requires managing information flow up the hierarchy to avoid bottlenecks.

4. *Redundancy and expendability*: Biological nervous systems tend to be redundant. Many nerve cells act as embedded sensors with overlapping ranges of interest. Since most nerve cells are not regenerative, the level of redundancy is finely tuned to match the expected loss of function over the lifetime of the creature.

5. *Self-healing*: While largely beyond the scope of this book, self-healing structural technologies can benefit some level of sensory input. Early applications of self-healing engineered structural systems include the self-sealing airplane fuel tanks that appeared early in the twentieth century and the use of automated check valves to protect fluid-hydraulic systems from rupture [182]. Mercier patented a self-healing material system in 1896 that used a layer of petroleum jelly to cause a partially vulcanized rubber layer to swell and seal punctures [183]. Dry developed a series of technologies primarily related to the encapsulation and embedment of repair epoxies in concrete and masonry materials [184]. White developed a similar microencapsulation technique for self-repair in polymer composites [185]. Inman et al. developed self-healing bolt systems where the bolts can sense a loosened state and attempt to retighten themselves [186]. Many biological systems execute self-healing by using vascular networks to sense injury, transport repair agents, and remove injured tissues. A major enabling technology for artificial self-healing materials could be to use vascular networks. Kim and Kim demonstrated the use of acrylonitrile polymers in a vascular self-healing system for concrete [187]. Huston and Esser patented a set of self-healing technologies for wiring systems that included microvascular and foaming techniques [188]. Duenas et al. demonstrated multifunctional polymers with sensing, morphing, and self-healing capabilities [189].

6. *Self-organization*: Biological sensing systems are self-organized. They grow from amorphous collections of cells into very specialized structures, seemingly with minimal central control and guidance. The self-assembly of *ad hoc* wireless sensor networks mimics some of these processes.

7. *Tight coupling and coordination of functions*: Biological systems are prime examples of systems that thrive by the coupling of functions. For example, bones not only provide structural support for vertebrate animals, but they are also key components of the vascular and immune systems. The coupling of structural systems with health monitoring apparatus remains a largely untapped opportunity. An example is the combination of sensors and strength elements in fiber composites.

Ecological systems exhibit highly complex interactions between competing and cooperating organisms and species. In a manner similar to biomimetic design, it may be possible to draw inspiration from ecological systems for the ecomimetic design of engineered systems such as those used in SHM. Interesting, but not well-understood properties of ecological systems are dynamism, efficiency, and robustness. Ecological systems are constantly changing and rarely close to a state of equilibrium. The closest that most ecological systems get to equilibrium is the fairly common occurrence of periodic cycling of populations.

Turbulent-type interactions are also common. Ecomimetic design can provide inspiration in at least two SHM applications. The first application is in energy harvesting systems. Established ecological systems tend to be efficient with interesting underlying thermodynamic principles of organization that waste very little waste of available energy. The second application is in the area of robustness and resilience. Ecological systems are usually very robust. They can sustain very large insults and then adapt, remodel, and recover. The ecomimetic design of systems and systems of systems that result in the robustness, inherent in some ecological systems, is an elusive, but desirable, goal.

10.9   Benchmarking of Designs

Benchmark comparisons establish the relative performance capabilities of various SHM systems. Perhaps due to the relative infancy of the field, only a handful of benchmarking studies appear in the SHM literature. Inspiration may be drawn from the medical radiology community that routinely uses commercially available objects, known as *phantoms*, to benchmark measurement validity and repeatability. It is anticipated that in the future, as SHM technology matures, similar sorts of phantoms may become widely available for SHM benchmarking [190].

Simple benchmark models have the advantage of being relatively low cost and easy to set up, but more importantly have the advantage of enabling a critical examination of the key issues underlying the strengths and weaknesses of various damage identification algorithms. More complicated laboratory benchmark test structures can capture more complicated dynamic behaviors that do not arise in the simpler models. Field tests of full-scale structures are even more complicated and difficult to reproduce. Field tests are notorious for introducing unexpected confounding effects. A major advantage of field testing is the credibility gained by successful methods.

Petitjean et al. describe a benchmark test for evaluating the endurance and sensitivity of SHM systems that detect damage in thick-stack aircraft joints [191]. The test specimen is a simple double lap shear joint with specified dimensions and bolting patterns. The first phase of the testing establishes the long-term survivability of the SHM system with an endurance test that subjects the splice joint to repeated fatigue loads in excess of 100,000 cycles. The intent is to mimic the loads experienced during the lifespan of an airplane. The second phase evaluates the ability of the SHM system to detect a growing crack with the same bolted lap joint. The test introduces small saw cuts defects into the specimen that grow into cracks under cyclic loading.

Vibration-based SHM methods may be particularly suitable for benchmarking studies. Models and test conditions are relatively easy to specify. Experimental and numerical results, that is, mode shapes, frequencies, damping ratios, and so on, are readily transferred in standard formats for analysis by different algorithms. Benchmarking models can range from being fairly simple in form to more complicated structures to full-scale structures.

An example of a relatively simple model for vibration-based damage detection benchmarking is a cantilever beam. Sun and Qin performed a benchmark comparison of four vibration-based damage detection, localization, and assessment algorithms with the cantilever beam made of A3 steel shown in Figure 10.3 [192]. The results of the study were that measuring natural frequency shifts to detect damage was largely ineffective due to the small size of the shifts. Mode shape analysis methods were also found to be relatively insensitive to damage. An analysis of modal flexibility curvature changes were more successful

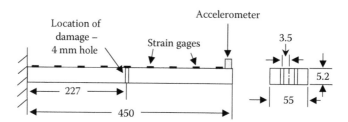

FIGURE 10.3   Cantilever A3 steel beam used in benchmarking studies of vibration-based damage detection, localization, and assessment algorithms by Sun and Qin (dimensions in mm). (Adapted from Sun X and Qin Q. 2006. *Proc 4th China–Japan–US Symp on Structural Control and Monitoring*, Hangzhou.)

in detecting and locating damage, but tended to identify some undamaged locations as being damaged. Modal strain energy methods proved to be the best of the four methods in detecting and locating the damage. Milanese et al. extended the cantilever benchmarking approach to include cubic stiffness nonlinearities at the clamped end and a gap with hard stops at the free end [193].

A shake table-mounted multistory building model forms the basis of a benchmark study of statistical methods of damage detection organized by the International Association for Structural Control (IASC)–ASCE Structural Health Monitoring Task Group [194]. The Phase I study examined a quarter scale four-story steel frame model building at the University of British Columbia. Johnson et al. simulated the multiple degrees of freedom (MDOF) response of the structure due to floor accelerations under various cross-brace removal damage scenarios [195]. The intent of the simulations was to provide a predictive numerical simulation of the effect of various damage types, such as the removal of diagonal bracing. Multiple researchers examined the data and applied different damage detection algorithms. Casciati detected damage imposed on the model with an ANOVA analysis of input–output data [196]. Ching and Beck used a Bayesian modal updating technique to detect damage in the form of weakened diagonal bracing [197]. Yuen et al. detected stiffness changes with a two-stage Bayesian approach [198]. The method incorporated a low-resolution 12 DOF model of the building into the analysis. In many of the cases, the technique identified the changes in stiffness due to the damage. However, the technique had difficulty with detecting small-scale damage and with the use of inaccurate lumped mass-inertia models. Adding the inertia as unknown identifiable parameters resolved some of these issues. Lam et al. used a similar statistical model updating approach based on frequency domain curve fitting [199]. Caicedo et al. identified modal vibration properties and then stiffness changes in a structure with a combination of the *natural excitation technique* and the *eigensystem realization algorithm* [200]. These methods could identify gross damage such as cross-brace removal, but had a bit of difficulty in identifying more minor damage, such as a loosened beam. Bernal and Gunes used a somewhat different approach [201]. The first step fits a state space model to the input and output data. The next step calculated the model mass and flexibility matrices within a multiplicative scalar. The final step identified and located the damage. These methods worked quite well, with the exception of the ability to locate damage from measurements at a reduced number of points. Lus et al. identified damage in the benchmark model with a Kalman filter technique [202]. Hera and Hou identified sudden changes in structural properties due to the simulated damage with wavelets [203]. Yang et al. used the benchmark data to evaluate the performance of a newly developed *recursive least-squares estimator for unknown inputs* (RLSE-UI) [204]. The RLSE-UI algorithm identified unknown structural parameters, changes in the parameters due to damage, and unknown inputs. Lin et al. used HHT methods with success [205].

10.10   Sensor Diagnostics

Virtually any part of a sensor system can fail. Since the data from SHM systems provides input into decisions with potentially large impacts, it is desirable to have systems in place to test and diagnose the health of the SHM system. Several approaches are possible.

1. Smart sensors with onboard diagnostics to detect sensor fault conditions.
2. Use checksums to verify the integrity of data storage and transmission.
3. Apply known stimuli to the structure and observe if the response is as expected.
4. Use redundant sensor systems with overlapping information content. Run statistical analyses, such as PCA, on the data to look for unexpected outliers [206,207]. In many respects this can be a matter of sensors voting to agree upon a correct answer.
5. Conduct routine and systematic analyses of the data as they are collected. This creates a database of observed behaviors for use in identifying anomalous behaviors.
6. Use automated systems to evaluate the integrity of wired networks with probing test signals [208].

10.11   New Structural Systems

New structural systems often offer excellent opportunities for the application of SHM technologies. One reason is that the detailed mechanics, particularly as they relate to damage and failure propensities, are largely unknown. Another is that it is possible to include SHM in early design stages for the structural system. Likewise, many new structural systems engage in some sort of high-performance activity that may open the possibility for increased risks of unexpected damage modes.

Offshore wind energy plants (OWEPs) are examples of novel high-performance structural systems that may benefit from SHM systems [209,210]. Possible uses of an OWEP SHM include (1) structural design verification and enhancement, (2) SHM, (3) environmental load monitoring, and (4) provide information for improved operational control of the OWEP. Possible principal design considerations for an OWEP SHM system are (1) damage detection, (2) sensor fault detection, (3) damage localization, and (4) synchronization of sensor signals. Rolfes et al. used wireless sensor modules with onboard data processing capabilities to validate models of OWEP dynamical behavior [211]. Fritzen and Klinkov developed an OWEP wind load measurement system that estimates the loads with an *unknown input observer* (UIO) combined with a finite element model of the structure [212]. Tests on a scale model OWEP verified the performance of the UIO.

References

1. Alampalli S and Ettouney M. 2003. *Proc of the Workshop on Engineering Structural Health*, New York State DOT, July.
2. Derriso MM, Pratt DM, Homan DB, Schroder JB, and Bortner RA. 2003. Integrated vehicle health management: The key to future aerospace systems. In *Structural Health Monitoring*, FK Chang (ed.). DEStech, Lancaster.
3. Kane AR and Aldayuz JL. 2004. A United States vision for total highway asset management. *Proc IABMAS'04 Bridge Maintenance, Safety, Management and Cost*, Kyoto. Taylor & Francis, London.
4. Buderath M. 2004. Review the process of integrating SHM systems into condition based maintenance as part of the structural integrity programme. *Proc 2nd European Workshop on Structural Health Monitoring*, DEStech, Munich.
5. Trego A, Akdeniz A, and Haugse E. 2004. Structural health management technology on commercial airplanes. *Proc 2nd European Workshop on Structural Health Monitoring*, DEStech, Munich.
6. Kress KP and Richter H. 2003. Overview on the AHMOS Project. In *Structural Health Monitoring*, FK Chang (ed.). DEStech, Lancaster.
7. Aktan E, Ciloglu K, Grimmelsman K, Pervizpour M, and Qin X. 2003. Monitoring the operations, security and structural health of major long-span bridge hyper-systems. In *Structural Health Monitoring*, FK Chang (ed.). DEStech, Lancaster.
8. Chase S and Ghasemi H. 2003. A vision for highway bridges for the 21st century. In *Structural Health Monitoring*, FK Chang (ed.). DEStech, Lancaster.
9. Peil U. 2003. Life-cycle prolongation of civil engineering structures via monitoring. In *Structural Health Monitoring*, FK Chang (ed.). DEStech, Lancaster.
10. Marotta SA, Ooi TK, Gilbert JA, and Bower MV. 2003. Remote readiness asset prognostic and diagnostic system (RRAPDS) near-term applications. In *Structural Health Monitoring*, FK Chang (ed.). DEStech, Lancaster.
11. Ikegami R. 2003. Structural health monitoring: Assessment of aircraft customer needs. *Proc. 2nd Intl Workshop on Structural Health Monitoring*, Stanford. Technomic, Lancaster.
12. Foote PD. 2003. Structural health monitoring: Tales from Europe. *Proc 2nd Intl Workshop on Structural Health Monitoring*, Stanford. Technomic, Lancaster.
13. Liu SC. 1999. Natural hazard mitigation: Exploring the technological frontiers. *Proc 2nd Intl Workshop on Structural Health Monitoring*, Stanford. Technomic, Lancaster.
14. Kudva JN, Grage MJ, and Roberts MM. 1999. Aircraft structural health monitoring and other smart structures technologies—Perspectives on development of future smart aircraft. *Proc 2nd Intl Workshop on Structural Health Monitoring*, Stanford. Technomic, Lancaster.
15. Hall SR. 1999. The effective management and use of structural health data. *Proc 2nd Intl Workshop on Structural Health Monitoring*, Stanford. Technomic, Lancaster.
16. Phares BM, Wipf TJ, Greiman LF, and Lee YS. 2005. *Health Monitoring of Bridge Structures and Components Using Smart Structure Technology*, Wisconsin Highway Research Program, WHRP 05-03, January.
17. Vandiver TL. 1997. Health monitoring of U.S. Army missile systems. In *Structural Health Monitoring Current Status and Perspectives*. F.-K. Chang (ed.). Technomic, Lancaster.
18. Bartelds G. 1997. Aircraft structural health monitoring, prospects for smart solutions from a European viewpoint. In *Structural Health Monitoring Current Status and Perspectives*. F.-K. Chang (ed.). Technomic, Lancaster.
19. Chong KP. 1997. Health monitoring of civil infrastructures. In *Structural Health Monitoring Current Status and Perspectives*. F.-K. Chang (ed.). Technomic, Lancaster.
20. Egawa K. 1997. A new concept of maintenance inspection. In *Structural Health Monitoring Current Status and Perspectives*. F.-K. Chang (ed.). Technomic, Lancaster.

21. Van Way CB, Marantidis C, and Kudva JN. 1994. Design requirements and system payoffs for an on-board structural health monitoring system (SHMS). *Smart Sensing, Processing, and Instrumentation*, SPIE 2191.

22. Moan T. 2005. Reliability-based management of inspection, maintenance and repair of offshore structures. *Struct Infrastruct Eng*, 1(1), 33–62.

23. Uckun S. 2005. Integrated systems health management for space exploration. In *Structural Health Monitoring*, FK Chang (ed.). DEStech, Lancaster.

24. Sundermeyer JN, D'Souza R, Townsend C, and Jones J. 2005. SHIELD (Structural health integrated electronic life determination). In *Structural Health Monitoring*, FK Chang (ed.). DEStech, Lancaster.

25. Ruderman GA. 2005. Health management issues and strategy for air force missiles. In *Structural Health Monitoring*, FK Chang (ed.). DEStech, Lancaster.

26. Kitajima H and Asakura K. 2005. Research and development policy of the advanced maintenance systems for the industrial structures and infrastructures founded by METI challenge to the future. In *Structural Health Monitoring*, FK Chang (ed.). DEStech, Lancaster.

27. Hochmann D and Duke D. 2005. Aircraft structural health monitoring, a vision. In *Structural Health Monitoring*, FK Chang (ed.). DEStech, Lancaster.

28. MacConnell JH. 2005. ISHM and design: A new capability perspective. In *Structural Health Monitoring*, FK Chang (ed.). DEStech, Lancaster.

29. Gaston B and Simmons K. 2007. Health monitoring to ensure fleet readiness. In *Structural Health Monitoring*, FK Chang (ed.). DEStech, Lancaster.

30. Mita A, Iwaswa O, and Ogawa S. 2007. Smart sensor network for biofication of living spaces. In *Structural Health Monitoring*, FK Chang (ed.). DEStech, Lancaster.

31. Egawa K. 1997. A new concept of maintenance inspection. In *Structural Health Monitoring Current Status and Perspectives*. F.-K. Chang (ed.). Technomic, Lancaster.

32. Van der Auweraer H and Peeters B. 2003. Sensors and systems for structural health monitoring. *J Struct Control*, 10, 117–125.

33. Staszewski W, Boller C, and Tomlinson G. 2004. *Health Monitoring of Aerospace Structures*. Wiley, Chichester.

34. Ullman DG. 2003. *The Mechanical Design Process*, 3rd Ed. McGraw-Hill, Boston.

35. Pahl G, Beitz W, and Wallace K. 1988. *Engineering Design—a Systematic Approach*. Springer, New York.

36. Hundal MH. 1997. *Systematic Mechanical Designing: A Cost and Management Perspective*. ASME, New York.

37. Habel WR. 2004. Fiber optic sensors for deformation measurements: Criteria and method to put them to the best possible use. *Smart Sensor Technology Measurement Systems*, SPIE 5384.

38. Hutcheson RS and Tumer IY. 2005. Function-based co-design paradigm for robust health management. In *Structural Health Monitoring*, FK Chang (ed.). DEStech, Lancaster.

39. Suh NP. 1995. Axiomatic design of mechanical systems. *Trans ASME, Special 50th Anniversary Design Issue*, 117(June), 2–10.

40. Perrow C. 1999. *Normal Accidents*. Princeton University Press, Princeton.

41. Braha D and Maimon O. 1998. The measurement of a design structural and functional complexity. *IEEE Trans Syst Man Cybern-Part A: Sys Human*, 28(4), 527–535.

42. Panahandeh M, Budiman ES, Rostam-Abadi F, and Kasper E. 2005. SMARTSIM: An integrated design tool for smart structures. In *Structural Health Monitoring*, FK Chang (ed.). DEStech, Lancaster.

43. Fang M, Liu W, and He Y. 2006. Basic design of structural health & safety monitoring system on Hangzhou Bay Bridge. *Proc 4th China–Japan–US Symp on Structural Control and Monitoring*, Hangzhou.

44. Vestroni F, Vidoli S, and dell'Isola F. 2004. Structural health monitoring based on dynamic measurements. *Proc IABMAS'04 Bridge Maintenance, Safety, Management and Cost*, Kyoto. Taylor & Francis, London.

45. Azzam H, Beaven F, Wallace M, Bryant K, Smith A, Hebden I, Foote P, and McFeat J. 2005. A certifiable approach to structural prognosis health management. In *Structural Health Monitoring*, FK Chang (ed.). DEStech, Lancaster.

46. Reinking JT, Sawyer BE, Corder A, and Ooi TK. 2007. Structural health monitoring of missile communications ground system. In *Structural Health Monitoring*, FK Chang (ed.). DEStech, Lancaster.

47. Hill AR. 2005. A systems engineering approach to structural health management system design. In *Structural Health Monitoring*, FK Chang (ed.). DEStech, Lancaster.

48. Kessler SS. 2005. Certifying a structural health monitoring system: Characterizing durability, reliability and longevity. *First Intl Forum on Integrated System Health Engineering and Management in Aerospace*, Napa.

49. Chambers JT, Wardle BL, and Kessler SS. 2007. Lessons learned from a broad durability study of an aerospace SHM system. In *Structural Health Monitoring*, FK Chang (ed.). DEStech, Lancaster.

50. Robeson E and Thompson B. 1999. Tools for the 21st century: MH-47E SUMS. *Proc 2nd Intl Workshop on Structural Health Monitoring*, Stanford. Technomic, Lancaster.

51. Rybak JM. 2006. Remote condition monitoring using open-system wireless technologies. *Sound Vib*, February, 16–20.

52. Bakker JD, Postema FJ, and Forster U. 2004. Web-based, fully automated remote monitoring of structures. *Proc IABMAS'04 Bridge Maintenance, Safety, Management and Cost*, Kyoto. Taylor & Francis, London.

53. Minser AJ, Schongar G, and O'Connor JS. 2002. Multi-year study to evaluate FRP wrapping of deteriorated concrete columns. In *Proc Structural Materials Technology V, an NDT Conference*, S Allampalli and G Washer, (eds.). American Society for Nondestructive Testing, Cincinnati.

54. Haney MA. 2005. Structural health monitoring of engine exhaust-washed structures. In *Structural Health Monitoring*, FK Chang (ed.). DEStech, Lancaster.

55. Isoe A, Kimoto J, Tsutui H, and Bandoh S. 2003. Conceptual study of the advance load and usage monitoring system using optical glass fiber for a helicopter rotor system. In *Structural Health Monitoring*, FK Chang (ed.). DEStech, Lancaster.

56. Brönnimann R, Nellen PM, and Sennhauser U. 1998. Application and reliability of a fiber optical surveillance system for a stay cable bridge. *Smart Mater Struct*, 7, 229–236.

57. Keller J, Gollhardt A, Vogel D, and Michel B. 2005. Nanoscale deformation measurements for reliability analysis of sensors. *Advanced Sensor Technologies for Nondestructive Evaluation and Structural Health Monitoring*, SPIE 5770.

58. Jemli ME, Karoumi R, and Lanaro F. 2003. Monitoring of the New Arsta Railway Bridge using traditional and fiber optic sensors. *Smart Systems and Nondestructive Evaluation for Civil Infrastructures*, SPIE 5057.

59. Teshigawara M, Isoda H, Kusunoki K, and Kitagawa Y. 2004. An application of structural health monitoring for existing buildings. In *Advanced Smart Materials and Smart Structures Technology*. FK Chang, CB Yun, and BF Spencer, Jr. (eds.). DEStech, Lancaster.

60. Del Grosso A, Torre A, Rosa M, and Lattuada BG. 2004. Application of SHM techniques in the restoration of historical buildings: The Royal Villa of Monza. *Proc 2nd European Workshop on Structural Health Monitoring*, DEStech, Munich.

61. Denton R. 2004. Sensor reliability impact on predictive maintenance program cuts. *Proc Vibration Institute National Technical Training Institute and 28th Annual Meeting*, Bloomingdale.

62. Kulcu E, Qin X, Barrish Jr RA, and Aktan AE. 2000. Information technology and data management issues for health monitoring of the Commodore Barry Bridge. *Nondestructive Evaluation of Highways, Utilities, and Pipelines IV*, SPIE 3995.

63. Barrish RA, Grimmelsman KA, and Aktan AE. 2000. Instrumented monitoring of the Commodore Barry Bridge. *Nondestructive Evaluation of Highways, Utilities, and Pipelines IV*, SPIE 3995.

64. Happold E. 1980. *Appraisal of Existing Structures*. Inst of Struct Eng, London.

65. Li Z, Koh BH, and Nagarajaiah S. 2007. Detecting sensor failure via decoupled error function and inverse input–output model. *J Eng Mech*, 133(11), 1222–1228.

66. Scott M, Bannister M, Herszberg I, Li H, and Thomson R. 2005. Structural health monitoring—the future of advanced composite structures. In *Structural Health Monitoring*, FK Chang (ed.). DEStech, Lancaster.

67. Wakha K, Majed MA, Dasgupta A, and Pines DJ. 2003. Multifunctional piezoelectric stiffness/energy sensor for monitoring the health of structures. *Smart Structures and Integrated Systems*, SPIE 5056.

68. Tayebi A and Ul Hoque MM. 2003. Design of experiments optimization of embedded MEMS sensors in composites for structural health monitoring. *Smart Systems and Nondestructive Evaluation for Civil Infrastructures*, SPIE 5057.

69. Kousourakis A, Mouritz AP, and Bannister MK. 2005. Effect of internal sensor galleries on the mechanical properties of aerospace composite laminates. In *Structural Health Monitoring*, FK Chang (ed.). DEStech, Lancaster.

70. ACI Committee. 2003. *In-Place Methods to Estimate Concrete Strength*. American Conference Institute, ACI 228.1R-03.

71. Kroening M, Berthold A, and Meyendorf N. 2005. Sensor modules for structural health monitoring and reliability of components. *Advanced Sensor Technologies for Nondestructive Evaluation and Structural Health Monitoring*, SPIE 5770.

72. French MJ. 1988. *Invention and Evolution Design in Nature and Engineering*. Cambridge University Press, Cambridge.

73. Ben-Haim Y, Cempel C, Natke HG, and Yao JT. 1993. Evaluation of diagnostic methods. In *Safety Evaluation Based on Identification Approaches*, HG Natke, GR Tomlinson, and JTP Yao (eds). Vieweg, Braunschweig.

74. Celebi M, Sanli A, Sinclair M, Gallant S, and Radulescu D. 2003. Real-time seismic monitoring needs of a building owner and the solution. In *Structural Health Monitoring*, FK Chang (ed.). DEStech, Lancaster.

75. de Silva V and Ghrist R. 2006. Coordinate-free coverage in sensor networks with controlled boundaries via homology. *Intl J Robot Res*, 25, 1205.

76. *Civil Engineering*. 1993. Dam rehab conference stresses safety issues. ASCE, 63(7), 16.

77. Lauridsen J. 2004. Bridge owner's benefits from probabilistic approached—experiences and future challenges. *Proc IABMAS'04 Bridge Maintenance, Safety, Management and Cost*, Kyoto. Taylor & Francis, London.

78. Sumitro S, Okarnoto T, and Inaudi D. 2004. Intelligent sensory technology for health monitoring based maintenance of infrastructures. *Smart Sensor Technology Measurement Systems*, SPIE 5384.

79. Frangopol M, Lin KY, and Estes AC. 1997. Life-cycle cost design of deteriorating structures. *J Struct Eng*, 123(10), 1390–1401.

80. Honjo Y, Ueki J, Sumitro S, Matsui Y, and Kato Y. 2001. Influences of quality and frequency of inspection on maintenance of civil engineering structures. *Proc JSCE Annual Conference*, Vol. 56, pp. CS6-004.

81. Koch GH and Thompson NG. 1991. Corrosion monitoring as a means to increase maintenance efficiency in aircraft. Paper no. 342, *Proc NACE Annual Conference*. Houston, TX.

82. Sanders T, Hess G, Davidson J, Ooi T, and Corder A. 2007. Multifunctional diamond sensor development for structural health monitoring. In *Structural Health Monitoring*, FK Chang (ed.). DEStech, Lancaster.

83. Hobbs GK. 2000. *Accelerated Reliability Engineering HALT and HASS*. Wiley, Chichester.

84. Callaway Jr EH. 2004. *Wireless Sensor Networks*. Auerbach, Boca Raton.

85. Arritt BJ, Buckley SJ, Ganley JM, Kumar A, Clayton E, Hannum R, Todd MD, Kennel MB, Welsh J, Beard S, Stabb MC, Xinlin Q, and Wegner P. 2007. Responsive satellites and the need for structural health monitoring. In *Structural Health Monitoring*, FK Chang (ed.). DEStech, Lancaster.

86. Stone AR. 1952. Stress indicating bolt or stud. US Patent 2,600,029.

87. Popenoe CH. 1990. Opti-mechanical displacement indicator with high sensitivity. US Patent 4,904,132.

88. Kibblewhite IE. 1990. Ultrasonic load indicating member, apparatus and method. US Patent 4,899,591.

89. McKee B, Shkarlet YM, Khatkate A, Banas T, Ingram R, and Perkins D. 1998. Electromagnetic smart washer for detecting bolthole cracking. *NASA Tech Briefs*, TSP MFS-26479.

90. Reichard K, Banks J, Conlon S, Swanson D, and Kozlowski J. 2005. Comparison of prognostic health monitoring system architectures and implementations. In *Structural Health Monitoring*, FK Chang (ed.). DEStech, Lancaster.

91. Mitchell J, Bond T, Bever K, and Manning N. 1998. MIMOSA—Four years later. *Sound Vib*, November, 12–21.

92. Cui Y, Gao RX, and Kazmer DO. 2005. Energy efficiency analysis of a self-powered pressure sensor using bond graph. *Sensors and Smart Structures Technologies for Civil, Mechanical, and Aerospace Systems*, SPIE 5765.

93. Deshmukh A, Pepyne D, Hanlon AB, and Hyers RW. 2005. Distributed sensor network for thermal protection systems of future spacecraft. *First Intl Forum on Integrated System Health Engineering and Management in Aerospace*, Napa, CA.

94. Medelius PJ, Eckhoff AJ, Angel LR, and Perotti JM. 2002. Advanced self-calibrating, self-repairing data acquisition system. US Patent 6,462,684.

95. Shuster LA. 2004. Redundancy pays off in positioning caissons for new Tacoma Narrows Bridge. *Civil Eng*, 74(3), 30–31.

96. Krishnamachari B. 2005. *Networking Wireless Sensors*. Cambridge University Press, Cambridge.

97. Braddee RW, Hause MG, and Pratt SG. 2000. Worker deaths by falls—a summary of surveillance findings and investigative case reports. DHHS (NIOSH) Publication no. 2000-116.

98. Hong CS, Kim CG, Kwon IB, and Park JW. 1996. Simultaneous strain and failure sensing of composite beam using an embedded fiber optic extrinsic Fabry–Perot sensor. *Smart Sensing, Processing and Instrumentation*, SPIE 2718.

99. Liang YC and Hwu C. 2001. On-line identification of holes/cracks in composite structures. *Smart Mater Struct*, 10, 599–609.

100. US Army Corps of Engineers. 1980. *Engineering and design instrumentation for concrete structures*. EM 1110-2-4300, Department of the Army, Office of the Chief of Engineers, Washington, DC.

101. Schaaf K, Cook B, Ghezzo F, Starr A, and Nemat-Nasser S. 2005. Mechanical properties of composite materials with integrated embedded sensor networks. *Sensors and Smart Structures Technologies for Civil, Mechanical, and Aerospace Systems*, SPIE 5765.

102. Kapoor C, Graves-Abe TL, and Pei JS. 2005. Development of an off-the-shelf field programmable gate array-based wireless sensing unit for structural health monitoring. *Sensors and Smart Structures Technologies for Civil, Mechanical, and Aerospace Systems*, SPIE 5765.

103. Padula SL and Kincaid RK. 1999. *Optimization Strategies for Sensor and Actuator Placement*. NASA/TM-1999-209126, NASA Langley Research Center, Hampton, Virginia.

104. Beck JL, Papadimitriou C, Au SK, and Vanik MW. 1998. Entropy-based optimal sensor location for structural damage detection. *Smart Systems for Bridges, Structures, and Highways*, SPIE 3325.

105. Riordan J. 2002. *Introduction to Combinatorial Analysis*. Dover Publications, Mineola, NY.

106. Wu HC, Warnemuende K, Yan A, and Mu B. 2003. Strategic sensor locations of FRP bridge decks. *Smart Nondestructive Evaluation and Health Monitoring of Structural and Biological Systems II*, SPIE 5047.

107. Palacz M, Ostachowicz W, Hoernlein H, Buderath M, and Boller C. 2004. Impacts of fatigue crack monitoring on structural optimisation using finite elements. *Proc 2nd European Workshop on Structural Health Monitoring*, DEStech, Munich.

108. Bayleyegn YS and Fanous FS. 2006. Field testing and finite element analysis of steel bridge retrofits for distortion-induced fatigue. Paper no. 06-1878, 86th Annual Meeting, TRB, Washington, DC.

109. Weis M, Bilgram R, Bautz B, Breu C, and Drechsler K. 2005. An investigation of strain distribution for fibre Bragg grating sensor applications. In *Structural Health Monitoring*, FK Chang (ed.). DEStech, Lancaster.

110. Olson SE, Leonard MS, and Malkin MC. 2007. Analytical modeling to develop SHM techniques for aircraft 'hot spots'. In *Structural Health Monitoring*, FK Chang (ed.). DEStech, Lancaster.

111. Facchini M, Botsis J, and Sorensen L. 2007. Measurements of temperature during fatigue of a thermoplastic polymer composite using FBG sensors. *Smart Mater Struct*, 16, 391–398.

112. Hosser D, Klinzmann C, and Schnetgoeke R. 2004. Optimization of structural assessment using reliability-based system assessment. *Proc 2nd European Workshop on Structural Health Monitoring*, DEStech, Munich.

113. Schnetgoeke R, Klinzmann C, and Hosser D. 2005. Structural health monitoring of a bridge using reliability based assessment. In *Structural Health Monitoring*, FK Chang (ed.). DEStech, Lancaster.

114. Kim SH, Kim BJ, and Yoon C. 1998. Optimum monitoring system design using artificial neural network and multi-level sensitivity analysis. *Structural Engineering Worldwide*, Elsevier, Amsterdam, T200-4.

115. Li DS, Li HN, and Guo XL. 2003. Computer simulation of optimal sensor locations in loading identification. *Modeling, Signal Processing and Control*, SPIE 4049.

116. Kang YK, Park HC, Hwang W, and Han HS. 1996. Optimum placement of piezoelectric sensor/actuator for vibration control of laminated beams. *AIAA J*, 34(9), 1921–1926.

117. Johnson RC. 1961. *Optimum Design of Mechanical Elements*. Wiley, New York.

118. Lasdon LS. 2002. *Optimization Theory for Large Systems*. Dover Publications, Mineola, NY.

119. Papalambros PY and Wilde DJ. 1988. *Principles of Optimal Design*. Cambridge University Press, Cambridge.

120. De Fonseca P, Sas P, and van Brussel H. 1999. A comparative study of methods for optimising sensor and actuator locations in active control applications. *J Sound Vib*, 221(4), 651–679.

121. Maghami PG and Joshi SM. 1993. Sensor/actuator placement for flexible space structures. *IEEE Trans Aerosp Electron Syst*, 29(2), 345–351.

122. Papadopoulos M and Garcia E. 1998. Sensor placement methodologies for dynamic testing. *AIAA J*, 36(2), 256–263.

123. Guratzsch RF. 2007. *Sensor placement optimization under uncertainty for structural health monitoring systems of hot aerospace structures*. Doctoral dissertation, Vanderbilt University.

124. Zhang G. 2005. *Optimum sensor localization/selection in a diagnostic/prognostic architecture*. Doctoral dissertation, Electrical Engineering, Georgia Institute of Technology.

125. Kammer DC. 1996. Optimal sensor placement for modal identification using system realization methods. *J Guid Control Dyn*, 19(3), 729–731.

126. Cherng AP. 2003. Optimal sensor placement for modal parameter identification using signal subspace correlation techniques. *Mech Syst Signal Process*, 17(2), 361–378.

127. Erisson-Jackson AJ, Bainum PM, and Xing G. 1997. Actuator/sensor displacement using degree of controllability and observability for digitally controlled orbiting platforms. *J Aeronaut Sci*, 45(1), 73–89.

128. Yan TH and Lin RM. 2006. General optimization of sizes or placement for various sensors/actuators in structure testing and control. *Smart Mater Struct*, 15, 724–736.

129. Meo M and Zumpano G. 2004. Optimal sensor placement on a large scale civil structure. *Health Monitoring and Smart Nondestructive Evaluation of Structural and Biological Systems III*, SPIE 5394.

130. Heo G, Wang ML and Satpathi D. 1997. Optimal transducer placement for health monitoring of long span bridge. *Soil Dyn Earthq Eng*, 16, 495–502.

131. Song Y, Wang JX and Liu LJ. 2006. Optimum sensor placement for damage detection of bridges. *Proc 4th China–Japan–US Symposium on Structural Control and Monitoring*, Hangzhou.

132. Gaitanaros S, Karaiskos G, Papadimitriou C, and Aravas N. 2007. Crack identification in structures using optimal experimental design. In *Structural Health Monitoring*, FK Chang (ed.). DEStech, Lancaster.

133. Gao RX, Wang C, and Sheng S. 2004. Optimal sensor placement strategy and sensor design for high quality system monitoring. *Sensors and Smart Structures Technologies for Civil, Mechanical and Aerospace Systems*, SPIE 5391.

134. Gawronski W. 1997. Actuator and sensor placement for structural testing and control. *J Sound Vib*, 208(1), 101–109.
135. Basseville M. 2004. On sensor positioning for structural health monitoring. *Proc 2nd European Workshop on Structural Health Monitoring*, DEStech, Munich.
136. Raich AM and Liszkai TR. 2003. Multi-objective genetic algorithm methodology for optimizing sensor layouts to enhance structural damage identification. In *Structural Health Monitoring*, FK Chang (ed.). DEStech, Lancaster.
137. Trendafilova I, Heylen W, and Van Brussel H. 2001. Measurement point selection in damage detection using the mutual information concept. *Smart Mater Struct*, 10, 528–533.
138. Sazonov E and Klinkhachorn P. 2005. Optimal spatial sampling interval for damage detection by curvature or strain energy mode shapes. *J Sound Vib*, 285, 783–801.
139. Worden K and Burrows AP. 2001. Optimal sensor placement for fault location. *Eng Struct*, 23, 885–901.
140. Azarbayejani M, El-Osery A, Choi KK, and Reda Taha M. 2007. Optimal sensor placement for efficient structural health monitoring. In *Structural Health Monitoring*, FK Chang (ed.). DEStech, Lancaster.
141. Fijany A and Vartan F. 2006. A new and efficient algorithm for analyzing and optimizing the system of sensors. *IEEE Aerospace Conference*, Big Sky.
142. Mishalani RG and Gong L. 2006. Optimal spatial sampling of infrastructure condition: A life-cycle-based approach under uncertainty. Paper no. 06-0637, 86th Annual Meeting, TRB, Washington, DC.
143. Miyamoto A and Matsukawa T. 2006. Development of a health monitoring system for deteriorating sewage pipe network. *Proc 4th China–Japan–US Symp on Structural Control and Monitoring*, Hangzhou.
144. Yang K, Araki H, Yabe A, Wu Z, and Li S. 2006. The optimum length of long-gage FBG sensors for structural health monitoring of flexure RC members. *Proc 4th China–Japan–US Symp on Structural Control and Monitoring*, Hangzhou.
145. Orlowska A, Kolakowski P, and Holnicki-Szulc J. 2004. Monitoring of delamination defects— dynamic case. *Proc 2nd European Workshop on Structural Health Monitoring*. DEStech, Munich.
146. Spillman Jr WB and Huston DR. 1998. Techniques for creating low-cost spatially weighted fiber optic sensors for structural health monitoring. In *Structural Engineering Worldwide*. Elsevier, Amsterdam, T200-4.
147. Hyland D and Fry G. 1999. A neural-genetic hybrid approach for optimizing structural health monitoring systems. *Proc 2nd Intl Workshop on Structural Health Monitoring*, Stanford. Technomic, Lancaster.
148. Nguyen VV, Smarsly K, and Hartmann D. 2007. A computational steering approach towards sensor placement optimization for structural health monitoring using multi-agent technology and evolutionary algorithms. In *Structural Health Monitoring*, FK Chang (ed.). DEStech, Lancaster.
149. Kang F, Li J, and Liu D. 2007. Strategy for optimizing sensor placement related to structural health monitoring. *Proc Intl Conf on Health Monitoring Structure Mater and Environment*, Nanjing.
150. Wang BT. 1996. Optimal placement of microphones and piezoelectric transducer actuators for far-field sound radiation control. *J Acoust Soc Am*, 99(5), 2975–2984.
151. Guo HY, Zhang L, Zhang LL, and Zhou JX. 2004. Optimal placement of sensors for structural health monitoring using improved genetic algorithms. *Smart Mater Struct*, 13, 528–534.
152. Reich GW and Sanders B. 2003. Structural shape sensing for morphing aircraft. *Smart Structures and Integrated Systems*, SPIE 5056.
153. Li J, Kapania WB, and Spillman Jr WB. 2004. Placement optimization of fiber optic sensors for a smart bed using genetic algorithms. *10th AIAA/ISSMO Multidisciplinary Analysis and Optimization Conference*, Albany, AIAA-2004-4334.
154. Spillman Jr W and Huston D. 1997. Detection, location and characterization of point perturbations over a two dimensional area using two spatially weighted distributed fiber optic sensors. *Smart Sensing, Processing, and Instrumentation*, SPIE 3042.

155. Worden K, Tomlinson GR, and Burrows AP. 1994. Fault detection employing transducer optimisation procedures. *Second European Conf on Smart Structures and Materials*, Glasgow, SPIE 2361.

156. Tumer IY and Stone RB. 2003. Mapping function to failure mode during component development. *Res Engrg Des*, 14, 25–33.

157. Sazonov E, Janoyan K, and Jha R. 2004. Wireless intelligent sensor network for autonomous structural health monitoring. *Smart Sensor Technology Measurement Systems*, SPIE 5384.

158. Gu H, Jin P, Zhao Y, Lloyd GM, and Wang ML. 2004. Design and experimental validation of a wireless PVDF displacement sensor for structure monitoring. *Nondestructive Detection and Measurement for Homeland Security II*, SPIE 5395.

159. Wu J, Yuan S, Zhao X, Yin Y, and Ye W. 2007. A wireless sensor network node designed for exploring a structural health monitoring application. *Smart Mater Struct*, 16, 1898–1906.

160. Shi ZY, Law SS, and Zhang LM. 2000. Optimum sensor placement for structural damage detection. *J Eng Mech*, 126(11), 1173–1179.

161. Eckhoff A, Lucena A, Medelius P, and Perotti J. 2001. Advanced data acquisition technology self-healing, self-calibrating signal conditioner. *NASA Tech Briefs*, TSP KSC-12301, January.

162. Beard RV. 1971. *Failure accommodation in linear systems through self-reorganization*. Doctoral dissertation, Aeronautics and Astronautics, MIT.

163. Alsum J and Fish P. 2002. NDT applications utilized in Hoan Bridge failure. *Proc Structural Materials Technology V, an NDT Conference*, Cincinnati.

164. Fisher JW, Wright W, Sivakumar B, Kaufmann EJ, Xi Z, Edberg W, and Tijang H. 2001. *Hoan bridge forensic investigation failure analysis*. Final Report, Wisconsin Department of Transportation and Federal Highway Administration.

165. Box GP, Hunter WG, and Hunter JS. 1978. *Statistics for Experimenters*. Wiley, New York.

166. Farrar CR and Doebling SW. 1997. Lessons learned from applications of vibration-based damage identification methods to a large bridge structure. In *Structural Health Monitoring Current Status and Perspectives*. F.-K. Chang (ed.). Technomic, Lancaster.

167. Felber A. 1997. Practical aspects of testing large bridges for structural assessment. In *Structural Health Monitoring Current Status and Perspectives*. F.-K. Chang (ed.). Technomic, Lancaster.

168. Hipley P and Huang M. 1997. CALTRANS/CSMIP bridge strong motion instrumentation. *Proc 2nd National Seismic Conf on Bridges and Highways*, Sacramento.

169. Walter PL. 2004. Air-blast and the science of dynamic pressure measurements. *Sound Vib*, December, 10–16.

170. Walter PL. 2007. Selecting accelerometers for mechanical shock measurements. *Sound Vib*, December, 14–18.

171. Choi MY, Kang KK, Kim JW, and Park HY. 2003. The real-time health monitoring system of a large structure based on non-destructive testing. *Smart Nondestructive Evaluation and Health Monitoring of Structural and Biological Systems II*, SPIE 5047.

172. Wahbeh M, Tasbihgoo F, Yun H, Masri SF, Caffery JP, and Chassiakos AG. 2005. Real-time earthquake monitoring of large-scale bridge structures. In *Structural Health Monitoring*, FK Chang (ed.). DEStech, Lancaster.

173. Gorinevsky D, Gordon GA, Beard SA, Kumar A, and Chang FK. 2005. Design of integrated SHM system for commercial aircraft applications. In *Structural Health Monitoring*, FK Chang (ed.). DEStech, Lancaster.

174. Yun H, Nayeri R, Tasbihgoo F, Wahbeh M, Caffrey J, Wolfe R, Nigbor R, Masri SF, Abdel-Ghaffar A, and Sheng LH. 2008. Monitoring the Collision of a Cargo Ship with the Vincent Thomas Bridge. *Struct Control Health Monit*, 15, 183–206.

175. Mijarez R, Gaydecki P, and Burdekin M. 2007. Flood member detection for real-time structural health monitoring of sub-sea structures of offshore steel oilrigs. *Smart Mater Struct*, 16, 1857–1869.

176. Beckerman M. 1997. *Adaptive Cooperative Systems*. Wiley, New York.

177. Weiss TF. 1996. *Cellular Biophysics Vol. 2: Electrical Properties*. MIT Press, Cambridge.

178. Argyrakis P, Cheung R, Hamilton A, Webb B, Zhang Y, and Gonos T. 2007. Fabrication and characterization of a biomorphic wind sensor for integration with a neuron circuit. *Microelectron Eng*, 84, 1749–1753.
179. Nolte J. 1999. *The Human Brain: An Introduction to Functional Anatomy*, 4th Ed. Mosby, St. Louis.
180. Inada H, Okuhara Y, and Kumagai H. 2004. Experimental study on structural health monitoring of RC columns using self-diagnosis materials. *Sensors and Smart Structures Technologies for Civil, Mechanical and Aerospace Systems*, SPIE 5391.
181. Kirikera G, Datta S, Schulz MJ, Ghosal A, Sundaresan MJ, Feaster J, and Hughes D. 2003. Recent advances in artificial neural system for structural health monitoring. *Nondestructive Evaluation and Health Monitoring of Aerospace Materials and Composites II*, SPIE 5046.
182. Eckelmeyer Jr EH. 1946. The story of the self-sealing tank. *US Naval Inst Proc*, 72(2), 205–219.
183. Mercier P. 1896. Material for protecting vessels, recepticles, etc. US Patent 561,905.
184. Dry C. 2003. Self-repairing reinforced matrix material. US Patent 6,527,849.
185. White SR, Sottos NR, Geubelle PH, Moore JS, Kessler MR, Sriram SR, Brown EN, and Viswanathan S. 2001. Autonomic healing of polymer composites. *Nature*, 409(6822), 794–797.
186. Inman DJ, Muntges DE, and Park G. 2001. Investigation of a self-healing bolted joint employing a shape memory actuator. *Smart Structures and Integrated Systems*, SPIE 4327.
187. Kim WJ and Kim IS. 2004. Development of self-diagnosis in addition to ability of repair concrete for damage. *Smart Sensor Technology Measurement Systems*, SPIE 5384.
188. Huston D and Esser B. 2007. Self-healing cable apparatus and method. US Patent 7,302,145.
189. Duenas T, Jha A, Lee W, Bortolin R, Mal A, Ooi TK, and Corder A. 2007. Structural health monitoring with self-healing morphing skins. In *Structural Health Monitoring*, FK Chang (ed.). DEStech, Lancaster.
190. Pherigo GL and Pherigo AL. 2000. Implanting flaws for NDT validation. In *Structural Materials Technology IV*, S Allampalli (ed.). Technomic, Lancaster.
191. Petitjean B, Simonet D, Choffy JP, and Barut S. 2004. SHM technology benchmark for damage detection. *Proc 2nd European Workshop on Structural Health Monitoring*. DEStech, Munich.
192. Sun X and Qin Q. 2006. Nondestructive damage detection methods and damage detection accuracy for beam-type structures. *Proc 4th China–Japan–US Symp on Structural Control and Monitoring*, Hangzhou.
193. Milanese A, Marzocca P, and Nichols JM. 2007. Modeling of randomly excited nonlinear structures for the benchmark of SHM techniques. In *Structural Health Monitoring*, FK Chang (ed.). DEStech, Lancaster.
194. Dyke SJ, Bernal D, Beck JL, and Ventura C. 2001. An experimental benchmark problem in structural health monitoring. *Proc 3rd Intl Workshop on Structural Health Monitoring*, Stanford.
195. Johnson EA, Lam HF, Katafygiotis LS, and Beck JL. 2004. Phase I IASC-ASCE structural health monitoring benchmark problem using simulated data. *J Eng Mech*, 130(1), 3–15.
196. Casciati S. 2004. Statistical models comparison for damage detection using the ASCE benchmark. *Proc 2nd European Workshop on Structural Health Monitoring*. DEStech, Munich.
197. Ching J and Beck JL. 2004. Bayesian analysis of the phase II IASC-ASCE structural health monitoring experimental benchmark data. *J Eng Mech*, 130(10).
198. Yuen KV, Au SK, and Beck JL. 2004. Two-stage structural health monitoring approach for phase I benchmark studies. *J Eng Mech*, 130(1), 16–33.
199. Lam HF, Katafygiotis LS, and Mickleborough NC. 2004. Application of statistical model updating approach on phase I of the IASC–ASCE structural health monitoring benchmark study. *J Eng Mech*, 130(1), 34–48.
200. Caicedo JM, Dyke SJ, and Johnson EA. 2004. Natural excitation technique and eigensystem realization algorithm for phase I of the IASC–ASCE benchmark problem: Simulated data. *J Eng Mech*, 130(1), 49–60.
201. Bernal D and Gunes B. 2004. Flexibility based approach for damage characterization: Benchmark application. *J Eng Mech*, 130(1), 61–70.
202. Lus H, Betti R, Yu J, and De Angelis M. 2004. Investigation of a system identification methodology in the context of the ASCE benchmark problem. *J Eng Mech*, 130(1), 71–84.

203. Hera A and Hou Z. 2004. Application of wavelet approach for ASCE structural health monitoring benchmark studies. *J Eng Mech*, 130(1), 96–104.

204. Yang JN, Pan S, and Lin S. 2007. Least-squares estimation with unknown excitations for damage identification of structures. *J Eng Mech*, 133(1), 12–21.

205. Lin S, Yang JN, and Zhou L. 2005. Damage identification of a benchmark building for structural health monitoring. *Smart Mater Struct*, 14, S162–S169.

206. Friswell MI and Inman DJ. 1999. Sensor validation for smart structures. *J Intel Mater Syst Struct*, 10, 973–982.

207. Kerschen G, De Boe P, Golinval JC, and Worden K. 2004. Sensor validation for on-line vibration monitoring. *Proc 2nd European Workshop on Structural Health Monitoring*, DEStech, Munich.

208. Furse C, Smith P, Lo C, Chung YC, Pendayala P, and Nagoti K. 2005. Spread spectrum sensors for critical fault location on live wire networks. *Struct Control Health Monit*, 12, 257–267.

209. Rohrmann RG, Rücker W, and Thöns S. 2007. Integrated monitoring systems for offshore wind turbines. In *Structural Health Monitoring*, FK Chang (ed.). DEStech, Lancaster.

210. Kraemer P and Fritzen CP. 2007. Concept for structural damage identification of offshore wind energy plants. In *Structural Health Monitoring*, FK Chang (ed.). DEStech, Lancaster.

211. Rolfes R, Zerbst S, Haake G, Reetz J, and Lynch JP. 2007. Integral SHM-system for offshore wind turbines using smart wireless sensors. In *Structural Health Monitoring*, FK Chang (ed.). DEStech, Lancaster.

212. Fritzen CP and Kinkov M. 2007. Online wind load estimation for the offshore wind energy plants. In *Structural Health Monitoring*, FK Chang (ed.). DEStech, Lancaster.

11

Case Studies

This chapter is a collection of some examples of SHM studies. Most of these are drawn from the experience of the author.

11.1 Shoring Load Measurements

Measuring the loads and the state of a structure while it is being built can provide information that leads to improvements in the quality, safety, and efficiency of construction processes. The following describes load measurements on a multistory reinforced concrete slab style building.

The construction of multistory reinforced concrete buildings typically requires the use of temporary formwork and vertical shores to support fresh concrete through curing solidification. Key issues are for the formwork to support the fresh and partially cured concrete without collapse and excessive deflection, and formwork removal as soon as the concrete has sufficient strength for self-support. Additionally, formwork and shoring should be lightweight to permit easy handling and economies of usage. Shoring and formwork failures can cause catastrophic structural collapses. The primary causes of formwork failures are (1) excessive loads, (2) premature removal of forms or shores, and (3) inadequate lateral support for shoring members [1]. Hadipriono and Wang reported that over a period of 25 years, more than 85 collapses in the United States were directly attributed to formwork or shoring failures, with 48% being due to inadequate vertical shores [2].

There are many factors that affect the behaviors of slab-formwork (or slab-shore) systems. A three-category classification is

1. *Design decisions*—concrete types, column and slab sizes, reinforcement arrangement, and so on.
2. *Construction schedules*—shoring and reshoring, construction cycle, method of concrete placement such as concrete bucket dropping, pumping, concrete placing paths, and so on.
3. *Environmental conditions*—ambient temperatures, wind conditions, and other possible natural excitations.

The development of rational design guidelines for safe construction should account for all of these factors and the associated uncertainties so that

$$\text{Resistance} \geq \text{Effect of factored loads.} \tag{11.1}$$

Loads on building formwork are primarily vertical gravity loads and applied lateral loads. The gravity loads consist of both dead and live loads. ACI 347 recommends a minimum design load for combined dead and live vertical loads of 100 psf (5.0 kPa) or 125 psf (6.25 kPa) when using buggies to transport the concrete. Dead loads tend to have less variation than live loads. Horizontal loads can be treated similarly as suggested by ACI 347 (1988) as either 100 lb/linear ft (1.5 kN/m) or 2% of the total superposed dead load, whichever is larger. Both ACI 347-88 and ANSI Standard A10.9 require a minimum design wind load of 15 psf (720 Pa).

An inherent limitation of computer or hand calculations is the difficulty in anticipating unusual or extreme circumstances, such as fires, or unanticipated local component failures. An additional complication is deviations from the shoring system design and/or planned construction sequence. A computer model may not correctly represent the loads present at the actual site [2]. Lee et al. examined the failure probability of time-dependent reinforced concrete structures [3]. They showed that a structure during construction is much more dangerous than one during its normal usage. Epaarachchi et al. extended these approaches to the reliability analyses of multistory buildings during construction [4].

In an effort to measure construction load variability, Huang measured dead and live loads at 20 construction sites in Taiwan during concrete pumping [5]. The average dead load was about 110 psf (5.5 kPa). This included slab and beam weights, as well as some part of the column weight. The coefficient of variation (COV) of dead load was 0.22 higher than that of the service dead load. The average live load was about 15 psf (0.72 kPa).

With this background in mind, an attempt was made to measure shoring loads during the construction of a multistory building. In the summer of 1997, a major concrete construction site—the *Boston Museum Towers* (BMT)—in Cambridge, MA, became available for testing. This site was a multiple building complex with two 25-story, one 5-story, and one 7-story reinforced concrete buildings.

The 25-story South Tower was the subject of this investigation. The construction plan called for each story to be virtually identical, with 8-in-thick floor slabs. Economic considerations, along with the desire to avoid cold-weather concreting, favored fast-pace construction. Approximately every 4 or 5 working days saw the placing of concrete for a new floor. Sustaining such a pace without resorting to the use of more expensive rapidly curing concrete required a multilevel shoring system.

An interesting aspect of the construction was the use of the Peri™ shoring system. This formwork system uses a drophead arrangement that allows for the removal of floor slab forms without removing the shoring. The drophead eliminates the need for dangerous reshoring operations. Figure 11.1 shows the Peri shores with the floor slab formwork stripped from the dropheads and with a three-axis load cell in the load path. Eliminating reshoring allows for a faster construction schedule. Using drophead shoring appears to be a cost-saving improvement that also promotes safety. A disadvantage of drophead shoring is that it requires more individual pieces and increases the complexity of managing the shoring equipment inventory.

The measurements collected multistory and multi-axis shoring load data from June 2, 1997 to August 29, 1997. The measurement locations for each floor comprised two positions in a large bay room. Figure 11.1 is a photograph of a load cell installed in the shoring. Figures 11.2 and 11.3 indicate the positions. A measurement cycle consisted of placing two load cells at locations *A* and *B*, as indicated in Figure 11.3, and recording the data during pouring and curing operations on that floor. The load cell placement on the next floor was at the same locations directly above the instrumented shores for the shores below. Repeated placement on subsequent floors allowed for the collection on up to five stories of shores in a single column. The reason for using five-story floor cycles was twofold. One was that the shoring system used four and five stories per cycle. The other was the finite length of

FIGURE 11.1   Peri™ shoring with formwork stripped from dropheads and an in-line three-axis load cell.

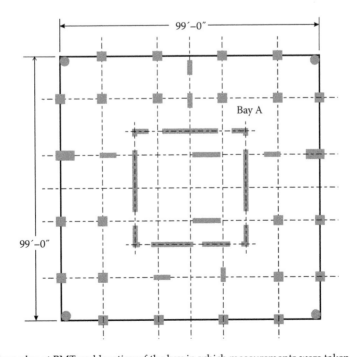

FIGURE 11.2   Floor plan at BMT and location of the bay in which measurements were taken.

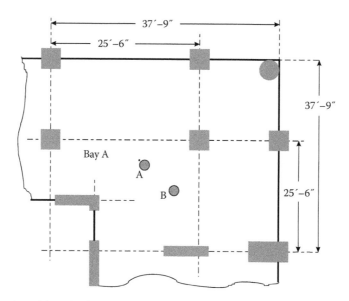

FIGURE 11.3   Location of shore load measurements at BMT.

the load cell cables. A rugged steel crate housed the centralized data acquisition system. Due to a possible excess of steel armor, the heavy data acquisition crate required a crane to lift it up to the higher floors.

Data collection started on the eigth floor and continued in a fairly continuous manner up to the 22nd floor. Interruptions in the data acquisition process occurred due to power losses, missed opportunities for load cell placement, and the water-induced shorting of load cell cables. The data acquisition system could run at a rate in excess of 20 kHz. In an effort to avoid collecting excessive amounts of data, the data acquisition rates were reduced to once every 10 min for the curing operations, and set at once every second during concrete pouring and placing operations. Nonetheless, the resulting size of the data files was substantial.

The first exploratory pass at data analysis separated the data into 24-h pieces. The next pass examined the data for any spurious signals. Spurious signals occasionally appeared as wild full-scale swings in the recorded data. The probable cause was the water hosing of the load cells and formwork before placing the concrete and the inadvertent inducement of electrical shorts in the cables.

Tables 11.1 through 11.3 show the 24-h maximum vertical and horizontal load for the months of June, July, and August. Consideration was given to (1) the relative magnitude of the horizontal loads and (2) the correlation of horizontal loads with vertical loads. Figure 11.4 shows the lateral load effects (in one direction) for the small pour area as a percentage of the vertical load acting at the same time. The pouring activities occurred approximately between 15 and 30 min. As Figure 11.4 suggests, when the majority of the vertical load is already applied, the lateral load is about ±10% of the corresponding vertical load. The reported percentage variations are somewhat higher earlier during the pour activity. This corresponds to points at which the actual vertical load is low; thus, the lateral load effect would not be large. Nearly identical trends were observed in the other lateral direction as well as for both lateral directions for the large pour area. This value of ±10% provides a simple guideline for approximating lateral loads on formwork.

TABLE 11.1

Maximum Z-Direction Vertical Loads Measured (N) at BMT in July 1997

Date	Comments	1st Story	2nd Story	3rd Story	4th Story	5th Story
7/1/97	13th story pouring	12,410	18,192	11,787	NA	NA
7/1/97		NA	20,928	22,965	13,117	NA
7/2/97		NA	24,010	24,927	13,064	NA
7/3/97		NA	22,436	24,664	14,376	NA
7/4/97		NA	20,536	25,136	13,798	NA
7/5/97		NA	19,509	23,770	13,099	NA
7/6/97		NA	19,602	24,366	13,558	NA
7/7/97		NA	18,437	22,280	11,943	NA
7/9/97	14th story pouring	NA	NA	NA	NA	NA
7/10/97		14,145	NA	22,507	17,125	NA
7/11/97		15,390	NA	22,329	14,100	NA
7/12/97		14,945	NA	21,261	13,833	NA
7/13/97		15,612	NA	22,151	23,396	NA
7/14/97		13,922	NA	20,016	20,906	NA
7/15/97	15th story pouring	NA	NA	NA	NA	NA
7/17/97		16,609	27,916	NA	25,016	29,201
7/18/97		17,490	24,348	NA	23,325	27,168
7/19/97		12,018	16,146	NA	12,930	15,733
7/20/97		12,748	17,227	NA	16,974	17,761
7/21/97		12,072	16,080	NA	15,648	NA
7/22/97	16th story pouring	11,293	14,398	NA	10,622	NA
7/22/97		NA	19,269	20,358	NA	15,421
7/23/97		NA	19,015	18,908	NA	13,411
7/24/97		NA	17,525	17,632	NA	15,261
7/25/97		NA	15,341	16,097	NA	16,382
7/26/97		NA	16,916	19,313	NA	24,010
7/27/97		NA	15,964	18,152	NA	22,115
7/28/97		NA	15,915	18,993	NA	22,707
7/29/97	17th story pouring	NA	14,189	16,031	NA	18,917
7/29/97		NA	NA	17,031	18,370	NA
7/30/97		NA	NA	18,521	18,908	NA
7/31/97		NA	NA	16,982	18,824	NA

Note: NA, not available.

Some conclusions from this study are as follows:

1. The observed lateral shore loads were typically on the order of ±10% of the corresponding (i.e., simultaneous) vertical load.

2. A magnification factor on the order of 2.0 may be appropriate to account for the (spatial) variability among a group of shores in a common pour area. The degree of variability shore-to-shore appears to be a result of the inherent static indeterminacy of shoring systems. Slight differences in shore length, structural geometry, and/or precompression may induce large variations in the in-service loads.

TABLE 11.2

Maximum *X*-Direction Horizontal Loads Measured (N) at BMT in July 1997

Date	Comments	1st Story	2nd Story	3rd Story	4th Story	5th Story
7/1/97	13th story pouring	NA	−276	658	NA	−125
7/1/97		NA	NA	−325	667	NA
7/2/97		NA	NA	−365	676	NA
7/3/97		NA	NA	−360	689	NA
7/4/97		NA	NA	−369	689	NA
7/5/97		NA	NA	−347	672	NA
7/6/97		NA	NA	−360	676	NA
7/7/97		NA	NA	−320	658	NA
7/9/97	14th story pouring	NA	NA	NA	NA	NA
7/10/97		−347	NA	NA	−302	NA
7/11/97		−342	NA	NA	−294	NA
7/12/97		−351	NA	NA	−285	NA
7/13/97		−351	NA	NA	−360	NA
7/14/97		−351	NA	NA	−325	NA
7/15/97	15th story pouring	NA	NA	NA	NA	NA
7/17/97		−262	−316	NA	NA	−440
7/18/97		271	−387	NA	NA	−414
7/19/97		67	−391	NA	NA	−294
7/20/97		76	−387	NA	NA	−302
7/21/97		62	−378	NA	NA	NA
7/22/97	16th story pouring	62	−374	NA	NA	NA
7/22/97		NA	89	−351	NA	NA
7/23/97		NA	80	−387	NA	NA
7/24/97		NA	53	−387	NA	NA
7/25/97		NA	76	−387	NA	NA
7/26/97		NA	76	−378	NA	NA
7/27/97		NA	58	−387	NA	NA
7/28/97		NA	53	−391	NA	NA
7/29/97	17th story pouring	NA	76	−387	NA	NA
7/29/97		NA	NA	80	−396	NA
7/30/97		NA	NA	80	−400	NA
7/31/97		NA	NA	80	−400	NA

Note: NA, not available.

3. The suggested ACI design load, including equipment, appears to be adequate for the slab areas considered in this study. The simpler tributary analysis design load well approximates the average shore load, whereas the more conservative ACI load (including equipment) well approximates the maximum shore load. However, an error in the installation of a shore may result in that shore being seriously overloaded.

4. The use of drophead shoring can eliminate highly dangerous reshoring operations, while possibly improving the efficiency of construction operations.

5. An area effect that can reduce the maximum shore load for a given pour area may exist (i.e., the ACI design load becomes more conservative as the effective slab area

TABLE 11.3

Maximum *Y*-Direction Horizontal Loads (N) Measured at BMT in July 1997

Date	Comments	1st Story	2nd Story	3rd Story	4th Story	5th Story
7/1/97	13th story pouring	NA	165	−209	NA	NA
7/1/97		NA	NA	182	−200	NA
7/2/97		NA	NA	182	−227	NA
7/3/97		NA	NA	187	−209	NA
7/4/97		NA	NA	187	−205	NA
7/5/97		NA	NA	187	−236	NA
7/6/97		NA	NA	191	−209	NA
7/7/97		NA	NA	169	−214	NA
7/9/97	14th story pouring	NA	NA	NA	NA	NA
7/10/97		−254	NA	NA	187	NA
7/11/97		−267	NA	NA	182	NA
7/12/97		−267	NA	NA	187	NA
7/13/97		−294	NA	NA	236	NA
7/14/97		−254	NA	NA	209	NA
7/15/97	15th story pouring	NA	NA	NA	NA	NA
7/17/97		449	−512	NA	NA	236
7/18/97		445	−485	NA	NA	231
7/19/97		44	−387	NA	NA	178
7/20/97		44	−409	NA	NA	187
7/21/97		53	−374	NA	NA	NA
7/22/97	16th story pouring	49	−320	NA	NA	NA
7/22/97		NA	67	−463	NA	NA
7/23/97		NA	44	−440	NA	NA
7/24/97		NA	−22	−414	NA	NA
7/25/97		NA	22	−369	NA	NA
7/26/97		NA	13	−467	NA	NA
7/27/97		NA	−27	−454	NA	NA
7/28/97		NA	−31	−471	NA	NA
7/29/97	17th story pouring	NA	−13	−391	NA	NA
7/29/97		NA	NA	31	−454	NA
7/30/97		NA	NA	−36	−471	NA
7/31/97		NA	NA	71	−480	NA

Note: NA, not available.

increases). A reduction in design load for construction may be a function of pour area, slab thickness, formwork arrangement, and concrete placement procedures.

6. Significant load variations in steel shores may result from large daily (or other) temperature variations. These variations were particularly evident during curing periods, even when there were no additional externally applied loads.

7. The shore load decreases by a small amount during the curing period. This decrease in shore load is likely a function either of creep effects in the concrete or of gain in strength and stiffness of the slab. The decrease in load appears to become greater with increasing effective slab area.

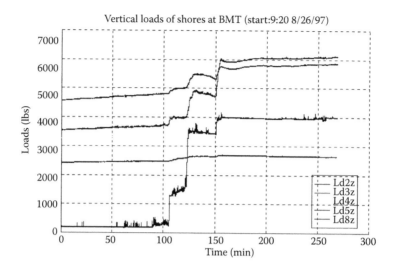

FIGURE 11.4 Typical multistory vertical shore load data collected during concrete placement on the top floor.

8. Shore removal can induce significant loads on the slab and shores, often of an impact nature. The magnitude of these loads is a function of both the amount of compression in the shore at the time of its removal and the removal procedure used. Shore removal loads take two forms: (1) additional compression due to the removal procedure and (2) load redistribution following the removal of a nearby shore. The amount of load redistribution is largely a function of the supporting formwork arrangement. Very large effective impact loads may be imparted to the (early-age) slab during formwork removal.

Acknowledgments: Tim Ambrose, Wai-Fah Chen, Peter Fuhr, Minhaj Kirmani, Matt Nelson, David Rosowsky, Weidlinger Assocs., and the U.S. National Institute of Occupational Safety and Health assisted and supported this project.

11.2 Cure Monitoring of Composite Laminates Used in the Manufacture of Snowboards

Snowboards are high-performance structures used in winter sports (Figure 11.5). The nature of the recreational sports business is such that snowboards need to be manufactured economically with a minimum of variation and defects and be able to survive demanding usage [6]. Competition to build the lightest, most responsive, and durable snowboard has prompted exploration into the fields of automated methods for monitoring the cure of composite snowboards during manufacture.

A typical snowboard is a multilayer laminate composite made of *ultrahigh molecular weight* (UHMW) plastic, *glass fiber-reinforced polymer* (GFRP), wood, and acrylonitrile butadiene styrene (ABS) plastic (Figure 11.6). The layers of importance for this study are two GFRP composite layers with thicknesses on the order of 1 mm. The GFRP layers bond the other layers together to add flexural strength along the major axes of a snowboard (tip to tail and width). The glass fibers have either a [0/90] orientation (biaxial) or a [0/+45/−45]

FIGURE 11.5 Top view of a snowboard.

orientation (triaxial) that depends on the design specifications. The epoxy-curing agent mix requires a heat-cure process (Figure 11.7).

The temperature versus time details of the cure cycle are a major consideration in efficient manufacturing and snowboard performance. Different cure cycles can result in parts with material properties that vary significantly. Controlling the cure process can be essential to ensuring end products with consistent material properties. Difficult to control variables include material properties, such as resins and fibers. Such variations can make it difficult to control the cure with an open-loop cycle. Using embedded cure sensors allows for eliminating or reducing end-product variability and production run setup costs.

The first step following part sizing impregnates glass fibers with a premixed two-part epoxy (Shell EPON® Resin 828 and EPI-CURE® Curing Agent 9552). The combination of the resin and curing agent in the premix initiates an exothermic cure reaction. The next steps lay up the uncured impregnated glass with the other material layers and place the stack into a heated press. The press is a hydraulic-driven system that controls the pressure (690 kPa—nominal) and temperature (85°C—nominal) during the cure cycle. The press cavity is preshaped to the desired board curvature. The end of the cure cycle sees the removal of the board from the mold. A fully cured board keeps the curvatures imposed by the mold. Final steps include shaping and finishing. Curing transforms the resin from a liquid to a solid state in a process that forms a 3-D network of covalently bonded polymer chains. During the first stage of cure, the resin has low viscosity and can impregnate the reinforcing fibers. As the cure reaction begins, molecular chains lengthen and form branches. The resin transforms into a vitrified state just prior to gelation. Next, the reaction continues into solidification and eventually reaches a plateau where the resin becomes insoluble and fully cured. Multiple physical properties change during this transformation and have the potential to form cure-monitoring measurands. These include dielectric coefficients, glass

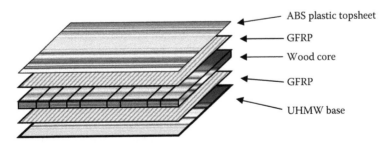

ABS plastic topsheet
GFRP
Wood core
GFRP
UHMW base

FIGURE 11.6 Schematic of the layers of a snowboard. (From Pelczarski N and Huston D. 2000. *Nondestructive Evaluation of Aging Materials and Composites IV*, SPIE 3993.)

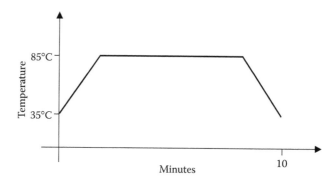

FIGURE 11.7 Typical cure cycle for a composite snowboard. (Adapted from Pelczarski N and Huston D. 2000. *Nondestructive Evaluation of Aging Materials and Composites IV*, SPIE 3993.)

transition temperature, thermal stability, tensile modulus, chemical resistance, strain to failure, viscosity, dynamic modulus, and fracture toughness.

There are several controllable variations in the cure schedule that may alter the end state of the snowboard. A higher-than-recommended cure temperature causes the resin to harden too quickly. This may cause *wet-out*—a state where there is not enough time for the resin to impregnate the fibers. The result is voids and other defects. Low-temperature cure cycles increase manufacturing time and risk pulling the board from the mold prematurely. This may adversely affect interlayer adhesion and final shape. Available *in situ* GFRP composite cure-monitoring techniques include dielectrometry, acoustic emissions, ultrasonics, fiber optics, and vibrational techniques [7–9]. The basis of dielectrometry is that most polymers exhibit dielectric macroscopic behaviors due to the molecular structure of polymer composite materials containing dipoles and ions that align with externally applied electric fields. The combination of ionic mobility and dipolar motion determines the dielectric constant [10]. Since ionic mobilities and dipolar motions change during the cure cycle of epoxies, capacitive measures of dielectric properties can indicate the state of cure [11]. For two parallel plate electrodes, the capacitance, C (F), is

$$C = \frac{\varepsilon_0 \varepsilon_r A}{d},$$ (11.2)

where ε_r is the relative dielectric constant, ε_0 is the permittivity of air, A is the area of the electrodes, and d is the distance between the electrodes.

The change in elastic properties of the resin enables using ultrasonic techniques for cure monitoring [12,13]. Most monitoring efforts measure the phase velocity and attenuation of ultrasonic waves as a function of cure. The phase velocity tends to increase with cure as the epoxy stiffens with the cross-linking of molecules. Typically the attenuation increases until the epoxy reaches the gel point, and then decreases as the epoxy hardens. Li and Menon monitored the cure of composite jackets used to retrofit concrete structures such as bridge columns with embedded waveguides that consisted of a *U*-shaped wire and an accompanying parallel wire [14]. The operational principle of the waveguides appears in Figure 11.8. An ultrasonic transducer launches gated sine waves into one end of the *U*-shaped waveguide. Ultrasound pickup transducers measure received signals at both the other end of the *U*-shaped waveguide and the parallel waveguide. While the epoxy is in a liquid state, there is an impedance mismatch and very little acoustic energy transfers from the waveguide to the epoxy. As the epoxy cures, it becomes stiffer. The impedance match

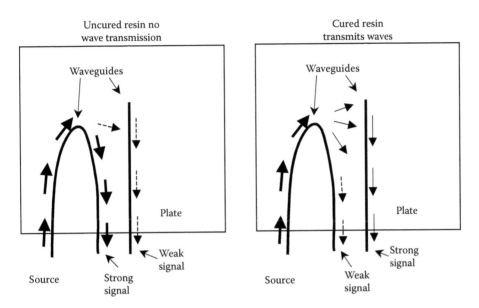

FIGURE 11.8 Ultrasound signal transmission between waveguides for cured and uncured polymer composites. (From Li Y and Menon S. 1998. *Sensors*, 15(2), 5–10.)

between the waveguide and epoxy is stronger and elastic waves leak out of the waveguide, especially around the curve. The wave leakage causes a reduced signal pickup at the end of the *U*-shaped waveguide and an increased signal at the parallel waveguide. In some respects the parallel waveguide provides redundant information.

A series of tests evaluated the effects of different cure parameters on capacitance and ultrasound measurements. The tests used a constant press-pressure of 689 kPa (100 psi). The design variables included three different temperatures: 66°C, 85°C (normal recommended temperature), and 100°C. The primary data collected were (1) capacitance versus time of cure and (2) wave attenuation versus time of cure (Figure 11.9). Embedding

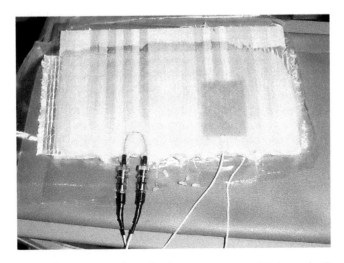

FIGURE 11.9 Waveguide and capacitance electrode placement on a cured test sample. (From Pelczarski N and Huston D. 2000. *Nondestructive Evaluation of Aging Materials and Composites IV*, SPIE 3993.)

cure-monitoring sensors into a snowboard requires accommodating constraints due to the thin geometry of the GFRP layers and the need for testing in a hot composite press. Two cure-monitoring systems satisfied the geometric and testing constraints. The first was a parallel plate capacitive transducer with the GFRP as the internal dielectric layer. The second system used two PZT ultrasonic transducers in a *U*-shaped waveguide (Valpey-Fisher VP-1093 Pinducers). A thin copper waveguide conducted the elastic waves from the transducers. An AC voltage excited the PZT into longitudinal oscillations. The waveguides geometry allows for the pinducer to remain outside of the curing epoxy and reusable. However, coupling of the copper waveguide and ultrasonic transducers required precise and consistent mechanical alignment.

An *inductance, current, and resistance* (LCR) meter measured the dielectric properties. The operating principle is to apply a constant-amplitude sinusoidal voltage waveform with a frequency of 10 kHz to a parallel plate capacitor sandwich with the curing composite in the center, and to measure the magnitude and phase angle of the response. The ultrasound measurements used a five-cycle sinusoidal burst at 0.8 Vp-p and 600 kHz. A pinducer converted the electrical test signal into mechanical oscillations that propagate as elastic waves through the waveguide. A second pinducer coupled to the opposite end of the waveguide changed the mechanical wave into an electrical signal.

The propagation of elastic waves in a waveguide is frequency and boundary condition dependent. Determining which modes are present in the copper waveguide can help to quantify stiffness values of the epoxy matrix during cure. A first step is to determine the time of flight of the wave through the waveguide from one pinducer to the other. The nondimensional frequency and wavenumber parameters are

$$\Omega = \frac{\omega a}{\delta c_2},\tag{11.3}$$

$$\bar{\xi} = \xi a,\tag{11.4}$$

where Ω is a nondimensional frequency parameter, ω is a radial wave frequency, $\delta = 3.832$ is the lowest nonzero root of the first-order Bessel function, ξ is the wavenumber (1/wavelength), a is the radius of the waveguide, and c_2 is the wave velocity. An evaluation of the typical test parameters above indicates that there is one longitudinal mode present in the waveguide for this frequency and testing configuration and that the group velocity is approximately equal to the compression wave phase velocity of copper of 4270 m/s [15].

The capacitance versus cure curves showed features of interest in the form of rounded corners (Figure 11.10). Since the electrode geometry remains constant during the test, capacitance changes are due to changes in the dielectric properties of the resin, which depends on dipolar and ionic molecular mobility. Initially, when the epoxy is in a low-viscosity state, the dipolar mobility and capacitance are high. As the gel state begins to form, the capacitance drops. This is attributed to the onset of cross-links in the resin, which limit dipolar mobility. The change in capacitance between Corners 1 and 2 suggests that the mobility of the dipoles is decreasing rapidly. At Corner 2 the rate of change slows. This suggests that the resin is no longer liquid-like. This correlates with the ultrasonic plot where the viscosity increase represents the fall in signal amplitude from Corner 2 to 3. During this period the capacitance changes slightly, since the resin is in a solid phase. Capacitance Corner 3 occurs consistently and can be a good indicator of the end of cure.

Figure 11.11 shows typical measurements of the relation between the ultrasonic waveguide source-to-barrier transmission and time of cure. Initially the transmission increases up to Corner 1. An explanation of the increase is the fact that the decreased

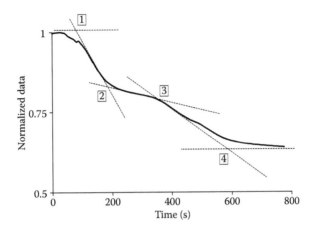

FIGURE 11.10 Capacitance plot with corner markers for a sample cured at 85°C. (Adapted from Pelczarski N and Huston D. 2000. *Nondestructive Evaluation of Aging Materials and Composites IV*, SPIE 3993.)

viscosity of the heated resin reduces dissipative elastic energy losses. It is believed that cross-linking begins between Corners 1 and 2. In the approach to Corner 2, the resin changes into a gel. At this point the resin surrounding the waveguide supports elastic waves and energy leaks into the surrounding medium. As the process continues, larger amounts of ultrasonic energy leak into the hardening resin. Corner 2 represents the start of this phase. The decreasing amount of soluble molecules in the resin likely causes the dramatic decrease in signal strength between Corners 2 and 3. Once solidified, the epoxy can support ultrasonic waves because of the increased stiffness due to cross-linking. At Corner 3, almost the entire wave energy leaks into the surrounding material. Fukuda and Osaka report similar results using a different transducer geometry [16]. The cross-linking slows down because of the lack of remaining soluble molecules, until the cure becomes essentially complete [14].

A comparison of the capacitance and ultrasound versus cure plots indicates that some of the features of interest occurred at the same time. For the tests at 85°C, Corners 1 and 2 occur at the same time. Similarly Capacitance Corner 3 is close to Ultrasound Corner 3

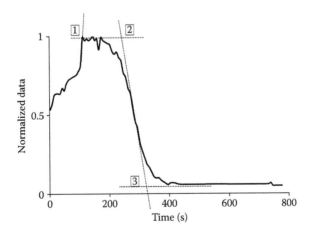

FIGURE 11.11 Ultrasound transmissibility versus curing time at 85°C with corners marked. (Adapted from Pelczarski N and Huston D. 2000. *Nondestructive Evaluation of Aging Materials and Composites IV*, SPIE 3993.)

on the ultrasonic plots. These corners represent the end of the cure reaction. This suggests that most of the changes in the capacitance and ultrasound data are due to the same phase transitions and, likewise, the end of cure. The ultrasonic and capacitance plots from samples cured at 66°C show a significant increase in the necessary time of cure. Corner 3 on the capacitance plots appears approximately 100 s later than it does in the 85°C plots. The ultrasonic transmission curves of the 85°C and 100°C samples differ from that of the 66°C sample. The latter two displayed a dramatic change in stiffness of the resin from Corners 2 to 3. However, the 66°C sample shows that this phenomenon occurs at a much slower rate. The slow rate of change of the ultrasonic plots suggests that the rate of cross-linking and chain extension is slower.

A conclusion is that ultrasound and capacitance measurements are possibly useful for monitoring the state of cure of snowboards during manufacture. Using such techniques for production runs requires additional design developments to optimize sensor configuration and to manage sensor cable ingress and egress issues. Therefore, embedding sensors in each snowboard is potentially costly and time-consuming. Nonembedded press-integrated systems may be more practical [17].

Acknowledgments: Noel Pelczarski conducted much of the data collection in this study. Burton Snowboards provided test materials. More details can be found in Refs. [6] and [18].

11.3 Building Vibration Measurements

It is desired for many buildings to have minimal levels of floor vibrations. Buildings that house precision scientific instruments are an example. The following describes a three-building vibration performance survey at the University of Vermont. Building 1 Health Sciences Research Facility (HSRF) was a newly constructed scientific research building that houses potentially vibration-sensitive scientific instruments. The HSRF building is five-storied with welded steel construction. Upon completion of the construction, the initial occupants anecdotally complained of excessive floor vibrations. This precipitated an effort to measure and reduce the vibrations. As a baseline, the study included vibration measurements at two older buildings. Building 2 is the Given Building (in the Medical Alumni wing). The building has a four-story, steel frame with masonry walls construction. Building 3 was the Votey Building with a three-story reinforced concrete slab frame. All three buildings appear in Figure 11.12.

The test protocol was to measure floor vibrations with piezoelectric accelerometers in selected rooms in each of the three buildings. The accelerometer configuration was a combination of a single triaxial set at one point and assorted uniaxial vertical sensors at other points. The accelerometers had a rated sensitivity of 0.001 g, which was barely adequate. A PC-based digital data acquisition system collected vibration data over a period of approximately 24 h at each location with a sampling frequency of 100 Hz. The amount of collected data was sufficiently large to complicate storage and analysis efforts. Statistical summaries identified the maximum acceleration for each minute, the maximum root mean square, and one-third octave spectra.

Figure 11.13 shows the maximum vertical accelerations per minute measured in the New Medical Research Building Room 407. Since the design of most scientific instruments uses passive techniques to suppress high-frequency vibrations, filtering the raw acceleration data prior to calculating extreme value statistics has the potential of providing additional useful information, especially for assessing potential problems with delicate instruments.

FIGURE 11.12 Three buildings at the University of Vermont: (a) HSRF building—steel frame with bricks; (b) Given building—steel frame; (c) Votey building—reinforced concrete. (Adapted from Huston DR et al. 2002. *Nondestructive Evaluation and Health Monitoring of Aerospace Materials and Civil Infrastructures*, pp. 237–245, 18 June 2002, SPIE 4704.)

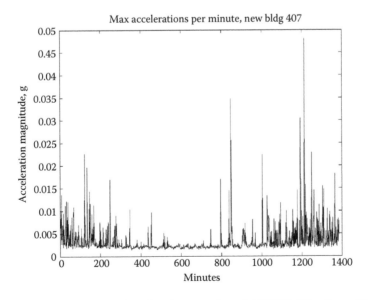

FIGURE 11.13 Maximum absolute vertical accelerations per minute in HSRF Building Room 407. (Adapted from Huston DR et al. 2002. *Nondestructive Evaluation and Health Monitoring of Aerospace Materials and Civil Infrastructures*, pp. 237–245, 18 June 2002, SPIE 4704.)

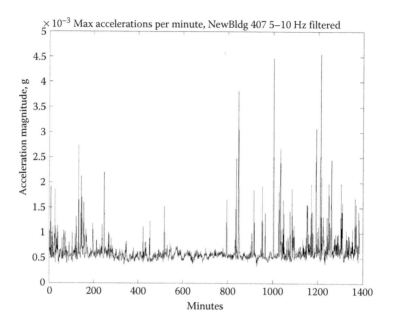

FIGURE 11.14 Maximum absolute vertical accelerations in HSRF Building Room 407 with 5–10 Hz band-pass filter. (Adapted from Huston DR et al. 2002. *Nondestructive Evaluation and Health Monitoring of Aerospace Materials and Civil Infrastructures*, pp. 237–245, 18 June 2002, SPIE 4704.)

To assess frequency-dependent effects, a 5–10 Hz band-pass filter conditioned the data. Figure 11.14 shows the maximum vertical accelerations in the New Medical Research Building Room 407 with a 5–10 Hz band-pass filter. The next step processed the acceleration data into a one-third octave spectrum and then plotted the result versus the BBN criteria (Figure 11.15). A histogram of the acceleration data appears in Figure 11.16.

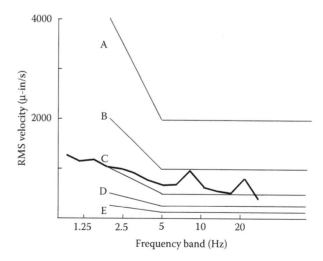

FIGURE 11.15 A one-third octave spectrum of acceleration with BBN criteria in HSRF Building Room 407. (Adapted from Huston DR et al. 2002. *Nondestructive Evaluation and Health Monitoring of Aerospace Materials and Civil Infrastructures*, pp. 237–245, 18 June 2002, SPIE 4704.)

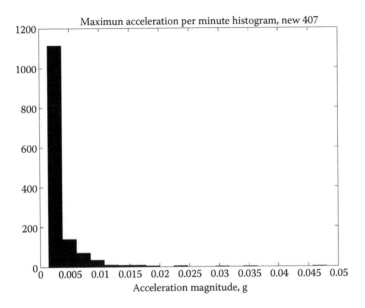

FIGURE 11.16 Histogram of vertical floor acceleration magnitudes in HSRF Building Room 407. (Adapted from Huston DR et al. 2002. *Nondestructive Evaluation and Health Monitoring of Aerospace Materials and Civil Infrastructures*, pp. 237–245, 18 June 2002, SPIE 4704.)

Similar analysis techniques processed the data collected in Given Building Room 165. Figure 11.17 shows the maximum acceleration per minute measured in Given Building Room 165. An examination of the data indicates that large acceleration spikes occur approximately every 7 min with quiescent periods between the spikes. A close-up examination of associated time histories revealed that the acceleration spikes were large enough to saturate

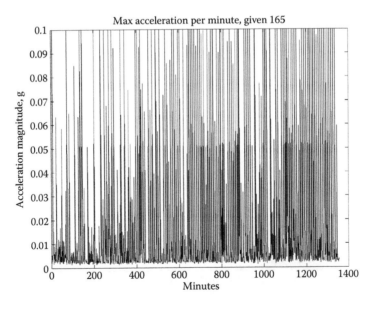

FIGURE 11.17 Maximum acceleration per minute in Given Building Room 165 with spikes due to on/off cycling of a nearby compressor. (Adapted from Huston DR et al. 2002. *Nondestructive Evaluation and Health Monitoring of Aerospace Materials and Civil Infrastructures*, pp. 237–245, 18 June 2002, SPIE 4704.)

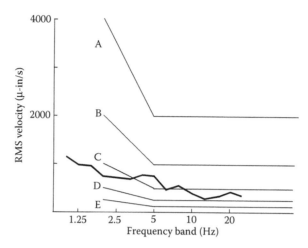

FIGURE 11.18 A one-third octave spectrum for vibrations between large spikes in Given Building Room 165. (Adapted from Huston DR et al. 2002. *Nondestructive Evaluation and Health Monitoring of Aerospace Materials and Civil Infrastructures*, pp. 237–245, 18 June 2002, SPIE 4704.)

the accelerometer amplifier, which prevented the acquisition of useful data during the spike and for a short period (~30 s) afterwards. The cause of these spikes was attributed to the on/off cycling of a cooling compressor for a nearby electron microscope. Data collected in between the initiation of the spikes and after the saturated amplifiers recovered appeared to be valid. Figure 11.18 shows the one-third octave spectrum for vibrations between large spikes. Figure 11.19 is a histogram of vibrations including the vibration spikes, which appear in the bin on the right side of the histogram. Figure 11.20 is the maximum acceleration per minute for Votey Building Room 116. The floor in this room was a relatively thick (280 mm) reinforced concrete slab on grade with a distance of approximately 100 m from a nearby

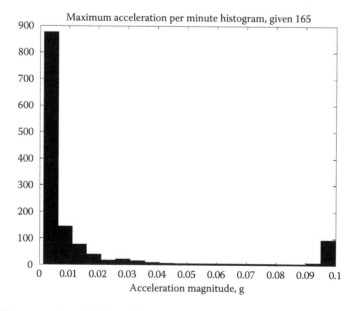

FIGURE 11.19 Histogram of vertical floor vibration magnitudes in Given Building Room 165. (Adapted from Huston DR et al. 2002. *Nondestructive Evaluation and Health Monitoring of Aerospace Materials and Civil Infrastructures*, pp. 237–245, 18 June 2002, SPIE 4704.)

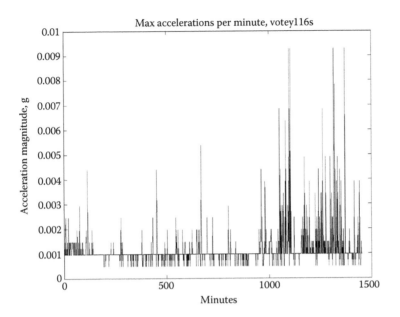

FIGURE 11.20 Maximum accelerations per minute in Votey Building Room 116. Accelerations are so small that ADC quantization is a concern. (Adapted from Huston DR et al. 2002. *Nondestructive Evaluation and Health Monitoring of Aerospace Materials and Civil Infrastructures*, pp. 237–245, 18 June 2002, SPIE 4704.)

heavily traveled road. An examination of these data indicated the presence of moderate quantization problems due to the low level of vibrations. Figure 11.15 shows the one-third octave spectrum of the accelerations versus the BBN criteria. Figure 11.21 is the histogram of the acceleration magnitudes. Note that the histogram has a Rayleigh-like appearance.

An examination of the data indicated that the reinforced concrete building (Votey) is the quietest—slightly larger than BBN level D. The histograms indicated that the data were

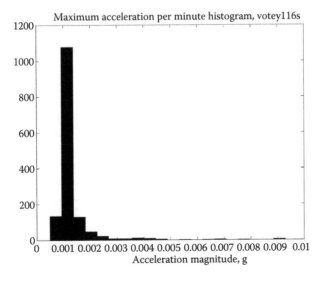

FIGURE 11.21 Histogram of vibrations in Votey Building Room 116. (Adapted from Huston DR et al. 2002. *Nondestructive Evaluation and Health Monitoring of Aerospace Materials and Civil Infrastructures*, pp. 237–245, 18 June 2002, SPIE 4704.)

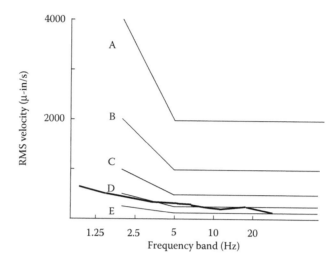

FIGURE 11.22 A one-third octave spectrum for vibrations between large spikes in Given Building Room 165. (From Huston DR et al. 2002. *Nondestructive Evaluation and Health Monitoring of Aerospace Materials and Civil Infrastructures.* pp. 237–245. SPIE 4704.)

fairly Gaussian. The steel-framed building with masonry walls (Given) had large acceleration spikes. The data were distinctly non-Gaussian. In between the spikes, the vibrations were relatively small—slightly larger than BBN level C. The welded steel building is the liveliest, with a BBN level B. However, anecdotal reports are that scientific instruments mounted on commercially available vibration isolation tables are generally able to operate successfully in this lively vibration environment (Figure 11.22).

Acknowledgments: Brian Esser, James Plumpton, and Xiangdong Zhao assisted with collecting and analyzing the data in these tests.

11.4 Frequency Shifting of a Highly Compressed Beam

This study examines the free vibrations of Bernoulli–Euler beams under axial loads that are large fractions of the Euler buckling values. Classical theory predicts that the fundamental frequencies decrease linearly with the axial load and equal zero at the buckling load. However, experiments on compressed columns exhibit different behavior. The natural frequency initially decreases with an increase in axial load, but increases rapidly with an approach to the buckling load. This phenomenon is explained qualitatively in terms of a model that incorporates initial imperfections in the beam shape. The sequel describes the mechanical model and results of simple experiments that corroborate the model.

The beam under consideration is simply supported and axially loaded with a force P that coincides with the nominal centroidal axis (Figure 11.23). The vibrations are small linear oscillations that can occur about a position of nonlinear static equilibrium [19]. LaGrange's equations with quadratic approximations of the virtual work of the axial force and of the kinetic and strain energies can describe linearized versions of the small oscillations.

The assumed beam geometry is nominally straight with slight imperfections so that the centroidal axis of the unloaded beam deviates slightly from that of a corresponding perfectly straight version. Compressing the beam causes it to bow out laterally due to initial

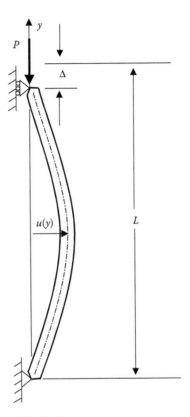

FIGURE 11.23 Geometry of a buckled slender beam.

imperfection [20]. This lateral deformation remains relatively small until the buckling load is approached. It is the vibrations about this position of static equilibrium that give rise to the observed frequency-shifting phenomenon.

The total lateral deflection of the beam measured at the centroidal axis, $u(y, t)$, is

$$u(y, t) = u_0(y) + u_b(y) + u_v(y, t), \tag{11.5}$$

where $u_0(y)$ is the initial crooked shape of the beam, $u_b(y)$ is the static deflection due to bending caused by the axial load, and $u_v(y, t)$ is the vibration about the deflected position. Since the beam is pin-ended, a Fourier sine series expansion of the deflection components satisfies the boundary conditions

$$u_0(y) = \sum_{n=1}^{\infty} a_n \sin\left(\frac{n\pi y}{L}\right), \tag{11.6}$$

$$u_b(y) = \sum_{n=1}^{\infty} b_n \sin\left(\frac{n\pi y}{L}\right), \tag{11.7}$$

$$u_v(y, t) = \sum_{n=1}^{\infty} c_n \sin\left(\frac{n\pi y}{L}\right). \tag{11.8}$$

Using a one-term truncated series approximation gives

$$u_0(y) \approx a_1 \sin\left(\frac{\pi y}{L}\right),$$ (11.9)

$$u_b(y) \approx b_1 \sin\left(\frac{\pi y}{L}\right),$$ (11.10)

$$u_v(y, t) \approx c_1(t) \sin\left(\frac{\pi y}{L}\right).$$ (11.11)

The principle of stationary potential energy provides a convenient means of calculating the static equilibrium configuration, $u_s(y) = u_0(y) + u_b(y)$. The total potential energy, U_T, is the sum of the strain energy in bending, U_b, and the potential energy due to the conservative axial compressive force U_P:

$$U_T = U_b + U_P$$ (11.12)

It is possible to ignore the strain energy due to axial compression because it does not contribute to the lateral deflections and does not alter the lateral vibration frequencies. The strain energy in bending, based on an assumption of small radii of curvature, is

$$U_b = \frac{1}{2} \int_0^L EI \left(u_b''\right)^2 \, dy.$$ (11.13)

The potential energy due to the axial load is

$$U_P = -P\Delta.$$ (11.14)

Δ is the axial deflection due to the bending of the beam, that is,

$$\Delta = \int_0^L \left\{\left[1 + (u')^2\right]^{1/2} - 1\right\} \, dy$$ (11.15)

or

$$\Delta = \int_0^L \left\{\left[1 + \frac{1}{2}(u')^2 - \frac{1}{8}(u')^4 + \cdots\right] - 1\right\} \, dy.$$ (11.16)

If the first two terms are retained in the binomial expansion of Equation 11.16, then

$$\Delta \approx \int_0^L \frac{1}{2}(u')^2 \, dy.$$ (11.17)

The total potential energy then takes the approximate form

$$U_T = -\frac{1}{2} \int_0^L \left\{EI\left[b_1\left(\frac{\pi}{L}\right)^2 \sin\left(\frac{\pi y}{L}\right)\right]^2 + P\left[(a_1 + b_1)\left(\frac{\pi}{L}\right)\cos\left(\frac{\pi y}{L}\right)\right]^2\right\} \, dy$$ (11.18)

or

$$U_T = \frac{1}{2}\left(\frac{L}{2}\right)\left(\frac{\pi}{L}\right)^2\left[EIb_1^2\left(\frac{\pi}{L}\right)^2 - P(a_1 + b_1)^2\right]. \tag{11.19}$$

Static equilibrium occurs when $\delta U_T = 0$,

$$\delta U_T = EIb_1\delta b_1\left(\frac{\pi}{L}\right)^4\left(\frac{L}{2}\right) - P(a_1 + b_1)\delta b_1\left(\frac{\pi}{L}\right)^2\left(\frac{L}{2}\right). \tag{11.20}$$

Solving for the static deflection coefficient b_1 yields

$$b_1 = \frac{\beta a_1}{1 - \beta}, \tag{11.21}$$

where

$$\beta = \frac{P}{EI(\pi/L)^2} = \frac{P}{P_{cr}}. \tag{11.22}$$

The parameter $\beta = P/P_{cr}$ is the ratio of the applied load to the Euler buckling load. The hyperbolic behavior in Equation 11.21 is a standard result [20]. The midspan deflection increases asymptotically as the axial load approaches the Euler buckling load (Figure 11.24). The initial crookedness in the beam causes the lateral deflection.

Retaining three terms in the binomial expansion of Equation 11.6 produces a higher-order approximation to the potential energy

$$U_T = \frac{1}{2}\left(\frac{L}{2}\right)\left(\frac{\pi}{L}\right)^2\left\{EIb_1^2\left(\frac{\pi}{L}\right)^2 - P\left[(a_1 + b_1)^2 - \frac{3}{16}(a_1 + b_1)^4\left(\frac{\pi}{L}\right)^2\right]\right\}. \tag{11.23}$$

Setting $\delta U_T = 0$ produces a cubic equation in b_1 for the static equilibrium position.

$$b_1 - \beta(a_1 + b_1) + \frac{3}{8}\left(\frac{\pi}{L}\right)^2\beta(a_1 + b_1)^3 = 0. \tag{11.24}$$

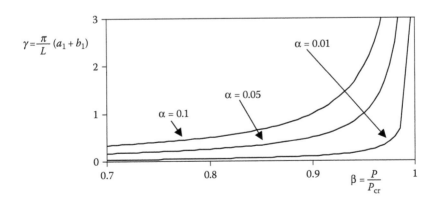

FIGURE 11.24 Static midspan deflection of a simply supported beam as a function of axial load, β, and initial crookedness, α, using two-term static binomial expansion.

Expanding the kinetic energy, T, and the potential energy, U_T, in a Taylor series, retaining only the terms that are quadratic or less, and then substituting into LaGrange's equation

$$\frac{d}{dt}\left(\frac{\partial T}{\partial \dot{c}_i}\right) - \frac{\partial T}{\partial c_i} + \frac{\partial U_T}{\partial c_i} = 0 \tag{11.25}$$

produces equations of motion for linear free vibrations about static equilibrium. The kinetic energy, T, of the vibrating beam with a mass per unit length of m, ignoring rotary and shear inertia, is

$$T = \frac{1}{2} \int_0^L m\dot{u}^2 \, dy, \tag{11.26}$$

which reduces to

$$T = \frac{1}{2} \int_0^L m \left[\dot{c}_1 \sin\left(\frac{\pi y}{L}\right)\right]^2 dy \tag{11.27}$$

and

$$T = \frac{1}{2}m\dot{c}_1^2 \left(\frac{L}{2}\right). \tag{11.28}$$

Retaining the first two terms in the binomial expansion of Equation 11.16 produces an approximate expression for the total potential energy of the beam:

$$U_T = \frac{1}{2}\left(\frac{L}{2}\right)\left(\frac{\pi}{L}\right)^2 \left[EI\,(b_1 + c_1)^2 \left(\frac{\pi}{L}\right)^2 - P(a_1 + b_1 + c_1)^2\right]. \tag{11.29}$$

Substituting Equations 11.28 and 11.29 into LaGrange's Equation 11.25 gives the equation of motion

$$m\ddot{c}_1 + EI\left(\frac{\pi}{L}\right)^4 (1 - \beta)c_1 = 0. \tag{11.30}$$

The shift in the square of the natural frequency due to the axial load is

$$\left(\frac{\omega}{\omega_0}\right)^2 = 1 - \beta, \tag{11.31}$$

where

$$\omega_0^2 = \left(\frac{\pi}{L}\right)^4 \frac{EI}{m}. \tag{11.32}$$

This indicates that the square of the natural frequency decreases linearly with respect to the axial load and equals zero at the buckling load. This is a standard result [21,22] (Figure 11.25).

Retaining the first three terms in the binomial expansion of Equation 11.16 results in a higher-order and presumably more realistic approximate model for the potential energy:

$$U_T = \frac{1}{2}\left(\frac{L}{2}\right)\left(\frac{\pi}{L}\right)^2 \left\{EI(b_1 + c_1)^2 \left(\frac{\pi}{L}\right)^2 - P\left[(a_1 + b_1 + c_1)^2 - \frac{3}{16}(a_1 + b_1 + c_1)^4 \left(\frac{\pi}{L}\right)^2\right]\right\}. \tag{11.33}$$

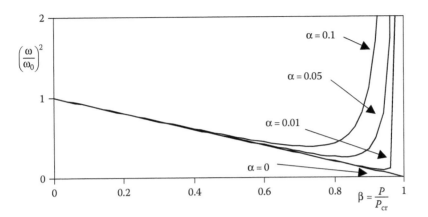

FIGURE 11.25 Shifting of the square of the fundamental vibration frequency as a function of axial load, β, using a two-term binomial expansion.

Expanding and retaining only terms that are quadratic in c_1 reduces U_T to

$$U_T = \frac{L}{4}\left(\frac{\pi}{L}\right)^2\left\{EIc_1^2 - Pc_1^2 + \frac{9P}{8}\left(\frac{\pi}{L}\right)^2(a_1 + b_1)^2 c_1^2\right\}. \tag{11.34}$$

Combining Equation 11.34 with Equations 11.25 and 11.28 results in the small-oscillation equation of motion

$$m\ddot{c}_1 + EI\left(\frac{\pi}{L}\right)^4\left\{1 - \beta\left[1 - \frac{9}{8}\left(\frac{\pi}{L}\right)^2(a_1 + b_1)^2\right]\right\}c_1 = 0. \tag{11.35}$$

The frequency shift due to the axial load is

$$\left(\frac{\omega}{\omega_0}\right)^2 = 1 - \beta\left(1 - \frac{9}{8}\gamma^2\right), \tag{11.36}$$

where

$$\gamma = \frac{\pi}{L}(a_1 + b_1). \tag{11.37}$$

Using the static solution to the two-term binomial expansion, Equation 11.21 for b_1, Equation 11.36 becomes

$$\left(\frac{\omega}{\omega_0}\right)^2 = 1 - \beta\left[1 - \frac{9}{8}\left(\frac{\alpha}{1 - \beta}\right)^2\right], \tag{11.38}$$

where

$$\alpha = a_1\left(\frac{\pi}{L}\right). \tag{11.39}$$

The frequency shifting described by Equation 11.38 indicates that initially the frequency decreases with an increase in β, but as β approaches 1, that is, as the axial load approaches the Euler buckling load, the frequencies increase asymptotically. The size of the initial crookedness parameter, α, determines the rate at which the frequency increases.

TABLE 11.4

Physical Properties of Two Steel Beams

Beam	Length (m)	E (MPa)	I (m^4)	m (kg/m)	f_0 (Hz)
1	0.606	2.07×10^5	5.15×10^{-10}	2.78	26.5
2	1.66	2.07×10^5	4.36×10^{-8}	3.43	29

An attempt to quantify the above results included tests on two different slender steel beams. The cross-section of the first beam (Beam 1) was flat and rectangular. The cross-section of the second beam (Beam 2) was a rectangular box. Table 11.4 lists the physical properties of these beams. The first set of tests measured the frequencies of vibration of Beam 1 when subjected to a sequence of increasing axial loads. A 22.5 kN servohydraulic material test system served as the test frame. A rubber sheet placed between the ends of the beam and the test machine platens simulated simply supported boundary conditions (Figure 11.26). A noncontacting cage placed around the beam prevented the beam from accidentally flying out of the test jig. Tapping the beam with a small hammer sets up free vibrations. A midspan magnetically mounted piezoelectric accelerometer measured the vibrations. The peaks of an autospectrum of the vibrations indicated the natural frequencies of the first two modes. Figure 11.27 shows the shifting of the frequencies of the first two modes with axial load, normalized by their respective unloaded values. The natural frequency of the first mode decreased initially with increasing axial load, but then began to increase at about 80% of the buckling load. The second and third natural frequencies shifted slightly lower as the load increased.

Similar testing procedures using a 225 kN screw-type material testing machine on Beam 2 provided a second set of frequency-shifting data. Placing steel cylinders between the ends of the beams and the test machine platens simulated simply supported boundary conditions

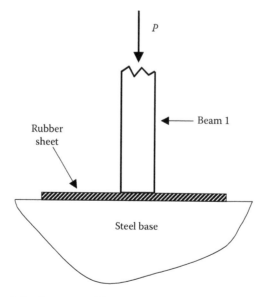

FIGURE 11.26 Simulation of simply supported boundary conditions on Beam 1 with a sheet of rubber placed between the beam and the test frame platen.

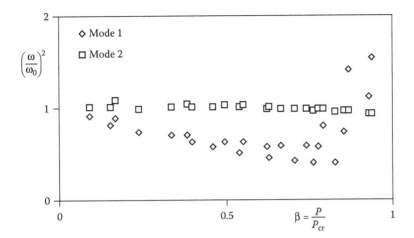

FIGURE 11.27 Shifting of the square of the normalized frequencies for the first two bending modes of Beam 1 as a function of axial load.

(Figure 11.28). The natural frequencies of Beam 2 shifted in a manner very similar to that of Beam 1 (Figure 11.29).

An assumption of the higher-order frequency-shifting model is that the vibration mode shapes do not change with the shift in axial load and frequency. Mode shape measurements on Beam 2 at 14% and 83% of the nominal Euler buckling load checked the validity of this assumption. The impact hammer technique with a spectrum analyzer and commercial modal analysis software measured the mode shapes [23]. The technique involves attaching an accelerometer to a fixed location on the beam, striking the beam at a series of locations that span the beam with an instrumented impact hammer, and processing the

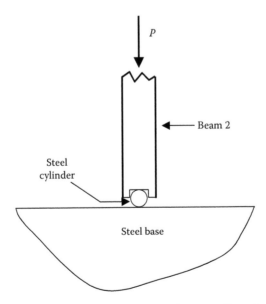

FIGURE 11.28 Simulation of simply supported boundary conditions on Beam 2 with a steel cylinder placed between the beam and the test frame platen.

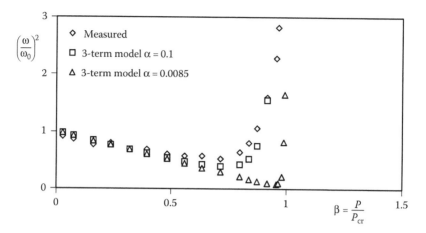

FIGURE 11.29 Shifting of the square of the normalized frequencies for the first mode of Beam 2 as a function of axial load, and with the three-term model for $\alpha = 0.1$ and $\alpha = 0.0083$.

dynamic data to estimate the mode shapes. Figure 11.30 shows the 10 impact locations. The accelerometer magnetically attached at location number 4. This configuration enabled the measurement of both bending and torsional modes. The lowest three modes were bending modes (Figure 11.31). The box-shaped cross-section of Beam 2 provided sufficient torsional rigidity to shift the torsional modes up and out of the frequency range of interest. The lower three vibration mode shapes did not change with the load, in spite of the large shift in the fundamental frequency.

Static measurements of midspan deflection with a dial indicator lead to an estimate of the initial crookedness parameter of $\alpha = 0.0085$. An examination of the frequency-shifting curves from the theoretical model (Figure 11.25), and the experimental measurements, indicates that the frequency is measured to shift upward at a load that is somewhat lower

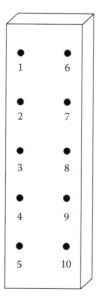

FIGURE 11.30 Impact test locations on Beam 2 for modal testing.

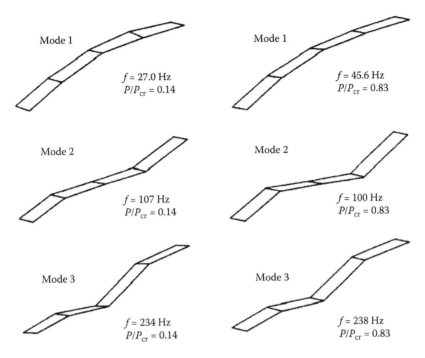

FIGURE 11.31 Vibration mode shapes and frequencies of Beam 2 as a function of axial load.

than that predicted by the model. Possible explanations for the discrepancy include the following: (1) the model is incorrect; (2) more terms need to be included in the series expansion (which does not necessarily converge quickly in the region of interest); and (3) the static equilibrium configuration estimate is inaccurate in shape and amplitude. The cubic Equation 11.24 should be solved for the static equilibrium position in Equation 11.36.

Acknowledgment: Edgardo Colon-Emeric, David Ogden, and Hai-Yan Zhang assisted with the data collection and analysis in this study.

References

1. Chen WF and Mosallam KH. 1991. *Concrete Buildings: Analysis for Safe Construction*. CRC Press, Boca Raton.
2. Hadipriono FC and Wang HK. 1986. Analysis of causes of falsework failures in concrete structures. *J Constr Eng Manag*, 112(1), 112–121.
3. Lee HM, Liu XL, and Chen WF. 1991. Creep analysis of concrete buildings during construction. *J Struct Eng*, 117, 10.
4. Epaarachchi DC, Stewart MG, and Rosowsky DV. 2002. Structural reliability of multistory buildings during construction. *J Struct Eng*, 128, 2.
5. Huang YL. 1994. *Design Considerations for Scaffold System in Concrete Building Construction*. School of Civil Engineering, Purdue University, West Lafayette, Indiana, November.
6. Pelczarski N and Huston D. 2000. Cure monitoring of composite laminates used in the manufacture of snowboards. *Nondestructive Evaluation of Aging Materials and Composites IV*, SPIE 3993.

7. Powell GR, Crosby PA, Fernando GF, France CM, Spooner RC, and Waters DN. 1996. Optical fiber evanescent wave cure monitoring of epoxy resins. *Smart Sensing, Processing and Instrumentation*, SPIE 2718.

8. Doo J, Zhang Y, Compton J, Kranbuehl D, and Loos AC. 2006. In-situ microfrequency dependent electric field sensors to verify model predictions of cure and aging. *J Mater Sci*, 41, 6639–6646.

9. Leng JS and Asundi A. 2002. Real-time cure monitoring of smart composite materials using extrinsic Fabry–Perot interferometer and fiber Bragg grating sensors. *Smart Mater Struct*, 11, 249–255.

10. Jang BZ. 1994. *Advanced Polymer Composites*. ASM International, Materials Park.

11. Nass KA and Seferis JC. 1989. Analysis of the dielectric response of thermosets during isothermal and nonisothermal cure. *Polymer Eng Sci*, 29(5), 315–324.

12. Biermann PJ, Cranmer JH, Lebowitz CA, and Brown LM. 1996. End-of-cure sensing using ultrasonics for autoclave fabrication of composites. *Nondestructive Evaluation for Process Control in Manufacturing*, SPIE 2948.

13. Shepard DD and Smith KR. 1997. A new ultrasonic measurement system for the cure monitoring of thermosetting resins and composites. *J Thermal Anal*, 49, 95–100.

14. Li Y and Menon S. 1998. Ultrasonic sensing of composite materials during the heat-cure cycle. *Sensors*, 15(2), 5–10.

15. Graff KF. 1973. *Wave Motion in Elastic Solids*. Ohio State University Press, Columbus.

16. Fukuda T and Osaka K. 1999. In situ strain measuring and cure monitoring of composite materials in autoclave molding. *Proc 2nd Intl Workshop on Structural Health Monitoring*, Stanford, Technomic, Lancaster.

17. Shepard DD and Smith KR. 1997. A new ultrasonic measurement system for the cure monitoring of thermosetting resins and composites. *J Thermal Anal*, 49, 95–100.

18. Pelczarski N. 1998. *Embedded sensor monitoring of composite curing*. MS thesis, University of Vermont, Mechanical Engineering.

19. Goldstein H. 2002. *Classical Mechanics*, 3rd Ed. Prentice-Hall, Englewood Cliffs, NJ.

20. Bleich F and Ramsey LB. 1952. *Buckling Strength of Metal Structures*. McGraw-Hill, New York.

21. Blevins RD. 1979. *Formulas for Natural Frequency and Mode Shape*. Van Nostrand Reinhold Co, New York.

22. Thompson JM. 1982. *Instabilities and Catastrophes in Science and Engineering*. Wiley, New York.

23. Allemang RJ and Brown DL. 1987. *Experimental Modal Analysis and Dynamic Component Synthesis*, Vols I–VI. AFWAL-TR-87-3069, Air Force Wright Aeronautical Labs, December.

24. Huston DR, Esser B, Plumpton JO, and Zhao X, 2002. Monitoring of microfloor vibrations in a new research building, *Nondestructive Evaluation and Health Monitoring of Aerospace Materials and Civil Infrastructures*, AL Gyekenyesi, SM Shepard, DR Huston, AE Aktan, and PJ Shull (eds.). pp. 237–245, 18 June 2002. SPIE 4704.

Appendix A

Waves

Wave propagation is the basis of many SHM systems and components. Acoustic, elastic, and EM waves are common tools that can both probe structural elements for condition assessment and transmit information. In spite of differences at the detail level, there is considerable commonality in the overall behavior and in the associated mathematical descriptions of wave phenomena. This appendix summarizes some of the descriptive highlights. More in-depth coverage can be found in standard references [1–8].

Waves are disturbances that propagate through distributed media. Different descriptive models of wave propagation provide solutions and insights in a variety of circumstances. These models vary in terms of mathematical simplicity, physical rigor, and the ability to describe specific physical phenomena. As an example, Figure A.1 shows a set of methods of EM wave analysis. The simplest models use geometric ray-tracing. Rays are directed lines/curves that are normal to wavefronts and point in the direction of propagation. Smooth changes in the refractive index and/or large smooth surfaces of refractive index discontinuity produce smooth wavefronts. Such situations lend themselves to ray-tracing analyses. Associated phenomena correspond to focusing, imaging, reflection, and refraction. When the index of refraction changes significantly over short distances, that is on the order of a wavelength, the wavefronts are not smooth enough to use ray-tracing analyses. Interference and diffraction are typical in these situations. More complicated scalar wave approaches can explain many interference and diffraction effects. Scalar waves, however, are inadequate for describing polarization effects. Instead, more complicated and more rigorous, vector wave models describe polarization phenomena. In turn, vector wave models cannot describe a variety of effects, such as fluorescence, lasing, holography, and optical-material interactions. Rational explanations for these phenomena require models based on quantum theory. In effect, models ranging from ray-tracing through quantum theory form a cascade of representations where the more complicated approaches, such as quantum theory, can explain all of the phenomena explained by the simpler models, such as ray-tracing, and more. It is often preferred to use the simplest viable model for instrument design and data interpretation.

A.1 Acoustic Waves

Fluids are a class of continuous materials that respond to shear stress with unrestrained deformation. Real fluids are massive with nonzero densities, and are compressible. The equation of state of many fluids expresses the pressure, p, as a function of the density, ρ:

$$p = p(\rho). \tag{A.1}$$

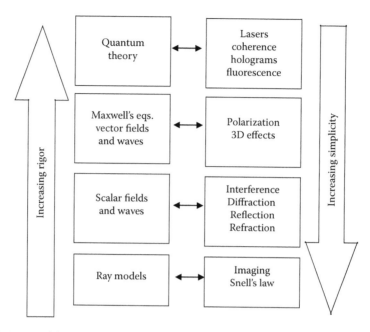

FIGURE A.1   Various models of electromagnetic wave progagation.

Small density fluctuations about an equilibrium state can propagate as waves of density fluctuations, commonly known as *acoustic waves* or *sound waves*. The 3-D differential scalar wave equation describes much of the linear behavior of acoustic waves [1]:

$$\nabla^2 p = \frac{1}{c^2}\frac{\partial^2 p}{\partial t^2},$$ (A.2)

where c is the acoustic wavespeed. Density and displacement fluctuations track with the pressure fluctuations. Density is a scalar and follows a wave equation also of the form as Equation A.2. The exception is that displacement is actually a vector wave equation. The displacement vectors are normal to the wavefronts, parallel to the direction of propagation, and are not polarized. Boundary conditions between different acoustic media require the continuity of pressure and displacement at interfaces. Modeling of coupled fluid–structure boundary conditions often involves integral equations [9].

Air has an equation of state that depends on both density and pressure. The wavespeed of air is 331 m/s at 0°C and 1 atm static pressure. If the fluctuations are adiabatic, the temperature dependence tends to stiffen the medium, but Equation A.2 still holds for small amplitude fluctuations. The wavespeed is

$$c^2 = \left(\frac{\partial p}{\partial \rho}\right)_{\text{adiabatic}}.$$ (A.3)

A.2   Electromagnetic Waves

Vector field equations describe many EM phenomena. The principal EM vector fields are the *electric field,* $\vec{E}$; the *magnetic field,* $\vec{H}$; the *electric displacement,* $\vec{D}$; and the *magnetic induction*, $\vec{B}$. In the absence of material discontinuities, these vector fields are normally finite, continuous, and differentiable functions of position and time. Heaviside's version of Maxwell's equations combines the four field vectors in a concise description of field interactions [10]:

$$\nabla \times \vec{H} = \frac{\partial \vec{D}}{\partial t} + \vec{J}_s + \vec{J}_c \tag{A.4}$$

$$\nabla \times \vec{E} = -\frac{\partial \vec{B}}{\partial t}, \tag{A.5}$$

$$\nabla \cdot \vec{H} = 0, \tag{A.6}$$

$$\nabla \cdot \vec{E} = \frac{\rho}{\varepsilon_0}, \tag{A.7}$$

where $\vec{J}_s$ is the source current density, and $\vec{J}_c$ is the conductivity current density ($\vec{J}_c = \sigma\vec{E}$), and ρ is the free charge density.

In addition to Maxwell's equations, EM wave descriptions also need constitutive material properties. In isotropic media, $\vec{D}$ is parallel to $\vec{E}$ and $\vec{H}$ is parallel to $\vec{B}$, that is,

$$\vec{D}(\vec{r}, \omega) = \varepsilon(\omega)\vec{E}(\vec{r}, \omega) \tag{A.8}$$

and

$$\vec{B} = \mu_0\vec{H}, \tag{A.9}$$

where ε and μ are the electrical permittivity and magnetic permeability, respectively. In a vacuum, or air at most wavelengths, $\varepsilon_0 = 8.854 \times 10^{-12}$ F/m and $\mu_0 = 4\pi \times 10^{-7}$ N/A^2. Equation A.8 includes frequency dependence in the constitutive description with the frequency variable ω. Frequency-dependent media are *dispersive*. Anisotropic materials generally require tensor-based constitutive models.

For nonconducting and nondispersive homogeneous media, Equation A.4 and Equation A.5 simplify to [11]

$$\nabla \times \vec{H} = \varepsilon\frac{\partial \vec{E}}{\partial t} \tag{A.10}$$

and

$$\nabla \times \vec{E} = -\mu\frac{\partial \vec{H}}{\partial t}. \tag{A.11}$$

Taking the curl of Equations A.10 and A.11 gives

$$\nabla \times \nabla \times \vec{H} = \varepsilon\frac{\partial(\nabla \times \vec{E})}{\partial t}, \tag{A.12}$$

$$\nabla \times \nabla \times \vec{E} = -\mu\frac{\partial(\nabla \times \vec{H})}{\partial t}. \tag{A.13}$$

For any arbitrary continuous vector field $\vec{A}$,

$$\nabla \times \nabla \times \vec{A} = \nabla \nabla \cdot \vec{A} - \nabla^2 \vec{A}. \tag{A.14}$$

Combining Equation A.6 with Equation A.14 produces the vector wave equations

$$\nabla^2 \vec{H} = \varepsilon \mu \frac{\partial^2 \vec{H}}{\partial t^2} = \frac{1}{c^2} \frac{\partial^2 \vec{H}}{\partial t^2}, \tag{A.15}$$

$$\nabla^2 \vec{E} = \varepsilon \mu \frac{\partial^2 \vec{E}}{\partial t^2} = \frac{1}{c^2} \frac{\partial^2 \vec{E}}{\partial t^2}, \tag{A.16}$$

where c is the wavespeed. The wavespeed, c_0, in a vacuum is

$$c_0 = \frac{1}{(\varepsilon_0 \mu_0)^{1/2}} = 2.998 \times 10^8 \, \text{m/s}. \tag{A.17}$$

EM waves propagate through a vacuum with no loss of energy or dispersion. The propagation of EM waves through all other media is lossy with a conversion of EM wave energy into other forms, for example heat. A macroscopic representation of wave transmission losses is a linear model with complex-valued permittivities and permeabilities, that is, $\varepsilon^* = \varepsilon' + j\varepsilon''$ and $\mu^* = \mu' + j\mu''$ [12]:

$$\nabla^2 \vec{E} = \varepsilon^* \mu^* \frac{\partial^2 \vec{E}}{\partial t^2}. \tag{A.18}$$

The propagation factor, γ, is

$$\gamma = j\omega[\varepsilon^* \mu^*]^{1/2} = \alpha + j\beta, \tag{A.19}$$

where α is the attenuation factor ($\alpha = 0$ for loss-free media) and β is the phase factor, which indicates the phase velocity of the wave. For most dielectric materials, it is reasonable to absorb the loss factor into the imaginary part of the permittivity ε^*. A useful and relatively easy-to-measure version of the loss factor is the *loss tangent*, $\delta = \varepsilon''/\varepsilon'$. Using complex-valued loss functions is generally appropriate only in the context of oscillations with stationary frequency domain content, for example, harmonic inputs and outputs. Using complex loss function models in transient analyses can cause anomalous noncausal behaviors [13]. Purely imaginary dielectric values cause strong reflections and are useful for modeling metallic surfaces [3].

Boundary conditions between media with differing EM properties generally require that the magnetic field vector be continuous across boundaries of dielectric nonmagnetic materials and that the electric field components parallel to the surface be continuous. The components of the electric displacement vector normal to the interface are continuous, but there may be a jump in the electric field vector due to a jump in dielectric properties.

A.3   Elastic Waves

Dynamic variations in elastic media gradients provide impetus for the propagation of elastic waves [14]. If the displacement gradients are small, the linear strain tensor, ε_{ij}, adequately describes most deformation gradient phenomena. The strain tensor in indicial notation is of the form

$$\varepsilon_{ij} = \frac{1}{2}\left(\frac{\partial u_i}{\partial x_j} + \frac{\partial u_j}{\partial x_i}\right), \tag{A.20}$$

where $u_i(x_1, x_2, x_3, t)$ are the displacements and x_i ($i = 1, 2,$ or 3) are the spatial (or material) coordinates. An isotropic elastic solid undergoing small gradient deformation has the stress–strain relation

$$T_{ij} = \lambda \varepsilon_{kk} \delta_{ij} + 2\mu \varepsilon_{ij}, \tag{A.21}$$

where T_{ij} is the stress tensor. The Lamé parameters λ and μ are elastic constants. Direct algebraic relations exist between the Lamé parameters and other elastic constants, such as Young's modulus, E, the shear modulus, G, and Poisson's ratio, v,

$$\lambda = \frac{vE}{(1-v)(1-2v)}, \tag{A.22}$$

$$\mu = G = \frac{E}{2(1+v)}. \tag{A.23}$$

Newton's second law and Cauchy's stress principle for a continuum with no body forces combine to produce the equation of motion

$$T_{ij,j} = \rho \ddot{u}_i. \tag{A.24}$$

Inserting Equation A.20 into Equation A.21 and then into Equation A.24 produces a displacement equation of motion

$$(\lambda + \mu)u_{j,ij} + \mu u_{i,jj} = \rho \ddot{u}_i. \tag{A.25}$$

Taking the divergence of Equation A.25 forms a scalar wave equation

$$\nabla^2 e = \frac{\rho}{(\lambda + 2\mu)}\ddot{e} = \frac{1}{c_p^2}\ddot{e}, \tag{A.26}$$

$$e = \nabla \cdot \vec{u}, \tag{A.27}$$

where c_p is the wavespeed. The waves corresponding to Equation A.26 are compression waves known as P-*waves* and are similar to acoustic waves. Taking the curl of Equation A.25 forms a vector wave equation

$$\nabla^2 \vec{W} = \left[\frac{\mu}{\rho}\right]\ddot{\vec{W}} = \frac{1}{c_s^2}\ddot{\vec{W}}, \tag{A.28}$$

where

$$\vec{W} = \nabla \times \vec{u}. \tag{A.29}$$

c_s is the wavespeed. The waves corresponding to Equation A.28 are shear or S-*waves*. The transverse vector nature of S-waves gives rise to polarization effects. Typical p-wavespeeds for steel and aluminum are 5.7×10^3 and 6.15×10^3 m/s, respectively. Typical s-wavespeeds for steel and aluminum are 3.2×10^3 and 3.1×10^3 m/s, respectively [6].

Boundaries between different elastic media reflect, refract, and scatter elastic waves. The boundary conditions between two different elastic media require that the displacement and traction conditions must be continuous. The strain fields may exhibit finite discontinuities at the boundaries. When a compliant fluid, such as air, forms a boundary with the solid, it is often reasonable to assume that the boundary conditions are traction-free.

A.4 1-D and Scalar Waves

Equations A.2 and A.26 are scalar wave equations. Equations A.15, A.16, and A.28 are vector wave equations. Vector waves carry more information than scalar waves and produce vector-based phenomena. Analyzing vector waves typically requires more mathematical sophistication than that of scalar waves. Aside from the notable exception of polarization, scalar wave surrogates can capture many salient features of phenomena described with vector waves. A generic scalar wave equation is

$$\nabla^2 \varphi = \frac{1}{c^2} \frac{\partial^2 \varphi}{\partial t^2}. \tag{A.30}$$

D'Alembert-type solutions to the scalar wave equation, Equation A.30, in a homogeneous isotropic medium, free of boundaries and active sources, are

$$\varphi_+ = f\left[h(x, y, z) - \omega t\right] \tag{A.31}$$

and

$$\varphi_- = g\left[h(x, y, z) + \omega t\right]. \tag{A.32}$$

The validity of this solution can be established by direct substitution of φ_+ or φ_- into Equation A.30. For φ_+, this reduces to

$$f''(\nabla h \cdot \nabla h) + f' \nabla^2 h = f'' \frac{\omega^2}{c^2}. \tag{A.33}$$

A special case arises when

$$\nabla h \cdot \nabla h = \frac{\omega^2}{c^2} \tag{A.34}$$

and

$$\nabla^2 h = 0. \tag{A.35}$$

These conditions automatically satisfy Equation A.33 and guarantee the validity of the D'Alembert solutions φ_+ and φ_-. If ω and c are constants, then the D'Alembert solutions use the constant wavenumber components K_x, K_y, and K_z, so that

$$h(x, y, z) = K_x x + K_y y + K_z z = \vec{K} \cdot \vec{r}, \tag{A.36}$$

$$K_x^2 + K_y^2 + K_z^2 = K^2 = \frac{\omega^2}{c^2}. \tag{A.37}$$

These solutions correspond to plane wavefronts propagating in the direction of the wavenumber vector $\vec{K}$. Rays are directed curves formed as tangents to wavenumber vectors. The special case of a plane wave with the normal to the propagation front aligned with the x-axis has the wavenumbers

$$K_x = K, \qquad K_y = 0, \qquad K_z = 0. \tag{A.38}$$

The representation of plane waves propagating along an axis simplifies to a 1-D form

$$\frac{\partial^2 \varphi}{\partial x^2} = \frac{1}{c^2} \frac{\partial^2 \varphi}{\partial t^2} \tag{A.39}$$

with 1-D D'Alembert solutions

$$\varphi_+ = \varphi_+(Kx - \omega t) = f\left(t - \frac{x}{c}\right) \tag{A.40}$$

and

$$\varphi_- = \varphi_-(Kx + \omega t) = g\left(t + \frac{x}{c}\right). \tag{A.41}$$

These solutions correspond to waves with the shape $f(\xi)$ propagating in the positive x-direction and waves with the shape $g(\xi)$ propagating in the negative x-direction. Both waves have a phase velocity of c.

The complex format of harmonic plane waves is

$$\varphi = \Phi_0 \exp[j(\vec{K} \cdot \vec{r} - \omega t)]. \tag{A.42}$$

Frequency and wavelength are inversely proportional. For harmonic waves with ω as the circular frequency and λ as the wavelength,

$$\lambda = \frac{2\pi}{K} = \frac{2\pi c}{\omega}. \tag{A.43}$$

An important feature of the above development is the assumption of constant material properties and no boundaries. These assumptions produce solutions that propagate at constant speed independent of the wavelength and without spatial distortion. Boundary conditions, spatially varying material properties, and frequency-dependent material properties can cause qualitative and quantitative changes in the nature of the solution. Possibilities include wavelength-dependent wavespeeds, wavelength-dependent attenuation, and the guiding of waves along certain geometric paths. Media with frequency-dependent (or wavelength-dependent) phase velocities are *dispersive*. Plots of frequency and wavelength dependence of phase velocities are *dispersion curves*. An additional consideration is that boundary conditions between different media can alter the form of the solutions. Guided waves are an interesting and often useful boundary-condition-dependent behavior.

Since the wave equation (Equation A.30) is linear, one approach to forming a general solution superposes an appropriate mix of plane waves of various directions, amplitudes, wavelengths, and phase angles [4]. Fourier's theorem enables representing arbitrary nonharmonic wave shapes as a superposition of harmonic waves. The propagation of nonharmonic waves in dispersive media causes different harmonic components to propagate at different speeds. The dispersion of the components causes the waves to change shape.

A spatially localized sinusoidal wave train often appears to travel at a *group velocity* $\vec{u}_g$ with the definition

$$\vec{u}_g = \frac{\partial \omega}{\partial \vec{K}}. \tag{A.44}$$

The direction of the group velocity vector corresponds to the direction of ray propagation in geometric optics [7]. Nonlinear wave-amplitude dependence combined with dispersive harmonic behavior can lead to isolated pulse waves known as solitons [15]. Solitons can travel long distances with a minimum of distortion.

It should be noted that not all frequency-dependent dispersion relations are physically realizable. The *Principle of Causality* states that all events must be preceded and caused by other events. Causality constrains the frequency domain behavior of linear dispersive media. The *Kramers–Kronig* relations are a Hilbert transform pair that expresses the constraint of causality. The format is typically a pair of integrals along the positive real axis.

Similar to the plane wave is the spherical wave. The Laplacian of ϕ in the spherical coordinates of Figure A.2 is [16]

$$\nabla^2 \varphi = \frac{1}{r^2} \frac{\partial}{\partial r} \left(r^2 \frac{\partial \varphi}{\partial r} \right) + \frac{1}{r^2 \sin \psi} \left(\sin \psi \frac{\partial \varphi}{\partial \psi} \right) + \frac{1}{r^2 \sin^2 \psi} \frac{\partial^2 \varphi}{\partial \theta^2}. \tag{A.45}$$

If φ is spherically symmetric, that is,

$$\frac{\partial \varphi}{\partial \psi} = \frac{\partial \varphi}{\partial \theta} = 0, \tag{A.46}$$

then

$$\nabla^2 \varphi = \frac{1}{r^2} \frac{\partial}{\partial r} \left(r^2 \frac{\partial \varphi}{\partial r} \right) = \frac{1}{r} \frac{\partial^2}{\partial r^2} (r\varphi). \tag{A.47}$$

Using $\xi = r\varphi$ enables writing Equation A.30 as a 1-D scalar wave equation

$$\frac{\partial^2 \xi}{\partial r^2} = \frac{1}{c^2} \frac{\partial^2 \xi}{\partial t^2}. \tag{A.48}$$

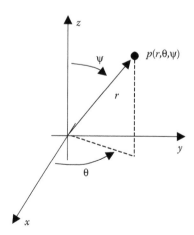

FIGURE A.2 Spherical coordinates.

The result is that D'Alembert-type spherically symmetric waves of arbitrary shape propagate at the speed c with an amplitude that decays as $1/r$, except at $r = 0$.

The hypothesis that each point in a wave propagating medium can be a source of spherical waves leads to *Huygens' principle*. One consequence is that each point in a wavefront generates secondary waves. The envelope of the secondary waves assembles to form the propagating wavefront. While Huygens' principle is simple in statement, it can be a useful tool in explaining many subtle and complicated phenomena, including diffraction, interference, scattering, and the propagation of waves through nonhomogeneous media [7,17].

When there are sources of wave motion embedded in the medium, then the wave equation (Equation A.30) becomes nonhomogeneous with the form

$$c^2 \nabla^2 \varphi - \frac{\partial^2 \varphi}{\partial t^2} = f(x, y, z, t). \tag{A.49}$$

Green's functions solve the nonhomogeneous differential equation (Equation A.49) with a convolution integral. An interpretation of the action of a Green's function is that it uses the superposition implied by Huygens' principle to invert the differential operator. The Green's function convolution integral for an infinite 3-D domain is [18]

$$\varphi(x, y, z, t) = -\iiint \frac{f[\xi, \eta, \zeta, t - (r/c)]}{4\pi c^2 r} \, d\xi \, d\eta \, d\zeta, \tag{A.50}$$

$$r = \left[(x - \xi)^2 + (y - \eta)^2 + (z - \zeta)^2\right]^{1/2}. \tag{A.51}$$

Finite domains with boundaries complicate the use of Green's functions.

Cylindrical coordinates are another set of orthogonal curvilinear coordinates that sustain wave motion (Figure A.3) [16]. The Laplacian in cylindrical coordinates is

$$\nabla^2 \varphi = \frac{1}{r} \frac{\partial}{\partial r} \left(r \frac{\partial \varphi}{\partial r} \right) + \frac{1}{r^2} \frac{\partial^2 \varphi}{\partial \theta^2} + \frac{\partial^2 \varphi}{\partial z^2}. \tag{A.52}$$

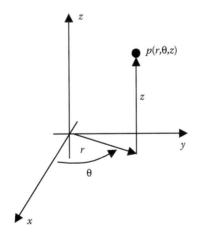

FIGURE A.3 Cylindrical coordinates.

Wave motions with cylindrical symmetry occur when

$$\frac{\partial \varphi}{\partial \theta} = \frac{\partial \varphi}{\partial z} = 0. \tag{A.53}$$

The 3-D wave equation simplifies to a 1-D differential equation in r and t:

$$\frac{1}{r}\frac{\partial}{\partial r}\left(r\frac{\partial \varphi}{\partial r}\right) = \frac{\partial^2 \varphi}{\partial r^2} + \frac{1}{r}\frac{\partial \varphi}{\partial r} = \frac{1}{c^2}\frac{\partial^2 \varphi}{\partial t^2}. \tag{A.54}$$

The 1-D cylindrical wave equation (Equation A.54) does not separate into D'Alembert-type wave solutions. The reason is that the $1/r$ and r terms on the left-hand side are frequency (or wavelength) dependent. Separation of variables gives a solution in terms of time harmonic wave components, with the assumption

$$\varphi(r, t) = \Phi(r)\,\exp(-j\omega t). \tag{A.55}$$

This produces a zero-order Bessel equation as the spatial component:

$$\frac{d^2 \Phi}{dt^2} + \frac{1}{r}\frac{d\Phi}{dr} + \left(\frac{\omega}{c}\right)^2 \Phi = 0 \tag{A.56}$$

with solution

$$\Phi(r) = A H_0^{(1)}\left(\frac{\omega r}{c}\right), \tag{A.57}$$

where A is a constant and $H_0^{(1)}$ is a Bessel function of the third kind, also known as a Hankel function [6]. For large values of r, solutions take the asymptotic form [16]

$$\varphi(r, t) \approx \frac{A}{\sqrt{r}}\,\cos\left[\omega\left(\frac{r}{c} \pm t\right)\right]. \tag{A.58}$$

The amplitude of the cylindrical wave attenuates approximately as $r^{-1/2}$.

The above wave solutions apply to homogeneous isotropic media. The wavespeed variation in nonhomogeneous media complicates the analysis and can produce reflections, refractions, and scattering phenomena. The scalar wave equation becomes

$$\nabla^2 \varphi = \frac{1}{c(x, y, z)^2}\frac{\partial^2 \varphi}{\partial t^2} = n(x, y, z)^2 \frac{\partial^2 \varphi}{\partial t^2}. \tag{A.59}$$

In general, analytic solutions to this variable coefficient partial differential equation are intractable. An approximate solution proceeds by assuming separation of variables as the product of spatial and time harmonic functions [19]:

$$\varphi(x, y, z, t) = \psi(x, y, z)\,T(t) = \psi(x, y, z)\,e^{j\omega t}, \tag{A.60}$$

and then uses an exponential form for the spatial term

$$\psi(x, y, z) = Q(x, y, z)\exp\left[j\omega h(x, y, z)\right] \tag{A.61}$$

with Q and h being real functions of position. The spatial component of the separation is

$$\nabla^2 \psi + \frac{\omega^2 \psi}{c^2} = 0, \tag{A.62}$$

which further separates into real and imaginary components

$$(\nabla h)^2 - \frac{\nabla^2 Q}{Q\omega^2} - \frac{1}{c^2} = 0 \tag{A.63}$$

and

$$\nabla^2 h + \frac{2(\nabla h \cdot \nabla Q)}{Q} = 0. \tag{A.64}$$

A simplification comes with the assumption that

$$\frac{\nabla^2 Q}{Q} \ll \frac{\omega^2}{c^2}, \tag{A.65}$$

then Equation A.63 reduces to

$$(\nabla h)^2 = \frac{1}{c(x,y,z)^2}. \tag{A.66}$$

Apart from the constant ω^2, Equation A.66 is the eikonal equation (Equation A.34). The approximation improves with an increase in frequency and a decrease in the spatial rate of variation of Q. This approximation forms the basis for the use of ray optics [7,20]. Rays are directed curves perpendicular to wavefronts that point in the direction of propagation. Rays can be very useful in the design of optical and other instruments based on wave propagation. An equivalent statement is *Fermat's principle*, which states that the length of the ray curve is stationary.

A point source emanates spherical waves. The corresponding ray pattern is a bundle of rays pointing radially outward and normal to the spherical wavefronts. A bundle of rays converges to an *image point* (Figure A.4). The image concept extends to sets of points, such as lines and surfaces and forms the basis of analyzing many forms of imaging optics [2,7].

When a bundle of rays travel along an axis with small angles of inclination and the rays interact with reflective and refractive objects, such as lenses, mirrors, and layers with different refractive indices, it is often possible to make use of the paraxial approximation.

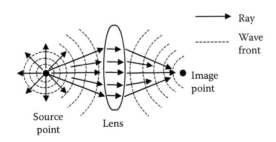

FIGURE A.4 Rays and wavefronts emanating from a point source and converging on an image point.

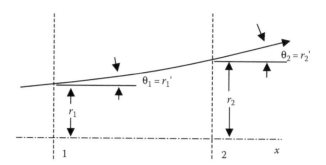

FIGURE A.5 Paraxial approximation of ray propagation. (Adapted from Verdeyen JT. 1989. *Laser Electronics*, 2nd Ed., Prentice-Hall, Englewood Cliffs, NJ.)

This specifies ray geometry as a function of position along an axis in terms of an offset, r, and slope, r' (Figure A.5). The small angles assumption gives rise to linear behavior where a matrix operator represents the change in offset and the direction between points 1 and 2 [21,22]:

$$\begin{bmatrix} r_2 \\ r_2' \end{bmatrix} = \begin{bmatrix} A & B \\ C & D \end{bmatrix} \begin{bmatrix} r_1 \\ r_1' \end{bmatrix}. \tag{A.67}$$

It is possible to represent the action of a cascade of multiple optical elements with a product of multiple matrix operators.

A simple case of propagation through nonhomogeneous media occurs by forming a joint along a planar surface of media with two different wavespeeds, c_1 and c_2. When the joint boundary is the y–z plane at $x = 0$ and the propagating waves are plane waves with normals pointing parallel to the x-axis, the analysis reduces to a 1-D problem. Each medium, 1 and 2, supports waves, $\varphi_1(x, t)$ and $\varphi_2(x, t)$, respectively. An assumption of continuous displacement at the joint requires that

$$\varphi_1(x_0, t) = \varphi_2(x_0, t). \tag{A.68}$$

When acting in isolation, each medium supports rightward and leftward traveling D'Alembert-type waves. Figure A.6 shows f_1 and g_1 as the rightward and leftward traveling waves in medium 1, respectively, and f_2 and g_2 as the rightward and leftward traveling waves in medium 2, respectively. Continuity of the displacement field variable at the interface results in the condition

$$f_1 + g_1 = f_2 + g_2. \tag{A.69}$$

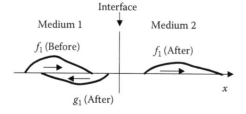

FIGURE A.6 Transmitted and reflected waves at an interface.

Another possible interface boundary condition is the continuity of the force (pressure, stress, electric displacement, etc.) that is driving the wave motion. For example, a string with the same tension and different mass density in each medium would have a force continuity boundary condition of the form

$$\frac{\partial \varphi_1(x,t)}{\partial x}\bigg|_{x=0^-} = \frac{\partial \varphi_2(x,t)}{\partial x}\bigg|_{x=0^+} \tag{A.70}$$

or

$$f_1' + g_1' = f_2' + g_2'. \tag{A.71}$$

Different wave mechanisms produce similar, but different, force continuity boundary conditions. The modeling of active surface effects requires additional terms.

A solution must satisfy the wave equation in medium 1 and 2, and the boundary conditions. Such a solution is

$$f_1 = F_1 \psi(\xi_{1+}), \quad g_1 = G_1 \psi(\xi_{1-}), \quad f_2 = F_2 \psi(\xi_{2+}), \quad g_2 = G_2 \psi(\xi_{2-}), \tag{A.72}$$

where F_1, G_1, F_2, and G_2 are constants and

$$\xi_{1+} = t - \frac{x}{c_1}, \quad \xi_{1-} = t + \frac{x}{c_1}, \quad \xi_{2+} = t - \frac{x}{c_2}, \quad \xi_{2-} = t + \frac{x}{c_2}. \tag{A.73}$$

When a rightward traveling wave in medium 1 interacts with the interface, waves transmit and reflect in both positive and negative directions. If $F_1 = 1$ and $G_2 = 0$, then G_1 is the amplitude of the reflected wave and F_2 is the amplitude of the transmitted wave, that is,

$$G_1 = \frac{c_2 - c_1}{c_1 + c_2}, \tag{A.74}$$

$$F_2 = \frac{2c_2}{c_1 + c_2}. \tag{A.75}$$

When the wave passes from a medium with a slow speed to that of a higher speed, that is, $c_2 > c_1$, then

$$0 < G_1 < 1, \quad 1 < F_2 < 2. \tag{A.76}$$

The reflected and transmitted waves have amplitude with a sign equal to that of the incident wave. If medium 2 has a very high wavespeed (low string mass density), that is, $c_2 \gg c_1$, then

$$G_1 \approx 1, \quad F_2 \approx 2. \tag{A.77}$$

If instead the wave passes from a fast to a slower medium, that is, $c_1 > c_2$, then

$$-1 < G_1 < 0, \quad 0 < F_2 < 1. \tag{A.78}$$

Table A.1 summarizes these results.

It is possible to superpose various forms of the two-medium cases to describe many of the often complicated and interesting interactions that form when waves impinge on multiple layers of different materials [23]. A relatively simple case occurs in Figure A.7 when a wave

TABLE A.1

Transmitted and Reflected Wave Amplitudes for String Wave
Traveling from Medium 1 to 2

Wavespeeds	$c_1 > c_2$	$c_1 < c_2$
Transmitted amplitude	$2c_2/(c_1 + c_2)$	$2c_2/(c_1 + c_2)$
Transmitted sign	Unchanged	Unchanged
Reflected amplitude	$(c_2 - c_1)/(c_1 + c_2)$	$(c_2 - c_1)/(c_1 + c_2)$
Reflected sign	Changed	Unchanged

travels from a region with a low wavespeed, c_1, into a thin layer with a higher wavespeed, c_2, and then into a third region with a low wavespeed, c_1. When a rightward traveling wave interacts with the thin layer, multiple reflections and transmissions occur. Two primary waves reflect and travel to the left in medium 1. The first wave has a reversed amplitude sign that reflects off the first interface. The first reflected wave has the same shape as the incoming wave, but has a negative sign and reduced amplitude. The second reflected wave reflects off of the second interface. The sign is the same as the incoming wave, but reduced in amplitude. This second wave travels leftward in medium 2 and interacts with the left interface. A portion of this wave transmits leftward through the interface. The net effect is an approximate differentiation of the wave (Figure A.7). The differentiation effect requires that the layer thickness be considerably less than the wavelength of the highest frequency component in the wave. If the intermediate layer has a thickness that is on the order of a wavelength, then the layer can act as a band-pass filter.

Figure A.8 is a graph that can elucidate some of the details associated with the transmission and reflection through multilayer media by plotting the timing versus depth and amplitude of wave propagations and reflections.

Oblique-angle intersections of plane waves with planar boundaries cause the additional complication of 2-D and 3-D effects. In many of these cases, it is simpler initially to perform Fourier decomposition on general waveforms, analyze each component individually, and to reconstruct the general waveform by superposition.

Figure A.9 shows the intersection of a harmonic plane wave with a planar joint between media with different wavespeeds. The waves reflect off of and transmit through the boundary. Aligning the coordinate system so that the wavefronts are parallel to an in-plane axis reduces the geometry from 3-D to 2-D. For the case of the plane waves being parallel to the y-axis, the y-components of the interactions vanish and the wavenumber vectors for the

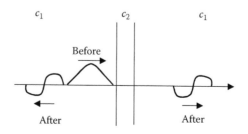

FIGURE A.7   Approximate differentiation effect of thin layer.

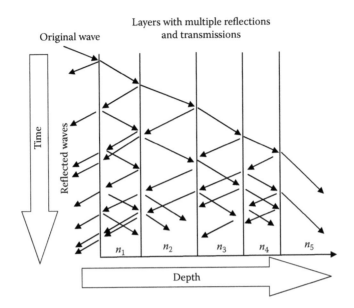

FIGURE A.8 Ray timing versus depth and amplitude representation of multilayer reflections.

incident ($\vec{K}_I$), reflected ($\vec{K}_R$), and transmitted ($\vec{K}_T$) waves are

$$\vec{K}_I = K_{Ix}\vec{i} + K_{Iz}\vec{k},$$
$$\vec{K}_R = K_{Rx}\vec{i} + K_{Rz}\vec{k}, \qquad (A.79)$$
$$\vec{K}_T = K_{Tx}\vec{i} + K_{Tz}\vec{k}.$$

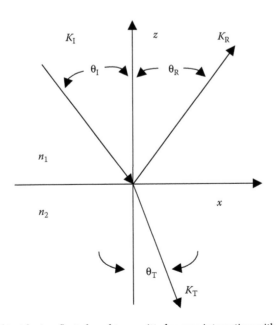

FIGURE A.9 Geometry of incident, reflected, and transmitted waves interacting with two different media.

Continuity requires equality of the wavenumber vector components in the x-direction at the interface, that is,

$$K_{Ix} = K_{Rx} = K_{Tx}. \tag{A.80}$$

Since the incident and reflected waves have the same wavespeed and frequency of oscillation, the wavenumber magnitudes are also equal. This implies that

$$\theta_I = \theta_R. \tag{A.81}$$

Similarly, the wavespeeds in the incident and transmitted media differ, whereas the waves retain the same frequency:

$$K_I^2 c_1^2 = K_R^2 c_1^2 = K_T^2 c_2^2. \tag{A.82}$$

Combining Equations A.79, A.80, and A.82 produces the relation

$$(K_{Ix}^2 + K_{Iz}^2)\, c_1^2 = (K_{Ix}^2 + K_{Tz}^2)\, c_2^2 \tag{A.83}$$

or

$$K_{Tz} = \left[K_{Ix}^2 \left(\frac{c_1^2}{c_2^2} - 1 \right) + K_{Iz}^2 \frac{c_1^2}{c_2^2} \right]^{1/2}. \tag{A.84}$$

The bracketed term of Equation A.84 is positive and K_{Tz} is real when

$$c_1 > c_2, \quad n_1 < n_2 \tag{A.85}$$

or

$$\sin \theta_I < \frac{c_1}{c_2} = \frac{n_2}{n_1}. \tag{A.86}$$

When either Equation A.85 or Equation A.86 is satisfied, and if K_{Tz} is real, then a combination of Equations A.80 and A.81 produces *Snell's law*:

$$\frac{\sin \theta_I}{\sin \theta_T} = \frac{c_1}{c_2} = \frac{n_2}{n_1}. \tag{A.87}$$

If $c_2 > c_1$ and $\theta_I > \theta_c$, where

$$\sin \theta_c = \frac{c_2}{c_1}, \tag{A.88}$$

then K_{Tz} is imaginary, that is,

$$K_{Tz} = \pm j\gamma. \tag{A.89}$$

The resulting waves have amplitudes that decay exponentially with distance from the interface. In elastodynamics, these surface waves are called *surface acoustic waves* or *Rayleigh waves*. In electromagnetism and optics, these surface waves are known as *evanescent waves*. The energy in these waves confines at the surface and does not propagate into the bulk material regions away from the surface. Energy confinement allows for the propagation of surface waves over long distances.

A.4.1 Interference and Interferometers

Interactions among waves produce many interesting and useful effects. Superposition methods can describe many of these interactions in linear media. *Interference* occurs when waves with similar harmonic content superpose to form local intensity fluctuations, that is, fringes. *Interferometers* use interference to transduce physical effects, such as changes in dimension or wavespeed. Many high-precision instruments use light-based interferometers. Some of the better ones have resolutions on the order of 1/1000 of a wavelength. The typical interferometer uses a single light source. Beam splitters segregate the source light into two separate paths of effective lengths L_1 and L_2. The separate light beams recombine at detectors and form interference patterns. Differential changes in the effective lengths of the light paths cause the interference patterns to shift. Classic interferometers include the Michelson, Fabry–Perot, and Mach–Zehnder variants. The Michelson interferometer uses a cross-geometry (Figure A.10). The Fabry–Perot interferometer uses an inline parallel beam splitter and mirror arrangement (see Chapter 2). The Mach–Zehnder interferometer uses ring geometry with counterpropagating paths and two detectors (Figure A.11). The configuration of these three interferometers is such that 1-D scalar interactions model most of the phenomena. Subwavelength curve fitting procedures can extend the resolution of these instruments down to the nanometer range.

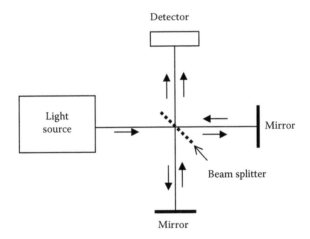

FIGURE A.10 Michelson's interferometer.

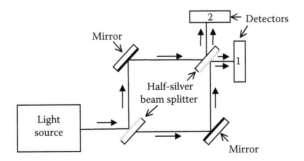

FIGURE A.11 Mach–Zehnder's interferometer.

An explanation of the behavior of an interferometer with a simple geometry and scalar intensity output considers the behavior of monochromatic light beams that oscillate at a single stable frequency, but may have a variable phase. Denoting φ_1 and φ_2 as the instantaneous amplitudes of two waves that arrive at the detector through paths 1 and 2, respectively, and denoting φ_D as the instantaneous amplitude of the combination of the waves at the detector, then

$$\varphi_1(t) = A_1 \exp[j(KL_1 - \omega t + \delta_1)] \tag{A.90}$$

and

$$\varphi_2(t) = A_2 \exp[j(KL_2 - \omega t + \delta_2)], \tag{A.91}$$

$$\varphi_D(t) = \varphi_1(t) + \varphi_2(t), \tag{A.92}$$

where δ_1 and δ_2 are phase angles. Since light oscillates at very high frequencies, most detectors measure an average intensity, I, with units of power per area:

$$I = \langle \varphi\varphi^* \rangle = \lim_{T \to \infty} \left[\frac{1}{T} \int_{-T/2}^{T/2} \varphi(t)\varphi^*(t)\, dt \right]. \tag{A.93}$$

The detected intensity, I_D, for the two combined beams is

$$I_D = \langle \varphi_D \varphi_D^* \rangle = \langle (\varphi_1 + \varphi_2)(\varphi_1^* + \varphi_2^*) \rangle, \tag{A.94}$$

$$I_D = A_1^2 + A_2^2 + 2A_1 A_2 \langle \cos[K(L_2 - L_1) - \delta_{12}(t)] \rangle, \tag{A.95}$$

$$I_D = A_1^2 + A_2^2 + 2A_1 A_2 \{\cos[K(L_2 - L_1)]\langle\cos[\delta_{12}(t)]\rangle + \sin[K(L_2 - L_1)]\langle\sin[\delta_{12}(t)]\rangle\}, \tag{A.96}$$

where δ_{12} is the phase angle difference $\delta_1 - \delta_2$.

Coherent light has a phase angle that remains constant, that is, the light appears as a perfect sinusoid. An alternate and less restrictive definition of coherence is that the phase angles δ_1 and δ_2 can fluctuate, but they do so in unison. The net effect is

$$\delta_1(t) - \delta_2(t) = \delta_{12}(t) = 0 \tag{A.97}$$

and

$$I_D = A_1^2 + A_2^2 + 2A_1 A_2 \cos[K(L_2 - L_1)]. \tag{A.98}$$

An examination of Equation A.98 indicates that the detected intensity is a constant offset plus a sinusoidal fluctuation with path length differences. The received intensity is always positive. The amount of intensity fluctuation is a maximum if the interferometer is balanced so that the amplitude of each arm A_1 and A_2 is equal.

The phase angles of light from incoherent sources shift randomly with time. The degree of statistical correlation between the phase angles $\delta_1(t)$ and $\delta_2(t)$ of the two segregated and then superposed light beams affects the behavior of the interferometer. Independent fluctuations in $\delta_1(t)$ and $\delta_2(t)$ can arise when two independent incoherent light sources form the separate light beams, or when the paths from a single light source are sufficiently different from $\delta_1(t)$ and $\delta_2(t)$ that appear to be independent. When the phase angles fluctuate independently,

$$\langle\cos[\delta_{12}(t)]\rangle = \langle\sin[\delta_{12}(t)]\rangle = 0 \tag{A.99}$$

and

$$I_D = A_1^2 + A_2^2. \tag{A.100}$$

If the light source is incoherent, but such that the effective path lengths L_1 and L_2 are equal, $\delta_1(t)$ and $\delta_2(t)$ can fluctuate in unison with the phase angle at the source. The detected intensity will reach a maximum value when the path lengths are equal and

$$I_D = A_1^2 + A_2^2 + 2A_1A_2. \tag{A.101}$$

An intermediate situation occurs when the two paths are nearly of the same length and the light is partially coherent. In this case, the maximum intensity occurs when the path lengths are equal, as in Equation A.101, but the intensity drops off at a more gradual rate as the two path lengths begin to differ [25]. This forms the basis of a white light interferometer. A somewhat more rigorous explanation of this effect is to note that ϕ_2 is a time lag of ϕ_1, based on the path length distance δ/c (Equation A.94). The detected signal is an autocorrelation function of the source amplitude, $\rho(\delta)$, where

$$\rho(\delta) = \left\langle\varphi(t)\varphi^*\left(t + \frac{\delta}{c}\right)\right\rangle. \tag{A.102}$$

For the case of an interferometer with an equal intensity split between light paths,

$$I_D = \frac{I_0}{2}\{1 + Re[\rho(\delta)]\}. \tag{A.103}$$

$I_D = I_0$ when the path lengths are equal.

Perfectly coherent light has a cosine-shaped autocorrelation function and a corresponding spectrum with a delta function shape. Conversely, perfectly incoherent light with a flat spectrum has a delta function at zero as an autocorrelation function. Partially coherent light with a band-limited spectrum will have an autocorrelation function that drops off from maximum to zero. The drop off rate depends on the coherence length.

The above description applies to scalar representations of interference effects at a point. Similar, but more geometrically complicated, interference phenomena occur over 2-D surfaces and inside 3-D volumes with fringe patterns and holograms being result phenomena [20,26].

A.4.2 Diffraction

Diffraction has many similarities to interference. Diffraction occurs when the index of refraction of a medium changes within a distance that is shorter than the wavelength of the propagating waves. Figure A.12 shows an example where a plane wave impinges on an absorber with a slit aperture.

According to Huygens' principle, the wave at $x = 0$ in the slit between $a/2$ and $-a/2$ acts as a continuum of point sources. The coherence of the plane wave causes all the point

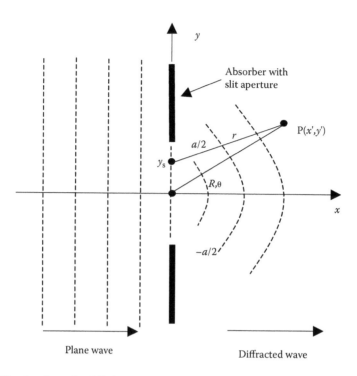

FIGURE A.12   Diffraction through a 2-D slit aperture in an absorber.

sources to be in phase. The waves that arrive at a point P downstream from the aperture, φ_P, are a superposition of the waves emanating from the point sources along the slit [1]:

$$\varphi_P = A\,e^{j\omega t} \int\limits_{-a/2}^{a/2} \frac{e^{-jkr}}{r}\,dy = A\,e^{j\omega t}\Phi(a,k,R),\qquad\text{(A.104)}$$

where

$$r = \left[x_P^2 + (y_P - y_s)^2\right]^{1/2}\qquad\text{(A.105)}$$

The time-averaged intensity at P is

$$I_P = \left\langle \varphi_P(t)\,\varphi_P^*(t)\right\rangle = A^2\Phi\Phi^*.\qquad\text{(A.106)}$$

Evaluating the integral for $\Phi(a, k, R)$ in Equation A.104 requires approximate analytical and/or numerical solutions. The *far-field* approximation assumes that P is sufficiently far from the aperture so that small angle geometric approximations apply. Expanding $r(y)$ in a power series gives

$$r(y_s) = R - y_s \sin\theta + (y_s^2/2R)\cos^2\theta + \cdots.\qquad\text{(A.107)}$$

The integral in Equation A.104 is only moderately sensitive to changes in the amplitude term $(1/r)$, but is more sensitive to changes in phase term (e^{-jkr}). When R is sufficiently large, then retaining the first term in Equation A.107 for the amplitude component and the first two terms for the phase component of the integral is the basis of the *Fraunhofer* approximation, that is,

$$\Phi(a,k,R,\theta) \approx a\, e^{-jkR}\frac{\sin\beta}{\beta}, \tag{A.108}$$

where

$$\beta = \frac{ka}{2}\sin\theta, \tag{A.109}$$

$$I_P(k,a,\theta) = I_P(\beta) = I_0\left(\frac{\sin\beta}{\beta}\right)^2. \tag{A.110}$$

I_P is an oscillating positive curve with values that range between 0 and $1/\beta^2$ (Figure A.13). The maximum intensity, I_0, occurs when $\theta = 0$. The intensity is zero at integral values of β/π.

Two limiting cases are as follows: (1) *The aperture length is much smaller than a wavelength*, that is $ka \ll 1$. In this case, β is small enough for $\sin(\beta) \approx \beta$ and

$$I_P(\theta) = I_0\left(\frac{\sin\beta}{\beta}\right)^2 \approx I_0. \tag{A.111}$$

The effect is that the aperture appears as a point source. The radiated intensity is independent of angle, in the far field. (2) *The aperture length is much larger than a wavelength*, that is, $ka \gg 1$. This causes $I_P \approx 0$, except when $\theta \approx 0$ and $I_P(\theta) \approx I_0$. This second case indicates that using an aperture to create a directed beam of collimated radiation requires that the aperture must be much larger than the wavelength of the radiation. This is a general result for directed radiation beams that also applies to other intensity distributions, such as Gaussians [21]. Shadows are a common manifestation of this effect. The above derivation of Fraunhofer's diffraction was for a 1-D slit aperture. Rectangular, circular, and more complex aperture geometries produce similar area diffraction integrals and behaviors.

The Fraunhofer approximation breaks down when the geometry of small angles approximations is inaccurate. This occurs when the distance between the aperture and the

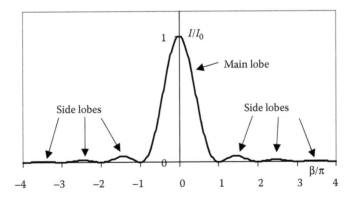

FIGURE A.13 Fraunhofer's slit intensity.

observation point decreases. Including an additional term in the series expansion in Equations A.104 and A.107 gives an improved model known as the *Fresnel* approximation [1,4,20]. The Fresnel approximation uses Fresnel integrals, $F(\alpha)$, with the form

$$F(\alpha) = \int_0^\alpha \exp\left(\frac{j\pi y^2}{2}\right) dy. \tag{A.112}$$

Fresnel's integrals do not have closed-form solutions, but are readily tabulated. One effect of the Fresnel approximation is the appearance of additional diffraction pattern intensity fluctuations that do not appear in the simpler Fraunhofer approximation.

Concepts from diffraction theory provide insights into the imaging of small features. The *Rayleigh criterion* is a standard method of defining the minimum resolvable distance between two objects. The criterion makes use of the notion that ray optics poses no limitations on the ability to image and discern small features. It is the overlapping of wave and diffraction effects, and not the close spacing of rays that limits the ability of an imaging system to discern closely spaced features. The diffraction pattern of an individual feature has a maximum value with a receiver point positioned directly on-axis, and drops to zero at a set of discrete nearby off-axis points. Placing a second feature at a position corresponding to a minimum in the diffraction pattern of the first feature produces a strongly detectable change in the combined intensity pattern. The closest position for this type of detection is the first minimum of the diffraction pattern of the first feature. This separation distance is the Rayleigh criterion (Figure A.14). The Rayleigh criterion also describes the resolution of capability instruments with lens systems, such as microscopes or telescopes, by the inclusion of the focal length. As indicated by Born and Wolf, the Rayleigh criterion has no fundamental physical basis other than that indicated in Figure A.14 [2]. Other similar criteria are also valid and useful [27]. Under certain circumstances, it is possible to resolve dimensions less than the Rayleigh limit, especially if imaging is not the goal. A non-imaging example of sub-Rayleigh limit optical measurements is the use of Vernier scales created by Moiré interference patterns to determine dimensions with nanometer-scale high resolutions [28].

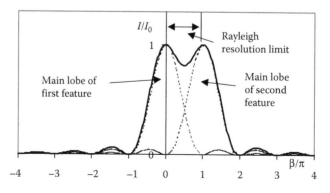

FIGURE A.14 Superposed diffraction patterns associated with Rayleigh resolution criterion for two closely spaced objects.

A.5 Vector Waves

Vector waves are propagating disturbances in vector fields. It is possible to describe a wider class of phenomena with vector wave fields than with scalar wave fields. In acoustic and elastic waves, the particle displacement vectors propagate as waves. For EM waves, it is the electric and magnetic field and displacement vectors that propagate as waves.

Polarization is a prime example of a vector wave phenomenon. Vector waves with vector components that point in directions transverse to the direction of propagation are polarized. Figure A.15 shows the geometric difference between polarized and unpolarized waves. Examples of polarized waves include transverse string vibrations, elastodynamic shear waves (S-waves), and EM waves. Waves with the direction of the vector parallel to the direction of propagation are unpolarized. Examples of unpolarized vector waves are the displacements associated with acoustic and elastodynamic compression waves (P-waves).

A generic vector wave equation in the absence of external forces and boundary conditions

$$\nabla^2 \vec{u} = \frac{1}{c^2} \frac{\partial^2 \vec{u}}{\partial t^2} \tag{A.113}$$

allows harmonic plane wave solutions of the form

$$\vec{u} = \vec{u}_0 \exp[j(\vec{K} \cdot \vec{r} - \omega t)]. \tag{A.114}$$

It is often convenient to represent vector field variables as derivatives of four potential functions (φ, A_x, A_y, and A_z) as the sum of a scalar gradient and vector curl of the potential functions to form the vector field [4]:

$$\vec{u} = \nabla \varphi + \nabla \times \vec{A}. \tag{A.115}$$

This representation provides a convenient segregation of the field equations. However, using four scalar potential functions to represent a three-component vector field introduces an excess degree of freedom. Formulating a unique solution requires the use of a constraint equation, known as a *gauge condition*. Setting the divergence of the vector potential field to zero is a common and convenient gauge condition, that is,

$$\nabla \cdot \vec{A} = 0. \tag{A.116}$$

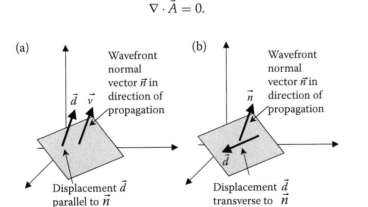

FIGURE A.15 Geometry of (a) unpolarized and (b) polarized vector waves.

The gage condition uncouples the scalar and vector potentials into separate wave equations:

$$\nabla^2 \varphi = \frac{1}{c_p^2} \frac{\partial^2 \varphi}{\partial t^2}, \tag{A.117}$$

$$\nabla^2 \vec{A} = \frac{1}{c_s^2} \frac{\partial^2 \vec{A}}{\partial t^2}. \tag{A.118}$$

In elastodynamics, the scalar gradient wave equation (Equation A.117) corresponds to volumetric changes that propagate as compression waves, that is, P-*waves*. The particle motions of P-*waves* are parallel to in the direction of propagation of planar wavefronts. The vector potential wave equation (Equation A.118) corresponds to volume-preserving shear waves, that is, S-*waves*. The particle motions lie in the plane of the planar wavefronts and are transverse with respect to the direction of propagation. The transverse vector nature of S-waves causes them to become polarized. P- and S-waves travel at different speeds and can propagate independently through bulk solids. Since both P- and S-waves are components of the same displacement field, interactions with boundaries and inhomogeneities can convert incident P- or S-waves into combined P- and S-waves [6]. SV-waves are a subset of S-waves that have a polarization such that the displacements lie in a plane that is normal to the boundary. SH-waves are a subset with a polarization such that the displacements are parallel to the boundary, and perpendicular to the SV-waves. SV-waves with an angle of incidence of 45° do not undergo a mode conversion upon reflection and reflect as pure SV-waves. These special waves are known as *Lamb waves*, and are of importance in wave propagation in plates and other NDE applications. An example is the use of polarized S-waves to evaluate surfaces with specific directional properties, including railroad wheels [29].

The vector field representation of EM waves uses four interacting vector fields (electric field, electric displacement, magnetic field, and magnetic displacement). Often the electric and magnetic field vectors lie in the plane transverse to the direction of propagation to form a *transverse electromagnetic* (TEM) wave. For a plane harmonic EM wave, both electric and magnetic fields are of the form as Equation A.114, with the directions transverse to the direction of propagation, that is,

$$\vec{E} = \vec{E}_0 \exp[j(\vec{K} \cdot \vec{r} - \omega t)], \tag{A.119}$$

$$\vec{H} = \vec{H}_0 \exp[j(\vec{K} \cdot \vec{r} - \omega t)]. \tag{A.120}$$

The representations in Equations A.119 and A.120 take advantage of the convenience of complex numbers to account for the phase angle differences between the electric and magnetic field vectors $\vec{E}_0$ and $\vec{H}_0$. Substitution of the complex harmonic plane wave solutions into the Maxwell equation (Equation A.10) produces

$$\vec{K} \times \vec{H}_0 = -\varepsilon \omega \vec{E}_0. \tag{A.121}$$

The cross product in Equation A.121 and the dot product in Equation A.6 indicate that the direction of propagation, the magnetic field vector, and the electric field vector are all mutually perpendicular (Figure A.16). The Poynting vector $\vec{S}$ gives the energy per unit area flow rate and direction as

$$\vec{S} = \vec{E} \times \vec{H}. \tag{A.122}$$

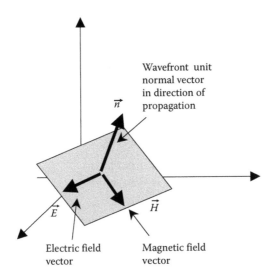

Wavefront unit
normal vector
in direction of
propagation

$\vec{n}$

$\vec{E}$

$\vec{H}$

Electric field
vector

Magnetic field
vector

FIGURE A.16   Geometry of field vectors in harmonic electromagnetic plane wave.

Two independent EM harmonic plane waves can propagate at the same frequency and in the same direction in a homogeneous isotropic medium. The decomposition of a generic harmonic wave into individual components is not unique. A standard decomposition resolves the wave into two components based on the magnitude and phase of the electric field vector with respect to orthogonal transverse axes. The directions and phasing of the magnetic and electric field vectors determine the polarization, which may be linear, circular, or elliptical [1]. When the phase of each orthogonal component is equal, the polarization is linear with the direction determined by the relative magnitude. When the phase angle differs by 90° and the amplitudes are equal, the polarization is circular. The general case corresponds to elliptical polarization. A mechanical analogy of polarization geometry occurs with the small angle motions of a spherical pendulum. The paths traced by the pendulum are the superposition of sinusoids in both transverse directions. These sinusoids have the same frequency but can differ in amplitude and phase. The superposed motion traces linear, circular, or elliptical paths.

Wave polarization effects are especially noticeable when the waves interact with material boundaries, material inhomogeneities, and the bulk properties of the transmissive medium. If the medium is dielectric and nonmagnetic, the direction of the electric field governs the type of interactions [3]. An example is a plane wave interacting with a boundary formed by the x–y plane between two dielectric media with indices of refraction n_1 and n_2 (Figure A.17). If the EM wave is linearly polarized so that the electric field vector lies in the x–y plane and the magnetic field vector lies in the x–z plane, then the wave is a transverse electric (TE) polarized wave. If the wave is linearly polarized so that the magnetic field vector lies in the x–y plane and the electric field vector lies in the x–z plane, then it is called a transverse magnetic (TM) polarized wave. TE waves and TM waves interact differently with a boundary, producing different amplitudes in the reflected and transmitted waves [30]. The reflection (R) and transmission (T) coefficients are

$$R_{\text{TE}} = \left(\frac{E_r}{E_i}\right)_{\text{TE}} = \frac{n_1 \cos \theta_i - n_2 \cos \theta_t}{n_1 \cos \theta_i + n_2 \cos \theta_t}, \tag{A.123}$$

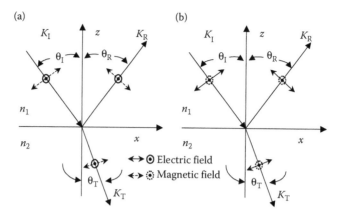

FIGURE A.17 Geometry of (a) transverse electric and (b) transverse magnetic waves interactions with a planar boundary between two different dielectric media. (Adapted from Udd E. 1991. *Fiber Optic Sensors*, Wiley, New York.)

$$T_{TE} = \left(\frac{E_t}{E_i}\right)_{TE} = \frac{2n_1 \cos \theta_i}{n_1 \cos \theta_i + n_2 \cos \theta_t}, \tag{A.124}$$

$$R_{TM} = \left(\frac{E_r}{E_i}\right)_{TM} = \frac{n_2 \cos \theta_i - n_1 \cos \theta_t}{n_1 \cos \theta_t + n_2 \cos \theta_i}, \tag{A.125}$$

$$T_{TM} = \left(\frac{E_t}{E_i}\right)_{TM} = \frac{2n_1 \cos \theta_i}{n_1 \cos \theta_t + n_2 \cos \theta_i}. \tag{A.126}$$

If the angle of incidence $\theta_i = 0$,

$$R_{TE} = -R_{TM} = \frac{n_1 - n_2}{n_1 + n_2}, \tag{A.127}$$

$$T_{TE} = T_{TM} = \frac{2n_1}{n_1 + n_2}. \tag{A.128}$$

An examination of the numerator of the reflection coefficients for the TE and TM waves indicates that there may be specific angles of incidence at which the reflection coefficient vanishes. A more detailed examination (using Snell's law) indicates that the case of zero reflections occurs only with TM waves. This specific angle of incidence is the *Brewster angle*, which occurs when

$$\theta_r + \theta_t = \frac{\pi}{2}. \tag{A.129}$$

Multilayer dielectric media with layer thickness dimensions similar to the wavelengths of propagating waves can produce strong reflecting (or antireflecting) effects. This phenomenon is the basis for quarter-wave mirrors, many optical antireflective coatings, and impedance matching ultrasound transducers. The characteristic matrix, $[M]$, is a useful tool for analyzing multilayer and/or multicomponent optical systems. $[M]$ accounts for changes in the TE (or magnetic) field components, $x(z)$ and $y(z)$, of a harmonic plane wave propagating in the z-direction by matrix multiplication, that is,

$$\begin{bmatrix} x(z_0) \\ y(z_0) \end{bmatrix} = [M(z_1 - z_0)]\begin{bmatrix} x(z_1) \\ y(z_1) \end{bmatrix} = \begin{bmatrix} M_{11}(z_1 - z_0) & M_{12}(z_1 - z_0) \\ M_{21}(z_1 - z_0) & M_{22}(z_1 - z_0) \end{bmatrix}\begin{bmatrix} x(z_1) \\ y(z_1) \end{bmatrix}. \tag{A.130}$$

The components of [M] depend on both the depth of the layers and the dielectric constants of the media. A determinant of [M] equal to 1 corresponds to lossless media. A multilayer stack of dielectrics forms an effective characteristic matrix $[M_s]$ as a product of the characteristic matrices of the individual layers:

$$\begin{bmatrix} x(z_0) \\ y(z_0) \end{bmatrix} = [M(z_1 - z_0)][M(z_2 - z_1)] \cdots [M(z_N - z_{N-1})] \begin{bmatrix} x(z_N) \\ y(z_N) \end{bmatrix}. \tag{A.131}$$

Born and Wolf attribute much of the original matrix analysis of multilayers to Abèles [2]. An example SHM application of matrix multilayer analysis is the quantitative analysis of fiber optic Bragg grating performance [31].

A.6 Guided Waves

Materials with spatially varying wavespeeds, such as boundaries, gradients, or inhomogeneities, distort and alter the directions of propagating waves. Occasionally, these effects combine to guide waves along specified paths. A common waveguiding technique embeds an elongated structure of one medium in a different medium, as in Figure A.18. This propagates waves in a manner that confines the waves to the inside of the elongated structure. The ray interpretation of many guided wave phenomena corresponds to repeated reflections off of the walls of the waveguide. Practical reasons of coupling efficiency and attempts to propagate signals over long distances with a minimum of dispersion and losses require reducing the waveguide cross-sectional dimensions to one or a few wavelengths. In these cases, a detailed understanding of the phenomena requires consideration of the wave nature of the propagation. Elastic Lamb waves in plates and fiber optic signal transmission are common SHM applications of guided waves.

The relative size of wavelength and plate thickness has a large effect on the nature of the propagation of elastic waves. Figure A.19 shows the geometry of a plate with lateral dimensions that are large relative to wavelength. When the wavelength is much larger than the plate thickness, the waves tend to be either transverse flexural waves or in-plane compression waves. When the wavelength is much shorter than the plate thickness, the waves

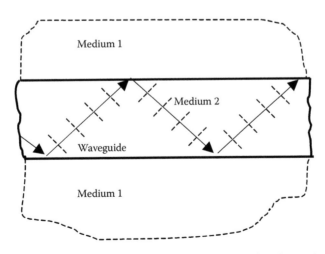

FIGURE A.18 Waveguide with ray-tracing and wavefront representation of guiding action.

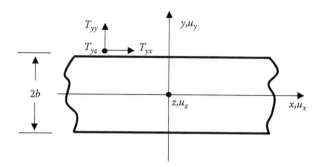

FIGURE A.19   Geometry of infinite plate of thickness 2b (z-axis is perpendicular to the page).

propagate as if the plate is infinite or semi-infinite elastic solid. When the wavelengths are of the same order as the plate thickness, then *Lamb* (a.k.a. *Rayleigh–Lamb*) waves propagate.

Lamb waves receive considerable interest as possible tools in SHM. The principal reasons include a combined ability to readily propagate elastic waves over long distances in plate-like structures and the strong coupling of the waves with inhomogeneities (delaminations, holes, inclusions, etc.). The following is an abbreviated description of results developed by Rayleigh, Lamb, Lamé, and Mindlin and collected by Graff [6]. For simplicity, it is assumed that a Lamb wave is a harmonic straight crested wave, with the crests parallel to the z-axis and the particle motions in planes parallel to the x–y plane, that is, $u_z = 0$. The boundary conditions at the top and bottom surface of the plate are assumed to be traction-free, that is, $T_{yx} = T_{yy} = T_{yz} = 0$ at $y = \pm b$.

A potential function description of the displacement field, in the form of Equation A.115 that satisfies the traction-free boundary conditions and forms straight-crested harmonic waves, is

$$\varphi(x,y,z,t) = f(y)\ \exp[j(\xi x - \omega t)],$$

$$A_x(x,y,z,t) = 0,$$

$$A_y(x,y,z,t) = 0,$$

$$A_z(x,y,z,t) = ja_z(y)\ \exp[j(\xi x - \omega t)].$$

(A.132)

Substituting these potentials into the elasticity Equations A.20 through A.25 and combining with Equation A.116 yields two sets of modal displacement fields as solutions. The first set, u_{xs} and v_{xs}, comprises the *symmetric* waves. The x-displacements are symmetric and the y-displacements are antisymmetric about the x-axis:

$$u_{xs} = j\left[B\xi \cos(\alpha y) + C\beta \cos(\beta y)\right] \exp[j(\xi x - \omega t)],$$

(A.133)

$$u_{ys} = \left[-B\alpha \sin(\alpha y) + C\xi \sin(\beta y)\right] \exp[j(\xi x - \omega t)].$$

(A.134)

The second set comprises the *antisymmetric* waves. The x-displacements are antisymmetric and the y-displacements are symmetric about the x-axis:

$$u_{xa} = j\left[A\xi \sin(\alpha y) - D\beta \sin(\beta y)\right] \exp[j(\xi x - \omega t)],$$

(A.135)

$$u_{ya} = \left[A\alpha \cos(\alpha y) + D\xi \cos(\beta y)\right] \exp[j(\xi x - \omega t)].$$

(A.136)

Boundary conditions along with the following characteristic frequency–wavenumber equations determine values for the coefficients A, B, C, and D in Equations A.133 through A.136:

$$\frac{B}{C} = -\frac{2\xi\beta\cos(\beta b)}{(\xi^2 - \beta^2)\cos(\alpha b)}, \tag{A.137}$$

$$\frac{A}{D} = -\frac{2\xi\beta\sin(\beta b)}{(\xi^2 - \beta^2)\sin(\alpha b)}, \tag{A.138}$$

with the wavenumbers α and β defined as

$$\alpha = \left[\frac{\omega^2}{c_p^2} - \xi^2\right]^{1/2}, \tag{A.139}$$

$$\beta = \left[\frac{\omega^2}{c_s^2} - \xi^2\right]^{1/2}. \tag{A.140}$$

Possible wavenumbers and frequencies that satisfy the Rayleigh–Lamb characteristic equation for free boundaries on both sides of the plate are

$$\frac{\tan(\beta b)}{\tan(\alpha b)} + \left[\frac{4\alpha\beta\xi^2}{(\xi^2 - \beta^2)^2}\right]^{\pm 1} = 0 \quad \begin{cases} +1 = \text{symmetric} \\ -1 = \text{antisymmetric.} \end{cases} \tag{A.141}$$

A continuum of possible wavenumber combinations—including those with imaginary components—are solutions to Equation A.141. A special case arises when $\xi = \beta$ and

$$\beta = \frac{n\pi}{2b}, \tag{A.142}$$

where $n = 1, 3, 5, \ldots$ corresponds to the symmetric modes and $n = 0, 2, 4, \ldots$ corresponds to the antisymmetric modes. The associated frequencies are

$$\omega_n^2 = 2(\xi c_s)^2 = \frac{1}{2}\left(\frac{\pi n c_s}{b}\right)^2, \quad n = 0, 1, 2, \ldots. \tag{A.143}$$

The waves corresponding to the solutions in Equations A.142 and A.143 are sometimes called *Lame modes* or *Lame waves*. It is also now common to call these waves Lamb waves. However, since the term "Lamb wave" describes a broader class of wave phenomena, the meaning of the term needs to be considered within the proper context. Common labels for the symmetric mode Lamb waves are $S_0, S_1, S_2, \ldots$. The displacements for the symmetric modes ($n = 1, 3, 5 \ldots$) are

$$u_{xs}(x, y, z, t) = \mathrm{j}\left(\frac{n\pi}{2b}\right) C \cos\left(\frac{n\pi y}{2b}\right) \exp\left[\mathrm{j}\left(\frac{n\pi x}{2b} - \omega_n t\right)\right], \tag{A.144}$$

$$u_{ys}(x, y, z, t) = \left(\frac{n\pi}{2b}\right) C \sin\left(\frac{n\pi y}{2b}\right) \exp\left[\mathrm{j}\left(\frac{n\pi x}{2b} - \omega_n t\right)\right]. \tag{A.145}$$

Common labels for the antisymmetric mode Lamb waves are $A_0, A_1, A_2, \ldots$. The displacements for the antisymmetric modes ($n = 0, 2, 4, \ldots$) are

$$u_{xa}(x, y, z, t) = -\mathrm{j}\left(\frac{n\pi}{2b}\right) D \sin\left(\frac{n\pi y}{2b}\right) \exp\left[\mathrm{j}\left(\frac{n\pi x}{2b} - \omega_n t\right)\right], \tag{A.146}$$

$$u_{ya}(x, y, z, t) = \left(\frac{n\pi}{2b}\right) D \cos\left(\frac{n\pi y}{2b}\right) \exp\left[\mathrm{j}\left(\frac{n\pi x}{2b} - \omega_n t\right)\right]. \tag{A.147}$$

It can be shown that Lamb waves correspond to the case of SV-waves propagating back and forth across the thickness and down the length of a plate at angles of 45°, with the harmonic wavelengths matching the diagonal length $(2b)\sqrt{2}$. As indicated in the previous section, SV-waves interacting with a free surface at an angle of 45° reflect into SV-waves without any mode conversion (Figure A.20). This distinctive property enables Lamb waves to propagate over large distances in plates with minimal losses. A complication is that independent forward and backward propagating waves with the same wavelength can coexist in the plate [32].

An important consideration is that the above derivation is for straight crested waves with geometrically linear sources. In practice, it is usually more convenient to generate waves with point transducers. The waves resulting from point sources are cylindrical with modified Bessel (Hankel) function geometries with an approximate $1/\sqrt{r}$ drop in amplitude versus distance from the source. Kohler et al. confirmed the presence of such cylindrical Lamb wave shapes with interferometric measurements and numerical calculations [33]. Liu and Giurgiutiu, and Martin and Jata demonstrated that the finite element method is capable of simulating Lamb waves in possibly more complicated geometries, especially when the geometry supports S_0 and A_0 waves [34,35]. Alleyne and Cawley found that S_i and A_i waves transmitted and reflected off of damage in plates in a manner that depends on the frequency-thickness products, mode symmetry, mode order, and damage geometry [36].

SHM Lamb wave applications typically involve forced excitation and response measurements [37]. Lamb wave launching and receiving techniques are critical details [38]. Neary et al. used a variable angle transducer to launch Lamb waves into composite plates for damage detection [39]. Applying the stimulus to only one side of the plate introduces a midplane asymmetry [40]. Giurgiutiu et al. and Wilcox et al. used asymmetric one-sided plate excitation to induce symmetric Lamb waves by exploiting sweet spots in the frequency response that favor symmetric over antisymmetric modes [41,42].

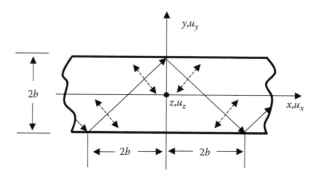

FIGURE A.20 Ray and wavefront representation of Lamb wave propagation of SV-waves at 45°. Solid arrows indicate rays. Dashed arrows indicate particle displacement wavefronts.

In addition to Lamb waves, elastic structures with elongated geometries support additional similar guided waves. These include Rayleigh waves that occur at the surface of the structure, compression waves in rods, and torsion waves in cylindrical pipes [6] [43].

Optical waveguides usually guide light with total internal reflection [30,44]. A common configuration is the cylindrical waveguide used in fiber optic cables (Figure A.21). Total internal reflection occurs when the wavespeed of the waveguide, c_1, is less than the wavespeed of the surrounding medium, c_2; or conversely for the refractive indices $n_1 > n_2$. Total internal reflection causes the guiding of light rays when the condition of Equation A.86 is exceeded, that is,

$$\sin \theta_1' = \cos \theta_1 > \frac{n_2}{n_1}. \tag{A.148}$$

Using air with a relative index of refraction $n_0 = 1$ gives the maximum off-axis input angle, θ_0, that permits guiding as

$$\frac{\sin \theta_0}{\sin \theta_1} = \frac{n_1}{n_0} = n_1. \tag{A.149}$$

The *numerical aperture* (NA) is a standard measure of the maximum off-axis input angle that depends on the relative refractive indices of the inner core and outer cladding [45]:

$$NA = n_0 \sin \theta_0 = \left[n_1^2 - n_2^2 \right]^{1/2}. \tag{A.150}$$

Since the index of refraction of the waveguide material, n_1, is greater than that of air, n_0, it is possible to use air as the cladding. A difficulty with air is that total internal reflection produces an evanescent wave in the outer material (cladding or air). Inhomogeneities, absorptions, and contact with mounting fixtures in the outer layer absorb energy and reduce the effective propagation distance of the waveguide. A low-loss cladding of sufficient thickness (several microns) eliminates most of the evanescent wave losses. Occasional applications of evanescent waves couple light in and out of waveguides to form the basis of fiber optic chemical sensors and optoelectronic devices. Ray-tracing models adequately describe waveguiding when the waveguide transverse dimensions are bigger than the length of the guided waves. Ray-tracing usually indicates that an infinite continuum of paths satisfies

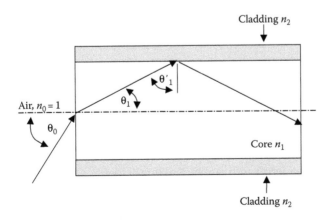

FIGURE A.21   Ray representation of light guiding in an optical waveguide.

the condition of total internal reflection in an optical waveguide (Figure A.21). The presence of multiple propagation paths, or modes, causes the light to travel at different axial speeds. Multipath propagation causes the smearing of signals and limits the maximum data transmission rate.

Reducing the transverse dimension of the waveguide to sizes on the order of a wavelength causes wave properties to be important and ray-tracing models to be inadequate. Instead, scalar and vector waves produce more realistic models of the waveguide behavior. Vector wave models based on harmonic solutions to Maxwell's equations indicate that only a few modes will propagate through a small diameter waveguide [45,46]. A sufficiently small transverse dimension allows for only a single mode with a direction of travel straight down the center of the waveguide. Early communication-grade fiber optic waveguides used a step change in refractive index to guide the light. Typical core diameters were on the order of 50–200 μm. These step-index fibers allow for the propagation of the hundreds (or more) modes and are usually referred to as *multimode* fibers. The dispersive nature of the multimode wave propagation in these fibers limited the high-speed digital communication performance. As manufacturing techniques improved, it became possible to mass-produce fiber optic waveguides with radially graded indices and transverse dimensions of about 5 μm. These fibers tend to allow only a single mode of propagation down the center. Data rates of 10 Gb/s are routine with single-mode optical fibers. A feature of circular-core single-mode optical fibers is that they allow for two virtually identical superposed modes with different polarizations. When necessary, elliptical core and other asymmetric geometry polarization-maintaining fibers can control such polarization effects [30].

A.7 Scattering

The interactions of waves with inhomogeneities alter the direction, wavefront shape, and frequency content of waves. Such processes are collectively referred to as *scattering*. *Elastic scattering* changes the direction and shape of the wavefront but does not alter the wavelength or frequency content. When the scattering is elastic and sufficiently weak so as not to affect the magnitude and nature of incident wave, the *first Born approximation* can simplify the analysis [2,47]. A rigorous derivation of the Born approximation uses the first term in a perturbation expansion of an integral representation of the combined incident and scattered fields. Mode conversion between P-waves and the different polarization angles of the S-waves in elastic materials is a further complication [48]. Combined global and local methods use analytical expressions for global representations of wave propagation, and finite element models of the interactions near to the scatterers are a compromise that avoids full finite element modeling [49]. *Inelastic* scattering changes the wavelength and frequency. *Coherent* scatterers have an ordered spatial arrangement. *Incoherent* scatterers have random spatial arrangements.

Figure A.22 shows an illustrative case of coherent scattering. The geometry is that of harmonic plane waves interacting with a coherent line of identical scatterers with an equidistant spacing pattern of pitch d and angle of incidence θ_i. Based on the assumptions of elastic scattering and the Born approximation, the incident harmonic wave interacts with the scatterers to produce spherical waves. Each scattered spherical wave has the same frequency and amplitude, but may have a different phase angle, φ_i. The phase angle depends on the

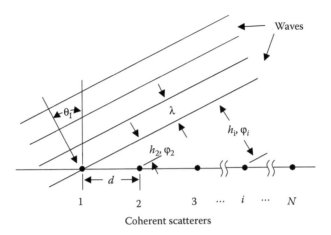

FIGURE A.22 Plane waves intersecting a linear array of coherent scatterers.

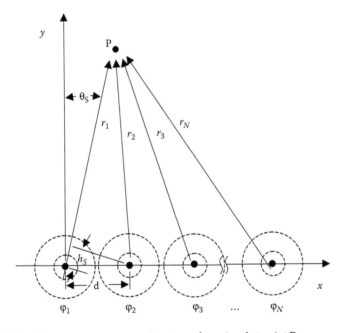

FIGURE A.23 Multiple coherent sources superposing to produce signal at point P.

position of the scatterer relative to the incoming wavefront and the distance to an observation point, P (Figure A.23). The superposition of the individual scattered waves with varied phase angles produces effects that include interference, diffraction, and beam steering.

The signal received at P is a harmonic wave with the same frequency as the incoming wave. The scattered waves, S_P, that arrive at P are the superposition of the waves from the individual scatterers, S_i:

$$S_P = \sum_{i=1}^{N} S_i = \sum_{i=1}^{N} f(r_i) \exp\left[j\left(\omega t - \varphi_i + \beta_i\right)\right], \tag{A.151}$$

where $f(r_i)$ is an amplitude that corresponds to the distance between the receiver and the source, φ_i is the phase angle due to the relative geometry of individual scatterers to incoming waves, and β_i is the phase angle term due to the relative geometry of individual scatterers and the receiver. The amplitude of the received signal depends both on the amplitude terms, $f(r_i)$, and on the superposition of the phase terms, φ_i and β_i. Sufficient distance between the receiver and scatterers allows for a small angle approximation to apply and $f(r_i) \approx f(r_1)$:

$$S_i \approx f(r_1) \exp[j(\omega t - \varphi_i + \beta_i)], \qquad (A.152)$$

where

$$\beta_i = \frac{(i-1)h_S}{\lambda} = \frac{(i-1)\,d\sin(\theta_S)}{\lambda} \qquad (A.153)$$

$$\varphi_i = \frac{(i-1)h_I}{\lambda} = \frac{(i-1)\,d\sin(\theta_I)}{\lambda}, \qquad (A.154)$$

and

$$S_P = f(r_1) \exp(j\omega t) \sum_{i=1}^{N} \exp[j(-\varphi_i + \beta_i)]. \qquad (A.155)$$

The intensity of the signal received at P depends on the scattering angle, θ_S. The intensity of the signal at P has a maximum when $\varphi_i = \beta_i$, that is, $\sin(\theta_I) = \sin(\theta_S)$. These maxima appear when $\theta_I = \theta_S + n\pi$, $n = 0, 1, 2, 3 \ldots$ and are the basis of a diffraction grating. Expanding the sum in Equation A.155 as a geometric series determines the resolution and scattering behavior [1]. 2-D and 3-D patterns and other more complicated organized arrangements of scatterers, for example, photonic crystals, produce similar effects [50]. Deformation of the patterns by strain fields alters the scattering effects and can form the physical basis of strain and force transducers [51]. Nonuniform and chirped diffraction gratings extend these effects. An example due to Fuhr et al. demonstrated noncontacting structural deflection measurements with a chirped diffraction grating and a remotely positioned laser [52].

Noncoherent scatterers collectively scatter coherent waves into a random noncoherent pattern. An amplitude histogram of the wave scattered from a coherent incoming wave can reveal the degree of coherence in the scatterers. When the source is a gated harmonic and the scatterers are coherent, the histogram of the scattered wave corresponds to that of a coherent superposition of harmonic signals and is non-Gaussian. Noncoherent scatterers tend to produce noncoherent scattered waves with Gaussian histograms (Figure A.24). A histogram of the envelope of a wave signal from incoherent scatterers can correspond to that of a random variable with a Rayleigh distribution. *Rician* and *Nakagami* statistical models can be appropriate tools for analyzing situations with combined coherent and noncoherent scattering [53,54].

The scattering of waves by defects in nonhomogeneous anisotropic materials is similar to that of isotropic materials, but requires consideration of the more complicated material structure in the analysis. For example, Banerjee et al. studied the scattering of elastic Lamb waves in nonhomogeneous anisotropic composite plates and found that the scattering was very sensitive to manufacturing and defect details [55].

Inelastic scattering effects shift frequencies and wavelength in a manner that depends on the properties of the scatterers and the surrounding medium. *Raman scattering* results from the interaction of the vibration energy of a molecule with that of a photon to shift the frequency of a scattered photon. The shift is usually to a lower frequency (Stokes), but

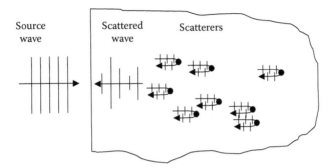

FIGURE A.24 Noncoherent scatterers.

upshifts (anti-Stokes) also occur [56]. The frequency shift is relatively insensitive to the excitation wave frequency, but does depend on the particular molecular species and can indicate the chemical makeup of a material. Many Raman scattering detection instruments are bulky and restricted to laboratory-based testing. Portable versions, including those with embeddable fiber optic probes, are becoming viable and may likely soon be a commonplace.

Brillouin scattering is an inelastic interaction of photons with phonons due to high-frequency elastic waves in solids [56]. Brillouin scattering shifts the frequency of the scattered wave by an amount that depends on the speed of the elastic waves, which in turn depend on temperature and elastic strain. A linearized form is [57]

$$\nu_B(\varepsilon, T) = \nu_B(0, T) + \alpha\varepsilon + \beta\Delta T, \tag{A.156}$$

where ν_B is the frequency of the Brillouin wave, ε is the strain, ΔT is a temperature shift, and α and β are constants.

A.8 Synthetic Methods

Instrument issues, such as bandwidth, resolution, and transducer number, and physical considerations, such as diffraction, limit the practical use of many wave phenomena. *Synthetic methods* attempt to overcome these limitations by synthesizing the results from a series of simpler tests. Most synthesis techniques use linearity. Examples include using *step frequency radar* to simulate impulse radar, SAR to simulate a large antenna array with a single antenna, *phased array* techniques in ultrasound and radar, and *optical proximity correction* (OPC) in photolithography. Figure A.25 is a schematic of the concept. Many of these

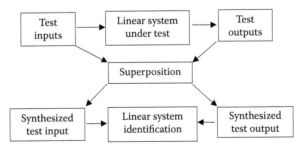

FIGURE A.25 Synthetic methods flow diagram.

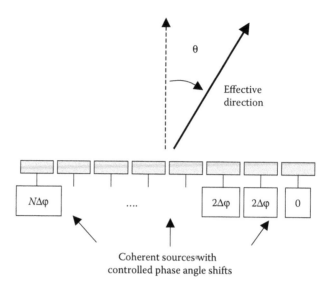

FIGURE A.26 Beam steering with phased source array. (Adapted from Moulin E et al. 2003. *Nondestructive Evaluation and Health Monitoring of Aerospace Materials and Composites II*, SPIE 5046.)

techniques now use digital implementations. Some, such as OPC, remain primarily analog techniques.

Synthetic beaming uses an array of same-frequency wave sources with controlled phases, that is, a phased array, to produce a steerable beam of waves. The operation requires the simultaneous synchronization of the frequency and phase shift of an array of transducers. Beam steering is similar to coherent scattering. The difference is that individual sources with controllable phase angles replace the incoming wave and scatterers. Phase adjustment enables the rapid steering and squinting of the beam. Figure A.26 shows a steerable linear array of sources [58]. A standard radar hardware implementation of beam steering uses *Butler matrices* (a network of phase-shifting circuit elements) [59,60]. A linear phased array is perhaps the easiest to analyze, but it has a directional bias. It is well suited for steering beams broadside to the array, but not in an in-line direction. Omni-directional steering is more practical with different geometries, such as circular arrays [61]. As an example application, Lozev et al. used a phased array of ultrasound transducers to steer focused beams inside of large steel pins in bridges [62]. Malinowski et al. developed a multichannel Lamb wave beam steering system for plate SHM [63]. Swift et al. demonstrated a laser thermoelastic ultrasound beam steering technique on an aluminum plate with a set of fiber optic wave guides that delivered the timed laser pulses [64]. Kramb and Nel used linear phased arrays for the ultrasonic inspection of aircraft engine parts [65,66]. Naito et al. developed an extraordinary tool for demonstrating phased array signal synthesis, known as the *advanced multiple organized experimental basin apparatus* (AMOEBA) [67]. The AMOEBA uses a circular array of wave generators and absorbers to form controlled wave patterns in a water tank for scale model testing of ship dynamics.

Phased arrays can operate in an inverse manner to help to identify the source direction of received waves [68]. Hirose and Kimoto developed methods for analyzing the phase array response of a linear ultrasound system based on the superposition of approximate Green's functions [69].

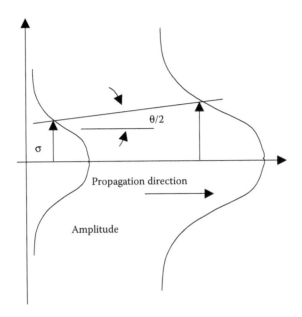

FIGURE A.27 Geometry of Gaussian beam spreading. (Adapted from Verdeyen JT. 1989. *Laser Electronics*, 2nd Ed. Prentice-Hall, Englewood Cliffs, NJ.)

While it is possible to create and control the characteristics of beams of waves in localized regions, physical constraints limit the amount of control that is possible with long-range propagation. Huygens' principle indicates that each point in a transmissive medium acts as a point source that generates spherical waves as part of wave propagation. The collective action of point sources causes localized beams to spread. Plane waves with infinite extent are an exception. They cannot spread any further. Mean square spread statistics are natural tools for quantifying beam spread. This allows for the use of the uncertainty principle. Based on a mean square measure of spread, the most spatially compact planar wave beam is one with Gaussian amplitude (Figure A.27). A Gaussian beam with a root mean square spread σ and wavelength λ diverges with an angle θ given by [21]

$$\theta = \frac{2\lambda}{\pi\sigma}. \tag{A.157}$$

Planar beams with initial spreads that are much larger than a wavelength have small divergence angles. This is the basis for the distinct beaming behavior of many lasers.

The costs (financial, weight, and complexity) required to build and operate a steerable phased array are often prohibitive. Superposition methods enable synthesizing the action of a phased array with a reduced count of transducers. Such techniques are common in the vibration modal analysis of structures. Moving excitation and motion transducers and repeating the data collection can simulate a phased array of sources and receivers [70,71]. As an example, Giurgiutiu et al. synthesized a phase array of Lamb wave sources in a plate with a dual channel switching and a superposition approach [72]. Gao et al. synthesized guided Lamb waves with a tuning technique that excited specific lower-order propagation modes [73]. Sundararaman et al. simulated a beamforming phased array of ultrasound transducers by physically moving a pair of transducers with repeated measurements [74].

Step frequency radar is a synthetic UWB radar technique. The method simulates the response of a system to EM pulses by superposing the response of the system to a sequence of harmonic test signals. The frequencies of the test signal sequence span a wide range. The multicycle repetition of the launch and received waveforms at each frequency allows for constructing a high-quality description of EM interactions with the material. The equivalent time domain version of the launch signal is that of an isolated pulse. Likewise, the equivalent time domain version of the received signal is that of a pulse response. Since time and frequency are conjugate Fourier variables, increasing the frequency span decreases the pulse width. Early analog versions of the technique swept continuously through frequency bands and went by the names of an FM (frequency-modulated) or CW (continuous wave) radar. The digital version of the technique uses finite Fourier transforms. Step or swept frequency approaches are also practical in ultrasonic testing. Particular frequencies corresponding to the wavelength depth multiples of subsurface features can form strong reflectors for use in specialized applications such as delamination detection [75].

The advantages of the step-frequency technique lie in the ability to create and analyze high-quality microwave signals without the need for the direct sampling of a transient waveform. Another advantage is the ability to avoid particular frequency bands that may interfere with other systems. Disadvantages include the need for longer processing times and distortions due to the frequency to time domain conversion. As an example of the timing issues, Eide describes a multi-antenna step-frequency GPR system that requires 25 ms to perform both a frequency sweep from 100 MHz to 2.4 GHz and an antenna sweep with an array of 25 antennas [76].

Many waveform synthesis techniques use the DFT and its inverse. Synthesis with the DFT can be quite convenient, but it runs the risk of introducing counterintuitive and sometimes erroneous results due to limitations of resolution and bandwidth in both the frequency and time domains. For N samples in time with a sampling interval Δt and frequency resolution, $\Delta f = 1/(N\Delta t)$, the DFTs are [77,78]

$$X_k = X(k\Delta f) = \Delta t \sum_{n=1}^{N} x_n \exp[-2\pi\, jkn/N], \quad k = 1, 2, \ldots, N, \qquad (A.158)$$

$$x_n = x(n\Delta t) = \Delta f \sum_{k=1}^{N} X_k \exp[2\pi\, jkn/N], \quad n = 1, 2, \ldots, N. \qquad (A.159)$$

The *magnitudes*, c_k, of the Fourier coefficients are

$$c_k = [\mathrm{Re}(X_k)^2 + \mathrm{Im}(X_k)^2]^{1/2}. \qquad (A.160)$$

The *phase angles*, φ_k, are

$$\varphi_k = atan2[\mathrm{Re}(X_k), \mathrm{Im}(X_k)], \qquad (A.161)$$

where $atan2$ is the two-argument inverse tangent function that avoids phase ambiguities. Standard time domain to frequency domain Fourier analysis initially measures the time domain x_n as real numbers and then calculates the frequency domain X_k as complex numbers with a DFT. Step-frequency methods do the inverse, that is, the initial measurement is the complex-valued X_k (c_k and φ_k) in the frequency domain following by a DFT calculation of the time domain x_n. Difficulties with this approach arise due to measurements being taken over a limited frequency range. Realizable physical systems and measurements use real signals in the time domain. The Fourier transform frequency domain equivalent has complex components. The constraint of realness in the time domain imposes frequency

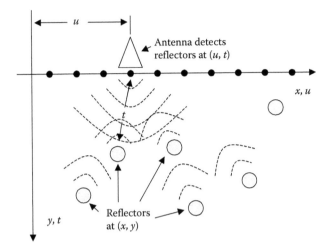

FIGURE A.28 A 2-D synthetic aperture radar geometry.

domain Fourier transform sign symmetry about the origin on the real components and sign antisymmetry on the imaginary components. If the X_k contain all of the information needed to reconstruct x_n, then the procedure proceeds smoothly. However, due to sampling constraints on frequency, time, quantization, and so on, the X_k may not contain all of the information needed to construct the x_n. Anomalous results, such as noncausal or nonreal-time domain behavior, can result.

SAR methods reconstruct images of radar reflectors from a series of multiple measurements. Similar reconstruction techniques applied to geotechnical radar or elastic waves are often called *migration* methods. Broadly, SAR and migration are identical methods, but actually comprise a family of methods that differ in details, such as the use of ray-tracing

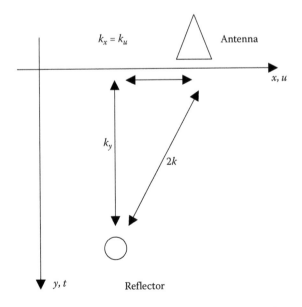

FIGURE A.29 Cross and downrange geometry that gives rise to a hyperbolic nonlinearity.

versus wave models and the treatment of wavespeed inhomogeneities. Both methods can accommodate many standard source waveform types, that is, pulse, sinusoidal, and gated sinusoids. The following presentation due to Soumekh is one of the simpler variants [79]. The concept is to synthesize the behavior and performance of a large aperture radar imaging system from a set of small aperture measurements taken across a range that spans the range of the equivalent large aperture system. A typical geometry appears in Figure A.28. Reflectors and scatterers located downrange at positions (x, y) form a reflection pattern $f(x, y)$. A single antenna positioned at u measures the reflections versus time to form the trace $s(t, u)$. Signals from off-axis reflectors arrive with a delay due to the increased diagonal distance. Stacking measurements from multiple positions in a single plot (usually with

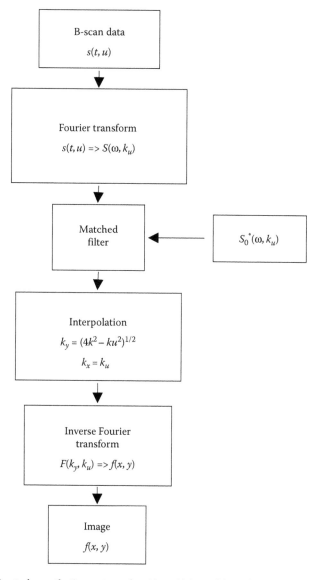

FIGURE A.30 Flowchart of a synthetic aperture algorithm. (Adapted from Soumekh M. 1999, *Synthetic Aperture Radar Signal Processing*, Wiley-Interscience, New York.)

pixel intensity equal to signal amplitude) form a B-scan. Off-axis downrange distortions form hyperbolae in B-scans. The goal of SAR methods is to construct an estimate of the scattering field $f(x, y)$ from a stack of $s(t, u)$ measurements.

Figure A.29 shows the geometry of the hyperbolic nonlinearity in wavenumber space. The wavenumber relations are

$$k_y^2 + k_u^2 = 4k^2, \tag{A.162}$$

$$k_y = (4k^2 - k_u^2)^{1/2}, \tag{A.163}$$

and

$$k_x = k_u, \tag{A.164}$$

where k_x is the downrange image wavenumber, k_y is the crossrange image wavenumber, k_u is the downrange wavenumber, and $k = \omega/c$ is the wavenumber. Figure A.30 shows a flowchart of a basic SAR algorithm. The algorithm removes the hyperbolic nonlinearity by algebra in the frequency and wavenumber domain. A complication is that the measured

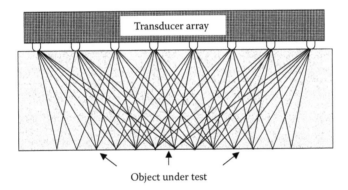

Object under test

FIGURE A.31 Ray representation of the MIRA synthetic aperture ultrasound system. (Diagram courtesy of Germann Instruments.)

FIGURE A.32 Arrays of shear wave dry-contact ultrasound transducers for synthetic aperture ultrasound inspection. (Photograph courtesy of Germann Instruments.)

FIGURE A.33 Concrete block with drilled holes that served as an ultrasound phantom. (Photograph courtesy of Germann Instruments.)

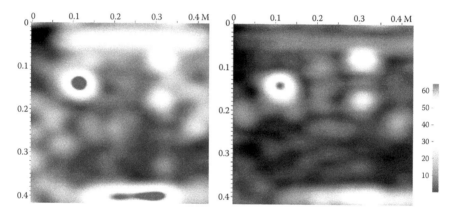

FIGURE A.34 Identification of drilled holes in the concrete block of Figure A.33 with synthetic aperture ultrasound using an array of shear wave dry contact transducers. (Photograph courtesy of Germann Instruments.)

data occur over finite spatial and temporal domains, whereas the Fourier transforms apply to infinite domains. One method to correct for finite domain mismatch is the matched filter algorithm [80]. The next step applies an inverse Fourier transform to form an image in the spatial x–y domain [81].

It may be possible to identify and locate scatterers, such as embedded defects, with synthetic time reversal methods [82]. Kim et al. reconstructed subsurface features in concrete with synthetic aperture single source and receiver continuous sinusoidal microwave signals [83]. Mast developed an SAR imaging method for use with layered media that accounts for different wavespeeds in layers [84,85].

Since the basis of most SAR methods is a scalar wave analysis, many similar techniques apply directly to other wave phenomena [86]. Elastic waves are a prime example. Applying migration methods to Mindlin-type flexural waves can lead to the imaging of damage in plates [87]. Direct numerical simulation of elastic waves interacting with damage is a possibility [88]. Figure A.31 shows a ray representation of the operation of the MIRA synthetic aperture ultrasound inspection system ("MIRA" is a contraction of the Spanish word "mirar"–to look and the English word "mirror."). This system uses multiple shear wave dry contact ultrasound transducers (Figure A.32). A concrete phantom test specimen with predrilled holes and synthetic aperture ultrasound test results on the phantom appear in Figures A.33 and A.34, respectively.

References

1. Towne DH. 1967. *Wave Phenomena*. Dover Publications, Mineola, NY.
2. Born M and Wolf E. 1999. *Principles of Optics*, 7th Ed. Cambridge University Press, Cambridge.
3. Feynman RP, Leighton RB, and Sands M. 1963. *The Feynman Lectures on Physics Mainly Mechanics, Radiation and Heat*, Vol. I. Addison-Wesley, Reading, MA.
4. Feynman RP, Leighton RB, and Sands M. 1963. *The Feynman Lectures on Physics Mainly Electromagnetism and Matter*, Vol. II. Addison-Wesley, Reading, MA.
5. Feynman RP, Leighton RB, and Sands M. 1963. *The Feynman Lectures on Physics Quantum Mechanics*, Vol. III. Addison-Wesley, Reading, MA.
6. Graff KF. 1973. *Wave Motion in Elastic Solids*. Ohio State University Press, Columbus.
7. Pauli W. 1973. *Optics and the Theory of Electrons*. Dover Publications, Mineola, NY.
8. Balanis CA. 1989. *Advanced Engineering Electromagnetics*. Wiley, New York.
9. Dowell EH, Crawley EF, Curtiss Jr HC, Peters DA, Scanlan RH, and Sisto FA. 1995. *Modern Course in Aeroelasticity*, 3rd Ed. Kluwer, Dordrecht.
10. Stratton J. 1941. *Electromagnetic Theory*. McGraw-Hill, New York.
11. Slater JC and Frank NH. 1969. *Electromagnetism*. Dover Publications, Mineola, NY.
12. Condon EU and Odishaw H. 1967. *Handbook of Physics*, 2nd Ed. McGraw-Hill, New York.
13. Scanlan RH and Mendelson A. 1963. Structural damping. *AIAA J*, 1, 938–939.
14. Malvern LE. 1969. *Introduction to the Mechanics of a Continuous Medium*. Prentice-Hall, Englewood Cliffs, NJ.
15. Bhatnagar PL. 1980. *Nonlinear Waves in One-Dimensional Dispersive Systems*. Oxford University Press, Oxford.
16. Hildebrand FB. 1976. *Advanced Calculus for Applications*, Prentice-Hall, Englewood Cliffs, NJ.
17. Courant R and Hilbert D. 1962. *Methods of Mathematical Physics V II Partial Differential Equations*. Wiley, New York.
18. Greenberg MJ. 1971. *Application of Green's Functions in Science and Engineering*. Prentice-Hall, Englewood Cliffs, NJ.
19. Freehafer JE. 1965. Geometrical optics. In *Propagation of Short Radio Waves*, DE Kerr, (ed.). Dover Publications, Mineola, NY.
20. Lizuka K. 1987. *Engineering Optics*, 2nd Ed. Springer, Berlin.
21. Verdeyen JT. 1989. *Laser Electronics*, 2nd Ed. Prentice-Hall, Englewood Cliffs, NJ.
22. Moore JH, Davis CC, and Coplan MA. 2001. *Building Scientific Apparatus*, 3rd Ed. Westview Press, Cambridge.
23. Bhattacharyya AK. 1994. *Electromagnetic Fields in Multilayered Structures*. Artech House, Boston.
24. Algernon D, Schubert F, and Wiggenhauser H. 2006. Impact-echo: Experimental and numerical study of acoustic wave propagation. *NDE Conf on Civil Engineering*, American Society for Nondestructure Testing, St. Louis.
25. Belleville C and Duplain G. 1993. White-light interferometric multimode fiber-optic strain sensor. *Opt Lett*, 18, 1.
26. Vanderlugt A. 1992. *Optical Signal Processing*. Wiley, New York.
27. Rai-Choudhury P. 1997. *Handbook of Microlithography, Micromachining, and Microfabrication*. SPIE, Bellingham.
28. Smith HL, Modiano AM, and Moon EE. 1995. On-axis interferometric alignment of plates using the spatial phase of interference patterns. US Patent 5,414,514.
29. Schramm RE, Clark AV, and Szelazek J. 1995. Safety assessment of railroad wheels by residual stress measurement. *Nondestructive Evaluation of Aging Railroads*, SPIE 2458.

30. Udd E. 1991. *Fiber Optic Sensors*. Wiley, New York.
31. Tai H. 2003. Theory of fiber optic Bragg grating—revisited. SPIE 48th Annual Meeting, San Diego.
32. Hillger W. 2007. Visualization and animation of the lamb wave propagation in composites. In *Structural Health Monitoring*, FK Chang (ed.). DEStech, Lancaster.
33. Kohler B, Schubert F, and Frankenstein B. Numerical and experimental investigation of lamb wave excitation, propagation and detection for SHM. *Proc 2nd European Workshop on Structural Health Monitoring*. DEStech, Munich.
34. Liu W and Giurgiutiu V. 2007. Finite element simulation of piezoelectric wafer active sensors for structural health monitoring. In *Structural Health Monitoring*, FK Chang (ed.). DEStech, Lancaster.
35. Martin SA and Jata KV. 2007. Finite element simulation of lamb wave generation with bonded piezoelectric transducers. In *Structural Health Monitoring*, FK Chang (ed.). DEStech, Lancaster.
36. Alleyne DN and Cawley P. 1992. The interaction of lamb waves with defects. *IEEE Trans Ultrasonics, Ferroelectrics, and Frequency Control*, 39(3), 381–397.
37. Raghavan A and Cesnik CE. 2004. Modeling of piezoelectric-based lamb-wave generation and sensing for structural health monitoring. *Sensors and Smart Structures Technologies for Civil, Mechanical and Aerospace Systems*, SPIE 5391.
38. Ende S and Lammering R. 2007. Investigation on piezoelectrically induced lamb wave generation and propagation. *Smart Mater Struct*, 16, 1802–1809.
39. Neary TE, Huston DR, Wu JR, and Spillman Jr WB. 1996. In-situ damage monitoring of composite structures. *Smart Sensing, Processing and Instrumentation*, SPIE 2718.
40. Greve DW, Neumann JJ, Nieuwenhuis JH, Oppenheim IJ, and Tyson NL. 2005. Use of lamb waves to monitor plates: Experiments and simulations. *Sensors and Smart Structures Technologies for Civil, Mechanical, and Aerospace Systems*, SPIE 5765.
41. Giurgiutiu V. 2003. Lamb wave generation with piezoelectric wafer active sensors for structural health monitoring. *Smart Structures and Integrated Systems*, SPIE 5056.
42. Wilcox PD, Dalton RP, Lowe MJ, and Cawley P. 1999. Mode selection and transduction for structural monitoring using lamb waves. *Proc 2nd Intl Workshop on Structural Health Monitoring*, Stanford. Technomic, Lancaster.
43. Kwun H, Kim SY, Choi MS, and Light GM. 2003. Guided-wave structural health monitoring of piping in processing plants. In *Structural Health Monitoring*, FK Chang (ed.). DEStech, Lancaster.
44. Lin C. 1991. Optical fiber transmission technology. In *Handbook of Microwave and Optical Components, V 4: Fiber and Electro-Optical Components*, K Chang (ed.). Wiley-Interscience, New York.
45. Handerek V. 1995. Foundations of optical fiber technology. In *Optical Fiber Sensor Technology*, K Grattan and B Meggitt (eds.). Chapman Hall, London.
46. Krohn DA. 2000. *Fiber Optic Sensors*, 3rd Ed. Instrument Society of American, Research Triangle Park.
47. Kino GS. 1978. The application of reciprocity theory to scattering of acoustic waves by flaws. *J Appl Phys*, 496, 3190–3199.
48. Gubernatis JE, Domany E, and Krumhansl JA. 1977. Formal aspects of the theory of the scattering of ultrasound by flaws in elastic materials. *J Appl Phys*, 48, 7.
49. Srivastava A, Bartoli I, Lanza di Scalea F, and Sabra K. 2007. Global-local ultrasonic method applied to the quantitative detection of defects in aircraft panels. In *Structural Health Monitoring*, FK Chang (ed.). DEStech, Lancaster.
50. Sheyka MP, Su MF, Reda Taha MM, and El-Kady I. 2007. Sub-micron damage identification using photonic crystals: Innovative simulation. In *Structural Health Monitoring*, FK Chang (ed.). DEStech, Lancaster.

51. Stomeo T, Grande M, Qualtieri A, Passaseo A, Salhi A, De Vittorio M, Biallo D, D'orazio A, De Sario M, Marrocco V, Petruzzelli V, and Prudenzano F. 2007. Fabrication of force sensors based on two-dimensional photonic crystal technology. *Microelectron Eng*, 84, 1450–1453.

52. Fuhr PL, Huston DR, and McPadden AJ. 1996. Noncontact deflection measurements in structures using diffraction gratings. In *Recent Developments in Deflection Evaluation of Concrete*, EG Nawy (ed.). ACI SP 161-6, American Concrete Institute, Farmington Hills, MI.

53. Wagner RF, Insana MF, and Brown DG. 1987. Statistical properties of radio-frequency and envelope-detected signals with applications to medical ultrasound. *J Opt Soc Amer A*, 4, 910.

54. Shankar PM. 2001. Ultrasonic tissue characterization using a generalized Nakagami model. *IEEE Trans Ultrason, Ferroelectr Freq Control*, 48(6), 1716–1720.

55. Banerjee S, Banerji P, Berning F, and Eberle K. 2003. Lamb wave propagation and scattering in layered composite plates. *Smart Nondestructive Evaluation and Health Monitoring of Structural and Biological Systems II*, SPIE 5047.

56. Meyers RA and Shore SN. 1990. *Encyclopedia of Modern Physics*. Academic Press, San Diego.

57. Zhang W, Shi B, Gao JQ, Ding Y, and Zhu H. 2005. Nondestructive testing of bridges strengthened by external prestressing using distributed fiber optic sensing. *Proc IABMAS'04 Bridge Maintenance, Safety, Management and Cost*, Kyoto. Taylor & Francis, London.

58. Moulin E, Bourasseau N, Assaad J, and Delebarre C. 2003. Lamb wave beam-steering for integrated health monitoring applications. *Nondestructive Evaluation and Health Monitoring of Aerospace Materials and Composites II*, SPIE 5046.

59. Hayashi H, Hitko DA, and Sodini GC. 2002. Four-element planar Butler matrix using half-wavelength open stubs. *IEEE Microw Wirel Compon Lett*, 12(3), 73–75.

60. Pattan B. 2004. The versatile Butler matrix. *Microwave J*, (November), 47(11), 126–149.

61. Fromme P, Wilcox P, Cawley P, and Lowe MJ. 2004. A guided wave array for structural health monitoring. *Proc 2nd European Workshop on Structural Health Monitoring*. DEStech, Munich.

62. Lozev MG, Grimmett B, Washer G, Fuchs P, and Spencer R. 2002. Ultrasonic phased-array inspection of bridge pins and eyebars. In *Proc Structural Materials Technology V, an NDT Conf*, S Allampalli and G Washer, (eds). Cincinnati, American Society for Nondestructive Testing, Columbus, OH.

63. Malinowski P, Wandowski T, and Ostachowicz W. 2007. Experimental application of signal processing algorithm for damage localization. In *Structural Health Monitoring*, FK Chang (ed.). DEStech, Lancaster.

64. Swift CI, Pierce SG, and Culshaw B. 2007. Generation of a steerable ultrasonic beam using a phased array of low power semiconductor laser sources and fiber optic delivery. *Smart Mater Struct*, 16, 728–732.

65. Kramb VA. 2004. Defect detection and classification in aerospace materials using phased array ultrasonics. *World Conference on Nondes Testing*, Montreal.

66. Nel MD. 2006. Ultrasonic phased arrays: An insight into an emerging technology. *Proc National Seminar Non-Destructive Evaluation*, Hyderabad.

67. Naito S, Minoura M, and Takeda M. 2006. Ship motion analyses in compact wave basin with absorbing wave-maker. *Intl J Offshore Polar Eng*, 16(1), 10–17.

68. Pena J, Kawiecki G, Ullate YG, Freijo F, and Melguizo CP. 2004. Low-cost, low frequency phased array system for damage detection in panels. *Proc 2nd European Workshop on Structural Health Monitoring*. DEStech, Munich.

69. Hirose S and Kimoto K. 2004. Ultrasonic wave radiation from a linear phased array transducer. *Proc IABMAS'04 Bridge Maintenance, Safety, Management and Cost*, Kyoto. Taylor & Francis, London.

70. Ewins DJ. 2001. *Modal Testing: Theory and Practice*, 2nd Ed. Research Studies Press Ltd, Baldock.

71. Allemang RJ and Brown DL. 1987. *Experimental Modal Analysis and Dynamic Component Synthesis*. AFWAL-TR-87-3069, Air Force Wright Aeronautical Laboratories, December.

72. Giurgiutiu V, Blackshire JL, Thomas DT, Welter JT, and Yu L. 2004. Recent advances in the use of piezoelectric wafer active sensors for structural health monitoring. *Proc 2nd European Workshop on Structural Health Monitoring*. DEStech, Munich.

73. Gao H, Rose JL, and Lissenden CJ. 2007. Ultrasonic guided wave mode selection and tuning in composites using a piezoelectric phased array. In *Structural Health Monitoring*, FK Chang (ed.). DEStech, Lancaster.

74. Sundararaman S, Adams DE, and Rigas EJ. 2003. Structural damage characterization through beamforming with phased arrays. In *Structural Health Monitoring*, FK Chang (ed.). DEStech, Lancaster.

75. Harmon LM, Gyekenyesi AL, Martin RE, and Baaklini GY. 2002. Investigation of delaminations with ultrasonic spectroscopy. *Nondestructive Evaluation and Health Monitoring of Aerospace Materials and Civil Infrastructures*, SPIE 4704.

76. Eide E. 2002. Ultra-wideband 3D-imaging ground penetrating radar using synthetic waveforms. *European Microwave Conference*, Milan.

77. Bendat JS and Piersol AG. 1980. *Engineering Applications of Correlation and Spectral Analysis*. Wiley-Interscience, New York.

78. Bloomfield P. 1976. *Fourier Analysis of Time Series: An Introduction*. Wiley, New York.

79. Soumekh M. 1999. *Synthetic Aperture Radar Signal Processing*. Wiley-Interscience, New York.

80. Papoulis A. 1968. *Systems and Transforms with Applications in Optics*. McGraw-Hill, New York.

81. Franceshetti G and Lanari R. 1999. *Synthetic Aperture Radar Processing*. CRC Press, Boca Raton.

82. Wang CH, Rose JT, and Chang FK. 2003. A computerized time-reversal method for structural health monitoring. *Nondestructive Evaluation and Health Monitoring of Aerospace Materials and Composites II*, SPIE 5046.

83. Kim YJ, Jofre L, De Flaviis F, and Feng MQ. 2004. Microwave subsurface imaging technology for damage detection. *J Eng Mech*, 130(7), 858–866.

84. Mast JE. 1998. Automated position calculating imaging radar with low-cost synthetic aperture sensor for imaging layered media. US Patent 5,796,363.

85. Warhus JP and Mast JE. 1998. Ultra wideband ground penetrating radar imaging of heterogeneous solids. US Patent 5,835,054.

86. Stolt RH. 1978. Migration by Fourier transform. *Geophysics*, 43(1), 23–48.

87. Lin X, Pan E, and Yuan FG. 1999. Imaging the damage in the plate with migration technique. *Proc 2nd Intl Workshop on Structural Health Monitoring*, Stanford. Technomic, Lancaster.

88. Schubert F, Koehler B, and Zinin P. 2005. Numerical time-domain simulation of wave propagation and scattering in acoustic microscopy for subsurface defect characterization. *Testing, Reliability, and Application of Micro- and Nano-Material Systems III*, SPIE 5766.

Appendix B

Abbreviations and Acronyms

Acronym	Equivalent Expanded Version
0-D	zero-dimensional
1-D	one-dimensional
2-D	two-dimensional
3-D	three-dimensional
A-scan	amplitude scan
AAR	alkali aggregate reaction
AASHTO	American Association of State Highway and Transportation Officials
ABS	acrylonitrile butadiene styrene
AC	alternating current
ACI	American Concrete Institute
ADC	analog-to-digital converter
ADPR	average driving point residue
AE	acoustic emission
AIC	Akaike information criterion
aka	also known as
AMOEBA	advanced multiple organized experimental basin apparatus
ANOVA	analysis of variance
AR	autoregressive
ARIMA	autoregressive integrated moving average
ARMA	autoregressive moving average
ARMAX	autoregressive moving average with exogenous input
ARV	vector autoregressive
ARX	autoregressive with exogenous input
ASCE	American Society of Civil Engineers
ASIC	application-specific integrated circuit
ASR	alkali silica reaction or alkali silica reactivity
ASTM	American Society for Testing and Materials
atm	standard atmospheric pressure
AURA	articulated ultrasound robot arm
AVS	ambient vibration survey
AWG	arrayed wavelength grating
B-scan	brightness scan
BAM	Bundesanstalt für Materialforschung und -prüfung
BBN	Bolt, Berenek, and Neuman
BMT	Boston Museum Towers
BOTDR	Brillouin optical time domain reflectometer
BSS	blind signal separation or blind source separation

C-scan	coronal plane scan
CA	California, USA or cellular automaton
CAN	controller area network
CDA	confirmatory data analysis
CFRP	carbon fiber-reinforced polymer
CMRR	common mode rejection ratio
CMSEM	cross-modal strain energy method
CO	Colorado, USA
COMAC	complimentary modal assurance criterion
COV	coefficient of variation
CVM	comparative vacuum monitoring
CW	continuous wave
CWT	continuous wavelet transform
DARPA	Defense Advanced Research Project Administration
DATA-SIMLAMT	dynamic across time autonomous—sensing, interpretation model learning, and maintenance
dB	decibels
dBA	decibels A-band
DBST	dual bond stress and temperature
DC	direct current or District of Columbia
DDF	dynamic damage factor
DFT	discrete Fourier transform
DOF	degree of freedom
DS	Dempster–Shafer
DSI	damage signature index
EDA	exploratory data analysis
EFIT	elastodynamic finite integration technique
EIT	electrical impedance tomography
EKBF	extended Kalman–Bucy filter
EM	electromagnetic
EMD	empirical mode decomposition
EMP	electromagnetic pulse
ESD	electrostatic discharge
ETDR	electric time domain reflectometer
EWF	electron work function
FBG	fiber Bragg grating
FCC	Federal Communications Commission
FDR	Fisher discriminant ratio
FFT	fast Fourier transform
FHWA	Federal Highway Administration
FM	frequency modulation
FMEA	failure mode effects analysis
FPGA	field programmable gate array
FRF	frequency response function
FRP	fiber-reinforced polymer
FRPC	fiber-reinforced polymer composite
FSM	fault signature matrix
FTIR	Fourier transform infrared
FWD	falling weight deflectometer
g	acceleration due to gravity

GA	genetic algorithm
GFRP	glass fiber-reinforced polymer
GHz	gigahertz (10^9 Hz)
GMR	giant magneto resistance
GPIB	general purpose interface bus
gpm	gallons per minute
GPR	ground-penetrating radar
GPS	Global Positioning System
GRIN	graded index
HALT	highly accelerated life testing
HART	highway addressable remote transducer
HASS	highly accelerated stress screening
HERMES	high-speed electromagnetic roadway measurement and evaluation system
HF	high frequency
HHT	Hilbert–Huang transform
HSRF	Health Sciences Research Facility
HUMS	Health and Usage Monitoring System
Hz	Hertz
I/O	input–output
ICA	independent components analysis
IE	impact echo
IEEE	Institute of Electrical and Electronics Engineers
IMS	inductive monitoring system
in	inch
IR	infrared
jpeg or jpg	Joint Photographic Experts Group
kb	kilobit (10^3 bit)
kg	kilogram (10^3 g)
kHz	kilo-Hertz (10^3 Hz)
km	kilometer (10^3 m)
kN	kilo-Newton (10^3 kg m/s^2)
kPa	kilo-Pascals (10^3 N/m^2)
lb	pound
LC	inductor capacitor
LCC	life cycle cost
LCR	inductance, current, and resistance
LED	light-emitting diode
LF	low frequency
LH$_2$	liquid hydrogen
LIBS	laser-induced breakdown spectroscopy
LLNL	Lawrence Livermore National Laboratory
LNA	low-noise amplifier
LOS	level of safety
LPR	linear polarization resistance
LVDT	linear variable displacement transformer or linear variable displacement transducer
m	meter
MA	Massachusetts or moving average

mA	milliampere (10^{-3} A)
MAC	modal assurance criterion
MADS	Modular Auxiliary Data System
Mb	megabits (10^6 bit)
ME	Maine
MEMS	microelectromechanical system
MFC	macrofibre composite
MHz	megahertz (10^6 Hz)
MIMO	multiple input multiple output
MIMOSA	Machinery Information Management Open Systems Alliance
MIR	micro-impulse radar
mm	millimeter (10^{-3} m)
MOORAD	magnetic On/Off robot attachment device
MVER	moisture–vapor emission rate
MW	megawatt
N	Newton
NA	numerical aperture or not available
NAT	normal accident theory
NBEMI	narrow band electromechanical impedance
NCAP	network capable application processor
NCHRP	National Cooperative Highway Research Program
Nd:YAG	neodymium-doped yttrium aluminum garnet
NDE	nondestructive evaluation
NDT	nondestructive testing
NFA	nonlinear factor analysis
NH	New Hampshire
NIR	near infrared
NN	neural network
ODE	ordinary differential equations
OFDR	optical frequency domain reflectometry
OPC	optical proximity correction
OSD	Office of Secretary of Defense
OTDR	optical time domain reflectometer
OWEP	offshore wind energy plant
p-p	peak to peak
P-wave	pressure wave or compression wave
Pa	Pascal (N/m^2)
PC	pretensioned concrete
PCA	principal components analysis
PCI	pavement condition index
PDA	pile driving analyzer
PDE	partial differential equation
PDF	probability density function or probability distribution function
PEEK	polyether ether ketone
PERES	Precision Electromagnetic Roadway Evaluation System
pH	potential of hydrogen, logarithmic measure of acid/alkali concentration
PKC	problem knowledge coupler
PMN–PT	lead magnesium niobate–lead titanate

POD	probability of detecting
PRC	polymer reinforced concrete
psf	pounds per square foot
psi	pounds per square inch
PSO	particle swarm optimization
PSPA	portable seismic pavement analyzer
PVDF	polyvinylidene fluoride
PZT	lead zirconate titanate
QDG	quantifiable directed graph
RC	reinforced concrete
RCA	root cause analysis
RCI	riding comfort index
RF	radio frequency
RFID	radio frequency identification
RLSE-UI	recursive least-squares estimator for unknown inputs
RMS	root mean square
RMSD	root mean square deviation index
ROC	receiver operating characteristic
RoHS	restriction of hazardous substances
ROI	region of interest
RS	recommended standard
RTD	resistance temperature detector
S	Strouhal number
S-wave	shear wave
SAMCO	structural assessment, monitoring, and control
SAR	synthetic aperture radar
SASW	spectral analysis of surface waves or surface acoustic stress wave
SAW	surface acoustic wave
Sc	Scruton number
SCI	structural condition index
SDI	structural damping index or surface distress index
SEI	Structural Engineering Institute
SH	shear horizontal
SHM	structural health monitoring
SIDI	single impact damage index
SMA	shape memory alloy
SMART	Stanford Material Receive and Transmit
SNR	signal-to-noise ratio
SOFO	Surveillance d'Ouvrage par Fibre Optique
SPR	statistical pattern recognition
SRI	skid resistance index
STFT	short-term Fourier transform or short-time Fourier transform
SVM	support vector machine
SWISS	Smart Wide Area Imaging Sensor System
TBC	thermal barrier coating
TDM	time division multiplexing
TDMA	time division multiple access
TDR	time domain reflectometer
TEDS	transducer electronic data sheet

TFHRC	Turner-Fairbanks Highway Research Center
TPS	thermal protection system
TRIP	transformation-induced plasticity
TRIZ	Teoriya Resheniya Izobretatelskikh Zadatch
Tx	transmit
UHF	ultrahigh frequency
UHMW	ultrahigh molecular weight
UIO	unknown input observer
US	United States
USB	universal serial bus
USD	United States Dollar
UV	ultraviolet
UWB	ultrawideband
V	volt
VA	Virginia, USA
VARTM	vacuum-assisted resin transfer molding
VDOT	Virginia Department of Transportation
VIEWS	visualizing impacts of earthquakes with satellites
VSWR	voltage standing wave ratio
VT	Vermont, USA
WDI	wire damage index
Wi-Fi	Wireless Fidelity
WIM	weigh-in-motion
WPAN	wireless personal area networks
WSN	wireless sensor network
WV	West Virginia
WVD	Wigner–Ville distribution
YSZ	yttria-stabilized zirconia
$\mu\varepsilon$	microstrain (10^{-6} Strain)
μA	micro-ampere (10^{-6} A)
μm	micrometer (10^{-6} m)
μs	microsecond (10^{-6} m)
μW	microwatts (10^{-6} W)

Index